TROPICAL RAINFOREST RESEARCH – CURRENT ISSUES

MONOGRAPHIAE BIOLOGICAE

VOLUME 74

Series Editors

H.J. Dumont and M.J.A. Werger

Tropical Rainforest Research – Current Issues

Proceedings of the Conference
held in Bandar Seri Begawan, April 1993

Edited by

D. S. EDWARDS

W. E. BOOTH

and

S. C. CHOY

University of Brunei Darussalam

KLUWER ACADEMIC PUBLISHERS

DORDRECHT / BOSTON / LONDON

Library of Congress Cataloging-in-Publication Data

Tropical rainforest research : current issues / edited by D.S. Edwards and W.E. Booth and S.C. Choy.
p. cm. -- (Monographiae biologicae ; v. 74)
Includes bibliographical references and index.

1. Rain forests. 2. Rain forest ecology. 3. Rain forest conservation. 4. Rain forests--Economic aspects. 5. Biological diversity--Tropics. I. Edwards, David S. II. Booth, W. E. (Webber E.) III. Choy, S. C. (Satish C.) IV. Series.
QH86.T76 1996
574.5'2642'0913--dc20 96-14529

ISBN-13: 978-94-010-7255-7 e-ISBN-13: 978-94-009-1685-2
DOI: 10.1007/978-94-009-1685-2

Published by Kluwer Academic Publishers,
P.O. Box 17, 3300 AA Dordrecht, The Netherlands.

Kluwer Academic Publishers incorporates
the publishing programmes of
D. Reidel, Martinus Nijhoff, Dr W. Junk and MTP Press.

Sold and distributed in the U.S.A. and Canada
by Kluwer Academic Publishers,
101 Philip Drive, Norwell, MA 02061, U.S.A.

In all other countries, sold and distributed
by Kluwer Academic Publishers Group,
P.O. Box 322, 3300 AH Dordrecht, The Netherlands.

Printed on acid-free paper

CONTENTS

RAINFOREST DYNAMICS

SOCIO-ECONOMICS

Preface

The Conference on Tropical Rainforest Research: Current Issues was organised by the University of Brunei Darussalam and The Royal Geographical Society, London, and held in Bandar Seri Begawan, Brunei Darussalam, over 8 days in April 1993. Over 160 participants from 22 countries attended the Conference, which was opened on the 9th April by the Brunei Darussalam Minister for Home Affairs, Yang Berhormat Pehin Orang Kaya Laila Setia Bakti Diraja Dato Laila Utama Haji Awang Isa bin Datu Perdana Menteri Dato Laila Utama Haji Awang Ibrahim.

The conference was initially intended to provide a forum to present the results of the expedition into the lowland mixed dipterocarp forest of the Temburong District of Brunei Darussalam, which had been jointly organised by the University of Brunei Darussalam and The Royal Geographical Society, London. The 15-month expedition, lasting from January 1991 to March 1992, was based at the then newly-completed Kuala Belalong Field Studies Centre, a research and education facility set up by the University with funds provided by the Brunei Government and Brunei-Shell Petroleum Sdn Bhd. The expedition, with over 70 scientists taking part, received financial support from a wide range of sponsors and Corporate Patrons, including Royal Brunei Airlines, the Baring Foundation, Daiwa-Dicam, Greencard Trust, the Hongkong Bank, Morgan Grenfell and Nomura-Nimco. The conference itself was supported by donations from Brunei-Shell Sdn Bhd, Royal Brunei Airlines and Standard Chartered Bank.

In the event, although papers resulting from work at the Field Centre formed a substantial part of the presentations, the theme of the conference was widened to include other aspects of research into tropical forests. Three main themes were identified - Biodiversity, Forest Dynamics and Socio-economic Aspects, with each theme being introduced by a keynote speaker; the proceedings have been organised in a similar manner. A total of 85 papers were presented over the 4 days of lecture sessions; the present volume contains the texts of 53 of these. Apart from the formal presentations at the Conference, a session was also devoted to discussing the future development of the Kuala Belalong Field Studies Centre, and the integration of its activities into the wider context of research into tropical forests within the south-east asian region as a whole; the resolutions drawn up to guide the University have not been included in these Proceedings.

The editors would like to thank the authors for their patience and understanding through what has been a long gestation period, as well as the many anonymous reviewers for their comments on the original manuscripts. The conference itself benefitted from the dedication and commitment of many of the staff of the University, far too numerous to mention here; the organisational abilities of Madam Pang Yoke Kiew (University of Brunei Darussalam) and Ms Catriona Prebble (Royal Geographical Society) were admirably demonstrated through its success. Publication of these Proceedings has been made possible through the generosity of grants from the Government of Brunei Darussalam and Brunei-Shell Petroleum Sdn Bhd.

Madam Pang Yoke Kiew and Mrs C.Fereday were most helpful in assisting in the preparation of the typescripts; the editors are also appreciative of the efforts of the staff at Kluwer Academic Publishers, particularly Mr. Wil Peters and Ir. Hendrik Prins, whose patience in the face of interminable delays was endless. We hope that this collection of papers will serve as a fitting epilogue to this phase of the productive collaboration between the University of Brunei Darussalam and The Royal Geographical Society, and will give an indication of the current state of tropical rain forest research in south-east Asia.

Finally, we would like to thank The Honourable Minister of Education, Yang Berhormat Pehin Orang Kaya Laila Wijaya Dato Seri Setia Awang Haji Abdul Aziz bin Begawan Pehin Udana Khatib Dato Seri Paduka Awang Haji Umar, and the Vice-Chancellor of Universiti Brunei Darussalam, Yang Dimuliakan Pehin Orang Kaya Putera Maharaja Dato Seri Laila Jasa Haji Abu Bakar bin Haji Apong, for the-r enthusiastic and wholehearted support for our efforts in developing the Kuala Belalong Field Studies Centre and in the preparation of these Proceedings.

David S.Edwards
Webber E.Booth
Satish C.Choy

Bandar Seri Begawan,
Brunei Darussalam.
May 1996

Tropical forest dynamics: the faunal components

N.E. STORK

Biodiversity Division, Department of Entomology, The Natural History Museum, Cromwell Road, London SW7 5BD, United Kingdom

ABSTRACT. The form, diversity and dynamics of tropical forests are influenced by both abiotic factors, such as the physical environment and climate, and biotic factors, such as interactions between organisms. To most people forests are principally comprised of trees, large mammals and birds. But, tropical forests are comprised of many other smaller life forms including arthropods, fungi, nematodes and bacteria, and these too have an important part to play in forest dynamics. The relative importance of different groups of animals in tropical forests, as measured through abundance, biomass and species richness, is presented here. Because of the sheer diversity of the life forms in tropical forests, only a few of the complex webs of interactions between organisms of the different kingdoms have been determined with any accuracy. In spite of this fact, some general patterns have been recognised and the major components of trophic systems in tropical forests have been determined. Many detailed studies of large mammals and birds have shown the significance of their roles in the dynamics of tropical forests. What is less clear is the precise contribution of many groups of invertebrates such as arthropods and nematodes. In part this is because the abundance, species richness and sometimes biology of these groups is poorly understood. Data are presented on the (i) distribution, abundance and biomass of arthropod groups in the main habitats of lowland rain forest in Seram, and (ii) the species richness of different arthropod groups in lowland forest in Sulawesi. Arthropods are particularly important in two contrasting strata of tropical forests, the canopy and soil, where herbivory and decomposition, respectively, are the principal trophic processes. The importance of arthropods in these forest strata and to forest dynamics are discussed through examples of two groups of arthropods, ants and termites. An understanding of tropical forest dynamics may help in determining how biodiversity and sustainability may sit more easily together. This is discussed with suggestions as to how we can best use available information toward this end, what new information is needed, and which ideas require examining.

KEY WORDS: diversity, biomass, biotic factors, abiotic factors, arthropods, ants, termites, Seram, Sulawesi.

INTRODUCTION

For obvious reasons much of the attention within studies of tropical forest dynamics has been on trees. The natural process of tree growth, death, gap creation and new tree growth has been well documented (Brokaw 1874, Grubb 1977. Whitmore 1984). The role of nutrient cycling in forest dynamics is also well documented but again largely focuses on botanical elements (DeAngelis 1992). However, tropical forests are not just comprised of trees or even just of plants. Other groups of organisms, such as insects and other invertebrates, are also essential to tropical forests, as they are to all ecosystems, and are often highly abundant and speciose. Prokaryote organisms (e.g. bacteria and viruses) and eukaryote plants (e.g. fungi) are clearly essential to ecosystem function but are beyond the scope of this paper and are not discussed here. The first aim of this paper is to provide a brief review of the eukaryote faunal components of tropical forests and their contribution to forest dynamics.

In reviewing faunal dynamics in tropical forests it is perhaps appropriate to consider this subject in the light of current scientific interest in both environmental change and biodiversity. Biodiversity, a contraction of 'biological diversity' (Wilson 1988), is the term used by many individuals, as diverse in interests as biologists and economists, to cover a very wide-ranging field of science. Most recently it has come to prominence through the

D. S. Edwards et al. (eds.), Tropical Rainforest Research – Current Issues, 1–20.

Biodiversity Convention agreed at the UNCED meeting in Rio de Janeiro in 1992 and signed by more than 160 countries. To most scientists biodiversity represents the diversity of life on Earth as expressed typically by genetic diversity, species diversity and ecosystem diversity (IUCN, UNEP & WRI 1990). In reality it involves many other aspects of nature including landscape diversity, biome diversity, and historical biogeography (Holloway & Stork 1991). Recent upward revisions of estimates of the number of species on Earth to around 5-15 million species (Gaston 1991, Hammond 1992, in press (a,b) May 1992, Stork 1993) and beyond have implicitly suggested that perhaps (i) more than 80% of species are undescribed, (ii) more than 80% of species of all living eukaryote organisms are insects, and (iii) perhaps 90% of all species are in tropical forests (although some marine biologists might argue this latter point - see Grassle & Maciolek 1992). The major faunal component of tropical forests is the arthropods, and their significance is vastly underestimated.

The second aim of this paper, therefore, is to examine more closely the contribution of arthropods to tropical forests. First, the abundance and biomass of different arthropod groups, and their distribution in the different habitats (canopy, soil etc.) are discussed using an example of a study in Seram. Second, the species richness of arthropods in tropical forests is examined using the example of an intensive study in Sulawesi. Although many arthropods are found in aquatic systems these are not considered here.

The third aim of this paper is to demonstrate the significance of arthropods to tropical forest dynamics. What we know of the contribution of arthropods to the dynamic processes in tropical forests is extremely variable. A few groups are better known than others and here I use two examples, ants and termites, to show the extent and complexity of their role in forest dynamics.

Scientists are faced with the task of providing answers to many pressing international problems. With limited budgets it is essential that efforts are coordinated to produce best value for money. Some answers are required more rapidly than others and therefore priorities need to be made if we are to contribute to limiting the effects of problems such as global climate change, forest destruction and pollution. Particularly important is the implementation of policies advocating sustainable utilisation of resources such as tropical forests. Understanding the natural processes of tropical forests and the changes in these processes when forests are affected by humans is essential. In the final section of this paper I briefly review some ideas that are either being discussed by international organisations or are being implemented.

CONCEPTS OF ECOSYSTEM DYNAMICS

What are the forces that drive tropical forest systems? The study of the dynamics of tropical forest ecosystems is extremely complex and no unifying theory exists. Numerous concepts overlap and sometimes conflict and no attempt is made in this paper to discuss these. However, perhaps the concept most fundamental to an understanding of tropical forest dynamics is that of scale and the balance between local and regional forces (see Figure 1, after di Castri 1991, and Delcourt & Delcourt 1987, and Figure 2 after Cornell & Lawton 1992). As humans we perceive nature in a limited range of scale and this has created bias in the way we examine it. As a result, large more obvious organisms such as birds, large mammals and trees have received more of our attention than smaller ones (Gaston & May 1992). Similarly, we have difficulty in looking at changes with respect to time when the timescale is either very short (less than a few seconds) or very long (more than a few years) with respect to the life-span of human. Cornell & Lawton (1992) noted that two groups of processes they identify, ecological and biogeographical, operate at the local and regional spatial scales, respectively. This is also true of the temporal scale: ecological processes

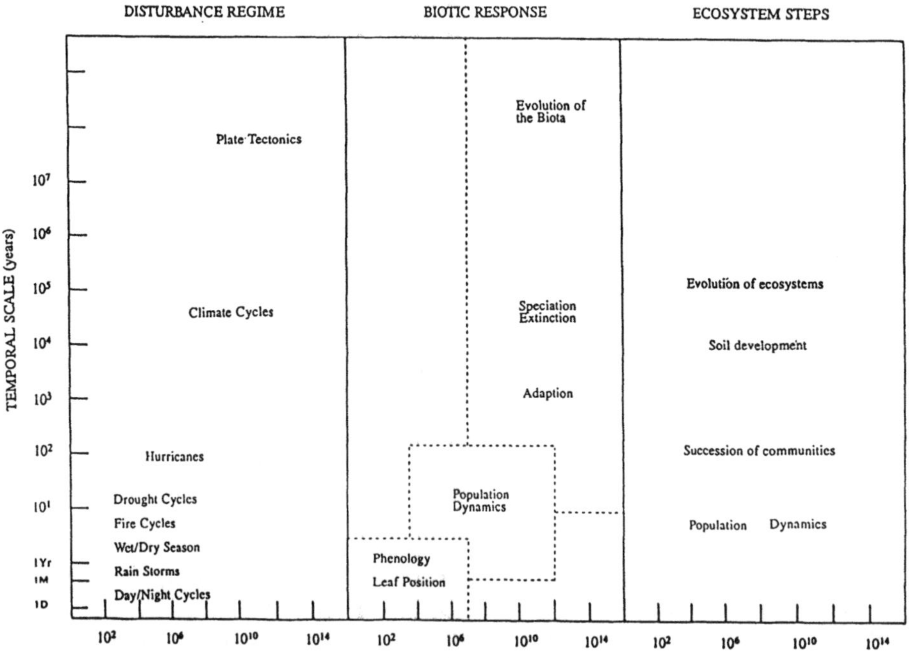

Figure 1. Spatial and temporal scales of disturbance forces and biotic responses (after di Castri 1991 and simplified from Delcourt & Delcourt 1987).

LOCAL FORCES	REGIONAL FORCES
population dynamics	long-distance dispersal
parasitism	speciation
predation	species extinction
predation	fluctuation in species distribution
abiotic fluctuations or disturbance	
= 'ecological forces'	= 'biogeographical forces'

Figure 2. The balance between local and regional forces and how these forces are represented in the biology of organisms.

operating more in the short-term and biogeographical processes operating in the long-term (Figure 2).

Of the abiotic forces affecting the balance of tropical forests, particularly important are the flows of water, carbon, nitrogen and other elements such as phosphorus. Changes in the balance of the cycling of these can considerably affect the nature and stability of the ecosystem (e.g. see Furley *et al.* 1992). Also extremely important is the stability of local and regional climatic conditions; hurricanes, for example creating immense natural damage to some forest systems (e.g. Weaver 1989). These abiotic forces can rapidly become controlling in some circumstances of forest change leading to erosion and desertification. The organisms that combine to form tropical forests are each physiologically adapted to a limited range of tolerances with respect to water and nutrient availability, sunshine, temperature, and relative humidity. Some species have sufficient genetic variability to be able to adapt to large and rapid changes in the environment whereas others are replaced by more adaptable species. The level of fluctuation in such abiotic factors will determine the degree of change in species composition, population structure, and biological disturbance, and the rapidity of recovery.

THE FAUNAL COMPONENTS OF TROPICAL FORESTS

What groups of animals are found in tropical forests? What are their abundances, species richness, biomass, and what are their roles in the dynamics of tropical forests? In practice, abundances and species richness of different organisms vary in time and space and no sites have been adequately sampled to present even a model view of their dynamics. Table 1 provides an overview of several measures of the faunal component of tropical forests and is discussed below. No attempt has been made to give a complete list of the higher taxa within the phyla, only those being listed which have a recognised role in one way or other in tropical forests.

Many phyla are entirely restricted to aquatic environments and only a few, such as Arthropoda, Chordata, Mollusca and Annelida have groups that have adapted to the rigours of life out of water. Many groups of small organisms found in soil, such as Protozoa, Rotifera and some Nemata, are still aquatic, surviving in the interstitial water between soil particles. Data for abundance, biomass, and species richness for these and many of the other groups figured in Table 1 are scarce. Furthermore, the term 'tropical forest' covers a wide range of forest types and hence data from one source are rarely directly comparable with those from another. To these extents, Table 1 is a conglomerate and provides a simplistic view of the faunal contribution to tropical forest dynamics. The actual contribution of different groups such as earthworms and termites in different forest types or parts of the world may vary considerably.

Nematodes are known from numerous studies to be superabundant in agricultural ecosystems but we are uncertain how diverse or how numerous they are in tropical forests. Preliminary results from studies in Brunei and Cameroon (Hodda *et al.*, in press) indicate they are very abundant and contribute large biomass in moist tropical forests. Evidence is so far lacking on the species richness of this group in tropical forests because of the enormous numbers of individuals per m^2. Lepidoptera can be very abundant particularly on short time scales and, as most are herbivores, they can have a dramatic affect on temperate forests of low tree species richness. Whether they have the same affect on similarly species poor tropical systems such as *Brachystegia* forests, or in more diverse tropical forests is uncertain. In Australian eucalyptus forests detritivorous Lepidoptera of the family Oecophoridae, numbering some 2,300 species, are essential in the nutrient cycle as they are major decomposers of eucalypt leaf litter (Common 1990).

It is evident from Table 1 that in terms of biomass, abundance and species richness, Arthropoda is the most important group of animals with respect to tropical forests. Many groups of arthropods have key roles in ecosystem function and the dynamic processes of tropical forests and examples of these are discussed below.

Site inventories

There is no single terrestrial site in the world, tropical or temperate, where we know all the living organisms and therefore not surprisingly, our understanding of the biological components of tropical forests is poor. Botanists have tackled this problem in a more systematic way (e.g. Dallmeier 1992) and lists of species for trees above 10 cm dbh are now available for many hectare units around the world. More intensive studies of the plants are

Table 1. Summary of the main parameters of the terrestrial faunal components of tropical forests. Arthropod data compiled from sources quoted in the text and from unpublished data. Vertebrate data compiled from (1) Inger 1979, (2) Inger 1979, (3) Davidson 1991, Johns 1992 and Woinarski 1991, (4) Chivers & Davis 1979, Curtin & Chivers 1978, Johns 1992, Medway 1983, Payne 1979 and Payne & Mahedi 1991. Groups in italics are exclusively marine. Classification following Parker (1982). Habitats: T - terrestrial (ie. not aquatic), G - ground, C - canopy; Key (key groups): P - predators, D - decomposers, H - herbivores.

Key to symbols:

	***	**	*	-
Abundance (individuals)	$>10^6$/ha	$>10^4$/ha	$>10^2$/ha	$<10^2$/ha
Biomass (gm)	$>10^4$/ha	$>10^3$/ha	$>10^2$/ha	$<10^2$/ha
Sp. richness (species)	$>10^3$/ha	$>10^2$/ha	>10/ha	<10/ha

	Abundance	Biomass	Sp. rich.	Habitat	Key
Subkingdom Protozoa	?	?	?	T:G	?
Subkingdom Phagocytellozoa	-	-	-	-	-
Subkingdom Parazoa	-	-	-	-	-
Subkingdom Eumetazoa					
Phylum Cnidaria	-	-	-	-	-
Phylum Ctenophora	-	-	-	-	-
Phylum Platyhelminthes	-	-	-	T:G	-
Phylum Nemertea	-	-	-	-	-
Phylum Gnathostomulida	-	-	-	-	-
Phylum Mesozoa	-	-	-	-	-
Phylum Gastrotricha	-	-	-	-	-
Phylum Rotifera	-	-	-	-	-
Phylum Kinorhyncha	-	-	-	-	-
Phylum Nemata	***	**	**	T:G	P,D,H
Phylum Nematomorpha	-	-	-	-	-
Phylum Acanthocephala	-	-	-	-	-
Phylum Priapulida	-	-	-	-	-
Phylum Mollusca	-	-	-	T:G	-
Phylum Annelida	***	***	*	T:G	D
Phylum Pogonophora	-	-	-	-	-
Phylum Echiura	-	-	-	-	-
Phylum Sipuncula	-	-	-	-	-
Phylum Onychophora	*	-	-	T:G	-
Phylum Tardigrada	-	-	-	-	-
Phylum Phoronida	-	-	-	-	-
Phylum Bryozoa	-	-	-	-	-
Phylum Entoprocta	-	-	-	-	-
Phylum Brachiopoda	-	-	-	-	-
Phylum Chaetognatha	-	-	-	-	-
Phylum Echinodermata	-	-	-	-	-
Phylum Hemichordata	-	-	-	-	-
Phylum Arthropoda					
Class Arachnida					
Order Scorpiones	*	-	-	T:G	-
Order Araneae	**	**	***	T:G,C	-
Order Pseudoscorpiones	**	*	*	T:G,C	-
Order Opiliones	**	*	*	T:G,C	-
Order Acari	***	*	***	T:G,C	P
Class Crustacea					
Order Isopoda	-	-	*	T:G,C	-
Class Hexapoda					
Order Collembola	***	*	**	T:G,C	D
Class Insecta					
Order Thysanura	*	-	-	T:G	-
Order Odonata	*	-	**	F	-
Order Isoptera	***	***	*	T:G	D
Order Orthoptera	**	**	*	T:G,C	-
Order Dermaptera	**	-	*	T:G,C	-
Order Hemiptera	**	*	**	T:G,C	-
Order Homoptera	**	*	**	T:G,C	H
Order Neuroptera	**	-	*	T:C	-
Order Coleoptera	***	**	***	T:G,C	D,H,P
Order Diptera	***	**	***	T:G,C	D
Order Lepidoptera	***	***	***	T:G,C	H
Order Hymenoptera	***	**	***	T:G,C	P
Phylum Chordata					
Subphylum Vertebrata					
Class Osteichthyes	-	-	-	-	-
Class Amphibia$_1$	*	**	*	T:G	P
Class Reptilia$_2$	*	**	*	T:G	P,H
Class Aves$_3$	*	**	*	T:G,C	P,H
Class Mammalia$_4$	*	***	*	T:G,C	P,H

far fewer although some have managed to identify and measure all trees above 1 cm DBH for 50 hectare plots (Foster & Hubbell 1990). Detailed censuses are also available for large mammals, birds, reptiles and amphibians. However, for the vast majority of groups such as the ferns, mosses and lichens on the plant side, and for arthropods, annelids, and protozoa on the animal side, we have very little idea of their abundance let alone their richness in species. Furthermore, the data that are available are from a variety of different sites and it is not always possible to make valid comparisons.

Two attempts, both in Indonesia, have been made specifically to document arthropod diversity in restricted areas of tropical forest, loosely comparable with the botanists' hectare and 50 hectare plots, and these are discussed here. The first examines arthropod abundance and biomass and the second largely examines species richness.

(i) *Arthropod abundance and biomass in Seram*

The relative abundance and biomass of the arthropods were estimated in a hectare of lowland tropical forest in Seram, Indonesia through collection of direct quantitative samples of arthropods from what were deemed to be the five most important habitats - canopy, tree trunks, ground herb-layer, leaf litter and soil (Stork 1988, Stork & Brendell 1993). Samples were collected using a variety of sampling methods (canopy - knockdown insecticide fogging; tree trunks - knockdown insecticide spraying; ground herb layer - DVac; leaf litter and soil - Tullgren funnels). Samples were sorted to orders and counted. In order to estimate biomass the samples were further sorted to log. size classes and the following body size/biomass relationship used:

$$B = N(0.0305L^{2.62}) \quad \text{(Rogers \textit{et al.} 1976)}$$

where B is biomass in mg, N is number of individuals, and L is body length in mm. The geometric mean of the two limits for each size class was used as the average body length of each size class in these calculations.

The number or biomass of arthropods from each of the five habitats in a hectare of Seram forest was calculated on the expected number of duplicate samples in a hectare. For example, since a 1 m^2 sampling unit was used for the canopy and a hectare is 10,000 m^2, the number of arthropods in the canopy in a hectare = 10,000 x M (where M = the mean number of arthropods /m^2). Multiples for the other habitats are: leaf litter - 10,000, herb-layer vegetation - 10,000/16, tree trunks - 800, soil - 650,000 (the figure for tree trunks is based on a guess of 800 'tree trunk equivalents' in a hectare, there being no data available for a more accurate assessment).

The results of this study indicated that in a hectare of forest (i) there are more than 42 million arthropods, equivalent to more than 30 kg dry weight, (ii) soil, canopy and leaf litter contribute the greatest abundance and biomass of arthropods, (iii) Collembola, Formicidae and Acarina are the most abundant groups, (iv) non-insect arthropods (primarily Diplopoda) comprise more than 50% of the biomass (Figure 3 and see Figures 7-8 in Stork 1988, Figures 1-3 in Stork & Brendell 1993).

The biases in the various sampling methods used in this study are well known and discussed in full by Stork & Brendell (1993). These and the results of other studies (Adis 1988, Adis & Schubart 1985, Collins 1983, André *et al.* 1992, 1994, Eggleton *et al.* in press, Eggleton & Bignell, in press) would suggest that these estimates for total arthropod abundance and biomass are low. Those for soil arthropods are particularly suspect and the abundance of Collembola and Acarina may be much higher. Termites are almost certainly a large component of the fauna of most tropical forest soils (see below) even if they were of genuinely low abundance or absent from the Seram study site. In spite of these anomalies, it seems probable that the relative significance of the five habitats studied, in terms of arthropod abundance and biomass, is correctly identified. For all five habitats examined

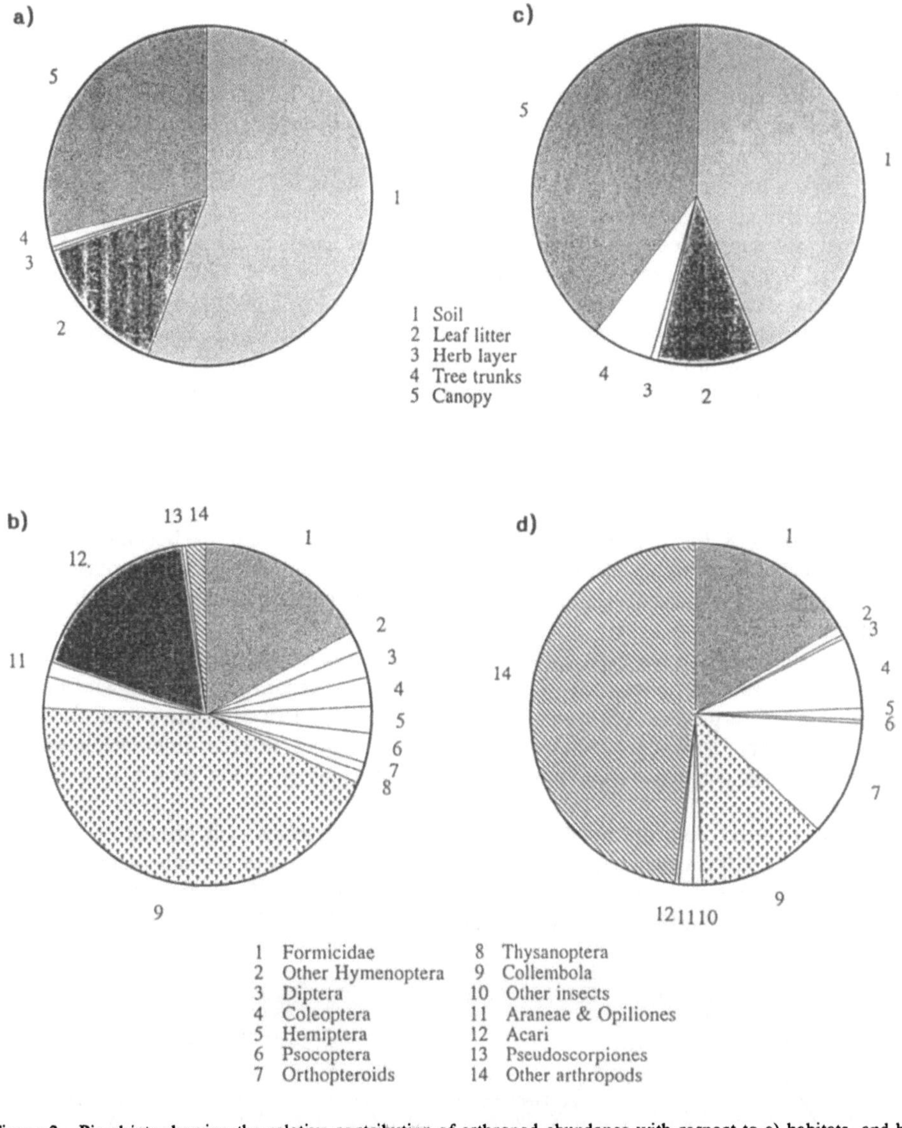

Figure 3. Pie charts showing the relative contribution of arthropod abundance with respect to a) habitats, and b) taxonomic groups, and of biomass with respect c) habitats, and d) taxonomic groups (note: Thysanoptera and Pseudoscorpiones biomass too low to include in Fig. 3d), in lowland tropical forest in Seram, Indonesia (after Stork & Brendell 1993).

arthropod abundance decreases and biomass increases with body size (Stork & Blackburn 1993), although the average body size increases vertically with habitat from soil to canopy. The modal body size for arthropod abundance is about 1.2 mm for above ground habitats and less than 0.5 mm for below ground habitats (soil and leaf litter).

Although some 45,000 arthropods were sampled in the Seram study this was insufficient to carry out an analysis of the species richness of the arthropods in a hectare of forest and the relative species contribution and overlap for each of the five habitats. A more intensive sampling programme, such as that described below, is required to determine arthropod species richness of tropical forest.

(ii) *Arthropod species richness and composition in N.Sulawesi*

The insect fauna of the Dumoga-Bone region of N.Sulawesi was intensively studied by over 100 entomologists during 1985 as part of the Royal Entomological Society's Project Wallace. The main results of this project are summarised in Knight and Holloway (1990). The largest single study was that of the insect fauna of lowland forest by a team from the Natural History Museum (London, UK). Their intensive sampling programme (see Hammond 1990, Stork & Brendell 1990), which involved Malaise, flight interception, light, yellow pan, and pitfall (baited and unbaited) trapping, knockdown insecticide canopy fogging, and litter sampling, was supplemented by additional sampling and hand collecting by others (Barlow & Woiwod 1990, Hodkinson & Casson 1987, 1991, Holloway *et al.* 1990, Noyes 1989a,b).

The most extensively sampled group were Coleoptera with over 1.1 million collected specimens being examined (Hammond 1990, Stork & Brendell, 1990). The vast majority of these were from a lowland forest area of about 500 hectares. Hammond identified 5,649 morphospecies from the whole region and 3,488 from the lowland forest area. He estimated that with additional unsorted materials the totals for the latter would be over 6,000. From his detailed analyses of these samples a number of observations can be made: (1) the most species rich groups in the samples are also those groups with the largest numbers of described species (see Gaston 1991); Staphylinoidea (26% of all beetle species), Curculionoidea (18.7%), Tenebrionoidea (12.9%), (2) only 17.4% of the beetle species feed on non-woody living higher plant materials (ie leaves, pollen etc.); 29% are predators and 23.4% fungivores (Figure 4). Hammond's (1990) cumulative species curves for both ground and aerial Malaise trap weekly catches throughout the year show no obvious asymptote. However, since none of the traps for which cumulative totals were presented collected more than about 400 species (ie less than 10% of the minimum number of species known from the area) this is not surprising. What would be interesting is to see a cumulative catch week by week for all trap methods combined. However, difficulties in cross-referencing species between samples and databasing this information probably preclude such a task for the moment.

Species richness data are available for several other species rich groups of Sulawesi insects. Noyes' (1989a,b) staggering collection of 4,109 species of Hymenoptera was from samples he and others collected in a three month period (April - June). Hodkinson and Casson's studies of the Hemiptera (Casson & Hodkinson 1991, Hodkinson & Casson 1987,

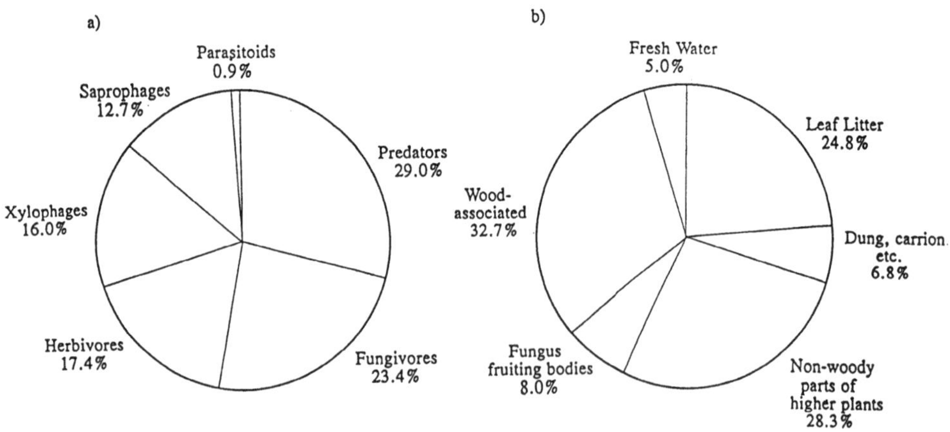

Figure 4. Pie charts showing a) the feeding guild associations and b) the habitat associations of 5,469 beetle species collected from N. Sulawesi, Indonesia (see text) (after Hammond 1990).

Table 2. Species richness of different arthropod groups for areas within Sulawesi: (i) for the lowland forest area of about 500 hectares including three hectare plots A-C at 200 m, 300 m and 400 m altitude, (ii) the Dumoga valley and surrounding regions, including upland sites up to 1760 m and over a geographical distance of 150 km, (iii) the whole of the island of Sulawesi.

	Plots A-C 500 ha	N. Sulawesi 15 km²	Sulawesi 150,000 km²	Author
Coleoptera	3500	6,250 (10,000)	?	Hammond (1990)
Diptera	?	?	?	
Hemiptera	?	1,690 (2,500)	?	Hodkinson & Casson (1991)
Hymenoptera	?	4,100 (10,000)	?	Noyes (1989a, b)
Lepidoptera				
Butterflies	300	350	470	R. Vane-Wright (pers. comm.)
Macro-Lepidoptera	?	1,600	?	Holloway *et al.* (1990) and J. Holloway & G. Robinson (pers. comm.)
Micro-Lepidoptera				
Tineidae	?	114	?	
Tortricidae	?	285	?	
Orthoptera	?	?	?	
Thysanoptera	?	?	?	
Psocoptera	?	?	?	
Estimated Totals	(30,000)	(50,000)	(75-100,000)	

1991) were also based on limited samples; most of their 1690 species were collected by knockdown insecticide fogging from trees (Stork & Brendell 1990).

In Table 2 an attempt has been made to compile very rough and preliminary species richness estimates for the insect fauna of the Dumoga-Bone area at three geographic scales: (i) the insect fauna of the lowland forests area of about 500 hectares including three hectare plots A-C at 200m, 300m and 400m altitude, respectively, (ii) the Dumoga valley and surrounding regions, including upland sites up to 1760m and over a geographical distance of 150 km, (iii) the whole of the island of Sulawesi. The Sulawesi project gives an approximate idea, perhaps for the first time, of the number of species in an area of moist tropical forest. It also indicates that unless levels of local endemism of species have been seriously underestimated, estimated local and regional species richness differ by less than an order of magnitude.

TROPICAL FOREST DYNAMICS: TWO EXAMPLES OF KEY GROUPS OF ARTHROPODS

The contribution of ants to tropical forest dynamics

(1) Ant abundance, diversity and biomass

A number of studies have shown that ants form a major part of the faunal component of tropical forests in terms of both abundance and biomass. Fittkau & Klinge (1973) suggested that ants, termites, bees and wasps comrise 75% of the insect biomass in Amazonian forest. In many knockdown insecticide studies of the canopy arthropod fauna of tropical forests in countries such as Brazil (Adis *et al.* 1984), Peru (Erwin 1988), Venezuela (Stork & Davies unpublished), and Seram (Stork 1988, Stork & Brendell 1993), ants comprise 50% of the individuals. Ants are low in abundance in only a few tropical forests, such as in Sulawesi and Brunei (Stork 1991, Stork & Brendell 1990). In the Seram study discussed above, ants comprised an estimated 17% of all arthropods in lowland tropical forest.

Table 3. Abundances of the eight species (from 98 in total) of ant with more than 100 individuals on ten trees sampled by canopy fogging in Borneo (Stork 1991, identifications provided by B. Bolton).

Species	1	2	3	4	5	6	Trees 7	8	9	10	Total
Crematogaster sp. B	0	15	522	0	0	0	214	38	0	0	789
Crematogaster sp. H	0	0	119	3	0	0	0	0	0	0	122
Crematogaster sp. I	6	0	21	0	75	0	17	0	5	0	124
Crematogaster sp. L	317	0	0	0	0	0	0	0	0	0	317
Myrmicaria sp. A	45	0	94	159	62	6	0	0	109	13	488
Myrmicaria sp. B	0	458	0	0	0	0	799	53	0	0	1310
Dolichoderus sp. D	0	9	0	52	59	0	0	0	0	0	120
Camponotus sp. C	0	0	0	0	0	0	0	0	0	199	199

In species terms, ants represent a relatively high diversity group. B. Bolton (personal communication) collected almost 400 species of ants in a hectare of lowland tropical forest in Sarawak. Hölldobler & Wilson (1990) quote other examples with lower species numbers from other ant workers collecting intensively in different tropical forest sites.

(2) Patterns of distribution

Ants have a broad geographical distribution being found throughout the world in temperate to tropical ecosystems. The well known social structure of ant colonies is discussed by Hölldobler & Wilson (1990) and elsewhere. Some species have interconnected nests forming colonies that may cover several km^2. One of the most unusual patterns that has been described for ants is the 'mosaic' distribution pattern. First described for ant species in cocoa plantations in Ghana by several authors (Leston 1973a,b, Majer 1976a,b,c, Room 1971), the mosaic pattern of distribution is now also recognised as a phenomenon of arboreal ants in natural forests (Hölldobler 1983, Majer 1993). Dominant species occupy large areas of single or several trees to the exclusion of other competing species. Some sub-dominant species appear to co-exist successfully in the territories of dominant species. In Table 3 the numbers of individuals are shown for the eight commonest species of the 98 collected from ten trees in Brunei by knockdown insecticide fogging (Stork 1991). This table demonstrates that individuals of the different species are not evenly distributed across all trees. The majority of individuals of a species are found on just one or two trees. The two commonest species of ant (both *Myrmicaria* spp) appear to be mutually exclusive. This would appear to demonstrate the dominance pattern typical of the ant-mosaic model.

(3) Contribution to forest dynamics

The importance of ants to tropical forest dynamics has been lauded by many including Wilson (1971, Hölldobler & Wilson 1990). Some of the more well recognised roles of ants are discussed here.

(i) Controlling levels of herbivory

Most species of ants have a significant role in herbivory either as predators of herbivores, tenders of, and hence protectors of, herbivores, or as herbivores themselves. The two main groups of herbivores, 'chewers' and 'suckers', are affected in different ways by some species of ants. Most sucking insects are Homoptera, are largely sedentary, and feed on plant phloem. Some species of ants feed on the honeydew excreted by Homoptera, and thus could be termed herbivores. These ants provide protection to Homoptera from predators and parasitoids, and by depredation of leaf-chewing herbivores (e.g. caterpillars) reduce herbivorous competition. Some species of *Monomorium* are predators of batches of Lepidoptera eggs thus reducing herbivory at the earliest stage (Way *et al.* 1989). Some reduce levels of herbivory through predation of leaf mining larvae (Memmott *et al.*, in press). Leaf-cutter ants of the genus *Atta* have been estimated to account for the removal of

12-17% of total leaf production in Neotropical forests (Cherrett 1986, 1989). Although these ants do not eat the leaves they cut (see below), they are nevertheless responsible for the leaf damage and hence can be considered herbivores. Their harvesting of leaves from trees is not arbitrary but follows clear patterns such that individual trees are not over-cropped. They maintain the correct degree of sunlight and shade over their extensive nests in the ground by pruning of the surrounding trees. Hence, these ants to some extent determine levels of plant growth in neotropical forests. Several studies have also shown how species of ant closely associated with a particular species of tree will prune other trees or destroy seedlings growing in the vicinity of their host tree (eg *Pseudomyrmex* with *Acacia*, Janzen 1967).

(ii) Control of other organisms by predation

Most ants are predominantly predatory and although many are generalist predators some are more specialised. Below are listed a few examples of such specialisms (B. Bolton, personal communication):

a) *Amblyopone*, *Leptanilla* - geophilomorph centipedes,
b) *Leptogenys* - Dermaptera, land isopods, termites, flat millipedes,
c) *Discothyrea*, *Proceratium* - arthropod eggs in leaf litter,
d) *Plectroctena*, *Gnamptogenys* - Diplopoda (eggs and adults),
e) Dacetini - mostly Collembola,
f) Many ants (e.g. *Dorylus*, *Centromyrmex*, *Decamorium*, *Carebara*) - termites,
g) *Cerpachys*, some *Tetramorium* - other ants,
h) *Myopopone*, *Aphaenogaster* - beetle larvae,
i) *Nomamyrmex* - leaf-cutter ants,
j) *Psalidomyrmex* - earthworms.

Insects which have been parasitised by other insects are often more prone to attack by ants as they sometimes lose some of their mobility and defensive escape mechanisms.

(iii) Plant adaptations

The benefits of reduced herbivory from hosting large populations of predatory ants, have resulted in many plants evolving morphological and physiological adaptations to promote the establishment of ant populations. Some have developed extra-floral nectaries which produce a sugary sap on which the ants feed. Removal of the nectaries results in marked increases in herbivory (see Hölldobler & Wilson 1990). In one study 60% of tree species in clearings and 30-40% of tree species in canopy samples showed mutualistic ant-plant associations (McKey 1992). Some plants produce structures called domatia which serve to protect ant colonies. These structures include hollow thorns, stems, or expanded leaf bases. Some domatia provide food bodies on which the ants feed (Hölldobler & Wilson 1990). One of the best known ant-plant associations is that between *Cercropia* trees and *Azteca* ants (see Davidson & Fisher 1991, Longino 1989). The ants vigorously defend the trees from herbivores, leaf-cutter ants and even vines. In return the plant provides shelter in the form of hollow stems and nutrition from food bodies at the bases of leaf petioles. If a *Cercropia* stem is cut, ants pour out and aggressively attack the damage-causing organisms. One study showed that young *Cercropia* trees have a reduced rate of survival when their *Azteca* colonies are removed (Schupp 1986).

One of the most unusual forms of domatia is that of the hugely expanded stems of epiphytic plants of the *Myrmecodia* (Rubiaceae). These domatia provide a honeycomb-like nest for the ant *Iridomyrmex cordatus* (Huxley 1978, 1980).

(iv) Seed dispersal

Ants play an important role in seed dispersal and seed growth in tropical forests, perhaps more so than any other group of animals. Many species of plants produce seeds bearing chemical attractants for ants and appendages which ease the task of carrying them (Beattie 1985). Ants harvesting seed falls do not manage to successfully carry all seeds back to their nests. Some seeds are dropped on the way, thus aiding seed dispersal. Some ants also provide a service to seedling growth by carrying seeds underground into the nutrient rich environment of ant nests. In this way they protect the seeds from predators and from the possible hazards of fire in dry forests (Hölldobler & Wilson 1990).

(v) Ants as a food source/centres of communities

With ants comprising such a high proportion of insect biomass in most tropical forests they naturally provide a rich food source for other animals. Well known examples of vertebrates associated with ants are many species of birds. Some 50 species of neotropical birds feed on ants or are associated with species of ants. Many of these prey on the insects flushed out by army ant colonies. Some parasitic species of insects, such as phorid flies, specialise on ants and have a rich fauna (Disney 1990). Many other groups of arthropods (including spiders, beetles, mites, flies, isopods, Collembola, Orthoptera, mites, bugs, and Lepidoptera) take advantage of the large food resource that ant nests represent and Hölldobler & Wilson (1990) list some of these and their relationships with ants. In terms of species, beetles predominate, particularly staphylinoid beetles. Some use the ant brood as a food source, many species developing structures and chemical secretions mimicking those of the ants to do so. The evolution of species rich groups, such as Paussidae and Pselaphidae (Coleoptera) and Lycaenidae (Lepidoptera), are strongly linked with those of the ants on which they depend (e.g. Pierce 1983, 1987). Many species of arthropods utilise the rich nutrient source of ant waste products.

Many plants also utilise the nutrient rich environment of ant nests for development, notably those species that develop in so-called ant gardens (Ule 1902). Seeds of epiphytes brought in by the ants develop in the ant carton nests which they build above ground in trees or bushes.

(vi) Nutrient cycling

In most tropical forests ants play an important role in soil creation and turnover, although probably to a lesser extent than termites, except perhaps in some neotropical forests. Here, leaf-cutter ants, *Atta* species, create huge nests many metres in diameter and depth. These ants carry leaves underground where they are decomposed by fungi cultured by the ants in specialised chambers. The ants feed on the fungi rather than the leaves. Thus the ants are instrumental in the decomposition process. The by-products from the decomposition of leaves are stored in dumping chambers and contribute considerably to nutrient cycling in neotropical forests.

The underground chambers and galleries of ants probably play an important but unmeasured role in the passage of rain through the soil, hence reducing water-logging, subsequent run-off, and soil erosion.

The contribution of termites to tropical forest dynamics

(1) Termite abundance, diversity and biomass

Termites are far less visible than ants and it is only through their nests and galleries that most people view their contribution to tropical forests. However, although not particularly speciose, they are one of the most abundant groups of organisms in this environment (Table 4). Estimating abundance and biomass is not easy and hence some of the figures in Table 4 may be gross underestimates. In dry tropical forests termites represent as much as 65% of the soil faunal biomass (Goffinet 1973). Some nests or galleries reach many metres into the soil and during the time taken to reach them by digging, many of the termites have long since disappeared. Similarly estimating termite abundance in dead wood is difficult. Numbers of species of termites recorded from tropical forest sites range from 25 to 100 (Constantino 1992, Eggleton *et al.*, in press, Eggleton & Bignell, in press, Wood *et al.* 1982).

(2) Patterns of distribution

Termites have a tropical/sub tropical distribution and few are found in temperate zones. Most termites are found under ground, in fallen timber on the ground, or in the lower parts

Table 4. Abundance and biomass of termites in tropical forests (after Wood *et al.* 1982).

Locality	Numbers m^2	Biomass g m^{-2}	Author
W. Africa			
Riparian forest, Rabba, Nigeria	2646	6.9	Wood *et al.* 1982
Semi-decid. forests, Ibadan, Nigeria	3163	8.0	Wood *et. al.* 1982
Riparian forest, Yangambi, Zaire	1000	11.0	Maldague 1964
S. & C. America			
Rain forest, Trinidad, W. Indies	4450		Strickland 1944
S.E. Asia			
Rain forest, Pasoh, Malaya	3280	10.9	Abe 1975
Sarawak:			
Lowland rain forest			
Mixed dipterocarp forest	1125		Collins 1979
Alluvial	254		Collins 1979
Kerangas	1405		Collins 1979
Lower montane forest	38		Collins 1980
Upper montane (1310-1860 m)	99-295		Collins 1980
Upper montane (1970-2376 m)	0		Collins 1980

of trees. Few are truly arboreal. They form large or small nests above or below the ground and some are connected to each other or to food sources by galleries in the soil or in dead and living wood. As Collins (1979, 1980) has shown, termites are predominantly a lowland forest phenomenon, their abundance and diversity decreasing with altitude.

(3) Contribution to forest dynamics

The role of termites in tropical forests is discussed in some depth by several authors (Lee 1983, Lee & Wood 1971, Wood 1988) and a summary is provided here.

(i) The decomposition process and nutrient cycling

Termites are an essential component of many tropical forests representing one of the major groups concerned with the decomposition of plant tissues aided by symbiotic microflora and microfauna (Collins 1981, Coventry *et al.* 1988). In breaking down these tissues they play a key role in nutrient cycling and hence maintenance of forest stability (Lee 1983, Lee & Wood 1971).

(ii) Soil generation, turnover and fertility

The construction of vast termite nests and complex galleries involves the movement of enormous quantities of soil and leaf litter (Collins 1983). Such termite constructions are composed of a mixture of termite faeces, saliva and soil particles of a particular size range. Thus the texture of the soil is affected by termites. Termitaria and galleries provide a rich source of nutrients for plant growth. C. Martius (personal communication) has shown that old arboreal termitaria falling to the ground provide nutrients for trees in varzea forest of Amazonia. In some dry tropical forests termites are responsible for the deposition of large amounts of soil on the surface over long periods of time (Bagine 1984).

(iii) Termitophilous communities

Many other organisms, particularly insects, live with termites. These associations are less well documented than those between ants and other organisms.

(iv) Termites as a food source

Because termites comprise such a large portion of the standing faunal biomass they offer an important food source to other organisms such as insects, birds and mammals. In E. Africa, even the mounds of termites are eaten by primates (G. Davies, personal communication).

(v) Gap creation and cycles of tree growth

Many tree falls are the result of termite attack and thus termites have a crucial role in gap creation and cycles of tree growth. Precisely how important this role is uncertain.

(vii) Hydrology

The galleries and chambers constructed by termites offer ready made channels for the through flow of rain and therefore termites, as many other insects, are important in the hydrology of tropical forests. Elkins *et al.* (1986) found that low vegetation cover areas treated with insecticide had significantly lower water infiltration rates than untreated areas. They suggested that the galleries of termites enhance water infiltration and, through reduced runoff, aid retention of topsoil.

(vi) The global carbon cycle

We are only just beginning to recognise the important role that termites may play in the global carbon cycle (Higashi *et al.* 1992, Jones 1990). Termites are believed to produce up to twice the global CO_2 generated by fossil fuel combustion and up to 40% of the global atmospheric CH_4 released each year (see Jones 1990, and references therein).

In the sections above we can see how at least two groups of insects, ants and termites, are essential to tropical forest dynamics. But what of other groups? The Sulawesi and Seram studies described above and Table 1 show that other arthropod groups such as Collembola, Acarina, Coleoptera, parasitic Hymenoptera, Lepidoptera, Diptera, and Hemiptera, are all either very abundant, have great biomass or are hyper-diverse groups with 10 times or more species per unit area than either termites or ants. The roles of some of these groups in tropical forest dynamics have received much less attention than ants and termites and are far less clear. One can see how the loss of some species of ants and termites could affect an ecosystem, but the effects of loss of many species of beetle, moth or fly are less apparent. The chains of interactions of these organisms with each other and with plants, fungi and microorganisms are extremely complex. Few such food webs have been examined for tropical forest systems and as a result the ecosystem function and value of many species of invertebrates are unclear. Separate studies of the biology of individual species or groups of species indicate that many are surprisingly essential for the survival of other organisms. However, there has been a real bias towards studies of the interactions of insects and living plants and of the processes of herbivory and insect parasitism. Far less attention has been paid towards insects and fungi and the decomposition process, and towards the role of animals in nutrient cycling.

THE WAY FORWARD

The purpose of this paper has been to highlight the importance of the faunal component and, in particular, the arthropod component of tropical forest dynamics and to clarify those areas where new research will provide a better understanding. Why we need this better understanding is discussed in the conclusions.

That some groups of organisms have received a disproportionate amount of attention or conversely, lack of attention, is clear. May (1988) showed that the number of scientific papers published per species for mammals, for example, is 100 times more than that for insects and 1000 times that for nematodes. The possible reasons for that are several. First, humans recognise features in mammals and birds that are also found in man. Human attributes, such as eyes, nose, mouth and ears, or gentleness, anger and kindness, are not always quite so readily identifiable in nematodes, mites, springtails or slugs! Second, some groups, such as butterflies, giant pandas and so on, appear to be more 'attractive' to the human eye than others (see Kellert 1993, Kim 1993). Third, we readily see and identify only a small size range of organisms. Many invertebrates and all microorganisms are beyond the vision of the human eye and optical aids are necessary for their study. Fourth, groups that are taxonomically intractable and species rich have been poorly examined or ignored.

The vast numbers of individuals and species of insects and other invertebrates, and the lack of synthesising taxonomic studies deter many from the study of these organisms. Finally, the important contribution that many of these groups of smaller organisms make to ecosystem function is not known or is poorly publicised within the scientific and public communities. The continuing disproportionate distribution of scientific resources maintains our level of ignorance; a greater proportion of resources needs to be applied to studies of microorganisms and invertebrates and their role is ecosystems.

These points are not simply of academic interest but show how, as often as not, selection of groups for study is not based on sound scientific criteria. The same is also true of what we know about the distribution and biology of those organisms that have received attention. The goal of most early naturalists and taxonomists was to maximise the richness of their collections. Far less attention was paid to the examination of geographical or biological patterns. Thus much of the data provided by museum specimens, although important, is difficult to use until such biases have been determined (Stork 1994).

It is clear from examination of present rates of description of species (circa 10,000 per year) and new estimates of 5-15 million species on earth, that unless remarkably fundamental changes occur in the way we describe species, the task is not going to be completed within the foreseeable future. Given the international concern for the environment, as exemplified the Biodiversity Convention and other international agreements signed at the United Nations Conference on the Environment and Development in Rio de Janeiro in 1992, drastic scientific measures are needed to assess the extent of biodiversity and the threats to it. Measures to inventory and monitor changes to the world's fauna and flora are, and have been, considered by international and national organisations. Some aspects are discussed briefly below, and in more detail elsewhere (di Castri *et al.* 1992, Stork 1994).

1) *Selection of indicator groups and key species based on sound rather than historical criteria*: It is impossible to fully inventory all organisms in the foreseeable future even with a massive injection of resources and yet it is important to have a better understanding of the distribution of life on Earth both for purely scientific reasons and for practical reasons concerned with the conservation of biodiversity. Similarly, it is impossible to monitor the effect of environmental change on all organisms. It is therefore necessary to select species and groups of species that act as indicators (*sensu lato*) for a wider group of organisms. In this respect no group can truly be an indicator for others as all species and groups of species have their own unique characteristics, but in practice compromises have to be made. A broad and biologically representative range of indicator groups need to be studied, their selection being based on sound ecological criteria such as species richness, body size, trophic group, and geographical distribution. There is a particularly urgent need for indicators in tropical regions and the proposed 'Tropical Areas Bioindicators System' (TABS)(Hawksworth & Ritchie, in press) should provide a method for early detection of environmental threats to forest disturbance as well as air pollution and water pollution in tropical forest ecosystems.

2) *Preparation of manuals/handbooks and 'sampling packages' for indicator groups*: Sampling of indicator groups needs to be evaluated and standardised to allow comparison between sites (e.g. see Hammond 1990). Where a range of indicator groups are to be examined, then the use of standard 'sampling packages' will be useful (Gadagkar *et al.* 1990, Stork 1994). The handbook for inventorying and monitoring changes in amphibian populations and species distributions is perhaps a model for other groups (Heyer *et al.* 1994).

3) *Selection of intensive and extensive study sites*: No single tropical or temperate site in the world has been *completely* inventoried. An intensive study of all the taxa at a single site - a so-called 'all-taxa biological inventory' (Janzen, personal communication) - could provide measured standards for less intensive studies at other sites, and a better understanding of ecosystem function (di Castri & Younès 1990). A full inventory of a single site using standardised sampling methods raises many experimental and procedural problems, such as

the problem of scale (e.g. compare the temporal and spatial problems of sampling bacteria and vertebrates). One good example of the benefits of an intensive study of a single site is the result of the long-term sampling of the components of the beech woodland soil fauna in Germany by Schaefer & Schauermann (1990). Through their studies they were able to give a complete or near-complete analysis of the trophic and energetic systems of the soil.

At a large number of 'extensive' study sites the species richness and abundance of species of the selected indicator groups should be determined using standardised sampling methods and sampling packages. These data could be used to determine local, regional and global patterns in diversity and to provide biological measures of the consequences of environmental change.

4) *Improvement of taxonomic and ecological resources in developing countries through training and development of regional and country centres*: The global species inventory task so far has largely failed because it has relied almost entirely on human resources in the developed world. If this process is to change, then the taxonomic community needs to be strengthened in developing countries. INBio in Costa Rica is seen by some as one model for developing countries (see Gamez 1991).

5) *Integration of international and national programmes for inventories and monitoring change*: Much of the funding for environmental and projects has been directed towards single countries, either through bilateral agreements or through international organisations. There has been a lack of coordination of research *between* countries with resultant duplication of effort and lack of information flow. There also has been no recognition of the need for funding of international programmes of research.

6) *Improve availability of information through networking*: There have been enormous advances in information technology such that networking and exchange of data has been made much easier and relatively cheap. Development of such resources in tropical countries will enhance the ability of tropical ecologists and taxonomists to make maximum use of data.

CONCLUDING REMARKS

In a recent editorial André *et al.* (1992) made a strong plea for more 'biodiversity data' particularly with respect to the African continent. I strongly endorse this call but with a warning. If we are to form a better understanding of the patterns of distribution of life on earth, the value of individual faunal and floral components to ecosystem function, and most important of all, develop methods of truly sustainable use of the environment, then we need to examine carefully what we really know about life on earth. We need to determine the biases and correct for them in future research programmes. Undoubtedly, one of the results such actions will be a better understanding of the faunal components of tropical forest dynamics.

ACKNOWLEDGEMENTS

I am grateful to the organisers of the Brunei Rain Forest Symposium for the opportunity to present this paper. I thank Victoria Best for assistance in compiling Table 1 and Martin Brendell and Victora for preparing the figures. I am particularly grateful to my colleagues Barry Bolton, Gaden Robinson and Dick Vane-Wright for provision of unpublished data. Constructive criticisms of the manuscript were provided by Barry Bolton, Paul Eggleton and Kevin Gaston.

LITERATURE CITED

ABE, T. 1975. *Studies on the distribution and ecological role of termites in a lowland rain forest of West Malaysia.* Doctoral Thesis, Kyoto University, Japan.

ADIS, J. 1988. On the abundance and density of terrestrial arthropods in central amazonian dryland forests. *Journal of Tropical Ecology* 4:19-24.

ADIS, J., LUBIN, Y. D. & MONTGOMERY, G. G. 1984. Arthropods from the canopy of inundated and terra firme forests near Manaus, Brazil, with critical considerations of the Pyrethrum-fogging technique. *Studies of the Neotropical Fauna & Environment* 19:223-236.

ADIS, J. & SCHUBART, H. O. R. 1985. Ecological research on arthropods in central amazonian forest ecosystems with recommendations for study procedures. Pp. 111-144 in Cooley, J.H. & Golley, F.B. (eds). *Trends in Ecological Research for the 1980's.* Nato Conference Series, Series 1: Ecology. Plenum Press, New York, London.

ANDRÉ, H.M., LEBRUN, P. & NOTI, M. 1992. Editorial. Biodiversity in Africa: a plea for more data. *Journal of African Zoology* 106:3-15.

ANDRÉ, H.M., LEBRUN, P. & NOTI, M. 1994. The soil fauna: the other last biotic frontier. *Biodiversity and Conservation* 3: 45-56.

BAGINE, R.K.N. 1984. Soil translocation by termites of the genus *Odontotermes* (Holmgren) (Isoptera: Macrotermitinae) in an arid area of Northern Kenya. *Oecologia (Berlin)* 64:263-266.

BARLOW, H.S. & WOIWOOD, I.P. 1990. Seasonality and diversity of macrolepidoptera in two lowland sites in the Dumoga-Bone National Park, Sulawesi Utara. Pp. 167-172 in Knight, W.J. & Holloway, J.D. (eds). *Insects and the Rain Forests of South East Asia (Wallacea).* Royal Entomological Society of London, United Kingdom.

BEATTIE, A.J. 1985. *The Evolutionary Ecology of Ant-Plant Mutualisms.* Cambridge University Press, New York.

BROKAW, N.V.L. 1984. Treefalls, regrowth, and community structure in tropical forests. In Pickett, S.T.A. & White, P.S. (eds). *Natural disturbance: the patch dynamics perspective.* Academic Press, New York.

CASSON, D.S. & HODKINSON, I.D. 1991. The Hemiptera (Insecta) communities of tropical rain forest in Sualwesi. *Zoological Journal of the Linnean Society* 102:253-275.

di CASTRI, F. 1991. Ecosystem Evolution and Global Change. Pp. 189-217 in Solbrig, O.T. & Nicolis G. (eds.). *Perspectives on Biological Complexity.* IUBS Monograph Series No 6, Paris.

di CASTRI, F. & YOUNèS, T. 1990. *Ecosystem Function of Biological Diversity.* Biology International, Special Issue 22.

di CASTRI, F., ROBERTSON VERNHES, J. & YOUNèS, T. 1992. *Inventorying and Monitoring Biodiversity.* Biology International, Special Issue 27, 28 pages.

CHERRETT, J.M. 1986. History of the leaf-cutting ant problem. Pp. 10-17 in Lofgren, C.S. & Van der Meer, R.K. (eds). *Fire Ants and Leaf-Cutting Ants; Biology and Management.* Westview Press, Boulder, Colorado.

CHERRETT, J.M. 1989. Leaf-cutting ants. Pp. 473-488 in Lieth, H. & Werger, M.J.A. (eds). *Tropical Rain Forest Ecosystems: Biogeographical and Ecological Studies.* Ecosystems of the World 14B. Elsevier, Amsterdam.

CHIVERS, D.J. & DAVIS, G. 1979. Abundance of primates in the Krau Game Reserve, Peninsular Malaysia. Pp. 9-32 in Marshall, A.G. (ed.). *Transactions of the Sixth Aberdeen-Hull Symposium on Malesian Ecology.* Dept. of Geography, University of Hull.

COLLINS, N.M. 1979. Comparison of the soil macrofauna of three lowland forest types. *Sarawak Museum Journal* 4, 27:267-282.

COLLINS, N.M. 1980. The distribution of the soil macrofauna on the west ridge of Gunung (Mount) Mulu, Sarawak. *Oecologia (Berlin)* 44:263-275.

COLLINS, N.M. 1981. The role of termites in the decomposition of wood and leaf litter in the southern Guinea savanna of Nigeria. *Oecologia (Berlin)* 51:389-399.

COLLINS, N.M. 1983. Termite populations and their role in litter removal in Malaysian rain forests. Pp. 311-325 in Sutton, S.L., Whitmore, T.C. & Chadwick A.C. (eds). *Tropical Rainforest: Ecology and Management.* Blackwell Scientific Publications, Oxford.

COMMON, I.F.B. 1990. *Moths of Australia.* Melbourne. vi + 535pp.

CONSTANTINO, R. 1992. Abundance and diversity of termites (Insecta: Isoptera) in two sites of primary rain forest in Brazilian Amazonia. *Biotropica* 24:420-430.

CORNELL, H.V. & LAWTON, J.H. 1992. Species interactions, local and regional processes, and limits to the richness of ecological communities: a theoretical perspective. *Journal of Animal Ecology* 61:1-12.

COVENTRY, R.J., HOLT, J.A. & SINCLAIR, F. 1988. Nutrient cycling by mound building termites in low fertility soils of semi-arid tropical Australia. *Australian Journal of Soil Research* 26:375-390.

CURTIN, S.H. & CHIVERS, D.J. 1978. Leaf-eating primates of Peninsular Malaysia. Pp. 441-464 in Montgomery, G.G. (ed.). *Ecology of Arboreal Folivores.* Smithsonian Institution, Washington D.C..

DALLMEIER, F. 1992. *Long-term monitoring of biological diversity in tropical forest areas: methods for establishment and inventory of permanent plots.* MAB Digest 11. UNESCO, Paris. 72pp.

DAVIDSON, D.W. & FISHER, B.L. 1991. Symbiosis of ants with *Cercropia* as a function of light regime. Pp. 289-309 in Huxley, C.R. & Cutler, D.F. (eds). *Ant-Plant Interactions*. Oxford University Press, Oxford.

DAVIDSON, G.W.H. 1991. Birds. Pp. 135-142 in Kiew, R. (ed.). *The State of Nature Conservation in Malaysia*. Malayan Nature Society, Malaysia.

DEANGELIS, D.L. 1992. *Dynamics of nutrient cycling and food webs*. Population and community biology series, 9. Chapman & Hall, London, 270pp.

DELCOURT, H.R. & DELCOURT, P.A. 1987. *Quaternary Ecology. A Paleoecological Perspective*. Chapman & Hall, London. 242pp.

DISNEY, R.H.L. 1990. Scuttle flies (Diptera: Phoridae) associated with ants and termites in Toraut Forest, Sulawesi. Pp. 301-304 in Knight, W.J. & Holloway, J.D. (eds). *Insects and the Rain Forests of South East Asia (Wallacea)*. Royal Entomological Society of London, United Kingdom.

EGGLETON, P. & BIGNELL, D.E. in press. Monitoring the response of tropical insects to changes in the environment: troubles with termites. In Harrington, R. & Stork, N.E. (eds). *Insects in a changing environment*. Academic Press, London.

EGGLETON, P., BIGNELL, D.E., SANDS, W.I., WAITE, B., WOOD, T.G. & LAWTON, J.H. in press. The species richness of termites (Isoptera) under differing levels of forest disturbance in the Mbalmayo Forest Reserve, southern Cameroon. *Journal of Tropical Ecology*.

ELKINS, N.Z., SABOL, G.V. WARD, T.J. & WHITFORD, W.G. 1986. The influence of subterranean termites on the hydrological characteristics of a Chihuahuan desert ecosystem. *Oecologia (Berlin)* 68:521-528.

ERWIN, T.L. 1988. Canopy arthropod biodiversity: a chronology of sampling techniques and results. *Revista Reruana de Entomologia* 32:71-77.

FITTKAU, E.J. & KLINGE, H. 1973. On biomass and trophic structure of the central Amazonian rain forest ecosystem. *Biotropica* 5:2-14.

FOSTER, R.B. & HUBBELL, S.P. 1990. The floristic composition of the Barro Colorado Island Forest. Pp. 85-98 in Gentry, A.H. (ed.). *Four Neotropical Forests*. Yale University Press.

FURLEY, P.A., PROCTOR, J. & RATTER, J.A. (eds) 1992. *Nature and Dynamics of Forest-Savanna Boundaries*. Chapman & Hall, London.

GADAGKAR, R., CHANDRASHEKARA, K. & NAIR, P. 1990. Insect species diversity in the tropics: sampling methods and a case study. *Journal of the Bombay Natural History Society* 87:337-53.

GAMEZ, R. 1991. Biodiversity conservation through facilitation of its sustainable use: Costa Rica's National Biodiversity Institute. *Trends in Ecology and Evolution* 6:377-8.

GASTON, K.J. 1991. The magnitude of global insect species richness. *Conservation Biology* 5:283-296.

GASTON, K.J. & MAY, R.M. 1992. Taxonomy of taxonomists. *Nature* 356:281-282.

GOFFINET, G. 1973. Contribution a l'étude de l'écosystème forêt claire (miombo). Note 11. Etude comparative des effectifs de quelques groupes arthropodiens du sol intercalique de quatre biotopes Katangais (Zaire). *Annuaire Universitaire Abidjan Série E* 6:251-256.

GRASSLE, J.F. & MACIOLEK, N.J. 1992. Deep-sea richness: regional and local diversity estimates from quantitative bottom samples. *American Naturalist* 139:313-41.

GRUBB, P.J. 1977. The maintenance of species-richness in plant communities: the importance of the regeneration niche. *Biological Reviews* 52:107-145.

HAMMOND, P. 1990. Insect abundance and diversity in the Dumoga-Bone National Park, N. Sulawesi, with special reference to the beetle fauna of lowland rain forest in the Toraut region. Pp. 197-254 in Knight, W.J. & Holloway, J.D. (eds). *Insects and the Rain Forests of South East Asia (Wallacea)*. Royal Entomological Society of London, United Kingdom.

HAMMOND, P.M. 1992. Species Inventory. Pp. 17-39 in Groombridge, B. (ed.). *Global Diversity. Status of the Earth's Living Resources*. World Conservation Monitoring Centre, Chapman & Hall, London.

HAMMOND, P.M. in press. Practical approaches to the estimation of the extent of biodiversity in speciose groups. *Philosophical Transactions of the Royal society, Series B*.

HAMMOND, P.M. in press. Described and estimated species numbers: an objective assessment of current knowledge. *Philosophical Transactions of the Royal society, Series B*.

HAWKSWORTH, D.L. & RITCHIE, M. in press. *Tropical Areas Bioindicators System (TABS)*. CAB International, Wallingford.

HEYER, W.R., DONELLY, M.A., McDIARMID, R.W., HAYEK, L.-A.C. & FOSTER, M.S. (eds) 1994. *Measuring and monitoring biological diversity. Standard methods for amphibians*. Smithsonian Institution, Washington D.C. 364pp.

HIGASHI, M., ABE, T. & BURNS, T.P. 1992. Carbon-nitrogen balance and termite ecology. *Proceedings of the Royal Society of London, Series Biological Sciences* 249:303-308.

HODDA, M., BLOEMERS, G.F., WANLESS, F.R. & LAMBSHEAD, P.J.D. in press. Nematodes and carbon flux in a tropical forest in Cameroon: preliminary results. In Nilsson, L.-O. & Mathy, P. (eds). *Nutrient Uptake and Cycling in Forest Ecosystems*. C.E.C., Brussels.

HODKINSON, I.D. & CASSON, D. 1987. A survey of the food plant utilization by Hemiptera (Insecta) in the understorey of primary lowland rain forest in Sulawesi, Indonesia. *Journal of Tropical Ecology* 3:75-85.

HODKINSON, I.D. & CASSON, D. 1991. A lesser predilection for bugs: Hemiptera (Insecta) diversity in tropical rain forests. *Biological Journal of the Linnean Society* 43:101-109.

HÖLLDOBLER, B. 1983. Territorial behaviour in the green tree ant (*Oecophylla smaragdina*). *Biotropica* 15:241-250.

HÖLLDOBLER, B, & WILSON, E.O. 1990. *The ants*. Belknap Press of Harvard University Press, Cambridge, Massachusetts.

HOLLOWAY, J.D., ROBINSON, G.S. & TUCK, K.R. 1990. Zonation in the Lepidoptera of northern Sulawesi. Pp. 153-166 in Knight, W.J. & Holloway, J.D. (eds). *Insects and the Rain Forests of South East Asia (Wallacea)*. Royal Entomological Society of London, United Kingdom.

HOLLOWAY, J.D. & STORK, N.E. 1991. The dimensions of biodiversity: the use of invertebrates as indicators of human impact. Pp. 37-62 in Hawksworth, D.L. (ed.). *The Biodiversity of Microorganisms and Invertebrates: Its role in Sustainable Agriculture*. CAB International, Wallingford.

HUXLEY, C.R. 1978. The ant-plants *Myrmecodia* and *Hydnophytum* (Rubiaceae), and the relationships between their morphology, ant occupants, physiology and ecology. *New Phytologist* 80:231-268.

HUXLEY, C.R. 1980. Symbiosis between ants and epiphytes. *Biological Reviews* 55:321-340.

INGER, R.F. 1979. Abundances of amphibians and reptiles in tropical forests of South-East Asia. Pp. 93-110 in Marshall, A.G. (ed.). *Transactions of the Sixth Aberdeen-Hull Symposium on Malesian Ecology*. Department of Geography, University of Hull.

IUCN, UNEP & WRI 1990. *The Biodiversity Conservation Strategy Programme, Mobilizing Worldwide action to Sustain the Living Resources of our Planet*. Gland, Switzerland.

JANZEN, D.H. 1967. Interaction of the bull's horn acacia (*Acacia cornigera* L.) with an ant inhabitant (*Pseudomyrmex ferruginea* F. Smith) in eastern Mexico. *University of Kansas Science Bulletin* 47: 315-558.

JOHNS, A.D. 1992. Vertebrate responses to selective logging: implications for the design of logging systems. *Philosophical Transactions: Biological Sciences of the Royal Society of London* 335(1275):443-457.

JONES, J.A. 1990. Termites, soil fertility and carbon cycling in dry tropical Africa: a hypothesis. *Journal of Tropical Ecology* 6:291-305.

KELLERT, S.R. 1993. Values and perspections of invertebrates. *Conservation Biology* 7:845-855.

KIM, K. C. 1993. Biodiversity, conservation and inventory: why insects matter. *Biodiversity and Conservation* 2:191-214.

KNIGHT, W. & HOLLOWAY, J.D. (eds) 1990. *Insects and the Rain Forests of South East Asia (Wallacea)*. Royal Entomological Society of London, United Kingdom.

LEE, K.E. 1983. The influence of earthworms and termites on soil nitrogen cycling. Pp. 35-48 in Lebrun, P., André, H.M., de Medts, A., Gregoire-Wibo, C. & Wauthy, G. (eds). *New Trends in Soil Biology*. Dieu-Brichart, Louvain-la-Neuve.

LEE, K.E. & WOOD, T. G. 1971. *Termites and Soil*. Academic Press, London.

LESTON, D. 1973a. Ecological consequences of the tropical ant mosaic. *Proceedings of the Seventh International Congress of the International Union for the Study of Social Insects, London* 19:311-341.

LESTON, D. 1973b. Ants and tropical tree crops. *Proceedings of the Royal Entomological Society of London, series C* 38:1.

LONGINO, J.T. 1989. Geographic variation and community structure in an ant-plant mutualism: *Azteca* and *Cercropia* in Costa Rica. *Biotropica* 21:126-132.

MAJER, J.D. 1976a. The maintenance of the ant mosaic in Ghana cocoa farms. *Journal of Applied Ecology* 13:123-144.

MAJER, J.D. 1976b. The ant mosaic in Ghana cocoa farms: further structural considerations. *Journal of Applied Ecology* 13:145-155.

MAJER, J.D. 1976c. The influence of ants and ant manipulation on the cocoa farm fauna. *Journal of Applied Ecology* 13:157-175.

MAJER, J.D. 1993. Comparison of the arboreal ant mosaic in Ghana, Brazil, Papua New Guinea and Australia - its structure and influence on arthropod diversity. Pp. 115-141 in LaSalle, J. & Gauld, I.D. (eds). *Hymenoptera and Biodiversity*. CAB International, Wallingford, United Kingdom.

MALDAGUE, M.E. 1964. Importance des populations de termites dans les sols equitoriaux. *Transactions of the 8th International Congress of Soil Science, Bucharest* 3:743-751.

MAY, R.M. 1988. How many species are there on earth? *Science* 241:1441-9.

MAY, R.M. 1992. How many species inhabit the earth? *Scientific American*, October, 18-24.

McKEY, D. 1992. Interactions between ants and plants: comparison of canopy, understorey and clearing environments. In: *Biologie d'une Canopée de fôret Équatoriale - II Rapport de Mission: Radeau des Cimes Octobre Novembre 1991, Réserve de Campo, Cameroun*.

MEDWAY, Lord, 1983. *The Wild Animals of Malaya (Peninsular Malysia and Singapore*. 2nd ed. Oxford Univ.Press,Kuala Lumpur.

MEMMOTT, J., GODFRAY, H.C.J. & BOLTON, B. 1993. Predation and parasitism in a tropical herbivore community. *Ecological Entomology* 18:348-352.

NOYES, J.S. 1989a. The diversity of Hymenoptera in the tropics with special reference to Parasitica in Sulawesi. *Ecological Entomology* 14:197-207.

NOYES, J.S. 1989b. A study of five methods of sampling Hymenoptera (Insecta) in a tropical rainforest, with special reference to the Parasitica. *Journal of Natural History* 23:285-298.

PARKER, S.P. (ed.) 1982. *Synopsis and classification of living organisms*. 2 Volumes. McGraw-Hill, New York.

PAYNE, J.B. 1979. Abundance of Diurnal Squirrels at the Kuala Lompat Post of the Krau Game Reserve, Pahang, Peninsular Malaysia. Pp. 37-53 in Marshall, A.G. (ed.). *Transactions of the Sixth Aberdeen-Hull Symposium on Malesian Ecology*. Department of Geography, University of Hull.

PAYNE, J. & MAHEDI, A. 1991. Large Mammals in Sabah. Pp. 177-183 in Kiew, R. (ed.). *The State of Nature Conservation in Malaysia*. Malayan Nature Society, Malaysia.

PIERCE, N. E. 1983. *The ecology and evolution of symbioses between lycaenid butterflies and ants*. Ph.D. dissertation. Harvard University x + 274pp.

PIERCE, N. E. 1987. The evolution and biogeography of associations between lycaenid butterflies and ants. Pp. 89-116 in Harvey, P.H. & Partridge, L. (eds). *Oxford surveys in evolutionary biology, volume 4*. Oxford University Press, Oxford.

ROGERS, L.E., HINDS, W. T. & BUSCHBOM, R.L. 1976. A general weight vs length relationships for insects. *Annals of the Entomological Society of America* 69:387-389.

ROOM, P.M. 1971. The relative distributions of ant species in Ghana's cocoa farms. *Journal of Animal Ecology* 40:735-751.

SCHAEFER, M. & SCHAUERMANN, J. 1990. The soil fauna of beech forests: comparison between a mull and a moder soil. *Pedobiologia* 34:299-314.

SCHUPP, E.W. 1986. *Azteca* protection of *Cercropia:* ant occupation benefits juvenile trees. *Oecologia (Berlin)* 70:379-385.

STORK, N.E. 1988. Insect diversity: facts, fiction and speculation. *Biological Journal of the Linnean Society* 35:321-337.

STORK, N.E. 1991. The composition of the arthropod fauna of Bornean lowland forest trees. *Journal of Tropical Ecology* 7:161-80.

STORK, N.E. 1993. How many species are there? *Biodiversity and Conservation* 2: 215-232.

STORK, N.E. 1994. Inventories of biodiversity: more than a question of numbers. Pp. 81-100 in Forey, P., Humphries, C.J. & Vane-Wright, R.I. (eds). *Systematics and Conservation Evaluation*. Systematics Association Special Volume no 50, Oxford University Press.

STORK, N.E. & BLACKBURN, T.M. 1993. Abundance, body size and biomass of arthropods in tropical forest. *Oikos* 67: 483-489.

STORK, N.E. & BRENDELL, M.J.D. 1990. Variation in the insect fauna of Sulawesi trees with season, altitude and forest type. Pp. 173-190 in Knight, W.J. & Holloway, J.D. (eds). *Insects and the rain forests of South East Asia (Wallacea)*. Royal Entomological Society of London, London.

STORK, N.E. & BRENDELL, M.J.D. 1993. Arthropod diversity studies in lowland rainforest of Seram, Indonesia. Pp in *The Natural History of Seram*. Edwards, I., Macdonald, A.A. & Proctor, J. (eds). Intercept, Andover.

STRICKLAND, A.H. 1944. The arthropod fauna of some soils. *Tropical Agriculture (University of the West Indies, Trinidad & Tobago)* 21:107-114.

ULE, E. 1902. Ameisengärten im Amazonasgebiet. *Botanische Jahrbucher für Systematik, Pflanzengeschichte und Pflanzengeographien* 30:45-52.

WAY, M.J., CAMMELL, M.E., BOLTON, B. & KANAGARATNAM, P. 1989. Ants (Hymenoptera: Formicidae) as egg predators of coconut pests, especially in relation to biological control of the coconut caterpillar, *Opisina arenosella* Walker (Lepidoptera: Xyloryctidae), in Sri Lanka. *Bulletin of Entomological Research* 79:219-233.

WEAVER, P.L. 1989. Forest changes after hurricanes in Puerto Rico's Luquillo Mountains. *Interciencia* 14:181-192.

WHITMORE, T.C. 1984. *Tropical forests of the Far East*. Clarendon Press, Oxford, 2nd Edition.

WILSON, E.O. 1971. *The Insect Societies*. Belknap Press of Harvard University Press, Cambridge, Massachusetts. 548pp.

WILSON, E.O. 1988. The current state of biological diversity. Pp 3-18 in Wilson, E. O. & Peter, F.M. (eds). *Biodiversity*, National Academy Press, Washington D.C.

WOINARSKI, J.C.Z. & TIDEMANN, S.C. 1991. The bird fauna of a deciduous woodland in the wet-dry tropics of northern Australia. *Wildlife Research* 18:469-500

WOOD, T.G. 1988. Termites and the soil environment. *Biology and Fertility of Soils* 6:228-236.

WOOD, T.G., JOHNSON, R.A., BACCHUS, S., SHITU, M.O., & ANDERSON, J.M. 1982. Abundance and distribution of termites (Isoptera) in a riparian forest in the southern Guinea savanna vegetation zone of Nigeria. *Biotropica* 14:25-39.

The biodiversity of arthropods from Australian rain forest canopies: some results on the role of the tree species

R. L. KITCHING *and* J. ZALUCKI

Australian School of Environmental Studies, Griffith University, Nathan, Brisbane, Qld 4111, Australia

ABSTRACT. Australia's modest tropical rainforest resources have been recognised both by measures taken to conserve them in recent years and by the establishment of a major Cooperative Research Centre to investigate their biodiversity and dynamics to underpin proper management as well as to generate biological understanding. Investigations of the arthropod biodiversity of the canopy strata have been underway for several years. Over the last two years the focus of this work has moved to include sites in the lowland forests of Cape Tribulation. These canopy studies take a 'top-down' approach to biodiversity definition generating, in sequence, ordinal, familial and specific profiles as well as guild structures for particular sites, samples or target trees. Recent work, following earlier hypotheses of Terry Erwin, has included investigations of the tree-species specificity of arthropod assemblages focussing in particular on the Coleoptera. Early results of these studies are presented, in particular relating to the canopy faunas of the very local endemic tree *Noahdendron nicholasii* and the more widespread *Ryparosa javanicum*. Difficulties associated with sampling, analysis and the identification of specificity at the level of the assemblage are discussed. Other work in progress and extension of the tree-specificity work under the aegis of the newly proposed Tropical Rainforest Research Centre is foreshadowed.

KEY WORDS: biodiversity, arthropods, Australia, Coleoptera, canopy sampling, tropical rainforest.

INTRODUCTION

Rainforest covers only 5% of the Australian continent but, in terms of biodiversity, public interest and global importance, this charismatic ecosystem looms much larger in the public consciousness than its area would suggest. The Australian biological community and the political forces at work in the country are keenly aware of the importance of our rainforests and the biological diversity they contain. The recent announcement of federal government support, to the tune of 13 million Australian dollars, for the establishment of a Cooperative Research Centre for Tropical Rainforest Ecology and Management, is perhaps the most tangible evidence of this awareness among scientists and their paymasters. The proposal for and approval of this research centre, scheduled to open in mid-1993, coincides with the development of a National Strategy for Biodiversity Conservation, the draft of which was released in late 1991 and the final version of which is due for final release shortly.

At the Kuala Lumpur conference in 1991 attention was drawn to the rainforest resources of Australia (Kitching 1992). At that stage the Government of Queensland was in dispute with the federal government concerning the future declaration of a substantial area of remaining tropical rainforest area as a World Heritage property. Between the presentation of that paper and the appearance of the printed version there was an unexpected change of government at the state level, the World Heritage Dispute was resolved and, in December 1990, the Wet Tropics of Queensland was inscribed as a World Heritage property. Since then a statutory body, the Wet Tropics Management Authority, has been established with both federal and state support and is now designing and supporting a coordinated programme of research to develop management strategies for the region in line with Australia's responsibilities nationally and internationally. This programme includes an extensive programme of biological inventory across all taxa.

D. S. Edwards et al. (eds.), Tropical Rainforest Research – Current Issues, 21–28.

Canopy Arthropod Studies

It is now generally accepted that the arthropods in the canopies of rainforest trees represent 'the heart of biodiversity' to paraphrase E. O. Wilson. Following pioneering studies by Erwin (1982, 1983) in the Neotropics, Stork and his co-workers carried out seminal studies in Brunei, in Sulawesi and in Seram (Stork 1987a, 1987b, Stork & Brendell, in press). These workers developed pyrethrum-knockdown techniques to obtain large samples from high in the forest canopy. These samples, duly sorted, have revealed a massive speciosity unrivalled in the animal world. Erwin (1982) in particular was so impressed by the results he obtained in Panama that he erected a series of testable hypotheses which, if upheld, lead to upward revision of our estimates of global arthropod biodiversity by more than an order of magnitude.

Parenthetically these new estimates have caused such a furore in some circles as to have engendered a whole secondary industry of comment and re-analysis which continues more than ten years after their original publication (Erwin 1992, Gaston 1991, Hodkinson & Casson 1988, May 1988, Stork 1988). What has happened, but rarely, is any attempt to test the sequential hypotheses on which Erwin's estimates rest. It is into this last partial vacuum that our present contribution falls!

METHODS

Canopy Studies in Australia

Since 1988 we have been using comparable pyrethrum-knockdown techniques to obtain data on the arthropod assemblages from Australian rainforest canopies. Our work initially focussed on the cool temperate and subtropical forests of northern New South Wales and south-east Queensland but, since 1991, has expanded to include lowland tropical sites in Australia's north. Our basic approach, methods, field sites and ordinal level results are presented by Kitching *et al.* (1993).

To date we have taken two separate approaches to the study: the collection of 'baseline' data from a set of 10x10m plots within which we carry out parallel surveys of woody plants; and, studies of the canopy faunas of individual trees of known species. The basic methods and some results of the baseline studies will appear elsewhere (Kitching & Arthur 1993, Kitching *et al.*, in press).

The tree species specificity project is designed to examine the first of Erwin's sequential hypotheses: that is that the average host specificity of Coleoptera on tropical rain forest trees is 13.5%. This is the first of the multipliers used in Erwin's calculation and is one to which the overall calculation is very sensitive: a 10% figure, for instance, will produce a global estimate 25% smaller than that arrived at by Erwin; a 20% figure one that is 50% greater.

We hasten to add that our aim is not to produce better global estimates of arthropod biodiversity, which must be regarded, at best, as an elusive goal, and at worst an unachievable diversion from the real business in hand! There can be no doubt, however, that the estimation of tree species specificity, which is fundamental to an understanding of how canopy faunas are assembled, presents formidable technical difficulties, but is nevertheless an attribute vital to any biodiversity conservation activities.

Tree-specificity Studies

Our work on tree specificity remains at an early stage. We have selected a number of understorey trees the growth forms of which facilitate sampling. Each of these is between six and 12 metres in height, growing more or less isolated (or easily isolatable!) from its neighbours. In each case a set of 0.5 m^2 collecting hoops are disposed around the base of the subject tree and a painters' scaffold erected adjacent to it. The canopy of the tree is then sprayed with a commercially available mixture of pyrethrins (see Kitching *et al.*, in press, for details of composition, dilution etc) for about six minutes using a backpack dispenser with the operative on the scaffold. All samples are taken in the morning to reduce variance associated with diel activity patterns. This somewhat simple approach allows a carefully controlled and thorough sampling of the particular canopy of interest but is less effective in dense forest where canopies are interdigitated.

Arthropod samples are collected three to four hours after the spraying event. Samples are sorted to Order at field laboratories and returned to the laboratory in ethanol. Subsequently particular Orders are further curated and sorted to family and morphospecies. We chose to focus further on the Coleoptera, an Order in which we are developing particular expertise and one which allows for direct comparison with the results of workers elsewhere in the world. Following removal from liquid storage specimens are pointed and separately labelled before further sorting. Other work, not discussed further here, examines the Collembola, Psocoptera, Lepidoptera and Araneida from our samples.

The results presented below represent the partial outcome from our 1992 field season during which, *inter alia*, we sampled six trees each of two known species: *Ryparosa javanica,* a widespread species in Australia and elsewhere, and *Noahdendron nicholasii*, a very restricted endemic found only in the Noah Creek and adjacent catchments (see below).

STUDY SITE

The results to be described all originate from sites located in the catchment of Noah Creek, a permanent waterway some 11 km south of Cape Tribulation within the greater Daintree area of the Wet Tropics World Heritage area. The vegetation of the region has been described by Tracey (1982) and species lists of the trees and climbing plants of Cape Tribulation have been published by Jessup and Guymer (1985) and McKenzie (1985). The vegetation of the actual sites sampled has been described (McIntyre, Kitching & Jessup, in press). The forest is a complex mesophyll vine forest following the structural-physiognomic classification of Webb (1959) and Webb *et al.* (1976). Palms are an important element of the vegetation, with *Calamus* spp. being abundant on all sites; *Licuala ramseyi*, *Linospadix* spp. and *Normanbya normanbyi* were also present although in lesser numbers. Numbers of tree and shrub species in each 10 x 10 m plot ranged from about 12 to 22.

The climate of the neighbouring village of Cape Tribulation is described by Kitching *et al.* (in press). Although there may be considerable falls of rain throughout the year, the area has a distinct wet season centred in February and March, with monthly falls of over 600 mm. The Noah Creek catchment is substantially wetter than this although not separately instrumented. Single rainfall events of over 225 mm are not unknown. There is a small annual temperature range with mean maxima ranging from 27°C in July to 34°C in January, and mean minima ranging from 21°C in July to 26°C in January.

This largely undisturbed catchment is one of a number that are widely recognised as containing relictual fauna and flora and almost certainly were refugia during the last glaciation when broad-leaved rainforest in Australia retreated to a number of such locations along the north-east coast. There is substantial interest in this and neighbouring catchments both as conservation objects and natural laboratories for the study of rainforest processes. It is one of the sites designated by the Wet Tropics Management Authority for which the only designated use is scientific research.

RESULTS

Figure 1 presents the ordinal profiles for insects and non-insectan arthropods from the two species of tree. The insects profile shows some key differences that warrant further attention: Collembola and Psocoptera are far more abundant on *Ryparosa* than *Noahdendron* whereas the reverse is true for the Orthoptera, Homoptera and Hymenoptera. The relative abundance of some Orders (eg. Blattodea, Coleoptera, Thysanoptera) appears unaffected by the tree species. For the non-insects all three groups present were somewhat more abundant on *Noahdendron* but the differences were not large.

Focussing on the Coleoptera, 135 individuals sorted to 59 species, 38 on *Noahdendron* and 35 on *Ryparosa* (14 being shared). These 59 species represent 22 families (Figure 2) dominated for both tree species by the Curculionidae, Coccinellidae and Corylophidae.

The species profiles for each species are presented separately in Figure 3. Essentially both assemblages show the highly skewed distributions characteristic of such data from virtually all ecosystems. Only a few species are present in any numbers, the greater part of the species diversity being contributed by 'rare' species. For *Ryparosa*, of 35 species only 3 occurred in the samples more than three times, 27 species occurred as singletons. For *Noahdendron* comparable figures are 6 and 29 respectively, out of 38 species.

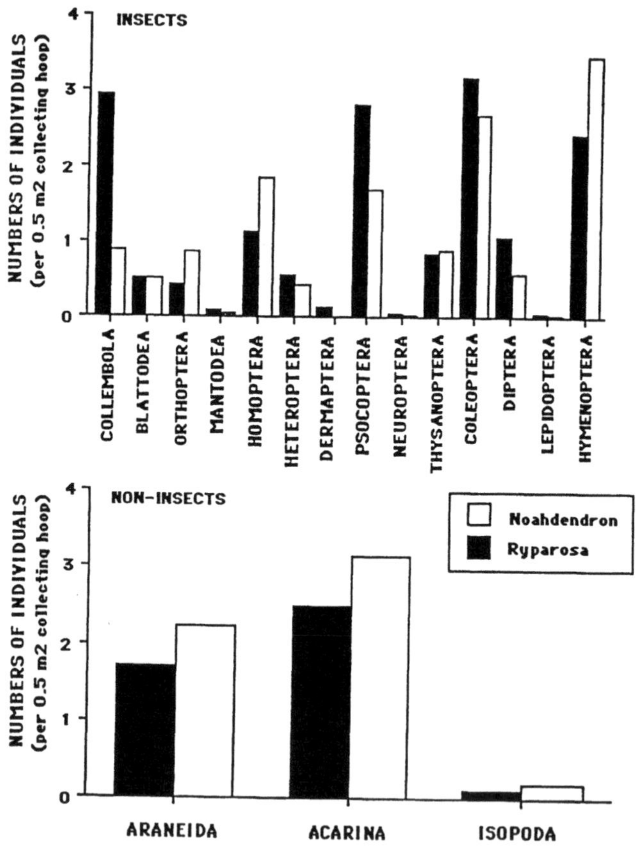

Figure 1. Distribution of insects and non-insects in samples from the canopies of *Ryparosa javanica* and *Noahdendron nicholasii* from Cape Tribulation, North Queensland.

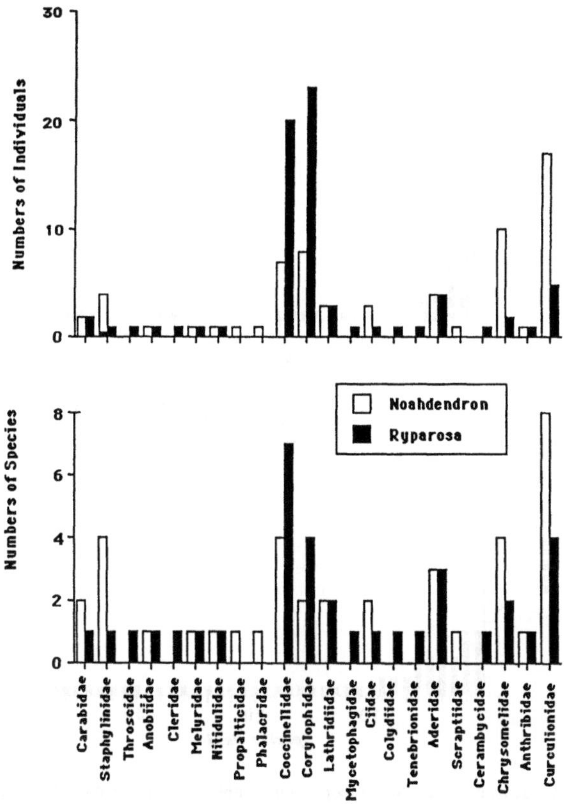

Figure 2. Distribution of Coleoptera into families in samples from the canopies of *Ryparosa javanica* and *Noahdendron nicholasii* from Cape Tribulation, North Queensland.

It is when these data are plotted together (Figure 4) that the overlap between the two assemblages become apparent and the question of 'specificity' can be addressed. Of the six 'commonest' species, only one is not shared by both tree species. That 'exclusive' species, a coccinellid, occurred on *Ryparosa*. Two species were almost 'exclusive' with just single individuals breaking the pattern. A coccinellid primarily on *Ryparosa* and a galerucine chrysomelid primarily on *Noahdendron* made up this set. More exclusivity was to be found among the rarer species, as indicated by asterisks in Figure 4, but the real contribution to beta diversity is made, by definition, by the singletons.

DISCUSSION

Top-down Approaches

The results we have presented can do little more than act as a vehicle for the posing of further questions at this stage. Attempts to define a figure for host tree specificity using the top-down approach we have taken can never do more than produce an indication of sample overlap - that is they can never be reliable guides as to whether or not any particular species is monophagous or polyphagous. Only detailed autecological studies can do that and they are not compatible, in general, with the sort of assemblage-level studies we have set ourselves to do. Nevertheless, in terms of defining beta, and subsequently gamma, diversity it is reliable measures of this overlap that are required.

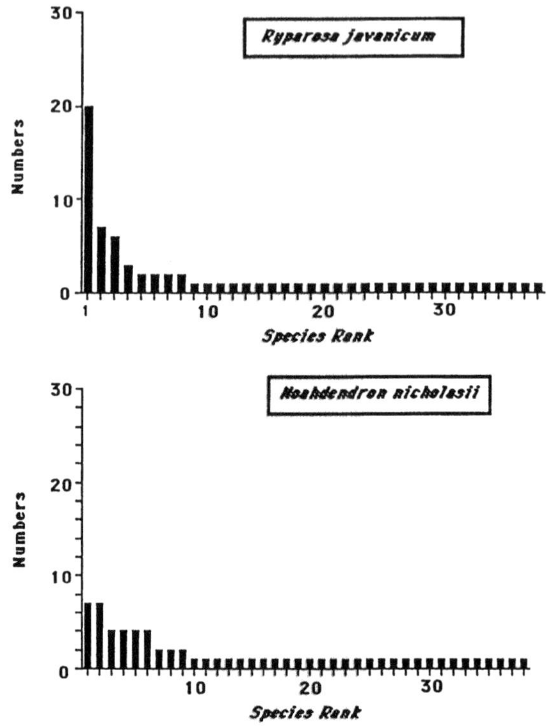

Figure 3. Numbers of individuals of species of Coleoptera in samples from the canopies of *Ryparosa javanica* and *Noahdendron nicholasii* from Cape Tribulation, North Queensland.

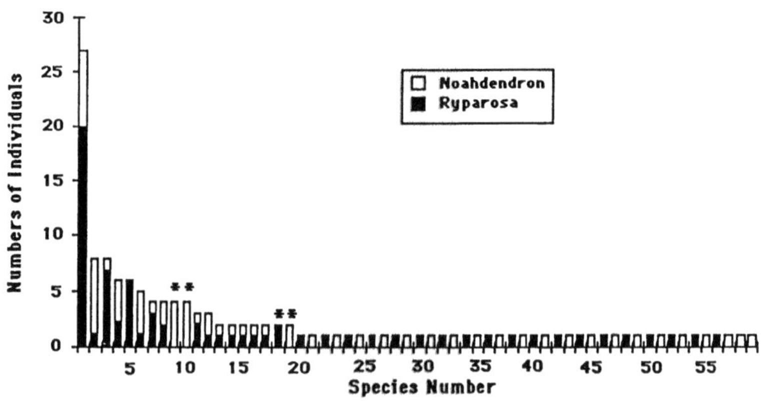

Figure 4. Overlap between the beetle fauna in samples from the canopies of *Ryparosa javanica* and *Noahdendron nicholasii*, from Cape Tribulation, North Queensland.

The Importance of Oligophagy

Erwin's specificity-multiplier is naive in the sense that it assumes that a beetle species is either tree species specific or not - that is, either monophagous or polyphagous. It does not allow for the various degrees of oligophagy - and yet these are probably more the rule than the exception. Tree species exhibit various sorts of phylogenetic relationship and it is entirely reasonable that two species of tree of the same genus, for instance, might be expected to share more species than two less closely related ones. The continuation and extension of our work in Australian forests will eventually allow quantification of the role of such factors in determining this level of overlap in diversity of the canopy faunas of co-occurring trees.

Current data shortcomings and Solow's analysis

Further analysis is required to give confidence that the numbers of specialists and generalists identified within our current data set is other than an artifact of small sample sizes and unequal distributions of species across replicates. However, M. P. Zalucki kindly carried out for us an analysis following that of Solow (1993) in which the distributions of beetle species and abundances within the assemblages from each tree species are compared with one constructed randomly. This suggested that the observed assemblages on *Ryparosa* and *Noahdendron* are significantly different from the one constructed on the null hypothesis that each is but a different random sample of a common assemblage.

The epiphyte problem

Problems inherent in our method will always restrict the power of biodiversity analyses of this kind. In particular, our selection for technical reasons of smaller, less mature trees means that these seldom have heavy loads of macroepiphytes. Less selective samples on high canopy trees inevitably include samples of the fauna of the rich epiphyte flora of these strata. These epiphytes undoubtedly play a large part in generating the very rich arthropod speciosity observed in canopy samples. At this stage separate sampling of epiphytes and their host trees on a sufficient scale to generate useful statistical estimates presents formidable technical difficulties with no obvious, affordable solutions.

Future Work

In July 1993 we are scheduled to spend an extended period in North Queensland continuing the tree specificity study. We shall repeat our studies of *Noahdendron nicholasii* and *Ryparosa javanica,* increasing replication and adding additional species including the endemic *Gymnostoma* and the widespread *Lindsayomyrtus racemoides*. We shall continue our 'baseline' studies obtaining dry season information for the first time and carry out light trap surveys of canopy Lepidoptera.

Our involvement in the Tropical Rainforest Cooperative Research Centre will also allow us to extend this and other work along an altitudinal base coordinated with extensive vegetation surveys and studies of the determinants of biodiversity in other animal groups, through the application of a centrally maintained database and associated geographical information systems facilities. The large databases on a few arthropod assemblages which we are accruing will also provide the context within which to define predictor-subsets of taxa which can be used statistically to predict the species richness of the whole set. Such techniques of Rapid Biodiversity Assessment, backed by adequate baseline data, will provide an essential and powerful tool for the assessment of impact and the efficacy of restoration, on key components of biological diversity. This aspect of our work may ultimately be its greatest significance.

ACKNOWLEDGEMENTS

The field work and subsequent analyses reported in this paper have been generously funded by the Wet Tropics Management Authority, the Ian Potter Foundation and Earthwatch

(Australia). We are grateful to these bodies for their patient and generous support. We thank Bill Upjohn for his provision of able technical support in the field, Heather Mitchell for sorting, mounting and generally assisting in the handling of very large samples and Graham Floater for assistance in the preparation of material for this paper. We are also grateful to John Lawrence, coleopterist with the Australian National Insect Collection, CSIRO Canberra, who has given freely of his time to advise and guide us during the sorting of our beetle specimens. Earlier work by Peter Hammond upon our collections has created a resource for which we remain grateful.

LITERATURE CITED

ERWIN, T. L. 1982. Tropical forests: their richness in Coleoptera and other species. *Coleopterists' Bulletin* 36:74-75.

ERWIN, T. L. 1983. Beetles and other arthropods of the tropical forest canopies at Manaos, Brasil, sampled with insecticidal fogging techniques. Pp. 59-75. in Sutton, S.L., Whitmore, T.C. & Chadwick, A.C. (eds). *Tropical Rain Forests: Ecology and Management.* Blackwell Scientific Publications, Oxford.

ERWIN, T. L. 1991. How many species are there?: revisited. *Conservation Biology* 5:330-333.

GASTON, K. J. 1991. The magnitude of global insect species richness. *Conservation Biology* 5:283-296.

HODKINSON, I. D. & CASSON, D. 1991. A lesser predilection for bugs: Hemiptera (Insecta) diversity in tropical rain forests. *Biological Journal of the Linnean Society* 43:101-109.

JESSUP, L. W. & GUYMER, G. P. 1985. Vascular plants recorded from the Cape Tribulation area. *Queensland Naturalist* 26:2-19.

MAY, R. M. 1988. How many species are there on Earth? *Science* 241:1441-1449.

MAY, R. M. 1990. How many species? *Philosophical Transactions of the Royal Society of London* B330:293-304.

KITCHING, R. L. 1992. Biodiversity, research and conservation in Australian rainforests. *Malayan Nature Journal* 45:31-54.

KITCHING, R. L., BERGELSON, J. M., LOWMAN, M. D., MCINTYRE, S. & CARRUTHERS, G., in press. The biodiversity of arthropods from Australian rain forest canopies: general introduction, methods, sites and ordinal results. *Australian Journal of Ecology* 18.

KITCHING, R. L. & ARTHUR, M. 1993. The biodiversity of arthropods from Australian rain forest canopies: summary and the impact of drought. *Selbyana* 14:29-35.

MCINTYRE, S., KITCHING, R. L. & JESSUP, L. W., in press. Structure and composition of rain forest plots at Cape Tribulation, Queensland. *Proceedings of the Royal Society of Queensland.*

MCKENZIE, F. 1985. Climbing plants from 'Pilgrim Sands', Cape Tribulation. *Queensland Naturalist* 26:20-25.

SOLOW, A. R. 1993. A simple test for change in community structure. *Journal of Animal Ecology* 62:191-193.

STORK, N. E. 1987a. Guild structure of arthropods from Bornean rain forest trees. *Ecological Entomology* 12:69-80.

STORK, N. E. 1987b. Arthropod faunal similarity of Bornean rain forest trees. *Ecological Entomology* 12:219-226.

STORK, N. E. 1988. Insect diversity: facts, fiction and speculation. *Biological Journal of the Linnean Society* 35:321-337.

STORK, N. E. & BRENDELL, M. J. D., in press. Arthropod diversity studies in lowland rain forest of Seram, Indonesia. In Proctor, J. & Edwards, I. (eds). *The Natural History of Seram.*

TRACEY, J. G. 1982. *The Vegetation of the Humid Tropical Region of North Queensland.* CSIRO Division of Plant Industry, Melbourne.

WEBB, L. J. 1959. A physiognomic classification of Australian rain forests. *Journal of Ecology* 47:551-570.

WEBB, L. J., TRACEY, J. G. & WILLIAMS, W. T. 1976. The value of structural features in tropical forest typology. *Australian Journal of Ecology* 1:3-28.

Describing and comparing high invertebrate diversity in tropical forest - a case study of small moths in Borneo

GADEN S. ROBINSON *and* KEVIN R. TUCK

Biodiversity Division, Department of Entomology, The Natural History Museum, Cromwell Road, London SW7 5BD, United Kingdom

ABSTRACT. The description and comparison of invertebrate diversity in tropical habitats poses enormous problems. Diversity indices, of which there is a plethora, are often seen as ends in themselves; this has led to much sterile phenomenology which is opaque to most naturalists. We mimic field samples by bootstrap sampling from computer-generated populations based on logarithmic series. This provides a simple way of comparing samples in terms of their shared species, defining confidence limits and describing diversity in easily understood terms. Field samples of Microlepidoptera from sites in Borneo, including two sites 1 km apart in the Batu Apoi Forest Reserve, are used to demonstrate how simple questions such as "Are the moth faunas of these habitats the same?", "How diverse are they?" or "How different are they?" may be answered. Extrapolations from these field samples to minimum estimates of the total number of species of a habitat or territory are attempted, and some of the problems and limitations encountered in sampling are examined.

KEY WORDS: Microlepidoptera, biodiversity, tropical rainforest, Brunei Darussalam.

INTRODUCTION

In this paper we discuss the diversity of a group of invertebrate animals in lowland Bornean tropical rainforest; specifically, we deal with Microlepidoptera - small moths - in the valley of the Sungai Belalong in the Batu Apoi Forest Reserve in Brunei.

The Lepidoptera, moths and butterflies, have traditionally been divided into three groups - the Macrolepidoptera, large moths, the Microlepidoptera, small moths, and the butterflies. Studies of the diversity of 'Lepidoptera' have almost invariably concentrated on the large moths. To complement previous studies we have focussed on the Microlepidoptera as they comprise more than 50% of the known species of moths and are a significant but neglected component of the invertebrate biodiversity of tropical forest. We include the Pyraloidea in our definition of Microlepidoptera. Although many pyraloids are robust and powerful flyers and are as large as many Macrolepidoptera, they have been omitted from most studies of 'Lepidoptera' diversity as being 'Microlepidoptera'. A noteworthy exception to this is the work of Barlow & Woiwod (1989). For further discussion of this separation of the Lepidoptera in studies of diversity see Robinson & Tuck (1993).

Diversity has come to mean many things to many people. Its measurement and significance has become an area of considerable debate. We do not intend contributing to this debate. Our approach is based on the principle that "if it ain't broke, don't fix it", tempered with common sense derived from considerable experience of working with Lepidoptera in the tropical forests of South-East Asia.

The diversity that we describe here falls in the category of 'alpha-diversity'. It is a measure of the species-richness of a particular group pertaining in a particular habitat or locality at a particular time. There are two ways to measure such species-richness. One is to enumerate all the species present in, say, the Belalong valley. The second is to measure the diversity of a sample or series of samples, compare these and extrapolate from them to

D. S. Edwards et al. (eds.), Tropical Rainforest Research – Current Issues, 29–42.

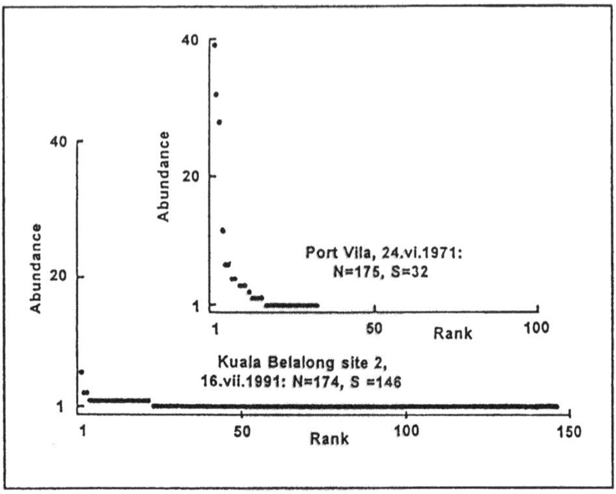

Figures 1 & 2 (inset). Rank abundance curve of (1) a high diversity sample of Microlepidoptera from lowland dipterocarp forest in Borneo, and (2) a low diversity sample of Macrolepidoptera from grassland and gardens in Vanuatu.

provide meaningful biological observations.

The first method is time-consuming: it necessitates the capture or at least recording of all the species present of the group of organisms under discussion. A full enumeration of Britain's Lepidoptera fauna (some 2500 species) has taken an army of naturalists some 200 years to achieve. It is impractical in the much more diverse tropical systems to contemplate such enumeration in the short term except with a few small groups of large vertebrates.

The sampling approach, while subject to the biases to which any sampling system is prone, has been applied widely in studies of tropical Lepidoptera with varying degrees of success. The measurement of diversity, or at least the method that we shall use, relies upon knowledge of the distribution of individual organisms among particular groups - in this case species. We thus need to sort our samples into their constituent species and count the numbers of each.

Measuring diversity

It is often obvious by inspection what is a high-diversity sample and what is a low-diversity sample (Figures 1 & 2). The first sample is of Microlepidoptera collected at light on 16 July 1991 on the eastern slope of the Belalong valley: it comprises 174 individuals representing 146 species. Figure 1 is a linear plot showing the number of individuals of each of these species in descending order of abundance. Only three species are represented by more than two individuals and most species are represented by just a single specimen; the resulting species-abundance curve is very long and flat.

The second sample is of Macrolepidoptera trapped on 24 June 1971 from the edge of a field in Port Vila, Vanuatu. It is of almost exactly the same size as the Belalong sample, consisting of 175 individuals, but contains only 32 species. Plotted in similar fashion to Figure 1, it can be seen that the rank-abundance curve (Figure 2) is very differently shaped; there are several abundant species and the curve, while hollow like that in Figure 1, is considerably steeper. These samples, in the context of tropical Lepidoptera, are respectively of high and low diversity.

Table 1. The number of species represented by different numbers of individuals in the entire (pooled) sample of Microlepidoptera from Kuala Belalong, with the number predicted from the log-series.

SPECIES WITH:	Observed	Expected (log series)
1 individual	320	309
2 individuals	116	116
3 individuals	49	58
4 individuals	33	32
5 individuals	15	20
6 individuals	19	12
7 individuals	6	8
8 individuals	4	5
9 individuals	2	3
10 individuals	0	3
RESIDUAL SPECIES (>10)	7	6

At its simplest, for a sample consisting of a given number of individuals, the diversity will be higher the greater the number of groups (species) in the sample.

In order to be able to understand, compare and manipulate samples we need a measure of the species-richness expressed in each that is independent of sample size. A species-abundance curve (such as that in Figure 1 or 2) may be summarised in terms of the number of species contributing to each abundance category starting with the rarest. As illustration, we take here the *total* sample of Microlepidoptera from the Belalong valley collected in 1991. It consists of 1230 individuals comprising 571 species (Table 1). The decreasing frequency of species in each class of (increasing) abundance is regular and predictable: it has long been known that such frequency distributions conform to a mathematical series known as the *logarithmic series*. The logarithmic series has been found to apply to an enormously wide range of samples of different organisms of different sizes taken over different periods of time (Fisher, Corbet & Williams 1943, Williams 1944, 1947, 1964).

Given the crucial elements N, the number of individuals, and S, the number of species in the sample, the applicable log-series can be calculated. In the case of the Kuala Belalong sample it will be seen by inspection that the expected (calculated) series matches closely the series observed in the field sample (Table 1). The log-series has particular properties that make it especially useful as a descriptor of samples:

1. It is relatively simple to generate log-series and manipulate them using a computer. Populations (*i.e.* mixed-species 'assemblages') of different sizes can be generated using parameters derived from samples so that samples can be 'magnified' or 'rarefied' and hypotheses explored.
2. One of the parameters of the logarithmic series, *alpha* (α), relates N and S directly and is a diversity index. Any particular log-series is defined by the three parameters N, S and α, or by a combination of any two. The three are related by the formula:

$$S=\alpha(\log_e(1+(N/\alpha)))$$

Alpha is most easily computed, given N and S, by iteration and substitution in this formula. The number of species represented by 1, 2, 3 (etc.) individuals is given by:

$$\alpha(N/(N+\alpha)),\quad \alpha((N/(N+\alpha))^2/2),\quad \alpha((N/(N+\alpha))^3/3),\ \text{etc.}$$

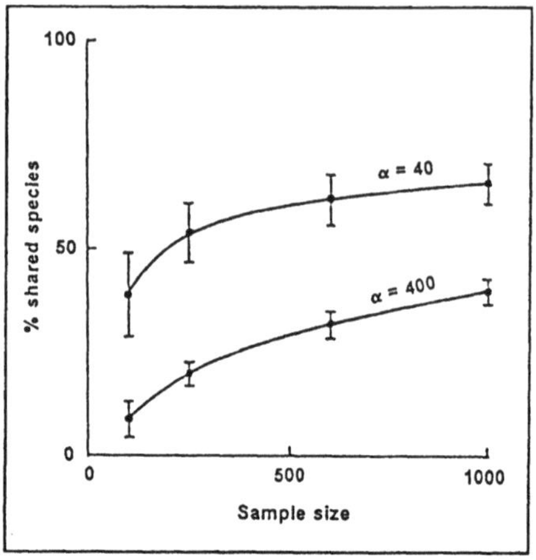

FIGURE 3: Shared species in pairs of similarly sized samples from the same mixed-species assemblage: variation in percentage with the different sample size and diversity. [Data from bootstrap sampling procedure from an assemblage of 10^7 individuals]

3. Random samples of a log-series are of themselves a log-series with the same α as the 'parent' assemblage (although, of course, N and S will be different).

The converse of the third point is that any sample must be from a putative mixed-species assemblage that has the same diversity (α) as the sample.

There are numerous other indices of diversity available; the reader is referred to Wolda (1981, 1983), Magurran (1988) and Taylor (1978) for further discussion. These authors provide additional justification for the use of Williams' α.

Comparing samples

Our next requirement is for a means to compare samples in terms of their species composition - do they represent the same or different assemblages?

One problem encountered in examining samples from high-diversity species-assemblages is that pairs of samples even from the same assemblage may have only a very small proportion of their species in common (Figure 3). Most indices of sample similarity or dissimilarity such as Preston's coefficient (Preston 1962) assume exhaustive sampling of species, rarely possible except in the case of a few small groups of large vertebrates. Wolda (1981) has provided a 'consumer guide' to similarity indices. Preston's coefficient was not included in his survey. We have used this index in the past (Holloway, Robinson & Tuck, 1990), and it is perhaps appropriate therefore that we demonstrate here its limitations in this context. Like many of the indices tested by Wolda, Preston's coefficient is sample-size and diversity-dependent (Figure 4) and is therefore not a good sample comparator.

This is not to condemn all such indices; Wolda (1981) found that modified versions of Morisita's index (Magurran 1988, Morisita 1959) for example, did not suffer defects of sample size- and diversity-dependence. However, computation of the Morisita index is difficult and we find it impossible to conceptualise what it is actually measuring.

Faced with the problem of comparing samples, we turned to Wolda's (1981, 1983) technique for randomly sampling computer-generated assemblages derived from the log-series. The concept of comparison that appeared to us to be statistically valid while at the same time remaining understandable was that of how many species are shared between our samples and how many we would *expect* to be shared.

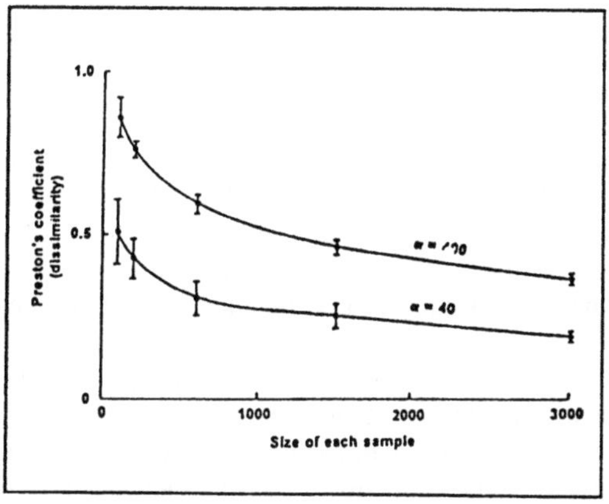

Figure 4. Variation in the value of Preston's coefficient of similarity between sample pairs (of the same size) from the same master population with different sample size and diversity. [Data derived as in Figure 3]

Given a log-series population, this can be sampled such that x individuals are randomly selected and the number and identity of species they represent can be recorded. Another sample can then be taken and the species common to the two compared. Replication of samples provides the means to evaluate the variance and confidence limits of the comparison.

Translated back into the field problem, we can therefore mimic the observed field situation by building a gigantic population (mixed-species assemblage) within the computer, the population of the same order of magnitude as that which might be present in, say, the Belalong valley. That population is generated at the same diversity as observed in the field samples. Pairs of samples of identical size to the field samples are then taken from the computer population and the number of species shared between sample pairs is compared with the number observed. The field samples are tested against the null hypothesis that they come from a common population. In practice we generate a population of 10^6 individuals; the diversity figure used is that of the pair of field samples combined, but there are other alternatives. Possibilities are the higher of the two diversities, the mean diversity, or the lower diversity (Robinson & Tuck 1993), but there are sundry other possibilities. Ten pairs of replicate samples are taken (with replacement) from the computer-generated log-series population; the species common to pairs of samples are enumerated, and means, variances and 2 x SD confidence limits calculated. The observed/expected ratio of shared species in the field samples is taken as a measure of similarity, with 1.00 indicating common origin - a match of observed to expected. Copies of the software are available from the authors.

The questions

Our sampling of Microlepidoptera in the forest adjacent to the Kuala Belalong Field Studies Centre addressed five questions:

1. How species-diverse is the forest of the Belalong valley?
2. How does the scale of this diversity compare with that of other habitats in Borneo or with temperate regions?
3. How many species are there likely to be in the area immediately adjacent to the Kuala Belalong Field Studies Centre - a minimum estimate?
4. Is diversity uniform across the Belalong valley?
5. Is species composition similar across the valley or between the Belalong valley and other sites in Borneo?

METHODS

Moths were attracted to a white sheet hung vertically and illuminated by a 125 watt mercury-vapour lamp suspended one metre from the sheet surface. Lamps were run from a portable generator.

Small specimens (wingspan <40 mm) were collected individually into corked glass tubes, the tubes kept cool overnight and the specimens killed, pinned and the wings spread (as described by Robinson 1984) the following day. Large specimens (>40 mm) were killed on site using conventional killing-bottles and pinned and stored the following morning.

Collecting began at dusk and usually ceased by about 2300 hrs when flight activity had become minimal. The capture of all Microlepidoptera and Pyraloidea on the sheet was attempted; on the few occasions when this proved logistically impossible a subsample was taken and it was attempted to make this random.

Pinned specimens were stored with a few crystals of chlorocresol to prevent growth of mould, dried, and returned to the laboratory. There they were labelled and sorted by appropriate specialists to family and to species. Individuals, species and higher taxonomic units (genera/subfamilies/families/superfamilies) were tabulated.

The localities studied

Kuala Belalong, Temburong, Brunei, altitude 125 m, sites 1 & 2 (KB1, KB2). These sites were on the sides of the valley of the Sungai (River) Belalong above the Kuala Belalong Field Studies Centre. Both were on steep slopes with closed-canopy primary lowland mixed dipterocarp forest. Site 1 was on the western side of the valley and site 2 diagonally opposite on the eastern slope about 1000 m from site 1. Collections were made from site 1 on 4, 6-8 and 10-11 July 1991 (6 nights) and from site 2 on 13-16 July (4 nights). Nightly samples were kept separate and specimens from each labelled with the precise night of collection.

Comparisons were made with two other localities in northern Borneo:

Crocker Range, Sabah, altitude 1500 m (CR). This site was in closed-canopy montane forest on a trail along a broad ridge crest at 5.45N, 116.19E, and appears to be in the transitional zone between lower and upper montane forest (Anderson & Chai 1983). Collections were made on the 6 nights from 7-12 August 1991. Nightly samples were pooled.

Kampong Kapok, Muara, Brunei, altitude 0 m (MR1 and MR2). This site was a white-painted house wall facing into undisturbed mangrove (*Rhizophora mucronata*) forest. Cleared land with sparse weed growth and a few ornamentals stretches 6-700 metres behind the house to the edge of disturbed secondary forest. The mangrove forest extends for more than 0.5 km before open water is reached. A more detailed description of this site is given by Robinson & Tuck (in press). Collections were made from 19-21 July 1991 (3 nights) and samples were pooled (MR1). Further samples were taken between 20 January and 10 February 1992 (ca. 14 nights' collections) and these were also pooled (MR2).

RESULTS

Numbers and diversity

The numbers of individuals and species and the diversity of each of the samples is summarised in Tables 2, 3 and 4. A more detailed summary is given by Robinson & Tuck (1993).

Values of alpha for the total sample at each of the two Kuala Belalong sites are very high - 355 and 413 (Table 4). In practical terms, samples of 10,000 specimens with the same diversities would contain 1197 and 1333 species respectively. A typical figure for

TABLE 2: Numbers of individuals of Microlepidoptera (sensu stricto - excluding Pyraloidea (MICRO), Pyraloidea (PYRAL) and Microlepidoptera including Pyraloidea (ALL) collected at thee sites in Borneo.

SITE	MICRO	PYRAL	ALL
KB1	372	377	749
KB2	166	315	481
KB1 + 2	538	692	1230
CR	386	127	513
MR1	128	57	185
MR2	323	362	685
MR1 + 2	451	419	870

TABLE 3: Numbers of species of Microlepidoptera (sensu stricto - excluding Pyraloidea) (MICRO), Pyraloidea (PYRAL) and Microlepidoptera including Pyraloidea (ALL) collected at three sites in Borneo.

SITE	MICRO	PYRAL	ALL
KB1	194	209	403
KB2	124	195	319
KB1 + 2	270	301	571
CR	198	70	268
MR1	40	35	75
MR2	99	113	212
MR1 + 2	118	135	253

TABLE 4: Diversity (Williams' α) of species of Microlepidoptera (sensu stricto - excluding Pyraloidea) (MICRO), Pyraloidea (PYRAL) and Microlepidoptera including Pyraloidea (ALL) collected at three sites in Borneo

SITE	MICRO	PYRAL	ALL
KB1	164±30	193±35	355±46
KB2	222±77	218±46	413±74
KB1 + 2	216±32	203±25	414±39
CR	163±29	64±20	226±35
MR1	20±6	39±19	47±11
MR2	49±9	56±9	105±13
MR1 + 2	52±8	69±11	120±13

Macrolepidoptera in Britain would be a diversity of 40, i.e. 221 species for 10,000 specimens.

The individual elements - Micro-lepidoptera and Pyraloidea - of the two Kuala Belalong samples have broadly comparable diversities at the two sites as do the total samples. However, the diversity (α) of Microlepidoptera at KB1 is notably lower than at KB2. The KB2 sample is small, however, and KB1 diversity falls within the 95% confidence limits of KB2 though not vice versa.

The total sample from the Crocker Range has a lower diversity than that from Kuala Belalong, and the sample confidence limits of the two do not overlap. This holds true if Microlepidoptera and Pyraloidea are considered separately.

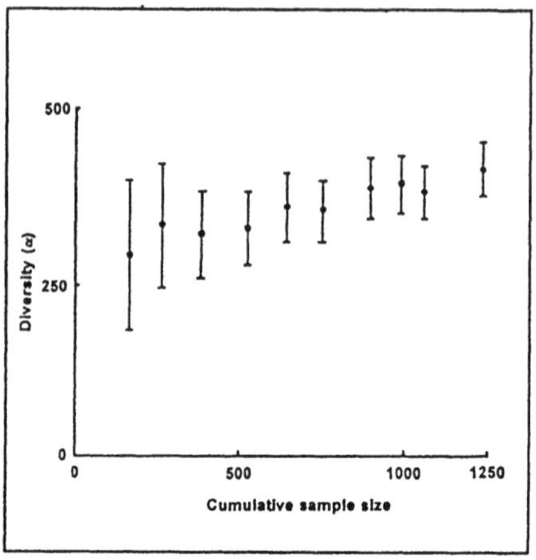

FIGURE 5: Trend in the index of diversity (Williams' α, with 95% confidence range) with the accumulation of individuals (N) of Microlepidoptera and Pyraloidea over ten days at Kuala Belalong, Brunei. Note that on 12 July (seventh point) the collecting site was changed from KB1 to KB2.

The mangrove forest samples from Muara are clearly of much lower diversity than those from the other sites and are little more diverse than samples from a mixed area of secondary vegetation in a temperate country.

If we enumerate the daily samples from Kuala Belalong and measure the diversity of the sample accumulated day by day it will be seen that there is a steady upward drift of the diversity index (Figure 5).

Comparison of the sites

The species content of the samples from KB, CR and MR was radically different. Kuala Belalong and Muara samples had more species in common than any other pair of sites, just 37 out of 787 (Table 5). However, expressed as a ratio of observed to expected similarity, this figure is not much different from that between Kuala Belalong and Crocker Range although much higher than the similarity between Muara and Crocker Range.

Comparison of the samples from the two Kuala Belalong sites, about one kilometre apart, shows that the total samples share species to within 15% of the figure expected, were the samples from a common assemblage. Examined separately, however, the Pyraloidea match the expected figure almost precisely whereas the Microlepidoptera do not (32% divergence). There is a remarkable disparity between the two Muara samples (MR1 and MR2) which were taken six months apart although the degree of difference between each of these and the samples from KB and CR is similar.

Commonness, rarity and taxonomic composition of the samples

Neither rainforest sample (KB or CR) contained markedly abundant species. Only seven species were represented in the KB sample by 10 or more individuals, the most abundant having 27. In the CR sample only three species had 10 or more individuals, the most abundant having 19. These "dominant" species comprise less than 3.75% of the total samples.

The pooled MR samples contained four numerous species with 30 or more individuals, the most abundant with 95, the four together comprising 28% of the sample; dominance in these samples is much more apparent.

TABLE 5: Differences between samples of Microlepidoptera (sensu stricto - excluding Pyraloidea) (MICRO), Pyraloidea (PYRAL) and Microlepidoptera including Pyraloidea (ALL) collected at three sites in Borneo in terms of their observed and expected shared species (expected figure derived by simulation method - see text).

SHARED SPP: SITES COMPARED	OBS.	EXP.	O/E
KB1 x KB2 MICRO	48	71±9	0.68
KB1 x KB2 PYRAL	103	100±13	1.03
KB1 x KB2 ALL	151	175±20	0.86
MR1 x MR2 MICRO	21	49±7	0.43
MR1 x MR2 PYRAL	13	32±6	0.40
MR1 x MR2 ALL	34	87±10	0.39
KB1+2 x CR ALL	27	223±13	0.12
KB1+2 x MR1 ALL	13	109±7	0.12
KB1+2 x MR2 ALL	28	266±22	0.11
KB1+2 x MR1+2 ALL	37	299±17	0.12
MR1 x CR	2	86±6	0.02

Further details, with notes on the taxonomic composition of the samples, are given elsewhere (Robinson & Tuck 1993).

DISCUSSION

How diverse?

The vegetation of the Belalong valley is mixed dipterocarp forest growing on slopes from the tame to the precipitous with a variety of aspects. This forest is a mosaic in which different trees and herbs dominate or are clumped in particular small areas and in which floristic diversity is remarkably high.

Most moth caterpillars feed on or in the leaves of plants and are to a greater or lesser extent host-specific. Thus moth diversity in a habitat reflects surrounding botanical diversity, and the high diversity of lowland dipterocarp forest in the Belalong valley is matched by high moth diversity. The mosaic nature of the forest is also matched by the moths but tempered by their ability to disperse by flight and thereby 'smooth out' the heterogeneity of the habitat.

We have taken a 'snapshot' of diversity of the smaller moths (Microlepidoptera and Pyraloidea) at Kuala Belalong. The α-value recorded, 414, is of a similar order of magnitude to that of about 300 for Macrolepidoptera found by Barlow & Woiwod (1989) in a mature secondary forest habitat in W. Malaysia and by Holloway (1984) in lowland dipterocarp forest in the Gunung Mulu National Park, Sarawak. Our diversity measure is probably an underestimate - see Figure 5, discussion below, and Robinson & Tuck (1993).

The scale of diversity?

This diversity is spectacularly high; mean α in the UK for Pyraloidea is 2.9; at Genting, W. Malaysia, it is 91 and at Kuala Belalong it is 203. We do not have data for the diversity of light-trapped Microlepidoptera in Britain but we suggest that the α value at any one site would be unlikely to exceed significantly the typical Macrolepidoptera value of 40 cited above.

The sample of Microlepidoptera described here from montane forest in the Crocker Range is less diverse than that from lowland forest. This matches the pattern found in Macrolepidoptera in SE Asia by Holloway (1984, 1987) and, in Sulawesi, in Macro- and Microlepidoptera by Holloway *et al.* (1990). These authors note that there is a peak of diversity at about 1000 meters. Holloway's (1984) site 11 at 150 metres altitude in mixed dipterocarp forest is broadly comparable to KB: his sample of Macrolepidoptera had an α of 309. His sites 1-5 in upper montane forest may be comparable to CR; α for Macrolepidoptera at these sites ranged from 73 to 118 whilst α for CR Microlepidoptera is 163.

Spectacularly high diversity is not universal in Bornean habitats, however. Floristically depauperate habitats such as mangrove forest, dry heath forest (kerangas), and *Shorea albida* peat-swamp forest have much lower diversities than mixed dipterocarp or montane forest. Species collected from the Muara site are from adjacent mangrove forest, cleared land with regenerating weeds and planted ornamentals, and (distant) secondary forest. Here the diversity (α) of the larger sample just exceeds 100. Our subjective assessment of diversity in dry kerangas and in peat-swamp forest, based on collecting in 1982 and 1988, is that these are no more diverse than mangrove forest.

How many species in the lowland forest?

There is no such thing as an absolute number of species present in a particular habitat except at a particular instant. Mobile organisms pass through, establish temporary populations, become locally extinct. Collecting and/or inventorying invertebrates in a lowland forest habitat such as the Belalong valley is an open-ended enterprise. However, the number of 'new' species yielded by each successive unit of collecting effort (usually measured as a number of individuals) becomes steadily less. The log-series formula can be used to predict the number of species obtained with progressively larger samples. This presupposes that the value of α remains constant. In the early stages of a sampling programme it is evident that it does not. Both at Kuala Belalong (Figure 5) and elsewhere (Barlow & Woiwod 1989) α rises with sampling effort. This phenomenon is also seen on 'bad' and 'good' collecting nights. A small sample invariably has a lower α than a large one (Robinson 1975). The effect is caused by heterogeneity of the surrounding habitat. On single good nights with abundant insect activity flight-ranges are considerable and the insects attracted to the light represent, metaphorically speaking, numerous tiles from the environmental mosaic. On a bad (perhaps cold and windy) night flight-ranges are restricted and the tiles sampled are fewer. Similarly, successive sampling will result in varied catches due to climatic factors and the random nature of some insect movement: the heterogeneity of the sampled habitat will increase steadily. Eventually some stability may develop, but superimposed now will be effects of seasonality and floristic phenology as well as longer-term environmental changes.

The values of α obtained by 'snapshot sampling' are therefore underestimates so that any prediction of species numbers from these will also be underestimates.

The basis for a 'total species estimate' can be anything that we want it to be. Here we propose the number of species that could be collected in two years. Likely values of α (from the figures above) for Macrolepidoptera, Pyraloidea and Microlepidoptera respectively are 310, 170 and 210 for lowland forest. The combination of a Rothamsted trap and MV light and sheet would allow a collector to collect on a nightly basis perhaps 100 Macrolepidoptera, 60 Pyraloidea and 60 Microlepidoptera. Extrapolating to a two-year catch and computing the number of species of each group from the proposed diversity value we get: Macrolepidoptera, 73000 individuals and 1694 species; Pyraloidea, 43800 individuals and 944 species; Microlepidoptera, 43800 individuals and 1122 species. The combined total is 160600 individuals and 3760 species. These represent no more than informed guesses of the number of Lepidoptera species that may be present at a single site in Bornean rainforest. We stress that they are likely to be underestimates.

Janzen (1988) has published inventory figures for the Lepidoptera of the Santa Rosa National Park, Costa Rica. This is a tropical dry forest and not necessarily comparable with the forest type described here. From 10 years' observations he recorded 2797 moth and 345 butterfly species; of the moths 1130 are Microlepidoptera and Pyraloidea and 1667 are Macrolepidoptera.

Elsewhere (Robinson & Tuck 1993) we have extended this method to derive a minimum figure for Lepidoptera in Borneo (excluding butterflies) of more than 8600. Holloway (1987) estimated that there are 4500 Macrolepidoptera species in Borneo; Robinson, Tuck & Shaffer (in preparation) have seen more than 6500 species of Microlepidoptera plus Pyraloidea from SE Asia (effectively Thailand + W. Malaysia + Borneo). These figures suggest to us that our extrapolation is of at least the right order of magnitude.

The question remains to what degree our 'snapshot' figures are underestimates of the 'true' diversity. This phenomenon requires further investigation to determine after what period of collection or trapping the diversity value stabilises if, indeed, it ever does. One approach that has been adopted to circumvent this problem is to define 'faunal elements' - associations of species characteristic of particular forest types - and to examine the diversities of these independently (Holloway 1989, Robinson 1975). However, large samples from a large number of sites are needed for this technique.

Uniform diversity?

The data presented above (Table 4) suggest that there is no significant difference in diversity between the two sites on either side of the Belalong valley. However, it must be stressed that the samples are small.

Similar species composition?

The extrapolation above assumes a uniformity in lowland Bornean forest types that may be illusory. We are able to speculate about habitat homogeneity in lowland dipterocarp forest only on the basis of the two KB sites, one kilometre (and a deep valley) apart. While the Pyralidae at the two sites appeared to be drawn from the same assemblage, the Microlepidoptera did not (Table 5). Many species in the latter group have limited dispersal powers. The mosaic nature of tropical forest may mean that Microlepidoptera appear more 'patchy' than the more strongly dispersive Pyralidae; further collecting of Microlepidoptera at the two sites should result in a convergence of the shared species with the number expected if that is the case. Further testing of habitat homogeneity of, particularly, lowland mixed dipterocarp forest is needed.

On a larger scale it is apparent that there are great faunistic differences between lowland and montane forest in Borneo (Table 5); few species are common to the two habitats. Elsewhere we describe differences in the proportion of families and superfamilies and indicate groups that are almost entirely restricted to one or other forest type (Robinson & Tuck 1993). The difference between the lowland forest samples and those from the edge of mangrove forest are of comparable magnitude. here is almost complete faunal dissimilarity between the Muara sample and that from the Crocker Range.

The two samples from Muara bear closer examination. Taken six months apart, the diversity of the Microlepidoptera is significantly different in the two as is the diversity of Microlepidoptera and Pyraloidea combined but not Pyraloidea alone. Sample comparison in terms of shared species (Table 5) indicates a much greater degree of difference in the temporally separated Muara samples than between the two spatially separated samples from Kuala Belalong. Combining the two Muara samples produces an increase in the index of diversity, suggesting that the two are drawn from putative parent assemblages that are different. There are several explanations.

It is possible that there is faunal change with season that is affecting the samples. Barlow & Woiwod (1989, 1990) have demonstrated temporal heterogeneity of samples of Macrolepidoptera from W. Malaysia and Sulawesi.

The samples may show the effects of succession. MR1 was taken shortly after completion of a block of houses built on bare infill soil. By the time MR2 was taken there was considerable growth of weeds, grass and ornamentals in the housing block which could add a particular specialised fauna to the second sample.

The samples may differ because of collector-bias; the two samples were taken by different entomologists.

The samples may differ because of their size. The mosaic nature of the habitat around the Muara site is obvious: it is at the intersection of mangrove forest and developing gardens, and mature secondary forest begins some 700 metres away. From samples taken elsewhere and from the work of Murphy (1990), we know that the majority of the 14 species with 10 or more individuals in the combined (MR1 + MR2) sample are mangrove specialists and have been found only in association with this habitat. Current work has identified other species that are also part of this specialised fauna. The remaining species in the Muara samples are characteristic of disturbed and open habitats and of secondary forest. We have given evidence above that heterogeneity within the sampled habitat contributes to an increase in diversity. It may be that in the case of the Muara site this heterogeneity is so acute as to affect faunal composition in samples of differing size to the extent that we have observed.

CONCLUSIONS

The rich lowland dipterocarp forest of the Batu Apoi Forest Reserve harbours a fauna of small moths that is spectacularly diverse. This moth fauna is quite different in terms of its species composition from that of montane forest and from a mixed secondary habitat with mangrove. It is probably no more diverse than the fauna of any other pristine tract of undulating lowland rainforest in northern Borneo, but such habitats are becoming increasingly rare. Whether they are uniform in terms of their species composition beyond a particular level of sampling remains to be investigated. However, first indications from Kuala Belalong suggest that over a distance of one kilometre there is apparent uniformity in a strongly-flighted group of moths but not in a poorly-dispersive group when comparatively small samples are used.

Projections from 'snapshot' measures of diversity suggest that such a forest contains upwards of 3750 Lepidoptera species excluding butterflies. Further extrapolations and a limited examination of small-scale turnover suggest that we are not looking at astronomical species numbers. However, the effect of turnover in increasing species diversity across a greater distance in this habitat now requires urgent investigation.

Apparently spectacular turnover figures for moths have been quoted from unpublished data by Erwin (1991) from two sites 500 km apart in the Amazon Basin. However, it is not made clear whether they are from the same forest type, and the total sample size is not given. Erwin recorded 3.2% of species common to the two sites; here 3.3% of species are common to our Kuala Belalong (lowland) and Crocker Range (montane) samples from sites 180 km apart.

Ultimately, figures such as these are intended to demonstrate that to inventory life on earth by naming and describing the planet's species may appear to be a Herculean task but is by no means impossible. As taxonomists we need to plan strategies to inventory biodiversity; strategies cannot be devised unless the scale of the problem is estimated. If the moth fauna of Borneo is 'only' four times the size of that of the United Kingdom there is no reason to fear that we cannot provide at least a top-down approach to describing the common, the conspicuous and the biologically important elements of the biodiversity that we talk so much about and seem to do so little to illuminate. Indeed, that is precisely the task that we as individuals are now engaged upon (Robinson, Tuck & Shaffer, in preparation).

ACKNOWLEDGEMENTS

We thank the many individuals and organisations who made our fieldwork in Borneo possible: work by Gaden Robinson at Kuala Belalong was undertaken as part of the Brunei Rainforest Project 1991-92 organised by Universiti Brunei Darussalam and the Royal Geographical Society. Kevin Tuck sampled in the Crocker Range with Operation Raleigh's Expedition Perwira, with project approval by the Social and Economic Research Unit of the Prime Minister's Department, Malaysia (SERU).

Fieldwork at Muara was made possible through the courtesy of Glenn and Maxine Ping; we are indebted to them and to Eric Classey for taking sample MR2.

Financial support for this project was provided by the Royal Society South-east Asian Research Committee, the Godman Fund and the Loke Wan Tho Memorial Foundation to whom we are most grateful.

We are indebted to our colleagues: Jeremy Holloway and Klaus Sattler reviewed early drafts of this manuscript; Pat Haynes provided painstaking technical support; Michael Shaffer gave us extensive help in sorting and identifying Pyraloidea; Peter Barnard gave unstinting assistance in overcoming serious software problems.

LITERATURE CITED

ANDERSON, J.A.R. & CHAI, P.P.K. 1983. Vegetation. Pp. 195-223 in Jermy, A.C. & Kavanagh, K.P. (eds). *Gunung Mulu National Park, Sarawak*. Sarawak Museum Journal Special Issue 2 (Part 1) 30.

BARLOW, H.S. & WOIWOD, I.P. 1989. Moth diversity of a tropical forest in Peninsular Malaysia. *Journal of Tropical Ecology* 5:37-50.

BARLOW, H.S. & WOIWOD, I.P. 1990. Seasonality and diversity of Macrolepidoptera in two lowland sites in the Dumoga-Bone National Park, Sulawesi Utara. Pp. 167-172 in Knight, W.J. & Holloway, J.D. (eds). *Insects and the rain forests of South East Asia (Wallacea)*. Royal Entomological Society, London. iv + 343 pp.

ERWIN, T. L. 1991. How many species are there?: revisited. *Conservation Biology* 5:330-333.

FISHER, R.A., CORBET, A.S. & WILLIAMS, C.B. 1943. The relation between the number of species and the number of individuals in a random sample of an animal population. *Journal of Animal Ecology* 12:42-58.

HOLLOWAY, J.D. 1984. The larger moths of the Gunung Mulu National Park; a preliminary assessment of their distribution, ecology, and potential as environmental indicators. *Sarawak Museum Journal* 30:149-190.

HOLLOWAY, J.D. 1987. Macrolepidoptera diversity in the Indo-Australian tropics: geographic, biotopic and taxonomic variations. *Biological Journal of the Linnean Society* 30:325-341.

HOLLOWAY, J.D. 1989. Moths. In: Lieth, H. & Werger, M.J.E. (eds). *Tropical Rain Forest Ecosystems. Ecosystems of the World*. 14B. Elsevier, Amsterdam. xvii + 713 pp.

HOLLOWAY, J.D., ROBINSON, G.S. & TUCK, K.R. 1990. Zonation in the Lepidoptera of northern Sulawesi. Pp. 153-166 in Knight, W.J. & Holloway, J.D. (eds). *Insects and the rain forests of South East Asia (Wallacea)*. Royal Entomological Society, London. iv + 343 pp.

JANZEN, D.H. 1988. Ecological characterization of a Costa Rican dry forest caterpillar fauna. *Biotropica* 20:120-135.

MAGURRAN, A.E. 1988. *Ecological diversity and its measurement*. Croom Helm, London. 179 pp.

MORISITA, M. 1959. Measuring of interspecific association and similarity between communities. *Memoirs of the Faculty of Science of Kyushu University, Series B, Biology* 3:65-80.

MURPHY, D.H. 1990. The natural history of insect herbivory on mangrove trees in and near Singapore. *Raffles Bulletin of Zoology* 38:119-203.

PRESTON, F.W. 1962. The canonical distribution of commonness and rarity. *Ecology* 43:185-215, 410-432.

ROBINSON, G.S. 1975. *Macrolepidoptera of Fiji and Rotuma: a taxonomic and biogeographic study*. Faringdon. vii + 363 pp.

ROBINSON, G.S. 1984. Microlepidoptera in Brunei. The fourth Ulu Temburong Expedition. *Brunei Museum Journal* 5(4):146-177.

ROBINSON, G.S. & TUCK, K.R. 1993. Diversity and faunistics of small moths (Microlepidoptera) in Bornean rainforest. *Ecological Entomology* 18:385-393.

TAYLOR, L.R. 1978. Bates, Williams, Hutchinson - a variety of diversities. In: Mound, L.A. & Waloff, N. (eds). *Diversity of Insect Faunas.* Royal Entomological Society, Oxford. x + 204 pp.

WILLIAMS, C.B. 1944. Some applications of the logarithmic series and the index of diversity to ecological problems. *Journal of Ecology* 32:1-44.

WILLIAMS, C.B. 1947. The logarithmic series and its application to biological problems. *Journal of Ecology* 34:253-272.

WILLIAMS, C.B. 1964. *Patterns in the balance of nature, and related problems of quantitative ecology.* Academic Press, London. 324 pp.

WOLDA, H. 1981. Similarity indices, sample size and diversity. *Oecologia (Berlin.)* 50:296-302.

WOLDA, H. 1983. Diversity, diversity indices and tropical cockroaches. *Oecologia (Berlin)* 58:290-298.

The herbaceous ground flora of the Batu Apoi Forest Reserve, Brunei Darussalam

AXEL DALBERG POULSEN

Department of Systematic Botany, Institute of Biological Sciences, Aarhus University, Nordlandsvej 68, DK-8240 Risskov, Denmark

ABSTRACT. The herbaceous ground flora was sampled in different habitats of the Batu Apoi Forest Reserve, Brunei, and analysed in more detail in two one-hectare plots on ridge slopes at about 250 m elevation in primary mixed dipterocarp forest. A total of 261 species were recorded, representing 26 families of angiosperms and 14 families of pteridophytes. Most of the species were proto-terrestrials (93%). The largest families of ground herbs in both the Forest Reserve generally and in the sampled plots were the Zingiberaceae and Araceae, although the collective contribution of all fern families is similar or greater. Grasses are not a significant component of ground herbs inside the forest; a major difference compared with results from a similar study in the Amazon. The species richness of vascular herbs was 92 in one plot (total of 6264 individuals with 191 m^2 cover) and 68 in the other plot (total of 2479 individuals with 72 m^2 cover). It is suggested that the rugged landscape and the dynamics of landslides may be important factors for the high species richness of ground herbs in the area.

KEY WORDS: diversity, ground herbs, inventory, mixed dipterocarp forest, Borneo.

INTRODUCTION

As early as in 1907 Raunkiaer (1934) noted that a vast number of more or less herbaceous plants of the phanerophytic type are found as ground cover, especially in forest regions with an almost constantly humid tropical climate. Their stems are less lignified than those of the trees and they are typically found under a closed tree canopy where the most favourable conditions prevail. Most inventories of tropical rain forests have concentrated on large trees and information of herbaceous plants is often not included. General descriptions of the ground herb flora of tropical lowland rain forests in Borneo mention most common plant families represented: Araceae, Begoniaceae, Cyperaceae, Gesneriaceae, Melastomataceae, and Zingiberaceae (Richards 1952), but information is often limited to local floras or checklists (Kiew 1978, 1987), and very few attempts have been made to measure quantitative parameters.

The general notion has been that ground herbs contribute insignificantly to the total diversity in the forest. However, Gentry & Dodson (1987) stressed the importance of non-woody species in contributing to total species richness in South America. The present paper includes a checklist of all ground herb species registered by the author in the scientific area of the Batu Apoi Forest Reserve. Other results were obtained from inventories of the ground herb vegetation in selected plots.

The ground herb life form is subordinate to trees and shrubs. The conditions on the ground in a closed-canopy forest are characterized by low light intensities and high humidity. Vegetative spread is important in ground herb reproduction (Burtt 1977) resulting in regenerative capabilities and mobility not found in trees. In some cases ground herbs depend on gaps to promote seed production (Smith 1987). Not surprisingly, ground herbs are often found to have a patchy distribution and large areas of the forest floor are often devoid of herbaceous cover (Poulsen & Balslev 1991).

D. S. Edwards et al. (eds.), Tropical Rainforest Research – Current Issues, 43–57.

THE STUDY AREA

The study was conducted in the Batu Apoi Forest Reserve which covers 45,000 hectares of Temburong District, Brunei Darussalam (115°10' E; 4°32' N). The topography is generally very hilly with steep slopes. The elevation ranges from 65 m at Kuala Belalong up to 913 m (2995 ft) at the highest peak in the area, Bukit (Bt.) Belalong. The climate is generally hot and humid. The total precipitation the first year of measurement was 4135 mm with less than 200 mm per month in January, February, July, and August. Periods with no rain for ten days occurred in these drier months. All soils are residuals from erosion of the shale rock of the Miocene Setap shale formation (Ashton 1964). The slopes are dominated by mixed dipterocarp forest. Human impact on the vegetation is low; Iban tribes have used the area for pig hunting exclusively. Bryophytes are rare on the forest floor in the lowland mixed dipterocarp forest as opposed to heath forest (kerangas).

One sample plot was established on the ridge west of the Kuala Belalong Field Studies Centre by scientists from the University of Aarhus and will be referred to as Plot A. Another plot, the second plot established by the Universiti Brunei Darussalam, on the ridge between Kuala Belalong and Bt. Belalong, will be referred to as Plot B. Each plot was of one hectare projected in primary mixed dipterocarp forest at approximately 250 m elevation on either side of the river Sungai Belalong (Figure 1). A flood stream was present in each plot but only after heavy rain. No major recent gaps existed in either of the plots. Plot A was much steeper than B - the differences between the highest and lowest points were about 60 and 30 m respectively (Figure 2). Because of its steepness, Plot A had 17.3% more surface area than a perfectly flat plot and Plot B had 6.4% additional surface area.

MATERIAL AND METHODS

Fieldwork was undertaken from April to June 1991 and from January to April 1992, a total of six months. Ground herb species were collected throughout the research area of the Reserve and notes were made of habitat and elevation. The ground herb vegetation was

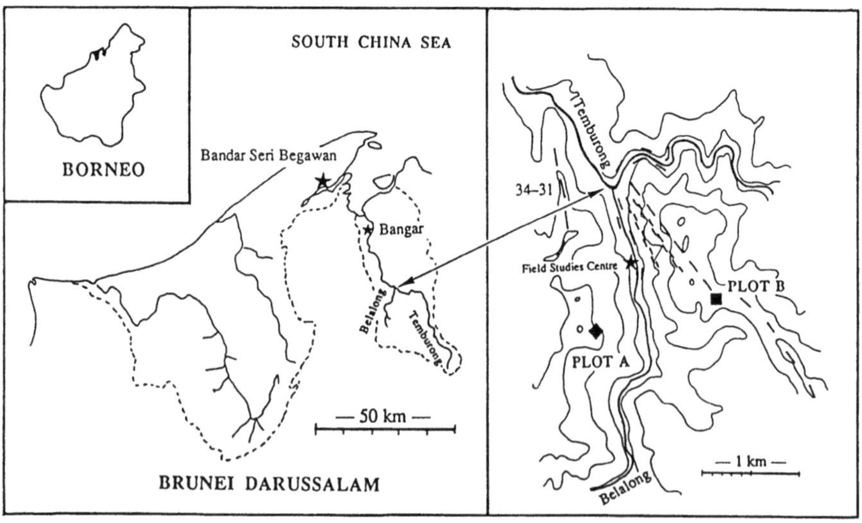

Figure 1. Map showing the position of the two plots in the Batu Apoi Forest Reserve, Brunei, Borneo.

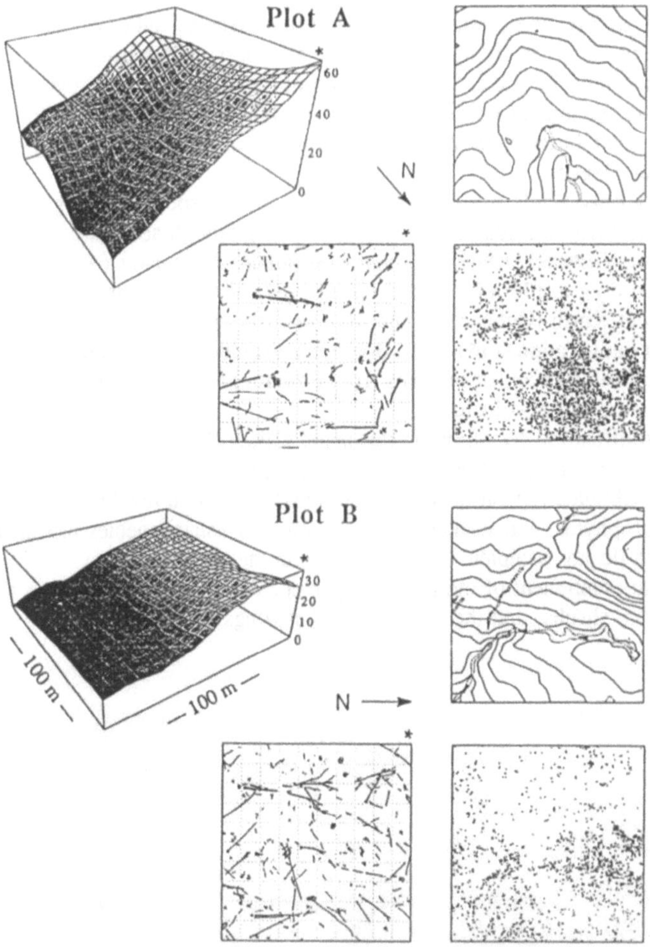

Figure 2. Topography, distribution of tree falls, and distributions of all herbaceous individuals on the ground in Plots A and B. The contours of Plot A are shown with 5 m intervals; Plot B with 2 m intervals. The position of floodstreams are shown on contour maps. The asterix indicates the position of the reference point (0;0).

studied in detail in the two quadratic one-hectare plots, A and B. Plot A was, at the same time, used for an inventory of trees (Poulsen *et al.*, this volume). The plots were subdivided into 400 5 x 5 m subplots with nylon strings. In each subplot the following data were recorded for all ground herb species: number of individuals, coordinates, and cover. Cover was estimated as the vertical projection of shoot area on a horizontal plane. Voucher specimens were collected for all species in the plots. The collections are deposited at the herbarium of Brunei (BRUN), with additional sets at Kew Herbarium (K), and Aarhus University (AAU).

Much influenced by botanical traditions at higher latitudes, a herb has been defined as a plant having stems which are not secondarily thickened and lignified and which die down annually (Lincoln *et al.* 1982). Herbaceous plants in tropical lowland rain forests are perennials and do not die back every year during the unfavourable season. Therefore tropical herbs may develop secondary lignified tissues in the basal part even if they are herbaceously derived. In the Batu Apoi Reserve this phenomenon is common in the Acanthaceae, Gesneriaceae, and Rubiaceae. The use of terms such as shrublet, undershrub, subshrub, suffrutex, and woody herb, noted from herbarium labels, indicates the difficulties that collectors experience when trying to categorize members of this transition group. It is not possible to come up with a clear cut definition of herbaceous plants, even after studies of anatomical cross sections of borderline species. Kiew (1988) stated that herbs may still be recognized as such because they have soft leaves, in contrast to shrubs and trees with typically leathery leaves.

In this study only a relatively small proportion of the species belongs to the transition group. The checklist includes all the Gesneriaceae although some are quite woody at the base (*Cyrtandra, Didymocarpus*). *Thottea penitilobata* Ding Hou (Aristolochiaceae) is an erect shrub according to Flora Malesiana (Ding 1984) and is excluded. Anatomical sections of *Neckia serrata* Korth (Ochnaceae) and *Myrioneuron cyaneum* Hallier (Rubiaceae) reveal well developed lignification (Figure 3). Both were excluded in this study despite the fact that the former occurs in a checklist of herbs from Peninsular Malaysia (Kiew 1987). Bamboos and tree ferns were not also included.

Life form stratification definitions in this paper follow Poulsen & Balslev (1991). Ground herbs are species that in certain developmental stages have stem parts at ground level rooted in the ground. Obligate terrestrial species are found on ground level only and include proto-terrestrials, saprophytes, and scandents. Facultative terrestrial species include climbers and epiphytes which do not complete their life cycle on the forest floor.

An important component of the ground layer not included in this study is that of woody plants like palms, shrubs, and the seedlings and saplings of trees.

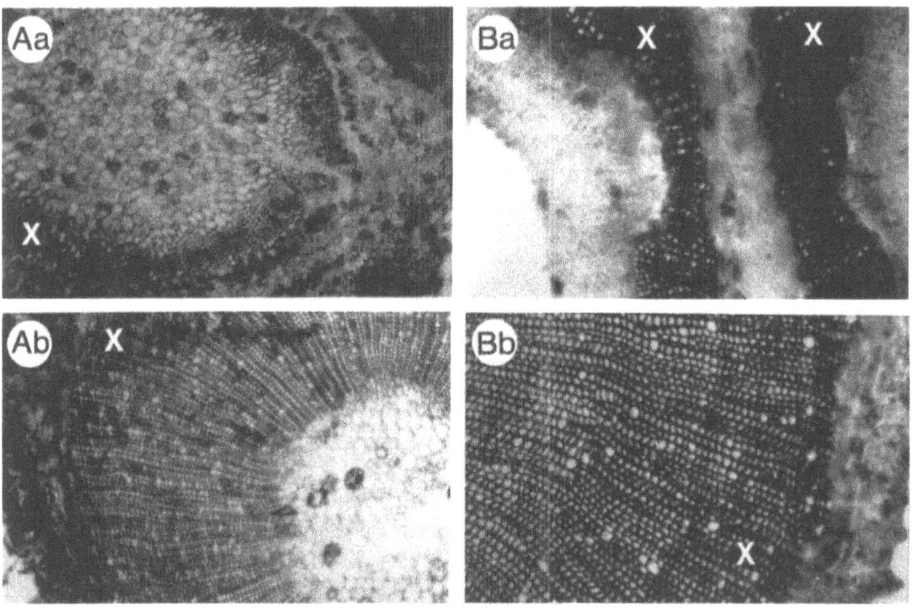

Figure 3. Anatomical cross-sections of A: *Neckia serrata* (Ochnaceae) and B: *Myrioneuron cyaneum* (Rubiaceae), at the apical (a) and basal part (b) of the stem showing secondary xylem (X). The material of Ba is compressed because it derives from a herbarium specimen.

RESULTS AND DISCUSSION

Species richness of the Reserve

Appendix 1 lists 261 species of ground herbs encountered by the author in the Batu Apoi Forest Reserve. It should be noted that this is not a complete account of the herbaceous ground flora. Other important collections of the ground herb life form in the area are those made by Argent & Mitchell, Boyce and Wong. The checklist nevertheless gives a good indication of the landscape-level (g) diversity in the Reserve. Kiew (1978) listed 105 species of the ground flora from the nearby Gunung Mulu National Park, including some woody species, but stressed that this figure probably is an underestimate.

The most species-rich family in the Reserve was the Zingiberaceae with 46 species, followed by Araceae (35), Gesneriaceae (16), Orchidaceae (16), and Cyperaceae (10); 21 families were found with less than 10 species (Figure 4). The group comprising the 14 families of pteridophytes (ferns and fern allies) is, however, more species-rich (72 species) than any of the angiosperm families.

The genera most rich in species were *Cyrtandra* (Gesneriaceae) and *Selaginella* (Selaginellaceae) each with 12 species, followed by *Diplazium* (10, Dryopteridaceae), *Tectaria* (10, Dryopteridaceae), *Amomum* (9, Zingiberaceae) and *Etlingera* (8, Zingiberaceae). Forty eight genera were only represented by one species.

Life form strategies

By far the most important life form strategy of ground herbs is the proto-terrestrial one (93%). The only scandent species found was the fern *Lygodium circinnatum*. Ten species of saprophytes were found in the area, with the genera *Epirixanthes* and *Sciaphila* being most

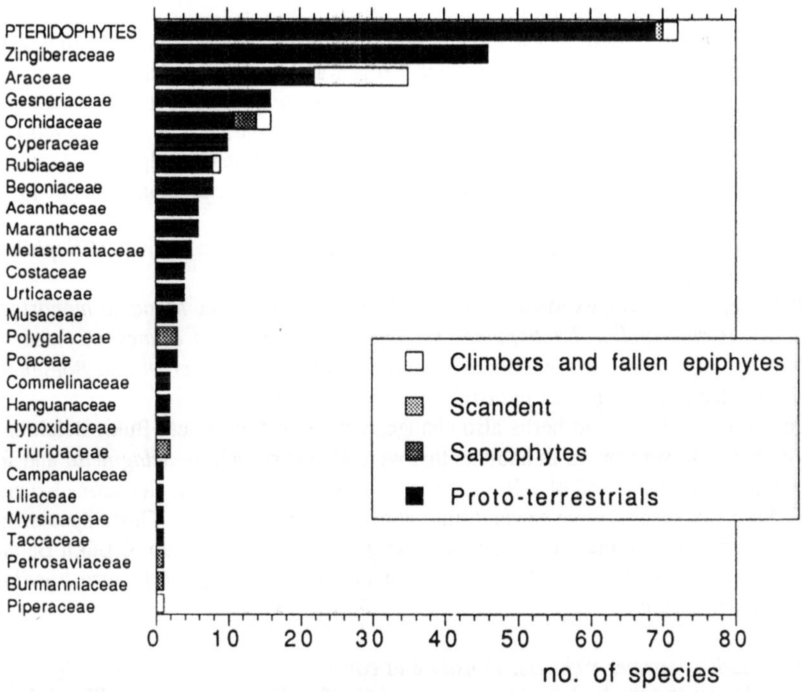

Figure 4. Number of species per family represented in the ground herb flora of the Batu Apoi Forest Reserve. The 14 families of pteridophytes are shown as one group.

common. Only 19 species of climbers and three epiphytes were rooted on the ground. Two of the most abundant species of the 13 aroid climbers were *Amydrium medium* and *Epipremnum falcifolium*. The genus *Rhaphidophora* was also found to have many juveniles contributing to the ground flora, and it is likely that more aroid species germinate on the forest floor than have been observed so far (e.g., *Pothos*). Even if climbers were omitted, Araceae would still be the second largest family in the area.

Habitats

In the nearby Gunung Mulu National Park, Kiew (1978) noted the association of ground layer species with different habitat types which she roughly divided into those associated with open canopy and others associated with closed canopy. Such habitat preferences were also observed in the Batu Apoi Forest Reserve.

About a quarter of the species were restricted to moist habitats near the rivers. Most strictly confined are the periodically flooded rheophytic herbs on shale rocks (e.g., *Bucephalandra motleyana, Piptospatha burbidgei, Tectaria lobii* var. *lobii*). On the river banks *Etlingera punicea* and *E. fimbribracteata* frequently occurred in dense stands in alluvial deposits. *Phacelophrynium maximum* was also abundant and attained a large size in the light open habitats near the river, but was also found on slopes in closed canopy forest. However, in the latter habitat no flowering was observed. This could be an example of the so called mass-effect (Shmida & Wilson 1985).

In areas with large gaps along the river created by landslides or major tree falls the herbaceous ground vegetation was dominated by giant herbs like bananas (*Musa*), and *Alocasia* spp. *Costus speciosus* was also typical of this habitat.

Alocasia lowii, A. guttata, Cyrtandra penduliflora, and many ferns grew on the moister slopes near the river but not in open gaps. Along small streams species of *Elatostema* and *Tectaria* were particularly common.

Within the closed canopy forest there was a difference between the ridge tops and the slopes. The ridge top community of ground herbs was less species-rich and characterized by *Didymocarpus lanceolatus* and *Diplazium crenatoserratum*. The proto-terrestrial orchid, *Apostasia elliptica*, of which only four individuals were found, was only observed on dry ridge tops. The ginger *Alpinia glabra* was also very abundant on dry ridge top areas - a fact that is in conflict with Kiew (1978), who states that this species dominates in areas of frequent waterlogging.

The saprophytic species were generally found in dark and moist habitats (*Sciaphila* spp. and *Epirixanthes* spp.).

On the steep slopes with exposed mineral soils, ground herbs like *Mapania monostachya, Ptyssiglottis psychotriifolia, Trichomanes singaporeanum, Begonia cyanescens* ined. were very abundant. On litter-covered slopes the gingers *Elettariopsis kerbyi* and *Boesenbergia burttiana* and the sedge *Mapania palustris* were more common.

The composition of ground herbs also changes with elevation. Near Bukit Belalong five species of *Sonerila* were found in habitats that were also very rich in *Selaginella* and ferns. These species of *Sonerila*, and 45 other species, were never observed at elevations lower than 600 m. Sixty-six species were never found above 100 m elevation. Other species had a distribution ranging from near the Sungai Belalong all the way to the top of Bukit Belalong, but never occurring on the dry ridge at intermediate elevations (e.g., *Etlingera brevilabris, Plagiostachys strobilifera*).

Vegetation studies, species richness, density and cover

In each plot seven pteridophyte and 13 angiosperm families were present: Plot A had one pteridophyte family and three angiosperm families which were not found in Plot B, and vice versa. In both plots the Zingiberaceae were represented by the highest number of species

(A:19; B:19), followed by Araceae (A:17; B:10). The remaining families were represented by less than 10 species. The pteridophyte families combined have more species than any angiosperm family in Plot A, and rank second in Plot B. This pattern of family contribution to species diversity based on small scale studies on habitat level (α-diversity) is similar to the results of the checklist (Figure 4) collected in a large area pooling various community types of the landscape (g-diversity).

Plot A had 92 species and Plot B 68 species of ground herbs. A total of 117 species were registered in the two plots and 38% were shared. Even when the larger surface area of the steep Plot A is compensated for, this plot has markedly more species. More microhabitats undoubtedly exist on the steep slope, thus enabling more species to coexist. Certainly a longer gradient from local ridge top and down slopes is represented in Plot A.

The number of tree species with dbh $\geq$ 10 cm in Plot A was 231 (Poulsen *et al.* this volume) and additional woody species not recorded (with dbh less than 10 cm) are present. Even though the total species richness in the plot is unknown, the species contribution of ground herbs is still significant.

Most of the individuals belong to a few species. In both plots 50% of the individuals are accounted for by 5–6% of the most common species of ground herbs. The pattern that a few species are common and many species rare is more pronounced for the herbs than for trees in the same plot (Figure 5).

Plot A had 6264 individuals of ground herbs whereas Plot B had 2479 individuals (Figure 2). It is noteworthy that ground herbs have a much more clustered distribution pattern than, for example, trees. Particularly in Plot A there is an evident density gradient correlated with topography and underlying edaphic factors. Plot B has a much lower overall density of ground herbs probably because the drier rich top constitutes a bigger part in this plot. Despite the high precipitation, a dry period of two weeks had a pronounced effect during which many ground herbs were observed to wilt.

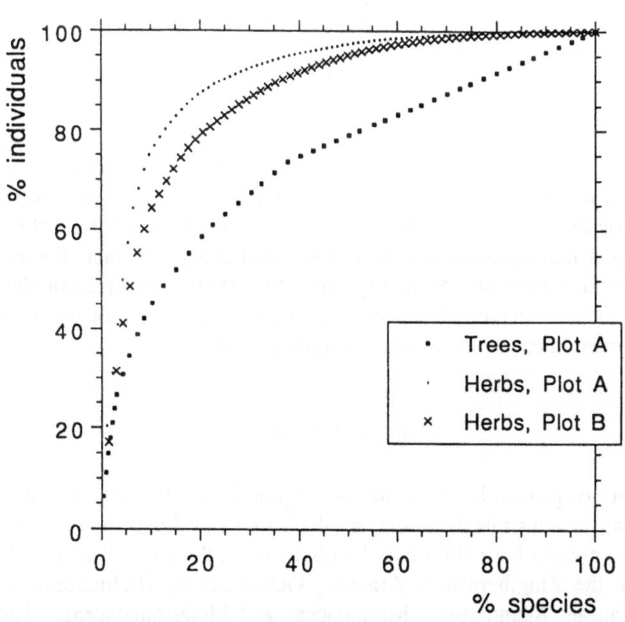

Figure 5. Species-abundance curve illustrating how a few of the species with highest density constitute the majority of the individuals.

Five species are found on both the lists of the 10 species with highest density for each plot: *Mapania monostachya, Mesophlebion falcatilobum, Elettariopsis kerbyi, Trichomanes singaporeanum*, and *Boesenbergia burttiana*. The density of these five species was in all cases more than 100 individuals per plot.

The total cover of ground herbs was 191.2 m^2 and 72.4 m^2 in Plot A and B respectively. The explanation for the markedly lower cover in Plot B may be the same as that for the density difference. In both plots, 50% of the cover is accounted for by 2–4% of the most common species of ground herbs.

Comparing the 10 species with highest cover in each plot, four species are the same: *Mapania monostachya, Mesophlebion falcatilobum, Elettariopsis kerbyi, Mapania palustris*. These species all had a cover greater than 2 m^2. In Plot A, *Stachyphrynium* cf. *jagorianum* had the greatest cover of almost 79 m^2. This species was found in a large, dense cluster, a feature probably explained by an historical event affecting the vegetation. The cluster was probably formed in a gap-phase situation, where vegetative spread as well as seed production for this species was optimal (Smith 1987). Established plants are then able to survive for some time after the canopy has closed. This is an example of a gap-phase species as opposed to the shade-loving species (Richards 1952), but the species may already have been present on the site before the gap situation.

Comparison with a Neotropical study

In a similar inventory of ground herbs in a one-hectare plot in lowland rain forest in the Amazon (Poulsen & Balslev 1991), the total number of species was 96, comprising 10,960 individuals and 2.5% cover. The Amazonian plot was not as steep as any of the Bornean ones; the additional surface area was only 2% more than a perfectly flat plot. The species richness of the Amazonian plot is, however, higher due to the larger contribution of facultative species, which have a high diversity in this region (Gentry 1982). If only proto-terrestrials are compared, the Amazonean plot has about the same number as the flattest Bornean Plot B (50/54). Despite the great variation in individual numbers associated with the plots, species richness of the Bornean and Amazonian plots is of the same magnitude, reflecting the similarities in conditions for herbs on the ground of tropical lowland rain forests in general.

The major floristic difference between the two studies is that grasses are a very prominent component of the Amazonian study. In contrast, in all inventories, only one individual of a grass (*Scrotochloa urceolata*) was found in the Bornean study. In both investigations the Araceae and pteridophytes contribute massively to the ground herb species richness, whereas Zingiberaceae plays a much greater role in the Bornean study. In fact, the Zingiberaceae seems to have a stable representation in the vegetation studies in terms of the number of species. The two most abundant ginger species in each Belalong plot were the same and their density and cover in each plot were also approximately equal.

CONCLUSIONS

The species diversity of ground herbs in the Batu Apoi Forest Reserve is relatively high in spite of the fact that the substrate is all clayish derivatives of Setap shale. Several habitats with a characteristic ground herb flora may be observed in the area. The 10 richest ground herb families were the Zingiberaceae, Araceae, Gesneriaceae, Orchidaceae, Cyperaceae, Rubiaceae, Begoniaceae, Acanthaceae, Marantaceae and Melastomataceae. These families were also among the highest ranking in respect of density and cover in vegetation studies in closed canopy forest. Pteridophytes, taken as a single group, contribute more to ground herb diversity than any of the angiosperm families. One of the reasons for the high number of species in the Batu Apoi Forest Reserve may very well be the diversity of habitats provided by the rugged landscape and the dynamics of landslides.

ACKNOWLEDGEMENTS

The author is indebted to P. Boyce, B. L. Burtt, J. Dransfield, D. S. Edwards, P. Goetghebeur, B. Hansen, K. U. Kramer, K. van der Meijden, M. Sands, G. Seidenfaden, D. Simpson R. M. Smith, E. de Vogel, B. Øllgaard for help with identification, and to the Royal Geographical Society and the Universiti Brunei Darussalam for permission to participate in the Brunei Rain Forest Project 1991–92. The study was conducted during a scholarship from the Faculty of Science, the University of Aarhus, and financed by the Danish Natural Science Research Council (Grant no. 11-9018).

LITERATURE CITED

ASHTON, P. S. 1964. Ecological studies in the mixed Dipterocarp forests of Brunei State. *Oxford Forestry Memoirs* 25:1–75.

BRUMMITT, R. K & POWELL, C.E. (eds) 1992. *Authors of plant names*. Royal Botanic Gardens, Kew.

BURTT, B. L. 1977. Notes on rain-forest herbs. *Gardens' Bulletin Singapore* 29: 73–80.

DING, HOU. 1984. Aristolochiaceae. Pp. 53–108 in *Flora Malesiana*, Ser. 1, Vol. 10.

GENTRY, A. 1982. Neotropical floristic diversity: phytogeographical connections between Central and South America, pleistocene climatic fluctuations, or an accident of the Andean orogeny?. *Annals of the Missouri Botanical Garden* 69:557–593.

GENTRY, A. & DODSON, C. 1987. Contribution of non-trees to the species richness of a tropical rain forest. *Biotropica* 19:149–156.

KIEW, R. 1978. Floristic components of the ground flora of a tropical lowland rain forest at Gunung Mulu National Park, Sarawak. *Pertanika* 1:112–119.

KIEW, R. 1987. The herbaceous flora of the Ulu Endau, Johore-Pahang, Malaysia, including taxonomic notes and descriptions of new species. *Malayan Nature Journal* 41:201–234.

KIEW, R. 1988. Herbaceous flowering plants. Pp. 56–76 in Cranbrook, Earl of (ed.). *Malaysia, key environments*. Pergamon Press, Oxford.

LINCOLN, R. J., BOXSHALL, G. A. & CLARK, P. F. 1982. *A dictionary of ecology, evolution and systematics*. Cambridge University Press, Cambridge.

POULSEN, A. D. & BALSLEV, H. 1991. Abundance and cover of ground herbs in an Amazonian rain forest. *Journal of Vegetation Science* 2:315–322.

RICHARDS, P. W. 1952. *The tropical rain forest*. Cambridge university Press, England.

RAUNKIAER, C. 1934. *The life-forms of plants and statistical plant geography*. Oxford University Press, London.

SHMIDA, A. & WILSON, M. V. 1985. Biological determinants of species diversity in temperate and tropical forests. *American Naturalist* 126:760–776.

SMITH, A. P. 1987. Respustas de hierbas del sotobosque tropical a claros ocasiondos por la caída de árboles. *Revista de Biología Tropical* 35 (Supl.1):111–118.

Appendix. Checklist of ground herb species collected or registered by the author in the scientific area of Batu Apoi Forest Reserve during the Universiti Brunei Darussalam and Royal Geographical Society's Rain Forest Project 1991–92, with notes on occurrence and habitat. Abbreviations: KBFSC = Kuala Belalong Field Studies Centre; K. = Kuala; Bt. = Bukit; S. = Sungai. Plot A was established by Aarhus University (AAU); plot B was the second plot established by Universiti Brunei Darussalam (UBD); C plots were established by Colin Pendry with specified occurrence as low (c 200 m elev.), middle (c 500 m elev.) or high (c 800 m elev.) in brackets. Voucher number is given in round brackets. Specimens are deposited at BRUN, with additional sets at K and AAU. Author names according to Brummitt & Powell 1992.

Acanthaceae

Cosmianthemum obtusifolium Bremek. Proto-terrestrial; S. Belalong, on bank above floodlevel; 70 m; (*Poulsen 117*).

Linariantha bicolor B.L.Burtt & R.M.Sm. Proto-terrestrial; small creeper along the ground, on trail along S. Belalong; 65 m; (*Poulsen 256*).

Ptyssiglottis psychotriifolia (Stapf) B.Hansen. Proto-terrestrial, rather woody at base; common on slopes particularly on exposed mineral soils; plot A, C (low); 150–250 m; (*Poulsen 40, 157, 158*).

Ptyssiglottis staminodifera B. Hansen. Proto-terrestrial, rather woody at base; S. Belalong, near river in dark, moist habitat; 65 m; (*Poulsen 67*).

Staurogyne jaherii Bremek. Proto-terrestrial; S. Belalong, near river; plot A; 60 m; (*Poulsen 28*).

Staurogyne spp. Proto-terrestrial; Bt. Belalong, steep rock near Pondok Belalong; 850 m; (*Poulsen 311*).

Araceae

Aglaonema simplex Blume Proto-terrestrial; ridge W. of KBFSC; plot A; 250 m; (*Poulsen 212*).

Alocasia denudata Engl. Proto-terrestrial; ridge W. of KBFSC; plot A, B, C (low, high); 150–850 m; (*Poulsen 79*).

Alocasia guttata N.E.Br.Proto-terrestrial; S. Baki, S. Belalong, and S. Temburong, on shale bank near river; 65–70 m; (*Poulsen 108, 235*).

Alocasia lowii Hook.f. Proto-terrestrial; S. Belalong and S. Enkiang, near river; 60 m; (*Poulsen 18, 264*).

Alocasia peltata M.Hotta Proto-terrestrial; ridge W. of KBFSC; plot A; 250 m; (*Poulsen 213*).

Alocasia sp. Proto-terrestrial, giant herb; in disturbed areas along S. Belalong; 65 m; (*Poulsen 43*).

Alocasia sp. Proto-terrestrial; ridge slope W. of KBFSC, gully between A and S. Belalong; 150 m; (*Poulsen 155*).

Alocasia sp. Proto-terrestrial; S. Esu, on vertical rock near creek; 80 m; (*Poulsen 296*).

Amydrium medium (Zoll. & Mor.) Nicolson Climber, juveniles germinate on the forest floor, flower about 2 m up a tree trunk; common; plot A, B, C (low, midlle, high); 150–850 m; (*Poulsen 173, 236, 364*).

Bucephalandra motleyana Schott. Rheophyte; on shale rocks in river; 70 m; (*Poulsen 109*).

Epipremnum falcifolium Engl. Climber, germinating on the forest floor; common; plot A, C (low, high); 65–850 m; (*Poulsen 172, 233, 363*).

Homalomena cordata (Houtt.) Schott. Proto-terrestrial; common; plot A, C (low, high); 250–850 m; (*Poulsen 25, 234*).

Homalomena geniculata M.Hotta Proto-terrestrial; Bt. Belalong; C (high); 850 m; (*Poulsen 322*).

Homalomena humilis Hook.f. Proto-terrestrial; S. Belalong, on rocks near river; 65 m; (*Poulsen 190*).

Homalomena ovata Engl. Proto-terrestrial; steep soil slopes near river and in forest; plot A, C (low); 70–250 m; (*Poulsen 168, 268*).

Homalomena vagans P.C.Boyce *ined.* Rheophyte; S. Belalong, on steep river banks close to K. Belalong; 65 m; (*Poulsen 47, 273*).

Homalomena sp. Proto-terrestrial; ridge slope W. of KBFSC; 150 m; (*Poulsen 181*).

Piptospatha burbidgei (N.E.Br.) M.Hotta Rheophyte; common on shale rocks in river; 60 m; (*Poulsen 19*).

Rhaphidophora korthalsii Schott. Climber, juvenile on ground; ridge W. of KBFSC; plot A, B; 250 m; (*Poulsen 211*).

Rhaphidophora minor Hook.f. Climber; ridge slope W. of KBFSC; plot A, B; 200 m; (*Poulsen 153*).

Rhaphidophora sp. Climber; ridge slope W. of KBFSC; plot B; 180 m; (*Poulsen 152*).

Rhaphidophora sp. Climber, fallen or juvenile on ground; ridge slope W. of KBFSC; plot A, B, C (low); 150–250 m; (*Poulsen 165*).

Rhaphidophora sp. Climber, juvenile on ground; ridge W. of KBFSC; plot A, B, C (low, high); 250–850 m; (*Poulsen 207*).

Rhaphidophora sp. Climber, juvenile on ground; ridge W. of KBFSC; plot A, C (high); 250–850 m; (*Poulsen 232*).

Rhaphidophora sp. Climber, juvenile on ground; on ridge between K. Belalong and Bt. Belalong; plot B; 250 m; (*Poulsen 362*).

Rhaphidophora sp. Climber, juvenile on ground; Bt. Belalong; plot C (high); 850 m; (*Poulsen 366*).

Rhaphidophora sp. Climber, juvenile on ground; plot C(low); 850 m; (*Poulsen 367*).

Schismatoglottis beccarianus Engl. Proto-terrestrial; Bt. Belalong; plot C (high); 850 m; (*Poulsen 365*).

Schismatoglottis brevicuspis Hook.f. Proto-terrestrial; S. Baki, on steep rock slope near stream; 70 m; (*Poulsen 139*).

Schismatoglottis calyptrata Zoll. & Mor. Proto-terrestrial; S. Belalong, on bank of small stream E. of S. Belalong; plot A; 65–250 m; (*Poulsen 192, 238, 285*).

Schismatoglottis modesta Schott Proto-terrestrial; ridge slope W. of KBFSC; plot A; 150–250 m; (*Poulsen 177, 237*).

Schismatoglottis sp. aff. *monoplacenta* M.Hotta Proto-terrestrial; ridge W. of KBFSC; plot A; 250 m; (*Poulsen 86, 239*).

Schismatoglottis sp. Proto-terrestrial; S. Belalong, near KBFSC, bank above floodlevel; 70 m; (*Poulsen 143*).

Scindapsus longipes Engl. Climber, fallen or juvenile on ground; ridge between K. Belalong and Bt. Belalong; plot A, B, C (low); 150 m; (*Poulsen 313*).

Scindapsus pictus Hassk.Climber, juvenile on ground; ridge W. of KBFSC; plot A, B, C (low,high); 250–850 m; (*Poulsen 206*).

Begoniaceae

Begonia baramensis Merr. Proto-terrestrial; Bt. Belalong; plot C (high); 850 m; (*Poulsen 316*). [= Martin Sands sp. # 1]

Begonia (cyanescens *ined.*) Proto-terrestrial; steep slope behind KBFSC; plot A, B, C (low, high); 70–850 m; (*Poulsen 6, 30*). [= Martin Sands #13]

Begonia sp. Proto-terrestrial; ridge W. of KBFSC; plot A; 250 m; (*Poulsen 31*). [= Martin Sands sp. # 9]

Begonia sp. Proto-terrestrial; S. Belalong, near waterfall; 65 m; (*Poulsen 45*). [= Martin Sands sp # 11]

Begonia sp. Proto-terrestrial; S. Temburong, on steep rocky bank near river; 70 m; (*Poulsen 58*). [= Martin Sands sp. # 10]

Begonia sp. Proto-terrestrial; S. Temburong, riverbank, in wet soil; 70 m; (*Poulsen 105*). [= Martin Sands sp. #18]

Begonia sp. Proto-terrestrial; trail along S. Belalong, above floodlevel, damp area; 65 m; (*Poulsen 258*). [= Martin Sands sp. # 4]

Begoniaceae (continued)

Begonia sp. Proto-terrestrial; Bt. Belalong; plot C (high); 850 m; (*Poulsen 317*). [= Martin Sands sp. # 17]

Burmanniaceae

Burmannia coelestis Don Saprophyte; ridge between Pondok Busiri and Bt. Belalong; 700 m; (*Poulsen 101*).

Campanulaceae

Pentaphragma prostratum Kiew Proto-terrestrial; trail along S. Belalong, above floodlevel, damp area; plot C (low, high); 65–800 m; (*Poulsen 254*).

Commelinaceae

Amischotolype mollissima (Bl.) Hassk. Proto-terrestrial; river bank and gully between A and S. Belalong; 65–150 m; (*Poulsen 156, 186*).

Pollia pumila Hallier Proto-terrestrial; S. Belalong, river bank; 65 m; (*Poulsen 187*).

Costaceae

Costus globosus Blume s. lat. sp. 1 Proto-terrestrial; ridge W. of KBFSC; plot C (low); 150 m; (*Poulsen 167*).

Costus globosus Blume s. lat. sp. 2 Proto-terrestrial; ridge slope W. of KBFSC; 150 m; (*Poulsen 175*).

Costus paradoxus K.Schum. Proto-terrestrial; S. Temburong and S. Belalong, river bank; 65 m; (no voucher).

Costus speciosus (Koenig) Sm. Proto-terrestrial; S. Temburong, near landslide in open and disturbed, rocky habitat; 75 m; (*Poulsen 54*).

Cyperaceae

Cyperus diffusus Vahl Proto-terrestrial; S. Temburong, in landslide area near river; 70 m; (*Poulsen 66*).

Mapania cuspidata (Miq.) Uittien Proto-terrestrial; S. Temburong, S. Belalong, above river, on bank of small flood stream near B; 75 m; (*Poulsen 135, 300, 369*).

Mapania graminea Uittien Proto-terrestrial; ridge W. of KBFSC; plot A, C (low, middle); 250 m; (*Poulsen 41*).

Mapania longiflora C.B.Clarke Proto-terrestrial; ridge between K. Belalong and Bt. Belalong; plot B; 250 m; (*Poulsen 318*).

Mapania monostachya Uittien Proto-terrestrial; common on ridge slopes particularly on steep, exposed soil; plot A, B, C (low, middle, high); 220-900 m; (*Poulsen 32, 39, 284*).

Mapania palustris (Hassk. ex Steud.) Fern.-Vill. Proto-terrestrial; common on ridge slopes; plot A, B, C (low, middle, high); 180-850 m; (*Poulsen 201*).

Mapania richardsii Uittien Proto-terrestrial; ridge W. of KBFSC; 200 m; (*Poulsen 36*).

Paramapania parvibracteata (C.B.Clarke) Uittien Proto-terrestrial; ridge W. of KBFSC; plot A, B, C (low, high); 250-850 m; (*Poulsen 53*).

Scleria motleyi C.B.Clarke Proto-terrestrial; Bt. Belalong, in clearing on the summit; plot C (high); 850–900 m; (*Poulsen 97*).

Scleria purpurascens Steud. Proto-terrestrial; S. Belalong, near river; 60 m; (*Poulsen 16*).

Gesneriaceae

Cyrtandra basiflora C.B.Clarke Proto-terrestrial, rather woody at base, about 1 m; on banks of flood stream; plot A, C (low, high); 75–750 m; (*Poulsen 288*).

Cyrtandra digitaliflora B.L.Burtt Proto-terrestrial; Bt. Belalong; plot C (high); 850 m; (*Poulsen 381*).

Cyrtandra penduliflora Kraenzl. Proto-terrestrial; on steep slope behind KBFSC; 75 m; (*Poulsen 49*).

Cyrtandra phoenicolasia Lauterb. Proto-terrestrial; S. Belalong, in rocky area near little stream; plot A, B, C (low) 70 m; (*Poulsen 188*).

Cyrtandra sarawakensis C.B.Cl. Proto-terrestrial, ascending woody stem with fleshy outer layer; S. Esu, wet bank above river; 70 m; (*Poulsen 297*).

Cyrtandra sp. aff. *lambirense* Proto-terrestrial; steep slopes on ridge top W. of KBFSC, near A; 250 m; (*Poulsen 76*).

Cyrtandra sp. Proto-terrestrial; ridge slope W. of KBFSC, gully between A and S. Belalong; 150 m; (*Poulsen 161*).

Cyrtandra sp. Proto-terrestrial; ridge W. of KBFSC; plot A, B, C (low); 250 m; (*Poulsen 119*).

Cyrtandra sp. Proto-terrestrial, rather woody at base; ridge slope W. of KBFSC; plot A, B, C (middle, high); 250–800 m; (*Poulsen 120*).

Cyrtandra sp. Proto-terrestrial, rather woody base; ridge W. of KBFSC; plot A; 250 m; (*Poulsen 220*).

Cyrtandra sp. Proto-terrestrial, rather woody base; ridge W. of KBFSC; plot A, B, C (low); 250 m; (*Poulsen 221*).

Cyrtandra sp. Proto-terrestrial; Bt. Belalong; plot C (high); 850 m; (*Poulsen 302*).

Didymocarpus amoenus C.B.Clarke s. lat. Proto-terrestrial; from river banks of S. Belalong to Bt. Belalong, below summit; 65–850 m; (*Poulsen 189, 283*).

Didymocarpus gracilipes C.B.Clarke s. lat. Proto-terrestrial; Bt. Belalong, steep rock near Pondok Belalong; 850 m; (*Poulsen 310*).

Didymocarpus lanceolatus C.B.Clarke Proto-terrestrial, woody base; ridge top W. of KBFSC; plot A, C (low); 280 m; (*Poulsen 85*).

Loxocarpus petiolaris (C.B.Clarke) B.L.Burtt Proto-terrestrial; Bt. Belalong; plot C (high); 850 m; (*Poulsen 321*).

Hanguanaceae

Hanguana malayana (Jack) Merr. Proto-terrestrial; short stem rather woody, leaves linear; S. Belalong, on river bank; 70 m; (*Poulsen 121*).

Hanguana sp. 1 Proto-terrestrial, stem rather woody, broad leaves; S. Belalong, on slope near river; plot B, C (high); 60–850 m; (*Poulsen 33*).

Hypoxidaceae

Curculigo latifolia Dryand. Proto-terrestrial; Bt. Belalong, S. Belalong, steep slope behind KBFSC; plot A, B, C (low, middle, high); 80–850 m; (*Poulsen 90; 295*).

Curculigo racemosa Ridl. Proto-terrestrial, occasional epiphyte; from S. Babi, steep bank above river, to ridge slopes near Bt. Belalong; plot C (high); 70–850 m; (*Poulsen 38, 60, 278*).

Liliaceae

Peliosanthes teta Andr. ssp. *humilis* (Andr.) Jessop. Proto-terrestrial; Bt. Belalong; plot C (high); 850 m; (*Poulsen 368*).

Marantaceae

Donax cannaeformis (G. Forst.) K. Scum. Proto-terrestrial, woody 4 m tall; on shale in gully between A and S. Belalong; 150 m; (*Poulsen 166*).

Phacelophrynium maximum (Blume) K.Schum. Proto-terrestrial; S. Belalong river bank and on steep slope behind KBFSC; plot A, C (low) 65-150 m; (*Poulsen 2, 44, 61*).

Phrynium cf. *placentarium* (Laur.) Merr. Proto-terrestrial; on bank near river; 65 m; (*Poulsen 191*).

Phrynium sp. aff. *inflatum* Merr. Proto-terrestrial; ridge W. of KBFSC; plot A, B, 250 m; (*Poulsen 209*).

Stachyphrynium borneense Ridl. Proto-terrestrial; ridge W. of KBFSC, ridgetop; plot A; 250 m; (*Poulsen 75*).

Stachyphrynium cf. *jagorianum* (K.Koch) K.Schum. Proto-terrestrial; ridge between K. Belalong and Bt. Belalong and ridge W. of KBFSC; plot A; 250 m; (*Poulsen 282*).

Melastomataceae

Sonerila begoniefolia Blume. Proto-terrestrial, very low with woody base; Bt. Belalong; plot C (high); 880 m; (*Poulsen 308*).

Sonerila borneensis Cogn. var. *setulosa* Stapf. Proto-terrestrial, occasionally epiphyte on mossy trunks; Bt. Belalong; plot C (high); 850 m; (*Poulsen 320*).

Sonerila heterophylla Jack. Proto-terrestrial; Bt. Belalong; plot C (high); 850 m; (*Poulsen 319*).

Sonerila obovata Schwartz. Proto-terrestrial; Bt. Belalong; plot C (high); 880 m; (*Poulsen 94*).

Sonerila pulchella Stapf. Proto-terrestrial; Bt. Belalong; plot C (high); 900 m; (*Poulsen 100*).

Musaceae

Musa 3 spp. Proto-terrestrial, giant herb; S. Belalong and S. Temburong, in large gaps near river; 65 m; (no voucher, Argent pers. comm.).

Myrsinaceae

Labisia pumila (Blume) Fer.-Vill. Proto-terrestrial, with woody base; ridge W. of KBFSC; plot B, C (low, high); 150–850 m; (*Poulsen 77*).

Orchidaceae

Aphyllorchis pallida Blume. Saprophyte, less than 20 cm; ridge between K. Belalong and Bt. Belalong; plot C (middle); 250–400 m; (*Poulsen 281*).

Apostasia elliptica J.J.Sm. Proto-terrestrial; ridge between K. Belalong and Bt. Belalong, near Pondok Busiri; plot B; 250–500 m; (*Poulsen 89*).

Apostasia wallichii R. Br. in Wall. Proto-terrestrial, 50 cm; ridge W. of KBFSC and Bt. Belalong; plot A, B, C (low, high); 250–800 m; (*Poulsen 122, 312*).

Claderia viridiflora Hook.f. Climber, on ground; ridge between K. Belalong to Bt. Belalong; plot B; 250–300 m; (*Poulsen 219, 371*).

Cystorchis variegata Blume. Proto-terrestrial; in moist, dark places in forest; plot B, C (high); 250–800 m; (*Poulsen 323*).

Dendrobium sp. Epiphyte, probably fallen, rooted in ground; ridge W. of KBFSC; plot A; 250 m; (*Poulsen 215*).

Dipodium sp. Climber, juvenile on ground; on ridge between K. Belalong and Bt. Belalong; plot B; 250 m; (*Poulsen 374*).

Eria sp. Epiphyte, probably fallen, rooted in ground; on ridge between K. Belalong and Bt. Belalong; plot B; 250 m; (*Poulsen 375*).

Lecanorchis multiflora J.J.Sm. Saprophyte, 50 cm; ridge between K. Belalong and Bt. Belalong; plot A, C (low, middle); 350 m; (*Poulsen 88*).

Liparis wrayi Hook.f. Proto-terrestrial; trail from K. Belalong to Bt. Belalong, on big, old well-decomposed log; plot C (low); 150–275 m; (*Poulsen 200*).

Malaxis sp. Proto-terrestrial; on trail along S. Belalong, above floodlevel, damp area; 65 m; (*Poulsen 255*).

Mischobulbum scapigerum (Hook.f.) Schltr. Proto-terrestrial; Bt. Belalong; plot C (high); 850 m; (*Poulsen 373*).

Plocoglottis sp. Proto-terrestrial; S. Belalong; 70 m; (*Poulsen 218*).

Tainia paucifolia (Breda) J.J.Sm. Proto-terrestrial; ridge W. of KBFSC; plot A; 250 m; (*Poulsen 214*).

Tropidia graminea Blume. Proto-terrestrial; ridge W. of KBFSC and Bt. Belalong; plot A, C (high); 250–850 m; (*Poulsen 216*).

Tropidia saprophytica J.J.Sm. Saprophyte; ridge between K. Belalong and Bt. Belalong, Bt. Belalong; plot B; 250–750 m; (*Poulsen 303*).

Petrosaviaceae

Petrosavia stellaris Becc. Saprophyte; damp flat area below Bt. Belalong; 850 m; (*Poulsen 305*).

Piperaceae

Piper cf. *abbreviatum* Opiz. Climber, juvenile on ground, ridge W. of KBFSC; plot A; 250 m; (*Poulsen 241*).

Poaceae

Paspalum conjugatum Berg. Proto-terrestrial; S. Belalong, near river; 60 m; (*Poulsen 17*).

Scrotochloa urceolata (Roxb.) Judz. Proto-terrestrial; valley West of KBFSC; plot B; 110–250 m; (*Poulsen 293*).

Thysanolaena maxima Kuntze Proto-terrestrial, 3-4 m long stems hanging a few meters above river; S. Temburong, steep rocky river bank in light gap; 70 m; (*Poulsen 274*).

Polygalaceae

Epirixanthes cylindrica Blume. Saprophyte; ridge between K. Belalong and Bt. Belalong, damp flat area below Bt. Belalong; plot C (high); 800 m; (*Poulsen 304*).

Epirixanthes elongata Blume. Saprophyte; ridge W. of KBFSC; plot A, B; 250 m; (*Poulsen 224*).

Epirixanthes pallida Wendt. Saprophyte; ridge W. of KBFSC; 80 m; (*Poulsen 3*).

Rubiaceae

Acranthera sp. aff. *frutescens* Valeton. Proto-terrestrial, with woody base; ridge slope W. of KBFSC; plot A, C (low, high); 250–800 m; (*Poulsen 52*).

Acranthera sp. aff. *involucrata* Valeton. Proto-terrestrial, with woody base; ridge slope W. of KBFSC; plot A, C (low); 250–800 (*Poulsen 51*).

Acranthera sp. aff. *longipetiolata* Bremek. Proto-terrestrial, woody at base; valley W. of KBFSC, near river; 100 m; (*Poulsen 286*).

Rubiaceae (continued)

Argostemma hameliaefolium Wernham in Gibbs. Proto-terrestrial; S. Belalong, river bank above floodlevel, damp area; 65 m; (*Poulsen 12, 253*).

Argostemma moultonii Ridl. Proto-terrestrial; Bt. Belalong; plot C (high); 850 m; (*Poulsen 309*).

Argostemma subfalcifolium Bakh.f. Proto-terrestrial; Bt. Belalong; plot C (low, high); 850 m; (*Poulsen 325*).

Hedyotis sp. Proto-terrestrial; ridge W. of KBFSC; plot A; 250 m; (*Poulsen 222*).

Ophiorrhiza sp. Proto-terrestrial; ridge W. of KBFSC; plot A; 250 m; (*Poulsen 223*).

Psychotria sp. Climber, germinating on ground, small plant; ridge W. of KBFSC; plot A, B, C (low, middle, high); 250–850 m; (*Poulsen 118*).

Taccaceae

Tacca integrifolia Ker Gawl. Proto-terrestrial; S. Temburong, bank above river; 65 m; (*Poulsen 240*).

Triuridaceae

Sciaphila densiflora Schltr. Saprophyte; ridge W. of KBFSC; plot A, C (low); 250 m; (*Poulsen 27*).

Sciaphila secundiflora Thw. ex Benth. Saprophyte; ridge W. of KBFSC, between K. Belalong and Bt. Belalong, and damp flat area below Bt. Belalong; plot A, C (high); 250–800 m; (*Poulsen 306, 307*).

Urticaceae

Elatostema sp. aff. *kabayense* (Gibbs) Schroter Proto-terrestrial, long creeping, also epiphytic(*193*); on trail along S. Belalong, above floodlevel, damp area; 65 m; (*Poulsen 193, 252*).

Elatostema sp. aff. *sesquifolium* (Bl.) Hassk. Proto-terrestrial; ridge slope W. of KBFSC, gully between A and S. Belalong; 150 m; (*Poulsen 159*).

Elatostema sp. Proto-terrestrial; S. Belalong and S. Temburong, river bank in wet soil; 60 m; (*Poulsen 10, 107*).

Elatostema sp. Proto-terrestrial; S. Baki, on shale rocks in river; 70 m; (*Poulsen 111*).

Zingiberaceae

Alpinia glabra Ridl. Proto-terrestrial; particularly common on dry ridges; plot B, C (middle, high); 180–800 m; (*Poulsen 29, 279*).

Alpinia ligulata K.Schum. Proto-terrestrial; creek adjoining S. Belalong and along trail to S. Esu, bank; 60–65 m; (*Poulsen 13, 329*).

Alpinia sp. aff. *capitellata* Jack. Proto-terrestrial; ridge W. of KBFSC; Ruth Levy's plot, A, B; 250 m; (*Poulsen 226*).

Alpinia sp. sect. Cenolophon. Proto-terrestrial, big dark green ginger over 2m; ridge slopes; plot B, C (high); 250–850 m; (*Poulsen 327*).

Alpinia sp. sect. Allughas. Proto-terrestrial; Bt. Belalong; plot C (high); 850 m; (*Poulsen 328*).

Amomum bicorniculatum K.Schum. Proto-terrestrial; ridge slopes; plot B, C (low, high); 250–850 m; (*Poulsen 326*).

Amomum coriaceum R.M.Sm. Proto-terrestrial; common from river banks to Bt. Belalong; plot A, B, C (low, high); 70–800 m; (*Poulsen 21; 277*).

Amomum laxisquarrosum K.Schum. vel aff. Proto-terrestrial; S. Belalong, steep soil; 70 m; (*Poulsen 267*).

Amomum sp. aff. *oliganthum* K.Schum. Proto-terrestrial; on bank of small stream from A adjoining S. Belalong; 70 m; (*Poulsen 261*).

Amomum pungens R.M.Sm. vel aff. Proto-terrestrial; S. Babi, on wet soil bank in shade; 70 m; (*Poulsen 276*).

Amomum sp. Proto-terrestrial; ridge W. of KBFSC; plot A; 250 m; (*Poulsen 20*).

Amomum sp. aff. *ridleyi* R.M. Sm. Proto-terrestrial; ridge between K. Belalong and Bt. Belalong; plot B, C (high); 300–850 m; (*Poulsen 87*).

Amomum sp. aff. *flavoalbum* R.M. Sm. Proto-terrestrial; very common on ridge slopes; plot A, B, C (low, middle, high); 250–800 m; (*Poulsen 280*).

Amomum sp. Proto-terrestrial; ridge W. of KBFSC; plot A; 250 m; (*Poulsen 24*).

Boesenbergia belalongensis A.D.Poulsen. Proto-terrestrial; from S. Belalong near river to Bt. Belalong, steep shale bank; plot C (middle, high); 65–800 m; (*Poulsen 182*).

Boesenbergia flavoalba R.M.Sm. Proto-terrestrial; common on ridge slopes in closed forest; plot A, B, C (low, high); 250–800 m; (*Poulsen 50*).

Boesenbergia gracilipes (K. Schum.) R.M. Sm. Proto-terrestrial ; S. Temburong, on flat bank 50 m from river; 75 m; (*Poulsen 137*).

Boesenbergia orbiculata R.M.Sm. Proto-terrestrial, creeping; on trail along S. Belalong, above floodlevel, damp area; 65 m; (*Poulsen 257*).

Boesenbergia urceoligena A.D.Poulsen.. Proto-terrestrial; on trail along S. Belalong, above floodlevel, damp area; 65 m; (*Poulsen 251*).

Elettariopsis kerbyi R.M.Sm. Proto-terrestrial; common species on ridges; plot A, B, C (low, middle, high); 250–850 m; (*Poulsen 74*).

Elettariopsis sp. aff. *curtisii* Baker Proto-terrestrial; on slopes; plot A, B, C (low); 250 m; (*Poulsen 227*).

Elettariopsis sp. Proto-terrestrial; S. Esu, wet bank near creek; plot B; 65–250 m; (*Poulsen 330*).

Etlingera brevilabris (Valeton) R.M.Sm. Proto-terrestrial; from S. Belalong, river bank to Bt. Belalong; plot A, B, C (low, high); 65–800 m; (*Poulsen 170*).

Etlingera fimbriobracteata (K. Schum.) R.M.Sm. Proto-terrestrial, giant ginger confined to river banks of S. Temburong and S. Belalong; 65 m; (*Poulsen 56; 376*).

Etlingera punicea (Roxb.) R.M.Sm. Proto-terrestrial, giant ginger; confined to river banks of S. Temburong and S. Belalong; iban name: topos; 65 m; (*Poulsen 46*).

Etlingera velutina (Ridl.) R.M.Sm. Proto-terrestrial; S. Baki, in wet area near creek, and on ridge slopes W. of KBFSC; plot A; 60–250 m; (*Poulsen 83*).

Etlingera sp. nov. Proto-terrestrial; ridge slopes; plot A, B, C (high); 80–800 m; (*Poulsen 35*).

Etlingera sp. nov. Proto-terrestrial; ridge W. of KBFSC; plot A, C (high) 220–800 m; (*Poulsen 130*).

Etlingera sp. aff. *muluensis* R.M.Sm. Proto-terrestrial; S. Belalong, on river bank; common in B; 65–250 m; (*Poulsen 169*).

Etlingera sp. Proto-terrestrial; ridge slopes; plot A, C (low, middle); 250–500 m; (*Poulsen 225*).

Globba atrosanguinea Teijsm. & Binn. Proto-terrestrial; S. Belalong; in moist areas near river; 65 m; (*Poulsen 171, 331*).

Globba tricolor Ridl. var. *gibbsiae* (Ridl.) R.M.Sm. vel. aff. Proto-terrestrial; S. Enkiang, S. Babi, common in moister places on slopes; plot A, C (low, high); 80–800 m; (*Poulsen 116, 263, 275*).

Zingiberaceae (continued)

Hedychium muluense R.M.Sm. Epiphyte, on fallen debris; S. Belalong, near KBFSC; 70 m; (*Poulsen 287*).
Hornstedtia scyphifera (Koenig) Steud. Proto-terrestrial; S. Temburong and S. Belalong, river bank; 70 m; (*Poulsen 55, 64*).
Hornstedtia tomentosa (Bl.) Bakh. f. Proto-terrestrial; Bt. Belalong; plot C (high); 850 m; (*Poulsen 332*).
Plagiostachys albiflora Ridl. Proto-terrestrial; very common on ridge slopes; plot A, B, C (middle, high); 150–800 m; (*Poulsen 22, 34, 176*).
Plagiostachys strobilifera (Baker) Ridl. Proto-terrestrial; very common species on ridge slopes, Bt. Belalong; Iban name: banjang; plot A, B, C (low, high); 180–800 m; (*Poulsen 4, 23, 95*).
Zingiber griffithii Baker. Proto-terrestrial; on steep slope behind KBFSC, Bt. Belalong; plot C (high) 70–800 m; (*Poulsen 202*).
Zingiber pseudopungens R.M.Sm. Proto-terrestrial; S. Baki, valley W. of KBFSC, in wet places; 70–110 m; (*Poulsen 110, 289*).
Zingiber cf. *albiflorum* R.M.Sm. Proto-terrestrial; ridge W. of KBFSC; plot A, B, C (high); 250–800 m; (*Poulsen 228*).
Zingiber sp. Proto-terrestrial; ridge slopes; plot A, B, 250 m; (*Poulsen 337*).
Genus indet. *Geocharis* or Alpinia. Proto-terrestrial; ridge slopes; plot A, B, C (low, middle); 250 m; (*Poulsen 229*).
Genus indet. Proto-terrestrial; Bt. Belalong; plot C (high); 850 m; (*Poulsen 333*).
Genus indet. Proto-terrestrial; .Bt. Belalong; plot C (high); 850 m; (*Poulsen 334*).
Genus indet. Proto-terrestrial; Bt. Belalong; plot C (high); 850 m; (*Poulsen 335*).
Genus indet. Proto-terrestrial; Bt. Belalong; plot C (high); 850 m; (*Poulsen 336*).

PTERIDOPHYTES

Blechnaceae

Blechnum finlaysonianum Wall. ex Hook. & Grev. Proto-terrestrial; S. Temburong, on bank in dark area; plot A; 70 m; (*Poulsen 136*).

Cyatheaceae

Cyathea moluccana R.Br. Proto-terrestrial; S. Temburong, river bank; 65 m; (*Poulsen 69*).

Dennstaedtiaceae

Lindsaea borneensis Hook. ex Baker. Proto-terrestrial; Bt. Belalong; plot C (middle, high); 500–850 m; (*Poulsen 93*).
Lindsaea cultrata (Willd.) Sw. Proto-terrestrial; ridge slope W. of KBFSC ; plot A, B, C (low, high); 250–850 m; (*Poulsen 128, 343*).
Lindsaea divergens Hook. & Grev. [=Isoloma divergens]. Proto-terrestrial; ridge between K. Belalong and Bt. Belalong, in small valley near B; 200 m; (*Poulsen 271*).
Lindsaea doryphora Kramer. Proto-terrestrial; on ridge between K. Belalong and Bt. Belalong; plot B; 250 m; (*Poulsen 341*).
Lindsaea integra Holttum. Proto-terrestrial; Bt. Belalong; plot B, C (high); 250–850 m; (*Poulsen 344*).
Lindsaea lucida Blume ssp. lucida. Proto-terrestrial; S. Temburong, muddy river bank; 65 m; (*Poulsen 72*).
Lindsaea parallelogramma Alderw. Proto-terrestrial; common from S. Temburong, muddy river bank to Bt. Belalong; plot A, B, C (high); 65–850 m; (*Poulsen 71, 342*).
Tapainidium pinnatum (Cav.) C.Chr. Proto-terrestrial; S. Temburong, muddy bank near river; 65 m; (*Poulsen 70*).
Tapeinidium luzonicum (Hook.) Kramer. Proto-terrestrial; Bt. Belalong; plot C (high); 850 m; (*Poulsen 350*).

Dipteridaceae

Dipteris lobbiana (Hook.) Moore. Rheophyte; S. Temburong, close to river; 70 m; (*Poulsen 63*).

Dryopteridaceae

Arachniodes cf. *tripinnata* (Goldm.) Sledge agg. Proto-terrestrial; Bt. Belalong; plot C (high); 850 m; (*Poulsen 353*).
Cyclopeltis crenata (Fée) C.Chr. Proto-terrestrial; ridge slope W. of KBFSC; 150 m; (*Poulsen 178*).
Diplazium asperum Blume. Proto-terrestrial, up to 2.5 m; S. Belalong, small stream adjoining S. Belalong; 65 m; (*Poulsen 184*).
Diplazium cordifolium Blume. Proto-terrestrial, often with irridescent leaves; common on moist slopes; plot A, B, C (low, high); 250–850 m; (*Poulsen 203*).
Diplazium crenatoserratum (Blume) Moore. Proto-terrestrial; on higher slopes near ridge tops; plot A, B, C (low, middle, high); 250–850 m; (*Poulsen 26, 73, 91*).
Diplazium esculentum (Retz.) Sw. Proto-terrestrial; young leaves eaten as vegetable; iban name: pakoh; K. Belalong, light open riverbank on accumulated organic material; 65 m; (*Poulsen 144*).
Diplazium mallacense C.Presl. Proto-terrestrial; Bt. Belalong; plot C (high); 850 m; (*Poulsen 348*).
Diplazium pallidum (Blume) Moore. Proto-terrestrial; leaves often irridescent; Bt. Belalong; plot C (high); 850 m; (*Poulsen 346*).
Diplazium porphyrorachis (Baker) Diels. Proto-terrestrial; common on steep soil slopes with exposed soil; plot A, B, C (low, high); 220–850 m; (*Poulsen 80*).
Diplazium riparium Holttum. Proto-terrestrial; near flood steram in forest; plot A, C (low); 250 m; (*Poulsen 123*).
Diplazium xiphophyllum (Baker) C.Chr. Proto-terrestrial; S. Baki, on clayish soil slope near river; 70 m; (*Poulsen 114*).
Diplazium sp. Proto-terrestrial; Bt. Belalong; plot C (high); 850 m; (*Poulsen 347*).
Pleocnemia irregularis (Presl) Holttum. Proto-terrestrial; valley and ridge slope W. of KBFSC, in wet area near pig wallow; 110 m; (*Poulsen 162, 294*).
Tectaria angulata (Willd.) Copel. Proto-terrestrial; ridge slope W. of KBFSC; plot A; 150–250 m; (*Poulsen 154, 163*).
Tectaria barberi (Hook.) Copel. Proto-terrestrial; ridge slopes in moist places; plot A, C (high); 250–850 m; (*Poulsen 125*).
Tectaria decurrens (Presl) Copel. Proto-terrestrial; S. Baki, on soil slope near stream; 70 m; (*Poulsen 140*).
Tectaria inopinata Holttum. Proto-terrestrial; Bt. Belalong, on rocks; plot C (high); 850 m; (*Poulsen 354*).
Tectaria lobii (Hook.) Copel var. *lobii*. Rheophyte; S. Belalong, river bank shale; 70 m; (*Poulsen 299*).
Tectaria ternata (Baker) Copel. Rheophyte; S. Baki, on shale rocks in river; plot A, C (low, high); 70–850 m; (*Poulsen 112, 132*).
Tectaria vasta (Blume) Copel. Proto-terrestrial; S. Temburong, soil slope on river bank, in shade; 70 m; (*Poulsen 138*).
Tectaria sp. Proto-terrestrial; ridge slope W. of KBFSC, between A and S. Belalong, gully; 150 m; (*Poulsen 164*).
Tectaria sp. Proto-terrestrial; Bt. Belalong; plot C (high); 850 m; (*Poulsen 349*).
Tectaria sp. Proto-terrestrial; Bt. Belalong on rocks; plot C (high); 850 m; (*Poulsen 355*).

Gleicheniaceae

Dicranopteris linearis (Burm.f.) Underw. ssp. *subpectinata*. Proto-terrestrial, becoming scandent with age; ridge W. of KBFSC, in light gap; plot C (high); 250–850 m; (*Poulsen 82*).

Gleichenia truncata Willd. Proto-terrestrial; from S. Temburong, overhanging river to Bt. Belalong, in clearing on the summit; 70–913 m; (*Poulsen 96, 104*).

Hymenophyllaceae

Trichomanes obscurum Blume. Proto-terrestrial; on ridge between K. Belalong and Bt. Belalong; plot B, C (high); 250–850 m; (*Poulsen 340*).

Trichomanes singaporeanum (Bosch) Alderw. Proto-terrestrial; common on ridge slopes, abundant on steep slopes with exposed soil; plot A, B, C (low, middle, high); 250–850 m; (*Poulsen 204*).

Lomariopsidaceae

Lomariopsis lineata (Presl) Holttum. Climber; juvenile on ground; plot A; 250 m; (*Poulsen 126*).

Teratophyllum rotundifoliatum (Bonap.) Holttum. Climber, juvenile with bathyphylls on ground; ridge slope W. of KBFSC; plot A; 250 m; (*Poulsen 131*).

Lycopodiaceae

Lycopodium cernuum L. Proto-terrestrial; Bt. Belalong; plot C (high); 850 m; (*Poulsen 356*).

Marattiaceae

Angiopteris sp. Proto-terrestrial; Bt. Belalong; plot C (high); 850 m; (*Poulsen 345*).

Christensenia aesculifolia (Blume) Maxon. Proto-terrestrial; S. Baki, soil slope near creek; 70 m; (*Poulsen 113*).

Polypodiaceae

Asplenium tenerum Forst. var. *retusum*. Proto-terrestrial; S. Esu, on rock; 70 m; (*Poulsen 259*).

Pteridaceae

Syngramma alismifolia (Presl) J.Sm. Proto-terrestrial; on ridgebetween K. Belalong and Bt. Belalong; plot B, C (high); 250–850 m; (*Poulsen 338*).

Taenitis blechnoides (Willd.) Sw. Proto-terrestrial; on ridge between K. Belalong and Bt. Belalong; plot B, C (low, high); 250–850 m; (*Poulsen 339*).

Schizaeaceae

Lygodium circinattum (Burm. f.) Sw. Scandent; ridge slope W. of KBFSC; plot A; 70–250 m; (*Poulsen 133*).

Schizaea dichotoma (L.) Sm. Proto-terrestrial; ridge slope W. of KBFSC; plot A, B; 250 m; (*Poulsen 127*).

Schizaea digitata (L.) Sw. Proto-terrestrial; on ridge near Pondok Busiri; 500 m; (*Poulsen 360*).

Selaginellaceae

Selaginella bluuensis Alderw. Proto-terrestrial; common on ridge slopes; plot A, B, C (low, high); 70–850 m; (*Poulsen 7*).

Selaginella dielsii Hieron. Proto-terrestrial; Bt. Belalong; in clearing on the summit; plot C (high); 800–900 m; (*Poulsen 99*).

Selaginella cf. *boschai* Hieron. Proto-terrestrial; on ridge W. of KBFSC; A ; 250 m; (*Poulsen 230*).

Selaginella subserpentina Alderw. Proto-terrestrial; Bt. Belalong, in clearing near the summit; plot C (high); 800–900 m; (*Poulsen 98*).

Selaginella cf. *brevipes* A.Br. Proto-terrestrial; on steep slope behind KBFSC; 70 m; (*Poulsen 8*).

Selaginella sp. Proto-terrestrial; S. Temburong, river bank; 70 m; (*Poulsen 231*).

Selaginella sp. Proto-terrestrial; ridge W. of KBFSC; 250 m; (*Poulsen 242*).

Selaginella sp. Proto-terrestrial; Bt. Belalong; plot C (high); 850 m; (*Poulsen 357*).

Selaginella sp. Proto-terrestrial; Bt. Belalong; plot C (high); 850 m; (*Poulsen 358*).

Selaginella sp. Proto-terrestrial; Bt. Belalong; plot C (high); 850 m; (*Poulsen 378*).

Selaginella sp. Proto-terrestrial; ridge slopes to Bt. Belalong; plot B, C (high); 250–800 m; (*Poulsen 379*).

Selaginella sp. Proto-terrestrial; Bt. Belalong; plot A, C (high); 850 m; (*Poulsen 380*).

Thelypteridaceae

Mesophlebion beccarianum (Ces.) Holttum. Proto-terrestrial; Bt. Belalong; plot C (middle, high); 850 m; (*Poulsen 92*).

Mesophlebion falcatilobum Holttum var. *falcatilobum*. Proto-terrestrial; one of the most common ferns in the area; plot A, B, C (low, middle); 250–800 m; (*Poulsen 134*).

Mesophlebion motleyanum (Hook.) Holttum. Proto-terrestrial, juvenile from ecological study [alternatively M. dulitense Holtt.]; Bt. Belalong; plot C (high); 850 m; (*Poulsen 351*).

Pronephrium nitidum Holttum. Proto-terrestrial; ridge slope W. of KBFSC; plot A; 250 m; (*Poulsen 129*).

Pronephrium cf. *menisciicarpon* (Blume) Holttum. Proto-terrestrial; valley W. of KBFSC, in wet flood stream; 110 m; (*Poulsen 291*).

Pronephrium sp. Proto-terrestrial; Bt. Belalong; plot C (high); 850 m; (*Poulsen 352*).

Sphaerostephanos pterosporus (Alderw.) Holttum var. *crenatus* Holttum. Proto-terrestrial; ridge slope W. of KBFSC, on rocks in floodstream; plot A; 220 m; (*Poulsen 124*).

Sphaerostephanos sp. Proto-terrestrial; K. Belalong, light open riverbank on accumulated organic material; 65 m; (*Poulsen 145*).

Sphaerostephanos sp. Proto-terrestrial, valley West of KBFSC, in wet flood stream; 110 m; (*Poulsen 290*).

Sphaerostephanos sp. aff. *trichoclamys* Holttum. Proto-terrestrial; on ridge NW of A, near floodstream; 250 m; (*Poulsen 361*).

Wild bees of Brunei Darussalam

DAVID W. ROUBIK

Smithsonian Tropical Research Institute, Unit 0948, APO AA 34002-0948, USA.

ABSTRACT. The wild bee fauna of Borneo and Brunei is singularly depauperate compared with other tropical areas but very rich in the two bee groups at the zenith of social insects. In the forests of Belalong I found 4 *Apis* and 22 stingless bee species; the latter, mostly *Trigona*, included 76% of all stingless bee species known from Borneo, and all three endemics. Other bee species probably number about 80 if the proportions of the highly social bees found in Indonesia apply to Belalong. A 'rapid assessment' of the highly social bee fauna was possible in Brunei, even during rainy weather in November, by spraying sugar and salt baits on vegetation. Virtually every species of *Trigona*, *Apis* and *Pariotrigona* came to these baits during a few days. Assays of various salt solutions revealed the following preferences by the stingless bees (predominantly *Trigona melina*) and honey bees (predominantly *Apis koschevnikovi* and *A. dorsata*) : Na > Mg > K > NH_4 > PO_4. Preference for sodium, magnesium, and potassium is possibly linked to neural transmission requirements, osmotic water balance and low levels of these elements in soil, expressed directly in the nectar and pollen that bees consume. *Apis koschevnikovi* was found to be similar to African and other tropical cavity-nesting *Apis* in showing the highest profitability (measured as energy harvest rate of sucrose solutions) at 45% sugar solutions, compared with solutions as low as 15% and as high as 65% sugar. Ground-level surveys of bee nests located 33 colonies of stingless bees along 12 km of trails. I calculate the density of stingless bee colonies to be approximately 3 per hectare - lower than the 4 to 6 colonies in each hectare found in some Neotropical forests. The sugar-salt baiting technique also revealed the first known association between an orchid and the genus *Apis*. *Apis koschevnikovi* was found with a small orchid pollinarium attached to its scutellum, thus it may have a pollinator role often played by *Trigona* in the neotropics for a few orchid genera. Other noteworthly bees were the large, nocturnal *Xylocopa myops*, megachilids and long-tongued *Amegilla*, all likely principal pollinators of obligately outcrossing tree species.

KEY WORDS: bee biodiversity, *Trigona, Pariotrigona, Apis,* feeding behaviour, sugar-salt baiting, Borneo.

INTRODUCTION

Recent treatments of animal or plant biodiversity place too much emphasis on species number, rather than on an actual assessment of their biological *diversity*, which includes ecological roles, behaviour, and many characteristics of living things that are seldom uncovered by comparative morphology and museum specimens (see also Williams 1993). In work described here I give results of a short survey of the bees at Belalong field station in Brunei, and an effort to characterize some of their feeding behavior. These results are compared to other equatorial and some temperate areas. In an attempt to make a rapid assessment of the richness of the socially-foraging colonial bees at Belalong, I used baiting techniques that generally provide results in any of the tall equatorial forests. By attracting bees to feeding stations, other studies can be made with the trained bees. At Belalong these were aimed at answering two questions: 1) What are some of the important non-floral resources that bees obtain by visiting such odd substrates as chili sauce containers, laundry, human skin, campfire ashes, and seepage areas along stream banks? 2) What kinds of nectar sources are colonial bees most likely to prefer, to judge from energy intake rate at different sugar solutions? (cf. Roubik & Buchmann 1984).

D. S. Edwards et al. (eds.), Tropical Rainforest Research – Current Issues, 59–66.

METHODS

Between 28 October until 25 November, 1991, when very little flowering took place in the forest understory, I set up monitoring stations for the social bees by spraying combinations of honey, sugar and table salt solutions with banana flavouring on small bushes 1-3 m high, near the laboratory clearing at Belalong. The amount of flavouring used was one or two drops per litre solution. Using a hand-held refractometer, I prepared solutions with sugar concentration of 28% by weight. A hand-held sprayer was used to make bait applications (after Wille 1962). General procedure was to spray 200-300 ml of solution at each site, beginning at 0640, and replenishing the solutions as they evaporated in direct sunlight or were depleted by foraging bees. A fine mist spray was used to enhance odour dispersal, while a coarser spray was used to cover leaf surfaces with the bait. After bees began to find the solutions and bring nest mates there, within the first morning in which weather was fair, bees were periodically collected with an aerial net. Additional survey and collecting were carried out by walking the forest trails searching for social bee nests in living trees. I covered about 12 km of hilly forest on either side of the Belalong River, monitoring a swath of forest 10 m wide (5 m on either side of the trail). Thus I surveyed roughly 12 ha of the forest for bee nests.

Baiting experiments were performed on 23-25 November at the laboratory clearing, under the corner of a building which provided shelter during intermittent rains. Bees which had not been given baits for more than two weeks immediately found baits when these were placed in their former positions on the afternoon of 23 November. Sugar solution feeders made of inverted, paired petri dishes were placed on one bush used for baiting, where bees visiting them were taken to the new site used for the feeding assays. This was done by carefully moving a group of bees on the dish they fed upon. Salt and sugar solutions for behavioural assays were made with distilled water as follows: sugar solutions of 15, 30, 45 and 65% were made with table sugar, their concentration determined using an Atago hand-held refractometer corrected for temperature. Five 0.5 M salt solutions were prepared of NaCl, $MgCl_2$, Na_3PO_4, NH_4Cl, and KCl, with the equivalent solute quantity per litre of 5.8 g, 20.3 g, 38 g, 5.4 g, and 7.5 g, respectively. Feeders for the salt solutions consisted of petri plate lids each with a moistened yellow paper square in the middle, and also of open vials 9 cm long by 2 cm in diameter, 1/3 filled with each solution.

The five salt solutions feeders were placed on a tray in a four-cornered arrangement with one feeder in the centre. Relative positions of feeders were changed periodically, and the tray was rotated 180^0 each 15 min. to prevent a tendency of forager site constancy from determining feeding choice. The total number of each bee species at the salt feeders was recorded each 15 min. during 5-min. periods between 0700 and 0930 on 23 and 24 November. During the latter part of the day on the 24th and all day on the 25th, the five open vials containing the solutions were left in the feeding arena. Vials served as traps, not allowing bees to leave once they had entered. Total bees captured on each day were combined with observed numbers feeding at the moistened paper patches. A feeding bout was scored when a bee extended its proboscis and fed >10 seconds.

Sugar solution feeding assays were made following Roubik & Buchmann (1984). Two sugar-solution reservoirs of 45% sugar were maintained in the feeding arena, where field bees fed freely during the study. Two smaller feeding platforms were maintained in the area for imbibement assays; a platform was carefully removed from the foraging area, thus avoiding disturbance by other bees, when an individual bee arrived at one. A 25 μL drop was measured using a calibrated micropipette and placed in a circular depression (the inverted base of a plastic film canister) for measurement of imbibement rate. Individual bees were timed at the depression from the start of feeding at the droplet until the end of a feeding bout, when the entire droplet had been removed. I terminated the stopwatch recording as soon as the bee started to move the proboscis laterally, as if seeking additional solution in the feeding depression. Imbibement assays were conducted on 25 November from 0930 until 1630, at ambient temperature of approximately 28°C, and 100% relative humidity. The lowest concentration of separate sets of the different solutions were presented first, progressing to the highest. At least 20 bees were recorded feeding at each type of resource, and no bee was monitored in more than one feeding assay at a particular concentration.

RESULTS

Species Richness and Nesting

The social bees recorded at feeders included all those found nesting in the area, and many whose nests were never located (Table 1). Some bees have Iban names that were provided by Awang Mergi (Table 1), others are mentioned by Beccari (1904). The stingless honey-making bees (Meliponinae, or 'clulut') were the richest group, having 22 species ranging in size from the minute *Pariotrigona pendleburyi* and *Trigona (Heterotrigona) fuscobalteata* (2.5 mm) to the relatively large *Trigona (Heterotrigona) thoracica* and *T. (Homotrigona) fimbriata* (8 mm). These constitute 76% of stingless bee species recorded for the island of Borneo (Sakagami *et al.* 1990). Further, the males of *Pariotrigona pendleburyi* were previously unknown (Michener 1990) but were found exiting from their nests at Belalong. There are four species of honeybees at Belalong, the most common being the giant

Table 1. Highly social bees and nests encountered near the Belalong field station.

Species	Nest (trees, No.)	Nest height (m)
Apis koschevnikovi 'ranyuan'	-	-
A. dorsata 'manyi' or 'lanyeh'	*Koompasia*, 1	40
A. cerana 'ranyuan tikung'	-	-
A. andreniformis 'nuang'	-	-
Pariotrigona pendleburyi	*Shorea*, *Dryobalanops*, 8	0.04-0.4
Trigona (Lepidotrigona) nitidiventris	-	-
T. (L.) terminata	1	4
T. (L.) ventralis	1	5.5
T. (Heterotrigona) melina	2	ground level
T. (H.) haematoptera	1	8
T. (H.) hobbyi	-	-
T. (H.) collina	11	ground level
T. (H.) thoracica	-	-
T. (H.) apicalis	3	1.5-3
T. (H.) drescheri	1	ground level
T. (H.) melanocephala	3	0.1-0.9
T. (H.) reepeni	-	-
T. (H.) itama	-	-
T. (H.) erythrogastra	-	-
T. (H.) canifrons	-	-
T. (H.) fuscobalteata	-	-
T. (H.) fuscibasis	-	-
T. (H.) rufibasalis	-	-
T. (H.) geissleri	2	0.2-0.5
T. (H.) laeviceps	-	-
T. (Homotrigona) fimbriata	-	-

honeybee *Apis dorsata* ('manyi' or 'lanyeh') and the Malesian honeybee *Apis koschevnikovi* ('ranyuan'), with the small *Apis andreniformis* ('nuang'), a bee about 8 mm long, and *Apis cerana* ('ranyuan tikung') seemingly less common.

Nests of the honeybees proved elusive, and only one small nest of *Apis dorsata* was seen on an individual of *Koompassia* (Fabaceae). Virtually all bee nests except those of *Lepidotrigona* were in or under the bases of large trees, 90 to 180 cm diameter at breast height (mean = 130 cm). The 33 nests of stingless bees listed in Table 1 were mostly within a few metres of ground level, and all were associated with living trees. All species but *T. apicalis* produced a resin and wax entrance tube that projected from a few cm to a few dozen cm from the substrate. Three species nested in the ground under large trees: *T. collina*, *T. drescheri* and *T. melina*. A nest of the last species was in the same tree base as a colony of the carpenter ant *Camponotus gigas*. *Pariotrigona pendleburyi* and *Trigona collina* were the most common colonies, amounting to over half of those seen. The former frequently had two nests separated by a few dozen centimetres in the same tree, and those of the latter were frequently found in groups of two to four, within root systems. The results of predator attacks (presumably civets and the sun bear, *Selenarctos*) were seen at the terrestrial nests of *T. drescheri* and *T. melanocephala*, while unmistakable claw marks of *Selenarctos* were seen going up the trunks of five trees, and a cavity in one tree had been chewed open, through 6 cm of living wood, at 3 m aboveground, exposing a hole 30 x 12 cm - very likely a former nest of *Apis* or *Trigona*.

Salt Selection

Three species were the primary foragers at salt solution feeders - *Apis dorsata*, *A. koschenikovi* and *Trigona melina*. Sodium salts appeared the most attractive to bees, with magnesium and potassium also sought to greater extent than phosphate or ammonium (Figure 1). The *Trigona* were substantially more numerous at magnesium and sodium solutions than at other mixtures, but there were only slight differences between potassium, ammonium and sodium for the two honeybees. Individual bees infrequently took full loads of these solutions, with the exception of *A. dorsata*, which repeatedly imbibed full loads of sodium solution. In one instance a worker of *Trigona melina* fought with one of *T. itama* by locking mandibles in a venter-to-venter position at a sodium bait. Combining all of the bee counts at feeders suggests that social bees were primarily interested in sodium salt (Figure 1).

Nectar Selection

The Malesian honeybee *Apis koschevnikovi* imbibed the dilute solutions more rapidly than viscous solutions, but the energy return rate was greatest at the 45% sugar solution and possibly at higher concentrations near 60% (see Roubik & Buchmann 1984), as it is for other honeybees of its size (Figure 2). For comparative purposes, I provide preliminary data, taken at ambient conditions of 26-31^{0}C and 80% humidity, for the other 'hive bees' of the world. Three other tropical varieties of *Apis mellifera*, two from the Yucatan peninsula in Mexico, where a tropically-adapted honeybee of European descent existed before colonization by Africanized honeybees, and Japanese *Apis cerana* - show nearly the same profile of energy harvest from sugar solutions. The most viscous solutions, 60-65% sugar, which provide 3.04 and 3.35 calories per microlitre, respectively, are nonetheless imbibed more slowly and are less energetically rewarding. The coefficients of variation at different solutions were between 10 and 26% for the 22 replicates of different solutions (N = 925 bees). However, *A. koschevnikovi*, with the smallest sample size (see legend for Figure 2), also had the smallest coefficient of variation, ranging between 11 and 14% for the time required to imbibe 25 microlitres of each of the four different solutions. *Apis koschevnikovi* was unusual in readily taking dilute solutions of 15% sugar, as do African

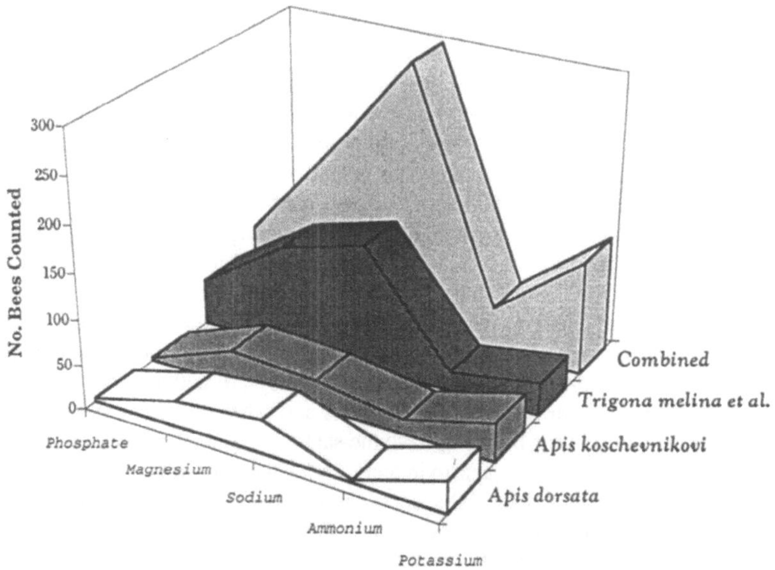

Figure 1. Salt Solution choice by highly social bees of the forest near Belalong field station.

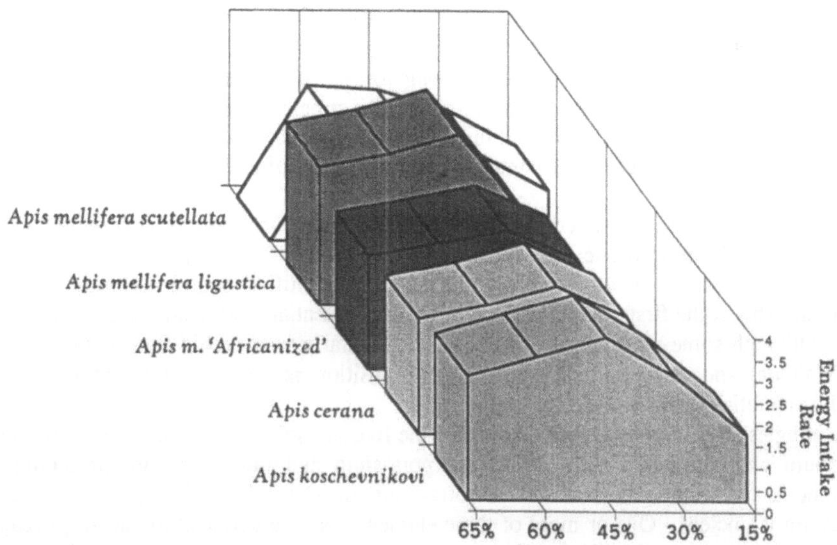

Figure 2. Sugar solution imbibement performance, shown as relative energy (calories) taken per second, by the multiple-comb or cavity-nesting *Apis*. Sample sizes at each concentration were usually equal for a given bee, N = 82, 142, 202, 249 and 250 for *Apis koschevnikovi*, *A. mellifera scutellata*, *A. cerana*, *A. mellifera ligustica* x *A. m scutellata*, and *A. m. ligustica*, respectively.

honeybees, *A. mellifera scutellata* (Figure 2). The other honeybees rarely completed feedin bouts at these solutions or rejected the solution after close approach.

DISCUSSION

In terms of biological diversity, temperate and tropical bee communities are difficult to compare, owing to the predominance of social species in the tropics and their near absence in the temperate zone (Michener 1979, Roubik 1979, 1989, 1990, 1992). For French Guiana and Sumatra, in which the total bee fauna in a given forest is approximately 250 and 125 species, respectively (Inoue *et al.* 1990, Roubik 1990), about 25% of the bees are stingless bees and honeybees, and 50% are to some degree social. Compared to Mexico and the United States, where there are 1600 and 3745 *recorded* bee species, respectively, only 12% and 2% are social, and in Mexico alone there are native highly eusocial species, comprising 3% of the fauna (Ayala 1992, Ayala *et al.* 1993). The number of genera, however, is 153 in Mexico and 122 in the United States (Ayala *et al.* 1993). There is thus higher generic and presumably biological diversity in areas such as Mexico that share tropical and temperate components. The ecological diversity of the stingless bees and honeybees, which are by definition perennial colonies, must be greater since they are active every day and have the opportunity to interact with far more plant species and other resources and natural enemies.

If the lowland forests near Belalong have the same component of honeybees and stingless bees as those of west Sumatra, approximately 25%, then the total number of forest bees should be somewhat higher than 100 species, exactly as it is in west Malesia (Inoue *et al.* 1990). The number of nests of 3 per hectare was perhaps low compared with forest of similar stature in the Neotropics, where between four and six nests might be found (Roubik 1989 & in press). The few other prominent bees that I observed in Belalong were the nocturnal carpenter bee, *Xylocopa (Nyctomelitta) myops*, *Amegilla (Glossamegilla)* sp., a large bee having among the longest tongues of any bee (Kato *et al.* 1991), and large megachilids, which along with *Amegilla* occasionally visited the salt solutions and laundry at the Belalong camp. Considering the stingless bees, all three of the species known only from Borneo (the other 26 are generally distributed from Thailand to Java [Sakagami *et al.* 1990]) were relatively common in the Belalong forest. These are *T. melanocephala*, *T. hobbyi* and *T. haematoptera*. The large complement of stingless bees and honeybees suggests that the bee fauna is intact and rich, and further study of solitary and other native bees would repay the effort for uncovering the composition and ecology of a lowland tropical forest bee assemblage.

Of all the species that recruited to baits at Belalong, none was notably aggressive toward other species and one in particular, *Apis koschevnikovi*, recruited very rapidly and quickly depleted baits. One *Apis koschevnikovi* carried an unidentified orchid pollinarium on its scutellum. This is the first known instance of an *Apis* potentially pollinating a tropical forest orchid, although some neotropical *Trigona* s. str. regularly visit orchid flowers (*Maxillaria* and *Xylobium*) and carry pollinaria in the same position as seen on *Apis* (Dodson 1967, Roubik, unpublished data).

The stingless bees and honeybee that visited the five salt solutions seemed to seek sodium, magnesium and potassium salts. Sodium, potassium and other ions are physiological requirements for water retention and neurotransmission (Nicholson 1990), but the role of magnesium is unclear. One or more of these elements may be deficient in the soil, because the pollen and nectar ordinarily forming bee diets often reflect the local soil conditions and nutrients (Kearns & Inouye 1993, Stanley & Linskens 1974). Previous studies with halictid bees, sometimes called 'sweat bees', had also shown a tendency to take sodium in solution, rather than pure water (Barrows 1974). The natural sources of such salts in the forest are both inorganic, where water leaches them from the subsoil or bedrock at river margins, or from animal feces and urine, particularly of carnivores. Whether any of the chemicals collected from these non-floral sources are critical in maintaining bee populations or species richness would be an interesting comparative study with no precedent.

The sugar contained in floral nectars used by bees ranges between 20 and 70% by weight, while that of fruit sap utilized by some bees, like the African honeybee in the neotropics, falls between 10 and 15% sugar (Roubik, unpublished data). Sugar composition of honey stored in nests of highly eusocial bees is generally between 70 and 80%, yet the most common nectars foraged by bees are usually less than 60% sugar, and often half of that (Roubik 1989). By demonstrating a peak efficiency at nectar of 45% sugar concentration, the cavity-nesting *Apis* (Figure 2) colonies would be expected to dominate these resources and prefer them to others, *all other factors being equal*. Given the predominance of these bees in their natural habitats, one might expect resource partitioning to encourage the solitary bees to seek nectars significantly different from 45% sugar, and to visit flowers presenting 45% sugar nectar only when floral morphology, abundance and distribution discourages recruitment by *Apis*. It is generally unknown if other components of nectar are favoured or rejected by some bees but not others. A study of trees and other flowering plants dependent on bees for outcrossing, which appears to be the predominant breeding system in tropical trees (Bawa 1990, 1992), might readily reveal whether or not they utilize these three *Apis* species. A plant providing 45% nectar, especially if nectar sugar increases during the day (Roubik & Buchmann 1984) would display a behaviour that would encourage visitation by *Apis mellifera*, *cerana* or *koschevnikovi*. Other patterns and concentrations of nectar would likely favour other pollinators, although the comparative study of energy harvest rate by the non-social bees that form the bulk of tropical species still awaits investigation.

ACKNOWLEDGEMENTS

I thank Marina Wong, Tamiji Inoue, Peter Becker, Gathorne Cranbrook, Catriona Prebble, David Edwards and collaborators at Universiti Brunei Darussalam and Royal Geographic Society in general for making this work possible. Special thanks are given to T. Inoue and R. Villanueva, and for funding from Smithsonian Scholarly Studies grants for assistance in studies of nectar use by *Apis*.

LITERATURE CITED

AYALA, R. 1992. *Revisión de las abejas sin aguijón de México (Hymenoptera: Apidae: Meliponinae)*. Thesis. National Autonomous University of Mexico, Mexico City. 73 pp.

AYALA, R., GRISWOLD, T.L. & BILLOCK, S.H. 1993. The native bees of Mexico. Pp. 179-228 in Ramamoorthy, T.P., Bye, R., Lot, A. & Fa, H. (eds). *Biological diversity in Mexico: Origins and distribution*. Oxford University Press, New York.

BARROWS, E. M. 1974. Aggregation behavior and response to sodium chloride in females of a solitary bee, *Augochlora pura* (Hymenoptera: Halictidae). *Florida Entomologist* 57:189-193.

BAWA, K. S. 1990. Plant-pollinatior interactions in tropical rain forests. *Annual Review of Ecology and Systematics* 21:399-422.

BAWA, K. S. 1992. Mating systems, genetic differentiation and speciation in tropical rain forest plants. *Biotropica* 24:250-255.

BECCARI, O. 1904. *Wanderings in the great forests of Borneo*. Archibald and Constable and Co., London [reprinted by Oxford University Press, Singapore, 1989]. 346 pp.

DODSON, C. H. 1967. Relationships between pollinators and orchid flowers. *Atlas do Simposio Biota Amazônica 5 (Zoologia)*:1-72.

INOUE, T., SALMAH, S., SAKAGAMI, S.F., YAMANE, S.K. & KATO, M. 1990. An analysis of anthophilous insects in central Sumatra. Pp. 201-218 in Sakagami, S.F., Ohgushi, R. & Roubik, D.W. (eds). *Natural history of social bees and wasps in equatorial Sumatra*. Hokkaido University Press, Sapporo, Japan.

KATO, M., ITINO, T., HOTTA, & INOUE, T. 1991. Pollination of four Sumatran *Impatiens* by hawkmoths and bees. *Tropics* 1:59-73.

KEARNS, C. A. & INOUE, D.W. 1993. *Techniques for pollination biologists*. University Press of Colorado, Boulder. 568 pp.

MICHENER, C. D. 1979. Biogeography of the bees. *Annals of the Missouri Botanical Garden* 66:277-347.

MICHENER, C. D. 1990. Classification of the Apidae. *University of Kansas Science Bulletin* 54:75-164.

NICHOLSON, S. W. 1990. Osmoregulation in a nectar-feeding insect, the carpenter bee *Xylocopa capitata*: water excess and ion concentration. *Physiological Entomology* 15:433-440.

ROUBIK, D. W. 1979. Africanized honey bees, stingless bees, and the structure of tropical plant-pollinator communities. Pp. 403-417 in Caron, D. (ed.). *Proceedings IV International Symposium on Pollination, Maryland. Agric. Exp. Sta. Misc. Publ. 1.*

ROUBIK, D. W. 1989. *Ecology and natural history of tropical bees.* Cambridge University Press, New York. 514 pp.

ROUBIK, D., W. 1990. Niche pre-emption in tropical bee communities: A comparison of neotropical and malesian faunas. Pp. 245-257 in Sakagami, S.F., Ohgushi, R. & Roubik, D.W. (eds). *Natural history of social bees and wasps in equatorial Sumatra.* Hokkaido University Press, Sapporo, Japan.

ROUBIK, D. W. 1992. Loose niches in tropical communities: Why are there so many trees and so few bees? Pp. 327-353 in Hunter, M.D., Ohgushi, T. & Price, P.W. (eds). *Resource Distribution and Animal-Plant Interactions.* Academic Press, San Diego.

ROUBIK, D. W. 1993. Direct costs of forest reproduction: bee-cycling and the efficiency of pollination modes. *Journal of Biosciences* 18:1-16.

ROUBIK, D. W. & BUCHMANN, S.L. 1984. Nectar selection by *Melipona* and *Apis mellifera* (Hymenoptera: Apidae) and the ecology of nectar intake by bee colonies in a tropical forest. *Oecologia (Berlin)* 61:1-10.

SAKAGAMI, S. F., INOUE, T. & SALMAH, S. 1990. Stingless bees of central Sumatra. Pp. 125-137 in Sakagami, S.F., Ohgushi, R. & Roubik, D.W. (eds). *Natural history of social bees and wasps in equatorial Sumatra.* Hokkaido University Press, Sapporo, Japan.

STANLEY, R. G. & LINSKENS, H.F. 1974. *Pollen: Biology, biochemistry, management.* Springer-Verlag, New York. 307 pp.

WILLE, A. 1962. A technique for collecting stingless bees under jungle conditions. *Insectes Sociaux* 9:291-293.

WILLIAMS, P. H. 1993. Choosing conservation areas: using taxonomy to measure more of biodiversity. Pp. 194-227 in Moon T.-Y. (ed.). *Colloquium on Biodiversity.* Korean Entomological Institute, Seoul.

Australian monsoon rainforest mammals and possible implications for the conservation of tropical rainforest biodiversity

D.M.J.S. BOWMAN *and* J.C.Z. WOINARSKI

Conservation Commission of the Northern Territory, PO Box 496, Palmerston, Northern Territory, 0831 Australia

ABSTRACT. Monsoon rainforests form an archipelago of small habitat fragments throughout the wet-dry tropics of northern Australia. According to one definition the current monsoon rainforest mammal assemblage contains only one rainforest specialist mammal species (restricted to Cape York Peninsula), and is dominated by eutherian habitat generalists (murids and bats) that mostly occur in surrounding savanna habitats. The mammal assemblages in monsoon rainforests across northern Australia (Cape York Peninsula, Northern Territory and the Kimberley) are essentially regional subsets of the local savanna and mangrove mammal assemblages, and consequently share only a limited number of species in common (most of which are bats). The lack of rainforest specialists in north western Australia is believed to be due to lack of large tracts (> 1000 ha) of monsoon rainforest habitat, the possible substantial contraction of these habitats in the past and the limited extent of gallery rainforests: gallery rainforests in South American savannas are important habitats for rainforest mammals. Unfortunately it is not possible to identify the threshold of habitat area required to maintain populations of rainforest specialist mammal species because of an impoverished fossil record pertaining to the past spatial distribution of monsoon rainforests. The implications of the lack of Australian monsoon rainforest specialist mammal fauna for the future of heavily fragmented tropical rainforests elsewhere in the world is briefly discussed. It is concluded that the analogy of habitat fragments to true islands is weak, that rainforest plant species are less vulnerable to local extinction than mammals, and that the loss of mammal rainforest specialists may not result in a dramatic loss of plant species.

KEY WORDS: biogeography, conservation, fragmentation, mammals, northern Australia, rainforests.

INTRODUCTION

It is widely accepted that the destruction of extensive rainforests will result in an archipelago of rainforest 'islands' within a 'sea' of anthropogenically modified vegetation. The consequences for biota specialized to the rainforest environment are uncertain (Simberloff 1990). Meave *et al.* (1991) have recently addressed this problem by analyzing tree species diversity in naturally created rainforest habitat islands that occur in large tracts of neotropical savanna.

Australian monsoon rainforests form an archipelago of small (typically less than 5 ha in extent) habitat fragments throughout the wet-dry tropics of northern Australia (Russell-Smith 1991). Given recent zoogeographic research these 'habitat islands' provide an opportunity to consider the impact of anthropogenic clearance upon rainforest mammal specialists.

ECOLOGICAL BIOGEOGRAPHY OF AUSTRALIAN MONSOON RAINFOREST MAMMALS

The mammal assemblages across northern Australia (Cape York Peninsula, Northern Territory and the Kimberley) are dominated by eutherian habitat generalists (murids and bats) and are essentially regional subsets of the local savanna and mangrove mammal assemblages

D. S. Edwards et al. (eds.), Tropical Rainforest Research – Current Issues, 67–69.

(Friend *et al.* 1991, Menkhorst & Woinarski 1992). There are only a few species (most of which are bats) that occur throughout northern Australia. The current monsoon rainforest mammal assemblage contains only one mammal species (*Rattus leucopus*) with exclusive dependence on rainforest habitats (Winter 1988, Winter & Atherton 1985).

Menkhorst & Woinarski (1992) suggest that the following features have prohibited colonization of north west Australian monsoon rainforests by non-volant mammal species that occur in the monsoon rainforests on Cape York Peninsula: the small area of monsoon rainforest and the typically large distances between patches; the extreme microclimatic differences between savanna and rainforest habitats (particularly in the dry season); the scarcity of gallery rainforests; and the existence of a major biogeographic barrier (the Gulf of Carpentaria and the Barkly Tablelands) that separates north western Australia from large tracts of tropical rainforest in eastern Australia.

In marked contrast to South American tropical rainforest mammal assemblages the majority of the Australian monsoon rainforest mammal assemblage have body weights below 0.1 kg (Redford 1992). In the South American seasonally dry tropics there are few mammals adapted to savanna conditions even though savanna dominates over 90% of the landscape (Mares *et al.* 1985, Redford & Da Fonseca 1986). Most of the mammals are restricted to gallery rainforests, which act as habitat corridors between Amazon and Atlantic rainforests (Redford & Da Fonseca 1986). Gallery rainforests are very restricted in northern Australia because the characteristically Australian genus *Melaleuca* dominates virtually all riverine habitats.

HISTORICAL BIOGEOGRAPHY OF AUSTRALIAN MONSOON RAINFOREST MAMMALS

Until the mid-Tertiary northern Australia supported a rich marsupial rainforest fauna (Archer *et al.* 1989). Currently Australian mammal species adapted to rainforest habitats have populations restricted to large tracts of rainforest in north eastern Australia (Winter 1988). It is assumed that the lack of rainforest specialists in south eastern Australia and north western Australia is due to reduction in the area of rainforest habitat in the past (Winter 1988). Localized extinction following recent European habitat destruction in the core area of Australian tropical rainforest mammals demonstrates the vulnerability of rainforest specialists to habitat fragmentation (Lawrence 1990). Patches less than 3000 ha in extent appear to be too small to support rainforest specialists, although mammal populations have probably not equilibrated to this latest episode of rainforest fragmentation. In this context it is significant that the largest tracts of monsoon forest in northern Australia are less than 1000 ha.

IMPLICATIONS FOR THE CONSERVATION OF TROPICAL RAINFOREST BIODIVERSITY

Given an extremely depauperate paleao-environmental record it is impossible to evaluate the hypothesis that past patch areas have fallen below a critical threshold to support monsoon rainforest specialists (cf. Winter 1988), or indeed if specialist mammals ever evolved to exploit large tracts of monsoon rainforest. Similarly, although both inter-patch distance and distance from core refugia appear to be important factors controlling the richness of monsoon rainforest mammal assemblages, uncertainty concerning the nature of past landscapes, including the extent of gallery forests, prohibits simple quantification of these variables.

The high diversity of monsoon rainforest plants and the low diversity of specialist rainforest mammals (following their mass extinction in the mid-Tertiary?) suggests that specialist mammals are not necessarily critical for the functioning of all rainforest ecosystems (cf. Da Fonseca & Robinson 1990, Dirzo & Miranda 1990, Howe 1984, Terborgh 1992). It appears that the lack of rainforest mammal specialists is compensated for by vagile winged vertebrates that exploit fruits and flowers from rainforest trees. Birds, frugivorous and nectarivorous bat species apparently play the most ecologically important role in the maintenance of the diversity of woody plants through their dispersal of pollen and seeds between isolated populations of plant species (Russell-Smith & Lee 1992).

The lack of rainforest specialists does not mean that the monsoon rainforests are of little significance to the regional mammal assemblage. The monsoon rainforests are opportunistically utilized by savanna adapted mammals in order to avoid climatic extremes, predators, and ubiquitous savanna fires. The use of monsoon rainforests by savanna animals supports Janzen's (1983) notion that "no park is an island" because small patches of rainforest will be exploited by biota in the surrounding habitat. North Australian monsoon rainforest biogeography also supports the notion that rainforest specialist vertebrates are far more vulnerable to extinction than plant species (Meave *et al.* 1991, Redford 1992). In this context the "empty" (*sensu* Redford 1992) north Australian monsoon rainforest fragments may be a window into the future of the world's tropical forests.

LITERATURE CITED

ARCHER, M., GODHELP, H., HAND, S.J. & MEGIRIAN, D. 1989. Fossil mammals of Riversleigh, Northwestern Queensland: Preliminary overview of biostratigraphy, correlation and environmental change. *Australian Zoologist* 25:29-65.

DA FONSECA, G.B.A. & ROBINSON, J.G. 1990. Forest size and structure: competitive and predatory effects on small mammal communities. *Biological Conservation* 53:265-294.

DIRZO, R. & MIRANDA, A. 1990. Contemporary Neotropical defaunation and forest structure, function, and diversity - A sequel to John Terborgh. *Conservation Biology* 4:444-447.

FRIEND, G.R., MORRIS, K.D., & McKENZIE, N.L. 1991. The mammal fauna of Kimberley rainforests. Pp. 393-412 in McKenzie, N.L., Johnson, R.B. & Kendrick P.G. (eds). *Kimberley Rainforests of Australia.* Surrey Beattie and Sons, Sydney.

HOWE, H.F. 1984. Implications of seed dispersal by animals for tropical reserve management. *Biological Conservation* 30:261-281.

JANZEN, D.H. 1983. No park is an island: increase in interference from outside as park size decreases. *Oikos* 41:402-410.

LAURANCE, W.F. 1990. Comparative responses of five arboreal marsupials to tropical forest fragmentation. *Journal of Mammalogy* 71:641-653.

MARES, M.A., WILLIG, M.R. & LACHER, T.E. 1985. The Brazilian Caatinga in South American zoogeography: tropical mammals in a dry region. *Journal of Biogeography* 12:57-69.

MEAVE, J., KELLMAN, M., McDOUGALL, A. & ROSALES, J. 1991. Riparian habitats as tropical forest refugia. *Global Ecology and Biogeography Letters* 1:69-76.

MENKHORST, K.A. & WOINARSKI, J.C.Z. 1992. Distribution of mammals in monsoon rainforest of the Northern Territory. *Wildlife Research* 19:295-316.

REDFORD, K.H. 1992. The empty forest. *BioScience* 42:412-422.

REDFORD, K.H. & Da FONSECA, G.A.B. 1986. The role of gallery forests in the zoogeography of Cerrado's non-volant mammalian fauna. *Biotropica* 18:126-135.

RUSSELL-SMITH, J. 1991. Classification, species richness, and environmental relations of monsoon rainforest vegetation in the Northern Territory, Australia. *Journal of Vegetation Science* 2:259-278.

RUSSELL-SMITH, J. & LEE, A. 1992. Plant populations and monsoon rain forest in the Northern Territory, Australia. *Biotropica* 24:471-487.

SIMBERLOFF, D. 1990. Species-area relationships, fragmentation and extinction in tropical forests. In Kheong, S. & Win L. (eds). *In Harmony with Nature.* Proceedings of the International Conference on Tropical Biodiversity. Malayan Nature Society, Kuala Lumpur.

TERBORGH, J. 1992. Maintenance of diversity in tropical forests. *Biotropica* 24:283-292.

WINTER, J.W. 1988. Ecological specialization of mammals in Australian tropical and sub-tropical rainforest: refugial or ecological determinism. *Proceedings of the Ecological Society of Australia* 15:127-138.

WINTER, J.W. & ATHERTON, R.G. 1985. *Survey of the mammals and other vertebrates of the Weipa region, Cape York Peninsula.* Queensland National Parks and Wildlife Service, Brisbane.

Presbytis species sympatry in Borneo versus allopatry in Sumatra: an interpretation.

DOUGLAS BRANDON-JONES

11 York Road, London SW11 3PX, United Kingdom

ABSTRACT. A review of the diversity of opinion on the number of Sumatran species recognizable for the genus *Presbytis* sensu stricto, indicates that an over-reliance on sympatry as a criterion for specific status, produces an underestimation of the species number in Sumatra, compared with Borneo. It is advocated that, for reasons of consistency, geographic variation in pelage colour must retain senior taxonomic status over geographic variation in adult male loud vocalization, but that the latter character corroborates the conspecificity rather than convergence of the Bornean and Sumatran grey-backed species, *Presbytis comata*. It is inferred that *Presbytis potenziani* was the sole species of its genus to survive an early stadial of the most recent glaciation, and that *Presbytis comata* derived from it during the subsequent interstadial and was approximately restricted to its modern distribution by a milder second stadial. The remaining species of the genus have derived from *Presbytis comata* since the glaciation. Allopatry in Sumatra is attributed to the northern part of the island forming a more active post-glacial dispersal centre than the comparable area in north Borneo. Only one Bornean species, *Presbytis frontata* has derived from *Presbytis comata* in Borneo. The other two, *Presbytis femoralis* and *Presbytis rubicunda*, had their origins in Sumatra.

KEY WORDS: dispersal, geographic variation, glaciation, pelage colour, *Presbytis melalophos* species group, vocalization.

The Asian representatives of the Old World monkey subfamily, Colobinae, consist of the proboscis monkey, the snub-nosed monkeys, the douc monkeys and twenty or so simple-nosed species (Brandon-Jones 1984). Most authors refer the simple-nosed species to a single genus, *Presbytis*, rejecting the opinion of Pocock (1935) that they constitute three genera. A growing body of evidence indicates that at least a bipartite generic division is warranted, and that the genus, *Presbytis* sensu stricto should include the Mentawai Island species, *Presbytis potenziani* (Brandon-Jones 1993). Bornean populations of this genus overlap each other in geographic distribution. Evidence of sympatry in Sumatra derives solely from speculation prompted by the absence of a river boundary between two distinct populations in the central part of the island (Borner 1974). This paper seeks to highlight the resulting imbalance in the number of species recognized for the two islands, and to propose an explanation for the zoogeographic contrast.

Miller (1934), dealing only with the "*Presbytis femoralis* group", recognized four species from Sumatra, *Presbytis batuana, P. catemana, P. percura* and *P. sumatrana*. Pocock (1935) segregated "*Presbytis thomasi*", but treated the remaining populations as respectively "blackish races", a "reddish race", or "whitish mutants" of "*Presbytis melalophus*". Chasen (1940), although acknowledging its earlier authorship, idiosyncratically treated *Simia melalophos* Raffles 1821, as a subspecies of "*Pithecus femoralis*". His recognition of "*Pithecus femoralis fusco-murinus*" [=*Presbytis melalophos mitrata]*, and his description of a new subspecies, "*Pithecus femoralis fluviatilis*", implied rejection of Pocock's (1935) opinion that they were whitish mutants of no taxonomic standing. Chasen (1940) innovatively recognized "*Pithecus aygula thomasi*" as conspecific with populations in Jawa and north Borneo, and showed remarkable insight in suggesting that, "taking an extremely broad view," *Presbytis potenziani* might "be regarded as a very distinct race" of the same

D. S. Edwards et al. (eds.), Tropical Rainforest Research – Current Issues, 71–76.

species which, for nomenclatural reasons, must now be referred to as *"Presbytis comata"* (Napier 1985, Napier & Groves 1983, Weitzel & Groves 1985). This relationship, appreciated only by Thomas (1895) in papers predating Chasen's, is corroborated by more recent research (Brandon-Jones 1993).

Hooijer (1962) and Kuhn (1967) recognized two Sumatran species, *Presbytis aygula* and *P. melalophos*. Medway (1970) considered Pocock's (1935) more conservative classification, implying convergence between the grey-backed populations in northwest Sumatra, Java and north and east Borneo, as more acceptable than Chasen's (1940) indication that *Presbytis comata* is disjunct in its present distribution and represents the initial colonizer of the Malay archipelago. Medway (1970) offered three alternative classifications. The most radical would recognize a single species *(Presbytis melalophos)* inhabiting Java, Sumatra, the Malay peninsula and west Borneo. The more orthodox would exclude Java from this distribution. The most cautious would also recognize *P. thomasi* as a distinct species in north Sumatra. Thorington & Groves (1970) opted for his most cautious interpretation.

Their extensive field survey of the primates of Sumatra led W.L. Wilson & C.C. Wilson (1975) to conclude that geographic variation in the adult male loud vocalization in Sumatran *Presbytis* (sensu stricto), differentiated it into three species, *P. melalophos, P. femoralis* and *P. thomasi*. C.C. Wilson & W.L. Wilson (1977) reported that the adult male three-part call of *Presbytis potenziani* "is as similar to that of *P. thomasi* as are the calls of the *P. melalophos* subspecies to one another. In fact, the Mentawai Island langur may prove to be a subspecies of *P. thomasi*" (*P. potenziani* is actually the earlier name). Their separate report (C.C. Wilson & W.L. Wilson 1975) that, based on vocalization, *"Presbytis aygula"* in east Kalimantan may have *"P. thomasi"* as its "nearest relative in Sumatra", corroborates Chasen's (1940) opinion that the distribution of *Presbytis comata* is disjunct, with *P. potenziani* possibly "as a very distinct race". Convergence both in pelage colour and vocalization between *P. potenziani* and two geographically isolated subspecies of *P. comata* would seem improbable.

Kawamura (1984) and Aimi *et al.* (1986) conducted further field surveys of Sumatran *Presbytis*, and endorsed the Wilsons' conclusion that there are three species. Aimi & Bakar (1992) claimed that their taxonomic arrangement was substantiated by the cranial morphology of specimens they collected in 1981 and 1983.

The classification adopted by Brandon-Jones (1984) was based on the appreciation that, for reasons of consistency, geographic variation in vocalization must remain subordinate in taxonomic status to geographic variation in pelage colour, the principal character on which *Presbytis* specific status has been allocated. Brandon-Jones (1984) reinterpreted as a natural group the cluster of pale-thighed populations in the southern Malay peninsula, Natuna Besar, the Riau archipelago and east central Sumatra, interpreted by Medway (1970) as convergent. The greyish white pelage colour of the thigh is exceptionally well developed in the east Sumatran subspecies, *Pithecus femoralis paenulatus* Chasen, 1940, which occurs in close proximity to surrounding dark-thighed populations. Unlike comparable populations in the Malay peninsula, there is no indication of subspecific pelage colour intergradation. It is this subspecies which is subject to speculation as to its sympatry with the adjacent species. Brandon-Jones (1984) accordingly recognized four species from Sumatra: the dorsally grizzled-grey species, *Presbytis comata;* the blackish brown species, *Presbytis femoralis;* the pale-thighed species, *Presbytis siamensis;* and the reddish and whitish species, *P. melalophos*.

These Sumatran populations, which Medway's (1970) radical and more orthodox classifications would reduce to one species, exceed all four unanimously recognized Bornean species together in pelage colour variation. The dorsally grizzled-grey populations in north and east Borneo are conspecific with, or convergent to, the similar populations in north

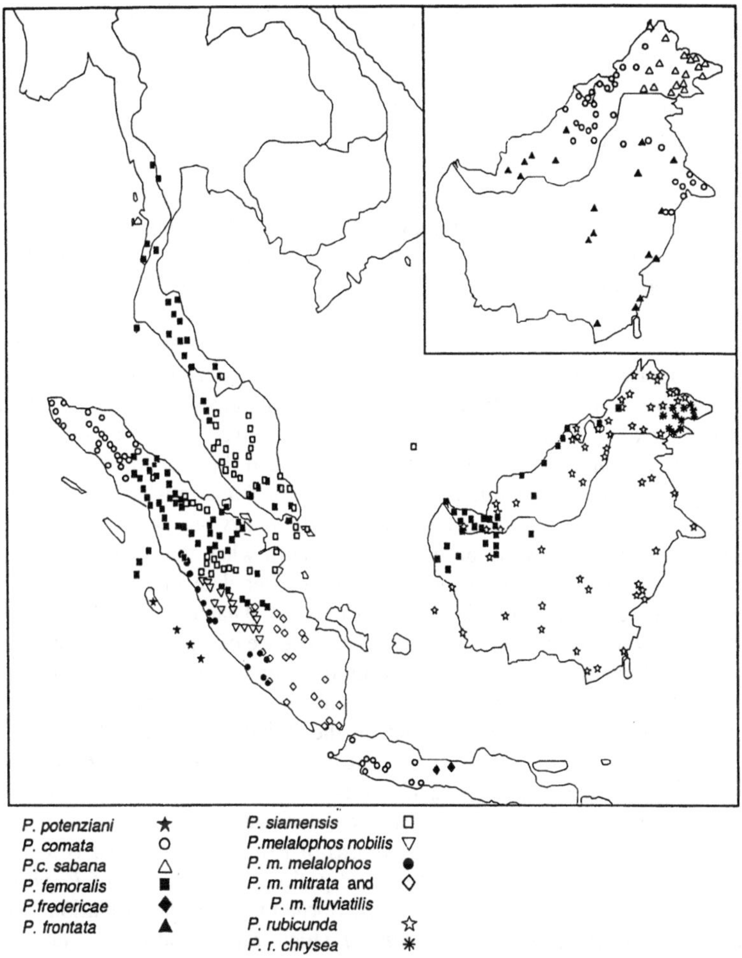

Figure 1. Geographic distribution of *Presbytis* (sensu stricto).

Sumatra. Even Pocock (1935) who granted it specific status, acknowledged that "*Presbytis chrysomelas*", inhabiting the northwest coastal region of Borneo, was "perhaps only racially distinct from *P. femoralis*". This view is now generally accepted. *P. femoralis* and the species, *Presbytis frontata*, endemic to east, southeast and central Borneo, are superficially so similar in pelage colour that museum skins have been subject to mistaken identity (e.g. Jentink 1898). *P. rubicunda*, which inhabits Karimata island and peripheral Borneo other than most of the area occupied by *P. femoralis*, is the counterpart of *P. melalophos* both in reddish pelage colour and, according to C.C. Wilson & W.L. Wilson (1975), in some aspects of behaviour and vocalizations. There is no Bornean equivalent of *P. siamensis*, nor of the southeast Sumatran whitish subspecies. Conversely, there is no Sumatran counterpart of the black, red and white pelage-coloured morph which has a high incidence in the eastern populations of Bornean *P. femoralis*.

Its geographic distribution leaves little doubt that *P. femoralis* originated in Sumatra and migrated to Borneo. Similarities in external characters and vocalization, and evidence of intergradation indicate that *P. melalophos* derived from *P. femoralis*. If it is possible for a predominantly reddish species to derive from a predominantly brownish species in Sumatra, then it is equally plausible for a similar species to have evolved in transit to Borneo. The similarities in vocalization between *P. rubicunda* and *P. melalophos* reported by the Wilsons have already been mentioned, but similarities in head hair disposition suggest that *P. rubicunda* actually originated from east Sumatran populations of *P. siamensis*.

On the face of it, the origins of *P. frontata* seem more obscure. Comparison reveals that the distinctive external cephalic features in *P. frontata* are only an exaggerated development of the similar features exhibited by *Presbytis comata sabana*. The two taxa also share a characteristic elongation of the nasal region of the skull. Their complementary geographic distribution and their pelage colour differentiation, show a remarkable parallel both to that between *P. comata* and *P. femoralis* in Sumatra, and between *P. comata* and *Presbytis fredericae* in Java.

Brandon-Jones (1978, & in preparation) has recounted how Asian colobine geographic distribution, supported by other biogeographic evidence, indicates that the Asian rainforest was fragmented by the cool and dry climate which accompanied the most recent glaciation. The approximate extent of three of these fragments is inferred from the modern geographic distribution of *P. comata*. The squirrel, *Callosciurus* and the gibbon, *Hylobates*, are both represented by a subspecies endemic to the area of north Sumatra where rainforest survived a period of glacial drought, but *Macaca* and *Semnopithecus* are not so represented. This implies that the latter genera were absent from Sumatra when that cold dry period began. They must however, have occurred there at some earlier date in order for the macaque, *Macaca pagensis*, to have colonized the Mentawai Islands, and the leaf monkey, *Semnopithecus auratus*, to have reached Java. The absence of a macaque and *Semnopithecus* endemic to north Sumatra therefore indicates the intervention of an earlier and more severe arid period. This earlier glacial drought was presumably also responsible for the eradication of the significant number of animals found on the Asian mainland and Java, but not on Sumatra (Dammerman 1929). It is deduced that *Callosciurus, Presbytis, Hylobates* and the orang-utan, *Pongo,* recolonized Sumatra after this drought and that their present distribution is the result of a second and less severe climatic deterioration.

The simplest explanation for the distinctiveness of *Presbytis potenziani* is that it was the sole member of its genus to survive the first arid period. The absence of *P. potenziani* on Borneo implies that the genus had not colonized Borneo prior to the glaciation. During the interstadial, *P. potenziani* gave rise to *P. comata* which migrated at least as far as Sumatra, Java and Borneo. The second cool and dry climatic period contracted the geographic distribution of *P. comata* probably to only the central areas of its modern distribution. With climatic remission, its distribution expanded initially to occupy the peripheral areas of its modern distribution. It then transmuted into *P. femoralis*, *P. fredericae* and *P. frontata* respectively, in Sumatra, Java and Borneo. *P. femoralis* gave rise to *P. melalophos* and *P. siamensis* and, through the latter, to *P. rubicunda*. This analysis would allow of the interpretation that *Presbytis* is a monotypic genus, were it not for the above discussed instances of sympatric distribution. The species display a chromatic succession in pelage and in some cases, skin pigmentation, comparable with those identified by Hershkovitz (1968, 1977).

Presbytis species diversity then, far from being the product of prolonged stability of the humid rainforest environment, is actually the result of its recent fragmentation. Its allopatric species distribution in Sumatra resulted from north Sumatra forming a recent centre of population dispersal, predominantly in a west to east direction, with one branch *(P. melalophos)* dispersing as far as the southern tip of the island, and another *(P. femoralis)* emigrating to the Malay peninsula and northwest Borneo. The only possible instance of sympatry resulted from a reverse dispersal of *P. siamensis* from the Malay peninsula. *Presbytis* sympatry in Borneo largely resulted from the greater productivity of the Sumatran dispersal centre, probably assisted by an inferred eastward ocean current between Sumatra and Borneo which favoured rafting in this direction. This colonization must have occurred by rafting because central Sundaland would have been climatically unable to sustain the requisite *Presbytis* habitat during its glacial emergence. Although *Presbytis comata canicrus* has probably supplanted part of the distribution of *P. frontata* in northeast Kalimantan, in general there has been decidedly less dispersal activity by indigenous Bornean *Presbytis*, possibly because expansion was blocked by the lingering of the cool dry climate for a longer period in central Borneo than occurred in central Sumatra.

Note added in proof: Fifteen kilometres south-west of Prapat near Lake Toba, during a visit to Sumatra in November-December 1993, J. Vermeer (*in litt.*) observed a troop of *Presbytis siamensis paentulata* whose identity was confirmed by the loud call. The locality is about 150 km north of the Barumun river, previously regarded as the northern limit of this subspecies (Aimi & Bakar 1992). It is difficult to interpret this evidence other than indicating significant sympatry between *P. femoralis* and *P. siamensis*.

LITERATURE CITED

AIMI, M. & BAKAR, A. 1992. Taxonomy and distribution of *Presbytis melalophos* group in Sumatera, Indonesia. *Primates* 33:191-206.

AIMI, M., HARDJASASMITA, H. S., SJARMIDI, A. & YURI, D. 1986. Geographical distribution of Aygula-group of the genus *Presbytis* in Sumatra. *Kyoto University Overseas Report of Studies on Asian Non-Human Primates* 5:45-58.

BORNER, M. 1974. *Progress report no. 4: The Sumatran rhinoceros (Dicerorhinus sumatrensis) in the provinces of Riau and West Sumatera.* World Wildlife Fund Project 884 - Sumatran Rhinoceros. 11 pp.

BRANDON-JONES, D. 1978. The evolution of recent Asian Colobinae. Pp. 323-325 in Chivers, D. J. & Joysey, K. A. (eds). *Recent Advances in Primatology, Vol. 3: Primate Evolution, Proceedings of the Sixth Congress of the International Primatological Society, 1976.* Academic Press, London.

BRANDON-JONES, D. 1984. Colobus and leaf monkeys. Pp. 398-408 in Macdonald, D. (ed.). *The Encyclopaedia of Mammals* Volume 1. Allen & Unwin, London. Pp. 102-112 in Macdonald, D. (ed.). *All the World's Animals: Primates.* Torstar Books, New York.

BRANDON-JONES, D. 1993. The taxonomic affinities of the Mentawai Islands sureli, *Presbytis potenziani* (Bonaparte 1856). *Raffles Bulletin of Zoology* 41:331-357.

CHASEN, F. N. 1940. A handlist of Malaysian mammals. *Bulletin of the Raffles Museum, Singapore* 15:i-xx, 1-209.

DAMMERMAN, K. W. 1929. On the zoogeography of Java. *Treubia* 11:1-88.

HERSHKOVITZ, P. 1968. Metachromism or the principle of evolutionary change in mammalian tegumentary colors. *Evolution* 22:556-575.

HERSHKOVITZ, P. 1977. *Living New World Monkeys (Platyrrhini): with an Introduction to Primates.* Volume 1. University of Chicago Press, Chicago

HOOIJER, D. A. 1962. Quaternary langurs and macaques from the Malay Archipelago. *Zoologische Verhandelingen uitgegeven door het Rijkmuseum van Natuurlijke Historie te Leiden* 55:1-64.

JENTINK, F. A. 1898. Zoological results of the Dutch scientific expedition to central Borneo. The mammals. *Notes from the Leyden Museum* 20:113-125.

KAWAMURA, S. 1984. Distribution and vocalization of *Presbytis melalophos* and *P. femoralis* varieties in westcentral Sumatra - a summarized report. *Kyoto University Overseas Report of Studies on Asian Non-Human Primates* 3:37-44.

KUHN, H.-J. 1967. Zur Systematik der Cercopithecidae. Pp. 25-46 in Starck, D., Schneider, R. & Kuhn, H.-J. (eds). *Progress in Primatology. First Congress of the International Primatological Society, Frankfurt am Main, 1966.* Fischer, Stuttgart.

MEDWAY, LORD. 1970. The monkeys of Sundaland: Ecology and systematics of the cercopithecids of a humid equatorial environment. Pp. 513-553 in Napier, J. R. & Napier, P. H. (eds). *Old World Monkeys: Evolution, Systematics and Behavior.* Academic Press, New York.

MILLER, G. S., Jr. 1934. The langurs of the *Presbytis femoralis* group. *Journal of Mammalogy* 15:124-137.

NAPIER , P. H. 1985. *Catalogue of Primates in the British Museum (Natural History) and elsewhere in the British Isles. Part III: Family Cercopithecidae, Subfamily Colobinae.* British Museum (Natural History), London.

NAPIER, P. H. & GROVES, C. P. 1983. *Simia fascicularis* Raffles, 1821 (Mammalia, Primates): request for the suppression under the plenary powers of *Simia aygula* Linnaeus, 1758, a senior synonym. Z.N. (S.) 2399. *Bulletin of Zoological Nomenclature* 40:117-118.

POCOCK, R. I. 1935. The monkeys of the genera *Pithecus* (or *Presbytis*) and *Pygathrix* found to the east of the Bay of Bengal. *Proceedings of the Zoological Society of London* 1934:895-961.

THOMAS, O. 1895. On some mammals collected by Dr. E. Modigliani in Sipora, Mentawei Islands. *Annali del Museo Civico di Storia Naturale di Genova* (2)14:660-672

THORINGTON, R. W., JR. & GROVES, C. P. 1970. An annotated classification of the Cercopithecoidea. Pp. 629-647 in Napier, J. R. & Napier, P. H. (eds). *Old World Monkeys: Evolution, Systematics and Behavior.* Academic Press, New York.

WEITZEL, V. & GROVES, C. P. 1985. The nomenclature and taxonomy of the colobine monkeys of Java. *International Journal of Primatology* 6:399-409.

WILSON, C. C. & WILSON, W. L. 1975. The influence of selective logging on primates and some other animals in East Kalimantan. *Folia Primatologica* 23:245-274.

WILSON, C. C. & WILSON, W. L. 1977. Behavioral and morphological variation among primate populations in Sumatra. *Yearbook of Physical Anthropology* 20:207-233

WILSON, W. L. & WILSON, C. C. 1975. Species-specific vocalizations and the determination of phylogenetic affinities of the *Presbytis aygula-melalophos* group in Sumatra. Pp. 459-463 in Kondo, S., Kawai, M. & Ehara, A. (eds). *Contemporary Primatology. Fifth International Congress of Primatology, Nagoya 1974*. Karger, Basel.

Diversity of sound-producing insects in a Bornean lowland rain forest

KLAUS RIEDE

Institut für Biologie I (Zoologie), Albert-Ludwigs Universität Freiburg, Germany

ABSTRACT. In the canopy of a lowland mixed dipterocarp forest in Mount Kinabalu Park (Sabah, Malaysia), animal sounds were recorded for several weeks during the dry and beginning of the rainy season. Spectrogram analysis revealed clearly identifiable sound patterns which in most cases could be classified at the order level (crickets, cicadas, katydids and vertebrate sounds) and attributed to certain 'ethospecies' which in several cases could be caught and identified. Song activity of cicadas and birds was diurnal, while crickets started singing at dusk, followed by strictly nocturnal katydids. This indicates pronounced temporal segregation at the order level by diel periodicity. During sunset, overall sound intensity increases due to several species engaging in a 'dusk chorus' composed of chorusing cicadas, crickets and frogs. A distinct 'acoustical guild' is observed at waterfalls, showing special adaptations of call structure to high background noise. The value of acoustical records as a tool to assess local and regional diversity is discussed.

KEY WORDS: bioacoustics, insects, Orthoptera, Cicadidae, sound recording, Sabah, Kinabalu N.P.

INTRODUCTION

Sound is an important medium for intermediate-range communication within rainforests, where optical communication is hampered by restricted visibility due to vegetation and the nocturnal life style of most rainforest creatures (Elton 1973). Consequently, as had been noted by 19th century naturalists (Wallace 1869), among the most exquisite sensory impressions when entering a tropical rainforest are the concertos produced by a great variety of songsters. The general notion of increased species numbers in the tropics holds also for those groups capable of sound production, and species-specific songs might well be an efficient way of mate location and species recognition for species of low population density within such complex habitats as rainforests. On the other hand, acoustic communication is limited by abiotic and biotic constraints: habitat structure affects sound propagation (Römer *et al.* 1989), and noise generated by abiotic sound sources like wind, rain and waterfalls, impairs efficient communication. Among the biotic constraints is the sound produced by con- and heterospecific animals. Therefore, it can be expected that signals will be shaped by competition for acoustic transmission channels. The aim of this paper is to sketch the composition of the 'acoustic commmunity' in a Bornean mixed dipterocarp forest and to outline a framework for detailed ecophysiological studies of singing animals.

MATERIAL AND METHODS

Study sites and periods

Investigations were made at Kinabalu National Park, Sabah, Malaysia, in a lowland mixed Dipterocarp forest at Poring Hot Springs (500 m asl), along Langanan river trail and Langanan waterfall (1100 m asl), and in mountain forest around the Park Headquarters (1200m asl). A canopy walkway system with a total length of 900 m was available at Poring. Observations were made from April to June 1991, March until May and August 1992, thereby covering different seasons and weather conditions.

D. S. Edwards et al. (eds.), Tropical Rainforest Research – Current Issues, 77–84.

Recording method

Recordings in the rainforest canopy were made by four pairs of electret microphones (EM-3) installed at four stations about 30 m apart at the walkway (Figure 1). Microphones were fed into two 4-channel mixers and their outputs were recorded by a stereo cassette recorder (Sony TCD5-PRO). Microphone stations could be selected by switching the respective mixer channels. Besides simple stereo recording at one site, simultaneous recording at two sites was possible by selecting microphones at different sites for the left and right channel, respectively. In addition, individual songsters were recorded and localized by a highly directional microphone consisting of an electret microphone capsule (type MCE-2000) mounted into a parabolic perspex reflector (PBR-330, Sony, diameter 33 cm).

Besides these recordings, sound intensity, light intensity, temperature and humidity were recorded. Overall sound intensity (dB[A]) was measured by a soundlevel meter (Noris NM-3).

Sound analysis and classification

Sounds were analyzed with a spectrum analyzer (Spectro 2000, MEDAV) which produced an on-line fast Fourier transform visible on a colour monitor (Figures 2-4). Sound parameters such as carrier frequency and pulse and chirp intervals were fed into a data base (Quattro Pro, Borland), together with information about locality and meteorological parameters. In most cases, sound patterns could be classified and attributed to certain 'ethospecies' (Riede 1993).

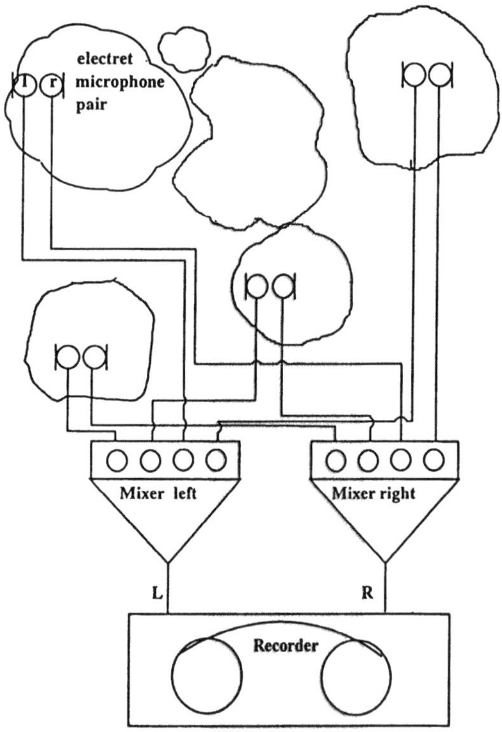

Figure 1. Wiring scheme of microphones. Two channels are recorded by a stereo cassette recorder; for each channel, microphones can be selected individually by changing the respective channel amplification by simply turning the mixer controls.

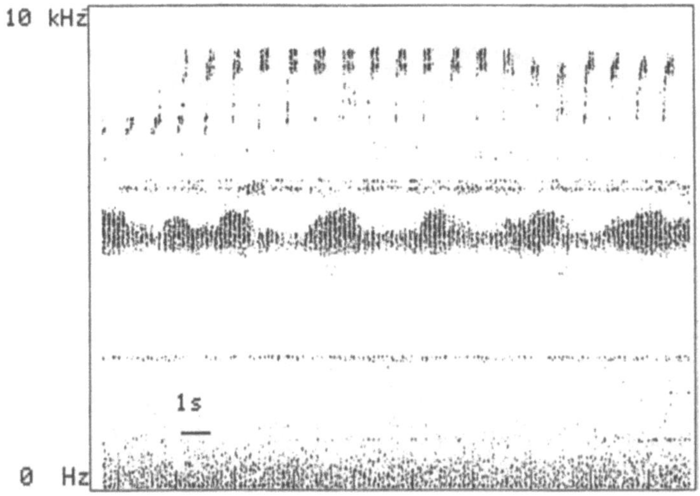

Figure 2. Sonagram of a 16 s record of the calling community in the canopy-region (Record canopy-walkway, central platform, 17.4.1991, 19:40) Ordinate: frequency [kHz]; Abscissa: time [s] .The broad frequency calls in the higher frequency range between 7 and 8 kHz might be calls of a katydid (Tettigoniidae), the narrow frequency bands at 6,3 kHz, 5,4 kHz, 4,7 kHz and 3,3 kHz could be cricket songs. Although the pulse structure of these songs cannot be recognized at this time-resolution, the different chirp structure appears clearly.

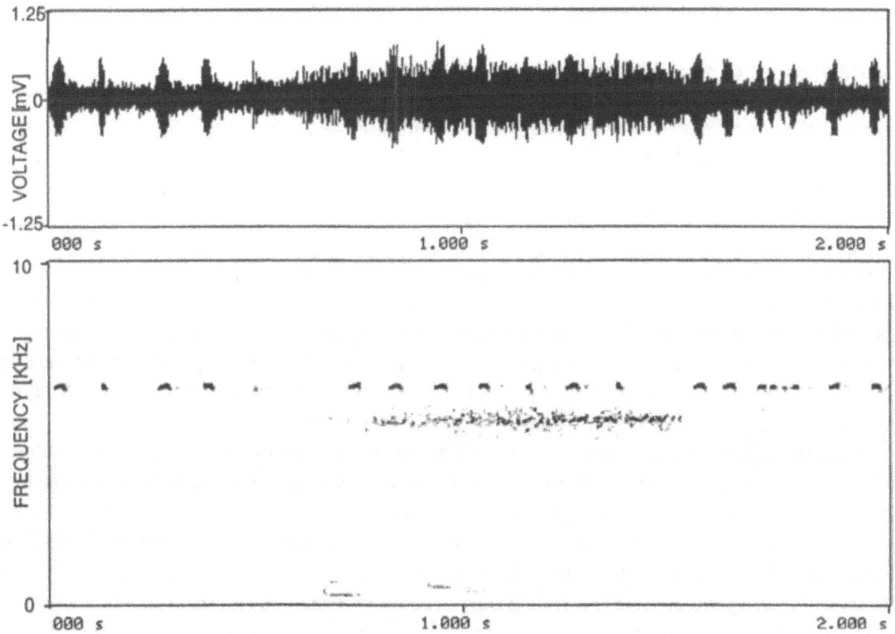

Figure 3. At a higher time-resolution, the pulse structure of a cricket species singing at 6,3 kHz emerges. In the upper half of the screen the time structure is represented by an oscillogram, the sonagram is plotted below. This call is clearly identifiable because of the shortening of pulse intervals at the end of the chirp, and could be heard at many sites during the whole time of the investigation. In addition, a second species with a deeper carrier-frequency and higher pulse rate appears. Record from canopy-walkway, central platform, 18.4.1991, 19:50.

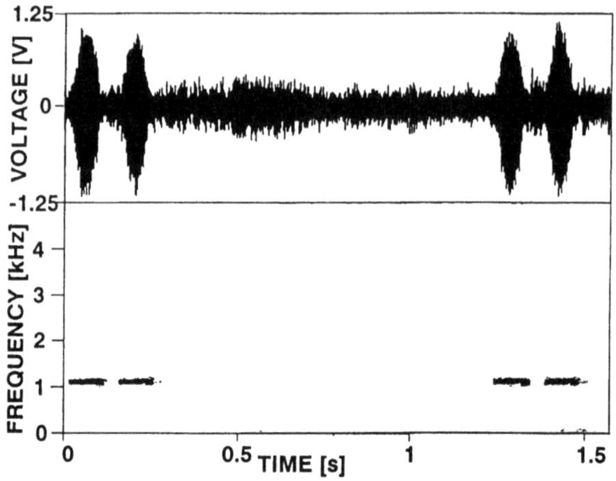

Figure 4. Euphonic low-frequency calls of *Tympanophyllum* sp. 1. The calls strongly resemble calls of the frog species *Metaphrynella sundana*. For call parameters, see Table 1.

RESULTS

Song activity of the numerous members of the 'acoustic community' depends on season, time of day and meteorological conditions. The aim of this paper is to analyze the diurnal variation of acoustic guild composition at the group level; more detailed analysis for within-group variation will be presented in later papers.

Characteristics of song types

Most of the songs can be attributed to vertebrates (mammals, birds and frogs) or different insect orders such as crickets and katydids (Orthoptera: Gryllidae and Tettigoniidae) (Figure 2) and cicadas (Cicadidae). In the following account, some general properties of sound producing mechanisms and song characteristics will be outlined.

Conspicuous, repetitive signals with a narrow-band carrier frequency between 4 and 9 kHz are generated by the rhythmic wing-strokes of male crickets (Gryllidae). During inward movement of the forewings, a file-and-scraper mechanism excites specialized resonance areas which radiate sound in a narrow carrier frequency band and at a pulse rate determined by the wing strokes (Figure 3). Besides these basic features, pulses can be arranged into complex secondary and tertiary structures called chirps. Carrier frequency, pulse rate and chirp structure are species-specific and are important cues for the innate releasing mechanism of searching conspecific females (Schildberger *et al.* 1989).

The related orthopteran family, Tettigoniidae, basically employs the same sound producing mechanism, but in general their songs consist of broader frequency spectra reaching up to the ultrasonic range, due to a highly damped alar resonator. However, species producing narrow carrier frequencies have been observed. Two species of the tettigoniid subfamily Pseudophyllinae (*Tympanophyllum* sp. Krauss 1903 and *Tympanophyllum* aff. *atroterminatum* [Brunner v. Wattenwyl 1895]) produce pure frequency songs of carrier frequencies between 1.4 and 2 kHz (Figure 4), which is the lowest frequency known for Tettigoniidae and strongly resembles vocalisations of the frog genus *Metaphrynella sundana* (Peters 1867) (Table 1). One could speculate that these Tettigoniidae acoustically mimic the well-protected frogs which live inaccessibly in tree and bamboo cavities. By shifting to frog-like low-frequency signals, *Tympanophyllum* could avoid falling into the scheme of a 'typical' high-

Table 1. Call structure of syntopic frogs and katydids (Pseudophyllinae: *Tympanophyllum* spp.)

Species	Frequency (kHz)	pulse length(ms)	pulses/chirp	chirp length (s)	chirp interval (s)
Metaphrynella sundana	1	125	1	125	>3
Tympanophyllum sp. 1	1.4	60	2	160	0.5
Tympanophyllum aff. *atroterminatum*	2	54	4-5	430	1

frequency tettigoniid song which might have been learned by vertebrate predators such as bats or tarsiers. The example shows that unequivocal attribution of ethospecies to higher systematic units is impossible in some cases, and that taxonomic identification of the 'ethospecies' can only be accomplished if the songster can be captured.

The cicadas (Cicadidae) generate their high intensity songs using tymbals situated at both sides of the first abdominal segments. Their characteristic, species-specific songs contain higher harmonics (Figure 5), and ethospecies can be identified reliably after some practice. This is reflected also in traditional classification systems where an onomatopoeic terminology describes each 'ethospecies' (e.g. "aisa-aisa" or "ngio-ngio" for certain cicada songs in the Rungus language).

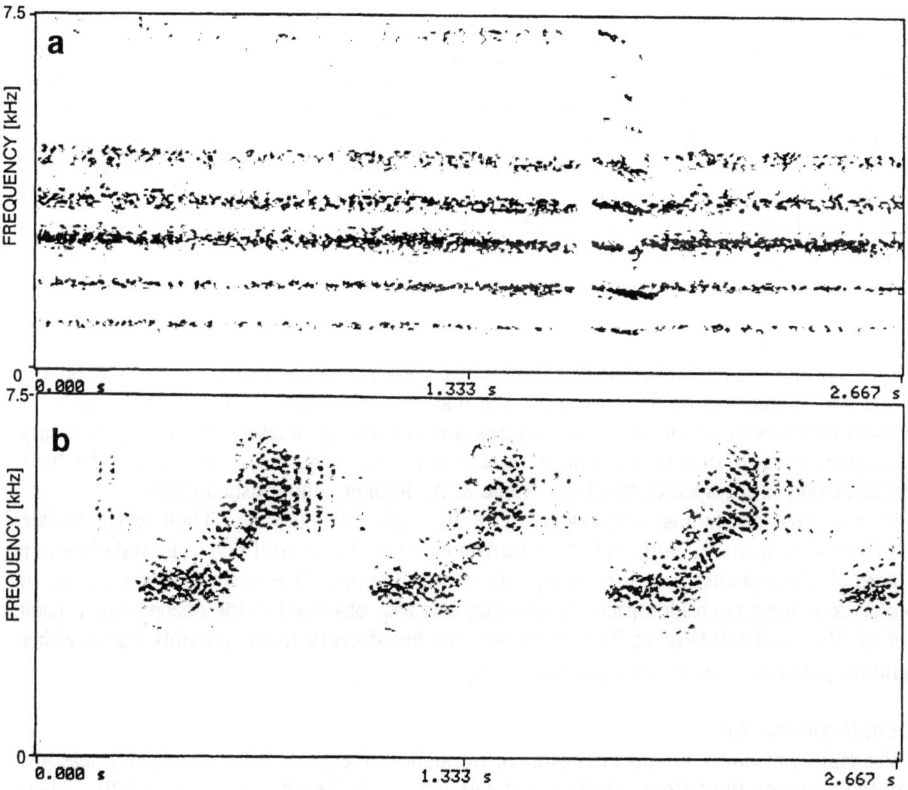

Figure 5. Cicada songs are characterized by their great proportion of higher harmonics. Left row: oscillograms and spectrograms of 2.7 s length. Right row: oscillograms of longer segments (20 s) of the respective songs. The first species (type I) is *Dundubia vaginata* (Fabricius 1787). None of the other 'ethospecies' have been caught yet; hence remain unidentified.

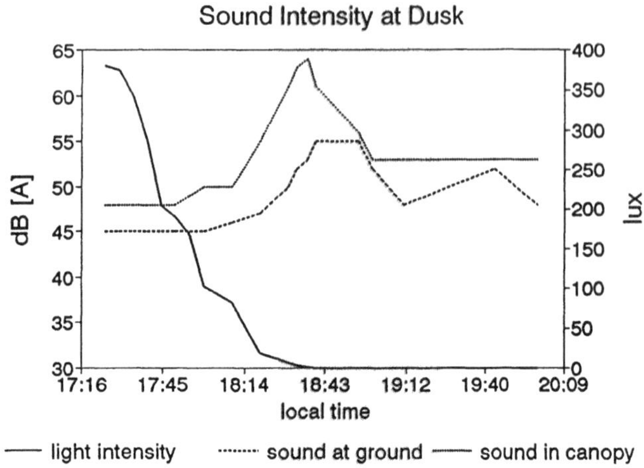

Figure 6. Sound intensity 1m above ground and in the canopy-region, recorded in the early evening hours together with light intensity. The 'dusk chorus' starts around 18:15, at a light intensity around 10 lx. The sound-intensity in the canopy-region is distinctly higher than at the ground.

Variation of overall sound intensity

Figure 6 shows the overall course of sound and light intensity between 17.00 and 20.00 at the Poring canopy walkway. Sound-intensity in the canopy is distinctly higher than at ground level. The first peak is due to cicadas and is more pronounced in the canopy sound track, while the second, more prolonged peak was measured at ground level and is generated by mole crickets. Increase in sound intensity in the evening is so pronounced that one can speak of a 'dusk chorus'. Dusk seems to be an efficient trigger for synchronisation of chorusing, and parameters such as change of light intensity or change of spectral composition of evening light might serve as cues for chorus members.

Analysis of the Chorus

A very pronounced temporal segregation is observed during the dusk chorus; this probably reflects strong competition between species at that time of the day. Spectral analysis shows that most of the noise of the early dusk chorus is generated by cicadas. Their songs are easy to recognize because of their content of higher harmonics (Figure 5). Several cicadas have been identified to the species level (K. Riede & A. Kroker, unpublished data).

When cicadas are fading, and exactly at 18:30, mole crickets start. Their songs contain one principal frequency. Although these burrowing animals are confined to humid places on the ground, their intense songs reach high up into the canopy. Two principal ethospecies of mole crickets have been identified. Chorusing was also observed in the eneopterine cricket *Itara* sp. Walker 1869 (Figure 7), but this was not heard every night, possibly due to either migration patterns or weather-triggered activity.

Waterfall communities

Waterfalls provide a humid environment throughout the year, even during dry seasons. A special community of frogs, crickets and katydids can be heard close to waterfalls. Their call structure is characterized by short duration and high intensity at rather high frequencies (Figure 8). This convergence must be interpreted as an adaptation to overcome the background noise of the waterfall, the power spectrum of which shows a linear decrease towards higher frequencies.

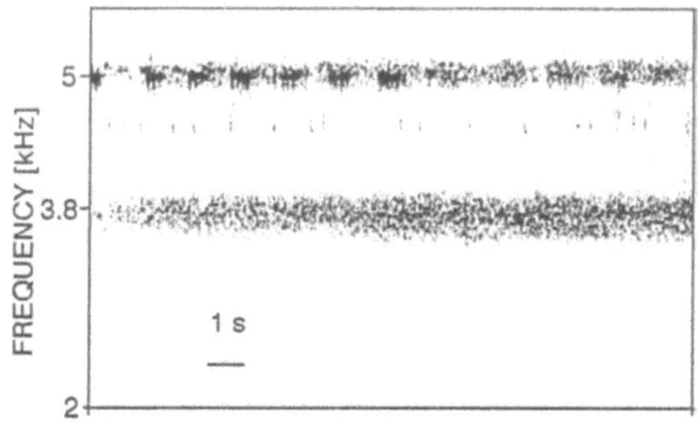

Figure 7. Two crickets forming the principal components of the late dusk chorus: a mole cricket (*Gryllotalpa* sp.) generates the lower frequency band between 3.7 and 3.95.1 kHz, and the eneopterine cricket *Itara* sp., sings at a higher frequency of 5 kHz (pulse intervals 76 ms, chirplength around 1 s, chirp intervals around 1.6 s). After several chirps, additional individuals of *Itara* sp. fall in forming a chorus, which results in a blurring of chirp structure.

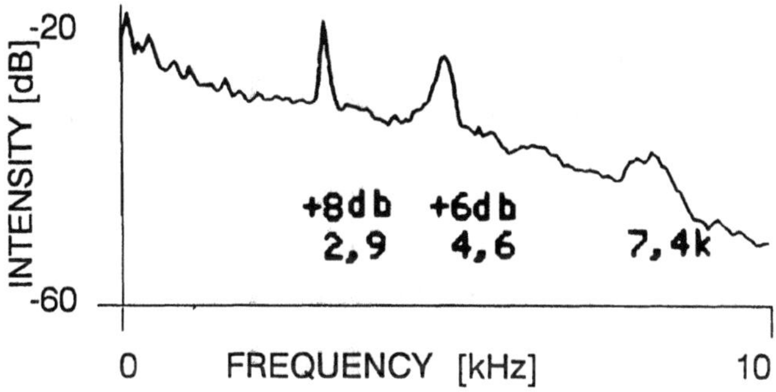

Figure 8. Amplitude spectrum of a record at Langanan-waterfall. Three species managed to call louder than the waterfall noise; these could be frogs or Tettigoniidae. Numbers below frequency maxima signify dB above waterfall noise and carrier frequency. Sound intensity of Langanan waterfall was around 65 dB[A] at 3 m distance, with a characteristic drop of -3 dB per frequency increase of 1 kHz.

DISCUSSION

In contrast to bird songs and vocalisations from higher vertebrates, the limited vocal repertoire of lower vertebrates and insects allows the recognition of species by analysing stereotyped, inborn (but by no means simple!) advertisement calls. Therefore, acoustical recording is a valuable tool for mapping and monitoring elusive, cryptic and nocturnal rainforest organisms (Riede 1993). The classification of 'ethospecies' presented here is preliminary and probably erroneous in some cases, but has the great advantage in as much as it allows a non-invasive study of singing animals.

The strong convergence between orthopteran and anuran song indicates that acoustic interference might occur across systematic units. Members from different phyla engage within the 'acoustic guild' and must be considered together in order to understand the constraints shaping acoustic communication among rainforest creatures. Members of the 'acoustic guild' interact within a soundspace defined by call parameters such as frequency and pulse intervals, and within a soundscape of abiotic sounds such as wind, rain and acoustical habitat properties. As with radio stations, 'senders' compete for the limited resource 'acoustic transmission channel'. One therefore should be able to observe competition, resource partitioning and niche segregation, and investigate ecological questions such as the saturation of a community. Acoustic transmission channels are a common, physically well-defined resource for a great variety of species, similar to light in a plant community. The observed diurnal resource partitioning points to strong competition within the acoustic guild which is especially pronounced during the dusk chorus where nocturnal and diurnal species overlap. Waning daylight might be attractive for many species for a variety of reasons. Dusk is well suited to trigger precisely timed, brief activity bouts between males and females, thereby reducing overall song time. Diurnal, visually oriented predators may stop hunting for want of light, but conspecifics would still have some residual light for orientation. In addition, members of a chorus are difficult to locate acoustically.

An interesting aspect of bioacoustic diversity is the comparison of acoustic guilds from different rainforests. A comparative analysis of acoustic partitioning in frogs showed strong similarities in anuran calling assemblies from three different neotropical forest habitats (Duellmann & Pyles 1983). The community described here indicates some basic differences with respect to a Neotropical community from Northwest-Amazonia (Riede 1993). In the Neotropical community, crickets (Gryllidae) sing during day and night, cicadas are much less dominant and the dusk chorus is less pronounced. The cricket community seems to be more densely packed, especially in the canopy, which may reflect differences in canopy architecture and/or epiphyte load.

In Southeast Asia, the diurnal acoustic dominance of loud cicada songs, at least at certain times of the year, may well form an important constraint displacing fainter orthopteran songsters to nocturnal activity.

ACKNOWLEDGEMENTS

I thank Sabah Parks for generous support of my field investigations, R. Malkmus, Wiesthal, for identification of frogs, and A. Kroker, Freiburg, for identification of cicadas. The study was conducted with the support of the Deutsche Forschungsgemeinschaft.

LITERATURE CITED

DUELLMAN, W. E. & PYLES, R. A. 1983. Acoustic resource partitioning in anuran communities. *Copeia* 3:639-649.

ELTON, C. S. 1973. The structure of invertebrate populations inside neotropical rain forest. *Journal of Animal Ecology* 42:55-104.

RIEDE, K. 1993. Monitoring biodiversity: Analysis of Amazonian rainforest sounds. *Ambio* 22:546-548.

RÖMER, H., BAILEY, W. & DADOUR, I. 1989. Insect hearing in the field. *Journal of Comparative Physiology* A164:609-620.

SCHILDBERGER, K., HUBER, F. & WOHLERS, D. 1989. Central auditory pathway: Neuronal correlates of phonotactic behaviour. Pp. 423-458 in Huber, F., Moore, T. E. & Loher, W. (eds). *Cricket behaviour and neurobiology*. Cornell University Press, Ithaca.

WALLACE, A. R. 1869. *The Malay archipelago*. Tynron Press, Scotland.

The arthropod community of bamboo internodes in Peninsular Malaysia: microzonation and trophic structure

DAMIR KOVAC[1] *and* BRUNO STREIT[2]

[1]*Forschungsinstitut Senckenberg, Entomologie I, Senckenberganlage 25, 60325 Frankfurt, Germany*
[2]*Zoologisches Institut, J. W. Goethe-Universität, Siesmayerstrasse 70, 60054 Frankfurt, Germany*

ABSTRACT. The hollow culm segments (internodes) of the Southeast-Asian bamboo *Gigantochloa scortechinii* Gamble were chosen as model systems for studying tropical animal communities. When the culm wall is punctured by boring insects or otherwise injured so that a connection to the exterior is established, the respective internode becomes filled with runoff rainwater. Thus a phytotelma - a small water body within a plant - is created. The phytotelma-containing internodes are colonized by numerous species of both aquatic and terrestrial animals. As a first step in the investigation of this poorly known community a survey was taken of the animal taxa occurring in living and dead internodes. In order to watch the interactions between the internode inhabitants more closely, artificial 'experimental internodes' were designed in which the internodes of horizontally positioned culm sections were prepared in such a way that their upper part formed a movable lid that could be opened for inspection. In such internodes the processes within the internode interior such as the sequence of colonization and the trophic interactions, could be observed. In all, several hundred animal taxa were found, many of them restricted to bamboo internodes, and many still undescribed. This makes bamboo internodes among the most species-rich of all phytotelmata. Several factors contribute to this extraordinary diversity: the nutrients that become available from the decomposition of the internode wall; the secluded and therefore well protected terrestrial zone above the water surface; and the occurrence of different types of internodes that provide a panoply of different ecological conditions.

KEY WORDS: food webs, phytotelmata, spatial distribution, bamboo internodes, *Gigantochloa scortechinii*.

INTRODUCTION

Various hypotheses have been proposed to explain the astonishing diversity and complexity of tropical communities. Our research group is studying this and other ecological questions in tropical ecosystems of West Malaysia. A microecosystem in the culms of the SE-Asian bamboo *Gigantochloa scortechinii* Gamble was chosen as a model system. Bamboo culms are usually subdivided into hollow segments, or internodes, whose interior is secluded from the outside environment. However, when the culm wall is penetrated by boring insects, the resulting holes not only allow other invertebrates to enter, but may also result in the internodes becoming filled with runoff-water during heavy rain. In this way a habitat is created which may be colonized by both aquatic and terrestrial organisms.

Other examples of such natural bodies of water in plants (=phytotelmata) in SE-Asia are water-filled tree cavities (Kitching 1987, 1990, Thienemann 1934) and the insectivorous pitcher plants (*Nepenthes*) (Beaver 1979a, 1979b, 1983, 1985, Thienemann 1932). These microhabitats have received increasing attention as methodologically amenable systems for testing hypotheses on the functioning of animal communities (Frank & Lounibos 1983, Kitching & Beaver 1990, Mogi & Yong 1992, Pimm & Kitching 1987). They offer several inherent advantages for the study of ecosystems: (1) they usually occur in sufficient number to allow statistical analyses; (2) they are colonized by a complex and specialized fauna which

D. S. Edwards et al. (eds.), Tropical Rainforest Research – Current Issues, 85–99.

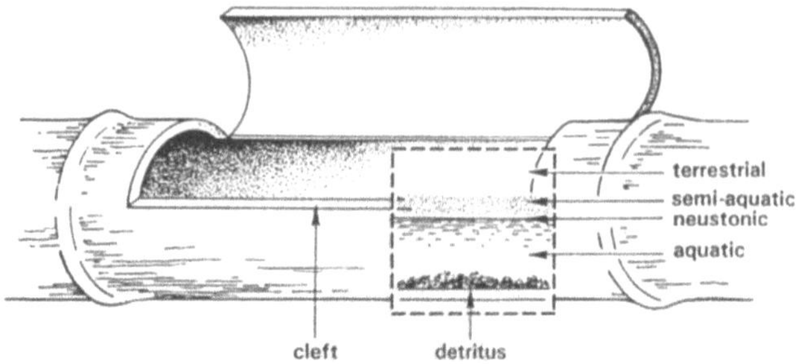

Figure 1. Schematic representation of a horizontal internode that has been sawed open in such a way that its upper half forms a movable lid that can be opened and closed ('experimental internode'). During rain the internode fills with water and several distinct microhabitats are created: water (aquatic), water surface (neustonic), wet transitional zone (semi-aquatic), and land zone (terrestrial). The cleft that remains between the top (lid) and the bottom of the internode also becomes a distinct habitat with a specialized community. The different zones can be subdivided further (detritus, free water, etc.), with gradual transitions between the individual zones.

can be sampled completely; (3) they are physically delimited from other habitats. Bamboo internodes offer even more advantages: (4) the process of colonization can be started experimentally at any time by boring artificial holes; (5) complex interactions between the aquatic and terrestrial fauna can be investigated; (6) the wide geographical distribution of bamboo makes possible comparative studies in different biogeographic regions and climatic zones.

As the fauna of bamboo internodes, with the exception of the mosquito larvae, was poorly known at the outset of this study (see Discussion), a faunal inventory of the internode inhabitants was the first step in our investigation. It soon became apparent that different types of internodes (living or dead, with holes or as stumps containing open water) lead to the formation of different animal communities within them. The second important step was taxonomic work on the internode inhabitants, many of which were still undescribed. Finally, an investigation of the structure and dynamics of the internode community was begun, with emphasis on trophic interrelationships and succession.

Among the different types of internodes the 'experimental internodes' were most thoroughly investigated. These horizontally positioned internodes of felled bamboo culms, whose upper halves could be swung open for inspection (Figure 1), allowed close monitoring of the internode fauna over an extended period of time without causing much disturbance. Moreover, the system could be experimentally altered by the removal of certain species. An additional advantage was the ease of observation of the animals in these internodes which, in contrast to open phytotelmata, do not fill up with fallen leaves. This allowed quantitative investigations of the trophic relationships under nearly natural conditions. This paper is intended to give a first, qualitative picture of the animal community occurring in the experimental internodes and of their trophic interactions.

METHODS

The study was conducted in 1991 in West Malaysia, Selangor Darul Ehsan, at the Ulu Gombak Field Studies Centre (University of Malaya), which is situated on the western slopes of the Main Range (location: 3°19'32''N, 101°45'16''E, altitude c. 250 m). The climate and geology of the area have been outlined by Marshall (1971) and Bishop (1973). The surrounding forest was logged selectively in 1956-1958. Clumps of the large bamboo *G. scortechinii* Gamble, which is indigenous to the Malay Peninsula (Holttum 1959), are abundant in the regenerating vegetation (Figure 3A). The culms grow to 25 m. In the lower section the length of the internodes measures c. 20-60 cm, the diameter c. 8-10 cm, and the thickness of the wall is c. 1 cm.

The faunas of different types of internode were investigated, including living internodes, dead internodes, bamboo stumps etc. Experimental internodes (Figure 1) were studied most intensively for methodological reasons. These were made by felling the bamboo culm, removing the upper section including all side branches, and placing the remaining lower section horizontally at a height of c. 1.5 m by wedging its terminal ends between other, erect bamboo culms. Ten internodes per suspended culm were sawn through halfway near the nodal septa at both ends of the internode, then the section between the two transverse cuts was split on one side with parang and hammer. The upper halves of the internodes could thus be bent upwards, with the only remaining connection between lower and upper side forming the hinge.

One hundred experimental internodes were checked daily during a six-months period (end of May through November 1991) to monitor the composition of the community and study its succession and trophic interactions. After the first heavy rain all internodes had filled with water (c. 200-500 ml per internode). With the exception of two which became temporarily dry, all internodes remained water-filled throughout the study period. Colonization of the internodes was possible via the two sawn gaps near the node septa, which were about 2 mm wide. Species that spend the adult stage outside the internodes were taken to the laboratory as larvae or pupae to be hatched. Freshly captured prey was taken from predators with a pair of forceps and identified in the laboratory. The animals were observed using a Zeiss[R] prism-headlens F (8x).

RESULTS

The 'experimental internode' habitat may be subdivided into several adjacent vertical zones: (1) aquatic zone, (2) water surface, (3) semiaquatic zone, and (4) terrestrial zone. The small gap between the movable upper and the lower part of the internode forms an additional microhabitat for a specialized fauna dominated by fly larvae. This latter microhabitat will not be considered in the present paper.

Aquatic microhabitat

The aquatic food chain of the internode community is fed by both water soluble nutrients from the inner bamboo wall and nutrient import from the outside, e.g. drowned animals. Once water has entered the internode, organic and inorganic substances of the nutritious internode wall become dissolved in it, stimulating the growth of bacteria and other microorganisms. Protozoa, e.g., ciliates, soon follow, feeding, among other things, on bacteria. Other members of the internode microfauna are small crustaceans (Copepoda and others) and water mites. Apart from these organisms, insect larvae also dominated the aquatic habitat. With the exception of *Scirtes* (Coleoptera), the insects involved all belonged to the order Diptera.

At the beginning of the succession, *Megaselia* sp. (Phoridae) larvae are numerous (see also Disney 1991, 1993). They move swiftly along the bottom, apparently scraping organic matter off the substrate. They are also capable of ingesting floating organic particles or of feeding on dead animals. In the course of time a substantial amount of detritus, including dead animals from the terrestrial zone of the internode, nesting material carried in by ants etc., accumulate at the internode floor.

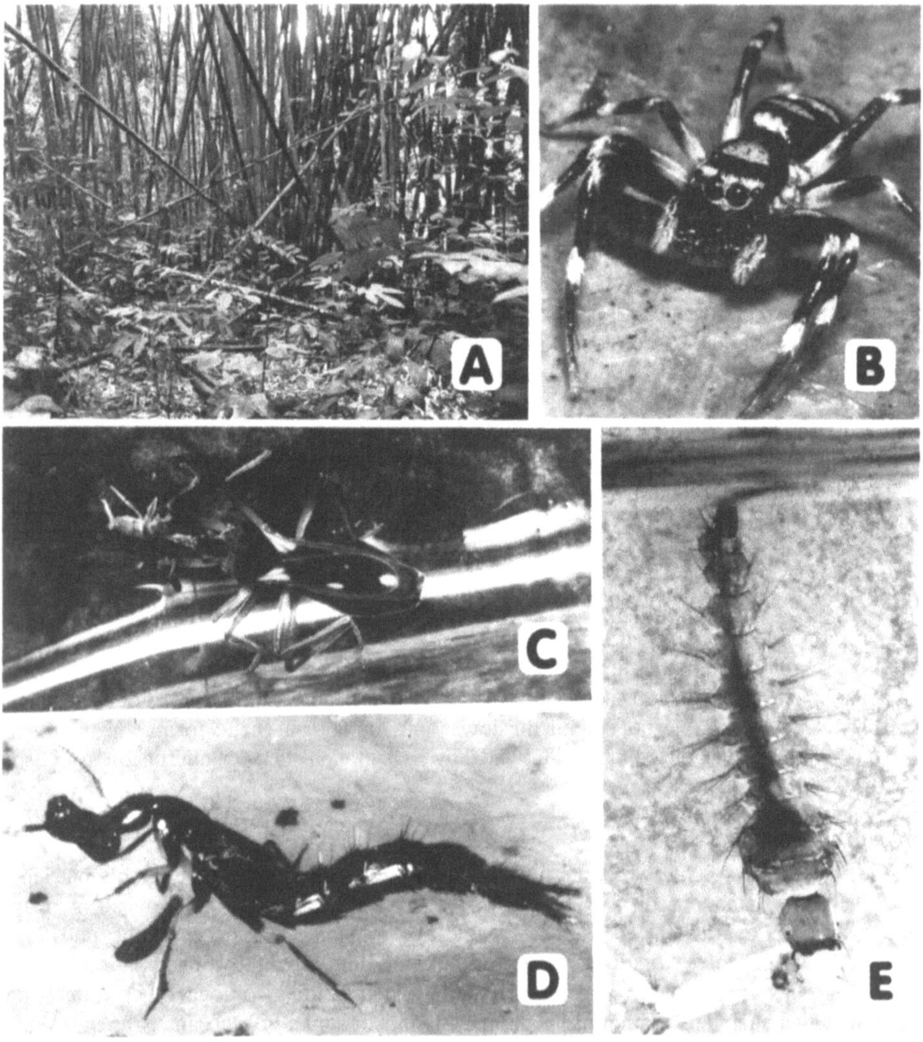

Figure 3. A. Stand of the bamboo species *Gigantochloa scortechinii* in Ulu Gombak, Peninsular Malaysia. There are several fallen bamboo culms in the foreground. In internodes of such fallen culms that have holes in their wall the predators shown in B-E occur. B. A jumping spider of the subfamily Spartaeinae (length c. 7 mm) has captured a predaceous mosquito larva, *Toxorhynchites metallicus*. C. A male of the water strider *Baptista* (length c. 5 mm) is sucking on a springtail (Collembola). D. The rove beetle *Acylophorus* sp. (length c. 6 mm) is holding a freshly captured mosquito pupa in its mandibles while standing on the water surface. E. The predaceous mosquito larva *Toxorhynchites metallicus* (length c. 7 mm) is feeding on the larva of a different species of mosquito. (With the exception of E, all photographs taken in the field).

The larvae of the ceratopogonids *Dasyhelea grata* (Johannsen) and *D. assimilis* (Johannsen) stay in or on the detritus layer and feed on dead animals. Beetle larvae of the genus *Scirtes* are found within the detritus layer, feeding on sizable organic particles. Larvae of a still undescribed *Forcipomyia* sp. (Ceratopogonidae), which also ingest organic particles, stay on the detritus layer or on the submerged part of the internode wall. A frequent and conspicuous inhabitant of the detritus layer are the larvae of the chironomid *Polypedilum (Pentapedilum) convexum* (Johannsen), which build vertical chimney-like structures out of faecal pellets and other detritus. With the help of their undulating body movements they create a water current through their tube and their fine-meshed silk web. They consume organic particles that get trapped in the web. Furthermore, on the detritus layer larvae of the cecidomyiid *Xylodiplosis* sp. are found which probably are not truly aquatic (see Discussion) but may nevertheless be an important dietary component for aquatic predators during the early succession stages.

In contrast to the other water inhabitants, the mosquito larvae, being able to swim or float, also colonize the open water. They feed mainly on minute organic particles and microorganisms which they strain from the water. During this activity some species move along the internode wall (*Uranotaenia* sp. A), others lie on their backs on the floor of the internode (*Culex* (*Lophoceraomyia*) *minor* (Leicester)), stay in the middle layer of the water (*Uranotaenia* sp. B), or hang down from the water surface (*Orthopodomyia anopheloides* (Giles)). *Anopheles asiaticus* Leicester strains food particles from the underside of the water surface. In addition, predatory behaviour may occur in the genera *Armigeres* (*Leicesteria*) and *Anopheles*. *Armigeres* may also be cannibalistic.

Two aquatic predators were found in the experimental internodes: mosquito larvae of the genus *Toxorhynchites* (*T. leicesteri* Theobald and *T. metallicus* Leicester, Figure 3E) and an undescribed midge larva of the subfamily Tanypodinae (Pentaneurini, near *Trissopelopia*). The midge larvae move slowly and inchworm-like along the bottom hunting small fly larvae. While searching for prey, the animals also climb onto raised objects (e.g. the tubes of *Polypedilum convexum*) then extend their body into the open water waiting for prey to swim by. In addition to live prey, dead animals are also taken. Several of the predaceous midges have often been observed feeding simultaneously on the same object. The *Toxorhynchites* spp. prey on all of the larger aquatic organisms. During the day, they tend to stay at the bottom of the water and during the night near the surface. In addition to aquatic prey, they also catch terrestrial animals which lie on the water surface or which have become trapped there. The majority of prey objects were other species of mosquito larvae. The prey spectra for both *Toxorhynchites* spp. were nearly identical. *T. leicesteri* reached its greatest abundance before *T. metallicus*, at the beginning of the succession. In both species fighting to the death was observed between the larvae until one or, rarely, two individuals per internode remained.

The water surface

Besides springtails (Collembola) and mites (Acari), water striders of the genus *Baptista* (Figure 3C) and rove beetles of the genus *Acylophorus* (Figure 3D) were the most conspicuous arthropods found on the water surface. *Baptista* sp. feeds to a large extent on Collembola (c. 45% of the prey of the nymphs and the adults, n=278) which it hunts actively on the water surface. The remaining prey objects were mainly terrestrial animals that had been trapped on the water surface. A small percentage were aquatic animals caught just below the water surface, e.g. small *Scirtes* or *Clogmia* larvae.

The adults of the rove beetle *Acylophorus* also stayed mostly on the water surface. They waited for mosquito larvae and pupae (c. 58% of the prey objects, n=51) to come to the surface for air, then seized the prey, dragged it out of the water, and ate it. Live and dead animals on the water surface were also caught by aquatic predators, including *Toxorhynchites* and the normally filter-feeding larvae of *Anopheles*. In two cases an *Anopheles* larva was seen to seize a live *Baptista* nymph by the hind leg and gradually swallow it entirely.

The semiaquatic microhabitat

Between the aquatic and the terrestrial zone there is a narrow band covered by a thin water film. This wet zone normally extends only a few millimeters vertically. The meniscus from the aquatic zone extends into the semiaquatic zone, and the semiaquatic zone in turn merges with the moist lower section of the terrestrial zone a little higher up.

The semiaquatic microhabitat is inhabited by various fly larvae which graze upon microorganisms growing on the internode wall. Some of the larvae are confined to this zone, while others occur also in the aquatic or the lower, moist part of the terrestrial zone. The larvae of *Forcipomyia* sp. (Ceratopogonidae), *Philosepedon* sp. (Psychodidae), and *Ptilona confinis* (Walker) (Tephritidae), for example, are restricted to the semiaquatic zone.

Anopheles asiaticus larvae enter the semiaquatic zone without ever leaving the meniscus region. The larvae of the psychodid, *Clogmia* sp., for the most part stay in the aquatic zone with their caudal spiracles in constant contact with the water surface, but they also enter the semiaquatic zone. The pupae of the mosquito, *Uranotaenia* sp. A, usually remain in the semiaquatic zone but flee into the water upon disturbance. Other inhabitants of the semiaquatic zone also occur in the moist part of the terrestrial zone, e. g., larvae of the phorid *Megaselia humida* Disney, larvae of the stratiomyid *Camptopteromyia fractipennis* de Meijere (Rozkosny & Kovac 1991), and larvae of several *Forcipomyia* species. Most larvae of the semiaquatic (and terrestrial) zone possess some passive anti-predator defenses such as protective secretions (*Forcipomyia* spp.), camouflage by covering the dorsal surface with detritus particles (the phorid *Megaselia humida*, the psychodid *Philosepedon* sp.), or surrounding themselves with threads of mucus (e.g., Mycetophilidae).

The inhabitants of the semiaquatic zone are preyed upon by spiders (Spartaeinae sp; Theridiidae spp., etc.) and by *Acylophorus* adults and, especially, larvae. The *Acylophorus* larvae remain completely in the semiaquatic zone. Unlike the adult beetles the larvae have no hydrophobous pubescence and therefore cannot walk on the water surface. Instead they incessantly walk along the internode wall just above the water surface hunting the fly larvae of the semiaquatic zone. They also catch *Baptista* nymphs and, like the *Acylophorus* adults, seize any mosquito larvae that come to the surface for air and pull them out of the water with their mandibles.

The terrestrial microhabitat

The terrestrial fauna is supplied with nutrients from the decomposition of the inner internode wall and from the import of organic substances from the outside, e. g., in the form of animals seeking protection within the internode. At the high ambient relative humidity, microorganisms and especially fungi begin to break down the internode wall. Various insect larvae graze on microorganismic growth on the internode wall. These include the flies, *Forcipomyia* species or, at later stages of succession, the larvae of Mycetophilidae and Sciaridae; the beetles, larvae of Endomychidae, Coccinellidae, and Tenebrionidae, and the moths, larvae of Arctiidae. Another type of decomposer are the various wood-boring beetles, e. g. *Conarthrus* and *Dinoderus*, that stay in the internode cavity for a brief period at the beginning of the succession, but then tunnel into the bamboo wall.

The terrestrial fauna also includes groups of detritivores which are normally abundant in the soil. These are species requiring high humidity, e. g. wood-lice, springtails, Psocoptera, and earwigs (see Srivastava & Kovac, in press). They have been observed to enter the internodes at night. In addition, numerous mites, true bugs, beetles and insects from other orders are found. In some internodes earthworms were found, mostly in the cleft between the upper and lower part of the internode, with part of the body submerged.

The predators among the terrestrial internode inhabitants belong to the true bugs, beetles, ants and spiders. The assassin bug *Emesopsis* sp. used its raptorial forelegs to catch mainly small, soft-bodied insects like Collembola and Ceratopogonidae, but also caught larger animals, e. g., adult mosquitoes. The staphylinid beetle *Stilicopsis* sp. had a similar prey spectrum. A number of additional predaceous and non-predaceous staphylinids were found in the internodes, but none of them was as common as *Stilicopsis* and *Acylophorus*. Ants were found only rarely and only a few prey objects were taken from them for identification (for example, larvae of Cecidomyiidae). As ants are the most common prey objects of the web-building internode spiders, however, they apparently do enter the internodes regularly, perhaps more frequently at night, when no regular checks were conducted.

The top predators in the experimental internodes were spiders. A species of salticid (Spartaeinae) (Figure 3B) fed on terrestrial and aquatic animals (c. 50% of the prey items each, n=84), especially mosquito larvae, that it caught at the water surface. The majority of the remaining spiders were web-builders of different families which constructed their webs in the upper region of the internode. The majority of species belonged to the Theridiidae, which build irregular, three-dimensional cobwebs. Ants and beetles were their main prey. To a lesser extent freshly emerged adults of insects with aquatic larvae were caught (Culicidae, etc.). In a few cases an undetermined Theridiid was seen to intrude into the webs of other spiders and prey on the occupants. The remains of the spider's prey organisms were dropped to the water surface after some time and thus became accessible to the aquatic fauna.

Most other species of animals found in the terrestrial zone were represented by few or even single specimens, for example various wasp and fly parasites that attack eggs, larvae and pupae (e.g., those of *Forcipomyia* and *Epicypta*) or animals which utilize the internodes for breeding (e.g., the crab spider *Smodicinodes kovaci* Ono (Thomisidae), see Ono 1993) or for temporary hiding places. Many of these animals certainly are not regular internode inhabitants. All of these frequently get trapped in spider webs or on the water surface and are therefore an important component of the nutrient import into the internode.

DISCUSSION

Comparison with earlier investigations on bamboo phytotelmata

The community in bamboo internodes has, with the exception of mosquito larvae, received little scientific attention to date, especially with respect to the terrestrial components of the community. Leicester (1903), who studied mosquitoes on the Malay Peninsula, was the first to mention this kind of habitat, notably the erect culms with holes. Thienemann (1932, 1934), who collected inhabitants from bamboo stumps and other phytotelmata in Sumatra, Java and Bali, was the next to make an important contribution. MacDonald & Traub (1960) conducted an extensive 3-year investigation on the ecology of the forest mosquitoes, mainly in the Ulu Gombak area. This study covered c. 100 species of mosquitoes that inhabit various internode types of *Gigantochloa scortechinii* and other bamboo species. In addition, information on the origin of the holes in live bamboo culms was given (MacDonald 1960). In the same area the bats *Tylonycteris pachypus* (Temminck) and *T. robustula* Thomas, which roost exclusively in live internodes (Medway 1972, Medway & Marshall 1970, 1972), and the parasitic batfly *Basilia hispida* Theodor (Nycterbiidae), which pupates on the internode wall (Marshall 1970, 1971) were investigated intensively.

Kitching (1987, 1990) investigated the fauna of bamboo internodes and other phytotelmata in Sulawesi and Papua New Guinea. He was the first to construct food webs for this habitat. Other investigations on the inhabitants of bamboo internodes were conducted in Japan. Most of these studies dealt with the ecology of culicid larvae in bamboo stumps and bamboo cemetery vases (see Kurihara 1983). Mogi & Suzuki (1983), in addition, investigated the complete fauna of live bamboo internodes with artificial holes made by the investigators. The most dominant groups were aquatic larvae of Culicidae and Ceratopogonidae. The most common animals in the terrestrial zone were mites, ants, collembolans and diplopods. Probably none of these animals were bamboo specialists, because there were no naturally occurring internodes with holes in this region and the animals occurred only in low numbers.

In the present investigation many previously undescribed internode-inhabiting species were found, especially in the terrestrial zone, which had never before been studied in SE-Asian bamboos. But even in the better known aquatic zone new animal taxa were discovered, for example, the semiaquatic larvae of Tephritidae. Another family which had not previously been found in bamboo internodes are the Cecidomyiidae. Aquatic larvae of this family (*Lestodiplosis syringopais* (Hering), Beaver 1979a) were known, however, to inhabit pitcher plants. The same genus also occurred in the experimental internodes (see Table 1), but only as adults. In the water we found, however, larvae of the cecidomyiid *Xylodiplosis*. The red, nearly fully grown larvae appeared in the water very suddenly, but then died within a few days both under natural and laboratory conditions. The larvae probably developed in the vascular tissue of the bamboo, as the larvae of other members of this genus do (Gagné 1985, Rock & Jackson 1985). *Xylodiplosis* females were regularly seen to deposit their eggs on the cut surfaces of the internode wall. Probably the larvae left the vessels to pupate, and the animals in the lower part of the internode got trapped in the water and died there or were caught by predators.

Comparison between different internode types

The kind of nutrient input and the accessibility of the bamboo internode have an important influence on the structure of the internode community. This was realized by Leicester (1903)

Table 1. Provisional list of arthropod species encountered in experimental internodes (families, genera and species in alphabetical order; list not exhaustive). Other macroscopic internode-inhabiting metazoans present were annelids and molluscs. The organisms designated * have not yet been described.

CRUSTACEA		
Copepoda		
ARACHNIDA		
Acarina		
Araneae	Clubionidae	*Clubiona* sp.*
		Heteropoda sp.
	Gnaphosidae	*Aphantaulax* sp.
	Mysmeridae	*Tamasesia* sp.
	Ochyroceratidae	*Merizocera* sp.*
	Oonopidae	*Gamasomorpha* sp.
	Pholcidae	*Spermaphora miser* Bristowe
		Uthina luzonica Simon
	Salticidae	Spartaeinae sp.*
	Theridiidae	*Coscinida* sp. A
		C. sp. B
		Euryopis sp. A
		Eu. sp. B
		Theridion sp. A
		T. sp. B
INSECTA		
Blattodea	Blattelidae	*Margattea* sp.*
Dermaptera	Chelisochidae	*Euenkrates* sp.*
		Hamaxas sp.
		Proreus sp.*
	Labiduridae	*Allostethella guttata* (Bormans)
	Spongiphoridae	*Chaetospania feae* (Bormans)
Psocoptera	Lepidopsocidae	*Soa* sp.
	Pachytroctidae	*Tapinella* sp.
Hemiptera	Anthocoridae	*Lasiochilus* sp.
	Aradidae	*Brachyrhynchus* sp.
	Ceratocombidae	
	Miridae	*Cylapomorpha* sp.
		Cylapus sp.
		Fulvius sp.
		n. gen., n. sp.
	Pentatomidae	

Table 1 (continued).

Hemiptera (cont.)	Reduviidae	*Emesopsis* sp.*
	Veliidae	*Baptista* sp.*
Coleoptera	Biphydidae	
	Brenthidae	
	Carabidae	*Pentagonica* sp.
	Cleridae	
	Coccinellidae	
	Corylaphidae	
	Cucujidae	
	Curculionidae	*Conarthrus* sp.
	Endomychidae	*Eumorphus vitalisi* Arrow
		Indalmus sp.
	Histeridae	
	Hydrophilidae	
	Lagriidae	
	Lathridiidae	
	Nitidulidae	
	Pselaphidae	*Batraxis* sp.
	Scirtidae	*Scirtes* sp. A
		S. sp. B
		S. sp. C
	Staphylinidae	*Acylophorus* sp.*
		Coproporus sp.
		Eleusis sp.
		Hesperus perakensis Cam.
		Pinophilus sp.
		Sepedophilus sp.
		Stilicopsis sp.
		Stiloderus sp.
		Thorakochirus sp.
	Tenebrionidae	*Leichrodes rufofulvus* Westw.
Diptera	Cecidomyiidae	*Clinodiplosis* sp.*
		Feltiella sp.*
		Lestodiplosis sp.*
		Xylodiplosis sp.*
	Ceratopogonidae	*Culicoides* sp.
		Dasyhelea assimilis (Johannsen)
		D. grata (Johannsen)
		Forcipomyia (Euprojoannisia) similans (Johannsen)
		F. (Forc.) bikanni Chan & LeRoux
		F. (Forc.) swezeyanaadfinis Chan & LeRoux
		further *F. (Forc.)* spp.*
	Chironomidae	*Compterosmittia* sp.
		Pentaneurini nr *Trissopelopia**
		Polypedilum (Pentapedilum) convexum Johannsen
	Culicidae	*Aedes (Christophersiomyia) gombakensis* Mattingly
		Ae. (Stegomyia) sp.
		Anopheles asiaticus Leicester
		Armigeres (Leicesteria) dolichocephalus (Leicester)
		A. (Leicesteria) flavus (Leicester)
		A. (L.) sp. A
		A. (L.) sp. B
		Culex sp.
		Orthropodomyia anopheloides (Giles)
		Topomyia (Suaymyia) spathulirostris Edwards
		Toxorhynchites leicesteri Theobald
		T. metallicus Leicester
		Tripteroides (T.) sp.
		Uranotaenia sp. A
		U. sp. B
	Mycetophilidae	*Allactoneura* sp.
		Epicypta sp. A
		E. sp. B
	Phoridae	*Megaselia animaloterga* Disney
		M. deningi Disney
		M. gombakensis Disney
		M. humida Disney
		M. kovaci Disney
	Psychodidae	*Clogmia* sp.
		Philosepedon sp.
	Sciaridae	*Bradysia* sp.
	Stratiomyidae	*Camptopteromyia fractipennis* de Meijere
		Ptecticus longipennis (Wiedemann)
	Syrphidae	*Graptomyza* sp.
	Tephritidae	*Ptilona confinis* (Walker)
	Tipulidae	
Lepidoptera	Arctiidae	
Hymenoptera	Formicidae	*Strumigenys* sp., *Pheidole* sp.
		many other ants and parasitic Hymenoptera

and MacDonald & Traub (1960). MacDonald & Traub (1960) classified the internodes intoseveral types: 1) artificial containers (bamboo pots placed in various situations), 2) bamboo stumps, 3) fallen and upright split bamboos, and 4) upright bamboos with holes. The latter group was further divided into three subgroups by the size of the holes: a) long axis less than 5 mm, b) long axis 5-10 mm, and c) long axis more than 10 mm.

We distinguish at a basic level between live and dead internodes:

(1) Access to live, upright internodes is possible through openings resulting from mechanical injury (e.g. fissures) or from the boring activities of animals. The most common types of holes are made by the larvae of the leaf beetle *Lasiochila goryi* (Guér.), by caterpillars of the moth family Pyralidae (3 species), and by at least one species of woodpecker, *Gecinulus viridis* Blyth (MacDonald 1960, and personal observations). Less common are openings made by adults of the long-horned beetle *Abryna regispetri* Paiva as they break out of the internode (Kovac & Yong 1992). During the first stage of succession, when the water has just entered, part of the surface tissue of the inner internode decomposes. The bamboo is not damaged in any recognizable way by this process and the water filled internodes, like tree cavities, may persist for many years. Our preliminary investigations indicate that the fauna of live internodes comprises fewer species than that of dead internodes, but there seems to be a higher proportion of specialists. The aquatic community of the live internodes has a similar structure to that of experimental internodes, but often there are differences at the species level. For example, *Toxorhynchites magnificus* (Leicester), the most common *Toxorhynchites* species of live internodes, is almost completely restricted to this internode type (MacDonald & Traub 1960, and personal observations). Some animals occur only in live internodes - the bats *Tylonycteris robustula* and *T. pachypus* for example (Medway & Marshall 1970).

Also in the terrestrial zone there are animals that probably live exclusively in live internodes. One of these is an undescribed ant, *Tetraponera* sp. near *attenuata* F. Smith, that inhabits only internodes with small holes, which may be made by pyralid or chrysomelid larvae or by the ants themselves (Klein *et al.* 1993, and personal observations). These ants get much of their nourishment from the honeydew of pseudococcids that inhabit the inner wall of the bamboo (Klein *et al.* 1992). The worker ants not only keep other animals out of the internode but they also keep the internodes dry by ingesting rainwater that has entered and regurgitating the water to the outside entrance (Klein *et al.* 1993).

Additional differences in the faunal composition of live internodes derive from the fact that some animals seem to be confined to the young shoots (e.g., the larvae of a syrphid species of the genus *Graptomyza*; see Kovac 1994). There is also evidence that some animals prefer internodes in the higher part of the culm while others prefer internodes in the lower part (MacDonald 1960, personal observations).

(2) We found that fallen or felled bamboo culms are broken down within 1-2 years by fungi and detritivorous animals (e.g. in the later stage by termites), with decomposition proceeding fastest for culms near the ground. The community within the experimental internodes corresponds closely to the fauna occurring naturally in fallen bamboo culms with fissures or small holes. Minor differences between the experimental internodes and internodes with small holes derive from the fact that the long, slit-like openings of the experimental internodes allow access also to larger animals if these are sufficiently flat (e.g. some spiders).

A different faunal composition results when the water surface of the bamboo phytotelmata is freely visible, as in longitudinally split internodes or in bamboo stumps. In that type of internode larger aquatic animals also occur, for example, tipulid larvae, water beetles (Dytiscidae: *Lacconectus krikkeni* Brancucci, Hydrophilidae: *Enochrus* sp. and others), the larvae of damselflies and dragonflies (Coenagrionidae: *Pericnemis stictica* Selys; Libellulidae: *Lyriothemis cleis* Brauer) and tadpoles of several amphibians (for example, *Kalophrynus pleurostigma* Tschudi). There seem to be more species of planktonic organisms such as small crustaceans than in closed internodes. Occasionally water measurers (*Hydrometra*), ripple bugs (*Microvelia*) and flies (Dolichopodidae: *Pelastoneurus* sp.) are found on the water surface. There are usually no animals in the freely visible terrestrial zone of these open internodes. 'Closed' internodes with a sufficiently large opening to admit sizable animals on the other hand, may be inhabited by both large aquatic inhabitants of open internodes and by terrestrial animals. In such internodes, crabs (*Johora johorensis intermedia* Ng) and large spiders that hunt mosquito larvae and dragonfly larvae may be found.

Differences in the community composition of closed and more or less open internodes result from the following factors: (1) nutrient import from the outside (plant and animal detritus) is more substantial in open internodes; (2) open internodes are colonized more easily by microorganisms and insects. There are a number of insects, e.g. water beetles, that use cues provided by the reflection of light at open water surfaces to find a small body of water. (3) Large animals require large openings to be able to colonize an internode; (5) fallen plant matter provides better hiding opportunities for the aquatic fauna.

The community composition of dead internodes depends not only on the size of the opening but also on the position of the internode. There are, for example, animals that prefer internodes lying on the ground (e.g. web-building spiders of the family Pholcidae) or others that colonize only internodes near a river (e.g. the crab *Johora johorensis intermedia*). Further, a 'mixed' internode fauna may form when a bamboo culm with holes topples over and the fauna of live internodes is gradually replaced by animals preferring dead internodes.

Comparison between bamboo internodes and other phytotelmata

The main differences between bamboo internodes and other phytotelmata derive from following special features of the internodes: (1) decomposition of part of the bamboo tissue adds to the nutrient supply of the community; (2) the space above the water surface is frequently secluded from the outside and thereby provides protection for a species-rich partially specialized terrestrial community; and (3) the occurrence of different internode types is equivalent to the availability of different habitats.

Other phytotelmata are usually open at the top and receive nutrient input from the outside in form of, for example, fallen leaves and animals. The phytotelma plant itself usually does not provide nutrients to the community. The plant tissue of bamboo internodes, in contrast, may contribute significantly to the nutrient supply. In live internodes plant-born nutrients play a significant role only in the beginning of the succession; in dead internodes the whole plant tissue is gradually broken down and decomposers and detritivores form an important species-rich segment of the community.

The aquatic fauna of the bamboo internodes is similar in structure to the fauna of other Southeast Asian phytotelmata (Beaver 1979a, 1979b, Kitching 1987, 1990). In open internodes, in particular, an aquatic community develops that shows great similarity to the fauna of tree cavities (Kitching 1990, MacDonald & Traub 1960, and personal observations). 'Closed' internodes (access only through small holes) on the other hand, are special in that they contain a species-rich terrestrial fauna. This terrestrial fauna interacts closely with the aquatic fauna, as is shown in the diagram of trophic interactions (Figure 2). The species-richness of the terrestrial fauna apparently depends on the presence of a protected space above the water surface which provides food in addition to protection. Relatively open and unprotected tree cavities, for example, do not have a distinct terrestrial fauna, whereas there are several - though not many - specialized terrestrial species in the better protected interior of pitcher plants, e.g. spiders of the genus *Misumenops* (Beaver 1979a).

In closed internodes spiders are the dominant predators. Spiders certainly also play a more important role as top predators in other tropical phytotelmata than generally acknowledged (see, for example, Schoenly *et al.* 1991). The senior author watched spiders at night, waiting for aquatic prey at the edge of the water surface of open phytotelmata, both in split bamboo

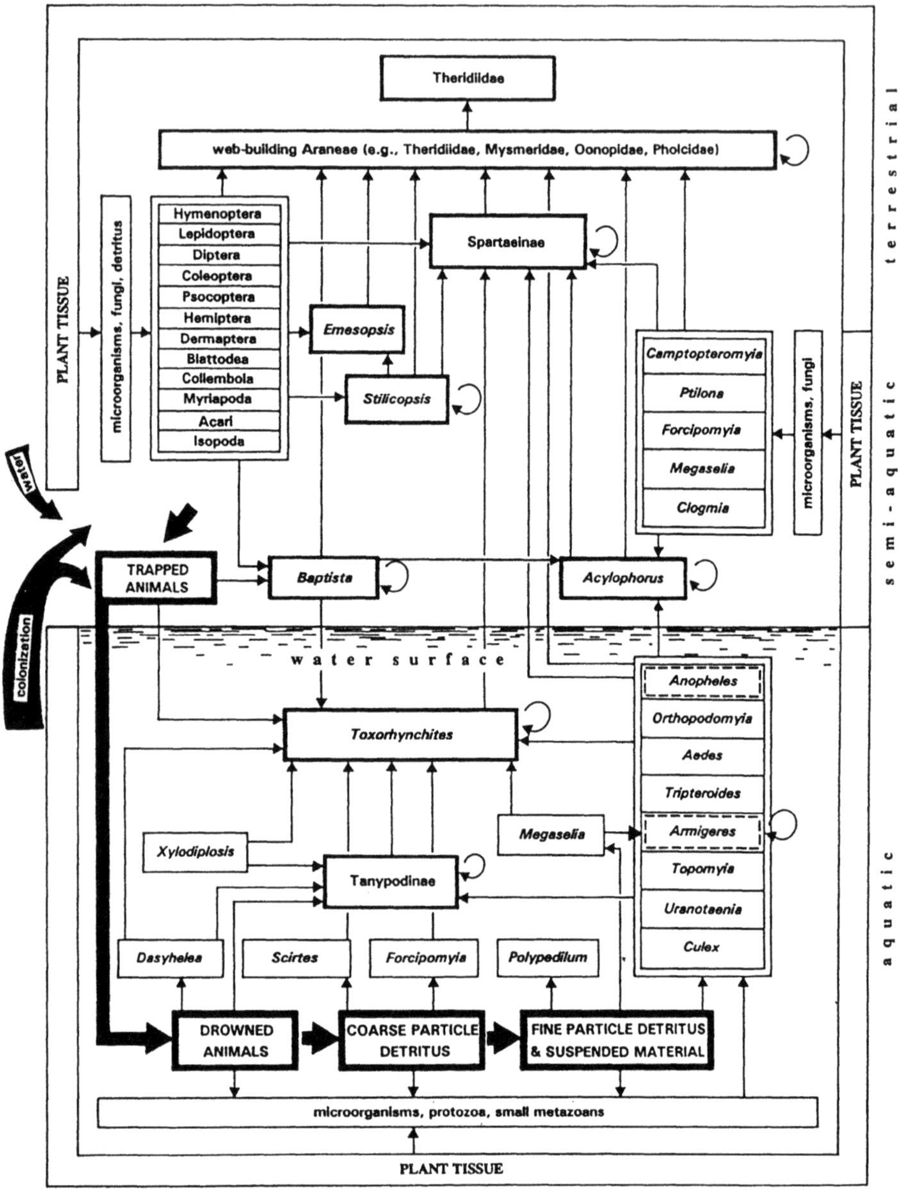

Figure 2. Simplified representation of the food web within the experimental internodes. The subdivision into different microhabitats is as shown in the framed section of Figure 1. Only the trophic interactions actually observed in the field were included. Slender arrows indicate from where individual organisms receive their nutrients. Circular arrows signify the occurrence of cannibalism. The broad arrows represent nutrient import from the outside or from the terrestrial to the aquatic part of the internode. Broad-framed boxes, in combination with broad arrows, indicate the process of decomposition of dead organic matter. The trophic interactions shown are not necessarily valid for all developmental stages. For example, *Toxorhynchites* preys only on the nymphs, not the adults of *Baptista*.

internodes lying on the ground, and in open tree cavities. Another visiting predator of phytotelma inhabitants was observed by D. H. Murphy in Singapore: The flying lizard, *Draco melanopogon* Boulenger entered water-filled bamboo poles to feed on mosquito larvae. Another interesting finding is that almost every predator will, on rare occasions, prey on conspecifics. Only in *Toxorhynchites* spp. and *Armigeres* (*Leicesteria*) spp., i.e. aquatic predators that can not escape to other internodes, cannibalism is common.

As is the case with other phytotelmata, a number of specialists that are probably restricted to this habitat occur in bamboo internodes. For example, many species of the mosquito genus *Armigeres*, subgenus *Leicesteria*, are closely tied in their distribution to certain species of bamboo (MacDonald 1960). Many other internode inhabitants are probably also bamboo specialists. The water strider *Baptista* was found exclusively in bamboo internodes (limnetic ecosystems in the Gombak region were investigated intensively by the senior author for their water strider fauna). In contrast to the specialists, there are a number of animals that inhabit a variety of habitats and have a wide distribution; *Megaselia deningi* Disney (distribution: Sri Lanka, Malaysia, Japan), for example, occurs also in *Nepenthes distallatoria* L. and even in steepers of Japanese garden radishes pickled with rice-bran and salt. In addition, this species has been reported in a case of urogenital myasis in a girl (see Disney 1991). Preliminary comparisons indicate that for many internode inhabitants the particular species of bamboo is not important - *Baptista* sp. for example, has been found in several species of bamboo in West and East Malaysia (Sabah) (*Gigantochloa scortechinii* Gamble, *G. ligulata* Gamble, *G. levis* (Blanco), *Gigantochloa latifolia* Ridl., *Schizostachyum grande* Ridl., *Dendrocalamus pendulus* Ridl.).

During the present investigation several hundred animal species were found in the different internode types of just one bamboo species, *Gigantochloa scortechinii*. Up to 26 different species (microorganisms and fungi not included) co-occurred in one single experimental internode. This makes the bamboo phytotelmata the most species-rich phytotelmata of Southeast Asia. There are probably several reasons why this was not recognized earlier. For one thing, very few long-term studies on bamboo internodes have been conducted. Furthermore, these earlier investigations were made in regions with fewer species (Japan) or areas where bamboo culms with holes are rare or absent. Although Thienemann (1934), who knew about the work of Leicester (1903), was looking for bamboo culms with holes on Sumatra, Java, and Bali, he did not find any. The senior author of this article was also unable to detect any holes in bamboo culms at several localities near settlements on Bali and in Sabah. The reason for that failure may be that either certain species of bamboo are never or only rarely attacked by boring insects (see Daniels 1908) or that the insect species able to attack certain introduced bamboo species do not occur in the area.

The results of this investigation endorse Leicester's (1903) closing sentence, "The full study of the bamboo as a breeding place for insects would, I feel sure, well repay careful and prolonged work".

ACKNOWLEDGEMENTS

This work is part of a research project funded by the Deutsche Forschungsgemeinschaft and supported by the Socio-Economic Research Unit of the Government of Malaysia and Universiti Malaya. Thanks are due to R. Klein for reviewing the manuscript and improving the English. Special thanks are due to I. Azarae for making available facilities at the Ulu Gombak Field Studies Centre. The authors are most grateful for the taxonomic assistance which was provided by the following taxonomists (in alphabetic order): M. Brancucci (Basel), A. Brindle (Nelson), Chan Kai Lok (National University of Singapore), P. Chandler (University of New Hampshire), P. S. Cranston (CSIRO, Canberra), C. Deeleman-Reinhold (Ossendrecht), H. Disney (University of Cambridge), W. W. MacDonald (Liverpool School of Tropical Medicine), D. Duckhouse (University of Adelaide), C. E. Dyte (University of Reading), R. J. Gagné (USDA, Washington), M. Hämäläinen (Espoo), P. Hammond (Natural History Museum, London), D. E. Hardy (University of Hawaii), E. Heiss (Innsbruck), J. K. H. Koh (Singapore), J. Murphy (London), T. R. New (La Trobe University, Bundoora), P. Ng (National University of Singapore), H. Ono (National Science Museum, Tokyo), L. M. Roth (The Agassiz Museum, Cambridge, Massachusetts), R. Rozkosny (Masaryk University, Brno), G. K. Srivastava (Zoological Survey of India, Calcutta), G. M. Stonedahl (Natural History Museum, London), J. R. Vockeroth (Centre for Land and Biological Research, Ottawa), W. W. Wirth (Gainsville), Yong Hoi Sen (Universiti Malaya, Kuala Lumpur).

LITERATURE CITED

BEAVER, R. A. 1979a. Biological studies of the fauna of pitcher plants (*Nepenthes*) in West Malaysia. *Annales de la Societé Entomologique de France (Nouvelle Série)* 15:3-17.

BEAVER, R. A. 1979b. Fauna and foodwebs of pitcher plants in West Malaysia. *The Malayan Nature Journal* 33:1-10.

BEAVER, R. A. 1983. The communities living in *Nepenthes* pitcher plants: Fauna and food webs. Pp. 129-159 in Frank, J. H. & Lounibos, L. P. (eds). *Phytotelmata: Terrestrial plants as hosts for aquatic insect communities*. Plexus, Medford, New Jersey.

BEAVER, R. A. 1985. Geographical variation in food web structure in *Nepenthes* pitcher plants. *Ecological Entomology* 10:241-248.

BISHOP, J. E. 1973. Limnology of a small Malayan river, Sungai Gombak. *Monographiae Biologicae* 22. Junk, The Hague.

DANIELS, C. W. 1908. Breeding grounds of Culicidae. *Studies from Institute for Medical Research, Federated Malay States* 3:1-7.

DISNEY, R. H. L. 1991. The aquatic Phoridae (Diptera). *Entomologica Scandinavica* 22:171-191.

DISNEY, R. H. L. 1993. New species of aquatic Phoridae (Diptera) from Malaysia. *Aquatic Insects* 15:149-158.

FRANK, J. H. & LOUNIBOS, L. P. (eds). 1983. *Phytotelmata: Terrestrial plants as hosts for aquatic insect communities*. Plexus, Medford, New Jersey.

GAGNÉ, R. J. 1985. Descriptions of new nearctic Cecidomyiidae (Diptera) that live in xylem vessels of fresh-cut wood, and a review of *Ledomyia* (s. str.). *Proceedings of the Entomological Society of Washington* 87:116-134.

HOLTTUM, R. E. 1958. The bamboos of the Malay Peninsula. *The Gardens' Bulletin, Singapore* 16:1-135.

KITCHING, R. L. 1987. A preliminary account of the metazoan foodwebs in phytotelmata from Sulawesi. *The Malayan Nature Journal* 41:1-12.

KITCHING, R. L. 1990. Foodwebs from phytotelmata in Madang, Papua New Guinea. *The Entomologist* 109:153-164.

KITCHING, R. L. & BEAVER, R. A. 1990. Patchiness and community structure. In Shorroks, B. & Swingland, I. (eds). *Living in a Patchy Environment*. Oxford University Press, Oxford.

KLEIN, R. W., KOVAC, D., SCHELLERICH, A. & MASCHWITZ, U. 1992. Mealybug-carrying by swarming queens of a Southeast Asian bamboo-inhabiting ant. *Naturwissenschaften* 79:422-423.

KLEIN, R., MASCHWITZ, U. & KOVAC, D. 1993. Flood control in ants: A Southeast Asian bamboo-dwelling *Tetraponera* (Formicidae: Pseudomyrmecinae) bails excess water from its internode nests. *Insectes Sociaux* 40:115-118.

KOVAC, D. 1994, in press. Die Tierwelt des Bambus: Ein modell für komplexe tropische Lebensgemeinschaften. *Natur und Museum* 124.

KOVAC, D. & YONG, H. S. 1992. *Abryna regispetri*: a bamboo long-horned beetle. *Nature Malaysiana* 17:92-98.

KURIHARA, Y. 1983. The succession of aquatic dipterous larvae inhabiting bamboo phytotelmata. Pp. 55-77 in Frank, J. H. & Lounibos, L. P. (eds). *Phytotelmata: Terrestrial Plants as Hosts for Aquatic Insect Communities*. Plexus, Medford, New Jersey.

LEICESTER, G. F. 1903. A breeding place of certain forest mosquitoes in Malaya. *Journal of Tropical Medicine and Hygiene* 6:291-292.

MARSHALL, A. G. 1970. The life cycle of *Basilia hispida* (Diptera: Nycterbiidae) in Malaysia. *Parasitology* 61:1-18.

MACDONALD, W. W. 1960. On the systematics and ecology of *Armigeres* subgenus *Leicesteria* (Diptera, Culicidae). *Malaysian Parasites XXXV-XLIX. Studies from the Institute for Medical Research, Federation of Malaya* 29:110-153.

MACDONALD, W. W. & TRAUB, R. 1960. An introduction to the ecology of the mosquitoes of the lowland dipterocarp forest of Selangor, Malaya. *Malaysian Parasites XXXV-XLIX. Studies from the Institute for Medical Research, Federation of Malaya* 29:79-109.

MEDWAY, LORD 1972. Reproductive cycles of the flat-headed bats *Tylonycteris pachypus* and *T. robustula* (Chiroptera: Vespertilioninae) in a humid equatorial environment. *Zoological Journal of the Linnean Society* 51:33-61.

MEDWAY, LORD & MARSHALL, A. G. 1970. Roost-site selection among flat-headed bats (*Tylonycteris* spp.). *Journal of Zoology, London* 161:237-245.

MEDWAY, LORD & MARSHALL, A. G. 1972. Roosting associations of flat-headed bats, *Tylonycteris* species (Chiroptera: Vespertilionidae) in Malaysia. *Journal of Zoology, London* 168:463-482.

MOGI, M. & SUZUKI, H. 1983. The biotic community in water-filled internodes of bamboos in Nagasaki, Japan, with special reference to mosquito ecology. *Japanese Journal of Ecology* 33:271-279.

MOGI, M. & Yong, H. S. 1992. Aquatic arthropod communities in *Nepenthes* pitchers: the role of niche differentiation, aggregation, predation and competition in community organization. *Oecologia* 90:172-184.

ONO, H. 1993. An interesting new crab spider (Araneae, Thomisidae) from Malaysia. *Bulletin of the National Science Museum, Tokyo, Series A* 19:87-92.

PIMM, S. L. & KITCHING, R. 1987. The determinants of food chain lengths. *Oikos* 50:302-307.

ROCK, E. A. & JACKSON, D. 1985. The biology of xylophilic Cecidomyiidae (Diptera). *Proceedings of the Entomological Society of Washington* 87:135-141.

ROZKOSNY, R. & KOVAC, D. 1991. First description of the male and the larva of *Camptopteromyia fractipennis* de Meijere from Malaysia (Diptera: Stratiomyidae). *Entomologica Scandinavica* 22:297-304.

SCHOENLY, K., BEAVER, R. A. & HEUMIER, T. A. 1991. On the trophic relations of insects: a food web approach. *The American Naturalist* 137:597-638.

SRIVASTAVA, G. K. & KOVAC, D., in press. Notes on some Dermaptera from Malaya with the description of two new species. *Records of the Zoological Survey of India.*

THIENEMANN, A. 1932. Die Tierwelt der *Nepenthes*-Kannen. *Archiv für Hydrobiologie, Supplement-Band* 11:1-54.

THIENEMANN, A. 1934. Die Tierwelt der tropischen Pflanzengewässer. *Archiv für Hydrobiologie, Supplement-Band* 13:1-91.

The avifauna of the Belalong forest, Brunei Darussalam

CLIVE F. MANN

122 Hartswood Road, London W12 9NG, United Kingdom

ABSTRACT. The areas explored were the full length of the East (Main) Ridge to the summit of Bukit Belalong and down to Sungai Temburong and Sungai Belalong, the length of the West Ridge from the Field Centre back to Sungai Belalong, much of Sungai Belalong from Kuala Belalong almost to the project boundary, and Sungai Temburong from Kuala Belalong to a little beyond Sungai Babi. 24-hour periods of mist netting were carried out, 10 periods at Pondok Bujat, one in an alluvial forest close to Sungai Babi, and another along Sungai Belalong at the Field Centre. Netting was carried out twice at the summit of Bukit Belalong. A total of 218 individuals of 58 species were captured, of which 214 of 57 species were ringed, the others being net casualties. Moult and weight were recorded, along with any changes in retraps, as were signs of breeding activity. 198 species were recorded in the project area, with three others identified to genus only. 25 species recorded are not forest dependent, and a further 15 are also likely to be found outside forest. During the expedition, 20 definite cases of breeding were discovered involving 16 species. Within the area three zones were recognised, namely riverside (Zone 1), slopes and ridges up to about 460 m (Zone 2) and slopes and ridges above 460 m (Zone 3). No true montane species were found. Bird species demonstrate vertical stratification in tall tropical forest, with four levels being recognised, with a fifth category for aerial feeders: ground level (leaf litter foragers) - 20 species, nine exclusive to this stratum; understorey - 76 species, 11 exclusive; mid levels - 95 species, 10 exclusive; canopy - 75 species, 44 exclusive; aerial - three forest-dependent species are exclusively aerial, whereas one other is also found in the canopy. Nine species are tied to waterways, of which two (*Enicurus* spp.) are also litter foragers on land, and another (*Ceyx erithacus*) may also be found some distance from water. 128 (116 forest dependent) species were found in the riparian zone (Zone 1); 139 (136 forest dependent) species were found on the lower slopes (up to c. 460 m, Zone 2) and 121 (114 forest dependent) on the higher slopes (above c. 460 m; ridge-top pole forest - Zone 3). 14 species of palaearctic migrants were recorded, together with two others, viz. *Ninox scutulata* and *Muscicapa daurica*, which have resident and migrant populations. Moult cycles were calculated to last from 100 to 220+ days.

KEY WORDS: bird diversity, tropical rainforet, population structure, breeding, distribution, Borneo.

INTRODUCTION AND METHODS

The aims of the study were as follows:
(a) to make an inventory of the birds, and as far as possible assess population numbers.
(b) to investigate the distribution of avian species in relation to habitat, altitude and vertical stratification.
(c) to record, through capture and ringing, the timing and rate of moult, weight changes, population structure and movements of species of the lower strata.
(d) to obtain data on breeding.

The area was first visited by the author on 6 August 1989, accompanying Lord Cranbrook and others on a reconnaissance trip. The next visits were during the expedition phase of the project, when a total of 68 days were spent by the author in the project area between 21 February and 13 October 1982. The areas explored during this period were the full length of the East Ridge to the summit of Bukit Belalong and down to Sungai Temburong and Sungai Belalong, the length of the West Ridge from the Field Centre back to Sungai Belalong, Sungai Belalong from Kuala Belalong almost to the project boundary, and Sungai Temburong from Kuala Belalong to a little beyond Sungai Babi (see Wills, this volume, Figure 1 for a map of the area). Apart from these main arteries some exploration was

D. S. Edwards et al. (eds.), Tropical Rainforest Research – Current Issues, 101–116.

carried out in intervening areas, but time prevented much of the area from being investigated.

Apart from netting, other ornithological activities consisted chiefly of walking trails and river banks, and observing from stationary posts and moving boats. As the main netting site, the northern end of the East Ridge (Bujat), received a disproportionate amount of attention, as did the area immediately surrounding the Field Centre. As well as making general observations on birds, Lord Cranbrook carried out netting (see below), as did D.R. Wells, who also produced a trip report on his visit to the area from 12 to 22 January 1992. These records, together with a report by C.M. Francis, have been made use of in the present work. A list of species recorded appears in Appendix 1. Mist netting was carried out during 10 x 24-hour periods (with nets closed during darkness) close to Pondok Bujat, in hill dipterocarp at the northern end of the East ridge; one period in an alluvial forest close to Sungai Babi, and another along Sungai Belalong at the Field Centre. After the author's study was completed, Lord Cranbrook netted twice at Pondok Bujat, as did D.R. Wells. The former also twice netted briefly at the summit of Bukit Belalong. The birds netted were ringed with uniquely numbered metal bands provided by Universiti Malaya, and with uniquely colour-coded plastic bands for individual identification in the field. The age and sex (where possible) was recorded for each individual ringed, and the weight, wing, tail and gape were measured. The state of moult was also recorded. Details of change of weight and moult status was noted for any bird retrapped on subsequent occasions. A total of 218 individuals of 58 species were captured, of which 214 of 57 species were ringed, the rest being net casualties (Appendix 3).

RESULTS

The Avifauna

The number of species recorded in the project area was 198 (two only recorded before the expedition), with three others identified to genus only (see Appendix 1). Since the end of the expedition phase, another species (Nieuwenhuis's Bulbul *Pycnonotus nieuwenhuisi*) has been recorded (R.S.R. Williams, personal communication). This species was previously known only from one specimen from the upper Kayan collected in 1900, and one specimen of a different subspecies from Sumatra (Smythies 1981). At the moment the record is *sub judice*. The cumulative totals of the species recorded in the area by month are shown in Figure 1, whereas Appendix 1 shows the months in which each species was recorded, and the number and percentage of days recorded during the author's presence. Within the area three zones were recognised, namely riverside/riparian forest (Zone 1), slopes and ridges up to about 460 m (Zone 2) and slopes and ridges above 460 m (Zone 3). Species recorded in each of these divisions are shown in Table 1. Appendix 1 shows the stratified and zonal distribution of each, their diets, and indicates those which are not forest dependent, those tied to waterways, and those which are Palaearctic migrants. The information on stratified distribution came from observations made during the project period, reinforced and supplemented by observations made elsewhere in Borneo and by published material such as Smythies (1957, 1981). Dietary data, summarised in Table 2, came mostly from published sources, e.g. Smythies (1957, 1981), supplemented by personal observations.

Although the majority of the resident species recorded undoubtedly breed in the project area, firm evidence of breeding in equatorial forest birds is notoriously difficult to obtain. However, during the expedition, 20 definite cases of breeding were discovered involving 16 species (Appendix 2). Breeding was detected by direct observations of active nests or recently fledged young, the carrying of nest material or the detection of brood patches on captured birds. Seven records are in January, three in February, five in March, one in May,

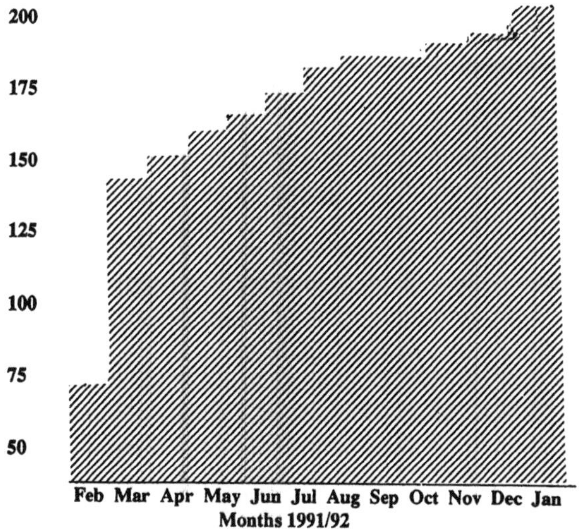

Figure 1. Cumulative total of species recorded by month.

Table 1. Numbers of species in each stratum. Levels: G = ground, U = Lower, M = Mid-levels, C = Canopy, A = Aerial; X = exclusive to that level, Non X = not exclusive to that level. Zone: 1 = riverside, 2 = slopes to 460 m, 3 = slopes above 460 m. [Since many species occur in more than level and more than one zone, the totals cannot be arrived at by simply totalling the numbers in a particular line or column]

Levels	G		U		M		C		A	TOTALS
	X	Non X	X	Non X	X	Non X	X	Non X		
Zones										
1	6	10	6	53	6	68	30	25	4	128
2	6	7	7	55	9	75	30	28	4	139
3	5	5	6	37	6	55	30	20	1	129
TOTALS	9	11	11	65	10	85	44	31	4	

one in June, two in July and one in August. Thus, the majority of records of breeding activity are from the early part of the year just after the Northeast monsoon rains, but the sample is too small to be considered significant, and covers a rather systematically disparate community.

About another 35 species could reasonably be expected to occur at times, so the present list is perhaps within 10-15% of the true total (see Figure 1), which will include some rarer residents, as well as palaearctic visitors. Some species, recorded from similar habitats, and at no great distance from the study area, in Borneo, were inexplicably absent. Whether or not this absence is real will hopefully be revealed by future field work in the area. Of the 198 species in the project area, 25 are not forest dependent, and a further 15 are also likely to be found outside the forest.

Bird species demonstrate vertical stratification in tall tropical forest, with four levels being recognised, viz. ground, understorey (shrub layer), mid-level (higher parts of small branches of tall trees) and canopy (including emergents). A fifth division accounts for aerial feeders (G, U, M, C and A in Appendix 1).

Table 2. Summary of diet preferences of birds in the study area. Numbers in brackets are non-forest dependent species.

	Exclusive	Non-Exclusive	Total
F : Fruit (+seeds, flowers)	11	61(+1)	73
N : Nectar (+pollen)	0	10	10
I : Invertebrates	94(+7)	65(+5)	171
P : Fish	3(+2)	3(+3)	11
C : Vertebrates (except fish)	5	14(+1)	20
F + N only	-	-	1
F + I only	-	-	57
I + N only	-	-	1
I + C only	-	-	7
I + P only	-	-	3
I + P + C	-	-	2

However, much of this stratification is preferential rather than absolute, with such canopy species as hornbills and barbets occasionally found in lower strata, and ground species such as pittas being found in the understorey, or higher, when singing. Preferences for each species in the study area are shown in Appendix 1 and a summary of stratified distribution in Table 1. These data came from observations in the study area and other parts of Borneo, and from published sources, particularly Smythies (1957, 1981). Data on zonal distribution from the study area are combined with these and discussed below.

Ground level (leaf litter foragers): Nine species are exclusively ground foragers, and 11 others use the ground for part of the time. The number of species in this guild diminishes with increasing altitude from 16 in Zone 1 (six exclusive to this stratum), to 13 in Zone 2 (six exclusive), to 10 in Zone 3 (five exclusive).

Understorey: Only 11 species feed exclusively in the understorey, with 65 other species which feed elsewhere as well. Greater numbers are found close to rivers and on slopes below 460 m, viz. 59 (six exclusive) and 62 (seven exclusive) as compared to 43 (six exclusive) above 460 m. These distributions reflect the attenuation of low vegetation with increasing altitude.

Mid-levels: Only 10 species feed exclusively in mid level vegetation, with 85 other species that also feed elsewhere. The figures for the three zones are 74 (six exclusive to this stratum), 84 (nine exclusive) and 61 (six exclusive).

Canopy: A total of 44 species feed exclusively in the canopy, with a further 31 that also feed elsewhere. 55 (30 exclusive) in Zone 1, 58 (30 exclusive) in Zone 2 and 50 (30 exclusive) above 460 m. The altitude and proximity to waterways appears to have little effect on the distribution of canopy feeders.

Aerial: Three forest-dependent species were identified as exclusively aerial, all occurring in the first two zones, and one in the last. Another species, occurring in the first two zones, was identified as both aerial and canopy.

In Bornean forests 900 m is generally found to be the altitude required for montane species. On mountains above this altitude these species may be found below 900 m, but they will not be found on mountains that do not reach this height (Banks 1937, Chasen 1935, Smythies 1960, C. F. Mann, personal observation). As was expected, no true montane species were found in the project area. Lowland species, such as *Rhinomyias umbratilis*, which are replaced by highland representatives at about 600-900 m on higher mountains, were found to the top of Bukit Belalong (approximately 900 m asl).

Alluvial forest: *Pitta baudi* was perhaps the only species confined to this forest type, as opposed to those in Zone 1 which are associated with rivers. However, little time was spent investigating alluvial forest which only exists in tiny pockets within the project area.

Waterways: A total of nine species are tied to waterways, of which two (*Enicurus*) are also litter foragers on land, and another (*Ceyx erithacus*) may also be found some distance from water. 128 (116 forest dependent) species were recorded in this zone.

Lower slopes (up to c. 460 m): 139 (136 forest-dependent) species were recorded in this zone.

Higher slopes (above c. 460 m; ridge-top pole forest): 121 species (114 forest-dependent) were found in this zone.

Palaearctic migrants: 14 species were recorded, and two others, viz. *Ninox scutulata* and *Muscicapa daurica*, which have resident and migrant populations. The former was believed to be of the resident population, whereas all records of the latter were almost certainly of migrant birds. Migrants, never numerous, were more frequently encountered during November and December, but most had moved on by January (Cranbrook, in preparation).

Species Diversity

This was based on totals of each species caught in each ringing session. These are recorded in Table 3. Figure 2 shows that a larger number of species is represented by a smaller number of individuals (18 singletons), and vice versa (one species represented by 22 individuals). Such a community structure has been found in other areas, e.g. Kenya (Mann 1985). Table 4 gives some species diversity indices for East African forests, Kenyan steppe/savanna, Southeast England woodland, and selectively-logged forest in Brunei (Lamunin) from Mann (*op. cit.*) for comparison.

Three indices were calculated, but only McIntosh's Index (D_m) could be calculated for sites other than Belalong due to the manner in which the data was recorded, and thus only this could be used in comparisons. Algorithms for these indices are found in Magurran (1988). The overall totals were perhaps rather too small in some cases to give a reliable species diversity index. There is a negative correlation between number of individuals in the catch and D_m, (r = -0.491, p = <5%). Therefore lower catches show a larger D_m.

The results clearly show a greater diversity at Belalong than in East African forests or in selectively-logged forest in Brunei. Some of the figures for savanna/steppe in Kenya are unexpectedly close, but it should be remembered that in such habitats there is no vertical stratification, and therefore the sample is closer to the whole avifauna of the area, rather than a segment.

Moult

Moult was recorded on trapped birds where it was detectable. The degree of moult of the primary wing feathers was recorded, and its progress monitored on birds retrapped. This is shown in Figure 3. The numbers in moult are shown in Figure 4. Individuals could be found moulting in every month, but greatest between May and August. A score of 50 indicates just completed primary moult. Extrapolation of figures on the graph indicate that a moult cycle would last from between 100 and 220+ days, assuming that most moult would be contained within the period required for primary moult. Two individuals (*Criniger bres* and *Malacopteron cinerum*) demonstrate interrupted moult, as shown by the shallow angles.

Age Structure of the Community

The numbers of adults and young in the trapped samples are shown in Figure 4. The numbers of young in all samples are very low, indicating a low rate of replacement, with

Table 3 : Species diversity and diversity indices of understorey community of birds in study area. D_m = McIntosh's index of diversity, range 0 - 1; E_m = McIntosh's index of diversity (modified) (range 0 - 1), 1/Ds = Simpson's index of diversity (reciprocal), range 1 - infinity.

Date	Total	No. of Species	Species diversity indices		
			D_m	E_m	$1/D_s$
Bujat					
28-29/03/91	15	11	0.86	0.92	15.10
02-03/04/91	14	10	0.89	0.95	18.2
03-04/05/91	29	14	0.82	0.92	13.1
21-22/06/91	33	20	0.92	0.95	24.8
19-20/07/91	25	14	0.87	0.95	17.7
21-22/08/91	28	17	0.89	0.96	23.6
31/10-01/11/91	30	18	0.91	0.97	27.1
13-14/01/92	14	8	0.81	0.92	10.1
20-21/01/92	22	13	0.85	0.91	13.0
Mean	23.3	13.9	0.87	0.94	18.1
Sg. Babi					
27-28/10/91	12	9	0.88	0.94	16.5

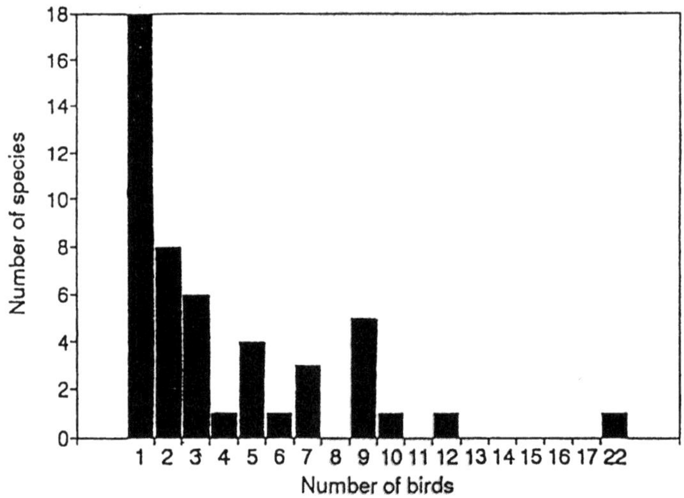

Figure 2. Species diversity among birds netted at Pondok Bujat, 1991-92.

greatest numbers between March and July, which shows some concurrency with observations on breeding. This is suggestive of greater longevity as found in other tropical forests, for example in Kenya (Mann *op.cit*), where one individual of *Andropadus* (= *Pycnonotus*) *curvirostris* had a minimum longevity of 19.5 years.

Abundance

No systematic study was carried out to estimate the abundance of species in the project area (with the exception of Operation Hornbill, Lord Cranbrook (personal communication)

Table 4 : Species diversity & diversity indices in other localities. D_m = McIntosh's index of diversity, range 0 -1.

Locality	No. of individuals	No. of species	D_m
East African mid-level/highland forests			
Kakamega Forest, Kenya (2 year average)	61.3	20.3	0.68
Mau forest, Kenya	45	15	0.76
Amani Forest, Tanzania (average of 2)	123	23.5	0.78
S.Nandi Forest, Kenya (average of 7)	56.4	22.7	0.81
Kenyan savanna/steppe			
Tot	44	19	0.74
Kaimosi	62	19	0.78
Kapsabet	80	19	0.61
Barsaloi	46	10	0.35
Archer's Post	59	15	0.49
Lokichar	46	14	0.79
Lake Hannington (Bogoria)	49	21	0.83
Lake Kisima	59	12	0.68
South Horr	61	23	0.82
Southeast England woodland (average of 4)	85.5	12.3	0.64
Lamunin Forest, Brunei (average of 2)	42	18.5	0.78

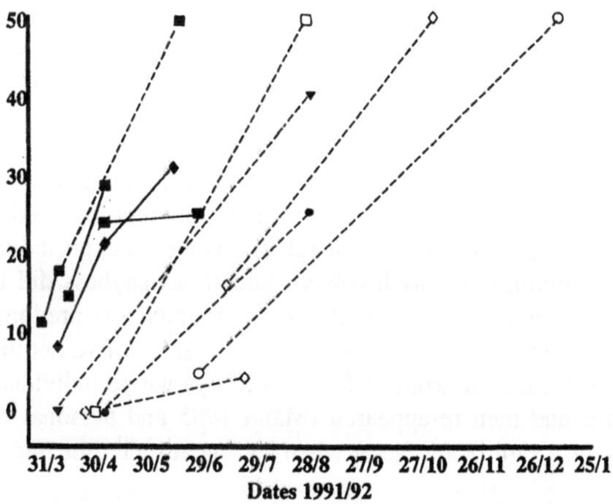

Figure 3. Primary moult scores for birds ringed in project area. Vertical axis shows primary moult score on a scale of 0 (no moult) to 50 (primary moult just completed). *Criniger bres* = ■; *Malacopteron magnum* = ♦; *Copsychus pyrrhopygus* = ○; *Arachnothera longirostris* = ▾; *Trichastoma malaccense* = ●; *Malacopteron cinereum* = ◇; *Arachnothera crassirostris* = □; --------- = minimum slope.

and reported elsewhere) due to time constraints and the difficulty of the terrain. Comments on numbers would be impressionistic only. Appendix 1 shows the percentage of the author's field days between February and October 1991, when each species was recorded.

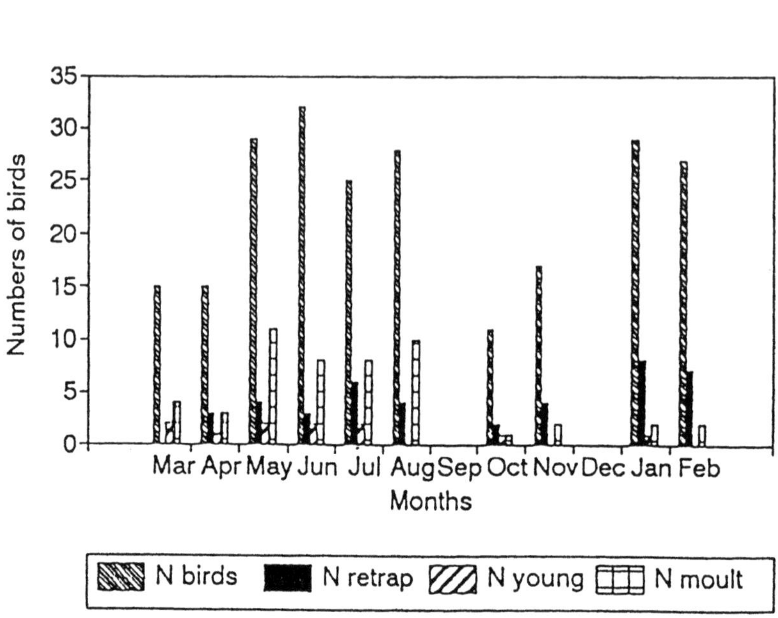

Figure 4. Statistics of understorey birds netted at Pondok Bujat, 1991-1992. N = number.

Despite the assumption that the majority of species were resident, only three were recorded on all days, and 22 on 50% or more days. All except five are species that are particularly conspicuous by their vocalisations. These figures are indicative of the difficulties encountered when trying to census birds in tropical forest. The large number of birds (79) seen on 5% or less days is made up of birds that are genuinely rare within the study area, or occasionally visit the area, or are difficult to detect. A few species, e.g. *Eumyias thalassina*, were only detected by netting.

Figure 4 and Table 5 show the percentage of retraps in the ringed population on each occasion of netting. Only figures from Bujat are meaningful, as Babi was only netted once, and Bukit Belalong summit twice, but with very small numbers involved. Over a period of time as short as this study, the percentage of retraps would be expected to increase steadily if a totally sedentary community was involved, and if net-shyness did not develop. The fluctuating results here suggest that although the community was predominantly sedentary, there is some movement into and out of the immediate area where netting was carried out. This phenomenon was detected in tropical forest in Kenya where individuals disappeared for up to a year or more, and then re-appeared (Mann 1985 and personal obserations). Net-shyness is unlikely to be a significant factor when netting at such time intervals (Mann 1985).

DISCUSSION

The Avifauna

D.R. Wells (personal communication) increased the numbers of species along river banks by 8.5%, on ridges to 460 m by 5.5%, and above 460 m by 34.4%. The overall species list was increased by 5.8%. The small increases for the first two areas are to be expected. It is not uncommon to add to the avifauna of lowland dipterocarp forest even after some years

Table 5. Ringing analysis.

Locality	Date	Total	No. of Species	% retraps
Bujat	28-29/03/91	15	11	0.0
Bujat	02-03/04/91	14	10	21.4
Bujat	03-04/05/91	29	14	13.8
Bujat	21-22/06/91	33	20	13.3
Bujat	19-20/07/91	25	14	24.0
Bujat	21-22/08/91	28	17	17.9
Sg. Belalong	24-25/08/91	8	6	0.0
Sg. Babi	27-28/10/91	12	9	0.0
Bujat	31/10-01/11/91	30	18	30.0
Bt. Belalong	05/11/91	6	3	0.0
Bujat	13-14/01/92	14	8	42.9
Bujat	20-21/01/92	22	13	22.7

of regular observation (personal observation). The larger figure for the higher groundreflects the less time found by the author to study that area prior to Wells's visit. Also, Wells's observations were made in January, a month not covered by other workers. Our present list is perhaps within 10 to 15% of the true total, which will include some rarer residents, as well as palaearctic visitors.

Cycles

In temperate species moult and breeding cycles generally do not overlap. Both events are a considerable drain on food resources, which at higher latitudes show marked seasonal fluctuations. Such fluctuations are also marked in the tropics, but much less so in equatorial forests. Seasonal breeding in humid tropical forests seems to be widespread, being recorded from Southeast Asia (Fogden 1972, Gibson-Hill 1952, Ward 1969, Voous 1950), East Africa (Mann 1985, Moreau 1936, 1950), New Hebrides (Baker *et al.* 1940) and Central and South America (Skutch 1950, Snow & Snow 1964). Due to both moult and breeding cycles being very protracted in some equatorial forest species (Mann 1985 and other references cited therein), they may overlap broadly. Generally there is a smaller number of young (and broods) which remain dependent on their parents (and in some cases related helpers) for extensive periods. Although annual cycles are common in equatorial bird species, cycles of less than, or more than, one year have been detected in Sarawak (Fogden 1972) and East Africa (Mann 1985),

Comparison with equatorial East African forests

The total number of forest-dependent species recorded in the study area (176) exceeds that of Kakamega Forest, Kenya (147), where the author carried out an extensive study from January 1973 to January 1975 (Mann 1985). Kakamega Forest lies on the equator at an average elevation of 1640 m above sea level; it is both less variable in altitude, and higher than Belalong (maximum altitude <900 m). The catch rate in mist nets at Kakamega was consistently higher (200-400%) than at Belalong for a shorter total length of netting (140 m compared to 213 m). Species Diversity Index (D_m) was lower at Kakamega than at Belalong (means 0.68 and 0.88 respectively).

ACKNOWLEDGEMENTS

Much assistance was received in the field from Jen Elkin and A. Conrad Ozog. Awang Samhan bin Nyawa, Dr. Joe Charles, Tom Watson and other expedition members contributed observations. Lord Cranbrook and Dr D.R. Wells contributed ringing data as well as their observations. Dr C.M. Francis and Dr. D.R. Wells contributed trip reports. Nigel de N. Winser, Catriona Prebble, Amanda Hardy and the staff of Kuala Belalong Field Studies Centre were helpful in ways too numerous to mention. John Banks and Lord Cranbrook helped with the preparation of materials. Above all, I would like to express my sincere appreciation to the Royal Geographical Society for its support of my field work and my attendance at the Tropical Rainforest Conference 1993 at Universiti Brunei Darussalam.

LITERATURE CITED

BAKER, J.R., MARSHALL, A.J. & HARRISON, T.H. 1940. The seasons in a tropical rain forest (New Hebrides). Part 5 Birds (*Pachycephala*). *Journal of the Linnean Society (Zoology)* 41:50-70

BANKS, E. 1937. The distribution of Bornean birds. *Sarawak Museum Journal* IV:453-496.

CHASEN, F.N. 1935. A handlist of Malaysian birds. *Bulletin of the Raffles Museum* 11:1-389.

FOGDEN, M.P.L. 1972. The seasonality and population dynamics of equatorial forest birds in Sarawak. *Ibis* 114:307-343.

GIBSON-HILL, C.A. 1952. The apparent breeding seasons of land birds in North Borneo and Malaya. *Bulletin of the Raffles Museum* 24:270-294.

MAGURRAN, A.A. 1988. Ecological diversity. Croom Helm.

MANN, C.F. 1985. An avifaunal study in Kakamega forest, Kenya, with particular reference to species diversity, moult and weight. *Ostrich* 56:236-262.

MOREAU, R.E. 1936. Breeding seasons of birds in East African evergreen forest. *Proceedings of the Zoological Society, London* 1936:631-653.

MOREAU, R.E. 1950. The breeding seasons of African birds. I Land birds. *Ibis* 92:223-267.

SKUTCH, A.F. 1950. The nesting season of Central American birds in relation to climate and food supply. *Ibis* 92:185-222.

SMYTHIES, B.E. 1960. *The birds of Borneo*, 1st edition. Oliver & Boyd, London & Edinburgh.

SMYTHIES, B.E. 1981. *The birds of Borneo*, 3rd edition. Sabah Society & Malayan Nature Society, Kota Kinabalu.

SNOW, D.W. & SNOW, B.K. 1964. Breeding seasons and annual cycles of Trinidad land birds. *Zoologica* 49:1-49.

VOOUS, K.H. 1950. The breeding seasons of birds in Indonesia. *Ibis* 92:279-287.

WARD, P. 1969. The annual cycle of the Yellow-vented Bulbul *Pycnonotus goiavier* in a humid equatorial environment. *Journal of Zoology, London* 157:25-45.

WILLS, J.T. (this volume). The creation of a Geographical Information System for the Kuala Belalong Field Studies Centre.

Appendix 1. Birds recorded in Belalong forest. 1 = alluvial forest/river banks; 2 = slopes and ridges upto c.460m; 3 = slopes and ridges above 460m. N = non forest; W = forested waterways; G = ground; U = understorey; M = mid layer; C = canopy; A = aerial; # = Palaearctic migrant. Diet [P = fish; C = other vertebrate; I = invertebrates; F = fruits, seeds, flowers; N = nectar, pollen]. Obs - Frequency of recording of species in project area. @ - recorded before 21 February 1991; # = only recorded after 13 October 1991; Arabic numerals = number of days recorded by author (max. 68) between 21 February and 13 October 1991. Roman numerals = months recorded.

	Habitat 1	2	3	Dist.	Diet	Obs	Months
Darter *Anhinga melanogaster*	+			N	P	2	IV,V
Little Green Heron *Butorides striatus*	+			N	P/I	9	II,IX,X
Osprey *Pandion haliaetus*	+			N#	C	#	
Blyth's Hawk-Eagle *Spizaetus alboniger*	+	+	+	C	C	2	VI
Wallace's Hawk-Eagle *S. nanus*	+			C	C	#	
Rufousbellied Eagle *Hieraaetus kienerii*	+	+		C	C	1	VIII @
Lesser Fish-Eagle *Ichthyophaga humilis*	+			W	P	5	II,V,VIII,IX
Greyheaded Fish-Eagle *I. ichthyaetus*	+			W	P	1	V
Accipiter sp. *Accipiter* (?*trivirgatus*)	+			C	C		
Crested Serpent Eagle *Spilornis cheela*	+	+	+	C	C/I	16	II-IV, VII-X
Blackthighed Falconet *Microhierax fringillarius*			+	C	C/I	1	III
Crested Wood Partridge *Rollulus roulroul*			+	G	F/I	@	
Great Argus Pheasant *Argusianus argus*	+	+	+	G	F/I	47	II-X, @
Common Sandpiper *Actitis hypoleucos*	+			N#	I	8	III,IX,X
Longtoed Stint *Calidris subminuta*	+			N#	I	1	VIII
Rednecked Phalarope *Phalaropus lobatus*	+			N#	I	#	
Thickbilled Green Pigeon *Treron curvirostra*		+		C	F	4	III,IV,VII
Little Green Pigeon *T. olax*	+			C	F	1	VIII
Mountain Imperial Pigeon *Ducula badia*	+		+	C	F	3	III, V
Emerald Dove *Chalcophaps indica*	+	+	+	G	F/I	3	II, III, V
Jambu Fruit Pigeon *Ptilonopus jambu*	+	+		G	F	1	V
Bluerumped Parrot *Psittinus cyanurus*		+	+	C	F	3	III, VIII
Bluecrowned Hanging Parrot *Loriculus galgulus*	+	+	+	C	F	5	IV, V, VII, VIII
Moustached Hawk-Cuckoo *Cuculus vagans*	+	+		U/M	I	32	III-V, VIII
Hodgson's Hawk-Cuckoo *C. fugax*	+	+	+	M/C	I	#	
Indian Cuckoo *C. micropterus*	+	+	+	C	I	12	II, III
Banded Bay Cuckoo *Cacomantis sonneratii*	+	+	+	M/C	I	43	II-X
Indonesian Cuckoo *C. sepulcralis*	+	+		M	I	1	V
Plaintive Cuckoo *C. merulinus*	+			N	I/F	3	III, IV
Violet Cuckoo *Chrysococcyx xanthorhynchus*	+	+		M/C	I/F	8	III-V,VII,VIII,X
Little Bronze Cuckoo *C. minutillus*			+	N	I	#	
Drongo-Cuckoo *Surniculus lugubris*	+	+	+	M/C	I	34	II-V
Raffles's Malkoha *Phaenicophaeus chlorophaeus*	+	+	+	M/C	I	4	II, V, VIII
Chestnutbreasted Malkoha *P. curvirostris*		+	+	M	I(C)	#	
Redbilled Malkoha *P. javanicus*	+	+	+	M	I	14	III, IV, VI-VIII
Blackbellied Malkoha *P. diardii*		+		M	I	#	
Shorttoed Coucal *Centropus rectunguis*	+			G/U	I	1	III
Greater Coucal *C. sinensis*	+			G/U	I/F(C)	1	VIII
Bay Owl *Phodilus badius*		+		U/M	C/I	1	VIII
Barred Eagle-Owl *Bubo sumatranus*		+	+	U/M	C	2	V, VIII
Buffy Fish-Owl *Ketupa ketupu*	+			N/W	P/I/C	1	VIII
Brown Wood-Owl *Strix leptogrammica*		+	+	U/M	I/C	3	III, VIII
Brown Hawk-Owl *Ninox scutulata*			+	U/M(#)	I	#	
Javan Frogmouth *Batrachostomus javensis*		+		U/?M	I	1	VIII
Grey Nightjar *Caprimulgus indicus*			+	U#	I	#	
Cave Swiftlet *Aerodramus* sp.	+	+	+	N/A	I	45	II-V, VII, VIII
Whitebellied Swiftlet *Collocalia esculenta*	+	+		N/A	I	30	II, III, V, VIII-X
Brown Spinetail *Hirundapus giganteus*	+			A	I	2	II, IV
Whitethroated Spinetail *H. caudacutus*			+	N/A	I#	2	III
Silverrumped Spinetail *Rhaphidura leucopygialis*	+	+	+	A	I	28	II-V, VII-X

Appendix 1 (continued). Birds recorded in Belalong forest.

	Habitat 1	2	3	Dist.	Diet	Obs	Months
Whiskered Tree-Swift *Hemiprocne comata*	+	+		C/A	I	3	II, III
Greyrumped Tree-Swift *H. longipennis*		+	+	?C/A	I	6	III, VII, VIII
Rednaped Trogon *Harpactes kasumba*	+	+	+	U/M	I	19	II,IV-VI, VIII,
Diard's Trogon *H. diardii*	+	+		U/M	I/F	3	III,VI,VIII
Scarletrumped Trogon *H. duvauceli*	+	+	+	U/M	I	18	II-VI,X
Banded Kingfisher *Lacedo pulchella*	+	+	+	M	I	12	III-IV,VIII
Storkbilled Kingfisher *Pelargopsis capensis*	+			N	P/I	18	V,VII-IX
Blueeared Kingfisher *Alcedo meninting*	+			W	P	1	II
Bluebanded Kingfisher *A. euryzona*	+			W	P/I/C	49	III,V-X
Rufousbacked Kingfisher *Ceyx erithacus*	+	+		W/U	P/I	3	VI,VIII,X
Redbearded Bee-eater *Nyctiornis amicta*	+	+	+	M/C	I	6	III, IV
Whitecrowned Hornbill *Berenicornis comatus*	+	+	+	C	F/C	6	II, III, VII
Black Hornbill *Anthracoceros malayanus*		+	+	C	F/I/C	1	VII
Bushycrested Hornbill *Anorrhinus galeritus*	+	+	+	C	F/I	25	II-VIII
Wreathed Hornbill *Rhyticeros undulatus*	+	+	+	C	F/I	20	II-V, VIII, X
Wrinkled Hornbill *R. corrugatus*	+	+	+	C	F/I	16	III, IV, VIII, X
Rhinoceros Hornbill *Buceros rhinoceros*	+	+	+	C	F/C	52	II-X
Helmeted Hornbill *Rhinoplax vigil*	+	+	+	C	F/C	34	II-X
Brown Barbet *Calorhamphus fuliginosus*	+	+	+	C	F/I	3	III
Goldwhiskered Barbet *Megalaima chrysopogon*	+	+	+	C	F	27	II-V, VII-VIII
Redthroated Barbet *M. mystacophanos*	+	+	+	M/C	F/I	44	II-X
Yellowcrowned Barbet *M. henricii*	+	+	+	C	F	33	II-VIII
Blueeared Barbet *M.australis*	+	+	+	C	F	44	II-VIII
Malaysian Honeyguide *Indicator archipelagicus*		+		?U/M	I	#	
Rufous Piculet *Sasia abnormis*	+	+		U/M	I	2	III/IV
Checkerthroated Woodpecker *Picus mentalis*	+	+	+	U/M	I	3	III, VIII
Crimsonwinged Woodpecker *P. puniceus*			+	M	I	#	
Greycapped Woodpecker *Picoides canicapillus*		+	+	M/C	I	1	III
Grey-&-Buff Woodpecker *Hemicircus concretus*	+		+	M/C	I/F	1	III
Rufous Woodpecker *Micropternus brachyurus*	+			U/M	I	7	II, III
Olivebacked Woodpecker *Dinopium rafflesii*	+	+		U/M	I	2	III
Buffnecked Woodpecker *Meiglyptes tukki*	+	+		U/M	I	2	VI, VIII
Maroon Woodpecker *Blythipicus rubiginosus*	+	+	+	U/M	I	7	III, IV, VII, IX
Whitebellied Woodpecker *Dryocopus javensis*		+	+	M/C	I	3	III, VII
Great Slaty Woodpecker *Mulleripicus pulverulentis*		+	+	M/C	I	6	III, VII, VIII
Orangebacked Woodpecker *Chrysocolaptes validus*		+	+	M/C	I	1	III
Green Broadbill *Calyptomena viridis*	+	+	+	U/M/?C	F	21	III, IV, VI-VIII
Black-&-Red Broadbill *Cymbirhynchus macrorhynchus*	+			M/C	I(F/P)	8	II, V, VIII, IX
Black-&-Yellow Broadbill *Eurylaimus ochromalus*	+	+	+	C	I/F	58	II-X
Banded Broadbill *E. javanicus*	+	+	+	M/C	I	46	II-VIII, X
Dusky Broadbill *Corydon sumatranus*			+	C	I(C)	1	III
Garnet Pitta *Pitta granatina*	+	+		G	I/F	8	II, V, VIII, IX
Banded Pitta *P. guajana*			+	G	I	@	
Blueheaded Pitta *P. baudi*	+			G	I	1	IX
Barn Swallow *Hirundo rustica*			+	N#	I	2	III#
Redrumped Swallow *H. daurica*			+	N#	I	#	
Pacific Swallow *H. tahitica*			+	N	I	#	
Asian House Martin *Delichon dasypus*			+	N#	I	#	
Grey Wagtail *Motacilla cinerea*	+			N#	I	2	II, III
Lesser Cuckoo-Shrike *Coracina fimbriata*	+		+	C	I/F	7	III, V, VII
Blackwinged Flycatcher-Shrike *Hemipus hirundinaceus*	+	+		C	I	3	III, VII
Large Wood-Shrike *Tephrodornis virgatus*	+	+	+	C	I	1	III?#
Scarlet/Fiery Minivet *Pericrocotus flammeus/igneus*	+		+	C	I/F	5	III, V-VII
Green Iora *Aegithina viridissima*		+	+	M/C	I	3	III, VII
Lesser Green Leafbird *Chloropsis cyanopogon*	+	+	+	M/C	F/I	8	III, V, VII, VIII
Greater Green Leafbird *C. sonnerati*	+	+	+	C	F/I	4	IV, IX
Blue-winged Leafbird *C. cochinchinensis*		+	+	C	F/I	4	III, VII, VIII
Fairy Bluebird *Irena puella*	+	+	+	C	F/I	10	II, III, VI-VIII, X

Appendix 1 (continued). Birds recorded in Belalong forest.

	Habitat 1	2	3	Dist.	Diet	Obs	Months
Greybellied Bulbul *Pycnonotus cyaniventris*	+		+	C	F/I	3	II, VII
Blackheaded Bulbul *P. atriceps*	+			C	F/I	1	VIII
Black-&-White Bulbul *P. melanoleucos*		+		M/C	F/I	5	IV-VI, VIII
Redeyed Bulbul *P. brunneus*	+	+	+	U/M	F/I	10	III, VII-IX
Spectacled Bulbul *P. erythrophthalmos*	+	+	+	U/M	F/I	31	II, III, V-VIII
Creamvented Bulbul *P. simplex*	+	+	+	U/M	F/I	3	VI, VIII
Puffbacked Bulbul *P. eutilotus*	+	+		M	F/I	15	VII-X
Strawheaded Bulbul *P. zeylanicus*	+			N/U	F/I	1	VIII
Greycheeked Bulbul *Criniger bres*	+	+	+	U/M	F/I	29	II-X
Yellowbellied Bulbul *C. phaeocephalus*	+	+		U/M	F/I	41	III, V-IX
Finsch's Bulbul *C. finschii*	+	+	+	U/M	F/I	7	II, III, VII-IX
Hairybacked Bulbul *Hypsipetes criniger*	+	+	+	U/M	F/I	26	II-VIII
Streaked Bulbul *H. malaccensis*		+	+	M/C	F/I	8	III, VI-VIII
Buffvented Bulbul *H. charlottae*	+	+		U/M/C	F/I	6	II, III, VIII
Hookbilled Bulbul *Setornis criniger*		+		U/M	F/I	#	
Siberian Blue Robin *Erithacus cyane*		+	+	G#	F/I	#	
Rufoustailed Shama *Copsychus pyrrhopygus*		+	+	G/U	I	25	II-VIII
Whiterumped Shama *C. malabaricus*	+	+	+	G/U	I	37	II-X
Chestnutnaped Forktail *Enicurus ruficapillus*	+			W/G	I/C	10	III, VII-IX
Whitecrowned Forktail *E. leschenaulti*	+			W/G	I	1	V
Blackcapped Babbler *Pellorneum capistratum*	+	+		G	I	13	II-V, VII, IX
Shorttailed Babbler *Trichastoma malaccense*	+	+	+	G/U	I	23	II-X
Whitechested Babbler *T. rostratum*	+	+		G/U	I	52	III-X
Ferruginous Babbler *T. bicolor*	+	+		?G/U	I	8	II-VI, VIII
Horsfield s Babbler *T. sepiarium*	+	+	+	G/U	I	10	II-V, VIII
Rufouscrowned Babbler *Malacopteron magnum*	+	+	+	U/M	I/F	68	II-X
Scalycrowned Babbler *M. cinereum*	+	+		U/M	I/F	22	II, III, V, VII-X
Moustached Babbler *M. magnirostre*	+	+	+	U/M	I	21	III-VIII
Sootycapped Babbler *M. affine*	+	+		U/M	I	10	II, III, V, IX
Whitethroated Babbler *M. albogulare*		+		?G/U	I	1	II
Chestnutbacked Scimitar-Babbler *Pomatorhinus montanus*	+	+	+	U/M	F/I	28	II-X
Striped Wren-Babbler *Kenopia striata*	+	+	+	G/U	I	18	II-IV, VI-IX
Fluffybacked Tit-Babbler *Macronous ptilosus*	+	+		U	I	59	II-X
Blackthroated Babbler *Stachyris nigricollis*	+	+		U	I	25	II-VI, VIII, IX
Chestnutrumped Babbler *S. maculata*	+	+	+	U/M	I	68	II-X
Chestnutwinged Babbler *S. erythroptera*	+	+	+	U/M	I	68	II-X
Whitenecked Babbler *S. leucotis*			+	U	I	1	III
Greyheaded Babbler *S. poliocephala*	+			U	I	4	III, IX, X
Rufousfronted Babbler *S. rufifrons*	+	+	+	M/C	I	6	IV, V, VII
Brown Fulvetta *Alcippe brunneicauda*	+	+	+	U/M	I	46	III-X
Whitebellied Yuhina *Yuhina zantholeuca*	+	+	+	M/C	I/F	10	III, IV, VII, VIII
Flyeater *Gerygone sulphurea*		+	+	C	I	1	III
Arctic Warbler *Phylloscopus borealis*		+	+	M/C#	I	1	IV#
Yellowvented Prinia *Prinia flaviventris*	+			N	I	1	III
Rufoustailed Tailorbird *Orthotomus sericeus*	+	+	+	U	I	57	II-X
Ashy Tailorbird *O. ruficeps*			+	U	I	2	III
Darknecked Tailorbird *O. atrogularis*	+	+	+	U	I	13	II,III,V,VII,VIII
Spotted Fantail *Rhipidura perlata*	+	+	+	?U/M/C	I	18	II-V, VII, VIII, X
Greyheaded Flycatcher *Culicicapa ceylonensis*	+	+	+	U/M	I	31	VI-IX
Asian Brown Flycatcher *Muscicapa dauurica*	+	+	+	M/C(#)	I	19	II-IV
Verditer Flycatcher *Eumyias thalassina*		+		U/M	I	1	V
Largebilled Blue Flycatcher *Cyornis caerulata*	+			U/M	I	31	III, V-VII, X
Bornean Blue Flycatcher *C. superba*		+		U/M	I	2	IV, V
Malaysian Blue Flycatcher *C. turcosa*	+			U/M	I	3	V, IX
Pale Blue Flycatcher *C. unicolor*			+	M/C	I	#	
Greychested Flycatcher *Rhinomyias umbratilis*	+	+	+	U/M	I	11	II, III, V-VIII
Olivebacked Flycatcher *R. olivacea*		+		U/M	I	1	VII
Rufoustailed Flycatcher *R. ruficauda*		+	+	U/M	I	3	III

Appendix 1 (continued). Birds recorded in Belalong forest.

	Habitat 1	2	3	Dist.	Diet	Obs	Months
Rufouswinged Monarch *Philentoma pyrrhoptera*	+	+	+	U/M	I	4	II, III
Maroonbreasted Monarch *P. velatum*	+	+	+	M	I	10	III, V-VIII
Blacknaped Monarch *Hypothymis azurea*	+	+	+	U/M	I	28	III-V, VII-IX
Asian Paradise Flycatcher *Terpsiphone paradisi*	+	+	+	M/C	I	42	II-VIII
Velvetfronted Nuthatch *Sitta frontalis*			+	C	I	2	III, VI
Yellowrumped Flowerpecker *Prionochilus xanthopygius*	+	+	+	U/M	F/I/N	15	III, V-VIII, X
Crimsonbreasted Flowerpecker *P. percussus*	+	+		U/M	F/I/N	1	VII
Yellowbreasted Flowerpecker *P. maculatus*	+	+		U/M	F/I/?N	43	II-VI, VIII
Yellowvented Flowerpecker *Dicaeum chrysorrheum*	+			U/M	F/I/?N	5	III, V
Brownbacked Flowerpecker *D. everetti*	+			U/M	F/I	1	IV
Orangebellied Flowerpecker *D. trigonostigma*		+		U/M	F/I	1	VIII
Plain Flowerpecker *D. concolor*			+	U/M	I	#	
Plain Sunbird *Anthreptes simplex*	+	+	+	U/M	I/N	13	III-VIII, X
Redthroated Sunbird *A. rhodolaema*	+			C	F/I/N	1	III
Rubycheeked Sunbird *A. singalensis*	+	+		M/C	F/I/N	6	III, VI-VIII
Scarlet Sunbird *Aethopyga mystacalis*		+	+	U/M	F/N(I)	3	III, VII
Crimson Sunbird *A. siparaja*	+			U/M	I	1	VIII
Purplenaped Sunbird *Hypogramma hypogrammica*	+	+	+	U	I	15	II, III, VI-VIII, X
Little Spiderhunter *Arachnothera longirostra*	+	+	+	U/M	I/N	25	II-X
Thickbilled Spiderhunter *A. crassirostris*	+	+	+	?U/M	I	3	V, VIII, X
Yelloweared Spiderhunter *A. chrysogenys*	+	+	+	C	F/I	2	III
Spectacled Spiderhunter *A. flavigaster*	+			C	?F/I	1	IX
Greybreasted Spiderhunter *A. affinis*	+	+	+	U/M/?C	F/I	2	VI, X
Everett s White-eye *Zosterops everetti*			+	C	F/I	1	VII
Bornean Bristlehead *Pityriasis gymnocephala*		+	+	C	I	1	III
Dusky Munia *Lonchura fuscans*	+			N	F	25	II-IV, VII, VIII
Philippine Glossy Starling *Aplonis panayensis*		+		N	F/I	2	VII
Hill Myna *Gracula religiosa*		+		C	F/I	2	III
Greater Rackettailed Drongo *Dicrurus paradiseus*	+	+	+	M/C	I	22	III, V-IX
Crowbilled Drongo *D. annectans*	+	+	+	M/C#	I	2	III, IV
Bronzed Drongo *D. aeneus*	+	+	+	M/C	I	3	III, VIII, X
Darkthroated Oriole *Oriolus xanthonotus*	+	+	+	C	F/I	22	III, V-X
Black Magpie *Platysmurus leucopterus*	+	+		M/C	F/I/C	8	II, III, V, IX, X
Crested Jay *Platylophus galericulatus*	+	+	+	U/M	I	2	III, VII
Slenderbilled Crow *Corvus enca*	+			C	C/I	1	V

Appendix 2. Breeding records from Belalong forest, Brunei Darussalam.

Species	Date	Locality	Comments
Spizaetus nanus	12-22/1/92	near KBFSC	on nest; ?young
Alcedo euryzona	26/3/91	near KBFSC	feeding young
Mulleripicus pulverulentus	19/7/91	below E. ridge	young in nest
Corydon sumatranus	25/3/91	below summit	fledgeling
Hypsipetes criniger	16/1/92	Busiri	nest material
Hypsipetes criniger	16/1/92	Bujat	brood patch
Pomatorhinus montanus	22/2/91	near Bujat	fledgeling
Pomatorhinus montanus	19/8/91	near summit	nest; 3 eggs
Stachyris erythroptera	23/2/91	KBFSC	fledgeling
Stachyris erythroptera	16/1/92	Bujat	brood patch
Alcippe brunneicauda	13/1/91	Bujat	just fledged
Orthotomus sericeus	23/2/91	KBFSC	2 fledgelings
Rhipidura perlata	16/1/92	Bujat	brood patch
Hypothymis azurea	31/3/91	W. ridge	on nest
Hypothymis azurea	18/7/91	KBFSC	on nest
Terpsiphone paradisi	12/6/91	W. ridge	building
Cyornis turcosa	16/5/91	Sg. Babi	nest material
Anthreptes simplex	8/3/91	E. ridge	young in nest
Arachnothera longirostra	16/1/92	Bujat	brood patch
Arachnothera chrysogenys	23/3/91	KBFSC	young in nest

Appendix 3. Birds netted in project area.

		Bujat	Total
Crested Serpent Eagle	*Spilornis cheela*	1	1
Rednaped Trogon	*Harpactes kasumba*	1	1
Banded Kingfisher	*Lacedo pulchella*	2	2
Rufousbacked Kingfisher	*Ceyx erithacus*	1	1
Brown Barbet	*Calorhamphus fuliginosus*	1	1
Goldwhiskered Barbet	*Megalaima chrysopogon*	1	1
Malaysian Honeyguide	*Indicator archipelagicus*	1	1
Rufous Piculet	*Sasia abnormis*	2	2
Buffnecked Woodpecker	*Meiglyptes tukki*	3	3
Green Broadbill	*Calyptomena viridis*	9	9
Blueheaded Pitta	*Pitta baudi*	0	1
Barn Swallow	*Hirundo rustica*	0	5
Redrumped Swallow	*H. daurica*	0	1
Asian Fairy Bluebird	*Irena puella*	1	1
Black-&-White Bulbul	*Pycnonotus melanoleucos*	7	7
Redeyed Bulbul	*P. brunneus*	1	1
Spectacled Bulbul	*P. erythrophthalmos*	10	10
Creamvented Bulbul	*P. simplex*	3	3
Puffbacked Bulbul	*P. eutilotus*	1	1
Greycheeked Bulbul	*Criniger bres*	9	9
Yellowbellied Bulbul	*C. phaeocephalus*	2	5
Hairybacked Bulbul	*Hypsipetes criniger*	9	9
Streaked Bulbul	*H. malaccensis*	3	3
Buffvented Bulbul	*H. charlottae*	1	1
Hookbilled Bulbul	*Setornis criniger*	2	2
Siberian Blue Robin	*Erithacus cyane*	1	1
Rufoustailed Shama	*Copsychus pyrrhopygus*	5	5
Whiterumped Shama	*C. malabaricus*	1	2
Chestnutnaped Forktail	*Enicurus ruficapillus*	0	2
Blackcapped Babbler	*Pellorneum capistratum*	2	3
Shorttailed Babbler	*Trichastoma malaccense*	4	5
Whitechested Babbler	*T. rostratum*	0	1
Ferruginous Babbler	*T. bicolor*	1	1
Horsfield s Babbler	*T. sepiarium*	0	1
Rufouscrowned Babbler	*Malacopteron magnum*	2	2
Scalycrowned Babbler	*M. cinereum*	9	9
Moustached Babbler	*M. magnirostre*	6	6
Sootycapped Babbler	*M. affine*	0	2
Fluffybacked Tit-Babbler	*Macronous ptilosus*	2	3
Blackthroated Babbler	*Stachyris nigricollis*	0	2
Chestnutwinged Babbler	*S. erythroptera*	12	12
Brown Fulvetta	*Alcippe brunneicauda*	3	3
Arctic Warbler	*Phylloscopus borealis*	1	1
Rufoustailed Tailorbird	*Orthotomus sericeus*	0	1
Spotted Fantail	*Rhipidura perlata*	5	5
Verditer Flycatcher	*Eumyias thalassina*	1	1
Bornean Blue Flycatcher	*Cyornis superba*	3	3
Greychested ,Flycatcher	*Rhinomyias umbratilis*	5	5
Rufouswinged Monarch	*Philentoma pyrrhoptera*	3	3
Maroonbreasted Monarch	*P. velatum*	1	1
Blacknaped Monarch	*Hypothymis azurea*	5	5
Yellowbreasted Flowerpecker	*Prionochilus maculatus*	7	8
Crimsonbreasted Flowerpecker	*P. percussus*	1	1
Yellowrumped Flowerpecker	*P. xanthopygius*	9	9
Plain Sunbird	*Anthreptes simplex*	7	7
Little Spiderhunter	*Arachnothera longirostra*	22	24
Thickbilled Spiderhunter	*A. crassirostris*	1	1
Greybreasted Spiderhunter	*A. affinis*	2	2
Total		192	218

Terrestrial Ants (Hymenoptera: Formicidae) of Poring, Kinabalu Park, Sabah

MARYATI MOHAMED, AZIZAH HUSSEIN *and* ARBAIN KADRI

Biology Department, Faculty of Science and Natural Resources, Locked Bag 62, 88996, Kota Kinabalu, Sabah, Malaysia

ABSTRACT. Collections of ants were carried out for a 7 month period in Poring, Kinabalu Park, Sabah. In total 46,999 individual ants were collected, representing 6 subfamilies. These subfamilies were represented by 52 genera and 125 morphospecies. Percentages of species in the various subfamilies were as follows: Myrmicinae 44.8%; Formicinae, 24.8%; Ponerinae, 14.4%; Dolichoderinae, 13.6%; Dorylinae, 1.6% and Pseudomyrmecinae, 0.8%. The 2 genera with the highest number of species were Pheidole and Dolichoderus with 12 each. Ninety-five species were collected by baited pitfall traps and 71 species were collected manually. Using a one way ANOVA no significant difference in the number of species was found between the 2 methods.

KEY WORDS: Borneo, ants, species richness, habitat quality, sampling techniques, diversity.

INTRODUCTION

The abundance and diversity of ants in the south east Asian tropical rain-forests have not been fully investigated. To understand these ecological aspects, as well as assist in the process of building up a reference collection of ants for Malaysia, we have been collecting ants since 1986, both in the peninsular as well as in Sabah, the North Borneo State of Malaysia. Previous collections have been concentrated in the lowland dipterocarp forests. The actual roles of ants in the tropical forest ecosystem are not fully understood. Their presence in all types of tropical forests are obvious. Burghout *et al.* (1992) and Wong (1984), working on 2 different locations found that ants were consistently the most abundant arthropods in the Malaysian forests. One of the roles played by ants thought to be highly plausible is the recycling of nutrient into the forest soil. Being abundant and diverse, ants are a potentially good bioindicator of the status of the forests. Greenslade & Greenslade (1984) emphasised the potential of invertebrates and in particular ants as bioindicators of habitat quality. This was emphasised by Majer (1987). It was our intention to determine the best method for collection of ants and the best ant species to be used as indicators not only to assess forest quality but also to serve as indicators of changes in the Malaysian environment.

METHODS

Our sampling plot was located at Poring Hot Spring in Kinabalu National Park, Sabah, at an altitude of 550 m (Figure 1). Rainfall for the area from 1981 to 1990 ranged from 1220.0 to 2757.8 mm per annum. Day temperatures ranged from 25-31^{0}C. A transect, parallel to the canopy walkway, was chosen for the location of five 10 m x 10 m plots. Vegetation along this transect was composed primarily of species of the families Euphorbiaceae, Arecaceae, Olecaceae, Meliaceae and Dipterocarpaceae. Seven 3-4 day collecting periods were spaced 3-4 weeks apart over a period of 7 months.

D. S. Edwards et al. (eds.), Tropical Rainforest Research – Current Issues, 117–123.

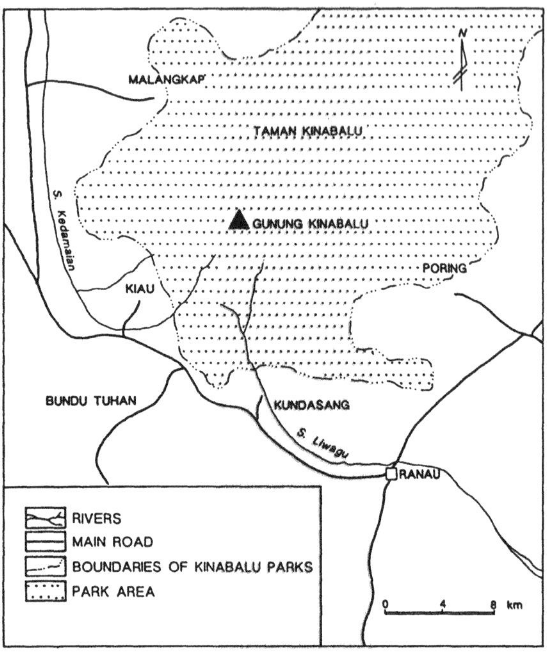

Figure 1. Map showing the location of Poring Hot Springs, Kinabalu Park, Sabah, Malaysia.

Two sampling techniques were employed. First, ants were collected manually from all the 5 plots. Collection was carried out by one person for about 5-6h from 9am to 4pm, for 2 consecutive days during each collection trip. Ants on the forest floor, in the forest litter and between crevices of rocks and bark, on tree trunks and undergrowth up to the eye level were collected and put in vials with 75% ethanol, and brought back to the laboratory for identification. The second method, baited pitfall trapping, was a modified version of the common pitfall traps used to collect forest floor arthropods. Plastic containers 7.5 cm diameter and 12 cm high were placed in holes dug into the ground so that the lip of the container was at ground level. Five containers were placed about 1 m apart at random in each plot. Five different baits were used in each plot, one in each container. The different baits used were canned tuna in oil (marine protein source), beefburger (Ramly brand, animal protein source), soft white tau-foo (plant protein source), sweet (Trebor, sour plum flavour), and pebbles as control. Each bait, about 1 cm square, was wrapped tightly in muslin cloth and secured with cotton thread. The baits were positioned about 2/3 of the way down into the containers from a wire rod. A soapy solution was put into the bottom of each container (ratio of water:washing liquid is 50:1). The addition of dish washing liquid was necessary to lower the surface tension of the water so that fallen ants would sink to the bottom of the container. The traps were left from 4pm until 7am the next morning.

Trapped ants were separated from the soap solution using a sieve. The ants were then rinsed in clean water, picked out and preserved in 75% ethanol and placed in coded vials. The trapping was repeated once during each visit. The two methods mentioned above were as in Maryati (1992).

Identifications were based on keys in Bolton (1990) and Holldobler & Wilson (1990). Ants were identified to generic level and designated as species 1, 2, 3, etc. Comparison was made in terms of species richness to see which collecting method was most successful in terms of numbers of species of ants caught.

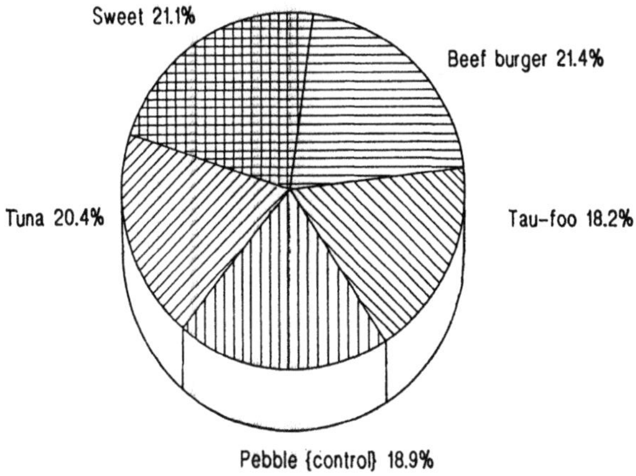

Figure 2. Percentage of species collected by various baits during the seven collecting periods.

RESULTS

Table 1 shows the overall results of the collection. Genera with 10 or more species were confined to *Pheidole* and *Dolichoderus*. Those with between 5 and 9 species included *Pheidologeton*, *Tetramorium* and *Polyrhachis*. As seen in Table 2, Myrmicinae included almost 45% of the species sampled at Poring, Formicinae 25%, and Ponerinae and Dolichoderinae each about 14%. Dorylinae and Pseudomyrmecinae made up only about 2% of the species sampled.

Table 3 shows species composition for subfamilies for each of the 7 collections, disregarding the collecting method. Myrmicinae had the highest representation of species followed by Formicinae or Dolichoderinae (Table 3 I-VII). For most of the sampling times the number of species of ponerines and dolichoderines were almost the same except for collections I, III and VII when the number of species of Ponerinae collected were about 1.5 to 2 times greater than species of Dolichoderinae. The ratio of Myrmecinae to Formicinae was about 1.5:1 to 3:1, while the ratio of Formicinae to Ponerinae or Dolichoderinae was about 1.4:1 to 2.7:1.

Table 4 compares the efficiency of the various baits in pitfall trapping programme. Using a one-way ANOVA it was found that there was no statistical difference between the efficiency of the baits. Even the control (baited with pebbles) was seen to be trapping as many ants as other kinds of baits, or even more (Figure 2).

DISCUSSION

Composition of terrestrial ants of Poring

Collections made in lowland forests such as Kuala Lompat, Pahang (Maryati 1991) and Gunung Danum, Sabah (Maryati 1989) showed a similar pattern of composition to that of the present study, with Myrmicinae and Formicinae being the 2 most highly represented subfamilies followed by Ponerinae or Dolichoderinae. The remaining subfamilies represent only 5-10% of the species composition or even less. Collections from Bukit-bukit Tawau, Sabah, showed a higher representation of species in the subfamily Formicinae compared with

Table 1. List of subfamilies and genera of ants collected by pitfall trapping and manually, at Poring. Numbers indicate the number of the morphospecies for each genus.

Subfamily	Genus	Pitfall traps	Manual	Total no. of morphospecies
PONERINAE	*Diacamma*	1	1	1
	Discothyrea		1	1
	Harpegnathos	1		1
	Hypoponera	1		1
	Leptogenys	1, 2, 3	1	3
	Mystrium	1, 2	3	3
	Odontomachus	1	1, 2	2
	Odontoponera	1	1	1
	Pachycondyla	1, 2, 3		3
	Platythyrea	1	1	1
	Prionopelta	1		1
	Sub-total	15	8	18
DORYLINAE	*Aenictus*	1, 2	1	2
	Sub-total	2	1	2
PSEUDOMYRMICINAE	*Tetraponera*	1	0	1
	Sub-total	1	0	1
MYRMICINAE	*Acanthomyrmex*	1, 2		2
	Aphaenogaster	1	1	1
	Cataulacus		1	1
	Crematogaster	1, 2, 3	1, 2, 3, 4	4
	Lophomyrmex	1, 2	1	2
	Mayriella	1, 2		2
	Meranoplus	1		1
	Myrmecina	1	2	2
	Oligomyrmex	1, 2, 3		3
	Paedalgus	1		1
	Paratopula	1		1
	Pentastruma	1, 2	1	2
	Pheidole	1-10	1-8, 11, 12	12
	Pheidologeton	1-7	1	7
	Proatta	1		1
	Quadrustruma	1		1
	Rhopalomastix	1		1
	Solenopsis	1		1
	Strumigenys	1, 2		2
	Tetramorium	1-6	1-3	6
	Vollenhovia	1		1
	Wasmannia	1		1
	Willowsiella	1		1
	Sub-total	51	23	56
DOLICHODERINAE	*Dolichoderus*	1-4	1-12	12
	Iridomyrmex	1	1	1
	Tapinoma	1, 2, 3	1-4	4
	Sub-total	8	17	17
FORMICINAE	*Acantholepsis*	1	1	1
	Acropyga	1		1
	Anoplolepis	1	2	2
	Camponotus	1, 2, 3	1, 2, 3	3
	Colobopsis	1	1,2	2
	Echinopla		1	1
	Euprenolepis	1, 2, 3	1	3
	Lasius		1, 2	2
	Myrmoteras	1		1
	Paratrechina	1-5	1, 2	5
	Plagiolepis		1	1
	Polyrhachis	1	1-6	6
	Prenolepis	1	2, 3	3
	Sub-total	18	22	31
TOTAL NO. OF SPECIES		95	71	125

Table 2. Species percentage composition for sub-families of ants collected at Poring.

Subfamily	Percentage
Ponerinae	14.4
Dorylinae	1.6
Pseudomyrmicinae	0.8
Myrmecinae	44.8
Dolichoderinae	13.6
Formicinae	24.8
Total	100

Table 3. Number of species of ants sampled during the seven collections (I-VII) in each subfamily.

	Collection						
	I	II	III	IV	V	VI	VII
Subfamily							
Ponerinae	9	10	6	5	9	8	11
Dorylinae	0	0	0	1	1	0	0
Pseudomyrmecinae	0	0	0	0	0	0	1
Myrmicinae	34	30	25	29	35	32	31
Dolichoderinae	4	7	6	6	10	7	7
Formicinae	13	13	18	13	17	13	18
Total	60	60	55	54	72	60	68

Table 4. Number of species of ant sampled for the seven collections based on baits used in the pitfall traps.

	Collection						
	I	II	III	IV	V	VI	VII
Kinds of baits							
Control (pebble)	34	45	21	40	37	23	25
Tau-foo	28	37	16	41	35	28	27
Beefburger	50	43	36	46	45	33	39
Tuna in oil	45	42	39	43	45	35	44
Sweet	30	43	33	45	47	33	41
Total	187	210	145	215	209	152	176

Myrmicinae, and a much lower representation of Dolichoderinae compared with Ponerinae (Maryati 1991). From a highland area such as Sayap-Kinabalu Park (Maryati 1993), Ponerinae had a higher representation of about 33%, a feature also associated with collections made from very disturbed forests such as Bangi forest, Selangor (Maryati 1990). These patterns suggest firstly that collections from undisturbed primary forest would give a different species composition from those collected from extreme habitats such as highland or very disturbed forests. Greenslade & Greenslade (1977) also noted that a habitat which was disturbed by human activities would have a lower diversity of ants compared to undisturbed habitats. Majer (1983) used ants as bioindicators of habitat quality. Secondly, in the tropical rain forests, subfamilies such as Myrmicinae, Formicinae, Ponerinae and Dolichoderinae are big subfamilies, whereas the remaining subfamilies are small.

Analysing the species present, *Dolichoderus* had a relatively high number of species (12). Our previous collections had never yielded this many species of *Dolichoderus*. Further work is needed to investigate the reason for such a high number of species. There may also be an advantage in looking into the biology and ecology of this group since one of the species, *Dolichoderus thoracicus*, has been successfully used in cocoa plantations to control infestations of the mirid *Helopeltis theobromae* (Khoo & Chung 1989).

Pheidole was represented by 12 species, a similar number to that found in many of the previous collections, especially those involving baited pitfall trapping. The genus *Polyrhachis* was also well represented, with 6 morphospecies collected. This genus is fairly common in tropical rain forest and shows high diversity. During the Bukit-bukit Tawau expedition 9 species of *Polyrhachis* were collected (Maryati 1991).

As shown in Table 3, the species composition of the ants collected did not vary much with time.

Comparison of collecting methods

Genera such as *Pheidole* could be collected efficiently by either of the two methods described. This was possible because members of this genus inhabit forest litter and the upper part of the forest soil. *Crematogaster* could also be caught efficiently by either method. This genus is generally arboreal, making nests on living plant parts such as branches or stems. Some even live inside plants such as within stems of *Macaranga* (Fiala *et al.* 1989). A species of *Tetraponera* was also collected from the pitfall traps in spite of this genus being largely arboreal. *Oligomyrmex*, *Tetramorium* and other small sized myrmicines such as *Myrmecina*, *Quadrustruma* and *Solenopsis* were mainly caught using baited pitfall traps, and not manually. On the other hand, as in previous studies, *Polyrhachis* could best be collected manually and perhaps by fogging (Maryati 1992).

These observations suggest that for inventory work, when as many species of ants as possible are required to be collected, all kinds of collecting techniques should be employed, as discussed by Maryati (1992).

There was no effect of the type of bait on the success of ant collection. According to Greenslade & Greenslade (1971), in areas with high ant diversity any advantage of a particular bait was marginal and the use of unbaited pitfall traps was recommended.

ACKNOWLEDGEMENTS

We wish to thank Sabah Parks for allowing this research to be conducted at Poring Hot Springs, Kinabalu National Park. This research was made possible by IRPA grant 04-07-03-007 managed by Prof. Dr. A. Latif Mohamad. The assistance of Mr.Sukup Akin of the Botany Section of Biology Department of the Faculty of Science and Natural Resources in the identification of plants around the sampling transect is also acknowledged.

LITERATURE CITED

BURGHOUT, T., ERNSTING, G., KORTHAL, G.W., & VRIES, T.H. 1992. Litterfall, leaf litter decomposition and litter invertebrates in the primary and selectively logged dipterocarp forest in Sabah, Malaysia. In Marshall, A.G. & Swaine, M.D. (eds). *Tropical Rain Forest: Disturbance and Recovery*. The Royal Society, London.

FIALA, B., MASCHWITZ, U., PONG, T.Y., & HELBIG, J. 1988. Studies of a South East Asian Ant-Plant Association: Protection of *Macaranga* trees by *Crematogaster borneensis*. *Oecologia* 7:463-470.

GREENSLADE, P. & GREENSLADE, P.J.M. 1971. The use of baits and preservatives in the pitfall traps. *Journal of the Australian Entomological Society* 10:253-260.

GREENSLADE, P. & GREENSLADE, P.J.M. 1977. Some effects of vegetation cover and disturbance on a tropical ant fauna. *Insectes Sociaux* 2:163-182.

GREENSLADE, J. & GREENSLADE, P. 1984. Invertebrates and environmental assessment In Environmental and Planning. *Department of Environment and Planning, Australia, Journal No.* 3.

HOLLDOBLER, B. & WILSON, E.O. 1990. *The Ants*. Belknap Press of Harvard University Press, Cambridge, Mass.

KHOO, K.C. & CHUNG, G.F. 1989. Use of black cocoa ant to control mirid damage in cocoa. *The Planter* 65:370-383.

MAJER, J.D. 1983. Ants: Bioindicator of Minesite Rehabilitation, Land Use and Land Conservation. *Environmental Management* 7:375-383.

MAJER, J.D. 1987. Invertebrates as indicators for management. In Saunders, D.A., Arnold, G.W., Burbridge, A.A. & Hopkins, A.J.M. (eds). *Nature Conservation: The Role of Native Vegetation*. Surrey Beatty & Sons, Pty. Ltd.

MARYATI, M. 1989. *The ants (Formicidae: Hymenoptera) of Gunung Danum, Lahad Datu, Sabah*. Report of Gunung Danum Expedition.

MARYATI, M. 1990. Fauna dan kelakuan semut (Formicidae: Hymenoptera). *Kawasan Ekologi Hutan Bangi Ekologi dan Ekologi Hutan Bangi, Kumpulan Kertas kerja 14*. Universiti Kebangsaan Malaysia.

MARYATI, M. 1991. *A preliminary list of ants (Formicidae: Hymenoptera) of Bukit-bukit Tawau, Sabah*. Report of Bukit-bukit Tawau Expedition.

MARYATI, M. 1992. *Penggunaan Kaedah yang sesuai bagi pengumpulan semut (Formicidae: Hymenoptera)*. Simposium Sumber Alam Kebangsaan Pertama.

MARYATI, M. 1993. *Terrestrial ants (Formicidae: Hymenoptera) of Sayap-Kinabalu Park, Sabah*. Report of Sayap-Kinabalu Park, Sabah Expedition.

WONG, M. 1984. Understorey Foliage Arthropods in the Virgin and Regenerating habitats of Pasoh Forest, West Malaysia. *The Malaysian Forester* 47:43-69.

Temporal and spatial patterns of butterfly diversity in a lowland tropical rainforest

A. G. ORR[1] *and* C. L. HAEUSER[2]

[1]*Department of Biology, Universiti Brunei Darussalam, Bandar Seri Begawan 2028, Brunei Darussalam*
[2]*Zoologisches Forschungsinstitut und Museum Alexander Koenig, Adenauerallee 150-164, D-53113 Bonn, Germany*

ABSTRACT. In nearly two years of regular sampling 324 species of butterfly were recorded from the lowland rainforest within one kilometer of the Kuala Belalong Field Studies Centre (KBFSC), Batu Apoi forest reserve, Temburong, Brunei Darussalam. This represents over one third of the Bornean fauna and perhaps over half the total fauna of the area. Records are biased towards the more conspicuous Papilionidae and Pieridae with relatively smaller proportions of Nymphalidae, Lycaenidae and Hesperiidae being represented. Samples taken at approximately monthly intervals indicated relatively slight variation in species richness. However there was a steady turnover in species composition, presumably the result of changing species abundance relationships as most of the more common species experienced substantial and irregular fluctuations in numbers. Some closely related species pairs showed phenological dissociation in abundance, but many were phenologically unrelated or were positively associated. There was no evidence for seasonal trends in community composition or in the phenology of any particular species. While many species ranged freely throughout the area, some were more or less restricted to either riverbanks and open areas, the forest understorey, or the canopy, subcanopy and tops of major ridges. Certain Lycaenidae and Hesperiidae were extremely localized, being sometimes restricted to an area of a few hundred m^2, presumably in the vicinity of their larval hostplants.

KEYWORDS: Borneo, butterflies, diversity, phenology, seasonality, species-richness, species-turnover.

INTRODUCTION

Approximately 950 butterfly species are recorded from the island of Borneo, and as the fauna is fairly well known, the true figure is probably not more than one thousand species (Corbet & Pendelbury 1992, Otsuka 1988, Seki *et al.* 1991a, 1991b). Of these, the great majority have been recorded in North Borneo, and with the exception of a handful of high altitude species, most could potentially occur in Brunei Darussalam, as it includes within its boundaries most of the forest types known from the island. The extent to which butterfly diversity is related to vegetational diversity is uncertain; previous studies suggest that most butterfly species are to be found in lowland mixed dipterocarp forest (Cassidy 1982, 1985, Holloway 1984), but even within this formation there is considerable floristic heterogeneity relating to edaphic and topographic factors (Ashton 1964), which might influence the composition of butterfly communities. The present study attempts to measure alpha diversity of butterflies within an edaphically and floristically uniform area of lowland rainforest, especially in an area of approximately one square kilometer surrounding the Kuala Belalong Field Studies Centre (KBFSC).

Overall diversity, and especially measurable or perceived diversity, is expected to have significant temporal and spatial components. We examine in particular the effects of seasonality and species turnover, and attempt to assess the importance of microdistributional patterns, within the habitat surveyed, on the measurement of diversity. The phenology of selected species and the extent of temporal association or disassociation between closely

D. S. Edwards et al. (eds.), Tropical Rainforest Research – Current Issues, 125–138.

related species is also examined, with the aim of testing the hypothesis that assemblages of closely related species will show temporal dissociation of phenologies, possibly in order to avoid competition for adult or larval resources, as has been suggested by several other studies (Clench 1967, Owen 1971, Owen *et al.* 1972) .

AREA AND METHODS

The study was conducted in the Batu Apoi Forest Reserve, Temburong, Brunei, in an area of primary lowland mixed dipterocarp forest of approximately one square kilometer in area and ranging from about 0-350m in elevation, in the vicinity of the KBFSC. The mean annual rainfall for the area was about 4000 mm (Pendry & Proctor, this volume) and the dry season was not strongly marked, with some rain falling on most days. On one occasion (14 September 1991) when helicopter transport was available a sample was also taken from nearby Bukit Belalong (elevation 932 m) in the space of four hours.

In the KBFSC area butterflies were sampled on 15 occasions over periods of two to five days at approximately monthly intervals between March 1991 and March 1993. There was a long discontinuity in data collection from December 1991 to April 1992. Specimens were netted along the east ridge trail as far as plot 2, and along the west ridge trail as far as the second summit (Figure 1). Collecting efforts were concentrated at fallen fruit, at sunspots and along the ridges, especially at the summits where various canopy species were hilltopping (see Shields 1967). Baits of fermenting banana were laid along the trail at regular intervals to attract certain fruit-feeding nymphalid species.

Every sampling period at least one estimation of species abundance patterns in the forest was made, using the transect walk technique of Pollard (1977). The area surveyed followed the west ridge trail to the second summit and back, and usually took place between 10.00h and 13.00h. Records were based mainly on visual sightings, hence included only those species which could be identified without capture, which excluded most Lycaenidae and Hesperiidae. Because of unevenness in sampling different taxonomic groups and species

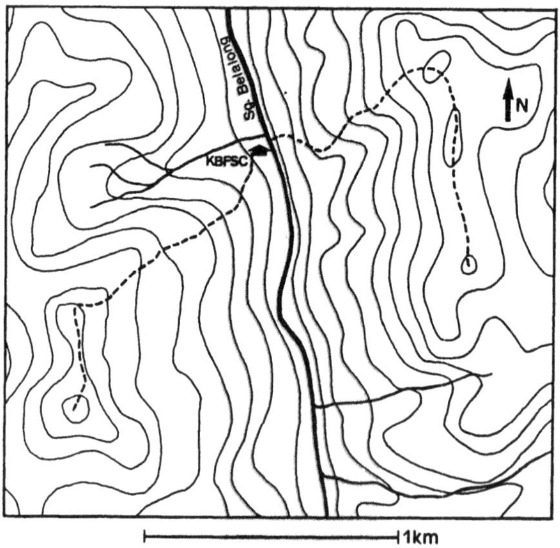

Figure 1. Area around the KBFSC showing the east and west ridge trails (dashed lines) along which sampling was done.

inhabiting different microhabitats, estimates of relative abundance are only considered comparable between equally conspicuous species of similar habits (e.g. the understorey fruit feeding guild), which usually meant members of taxonomically related assemblages.

Every day butterflies which came to 'mud-puddle' (Arms *et al.* 1974) at the sewage outlet by the river were censused. Numbers visiting the soak between 1400h and 1430h were recorded. A majority of papilionid and pierid species came to the sewage, and as most species could be identified without capture this yielded the most extensive and reliable data on relative abundance. As only newly emerged males visited the soak, recruitment rates rather than abundance are estimated, but if the rate of recruitment is reasonably steady over periods approximating the life expectancy of the butterflies it should also reflect overall abundance. Because the quality of the sewage source varied with the level of occupancy of the buildings, estimations of relative abundance of a species from one sampling period to another may be unreliable. However this is not expected to bias the measurement of the relative proportions of different species.

RESULTS

Species richness - general trends

Three hundred and twenty five species were recorded from the lowland area described above and an additional 24 sp were collected from Bukit Belalong. A full list of species with summary information on abundance and distribution is provided in the Appendix. One hundred and thirty-five species from our list are previously unrecorded in Brunei (Cassidy 1982, 1985) bringing the national tally to 492 species.

Together these samples represent about 37% of the total Bornean Fauna (Otsuka 1988, Seki *et al.* 1991a, 1991b). As is to be expected the more conspicuous Papilionidae, Pieridae and Nymphalidae are much better represented than are the Lycaenidae or Hesperiidae (Table 1). After 42 man days collecting the cumulative number of species recorded was still increasing at a rate of about one species every two days (Figure 2), hence there is obviously a substantial number of species in the area yet to be recorded.

Table 1. Proportional representation of families relative to the Bornean fauna (including sample from Bukit Belalong).

Family	Total species recorded	Number in Borneo	Propotion of Borneo total
Papilionidae	26	44	59%
Pieridae	31	41	76%
Nymphalidae	128	236	54%
Lycaenidae	112	393	28%
Hesperiidae	51	217	24%

Using the method of Clench (1979, see also Ragusa & Llorente-Bousquets 1990) species richness was estimated to be 454 species in total. Overall sampling is considered too uneven to describe species abundance relations with any degree of precision. However 54 species can be considered common (greater than 50 records overall), 112 were moderately common, at least sporadically (between 10 and 50 records) and the remaining 158 species were rare (less than 10 records, often only 1).

Phenology, species turnover and seasonality

Excluding very rainy days, the number of species recorded per day ranged from 59 to 102

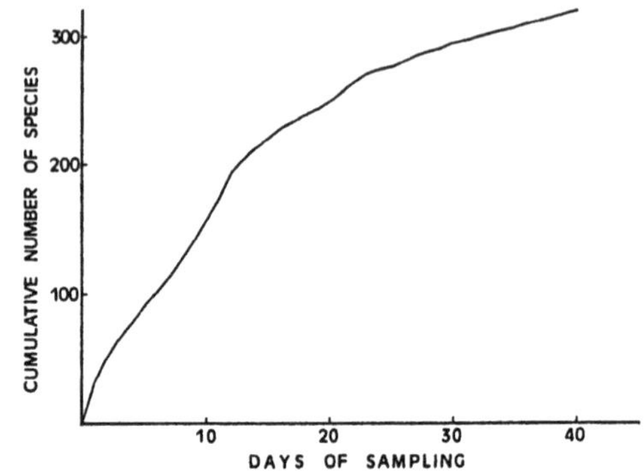

Figure 2. Cumulative species total against number of days spent sampling.

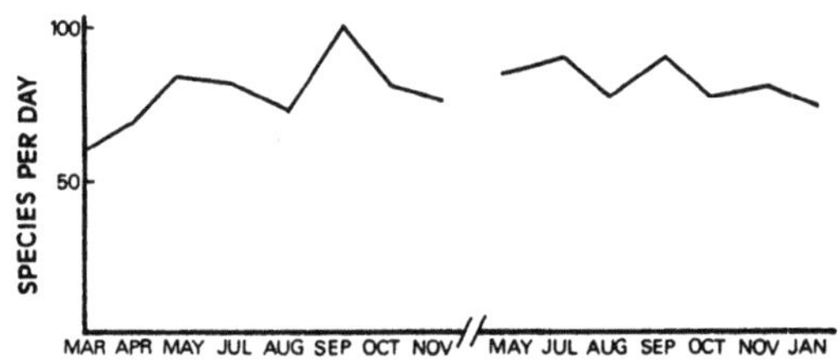

Figure 3. Average total number of species recorded per day for the various sampling periods from March 1991 to January 1993.

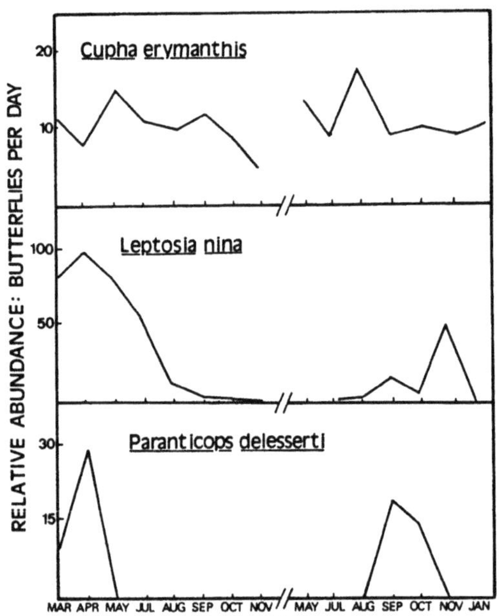

Figure 4. Three different patterns of phenology exhibited by (a) *Cupha erymanthis* - even, (b) *Leptosia nina* - fluctuating, and (c) *Paranticops delesserti* - sporadic, between March 1991 and January 1993.

with a mean of 78 (Figure 3). No regular seasonal trends in species richness were apparent, nor did any species show regular fluctuations in abundance. Certain species were always fairly common e.g. *Cupha erymanthis*, a species found mainly in open areas around the

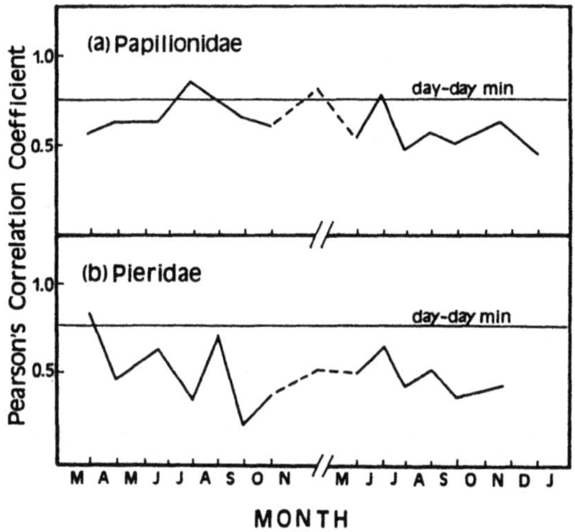

Figure 5. Month to month similarity (measured by Pearson's correlation coefficient) in species composition for mudpuddling (a) Papilionidae, and (b) Pieridae. Horizontal lines mark the minimum similarity recorded on consecutive days (asssumed to be zero), and any figure below this line probably represents a significant change in species composition.

centre (Figure 4a), whereas other very common species fluctuated markedly and apparently irregularly in numbers, e.g. *Leptosia nina* (Figure 4b). Some species were common only for one or two sampling periods and were not seen in between, e.g. *Paranticops delesserti* (Figure 4c).

Phenologies were classified as even, fluctuating or sporadic by the following scheme:

90% observations in 10-15 sampling periods - even
90% observations in 5-9 sampling periods - fluctuating
90% observations in 1-4 sampling periods - sporadic

By these criteria 43 species can be considered even in their phenology, 86 species fluctuating, and 37 species sporadic. The remaining species were represented by fewer than ten records and do not provide adequate data.

The prevalence of species with fluctuating or sporadic phenologies is reflected in the approximate monthly turnover of papilionid and pierid species visiting soaks (Figures 5a, 5b). Relative proportions of the various species are compared between successive sampling periods, using a Pearson product moment correlation coefficient which ignores variation in absolute abundance and hence variations resulting from varying attractiveness of the soak. It was found that similarities between consecutive months were almost always lower than the lowest correlation recorded for samples taken on consecutive days (in which turnover was assumed to be zero). This is true of both families, analysed independently, and indicates a continually changing community composition.

There was little evidence to suggest that species abundance patterns are repeated on a seasonal or yearly basis. High correlations were recorded for mudpuddling Papilionidae between May 1990 and September 1991 ($r = 0.85$) and for mudpuddling Pieridae between July 1990 and October 1990 ($r = 0.78$). There are no environmental correlates to suggest an explanation for these convergences. On the other hand there was no evidence of species abundance patterns repeating themselves on a yearly seasonal basis (Figure 6), indicating that the greatest similarity was between consecutive months. Samples from the same months in successive years tended to be dissimilar, suggesting a lack of any annual cycle.

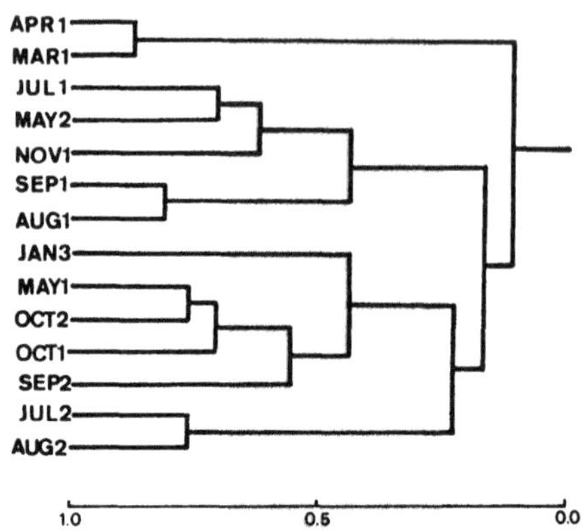

Figure 6. Relative similarity in species composition for different months throughout the sampling period March 1991 to January 1993. Numerals afer months indicate year (1=1991, 2=1992, 3=1993). Average linkage cluster analysis of Pearson correlation coefficients.

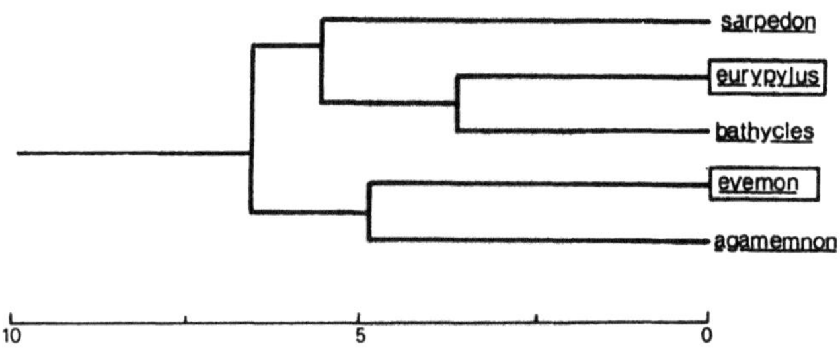

Figure 7. Similarity of phenologies of *Graphium* species: The sister species *eurypylus* and *evemon* group into different clusters. Average linkage cluster analysis of Pearson correlation coefficients.

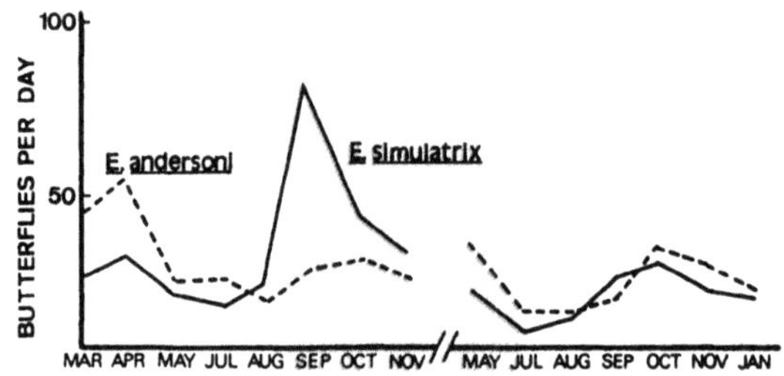

Figure 8. Positive association of phenologies in two species of *Eurema*, *E. andersoni* and *E. simulatrix*, mudpuddling at sewage (tau = 0.43, $p<0.05$), between March 1991 and January 1993.

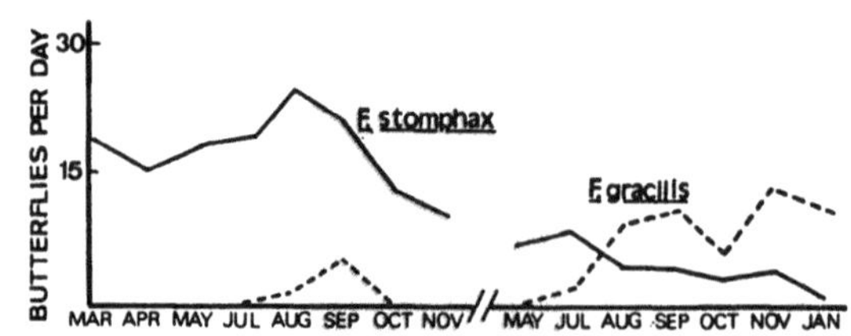

Figure 9. Dissociation of phenologies of two species of *Faunis*, *F. stomphax* and *F. gracilis*, feeding at fermenting banana baits (tau= -0.44, $p< 0.05$), between March 1991 and January 1993.

Table 2. Association/dissociation of phenologies of close species pairs. Only figures for *Eurema* and *Faunis* are significant at 0.05 level.

SPECIES PAIR	KENDAL'S TAU
Graphium evemon/G. eurypylus	-0.15
Eurema andersoni/E. simulatrix	0.43*
Cepora iudith/C. pactolicus	-0.02
Prioneris cornelia/P. philonome	-0.05
Erites elegans/E. argentata	-0.27
Lexias dirtea/L. pardalis	-0.20
Faunis gracilis/F. stomphax	-0.44*
Cirrochroa tyche/C. emalea	-0.17
Paralaxita orphne/P. telesia	0.23

Phenological relationships between closely related species were variable. A significant positive association was found between *Eurema simulatrix* and *E. andersoni* (Pieridae) (Figure 7), and a significant negative association was found between *Faunis stomphax* and *F. gracilis* (Nymphalidae) (Figure 8). However, most commonly, when the phenologies of closely related species pairs were compared there was no significant association or dissociation (Table 2). When the degree of similarity of phenologies was calculated for five *Graphium* species (Papilionidae) (Figure 9), the sister species *G. evemon* and *G. eurypylus* fell into different clusters, indicating at least a lack of similarity in their phenologies. However there was no significant negative correlation between them (Table 2).

Spatial patterns of diversity

Many species appear restricted to particular microhabitats, here divided into riverbanks and open areas, forest understorey and ridges (Table 3). Many of the 'ridge' species were probably found mainly in the canopy and subcanopy, but were either visible from the ridge or descended lower than usual along the crest. Information is available for about half the species recorded, the remainder being represented by too few records for conclusions to be drawn. About half of these are recorded as being in some way restricted in their distribution. Major trends observed were: 1. Many nymphalids were restricted to open habitats - these were mostly danaines and nymphaline species which also occur in secondary forest; 2. Many nymphalids were restricted to the understorey - these included most satyrines, amathusiines and the euthaliine nymphalines; 3. Many lycaenids were restricted to the ridge - these were mostly canopy dwelling theclines. Spatial separation of species in different microhabitats is a major source of unevenness in sampling, certain situations being much easier to sample than others.

About 12% of species were classified as local. Many of these species were restricted to very small areas, sometimes of only a few hundred square metres (Figure 10), and were found in only one place along the 1.5 km transects on either side of the KBFSC. Presumably these species, mostly Lycaenidae, were restricted to the immediate vicinity of their larval

Table 3. Microdistribution and habitat preference by family (including 176 species, representing 54% of all species).

Family	General	Open/Riverine	Ridge	Understorey	Local
Papilionidae	15	2	-	-	-
Pieridae	25	1	5	-	-
Nymphalidae	62	21	5	31	14
Lycaenidae	23	1	17	4	27
Hesperiidae	7	1	3	-	3

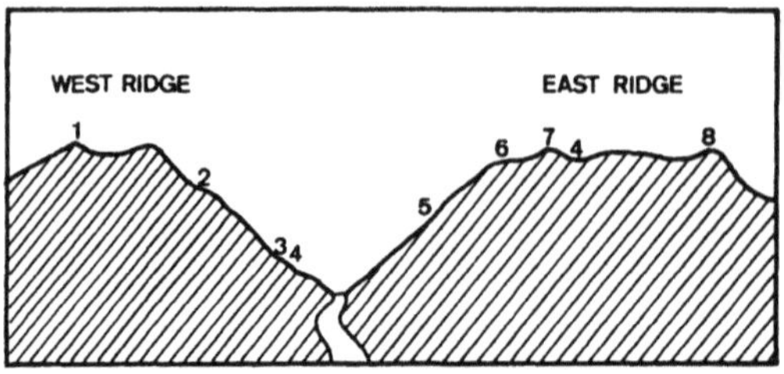

Figure 10. Sites along the east and west ridge transects where highly localized species occur (the transects have been transformed onto a vertical plane at right angles to the river). 1. *Trix scopula*, 2. *Paralaxita orphne*, 3. *Paralaxita telesia*, 4. *Dacalana* sp. 5. *Laxita tenata*, 6. *Poritia erycinoides*, 7. *Horaga syrinx*, 8. *Flos morphina*.

hostplant. Several of these species, such as *Flos morphina, Laxita tenata, Paralaxita telesia, Thrix scopula* and *Horaga syrinx*, were common in a very limited area and almost always present. Typically one male was visible, maintaining a territory in a sunspot. When it was removed it was very shortly replaced by another and this was in turn replaced when captured, and so on until at least three or four individuals had been recorded. Some species were slightly less localized but still restricted in their distribution. For example, the powerfully flying *Charaxes durnfordi* was found only on the east ridge between Pondok Bujak and Plot 2. Of 23 species with a highly restricted distribution, 19 were found in only one place. This suggests that the area covered was too small to provide an adequate sample of localized species.

DISCUSSION

The number of species recorded from Kuala Belalong in this study substantially exceeds the totals recorded for Mulu National Park, Sarawak, (276 species, Holloway 1984), and Endau Rompin, in Peninsular Malaysia (232 species, Tan *et al.* 1990) and is similar to the figure recorded for Kinabalu National Park and other localities in Sabah (340 species, Barlow *et al.* 1971). To put this result in perspective, the difficulty of comparing sampling efforts in different localities with different personnel must be taken into account, as well as the time frame within which the collections were made. The longest period of sampling for any of the other studies above was five months (continuous) at Mulu by Holloway and colleagues who collected 276 species from a range of habitat types. After five months periodic sampling at Kuala Belalong, 212 species had been recorded from a single forest type but overall sampling effort was probably considerably less.

It is well recognized, at least in seasonal environments, that measurement of species richness must take into account variation in species' phenologies, the so called 'temporal component of diversity' (Hill 1988, Shapiro 1975). This is especially true of temperate environments, or of tropical environments with well defined wet and dry periods, where species occur in a predictable sequence throughout the year (Owen 1971, Shapiro 1975). Within climatic constraints, species phenology is thought to be a function mainly of larval food availability (Slansky 1974), adult food availability (Clench 1967, Hill 1992) and mutual avoidance by species competing for either resource (Clench 1967, Owen 1971). In tropical areas, even where there is pronounced seasonality, there may be considerable variation in the phenologies of different butterfly families (Owen 1971, Owen *et al.* 1972), which tends to smooth out overall variation throughout the year. Some studies have found slightly greater diversity and/or abundance during the wet season (Owen 1971), whereas others report greater diversity during the dry season (Emmel & Leck 1970, Fox *et al.* 1965), but given the different family level trends, all such claims should be treated with caution because of the high probability of sampling bias in favour of particular groups.

At Kuala Belalong, which is without pronounced seasonality in rainfall and temperature, species phenology is mostly unpredictable. Apart from a small proportion of very abundant or very predictable species which are always reasonably common, the majority of species appear to undergo substantial and unpredictable fluctuations in abundance, resulting in a continual and fairly steady turnover in species abundance relationships and, hence in species actually sampled. Probably the main reason why this study has recorded substantially more species than comparable surveys such as Holloway (1984), is because it has continued over a longer period of time.

When considering the phenologies of closely related species three types of relationship are possible:

1. No interaction - phenologies of different species are independent.
2. Temporal association - similar species have similar phenologies because they are tracking the same set of environmental factors.
3. Temporal dissociation - similar species have dissimilar phenologies either because they are avoiding competition for larval or adult resources, or because they are adapted to different conditions.

Because of the difficulty in obtaining adequate sample sizes, phenological relationships among species assemblages are often unclear. Therefore a lack of correlation, positive or negative, between the phenologies of species pairs may just be a result of inadequate sample sizes of one or both species. This is particularly true when sister species are compared, since as a rule one species is always much rarer than the other in a given area. Nevertheless cases of dissociation in *Faunis* species and of positive association in *Eurema* species were evident, and it seems probable that numerous other cases of temporal dissociation occur. Previous studies emphasize the role of competition in promoting temporal dissociation (Clench 1967, Owen 1971, Owen *et al.* 1972), but in none of these studies can other explanations be ruled out. In general, species abundance and phenology seems likely to be influenced by a complex set of biotic interactions, including the phenology of resource availability, levels of predation and parasitoid attack, and within this set of potential factors interspecific competition might sometimes be the most significant factor.

Beta diversity, or turnover in space, is probably a less important component of total diversity in butterflies than in poorly dispersing groups such as Coleoptera (Mawdsley, this volume) or Microlepidoptera (Robinson & Tuck, this volume). However Pinheiro & Ortiz (1992) report high beta diversity in fruit feeding butterfly species within the understorey in gallery forest in central Brazil; at Kuala Belalong, greater turnover than was evident may be revealed by a more rigorous sampling technique. However, assessing microdistributional patterns is fraught with difficulties; for example many mudpuddling species are essentially canopy dwellers and only young males come to drink salt. In many species the microdistribution of males and females is apparently quite different. Many species are too rare to provide reliable data. Moreover within the rainforest there is always a characteristic vertical stratification (DeVries 1988, Papageorgis 1975) which inevitably introduces bias into any sampling proceedure. At Kuala Belalong the localization of so many species to one site only, suggests that the forest was inadequately sampled.

The most important single finding to emerge from this study is that the lowland forest in the Batu Apoi Forest Reserve supports an exceptionally rich diversity of butterflies, with in all probability nearly half the entire Bornean fauna occurring within a single square kilometre. Presumably many additional species would have been recorded had more sampling been conducted on Bukit Belalong. Therefore the area certainly deserves special conservation efforts on the part of the global community.

ACKNOWLEDGEMENTS

We would like to thank the organizers of the 1991/92 UBD/RGS Brunei Rainforest expedition, especially Catriona Prebble and Amanda Simpson, and the staff of the KBFSC, including Samhan bin Nyawa and Kamariah Abu Salim for their practical assistance en route to and while at the KBFSC. AGO was supported by a Universiti Brunei Darusalam research grant. CLH was supported by a grant from the Deutsche Forschungsgemeinschaft.

LITERATURE CITED

ARMS, K., FEENY, P. & LEDERHOUSE, R.C. 1974. Sodium: stimulus for puddling behaviour by tiger swallowtail butterfly butterflies, *Papilio glaucus*. *Science, N.Y.* 185:372-374.

ASHTON, P.S. 1964. *Ecological Studies in the Mixed Dipterocarp Forests of Brunei State*. Oxford Forestry Memoirs 25.

BARLOW, H.S., BANKS, H.J. & HOLLOWAY, J.D. 1971. A Collection of Rhopalocera (Lepidoptera) from Mt. Kinabalu, Sabah, Malaysia. *Oriental Insects* 5:269-296.

CASSIDY, A.C. 1982. An annotated checklist of Brunei butterflies, including a new species of the genus *Catapaecilma* (Lycaenidae). *The Brunei Museum Journal* 5:202-272.

CASSIDY, A.C. 1985. An enlarged checklist of Brunei butterflies (Lepidoptera: Rhopalocera) including descriptions of one new species and two new subspecies. *Brunei Museum Journal* 6:135-168.

CLENCH, H.K. 1967. Temporal dissociation and population regulation in certain hesperiine butterflies. *Ecology* 48:1000-1006.

CLENCH, H.K. 1979. How to make regional lists of butterflies: some thoughts. *Journal of the Lepidopterists' Society* 33:215-231.

CORBET, A.S. & PENDELBURY, H.M. 1992. *The butterflies of the Malay Peninsula*. Fourth edition revised by J.N. Eliot. Malayan Nature Society, Kuala Lumpur. xiv + 577 pp.

DEVRIES, P.J. 1988. Stratification of fruit feeding nymphalid butterflies in a Costa Rican rainforest. *Journal of Research on Lepidoptera* 26:98-108.

EMMEL, T.F. & LECK, C.F. 1970. Seasonal changes in organization of tropical rainforest butterfly populations in Panama. *Journal of Research on Lepidoptera* 8:133-152.

FOX, R.M., LINDSEY, A.W. Jr., CLENCH, H.K. & MILLER, L.D. 1965. The butterflies of Liberia. *Memoirs of the American Entomological Society* 19:1-438.

HILL, C.J. 1988. The temporal distribution of butterfly species richness in a subtropical region of Australia. *Proceedings of the Ecological Society of Australia* 15:115-126.

HILL, C.J. 1992. Temporal changes in abundance of two lycaenid butterflies (Lycaenidae) in relation to adult food resources. *Journal of the Lepidopterists' Society* 46:173 -181.

OTSUKA, K. 1988. *Butterflies of Borneo*, Volume 1. Tobishima Corp., Tokyo.

OWEN, D.F. 1971. *Tropical Butterflies*. Clarendon Press, Oxford. 214 pp.

OWEN, D.F., OWEN, J. & CHANTER, D.O. 1972. Seasonal changes in the relative abundance and estimates of species diversity in a family of tropical butterflies. *Oikos* 23:200-205.

PAPAGEORGIS, C. 1975. Mimicry in neotropical butterflies. *American Scientist* 63:522-532.

PINHEIRO, C.E.G. & ORTIZ, J.V.C. 1992. Communities of fruit-feeding butterflies along a vegetation gradient in central Brazil. *Journal of Biogeography* 19:505-511.

POLLARD, E. 1977. A method for assessing changes in abundance of butterflies. *Biological Conservation* 12:115-134.

RAGUSO, R. A. & LLORENTE-BOUSQUETS, J. 1990. The butterflies of the Tuxlas Mountains, Veracruz, Mexico, Revisited: Species-Richness and Habitat Disturbance. *Journal of Research on Lepidoptera* 29:105-133.

ROBINSON, G.S. & TUCK, K.R. 1993. Diversity and faunistics of small moths (Microlepidoptera)in Bornean rainforest. *Ecological Entomology* 18:385-393.

SEKI, Y., TAKANAMI, Y. & OTSUKA, K. 1991a. *Butterflies of Borneo*, Volume 2, No. 1: Lycaenidae. (In Japanese and English). Tobishima Corp. 252 pp., 72 pl.

SEKI, Y., TAKANAMI, Y. & OTSUKA, K. 1991b. *Butterflies of Borneo*, Volume 2, No. 2: Hesperiidae. (In Japanese and English). Tobishima Corp. 172 pp., 48 pl.

SHAPIRO, A.M. 1975. The temporal component of butterfly species diversity. Pp. 181-195 in Cody, M.L. & Diamond, J.M. (eds). *Ecology and evolution of communities*. Harvard University Press, Cambridge.

SHIELDS, O. 1967. Hilltopping. *Journal of Research on Lepidoptera* 6:69-178.

SLANSKY, F. 1974. Relationship of larval food-plants and voltinism patterns in temperate butterflies. *Psyche* 81:243-253.

TAN, M.W., KIRTON, L.G. & KIRTON, C.G. 1990. Composition and distribution of butterflies in Rompin-Endau, especially along Sungei Kinchin and its vicinity. Pp. 193-212 in *Proceedings of the International Conference on Tropical Biodiversity 'In Harmony with Nature', 12-16 June 1990, Kuala Lumpur, Malaysia*. Malaysian Nature Society, Kuala Lumpur.

Appendix. Checklist of the butterflies of Belalong. Abundance classes (ABUND.): 1 - common (+ moderately common, 10-50; ++ very common >50), 2 - sporadic (>10 records limited to one or two recording periods), 3 - rare (<10, frequently one record only). Distribution classes (DIST.): 1 - general (within 1.5 km radius of KBFSC); 2 - localised (+ localised, ++ very localised); 3 - Bkt Belalong (submontane). Habitat preference (H): rv - open areas or riverine, u - understorey, rg - ridge.

SPECIES	ABUND.			DIST.			H.
	1	2	3	1	2	3	
Papilionidae							
Trogonoptera brookiana	++			+			
Troides miranda	+			+			
Troides amphrysus			+	?			
Pachliopta aristolochiae		+		+			rv
Atrophaneura nox			+	+			
Atrophaneura neptunus			+				
Papilio karna			+	+			
Papilio demolion	++			+			
Papilio nephelus	+			+			
Papilio helenus			+	?			
Papilio fuscus	++			+			
Papilio iswara			+	+			
Papilio memnon		+		?			rv
Chilasa slateri			+			+	
Chilasa paradoxa			+	?			
Meandrusa payeni			+	?			
Graphium sarpedon	++			+			
Graphium evemon	++			+			
Graphium eurypylus	++			?			
Graphium bathycles	++			?			
Graphium agamemnon	++			+			
Paranticops ramaceous			+	+			
Paranticops delessertii		+		+			
Pathysa antiphates	++			+			
Lamproptera curius			+	?			
Lamproptera meges			+	?			
Pieridae							
Leptosia nina	++			+			rv
Delias ninus			+	+			rg
Delias eumolpe	?					+	
Delias hyparete	++			+			rg
Delias heningia			+	+		+	rg
Delias singhapura			+	?			
Prioneris philonome	+			+			
Prioneris cornelia	+			+			
Cepora pactolicus	++			+			
Cepora iudith	++			+			
Appias lyncida	++			+			
Appias paulina	+			+			
Appias indra	++			+			rg
Appias nero	+			+			
Appias cardena	++			+			
Saletara liberia	++			+			
Ixias pyrene	+			+			
Hebomoia glaucippe		+		+			
Pareronia valeria		+		+			rg
Dercas verhueli		+		+			
Catopsilia pomona			+	?			
Gandaca harina	++			+			
Eurema hecabe			+	?			
Eurema ada	++			+			
Eurema lacteola	+					+	
Eurema blanda	++			+			
Eurema simulatrix	++			+			
Eurema andersonii	++			+			
Eurema sari		+		+			

SPECIES	ABUND.			DIST.			H.
	1	2	3	1	2	3	
Eurema alitha			+	?			
Eurema tilaha	++			+			
Eurema tominia			+			+	
Nymphalidae - Danainae							
Tiramula septentrionis			+	+		+	
Parantica aspasia	+			+			rv
Parantica agleoides	+			+			rv
Ideopsis gaura	++			+		+	
Ideopsis vulgaris	+			+		+	rv
Idea lynceus			+	+			
Idea stolli	++			+			
Euploea modesta			+	?			rv
Euploea camaralzeman	+			+			rv
Euploea sylvester			+	+			rv
Euploea mulciber	++			+		+	rv
Euploea midamus			+	?			
Euploea phaenarete			+	?			
Euploea radamanthus	++			+			
Euploea algea		+		+			rv
Euploea eyndhovii			+	?		+	
Nymphalidae - Satyrinae							
Melanitis leda			+	?			u
Coelites epimintha			+	+			u
Coelites euptychoides			+	+			u
Neorina lowii		+		+			rv
Erites elegans	++				+		u
Erites argentina		+			+		u
Ragadia makuta	++			+			u
Mycalesis maianeus	+			+			
Mycalesis orseis		+			+		rv/u
Mycalesis fuscum		+		+			
Mycalesis horsfieldi			+	?			u
Mycalesis anapita	++			+			
Mycalesis kina	+				+		rg/u
Mycalesis amoena			+	?			
Ypthima fasciata			+		?		rv
Ypthima hanburyi		+			+		rv
Elymnias hypermnestra			+		?		
Elymnias nesaea		+			+		rg
Elymnias kuenstleri			+		?		
Elymnias penanga			+		?		
Nymphalidae - Amathusinae							
Faunis kirata		+		+			u
Faunis gracilis		+		+			u
Faunis stomphax	++			+			u
Xanthotaenia busiris	++			+			u
Amathusia phidippus			+	+			u
Amathusia masina			+	?			
Amathuxidia amythaon		+			+		u
Zeuxidia amythystus		+			+		u
Zeuxidia doubledayi			+		+		u
Zeuxidia aurelia			+		?		u
Thaumantis klugius			+		+		u
Thaumantis odana			+		+		u
Thauria aliris	++			+			u
Discophora sondaica			+	?			
Discophora necho			+	?			

Appendix. Checklist of the butterflies of Belalong. Abundance classes (ABUND.): 1 - common (+ moderately common, 10-50; ++ very common >50), 2 - sporadic (>10 records limited to one or two recording periods), 3 - rare (<10, frequently one record only). Distribution classes (DIST.): 1 - general (within 1.5 km radius of KBFSC); 2 - localised (+ localised, ++ very localised); 3 - Bkt Belalong (submontane). Habitat preference (H): rv - open areas or riverine, u - understorey, rg - ridge.

SPECIES	ABUND.			DIST.			H.
	1	2	3	1	2	3	
Nymphalidae - Nymphalinae							
Cupha erymanthis	++			+		+	rv
Phalanta alcipe	+			+		+	
Vagrans egista		+		+		+	
Vindula erota	++			+			rv
Paduca fasciata			+	?			
Cirrochroa tyche	+			+		+	
Cirrochroa emalea	++			+			
Cirrochroa malaya	+			+		+	
Cirrochroa satellita	+			+			
Cirrochroa orissa			+	?			
Terinos terpander	++			+			u
Terinos clarissa			+	?			
Cethosia hypsea			+	+			rv
Rhinopalpa polynice			+	?			
Doleschallia bisaltide	++			+		+	rv
Dichorragia nesimachus		+		+			
Chersonesia rahria	+			+			rv
Chersonesia peraka		+		+		+	
Chersonesia risa			+	?			
Cyrestis nivea		+		+			
Cyrestis theresae			+	?			
Pantoporia hordonia			+	?			
Pantoporia dindinga			+	?			
Pantoporia paraka			+			+	
Pantoporia aurelia	++			+			
Lasippa tiga		+		+			
Lasippa viraja			+	?			
Neptis hylas			?			+	
Neptis duryodana	+			+			
Neptis nata			+	?			
Neptis leucoporos	++			+		+	rv
Neptis miah			?			+	
Neptis clinia			?			+	
Neptis harita			+	?			
Neptis omeroda		+		+			rv
Neptis anjana			?			+	
Athyma pravara			+	?			
Athyma asura			+	?			
Athyma kanwa	+			+		+	
Athyma reta			+	?			
Athyma adunora			?			+	
Athyma selenophora			?			+	
Athyma nefte			+	?		+	
Sumalia daraxa			+				
Moduza procris	+			+			
Pandita sinope		+		+			
Lebadea martha		+					
Parthenos sylvia	+			+			rv
Tanaecia munda			+	?			
Tanaecia aruna			+	?			
Tanaecia orphne	++?			+			u
Tanaecia pelea	++?			+			u
Tanaecia godartii			+	?			u
Tanaecia iapis	++			+			u
Dophla evelina	+			+			u
Bassarona dunya	++			+			u
Bassarona teuta	+			+			*u*
Euthalia adonia			+			+	
Euthalia whiteheadi			+			+	
Euthalia aconthea			+	?		+	
Euthalia monina	+			+			
Euthalia kanda			+	?			
Lexias dirtea	++			+			u
Lexias pardalis	+			+			u
Lexias canescens		+		+			
Eulaceura osteria			+	?			
Herona sumartrana			+	?			
Agatasa calydonia	+				+		rg
Prothoe franck	+				+		rg
Polyura schreiber			+	?			
Polyura athama	+			+			rv
Polyura moori			+	?			
Polyura delphis			+	+			
Charaxes durnfordi	+				+		rg
Charaxes distanti			+	?			
Charaxes borneensis			+	?		+	
Charaxes bernardus	+			+			
Lycaenidae - Riodininae							
Zemeros emesoides	+				+		rv
Dodona elvira			+			+	
Abisara geza	+				+		rg
Abisara kausambi			+		?		
Paralaxita telesia	+				++	+	u
Paralaxita orphna			+		++		u
Laxita teneta	+				++		u
Lycaenidae - Lycaeninae							
Poritia erycinoides	+				++		rg
Poritia sumatrae			+		?		
Poritia plateni			+		?		
Simiskina pharyge			+		++		
Simiskina phalena			+		++		
Deramas nelvis			+		++		
Deramas yasoda			+		++		
Miletus gopara	++				+		rg
Miletus cellarius	+				+		rg
Allotinus subviolaceous	+			+			
Allotinus borneensis	+			+			
Allotinus horsefieldi			+	?			
Allotinus sarrastes			+	?			
Allotinus nicholsi			+	?			
Allotinus substrigatus			+	?			
Logonia regina	+				+		rg
Logonia malayica	++				+		rg
Logonia marmorata			+	?			
Discolampa ethion	++			+			
Caleta elna	++			+			
Neopithecops zalmora	+			+			u
Acytolepis puspa	++			+			
Acytolepis ripte	+			+			
Plautella cossaea			+			+	
Celastrina lavendularis			+			+	
Callenya lenya			+	?			
Cebrella pellecebra			+	?			

Appendix. Checklist of the butterflies of Belalong. Abundance classes (ABUND.): 1 - common (+ moderately common, 10-50; ++ very common >50), 2 - sporadic (>10 records limited to one or two recording periods), 3 - rare (<10, frequently one record only). Distribution classes (DIST.): 1 - general (within 1.5 km radius of KBFSC); 2 - localised (+ localised, ++ very localised); 3 - Bkt Belalong (submontane). Habitat preference (H): rv - open areas or riverine, u - understorey, rg - ridge.

SPECIES	ABUND.			DIST.			H.
	1	2	3	1	2	3	
Jamides virgulatus			+	?			
Jamides bochus			+	?			
Jamides caeruleus		+		+			
Jamides pura	++			+			
Jamides aratus	+			+			
Jamides elpis	+			+			
Jamides lugine	+			+			
Jamides zebra			+	?			
Jamides celeno			+	?			
Nacaduba subperusia	+				?		rg
Nacaduba hermus	+				?		rg
Nacaduba calauria			+	?			
Nacaduba beroe	+			?			
Nacaduba kurava	+			?			
Prosotas aluta			+	?			
Prosotas nora	++			?			
Prosotas pia	+			?			
Anthene emolus	++			+			
Anthene lycaenina	++			+			
Arhp. pseudocentaurus			+	?			
Arhopala kinabala			+			+	
Arhopala hypomuta			+	?			
Arhopala aedias			+	?			
Arhopala epimuta	+				+		rg
Arhopala lurida (?)			+	?			
Arhopala delta (?)			+	?			
Arhopala tropaea			+	?			
Arhopala moorei	+			+			
Arhopala antimuta			+	?			
Arhopala inornata (?)			+	?			
Arhopala democritus	+				+		rg
Arhopala denta (?)			+	?			
Arhopala elopura (?)			+	?			
Arhopala alitaeus (?)			+	?			
Arhopala zambra	+				+		rg
Arhopala borneensis			+		+		rg
Arhopala vihara			+	?			
Arhopala arvina (?)			+	?			
Arhopala labuana			+	?			
Flos fulgida			+			+	
Flos morphina	+				++		rg
Iraota distanti			+		?		
Iraota rochana			+		?		
Catapaecilma elegans			+		?		
Loxura cassiopeia			+		?		
Exooylides tharis		+		+			
Horaga syrinx		+			++		rg
Cheritra freja		+		+			
Ritra aurea			+		?		
Ticherra acte			+		?		
Drupadia ravindra			+	?			
Drupadia rufotaenia			+	?			
Drupadia theda	++			+			
Drupadia niasica			+	?			
Drupadia cinesia			+	?			
Drupadia cineus	+				+		rg
Dacalana vidura		+			+		

SPECIES	ABUND.			DIST.			H.
	1	2	3	1	2	3	
Dacalana sinhara			+		+		
Dacalana lowii			+		+		
Thrix scopula	+				++		rg
Mantoides gama			+			+	
Suasa lisides			+		?		
Remelana jangala	+			+		+	
Hypolycaena amasa	+			+			
Hypolycaena thecloides			+	?			
Hypolycaena amabilis			+		+		
Hypolycaena erylus	+			+			
Deudorix epijarbas			+		?		
Deudorix staudingeri			+		?		
Rapala pheretima	+				++		rg
Rapala varuna			+		?		
Rapala domitia			+		?		
Virachola subguttata		+		+		+	
Curetis santana			+	?			
Curetis felderi			+	?			
Curetis sperthis			+	?			
Curetis tagalica	+			+			
Hesperiidae							
Bibasis sena			+	?			
Hasora proxissima			+	?			
Hasora schoenherr	+				+		rv
Hasora chromus			+	?			
Hasora quadripunctata			+	?			
Hasora vitta	+				+		
Choaspes plateni			+	?			
Capila phaneus			+	?			
Charmion ficulnea	+			+			
Charmion species			+	?			
Celaenorrhinus bazilanus			+	?			
Tapena thwaitesi			+	?			
Darpa striata	+			+			rg
Mooreana trichoneura			+	?			
Tagiades japetus	+			+			
Tagiades gana			+	+			
Tagiades lavatus			+	?			
Tagiades parra			+	?			
Tagiades toba	+			+			
Tagiades waterstradti			+	?			
Odontoptilum pygela			+	?			
Arnetta verones			+	?			
Halpe sikkima			+	?			
Iambrix stellifer/salsala			+	?			
Koruthaialos rubicula	+			+			
Koruthaialos sindu			+	?			
Psolos fuligo			+	?			
Ancistroides nigrita			+	?			
Ancistroides armatus			+	?			
Notocrypta paralysos	+			+			
Notocrypta clavata			+	?			
Cupitha purreea			+	?			
Zographetus kutu			+	?			
Oerane microthyrus			+	?		+	
Hyarotus adrastus			+	?			
Isma kuala			+	?			

Appendix. Checklist of the butterflies of Belalong. Abundance classes (ABUND.): 1 - common (+ moderately common, 10-50; ++ very common >50), 2 - sporadic (>10 records limited to one or two recording periods), 3 - rare (<10, frequently one record only). Distribution classes (DIST.): 1 - general (within 1.5 km radius of KBFSC); 2 - localised (+ localised, ++ very localised); 3 - Bkt Belalong (submontane). Habitat preference (H): rv - open areas or riverine, u - understorey, rg - ridge.

SPECIES	ABUND.			DIST.			H.	SPECIES	ABUND.			DIST.			H.
	1	2	3	1	2	3			1	2	3	1	2	3	
Pyroneura flavia			+		?		rg	*Gangara lebadea*		+		?			
Pyroneura latoia			+		?			*Erinota thrax*		+		?			
Pyroneura niasana			+		?			*Erinota sybirita*			+	?			
Pyroneura aurantiaca		+			+		rg	*Hidari dosoena*			+	?			
Lotongus calathus			+	?				*Eetion elia*			+	?			
Zela excellans			+	?				*Acerbus anthea*			+				
Gangara thyrsis			+	?				*Pirdana hyela*			+	?			
Gangara sanguinocculus		+		?											

A quantitative inventory of trees in one hectare of mixed dipterocarp forest in Temburong, Brunei Darussalam

AXEL DALBERG POULSEN[1], IVAN C. NIELSEN[1], SYLVESTER TAN[2] *and* HENRIK BALSLEV[1]

[1]*Department of Systematic Botany, Institute for Biological Sciences, Aarhus University, Nordlandsvej 68, DK-8240 Risskov, Denmark.*
[2]*Forest Department Headquarters, Kuching, Sarawak, Malaysia.*

ABSTRACT. One hectare of mixed dipterocarp forest in Temburong, Brunei, was studied in a square one hectare plot. A 5 x 5 metres grid (horizontal projection) was laid out and marked with stakes and strings. All trees ≥ 10 cm dbh were marked with numbered aluminium plates and, for each tree, its coordinates in the plot were marked, its dbh and height were measured, and a voucher was collected for botanical identification. For each species the total and relative densities, dominance, and frequencies, and Importance Value Index (IVI) were calculated. For each family the total and relative values for density, diversity, and dominance, and Family Importance Value (FIV) were calculated. The plot contained 550 individual trees with dbh ≥ 10 cm and these were identified as belonging to 231 species divided among 43 families. The total basal area of trees with dbh ≥ 10 cm dbh was 40.79 square metres. The average height of the trees was 22.5 m, the tallest was 62 m, and the closed canopy was found at 30-40 metres. Dipterocarpaceae ranked highest in FIV with a score of 91.0 (out of 300 possible) followed by Euphorbiaceae with an FIV score of 34.5. The remaining 41 families all had FIV scores of less than 15, the lowest being Dilleniaceae with FIV score of 0.6. *Shorea parvifolia* (Dipterocarpaceae), *Elateriospermum tapos* (Euphorbiaceae), *Shorea laevis* (Dipterocarpaceae) and *Gluta laxiflora* (Anacardiaceae) were the most important species with IVI values between 18.2 and 12.9 (of 300 possible).

KEY WORDS: diversity, density, dominance, frequency, importance value, Borneo.

INTRODUCTION

The aim of the present study was to establish a permanent plot that could serve as an area of reference for future studies in the floristic diversity in both Brunei Darussalam and adjacent areas of NW Borneo. An equally important purpose was to create a framework for studies in other biological disciplines. The mixed dipterocarp forest at Belalong was subject to the classical study by Ashton (1964), where the study area chosen was "an ill-defined area at the meeting place of the Temburong and Belalong rivers". Ashton included trees above 12 inches girth (approximately 9.7 cm dbh). The present plot was placed near acre plots no. 31–34 of Ashton (op. cit. Figure 1). In the nearby Gunung Mulu National Park (Sarawak) Proctor *et al.* (1983) surveyed a one hectare plot in a dipterocarp forest site at 200 to 250 m altitude on "heavy-textured red-yellow podzolics with a surface layer (up to about 15 cm thick) of reddish-brown fibrous organic matter" (op. cit., p. 242). In that study, all trees above 10 cm dbh were enumerated.

THE STUDY AREA

The study was conducted in Batu Apoi Forest Reserve, Temburong District, Brunei (115°10' E; 4°32' N) in Northern Borneo. The soil is derived from the Setap Shale Formation. Precipitation is above 4000 mm per year with less than 200 mm in January–February and

D. S. Edwards et al. (eds.), Tropical Rainforest Research – Current Issues, 139–150.

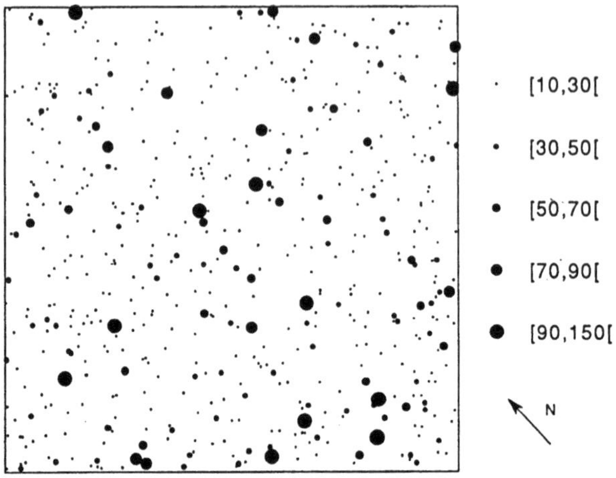

Figure 1. Map of the 550 trees ≥ 10 cm dbh on a 1-ha plot (100 x 100 m). The dot size represent trees of different dbh classes (cm).

July–August. A one hectare plot was laid out on the ridge west of Kuala Belalong Field Studies Centre on a steep slope in primary, mixed dipterocarp forest at about 250 m elevation. The difference between highest and lowest point of the sample plot was about 60 m.

MATERIAL AND METHODS

A grid of 400 squares, each one measuring 5 x 5 metres in horizontal projection was laid out, using measuring tape and compass. The corners of each grid were marked with 40 cm aluminium tubes which were then connected with nylon string. The total study plot of one hectare was then surveyed for trees with dbh ≥ 10 cm which were marked consecutively with numbered aluminium tags. Dbh was measured using a tree caliper. For stems with irregular circumference (i.e., not circular) the average of the largest and smallest diameter was used. For buttressed trees the diameter was measured just above the buttresses (if they reached over breast height). Tree height was measured using a clinometer, based on the angles to the top and base of the tree and the distance from observer to tree. Voucher specimens of all individual trees were collected whenever possible. Duplicates are deposited at BRUN and AAU (acronyms according to Holmgren *et al.* 1990) where identifications were carried out. Unvouchered trees were identified in the field by comparison with vouchered individuals.

For calculations the following definitions were used:

Density: Number of individuals per hectare.

Basal Area (Dominance): The area occupied by a cross-section of the stem at breast height.

Frequency: The number of sample units (5 x 5 m) in which a species is found.

Diversity: The number of species of a family encountered in the sample.

Relative density, dominance, frequency, and diversity are calculated as the percentages of the sum of values of all species or families.

Importance Value Index (IVI) for a species is the sum of its relative density, dominance and frequency (Curtis & Macintosh 1951).

Family Importance Value (FIV) for a family is the sum of its relative density, diversity and dominance (Mori *et al.* 1983).

RESULTS AND DISCUSSION

Density

The total density in the sample plot was 550 trees. This was unequally divided among the 231 species encountered (Appendix). The highest densities found for a species were those of *Elateriospermum tapos* (Euphorbiaceae) with 35 individuals per hectare, *Gluta laxiflora* (Anacardiaceae) with 26 individuals, and *Vatica odorata* (Dipterocarpaceae) with 21 individuals per hectare. Three species had densities between 10 and 20 individuals/hectare (*Shorea parvifolia, Moulthonianthus leembrugianus, Hydnocarpus sumatrana*). The remaining 225 species all had densities below 10 individuals per hectare, 143 species (62%) being represented by only a single individual in the plot.

Densities for families were also unequally divided among the 43 families represented (Table 1). Dipterocarpaceae were represented by 133 individuals and Euphorbiaceae by 93 individuals. Anacardiaceae, Ebenaceae, Flacourtiaceae, and Myristicaceae all had between 25 and 34 individuals, whereas the remaining families all had less than 25 individuals. Eight families (Dilleniaceae, Ixonanthaceae, Linaceae, Loganiaceae, Myrsinaceae, Olacaceae, Sabiaceae, and Thymelaeaceae) were very rare, each being represented by a single individual in the study plot. The distribution pattern of trees reveals no obvious clustering.

Dominance

The total basal area of trees ≥ 10 cm dbh was 40.79 square metres. The most dominant species were *Shorea parvifolia* and *Shorea laevis*, representing 12.3% and 14.7%, respectively, of the basal area. Eighteen species had more than 1% relative dominance; 13 of them belong to Dipterocarpaceae (see Appendix). The distributions of trees in specific size classes appear to be random (Figure 1.)

At the family level Dipterocarpaceae account for over half of the basal area at the study site, followed by Euphorbiaceae with 8.5% of the basal area. The remaining 41 families each have less than 5% of the basal area, 30 of them less than 1% (Table 1).

Frequency

Frequency conveys information about how evenly a group of individuals is distributed over the study plot. It therefore only makes sense to discuss this parameter for taxa with several individuals in the sample. If a taxon has many of its individuals located in a few sample units, i.e. it has a clumped distribution, then its relative frequency would be markedly lower than its relative density. Six species (*Gluta laxiflora, Shorea parviflora, Vatica odorata, Elateriospermum tapos, Moultonianthus leembruggianus, Hydnocarpus sumatrana*) all have over 10 individuals in the sample. For each them there is no dramatic difference between their relative density and their relative frequency (see Appendix), indicating that none of them have a clumped distribution at the study site.

Diversity

The most diverse family is Dipterocarpaceae which is represented by 28 species, Euphorbiaceae coming a close second with 21 species. Myristicaceae (17 sp.), Burseraceae (11 sp.), Ebenaceae (11 sp.), and Myrtaceae (11 sp.) are also species rich (Table 1). The total number of species with dbh ≥ 10 cm in the plot of 231 is comparable to the estimate of about 215 using the species-area curve of a dipterocarp forest from Mulu (Proctor *et al.* 1983). Ashton's (1964) species-area curve for the Belalong area gives a figure of only 90 species per hectare for trees exceeding 9.7 cm diameter even though this number is obtained lumping rectangular 1-acre subplots over a larger area. Although Ashton's plots were more strictly confined to a single habitat (ridge or hillside) thereby minimising beta-diversity, the extreme difference between his and our results can only be interpreted as an underestimation by him (the more likely) or overestimation by us (less likely to be the major cause) of the number of rare species, often based on inadequate fragmentary samples (Ashton, personal communication).

Table 1. Families of trees represented at Belalong in the plot west of Kuala Belalong Field Studies Centre showing the number of species, individuals, basal area, relative diversities (rel. div.), densities (rel. dens.), dominance (rel. dom.), and Family Importance Value (FIV).

Family	sp.	indiv.	basal area	rel. div.	rel. dens.	rel. dom.	FIV
Alangiaceae	1	3	0.05	0.43	0.55	0.12	1.10
Anacardiaceae	6	34	1.90	2.60	6.18	4.65	13.43
Annonaceae	8	14	0.26	3.46	2.55	0.65	6.66
Apocynaceae	4	7	1.77	1.73	1.27	4.33	7.34
Arecaceae	1	2	0.05	0.43	0.36	0.12	0.91
Bombacaceae	3	4	0.08	1.30	0.73	0.18	2.21
Burseraceae	11	13	0.52	4.76	2.36	1.27	8.40
Celastraceae	1	3	0.08	0.43	0.55	0.19	1.17
Chrysobalanaceae	5	5	0.10	2.16	0.91	0.25	3.32
Clusiaceae	9	13	0.29	3.90	2.36	0.70	6.96
Crypteroniaceae	1	2	0.02	0.43	0.36	0.05	0.84
Dilleniaceae	1	1	0.01	0.43	0.18	0.03	0.64
Dipterocarpaceae	28	133	22.31	12.12	24.18	54.70	91.00
Ebenaceae	11	25	0.92	4.76	4.55	2.25	11.56
Euphorbiaceae	21	93	3.48	9.09	16.91	8.54	34.54
Fabaceae	3	3	0.21	1.30	0.55	0.52	2.37
Fagaceae	6	7	0.80	2.60	1.27	1.97	5.84
Flacourtiaceae	9	37	1.54	3.90	6.73	3.77	14.39
Icacinaceae	2	6	0.13	0.87	1.09	0.31	2.27
Ixonanthaceae	1	1	0.32	0.43	0.18	0.79	1.40
Lauraceae	7	7	0.44	3.03	1.27	1.09	5.39
Lecythidaceae	2	2	0.05	0.87	0.36	0.13	1.36
Linaceae	1	1	0.09	0.43	0.18	0.23	0.84
Loganiaceae	1	1	0.06	0.43	0.18	0.14	0.76
Melastomataceae	5	6	0.29	2.16	1.09	0.72	3.97
Meliaceae	9	9	0.26	3.90	1.64	0.63	6.16
Moraceae	7	9	0.63	3.03	1.64	1.54	6.21
Myristicaceae	17	26	1.04	7.36	4.73	2.55	14.64
Myrsinaceae	1	1	0.03	0.43	0.18	0.08	0.69
Myrtaceae	11	16	0.67	4.76	2.91	1.63	9.30
Olacaceae	1	1	0.06	0.43	0.18	0.14	0.75
Polygalaceae	6	18	0.63	2.60	3.27	1.54	7.41
Proteaceae	1	2	0.06	0.43	0.36	0.15	0.95
Rosaceae	3	3	0.29	1.30	0.55	0.72	2.56
Rubiaceae	8	10	0.24	3.46	1.82	0.59	5.88
Sabiaceae	1	1	0.01	0.43	0.18	0.03	0.64
Sapindaceae	1	2	0.05	0.43	0.36	0.13	0.92
Sapotaceae	7	10	0.33	3.03	1.82	0.82	5.67
Sterculiaceae	5	10	0.59	2.16	1.82	1.45	5.43
Thymelaeaceae	1	1	0.02	0.43	0.18	0.04	0.66
Tiliaceae	2	2	0.04	0.87	0.36	0.09	1.32
Ulmaceae	1	2	0.02	0.43	0.36	0.05	0.85
Verbenaceae	1	4	0.05	0.43	0.73	0.13	1.29

Importance Value Index (IVI)

The Importance Value Index (IVI) lumps three ecological parameters (density, dominance, frequency) that measure quite different features of a forest. In itself it therefore hardly has an ecological meaning, and ranking species by it must be taken only as a way of seeing which species come out as important elements of the forest even if they are looked upon from quite different viewpoints in ecological terms. Dipterocarpaceae is conspicuous in the IVI ranking, accounting for five (*Shorea parvifolia, S. laevis, S. pauciflora, Vatica odorata, V. endertii*) of the top 10 species, but Euphorbiaceae is also conspicuous with three (*Elateriospermum tapos, Hydnocarpus sumatrana, Moultonianthus leembruggianus*) species among the top 10, and Anacardiaceae (*Gluta laxiflora*) and Apocynaceae (*Dyera costulata*) are also represented (see Appendix).

Family Importance Value (FIV)

The Family Importance Value (FIV) is another index of questionable ecological meaning, mixing the different parameters of diversity, density and dominance. Nevertheless it gives some indication of importance regardless of one particular parameter. Dipterocarpaceae is by far the most important family accounting for 91.0 (of 300 possible) points in FIV (Table 1). If individuals less than 30 cm dbh are omitted 59 species remain, 22 of these belonging to the Dipterocarpaceae; this result emphasizes the importance of the family among the canopy trees. Euphorbiaceae again is important with a FIV of 34.5. A group of five families (Myristicaceae, Flacourtiaceae, Anacardiaceae, Myrtaceae, Ebeneaceae) have FIV's between 10 and 15, and the remaining 36 families all have a FIV less than 10. Ashton (1964) also found Dipterocarpaceae followed by Euphorbiaceae to be the two most important families in his plots in the same area. In the nearby Gunung Mulu National Park (Sarawak) Proctor *et al.* (1983) found that the five most important families, based on percent basal area and actual numbers of individuals, were: Dipterocarpaceae (43.2%/114), Leguminosae (5.1%/15), Sapotaceae (4.8%/52), Myrtaceae (4.05%/65), and Euphorbiaceae (2.8%/70) out of a total for all families of 57 m^2 and 778 individuals at or above 10 cm DBH.

Height

In the sample 87 (16%) trees were over 30 m high, 25 (5%) were over 40 metres, and five (1%) were over 50 metres. The tallest tree, a *Shorea laevis*, measured close to 62 metres and had a dbh of 110 cm. This species was also the tallest emergent in Ashton's profile diagrams (1964) from the same area. Of the 87 trees $\geq$ 30 m dbh, 47 were Dipterocarpaceae. The closed canopy of the forest was at 30-40 metres above the ground and appears irregular (Figure 2).

CONCLUSIONS

With 231 species, the plot is among the most species-rich recorded in Northern Borneo for trees $\geq$ 10 cm dbh in a one hectare plot of mixed dipterocarp forest. This high number may partially be attributed to the inclusion of both ridge top and hill side habitats. The number of species may also be slightly overestimated because a rather narrow concept of morpho-species was applied compared with, for example, that used by Ashton (1964). The number of species recorded increases markedly when 10 cm dbh is used as a lower limit, rather than 30 cm.

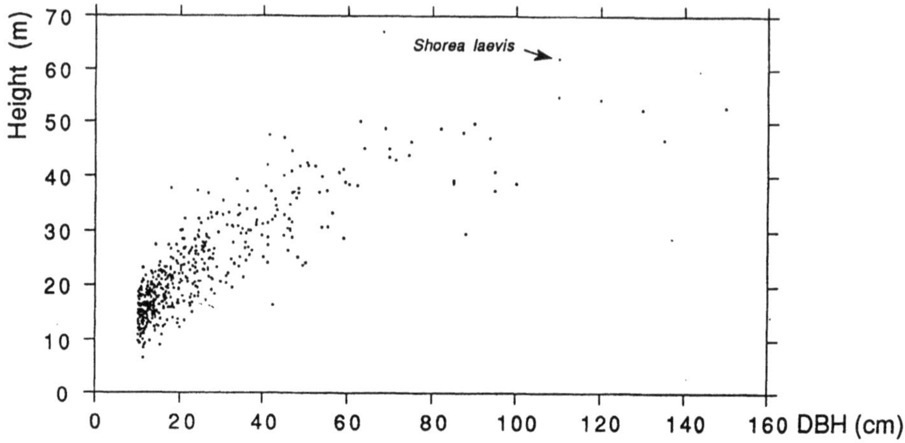

Figure 2. Relation between dbh and height of trees ≥ 10 cm dbh. Note the emergent *Shorea laevis*.

ACKNOWLEDGEMENTS

We are indebted to P. S. Ashton, M. M. J. van Balgooy, C. Kalkman, and W. J. J. O. de Wilde for help with identification. The study was financed by the Danish Natural Science Research Council (Grant no. 11-9018).

LITERATURE CITED

ASHTON, P. S. 1964. Ecological studies in the mixed Dipterocarp forests of Brunei State. *Oxford Forestry Memoirs* 25:1–75, 17 tables, 70 figures, 2 appendices.

CURTIS, J. T. & MACINTOSH, R.P. 1951. An upland forest continuum in the praire-forest border region of Wisconsin. *Ecology* 32:476-496.

HOLMGREN, P. K., HOLMGREN, N.H. & BARNETT, L.C. 1990. *Index Herbariorum. Part I: The Herbaria of the World*. New York Botanical Garden, Bronx.

MORI, S. A., BOOM, B.M., CARVALINO, A.M. & DOS SANTOS, T.S. 1983. Ecological importance of Myrtaceae in an eastern Brazilian wet forest. *Biotropica* 15:68-70.

PROCTOR, J, ANDERSON, J.M., CHAI, P. & VALLACK, H.W. 1983. Ecological studies in four contrasting lowland rain forests in Gunung Mulu National Park, Sarawak. *Journal of Ecology* 71:237–260.

Appendix 1. Tree species sampled at Belalong in the plot west of Kuala Belalong Field Studies Centre arranged alphabetically in families. One voucher number is given after the name. Number of individuals, basal area, frequency, relative density (rel. dens.), dominance (rel. dom.), frequency (rel. fre.) and Importance Value Index (IVI) are given for each species.

Species/Voucher	Ind.	bas.a	fre.	rel.den.	rel.dom.	rel.fre.	IVI
Alangiaceae							
Alangium javanicum (Bl.) Wang 50	3	0.05	3	0.55	0.12	0.58	1.25
Anacardiaceae							
Gluta laxiflora Ridl. 21	26	1.58	22	4.73	3.88	4.27	12.88
Gluta wallichii (Hk. f.) Ding Hou 146	1	0.02	1	0.18	0.05	0.19	0.42
Gluta sp. 1 367	2	0.05	2	0.36	0.11	0.39	0.86
Semecarpus heterophylla Bl. 72	3	0.04	3	0.55	0.09	0.58	1.22
Swintonia foxworthyi Elmer 8	1	0.03	1	0.18	0.08	0.19	0.46
Swintonia sp. 1 248	1	0.18	1	0.18	0.44	0.19	0.82
Annonaceae							
Alphonsea elliptica Hk. f. & Thoms. 288	2	0.06	2	0.36	0.15	0.39	0.90
Monocarpia marginalis (Scheff.) Sinclair 330	2	0.03	2	0.36	0.08	0.39	0.83
Polyalthia hypoleuca Hk. f. & Thoms. 46	3	0.06	3	0.55	0.14	0.58	1.27
Polyalthia sp. 104	1	0.03	1	0.18	0.08	0.19	0.46
Xylopia malayana Hk. f. & Thoms. 39	2	0.02	2	0.36	0.06	0.39	0.81
Xylopia sp. 251	1	0.02	1	0.18	0.04	0.19	0.42
genus indet sp. 1 282	1	0.01	1	0.18	0.03	0.19	0.41
genus indet sp. 2 131	2	0.02	2	0.36	0.06	0.39	0.81
Apocynaceae							
Alstonia scholaris (L.) R. Br. 153	1	0.16	1	0.18	0.39	0.19	0.77
Dyera costulata (Miq.) Hk. f. 87	4	1.58	4	0.73	3.88	0.78	5.38
Kibatalia sp. 510	1	0.01	1	0.18	0.03	0.19	0.40
Tabaernaemontana corymbosa Roxb. 75	1	0.02	1	0.18	0.04	0.19	0.41
Arecaceae							
Oncosperma horridum Bl. 53	2	0.05	2	0.36	0.12	0.39	0.87
Bombacaceae							
Durio excelsus (Korth.) Bakh. 349	1	0.01	1	0.18	0.02	0.19	0.40
Durio kutejensis (Hassk.) Becc. 427	1	0.02	1	0.18	0.05	0.19	0.43
Durio sp. 1. sect. bastia 269	2	0.04	2	0.36	0.11	0.39	0.86
Burseraceae							
Canarium sp. 538	1	0.10	1	0.18	0.24	0.19	0.62
Dacryodes costata (Benn.) Lam 83	3	0.07	3	0.55	0.18	0.58	1.31
Dacryodes rostrata (Bl.) Lam 448	1	0.05	1	0.18	0.12	0.19	0.49
Dacryodes rugosa (Bl.) Lam 267	1	0.01	1	0.18	0.02	0.19	0.40
Dacryodes sp. 1 10	1	0.01	1	0.18	0.03	0.19	0.40
Dacryodes sp. 2 71	1	0.02	1	0.18	0.04	0.19	0.41
Dacryodes sp. 3 302	1	0.14	1	0.18	0.34	0.19	0.72
Dacryodes sp. 4 417	1	0.05	1	0.18	0.12	0.19	0.50
Dacryodes sp. 5 501	1	0.01	1	0.18	0.03	0.19	0.40
Santiria megaphylla Kalkman 37	1	0.01	1	0.18	0.03	0.19	0.40
Triomma malaccensis Hk. f. 171	1	0.05	1	0.18	0.13	0.19	0.50
Celastraceae							
Lophopetalum sp. 1 480	3	0.08	3	0.55	0.19	0.58	1.32
Chrysobalanaceae							
Atuna racemosa ssp. *racemosa* Raf. 531	1	0.03	1	0.18	0.08	0.19	0.46
Licania splendens (Korth.) Prance 32	1	0.02	1	0.18	0.04	0.19	0.42
Parastemon grandifructus Prance 155	1	0.01	1	0.18	0.03	0.19	0.41
Parinari canarioides Kosterm. 259	1	0.03	1	0.18	0.06	0.19	0.44
Parinari oblongifolia Hk. f. 324	1	0.01	1	0.18	0.03	0.19	0.41
Clusiaceae							
Calophyllum nodosum Vesque 69	1	0.01	1	0.18	0.02	0.19	0.40
Garcinia penangiana Pierre 399	1	0.01	1	0.18	0.03	0.19	0.41
Garcinia sp. 1 481	1	0.01	1	0.18	0.02	0.19	0.40

Appendix (continued). Tree species sampled at Belalong in the plot west of Kuala Belalong Field Studies Centre arranged alphabetically in families. One voucher number is given after the name. Number of individuals, basal area, frequency, relative density (rel. dens.), dominance (rel. dom.), frequency (rel. fre.) and Importance Value Index (IVI) are given for each species.

Species/Voucher	Ind.	bas.a	fre.	rel.den.	rel.dom.	rel.fre.	IVI
Clusiaceae (continued)							
Garcinia sp. 2 49	2	0.02	2	0.36	0.05	0.39	0.80
Garcinia sp. 3 98	2	0.07	2	0.36	0.16	0.39	0.92
Garcinia sp. 4 517	1	0.01	1	0.18	0.03	0.19	0.40
Mesua beccariana (Baill.) Kosterm. 15	2	0.08	2	0.36	0.20	0.39	0.95
Mesua grandis (King) Kosterm. 405	1	0.06	1	0.18	0.14	0.19	0.52
Mesua oblongifolia (R.Sl.) Kosterm. 390	2	0.02	2	0.36	0.05	0.39	0.80
Crypteroniaceae							
Crypteronia paniculata Bl. 303	2	0.02	2	0.36	0.05	0.39	0.80
Dilleniaceae							
Dillenia beccariana Martelli 412	1	0.01	1	0.18	0.03	0.19	0.40
Dipterocarpaceae							
Anisoptera laevis Ridl. 224	3	0.21	3	0.55	0.52	0.58	1.65
Dipterocarpus caudatus ssp. *penangianus* (Foxw.) Ashton 13	6	1.28	6	1.09	3.13	1.17	5.39
Dryobalanops beccarii Dyer 4	3	0.23	2	0.55	0.56	0.39	1.50
Dryobalanops lanceolata Burck 135	1	0.15	1	0.18	0.36	0.19	0.73
Hopea dryobalanoides Miq. 143	3	0.38	3	0.55	0.94	0.58	2.06
Hopea sangal Korth. 512	1	0.01	1	0.18	0.02	0.19	0.40
Parashorea smythiesii Wyatt-Smith ex Ashton 148	2	0.24	2	0.36	0.59	0.39	1.35
Shorea amplexicaulis Ashton 60	7	0.64	6	1.27	1.58	1.17	4.01
Shorea confusa Ashton 178	1	0.14	1	0.18	0.33	0.19	0.71
Shorea domatiosa Ashton 211	6	1.12	4	1.09	2.76	0.78	4.62
Shorea faguetiana Heim 85	8	0.67	8	1.45	1.65	1.55	4.66
Shorea fallax Meijer 245	3	0.42	2	0.55	1.02	0.39	1.96
Shorea ferruginea Dyer ex Brandis 5	6	0.77	6	1.09	1.89	1.17	4.14
Shorea foxworthyi Sym. 149	1	0.01	1	0.18	0.04	0.19	0.41
Shorea hypoleuca Meijer 34	2	0.15	1	0.36	0.37	0.19	0.93
Shorea laevis Ridl. 11	7	5.98	7	1.27	14.66	1.36	17.29
Shorea leprosula Miq. 23	1	0.07	1	0.18	0.18	0.19	0.55
Shorea macroptera ssp. *macropterifolia* Dyer 95	6	0.42	5	1.09	1.03	0.97	3.09
Shorea maxwelliana King 467	3	0.30	3	0.55	0.74	0.58	1.87
Shorea obscura Meijer 204	1	0.44	1	0.18	1.07	0.19	1.44
Shorea parvifolia Dyer 6	17	5.03	14	3.09	12.34	2.72	18.15
Shorea pauciflora King 45	4	1.83	4	0.73	4.48	0.78	5.99
Shorea scaberrima Burck 126	1	0.01	1	0.18	0.03	0.19	0.41
Shorea scrobiculata Burck 444	1	0.05	1	0.18	0.12	0.19	0.49
Shorea sp 400	1	0.03	1	0.18	0.08	0.19	0.45
Vatica endertii Sloot. 62	16	0.78	15	2.91	1.91	2.91	7.74
Vatica odorata (Griff.) Sym. 28	21	0.90	16	3.82	2.20	3.11	9.12
Vatica sp. 1 397	1	0.04	1	0.18	0.11	0.19	0.48
Ebenaceae							
Diospyros borneensis Hiern 59	6	0.09	6	1.09	0.22	1.17	2.48
Diospyros sp. 1 48	6	0.23	6	1.09	0.57	1.17	2.83
Diospyros sp. 2 25	1	0.01	1	0.18	0.02	0.19	0.40
Diospyros sp. 3 511	1	0.03	1	0.18	0.08	0.19	0.46
Diospyros sp. 4 525	1	0.02	1	0.18	0.05	0.19	0.42
Diospyros sp. 5 31	1	0.17	1	0.18	0.41	0.19	0.79
Diospyros sp. 6 152	2	0.09	2	0.36	0.21	0.39	0.96
Diospyros sp. 7 375	1	0.02	1	0.18	0.05	0.19	0.43
Diospyros sp. 8 317	1	0.02	1	0.18	0.05	0.19	0.42
Diospyros sp. 9 362	1	0.17	1	0.18	0.43	0.19	0.80
Diospyros sp. 10 278	4	0.07	4	0.73	0.17	0.78	1.67
Euphorbiaceae							
Aporusa sp. 1 81	4	0.05	4	0.73	0.13	0.78	1.63
Aporusa sp. 2 292	1	0.02	1	0.18	0.05	0.19	0.43
Aporusa sp. 3 319	1	0.02	1	0.18	0.05	0.19	0.43

Appendix (continued). Tree species sampled at Belalong in the plot west of Kuala Belalong Field Studies Centre arranged alphabetically in families. One voucher number is given after the name. Number of individuals, basal area, frequency, relative density (rel. dens.), dominance (rel. dom.), frequency (rel. fre.) and Importance Value Index (IVI) are given for each species.

Species/Voucher	Ind.	bas.a	fre.	rel.den.	rel.dom.	rel.fre.	IVI
Euphorbiaceae (continued)							
Baccaurea latifolia King ex Hk. f. 252	1	0.01	1	0.18	0.02	0.19	0.40
Baccaurea sp. 1 80	1	0.01	1	0.18	0.03	0.19	0.40
Baccaurea sp. 2 212	1	0.04	1	0.18	0.09	0.19	0.47
Baccaurea sp. 3 455	1	0.03	1	0.18	0.08	0.19	0.45
Baccaurea sp. 4 477	1	0.01	1	0.18	0.02	0.19	0.40
Blumeodendron calophyllum Airy Shaw 120	1	0.01	1	0.18	0.03	0.19	0.40
Drypetes kikir Airy Shaw 272	2	0.15	2	0.36	0.38	0.39	1.13
Drypetes sp. 1 99	1	0.01	1	0.18	0.02	0.19	0.40
Elateriospermum tapos Bl. 55	35	2.36	28	6.36	5,78	5.44	17.58
Koilodepas longifolium Hk. f. 356	3	0.04	1	0.55	0.10	0.19	0.84
Macaranga kingii var *platyphylla* Hk. f. 381	1	0.01	1	0.18	0.02	0.19	0.40
Macaranga repando-dentata Airy Shaw 280	8	0.13	8	1.45	0.32	1.55	3.33
Macaranga trachophylla Airy Shaw 326	1	0.01	1	0.18	0.02	0.19	0.40
Mallotus sp. 1 74	7	0.07	7	1.27	0.16	1.36	2.79
Moultonianthus leembruggianus (Boerl. & Koord.) v. Steenis 105	16	0.40	13	2.91	0.99	2.52	6.42
Neoscortechinia kingii (Hook. f.) Pax & Hoffm. 167	2	0.02	2	0.36	0.04	0.39	0.79
Pimelodendron griffithianum (Muell. Arg.) Benth. 30	4	0.08	4	0.73	0.19	0.78	1.69
Ptychopyxis javanica (J.J. Smith) Croiz. 114	1	0.01	1	0.18	0.02	0.19	0.40
Fagaceae							
Castanopsis megacarpa Gamble 29	1	0.05	1	0.18	0.13	0.19	0.51
Castanopsis sp. 1 325	1	0.24	1	0.18	0.59	0.19	0.97
Lithocarpus sp. 1 16	2	0.21	2	0.36	0.52	0.39	1.27
Lithocarpu sp. 2 117	1	0.28	1	0.18	0.67	0.19	1.05
Quercus chrysotricha A. Camus 493	1	0.01	1	0.18	0.03	0.19	0.40
Quercus sp. 1 193	1	0.01	1	0.18	0.03	0.19	0.40
Flacourtiaceae							
Hydnocarpus borneensis Sleum. 271	3	0.08	3	0.55	0.19	0.58	1.32
Hydnocarpus pinguis Sleum. 184	2	0.29	2	0.36	0.70	0.39	1.45
Hydnocarpus sumatrana (Miq.) Koord. 17	15	0.61	13	2.73	0.11	0.39	0.86
Hydnocarpus woodii Merr. 82	3	0.19	3	0.55	0.08	0.58	1.21
Hydnocarpus sp. 1 36	2	0.04	2	0.36	0.08	0.19	0.46
Hydnocarpus sp. 2 432	3	0.03	3	0.55	1,50	2.52	6.75
Hydnocarpus sp. 3 513	1	0.03	1	0.18	0.47	0.58	1.59
Ryparosa baccaureoides Sleum. 350	1	0.02	1	0.18	0.06	0.19	0.44
Ryparosa kostermannsii Sleum. 77	7	0.24	6	1.27	0.58	1.17	3.02
Icacinaceae							
Gonocaryum minus Sleum. 12	3	0.09	3	0.55	0.23	0.58	1.36
Stemonurus umbellatus Becc. 187	3	0.03	3	0.55	0.08	0.58	1.21
Ixonanthaceae							
Irvingia malayana Oliv. et Benn. 130	1	0.32	1	0.18	0.79	0.19	1.16
Lauraceae							
Beilschmedia sp. 107	1	0.04	1	0.18	0.11	0.19	0.48
Endiandra clavigera Kosterm. 257	1	0.04	1	0.18	0.10	0.19	0.48
Endiandra sp. 1 44	1	0.16	1	0.18	0.39	0.19	0.77
Endiandra sp. 2 547	1	0.01	1	0.18	0.03	0.19	0.41
Phoebe sp. 86	1	0.05	1	0.18	0.12	0.19	0.50
genus indet sp. 1 181	1	0.12	1	0.18	0.30	0.19	0.68
genus indet sp. 2 306	1	0.01	1	0.18	0.02	0.19	0.40
Lecythidaceae							
Barringtonia sp. 1 449	1	0.01	1	0.18	0.03	0.19	0.40
Barringtonia sp. 2 127	1	0.04	1	0.18	0.10	0.19	0.48

Appendix (continued). Tree species sampled at Belalong in the plot west of Kuala Belalong Field Studies Centre arranged alphabetically in families. One voucher number is given after the name. Number of individuals, basal area, frequency, relative density (rel. dens.), dominance (rel. dom.), frequency (rel. fre.) and Importance Value Index (IVI) are given for each species.

Species/Voucher	Ind.	bas.a	fre.	rel.den.	rel.dom.	rel.fre.	IVI
Leguminosae							
Archidendron microcarpum (Benth.) Ivan Nielsen 494	1	0.04	1	0.18	0.10	0.19	0.47
Crudia sp. 431	1	0.11	1	0.18	0.26	0.19	0.64
Koompassia excelsa (Becc.) Taubert 458	1	0.07	1	0.18	0.16	0.19	0.54
Linaceae							
Ctenolophon parvifolius Oliv. 96	1	0.09	1	0.18	0.23	0.19	0.60
Loganiaceae							
Norrisia maior Soler. 469	1	0.06	1	0.18	0.14	0.19	0.52
Melastomataceae							
Memecylon sp. 1 35	1	0.02	1	0.18	0.05	0.19	0.43
Memecylon sp. 2 343	1	0.01	1	0.18	0.02	0.19	0.40
Memecylon sp. 3 177	1	0.04	1	0.18	0.11	0.19	0.49
Pternandra sp. 1 215	2	0.21	2	0.36	0.51	0.39	1.27
Pternandra sp. 2 347	1	0.01	1	0.18	0.02	0.19	0.40
Meliaceae							
Aglaia sp. 1 304	1	0.02	1	0.18	0.04	0.19	0.41
Aglaia sp. 2 333	1	0.02	1	0.18	0.04	0.19	0.41
Aglaia sp. 3 456	1	0.05	1	0.18	0.13	0.19	0.51
Aglaia sp. 4 457	1	0.01	1	0.18	0.02	0.19	0.40
Aglaia sp. 5 487	1	0.06	1	0.18	0.15	0.19	0.52
Azadirachta excelsa (Jack) Jacobs 162	1	0.06	1	0.18	0.14	0.19	0.52
Disoxylum cauliflorum Hiern 264	1	0.01	1	0.18	0.02	0.19	0.40
Sandoricum borneense Miq 479	1	0.01	1	0.18	0.03	0.19	0.40
genus indet (alive no leaves) 465	1	0.02	1	0.18	0.06	0.19	0.43
Moraceae							
Artocarpus lanceifolius Roxb. 373	1	0.02	1	0.18	0.05	0.19	0.43
Artocarpus nitidus Trec. 2	2	0.11	2	0.36	0.26	0.39	1.02
Artocarpus odoratissima Blanco 232	2	0.10	2	0.36	0.26	0.39	1.01
Artocarpus tamaran Becc. 520	1	0.17	1	0.18	0.02	0.19	0.40
Artocarpus sp. 1 537	1	0.01	1	0.18	0.43	0.19	0.80
Ficus sp. 1 175	1	0.20	1	0.18	0.49	0.19	0.86
Ficus sp. 2 433	1	0.02	1	0.18	0.04	0.19	0.41
Myristicaceae							
Gymnacranthera contracta Warb. 144	1	0.08	1	0.18	0.21	0.19	0.58
Gymnacranthera farquhariana (Hook. f. & Th.) Warb. 374	1	0.05	1	0.18	0.13	0.19	0.50
Horsfieldia borneensis de Wilde 70	1	0.10	1	0.18	0.26	0.19	0.63
Horsfieldia polyspherula (Hook. f. emend. King) Sinclair 378	3	0.04	3	0.55	0.10	0.58	1.22
Horsfieldia reticulata Warb. 509	1	0.01	1	0.18	0.04	0.19	0.41
Horsfieldia wallichii (Hook. f. & Th.) Warb. 328	1	0.06	1	0.18	0.15	0.19	0.52
Knema latericia Elmer ssp. *ridleyi* (Gandoger) de Wilde 18	2	0.02	2	0.36	0.06	0.39	0.81
Knema latifolia Warb. 19	2	0.03	2	0.36	0.08	0.39	0.83
Knema linguiformis (Sinclair) de Wilde 100	2	0.08	2	0.36	0.19	0.39	0.94
Knema membranifolia H. Winkler in Engl. 546	1	0.01	1	0.18	0.02	0.19	0.40
Knema oblongata Merr. ssp. *oblongata* 290	1	0.01	1	0.18	0.03	0.19	0.41
Knema percoriacea Sinclair forma *sarawakensis* de Wilde 411	1	0.01	1	0.18	0.02	0.19	0.40
Knema sp. 1 woodii/korthalsii/pulchra 408	1	0.06	1	0.18	0.15	0.19	0.52
Knema sp. 2 glauca group 535	1	0.02	1	0.18	0.04	0.19	0.41
Myristica borneensis Warb. 3	3	0.07	3	0.55	0.18	0.58	1.31
Myristica iners Bl. 201	2	0.35	2	0.36	0.85	0.39	1.60
Myristica maxima Warb. 102	2	0.03	2	0.36	0.07	0.39	0.82

Appendix (continued). Tree species sampled at Belalong in the plot west of Kuala Belalong Field Studies Centre arranged alphabetically in families. One voucher number is given after the name. Number of individuals, basal area, frequency, relative density (rel. dens.), dominance (rel. dom.), frequency (rel. fre.) and Importance Value Index (IVI) are given for each species.

Species/Voucher	Ind.	bas.a	fre.	rel.den.	rel.dom.	rel.fre.	IVI
Myrsinaceae							
Ardisia sp. 464	1	0.03	1	0.18	0.08	0.19	0.45
Myrtaceae							
Eugenia sp. 1 52	1	0.01	1	0.18	0.02	0.19	0.40
Eugenia sp. 2 353	2	0.03	2	0.36	0.07	0.39	0.82
Eugenia sp. 3 205	2	0.03	2	0.36	0.06	0.39	0.81
Eugenia sp. 4 179	4	0.31	4	0.73	0.77	0.78	2.27
Eugenia sp. 5 156	1	0.04	1	0.18	0.09	0.19	0.46
Eugenia sp. 6 462	1	0.03	1	0.18	0.07	0.19	0.45
Eugenia sp. 7 112	1	0.15	1	0.18	0.36	0.19	0.74
Eugenia sp. 8 287	1	0.01	1	0.18	0.02	0.19	0.40
Eugenia sp. 9 305	1	0.01	1	0.18	0.03	0.19	0.41
Eugenia sp. 10 233	1	0.03	1	0.18	0.07	0.19	0.44
Eugenia sp. 11 274	1	0.03	1	0.18	0.07	0.19	0.45
Olacaceae							
Strombosia sp. 7	1	0.06	1	0.18	0.14	0.19	0.51
Polygalaceae							
Xanthophylum adenotus Miq. 484	1	0.01	1	0.18	0.03	0.19	0.41
Xanthophyllum sp. 1 40	6	0.38	6	1.09	0.94	1.17	3.20
Xanthophyllum sp. 2 65	3	0.04	3	0.55	0.10	0.58	1.23
Xanthophyllum sp. 3 110	2	0.06	2	0.36	0.14	0.39	0.90
Xanthophyllum sp. 4 54	3	0.05	3	0.55	0.11	0.58	1.24
Xanthophyllum sp. 5 64	3	0.08	3	0.55	0.21	0.58	1.33
Proteaceae							
Helicia sp. 41	2	0.06	2	0.36	0.15	0.39	0.91
Rosaceae							
Maranthes corymbosa Bl. 406	1	0.20	1	0.18	0.50	0.19	0.87
Maranthes corymbosis Bl. 168	1	0.03	1	0.18	0.06	0.19	0.44
Prunus beccarii Ridl. 370	1	0.07	1	0.18	0.16	0.19	0.54
Rubiaceae							
Adina sp. 413	1	0.01	1	0.18	0.03	0.19	0.40
Canthium sp. 1	1	0.01	1	0.18	0.03	0.19	0.40
Neonauclea sp. 1 489	1	0.01	1	0.18	0.03	0.19	0.40
Neonauclea sp. 2 387	1	0.01	1	0.18	0.02	0.19	0.40
Psydrax sp. 1 388	1	0.01	1	0.18	0.03	0.19	0.40
Psydrax sp. 1 507	1	0.04	1	0.18	0.11	0.19	0.48
Urophyllum sp. 73	3	0.13	3	0.55	0.32	0.58	1.45
genus indet 172	1	0.02	1	0.18	0.04	0.19	0.41
Sabiaceae							
Meliosma sp. 460	1	0.01	1	0.18	0.04	0.19	0.41
Sapindaceae							
Nephelium sp. 357	2	0.05	2	0.36	0.13	0.39	0.88
Sapotaceae							
Madhuca sp. 1 318	1	0.05	1	0.18	0.12	0.19	0.49
Madhuca sp. 2 227	1	0.02	1	0.18	0.05	0.19	0.43
Palaquium sp. 1 111	2	0.02	2	0.36	0.06	0.39	0.81
Palaquium sp. 2 527	1	0.04	1	0.18	0.09	0.19	0.47
Payena sp. 1 43	2	0.09	2	0.36	0.23	0.39	0.98
Pouteria sp. 2 159	1	0.05	1	0.18	0.13	0.19	0.50
genus indet. 331	2	0.06	2	0.36	0.14	0.39	0.89

Appendix (continued). Tree species sampled at Belalong in the plot west of Kuala Belalong Field Studies Centre arranged alphabetically in families. One voucher number is given after the name. Number of individuals, basal area, frequency, relative density (rel. dens.), dominance (rel. dom.), frequency (rel. fre.) and Importance Value Index (IVI) are given for each species.

Species/Voucher	Ind.	bas.a	fre.	rel.den.	rel.dom.	rel.fre.	IVI
Sterculiaceae							
Heritiera sumatrana (Miq.) Kosterm. 124	1	0.01	1	0.18	0.02	0.19	0.40
Scaphium linearicarpum (Mast.) Pierre 285	5	0.28	5	0.91	0.68	0.97	2.56
Scaphium macropodum (Miq.) Blume ex Heyne 47	1	0.25	1	0.18	0.61	0.19	0.99
Sterculia scortechinii King 277	2	0.03	2	0.36	0.08	0.39	0.83
Sterculia sp. 226	1	0.02	1	0.18	0.05	0.19	0.43
Thymelaeaceae							
Gonostylus sp. 27	1	0.02	1	0.18	0.04	0.19	0.42
Tiliaceae							
Grewia cinnamomifolia (Burrett) Stapf ex Ashton 473	1	0.01	1	0.18	0.03	0.19	0.40
Pentace laxiflora Merr. 355	1	0.03	1	0.18	0.07	0.19	0.44
Ulmaceae							
Gironniera nervosa Planch. 254	2	0.02	2	0.36	0.05	0.39	0.80
Verbenaceae							
Teijsmanniodendron sarawakanum (H.H.W. Pearson) Kosterm. 323	4	0.05	4	0.73	0.13	0.78	1.64

Forest bird diversity in Oceania

JIRO KIKKAWA

Centre for Conservation Biology, the University of Queensland, Brisbane, Qld 4072, Australia

---

Forest bird diversity in Oceania reflects biogeographical patterns and historical factors of distribution rather more than the degree of niche differentiation that the forest birds are expected to have undergone. There are five main gradients of diversity found to the east and south of Wallace's Line, and these are summarised below in terms of species richness.

1) On the Australian continent climatic cycles of the past produced more dry-adapted forms in wetter periods than wet-adapted forms in drier periods (Brereton & Kikkawa 1963). This is because the continent has primarily been a dry continent dominated by sclerophyllous vegetation since the late Tertiary. Thus a prominent, biogeographical, gradient is found from the greatest diversity in open eucalypt forests (45 species per 8 ha and a high beta-diversity) to coastal closed forests (38 species per 8 ha and a low beta-diversity) and to inland savannahs and woodlands (25 species per 8 ha and a high beta-diversity) in the mid-latitude range (Kikkawa 1968).

2) The second gradient on the continent is developed in rainforest vegetation in which the latitudinal gradient associated with the resources of the forest is seen in forest birds (Kikkawa 1991). This gradient parallels the size of the area and in the wet tropical region, where the rainforest area is the largest on the continent, the diversity is greatest with 100 species, decreasing to 80 species in subtropical, and 42 species in temperate rainforests.

3) The third gradient on the continent is seen along the moisture gradient associated with seasonality of rainfall. From the wet tropics of the Cardwell-Cooktown region of north Queensland the forest bird fauna is attenuated towards the wet-dry region of Cape York Peninsula (Kikkawa *et al.* 1981a) and the Northern Territory (Kikkawa *et al.* 1981b). On the drier side rainforest vegetation occurs in smaller patches of monsoon forests containing many dry-defoliating tree species. Here the number of forest bird species declines to 85 with many replacements with dry-adapted forms.

4) In contrast to Australia, New Guinea is topographically very complex for its area, with great opportunities for speciation among the wet-adapted forms. Many of them exhibit conspicuous altitudinal zonation (Diamond 1972). An altitudinal gradient of diversity is found from the lowland rainforests (184 species at the Brown River, 147 species at the Trans-Gogol) to highlands (144 species in the Central Highlands) and mountain tops (8 species) (Driscoll & Kikkawa 1989, Kikkawa & Williams 1971b). This pattern is resource-related but, when the entire island is considered, the altitudinal gradient is even more pronounced, reflecting allopatry due to geographical isolation of rainforest areas in different altitudinal zones (Kikkawa & Williams 1971a).

5) The oceanic islands of the South Pacific have been colonised largely by birds of the Western Pacific region (Mayr 1945). The underlying diversity is therefore related to the distance from this source of colonisation. This gradient is shown for many plant and animal groups (Stoddart 1992). The number of forest bird species decreases from New Guinea (445) to Solomon Islands (127), New Caledonia (52), Central Polynesia (34) and south-eastern Polynesia (40). The only exception is found in the Hawaiian Islands where most of the forest birds are derived from the New World (Diamond 1984). These include the endemic family

D. S. Edwards et al. (eds.), Tropical Rainforest Research – Current Issues, 151–152.

of honeyceepers which have undergone spectacular radiation on the islands. On a much smaller scale speciation is evidenced among the white-eyes (*Zosterops*) in Micronesia and pigeons and parrots in Polynesia and Fiji. The fauna is impoverished on smaller and remote islands of Central Pacific (e.g. 12 species on Marshall, Nauru, Kiribati, Tokelau Islands (Pratt *et al.* 1987)) (Whiffin & Kikkawa 1992).

In historical times many exotic species have been introduced into islands either accidentally or on purpose. Some of them have established themselves in indigenous forests (e.g. New Zealand). Indigenous species were vulnerable to introduced influences on the islands and many forest birds became extinct in the past hundred years. These influences include habitat destruction caused by forest clearing, heavy grazing and disturbance by introduced mammals, competition from newly established species, predation by introduced predators (e.g. black rats, stoats, cats) and diseases (e.g. bird malaria).

These trends are significant modifiers of diversity gradients seen in the forest birds of the South Pacific region.

LITERATURE CITED

BRERETON, J. le G. & KIKKAWA, J. 1963. Diversity of avian species. *Australian Journal of Science* 26:12-14.

DIAMOND, J.M. 1972. *Avifauna of the Eastern Highlands of New Guinea.* The Nuttall Ornithological Club, Cambridge, Massachusetts. 438 pp.

DIAMOND, J.M. 1984. Biogeographic mosaics in the Pacific. Pp. 1-14 in Radovsky, F.J., Raven, P.H. & Sohmer, S.H. (eds). *Biogeography of the tropical Pacific.* Bishop Museum, Honolulu.

DRISCOLL, P.V. & KIKKAWA, J. 1989. Bird species diversity of lowland tropical rainforests of New Guinea and Australia. Pp. 123-152 in Harmelin-Vivien, M. & Bourliere, F. (eds). *Vertebrates in complex tropical systems.* Springer-Verlag, New York.

KIKKAWA, J. 1968. Ecological association of bird species and habitats in eastern Australia: similarity analysis. *Journal of Animal Ecology* 37:143-165.

KIKKAWA, J. 1991. Avifauna of Australian rainforests. Pp. 187-196 in Werren, G. & Kershaw, P. (eds). *The rainforest legacy: Australian national rainforests study.* Volume 2. Australian Heritage Commission, Canberra.

KIKKAWA, J., MONTEITH, G.B. & INGRAM, G.J. 1981a. Cape York Peninsula: a major region of faunal interchange. Pp. 1695-1742 in Keast, A. (ed.). *Ecological biogeography of Australia.* W. Junk, The Hague.

KIKKAWA, J., WEBB, L.J., DALE, M.B., MONTEITH, G.B., TRACEY, J.G. & WILLIAMS, W.T. 1981b. Gradients and boundaries of monsoon forests in Australia. *Proceedings of the Ecological Society of Australia* 11:39-52.

KIKKAWA J. & WILLIAMS, W.T. 1971a. Altitudinal distribution of land birds in New Guinea. *Search* 2:64-65.

KIKKAWA, J. & WILLIAMS, W.T. 1971b. Ecological grouping of species for conservation of land birds in New Guinea. *Search* 2:66-69.

MAYR, E. 1945. *Birds of the Southwest Pacific.* Macmillan, New York. 316 pp.

PRATT, H.D., BRUNER, P.L. & BERRETT, D.G. 1987. *The birds of Hawaii and the tropical Pacific.* Princeton University Press, Princeton. 409 pp. + 45 plates.

STODDART, D.R. 1992. Biogeography of the tropical Pacific. *Pacific Science* 46:276-293.

WHIFFIN, T. & KIKKAWA, J. 1992. The status of forest biodiversity in Oceania. *Journal of Tropical Forest Science* 5:155-172.

The fauna and flora of a semi-deciduous forest in Perlis, Peninsular Malaysia

DIONYSIUS SHARMA[1], HYDEIR KAMARUDIN[2], HALIJAH IBRAHIM[3] *and* ONG HEAN CHOOI[3]

[1]*WWF Malaysia, Locked bag No. 911, Jln Sultan P.O., 46990 Petaling Jaya, Selangor, Malaysia*
[2]*Malayan Nature Society, No. 485, Jln 5/53, 46000 Petaling Jaya, Selangor, Malaysia*
[3]*Department of Botany, University of Malaya, 59100 Kuala Lumpur, Malaysia*

ABSTRACT. A study was conducted to survey the fauna and flora at Mata Ayer Forest Reserve (3176 hectares) and the proposed Bukit Wang Mu Forest Reserve (2768 hectares), Perlis, Peninsular Malaysia. Thirty seven mammal species were recorded, including five primates and five ungulates. The sighting of the Stump-tailed macaque (*Macaca arctoides*) at Mata Ayer confirms its existence in Peninsular Malaysia. Thirty five species of reptiles were recorded. Eighty eight species of birds, including six species of hornbills were observed. One hundred and forty three non-timber plant species (herb, climber, shrub, sedge, grass, rotan) were identified. The study sites are important as water catchment areas. The establishment of a transfrontier reserve is recommended.

KEY WORDS: Mata Ayer Forest Reserve, Bukit Wang Mu Forest Reserve, transfrontier reserve.

INTRODUCTION

Perlis (6°15'- 6°43'N, 100°6'-100°23'E) is the smallest and northernmost State of Peninsular Malaysia, covering an area of some 81,063 hectares. It experiences a strongly monsoonal climate with an average annual rainfall of 1,735 mm (WWFM 1984) and a 3-4 month dry season. The original forest over most of Perlis (semi-deciduous with Burmese affinities) has been cleared for cultivation, or converted by cutting and fire into secondary *Schima*-bamboo forest.

The main forested areas (unalienated forested State land and forest reserves) left in Perlis lie in a continuous belt along the western border (Figure 1). The forests lie mainly on hilly terrain of limestone with numerous caves. The limestone forms part of the Setul Formation. Within this forested belt is the Mata Ayer Forest Reserve (3176 hectares) and the proposed Bukit Wang Mu Forest Reserve (2768 hectares). Both these areas lie adjacent to Thaleban National Park (NP) in Thailand and are important water catchment areas for the Timah-Tasoh Dam, situated in the south. Forest clearance projects, mainly for agriculture, have occurred in most of the valley-floor compartments of Mata Ayer. Due to this, the IUCN Commission on National Parks and Protected Areas felt compelled to place Thaleban NP on their list of 'Threatened Protected Areas of the World' (IUCN 1988). Since the forest at Mata Ayer and Bukit Wang Mu is unique to Peninsular Malaysia, and in view of the importance of the area for water catchment, the potential clearance of the forests and consequent threat to Thaleban NP, surveys were conducted to investigate the fauna and non-timber flora. The objective of the study was to collate the scientific information with the intention of developing a viable management plan for the area.

D. S. Edwards et al. (eds.), Tropical Rainforest Research – Current Issues, 153–161.

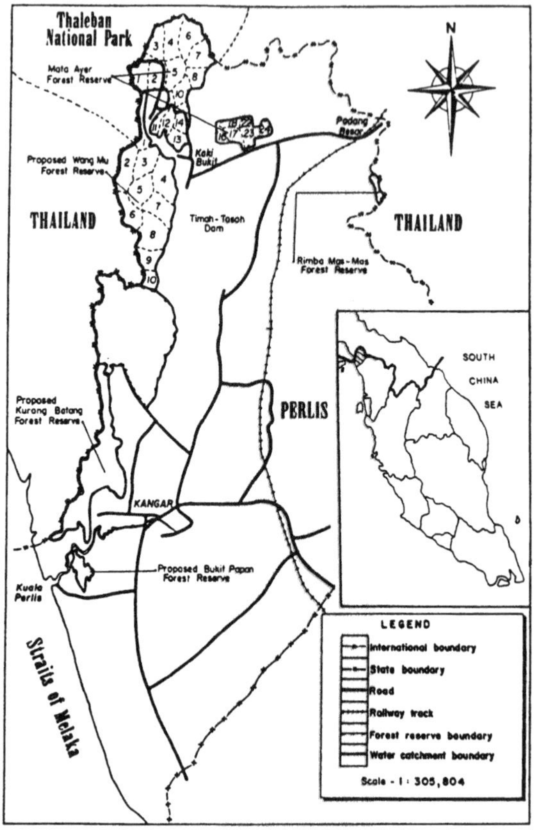

Figure 1. Existing and proposed nature reserves in Perlis, Penninsular Malaysia.

METHODS

Surveys were carried out at Mata Ayer between May and August 1992 and at Bukit Wang Mu in October and November 1992. Three line transects, each a kilometre long were cut in Mata Ayer and Bukit Wang Mu respectively. These were walked daily between 0700-0930 hours. Any mammal, bird or reptile observed along the transect was recorded. Calls of known animals (e.g. gibbons, dusky langurs, barbets, argus pheasants) were recorded as well. Animal tracks (especially of ungulates) and remains (e.g. skull, bones, quills, scales) were recorded when encountered. Night transect walks were conducted between 2100-2330 hours to investigate the nocturnal fauna. One hundred small mammal cage traps were set along the transects to investigate the squirrel, treeshrew and rat diversity. Traps were placed on the ground and up in the trees (two metres off the ground) at 25 metre intervals. At each locality surveyed, plants were collected for herbarium specimens and wherever possible, samples were selected for planting in the Botanic Garden, Department of Botany, University of Malaya. Specimens were identified with the aid of reference books (e.g. Henderson 1974, Holttum 1966, Keng 1969, Whitmore 1977) and by reference to herbarium specimens lodged in the Department of Botany, University of Malaya.

RESULTS

Thirty eight mammal species from twenty one different families were confirmed in the area (Appendix 1). In addition to this, several flying squirrels were seen but could not be identified to species level. Five species of primates and five ungulates were recorded. The most commonly seen primate was the White-handed gibbon (*Hylobates lar*). A troop of wild Stump-tailed macaques (*Macaca arctoides*) was observed in compartment 6, Mata Ayer FR. The group was estimated to include at least eight individuals. Two individuals were observed feeding on the ripe fruits of *Baccaurea kunstleri* (Sharma, in preparation). The troop was approximately 500 metres from the Malaysia/Thailand border.

Eighty eight species of birds from 34 families were recorded during the surveys (Appendix 1). The six species of hornbill recorded were the White-crowned hornbill (*Berenicornis comatus*), Bushy-crested hornbill (*Anorrhinus galeritus*), Wreathed hornbill (*Rhyticeros undulatus*), Rhinoceros hornbill (*Buceros rhinoceros*), Great hornbill (*Buceros bicornis*) and Helmeted hornbill (*Rhinoplax vigil*). The most common was the Great hornbill, frequently seen in pairs (n=12), although small flocks of three were observed as well (n=3). Two species of threatened Malaysian birds, the Chestnut-necklaced partridge (*Arborophila charltoni*) and Helmeted hornbill (*Rhinoplax vigil*), were recorded (see Collar & Andrew 1988).

Thirty five species of reptiles, including 21 species of snakes, five species of batagurid turtles, one trionychid and a tortoise, were recorded (Appendix 1). The striped racer (*Elaphe taeniura*) was often seen in caves, and together with the Oriental whip snake (*Ahaetulla prasina*), were the two most commonly encountered snakes. The Asian leaf turtle (*Cyclemys dentata*) was frequently encountered near slow-flowing streams, and three specimens were collected from inside caves. Juvenile specimens of the Malayan softshell (*Dogania subplana*) (carapace length 50mm) and Malayan flat shell (*Notochelys platynota*) (carapace length 88mm) were collected from stagnant pools.

One hundred and forty three non-timber plant species from 39 families were identified (Appendix 2). These include 17 species of orchids and seven species of palms. The Bukit Rongkit limestone hill flora in the Mata Ayer FR was interesting in that it includes seven species of orchids, two species of *Begonia* (*B. guttata* and *B. phoeniogramma*) and *Maxburetia rupicola*, a palm which is endemic to Peninsula Malaysia and of limited distribution. Of the 23 zingiberaceous species recorded, 36% were in flower, 33% sterile, 9% in bud stage, 9% were over flower and 5% were in fruit. Three zingiberaceous species, *Zingiber zerumbet, Globba sp.* and *Camptandra sp.*, found on the slope of Bukit Rongkit Utara, Mata Ayer FR, have not previously been recorded in limestone habitats.

DISCUSSION

The Asian elephant (*Elephas maximus*) and the Asian two-horned rhinoceros (*Dicerorhinus sumatraensis*) no longer occur in the Mata Ayer FR as a result of habitat loss (large scale conversion of lowland forests to agricultural land). The remaining forested areas are on hilly terrain (mainly limestone) and are not suitable habitats for these species. A confirmed sighting of the Stump-tailed macaque in northern Peninsular Malaysia was made by Syed Abdullah Idid, then Deputy Game Warden for Perlis/Kedah (Medway 1978). During their primate survey of Peninsular Malaysia, Marsh and Wilson (1981) spent several days at Mata Ayer but did not encounter these macaques. They were, however, informed by local people that the macaques may sometimes still be seen. In addition to this, a group of these macaques were reportedly seen in the State of Kelantan (Charles, J.K., personal communication). The Stump-tailed macaque was described as being 'possibly extinct' in

Peninsular Malaysia (Bennett 1991) as there were no confirmed sightings for the last fifteen years. The sighting of *M. arctoides* at Mata Ayer therefore further substantiates previous observations and confirms that this macaque still occurs in Peninsular Malaysia. Populations may wander to and fro across the international border. In Thailand, *M. arctoides* is now uncommon throughout its range and occurs only in isolated populations. It is nearing extinction in Peninsular Thailand, where it was once abundant (Roonwal & Mohnot 1977, Southwick & Siddiqi 1970).

Several other mammals recorded at Mata Ayer and Bukit Wang Mu are important in conservation terms. Mammals likely to be threatened include the Moonrat (*Echinosorex gymnurus*) and the Pencil-tailed tree mouse (*Chiropodomys gliroides*). Mammals which are already threatened include Low's squirrel (*Sundasciurus lowii*), Panther (*Panthera pardus*) and Serow (*Capricornis sumatraensis*).

The valley-floor compartments of Mata Ayer FR which have been cleared for agriculture, were regarded as being important for several bird species including the Streaked-breasted woodpecker (*Picus viridanus*), Dusky eagle owl (*Bubo coromandus*), Brown fish owl (*Ketupa zeylonensis*), Hooded pitta (*Pitta sordida*) and the extremely rare Green-backed Narcissus flycatcher (*Ficedula narcissina elizae*) (Wells 1987). Only the Hooded pitta was observed during the present surveys restricted to the remaining valley compartments.

The Wildlife at Mata Ayer FR and Bukit Wang Mu is threatened by hunting and trapping. Snare traps for Wild pig and Serow were seen, as well as bamboo traps for catching small rodents. Remains of wildlife found (with signs that it had been eaten) include Pangolin (*Manis javanica*), Malayan porcupine (*Hystrix brachyura*), Dusky langur (*Presbytis obscura*), Lesser mousedeer (*Tragulus javanicus*), Greater mousedeer (*Tragulus napu*), Spiny turtle (*Heosemys spinosa*), Giant pond turtle (*Heosemys grandis*), Malayan flat-shelled turtle (*Notochelys platynota*) and Elongated tortoise (*Indotestudo elongata*).

The limestone flora of Peninsular Malaysia has been well studied (e.g. Chin 1977, 1979, 1983a, 1983b, Henderson 1939, Kiew 1991). Several herbaceous plants found on limestone have great potential as ornamentals, especially those from the Balsaminaceae, Begoniaceae and Gesneriaceae. The species collected and identified at Mata Ayer FR to date are only a small fraction of the total number that may exist in the area. A systematic resurvey is required to fully document the flora.

The retention of natural forest cover at Mata Ayer and Bukit Wang Mu is crucial as these areas play a vital role in the hydrological cycle for the State of Perlis and are important catchment areas for the Timah-Tasoh Dam (Kumari & Ariffin 1988, WWFM 1984). During the drier months, demand for water for domestic consumption, industry and agriculture will have to be met. The loss of surface water, stagnant or flowing, will have immediate consequences on both the terrestrial and aquatic fauna.

On an international level, the establishment of a transfrontier reserve will serve to protect the fauna and flora along the common border. Any degradation or reduction of the forests at Mata Ayer and Bukit Wang Mu will not only extirpate Malaysian fauna and flora but also exert pressure on the biodiversity existing in Thaleban NP, Thailand.

ACKNOWLEDGEMENTS

We would like to thank the then Director of the Perlis Forestry Department, En. Borhanudin bin Haji Arshad, and the Assistant Director, En. Zainal Abidin, for permission to carry out field work. Thanks also to the staff of the Police Field Force, staff of the Balai Polis Kaki Bukit, Raleigh International, Caving Group of the Malayan Nature Society, Dr. Mikaail Kavanagh and Dr. Isabelle Louis for their assistance and support. This work was funded by the World Wide Fund for Nature, Malaysia.

LITERATURE CITED

BENNETT, E.L. 1991. Diurnal Primates. Pp. 150-172 in Kiew, R. (ed.). *The State of Nature Conservation in Malaysia*. Malayan Nature Society, Kuala Lumpur.

CHIN, S.C. 1977. The limestone flora of Malaya 1. *Gardens' Bulletin Singapore* 30:165-219.

CHIN, S.C. 1979. The limestone flora of Malaya 2. *Gardens' Bulletin Singapore* 32:64-203.

CHIN, S.C. 1983a. The limestone flora of Malaya 3. *Gardens' Bulletin Singapore* 35:137-190.

CHIN, S.C. 1983b. The limestone flora of Malaya 4. *Gardens' Bulletin Singapore* 36:31-91.

COLLAR, N.J. & ANDREW, P. 1988. *Birds to watch. The ICBP world check list of threatened birds*. ICBP Technical Publications No.8. ICBP, Cambridge.

HENDERSON, M.R. 1939. The flora of limestone hills of the Malay Peninsula. *Journal of the Malayan Branch of the Royal Asiatic Society* 17:13-87.

HENDERSON, M.R. 1974. *Malayan Wild Flowers*. 2 Volumes. The Malayan Nature Society, Kuala Lumpur.

HOLTTUM, R.E. 1966. *A Revised Flora of Malaya. Volume II. Ferns of Malaya*. Govt. Printing Office, Singapore.

IUCN 1988. *IUCN Commission on National Parks and Protected Areas. Newsletter No.45*. Gland, Switzerland.

KENG, H. 1969. *Orders and Families of Malayan Seed Plants*. University of Malaya Press, Kuala Lumpur.

KIEW, R. 1991. *Botanical survey of limestone hills in Perak*. Unpublished Report.

KUMARI, K. & ARRIFIN, I. 1988. *Conservation Issues Critical to the Timah-Tasoh Dam and its Catchment at Mata Ayer Forest Reserve in Perlis*. WWF Malaysia, Kuala Lumpur.

MARSH, C.W. & WILSON, W.L. 1981. *A Survey of Primates in Peninsular Malaysian Forests*. Universiti Kebangsaan Malaysia, Kuala Lumpur.

MEDWAY, LORD 1978. *The Wild Mammals of Malaya (Peninsular Malaysia) and Singapore*. Oxford University Press, Kuala Lumpur. 131 pp.

ROONWAL, M.L. & MOHNOT, S.M. 1977. *Primates of South Asia. Ecology, Sociobiology and Behaviour*. Harvard University Press, London. 421 pp.

SOUTHWICK, C.H. & SIDDIQI, M.F. 1970. Primate population trends in Asia, with specific reference to the rhesus monkeys of India. *Papers and Proceedings of the 11th Technical Meeting of the IUCN (New Delhi, Nov. 1969)* 1:135-147.

WELLS, D.R. 1987. A letter to the President, Chairman, Trustees and Scientific Advisory Committee, Wordl Wide Fund for Nature, Malaysia, entitled '*Water and Wildlife Conservation in Perlis - A Crisis*'.

WHITMORE, T.C. (ed.) 1977. *Tree Flora of Malaya*. 4 Volumes. Longman, Malaysia.

WWFM 1984. *Proposals for a Conservation Strategy for Perlis*. A paper submitted to the State Government of Perlis. World Wide Fund for Nature, Malaysia, Kuala Lumpur.

APPENDIX 1. Checklist of the fauna occurring at Mata Ayer Forest Reserve and the proposed Bukit Wang Mu Forest Reserve, Perlis, Peninsular Malaysia.

Family	Species	Common name
MAMMALS		
Erinaceidae	*Echinosorex gymnurus*	Moonrat/Gymnure
Soricidae	*Crocidura fuliginosa*	South-east Asian white-toothed shrew
Cynocephalidae	*Cynocephalus variegatus*	Colugo
Pteropodidae	*Cynopterus sphinx*	Short-nosed fruit bat
Rhinolophidae	*Rhinolophus affinis*	Intermediate horseshoe bat
	Hipposideros diadema	Diadem roundleaf horseshoe bat
Vespertilionidae		Greater flat-headed bat
Tupaiidae	*Tupaia glis*	Common treeshrew
	Tupaia minor	Lesser treeshrew
Cercopithecidae	*Presbytis obscura*	Dusky langur
	Presbytis melolophos	Banded langur
	Macaca fascicularis	Long-tailed macaque
	Macaca arctoides	Stump-tailed macaque
Hylobatidae	*Hylobates lar*	White-handed gibbon
Manidae	*Manis javanica*	Malayan pangolin
Sciuridae	*Ratufa bicolor*	Black giant squirrel
	Callosciurus notatus	Plaintain squirrel
	Callosciurus caniceps	Grey-bellied squirrel
	Callosciurus prevostii	Prevost's squirrel
	Sundasciurus lowii	Low's squirrel
	Lariscus insignis	Three-striped ground squirrel
Petauristinae	*Unidentified*	-
Muridae	*Chiropodomys gliroides*	Pencil-tailed tree mouse
	Ratufa tiomanicus	Malaysian wood rat
	Rattus bowersii	Bower's rat
	Rattus rajah	Brown spiny rat
	Leopoldamys sabanus	Long-tailed giant rat
Hystricidae	*Hystrix brachyura*	Malayan porcupine
Ursidae	*Helarctos malayanus*	Malayan sun bear
Viverridae	*Paradoxurus hermaphroditus*	Common palm civet
	Paguma larvata	Masked palm civet
	Herpestes urva	Crab-eating mongoose
Felidae	*Panthera pardus*	Leopard
Rhinocerotidae	*Dicerorhinus sumatrensis*	Asiatic two-horned rhinoceros
Suidae	*Sus scrofa*	Wild pig
Tragulidae	*Tragulus javanicus*	Lesser mousedeer
	Tragulus napu	Greater mousedeer
Cervidae	*Muntiacus muntjak*	Barking deer
Bovidae	*Capricornis sumatrensis*	Serow
BIRDS		
Accipitridae	*Elanus caeruleus*	Black-shouldered kite
	Spilornis cheela	Crested serpent-eagle
	Spizaetus cirrhatus	Changeable hawk-eagle
	Spizaetus alboniger	Blyth's hawk-eagle
Phasianidae	*Arborophila charltoni*	Scaly-breasted partridge
	Gallus gallus	Red junglefowl
	Argusianus argus	Great argus
Turnicidae	*Turnix suscitator*	Barred buttonquail
Gruidae	*Rallina fasciata*	Red-legged crake
Columbidae	*Treron vernans*	Pink-necked pigeon
	Chalcophaps indica	Green-winged pigeon
Psittacidae	*Loriculus galgulus*	Malay lorikeet
Cuculidae	*Cuculus micropterus*	Indian cuckoo
	Cacomantis merulinus	Plaintive cuckoo
	Chrysococcyx xanthorhynchus	Violet cuckoo
	Phaenicophaeus diardi	Black-bellied malkoha
	Phaenicophaeus chlorophaeus	Raffles's malkoha
	Centropus sinensis	Greater coucal
Caprimulgidae	*Eurostopodus temminckii*	Malaysian eared nightjar
	Caprimulgus macrourus	Large-tailed nightjar
Appodidae	*Collocalia esculenta*	White-bellied swiftlet
	Rhaphidura leucopygialis	Silver-rumped swift
	Apus affinis	House swift
	Cypsiurus batasiensis	Asian palm swift
Trogonidae	*Harpectes duvaucelii*	Scarlet-rumped trogon
Alcedinidae	*Lacedo pulchella*	Banded kingfisher
Meropidae	*Merops leschenaulti*	Chestnut-headed bee-eater
	Nyctyornis amictus	Red-bearded bee-eater

APPENDIX 1 (Continued). Checklist of the fauna occurring at Mata Ayer Forest Reserve and the proposed Bukit Wang Mu Forest Reserve, Perlis, Peninsular Malaysia.

Family	Species	Common name
Bucerotidae	*Berenicornis comatus*	White-crowned hornbill
	Anorrhinus galeritus	Bushy-crested hornbill
	Rhyticeros undulatus	Wreathed hornbill
	Buceros rhinoceros	Rhinoceros hornbill
	Buceros bicornis	Great hornbill
	Rhinoplax vigil	Helmeted hornbill
Capitonidae	*Megalaima chrysopogon*	Gold-whiskered barbet
	Megalaima rafflesii	Many-colored barbet
	Megalaima mystacophanos	Red-throated barbet
	Megalaima australis	Blue-eared barbet
Picidae	*Picoides canicapillus*	Grey-capped woodpecker
Eurylaimidae	*Corydon sumatranus*	Dusky broadbill
	Eurylaimus javanicus	Banded broadbill
	Eurylaimus ochromalus	Black and yellow broadbill
	Calyptomena viridis	Green broadbill
Pittidae	*Pitta sordida*	Hooded pitta
	Pitta guajana	Banded pitta
Hirundinidae	*Hirundo daurica*	Red-rumped swallow
Campephagidae	*Hemipus picatus*	Bar-winged flycatcher shrike
Chloropseidae	*Chloropsis sonnerati*	Greater green leafbird
Pycnonotidae	*Pycnonotus atriceps*	Black-headed bulbul
	Pycnonotus melanicterus	Black-crested bulbul
	Pycnonotus eutilotus	Puff-backed bulbul
	Pycnonotus brunneus	Red-eyed bulbul
	Pycnonotus erythropthalmos	Spectacled bulbul
	Criniger finschii	Finsch's bulbul
	Criniger bres	Grey-cheeked bulbul
	Hypsipetes malaccensis	Streaked bulbul
Dicruridae	*Dicrurus aeneus*	Bronzed drongo
Oriolidae	*Irena puella*	Asian fairy-bluebird
Paridae	*Melanochlora sultaneae*	Sultan tit
Tamiliidae	*Trichastoma malaccense*	Short-tailed babbler
	Trichastoma rostratum	White-chested babbler
	Malacopteron magnirostre	Moustached babbler
	Alcippe brunneicauda	Brown fulvetta
	Stachyris erythroptera	Chestnut-winged babbler
	Stachyris ruficeps	Rufous-capped babbler
	Macronous gularis	Striped tit-babbler
	Macronous ptilosus	Fluffy-backed tit-babbler
	Yuhina zantholeuca	White-bellied Yuhina
Turdidae	*Copsychus malabaricus*	White-rumped shama
	Enicurus ruficapullus	Crested-naped forktail
	Myophonus caeruleus	Blue whistling trush
Sylviidae	*Phylloscopus coronatus*	Eastern crowned warbler
	Orthotomus sutorius	Common tailorbird
	Orthotomus atrogularis	Dark-necked tailorbird
	Orthotomus ruficeps	Ashy tailorbird
Muscicapidae	*Culicicapa ceylonensis*	Grey-headed flycatcher
	Hypothymis azurea	Black-naped monarch
	Trepsiphone paradisi	Asian paradise flycatcher
Motacillidae	*Dendronanthus indicus*	Forest wagtail
Laniidae	*Lanius cristatus*	Brown shrike
Sturnidae	*Acridotheres tristis*	Common myna
Nectariniidae	*Hypogramma hypogrammicum*	Purple-naped sunbird
	Arachnothera longirostra	Little spiderhunter
	Arachnothera affinis	Grey-breasted spiderhunter
Dicaeidae	*Prionochillus maculatus*	Yellow-breasted flowerpecker
	Dicaeum trigonostigma	Orange-bellied flowerpecker
	Dicaeum cruentatum	Scarlet-backed flowerpecker
Ploceidae	*Lonchura striata*	White-rumped munia
REPTILES		
Typholopidae	*Typhlops diardi muelleri*	Diard's blind snake
Colubridae	*Elaphe taeniura*	Striped racer
	Zaocys caranatus	Keeled rat snake
	Oligodon purpurascens	Brown kukri snake
	Xenelaphis hexagonotus	Malayan brown snake (?)
	Calamaria lumbricoidea	Variable reed snake
	Chrysopelea ornata	Golden tree snake
	Boiga jaspidea	Jasper cat snake
	Boiga drapiezii	White-spotted cat snake
	Psammodynastes pulverulentus	Mock viper (?)

APPENDIX 1 (continued). Checklist of the fauna occurring at Mata Ayer Forest Reserve and the proposed Bukit Wang Mu Forest Reserve, Perlis, Peninsular Malaysia.

Family	*Species*	Common name
Colubridae	*Ahaetulla prasina*	Oriental whip snake
	Dryophios rubescens	Keel-bellied whip snake
	Xenochrophis trianguligerus	Triangle keelback
	Xenochrophis piscator	Chequered keelback
	Macropisthodon rhodomelas	Blue-necked keelback
Elapidae	*Bungarus candidus*	Malayan krait
	Maticora bivirgata	Blue Malayan coral snake
	Naja naja sputatrix	Black spitting cobra
	Ophiophagus hannah	King cobra
Viperidae	*Agkistrodon rhodostoma*	Malayan pit viper
	Trimeresurus sumatranus	Sumatran pit viper (?)
Bataguridae	*Cyclemys dentata*	Asian leaf turtle
	Heosemys grandis	Giant pond turtle
	Heosemys spinosa	Spiny turtle
	Notochelys platynota	Malayan flat shell
Testudinidae	*Indotestudo elongata*	Elongated tortoise
Trionychidae	*Dogania subplana*	Malayan softshell
Varanidae	*Varanus salvator*	Malayan water monitor
	Varanus nebulosus	Clouded monitor
Scincidae	*Mabuya multifasciata*	Common sun skink
Agamidae	*Bronchocela cristatellus*	Green crested lizard
	Calotes versicolor	Changeable lizard
	Draco melanopogon	Black-bearded flying dragon
	Draco volans	Common flying dragon
Gekkonidae	*Gekko smithi*	Large forest gecko
AMPHIBIANS		
Pelobatidae	*Leptobrachium hendriksoni*	Red-eyed ground toad
	Leptobrachium nigrops	Black-eyed ground toad
	Megophrys aceras	Short-horned toad
Bufonidae	*Bufo asper*	Giant toad
Ranidae	*Rana hosei*	Poisonous rock frog
	Rana limnocharis	Grass frog
	Rana blythi	Malayan giant frog
	Rana chalconota	Copper-cheeked frog
	Rana nigrovittata	-
Rhacophoridae	*Polypedates leucomystax*	Malayan house frog
Microhylidae	*Microhyla heymonsi*	Answering froglet
	Microhyla borneensis	Rufous froglet
	Kaloula pulchra	Banded bullfrog
	Kalophrynus pleurostigma	Red-sided sticky frog

APPENDIX 2. Checklist of the non-timber flora recorded at Mata Ayer Forest Reserve, Perlis, Peninsular Malaysia.

Family	Species
Annonaceae	*Anaxagorea javanica*
Aquifoliaceae	*Ilex cymosa*
Araceae	*Alocasia lowii*
	Arisaema roxburghii
	Arisaema wrayi
	Homalomena pendula
	Raphidophora sp.a
	Raphidophora sp.b
	Schismatoglottis wallichii
	Scindapsus hederaceus
	Scindapsus scortechini
	Scindapsus sp.a
	Scindapsus sp.b
	Scindapsus sp.c
	Typhonium sp.
Araliaceae	*Brassaiopsis mitis*
Aristolochiaceae	*Apama tomentosa*
	Thottea sp.
Asclepiadaceae	*Hoya diversifolia*
Begoniaceae	*Begonia gutta*
	Begonia phoeniogramma
Chloranthaceae	*Chloranthus officinalis*
Commelinaceae	*Aneilema sp.*
	Forrestia griffithii
	Pollia sorzogonensis
	Pollia thyrsiflora
Compositea	*Erechthites valeriannifolia*
Costaceae	*Costus globosus var.a*
	Costus globosus var.b
	Costus speciosus
Cucurbitaceae	*Trichosanthes wallichiana*
Cyperaceae	*Cyperus diffusus*
	Hypolytrum nemorum
	Mapania sp.
	Scleria lithosperma
Dennstaedtiaceae	*Microlepia speluncae*
Dilleniaceae	*Acrotrema costatum*
Dioscoreaceae	*Dioscorea sp.*
Euphorbiaceae	*Antidesma velutinosum*
	Antidesma sp.
	Chaetocarpus castonocarpus
	Croton cascanillioides
	Macaranga trichocarpa
	Mallotus porterianus
Flagellariaceae	*Hanguana malayana*
Gesneriaceae	*Didymocarpus corchorifolia*
	Paraboea sp.
Gramineae	*Leptaspis urceolata*
Hypoxidaceae	*Curculigo sp.*
Liliaceae	*Dracaena umbritica*
	Peliosanthes violacea
Marantaceae	*Donax grandis*
	Donax parviflora
	Phrynium terminale
Moraceae	*Ficus pisocarpa*
Musaceae	*Musa sp.*
Myrtaceae	*Eugenia sp.*
Orchidaceae	*Bulbophyllum membranaceum*
	Claderia viridiflora
	Cymbidium pubescens
	Dendrobium bifarium
	Dendrobium flabellum
	Dendrobium germinatum
	Dendrobium leonis
	Dendrobium pandaneti
	Dendrobium secundum
	Habenaria sp.
	Hetaeria sp.
	Luisia sp.
Orchidaceae	*Oberonia sp.*
	Paphiopedilum barbatum
	Phalaenopsis cornu-cervi
	Renanthera histrinica
	Sarcochilus stenoglottis
Palmae	*Calamus sp.*
	Caryota mitis
	Daemonorops longiper
	Licuala sp.
	Maxburretia rupicola
	Pinanga malaina
	Plectocomia griffithii
Pandanaceae	*Pandanus sp.*
Pentaphragmataceae	*Pentaphragma albiflora*
Piperaceae	*Piper kurzii*
	Piper ramipilum
Polypodiaceae	*Drymoglossum piloselloides*
Rubiaceae	*Argostemma pictum*
	Hedyotis congesta
	Heydotis philippensis
	Ixora javanica
	Ixora nigricans
	Ixora pendula
	Lasianthus cyanocarpus
	Lasianthus pterospermus
	Morinda sp.
	Mussaenda sp.a
	Mussaenda sp.b
	Mussaenda glabra
	Mycetia lanceolata
	Mycetia malayana
	Oldenlandia sp.
	Ophiorrhiza discolor
	Pavetta indica
	Uncaria pedicellata
	Unidentified species
Rutaceae	*Zanthoxylum nitidum*
Solanaceae	*Solanum ferox*
Taccaceae	*Tacca integrifolia*
	Tacca leontopetaloides
Verbenaceae	*Clerodendrum laevifolium*
Vitaceae	*Leea indica*
	Vitis sp.a
	Vitis sp.b
	Vitis sp.c
Zingiberaceae	*Alpinia sp.*
	Ammomum biflorum
	Ammomum sp.a
	Ammomum sp.b
	Ammomum sp.c
	Ammomum sp.d
	Boesenbergia curtisii
	Boesenbergia plicata
	Camptandra sp.
	Elettariopsis sp.a
	Elettariopsis sp.b
	Elettariopsis sp.c
	Elettariopsis cf smithiae
	Etlingera sp.a
	Etlingera cf littoralis
	Etlingera maingayi
	Globba sp.
	Hedychium sp.
	Kaempferia pulchra
	Zingiber zerumbet
	Zingiber sp.a
	Zingiber sp.b
	Zingiber sp.c

Brunei and biodiversity - the Kew-Brunei checklist project

M.J.E. COODE, J. DRANSFIELD *and* D.W. KIRKUP

The Herbarium, Royal Botanic Gardens, Kew, Richmond, Surrey TW9 3AE, United Kingdom

The small state of Brunei Darussalam, lying on the north-west coast of Borneo, is of botanical importance and diversity out of all proportion to its size. The biological richness of Brunei's rainforests is very well appreciated by the Brunei authorities. But, before the potential of the resource can be realised, it is essential to know what actually occurs within the forest.

The flora of Borneo is covered to some extent (c. 20%) by Flora Malesiana and other specialist accounts (e.g. Dipterocarpaceae - Ashton 1964, *Rhododendron* - Argent *et al.* 1988). The trees of Brunei itself are listed by Hassan bin Pukul & Ashton (1966, reprinted recently) which was based on Ashton's years in Brunei in the fifties. The trees of Sarawak are listed by Anderson (1980). Some tree groups are also covered by the two volumes of the Trees of Sabah (Cockburn 1976, 1980). Whitmore *et al.* (1990) have recently published several parts of a Flora of Indonesia (trees of timber size), including a checklist for Kalimantan.

So why is Kew involved? The reasons, not surprisingly, are partly historical and partly pragmatic.

In 1986 Wong Khoon Meng arrived in Brunei with a brief to rejuvenate and expand the Forest Herbarium. He realised that what was needed above all was a checklist of all species of vascular plants in Brunei which would not only be essential and indeed fundamental for proper development planning in Brunei itself but also would be valuable for the surrounding areas of Sarawak, Sabah and Kalimantan. He could not hope to accomplish such a task in the relative isolation of Brunei, without ready access to material from those surrounding areas, without types and without literature. Kew is blessed with these things.

Meanwhile, John Dransfield was seeking a focus for rejuvenating South-East Asian interests at Kew. A joint Kew-Brunei Checklist would serve both purposes; they began planning and, in March 1990, an accord between the then Director of Forestry Haji Yasin and the Keeper of the Herbarium at Kew, Gren. Lucas, was signed. In brief, this accord envisaged a programme of joint exploration of Brunei for 5 years during which the Forestry Department would provide all logistic support in Brunei while Kew would send specialists at regular intervals; Kew would also mastermind the production of the checklist of vascular plants itself and a database at the project's end as complete as possible as regards taxa. The checklist would be at least one step 'ahead' of the existing lists - it would contain specimen citations and, where practicable, keys and short descriptions; it was agreed that it would be unrealistic to expect a comparable level of detail throughout. Taxa will be separated out to the best of the specialist's abilities and named where possible, but in certain of the intractable groups - Rubiaceae and Ebenaceae, for instance, which need basic revision - it will not be possible to provide determinations other than species a, species b etc.

The expertise of Robert Johns, with experience in New Guinea before he came to Kew, and David Edwards, with experience in Brunei, will permit the ferns and fern-allies to be treated separately and in more detail.

Part of the agreement was the understanding that the top set of any collection would remain in the Forestry Department Herbarium at Sungai Liang (BRUN) and the second set

D. S. Edwards et al. (eds.), Tropical Rainforest Research – Current Issues, 163–168.

would come to Kew. Six other institutes (in alphabetical order: A, KEP, L, SAN, SAR, SING) were earmarked to receive the next 6 duplicates, while it was agreed that collaborating specialists would receive the third set, whether in that list or not.

Obviously, both sides have already benefitted from the collaboration. Both 'lead' herbaria have been enormously enriched. Kew botanists have been able to collect in tropical rainforest under near-ideal conditions. Two staff from the Brunei Herbarium have received training at Kew and a third is expected to begin training in October. Friendship and trust has been built up on both sides.

At Kew, much of the initial work was done by Leonard Forman who retired in 1991. Mark Coode was put in charge of the 'Kew end' towards the end of 1991 while John Dransfield remains responsible for overall policy and for liaison at all levels in Brunei. the team has been joined for some of his time by Don Kirkup, who manages the data-base. Happily, Leonard Forman has carried on working, in his retirement, on the identification of the collections. In 1992, Wong Khoon Meng, having set the system up in Brunei, moved to Sandakan and his place has been taken recently by Dr Idris Said.

THE DATABASE

Recent development of botanical databases has seen a division into two different but overlapping areas; specimen-based databases which contain information relating to individual specimens and taxon-based databases which deal with plant names and associated references to the literature. Once up and running, databases of the two types may be related through classes of information which both share (such as that of plant name) and represent a powerful taxonomic tool.

The Brunei database is a specimen-based database. The range of information held in such databases depends on its end use and so designs can vary greatly. A highly specialised design might involve exhaustive and detailed descriptors of specimens for the purpose of classification, on-line identification, generating 'automatic' descriptions or 'automatic' key writing. The Brunei database is of a more 'general purpose' design than this (Table 1), incorporating a more limited range of plant descriptors together with locality and habitat data, date of collection and ancillary information such as local names and uses. The individual specimen records are retrievable through the unique combination of collector name and number (or collector and date where *s.n.*).

It is not envisaged that the finished checklist should be generated automatically from the database; rather, a skeleton listing family by family be produced which summarises all the available information for each species. This will provide a framework into which other information (from the literature or re-examination of the specimens) can be incorporated.

The utility of a database like this increases with the number of records added. Interim listings include a gazetteer and a list of taxa, *ad hoc* queries include a listing of rheophytes.

Of course, at the end of the project Kew would expect to hand over control of the database to Brunei for Bruneian botanists to expand and use.

The database has been built up so far in two main phases.

Phase 1

As part of Kew's response to the agreement, three 'sandwich course students' worked consecutively over a period of nearly three years, combing the Kew Herbarium systematically for collections of flowering plants from Brunei. Label details were captured and in the end about 2900 collections were recorded, almost all of these Ashton collections in the BRUN series (undoubtedly, there will be more at BRUN).

Table 1. Structure of the Brunei Checklist database. All numeric fields are integers, apart from numbers 23 & 25 which have two decimal places.

Field	Field Name	Type	Width	Field	Field name	Type	Width
1	SPNUM	Numeric	6	20	LOCALTEXT	Character	40
2	FAMILY	Character	30	21	GEOLOGY	Character	40
3	GENUS	Character	25	22	HABITAT	Character	40
4	SP1	Character	25	23	LAT	Numeric	5
5	SPAUTH1	Character	30	24	NS	Character	1
6	SSPVAR	Character	3	25	LONG	Numeric	6
7	SP2	Character	25	26	EW	Character	1
8	SPAUTH2	Character	30	27	ALT	Numeric	6
9	COLLECTOR	Character	30	28	DETBY	Character	30
10	NUMBER	Character	15	29	DETYEAR	Numeric	2
11	ADDCOLL	Character	40	30	HABIT	Character	40
12	COLLDD	Numeric	2	31	FLOWEROBS	Character	40
13	COLLMM	Numeric	2	32	FRUITOBS	Character	40
14	COLLYY	Numeric	2	33	SIZE	Character	15
15	CENTURY	Numeric	2	34	NOTES	Memo	10
16	COUNTRY	Character	20	35	USES	Character	40
17	MAJORAREA	Character	30	36	VERNACULAR	Character	40
18	MINORAREA	Character	30	37	WILDCULT	Numeric	6
19	LOCALITY	Character	30				

The data were captured into a single DBASE3+ table on a laptop computer; the portability of the machine allowed the operator to move from site to site around the herbarium. The students were 'biologically oriented' and were competent to make decisions regarding the extraction of information from the specimens into the database without involving large amounts of specialists' time. Printouts were checked for errors by senior staff.

Phase 2

This is a continuing process involving the addition of collections new to the database. Collections made under the Collaborative Agreement have also been recorded - Wong's 2000 collections and Kew's 4850 (with more to be processed). Together with collections made by other members of the Forest Department staff and by colleagues visiting from overseas the total now stands at about 10,000.

At Kew, the project collectors all input their own data. Originally, a DBASE3+ programme, LABELS 3, was used for data input and herbarium specimen label production. Although somewhat limited in its original form, it was uncompiled and therefore easily customised by the project to produce acceptable labels. LABELS 3 has now been superseded by a revised database structure compatible with the BRAHMS programme being developed at the Oxford Forestry Institute, a structure gaining wider currency. Data entry is direct into a single DBASE4 table and label production is via an in-house programme (essentially a mail merge process).

We also need to compare the Kew listings with the BRUN card index and add those records only at BRUN and we hope also to include at least the fern component of the Herbarium here in the University of Brunei Darussalam. There are also van Niel's collection of c. 1500 numbers from the early sixties, lodged in Leiden but never systematically written up.

Database Management: Following the completion of phase 1 came a shift in requirements, from data capture to both data capture and manipulation, the additional workload involving updating determinations as the newly collected material was named; producing labels and listings; editing, cleaning up the data; routine housekeeping tasks.

The replacement of the laptop computer with a faster machine with a larger storage capacity (486/33mhz 120mb HDD) was essential at this juncture. At the same time the less easily solved problem of maintaining the database integrity had to be addressed. At present this is achieved by restricting access to computer-literate users only. New records and name changes are entered into temporary files which are used to update the main database. All queries are channelled through trained users.

A longer-term solution is for some sort of database management program, e.g. BRAHMS, which offers verification of data at entry and has built-in security procedures, safely allowing unskilled users access to the database for basic queries and edits. The BRAHMS program is still evolving and the Checklist Project has been closely involved in its recent development and evaluation. In addition, BRAHMS has recently been adopted by the Tree Flora of Sabah and Sarawak. Future developments of BRAHMS will include its own GIS.

Kew will also deliver the checklist in book form, for normal publication.

ASSESSMENT

The question has often been asked: Why does Kew put so much effort into Brunei? The flora's well-known already; Ashton was there for years. Perhaps the following figures will show that the effort has not been misplaced, that the flora was not well-known, and that, if we thought Brunei was rich before, now we can say it has twice the genera (Table 2) and at least twice the number of species (Table 3) we knew it had.

As an example: for Palms, fifteen species were known in the collections at the start of the project; there are now 125. For *Elaeocarpus*, Ashton listed 15 of which ten can be accepted with the names he applied to them; one was added in the seventies and the tally now is 12 of his original 15 and 16 more at least (including four from our recent trip to Gunung Pagon) out of a total for Borneo of about 66. So the increase in known taxa, while due mostly to recent collections of non-timber species, also includes many trees which Ashton missed. This

Table 2. Figures from the database. Figures for genera collected since 1988 will be slightly inflated by uncorrected errors being read as 'new' genera - but we expect to 'replace' as many 'genera' as are lost when errors are eliminated by genuinely new generic records amongst specimens currently recorded to family only.

Genera, pre-1988, in Kew	Genera collected since	Total known genera
478	529	1007

Table 3. Numbers of determined specific epithets for Brunei. The same proviso applies to this table as to Table 2. Note - no account taken of specimens named simply to genus, which outnumber those so far identified to species.

Species pre-1988 in Kew	Species collected since	Total named species
1425	1626	3051

Table 4. Approximate numbers of species in a range of groups in which the taxonomy is reasonably well-known. For Bamboos, we are indebted to S. Dransfield; Ferns - R. Johns; Aroids - P. Boyce; Palms - J.D.; *Elaeocarpus* - M.C.; *Begonia* - M. Sands; Loranthaceae - D.K.; Dipterocarps - Ashton (1982) and Forestry Department figures; *Mapania* - D. Simpson (1992).

	Borneo total	Brunei total	Rough Percentage
Bamboos:	50-60	c. 10	20
Ferns:	1000-1200		25-30
Aroids:	c.220	c. 110	50
Palms:	302	125	45
Elaeocarpus	65-70	c. 28	40
Begonia	120	17	14
Loranthaceae	77	27	35
and from published accounts:			
Dipterocarps	267	171	55
Mapania	26	13	50

is not to lessen the importance of Ashton's achievement - but to emphasise rather the richness of the flora. For instance, let us consider the easily accessible locality known as 'Bukit Teraja', visited by Ashton and subsequently by at least 5 teams under the collaborative venture. Ashton found *Sloanea javanica* at Bukit Teraja, but later visits by others did not. D. Kirkup and M. Coode found at least one *Elaeocarpus* and a very odd member of the Euphorbiaceae there, both new to Brunei, which none of the others found. J. Dransfield found *Pterisanthes grandis*, a new species of *Calamus* and a curious *Monocarpia* (Annonaceae). In fact, each trip to Bukit Teraja has resulted in finding species that all other teams have missed.

For those groups we at Kew have looked at more closely, a few figures will emphasise the increase in knowledge (Table 4). From these we can hazard a guess that either about one third of all Borneo's species are present in this one small area of less than 1% of the land surface (probably nearer one half of Borneo's flora below 2000 m) or that hundreds, even thousands, of new species still await discovery in the surrounding areas, in numbers sufficient to bring the figures to more reasonable percentages area for area. While undoubtedly new species will be found elsewhere in Borneo, we do not believe that their number will materially alter these figures and, for us, the first conclusion is the more likely. Finally, and most importantly, each new record increases the known value of Brunei's forests as a 'gene-pool' island for the future.

LITERATURE CITED

ANDERSON, J.A.R. 1980. *A Checklist of the trees of Sarawak*. Dewan Bahasa dan Pustaka Lawangan Sarawak, for Forest Department, Sarawak.

ARGENT, G., LAMB, A., PHILLIPS A. & COLLENETTE, S. 1988. *Rhododendrons of Sabah*. Sabah Parks Publication number 8. Kota Kinabalu.

ASHTON, P.S. 1964. *A Manual of the dipterocarp trees of Brunei State*. Oxford University Press, London.
ASHTON, P.S. 1982. *Dipterocarpaceae*. In Flora Malesiana Ser. I, Vol. 9, Pt. II:237-552. M. Nijhoff, The Hague.
COCKBURN, P.F. 1976. *Trees of Sabah*. Vol. 1. Sabah Forest Record 10. Borneo Literature Bureau for Forest Department, Sabah. Kuching, Sarawak.
COCKBURN, P.F. 1980. *Trees of Sabah*. Vol. 2. Sabah Forest Record 10. Borneo Literature Bureau for Forest Department, Sabah. Kuching, Sarawak.
HASAN BIN PUKUL & ASHTON, P.S. 1965, reprinted 1988-9. *A Check-List of Brunei Trees*. Forest Department, Brunei Darussalam.
SIMPSON, D.A. 1992. *A revision of Mapania (Cyperaceae)*. Royal Botanic Gardens, Kew.
WHITMORE, T.C., TANTRA, I.G.M. & SUTISNA, U. 1990. *Tree Flora of Indonesia. Checklist for Kalimantan*. Parts I, II.1, II.2. Forest Research and Development Centre. Bogor, Indonesia.

Stalk-Eyed Flies: Caprices of Evolution?

DIETRICH BURKHARDT *and* INGRID DE LA MOTTE

Institut für Zoologie der Universität, D 93040 Regensburg, Germany

ABSTRACT. A report is given on the species of stalk-eyed flies found around the Kuala Belalong Field Studies Centre (Brunei Darussalam, Borneo) during the Brunei Rain Forest Project 1991-92 of the University of Brunei and the Royal Geographical Society. Twelve of thirteen species of stalk- eyed flies found belong to the family Diopsidae. Ten of these are new species which are yet to be described. Local diversity of that family is much higher than hitherto assumed. Some of the species closely resemble their sister species present on the Malay Peninsula but do not interbreed with them. Allometric data of body length and eyespan are given for two species; some features of their behaviour are described. The only non diopsid stalk-eyed fly found, *Themara hirtipes* (Rondani), belongs to the family Tephritidae. Here only males have eyestalks. Allometric data of males are presented, together with field observations on territoriality, mating behaviour and reproductive behaviour.

KEY WORDS: Diopsidae, eyestalks, male competition, Tephritidae, Borneo, Brunei Darussalam.

INTRODUCTION

Stalk-eyed flies have their compound eyes on the tips of long, rigid lateral extensions of the head. Eyestalks are a typical male feature in a few species of several dipteran families, including the Tephritidae, Platystomatidae, and Drosophilidae. In the family Diopsidae, however, both sexes of all species have eyestalks (Burkhardt & de la Motte 1987a, Feijen 1989, Grimaldi & Fenster 1989, Shillito 1971). In large diopsid species the approximately 2500 facets of each compound eye have small receptive angles. These provide a high visual acuity, often well below one degree in the frontal direction (Burkhardt & de la Motte 1983). Comparable high visual acuity is mostly found in much larger insects such as honey bees and dragonflies. The eyestalks of diopsids provide several selective advantages by improving visual capacities: the visual field of each compound eye exceeds the perimeter of its respective hemisphere. Thus both eyes together provide panoramic vision and additionally binocular vision for two thirds of the entire sphere, leaving monocular vision to only the lateral parts. Eyestalks also allow them to look out from hiding places and, most importantly, a wide eye separation allows a long range estimate of an object's size. This last ability is crucial for recognizing and sizing up conspecifics. In species with sexual dimorphism in eyestalk length, the male's eyespan is strongly and steeply correlated with its body size: 25% increase of body length corresponds to 75% increase of eyespan (Burkhardt & de la Motte 1987b). Thus eyespan signals a male's strength in a much exaggerated way. This facilitates male competition and female choice. With the sensory advantages of eyes on long stalks a male is better equipped to size up a competitor's strength, yet simultaneously the eyespan is an enhanced signal of the male's own strength.

RESULTS

Diopsidae

In South-East Asia diopsids are found close to creeks and rivers. They form a substantial portion of the dipteran fauna along these waterways in areas where the vegetation is

D. S. Edwards et al. (eds.), Tropical Rainforest Research – Current Issues, 169–174.

undisturbed. Diopsid larvae feed on decaying plant material and, in rainforests, aid in the decomposition of organic matter, as do other dipteran larvae. For Borneo seven diopsid species have been listed (Steyskal 1975). During two short stays with the Brunei Rainforest Project in the Belalong area we found twelve species most of which are yet to be described (H. Feijen, personal communication). This indicates that local diversity of this family is much higher than has been assumed hitherto. Some of the Bruneian species show a close resemblance with sister species from the Malay Peninsula, but speciation seems to have progressed since, in our cultures only one of these species, *Megelabops sexguttata* (Brunetti), crossbreeds with its counterpart from the Malay Peninsula.

In some diopsid species, males and females have eyestalks of about the same length and are monomorphic in that character. The two largest species from our Brunei samples, *Eurydiopsis argentifera* (Bigot) and a *Cyrtodiopsis* species (sp. nov., close to *C. quinqueguttata* (Walker)), belong to this group. In the monomorphic species eyespan measurements have a standard deviation of 3% to 5% of the mean, which corresponds to the typical variability of such insect body measurements. In other species, especially in the genus *Teleopsis*, males have much longer eyestalks than females. In these species variability of eyespan is extremely high and the standard deviation may reach 12% of the mean in females and 30% of the mean in the males with their longer eyestalks.

Of all the diopsids we have found in the Belalong area the smallest (two species of the *Sphyracephala* group) showed the closest affinity to water. We found them on emergent boulders and logs in the river or close to the river banks. The two largest species *Eurydiopsis argentifera* and a *Cyrtodiopsis* species (sp. nov., close to *C. quinqueguttata*), were sometimes found distant from the rivers, several hundred metres up the slopes. The predominantly dimorphic *Teleopsis* species kept to the river banks or fairly close to smaller creeks uphill.

Among the new Bruneian diopsids a *Teleopsis* (sp. nov., close to *T. belzebuth* (Bigot)) sets a new record: the eyespan of large males may reach 20 mm, with a body length of only 8 mm. This large eyespan makes eye cleaning a difficult task. We propose the name *Teleopsis belalongensis* in the description to come.

In some of the sexual dimorphic species females outnumber males. Typically the flies gather at sunset in sleeping societies, usually on an aerial rootlet, a dried tendril or a dry vein of a rotten leaf. In most cases a large male is the first to settle on a sleeping site, followed by smaller males and females. Given a choice, females favour a sleeping site already occupied by a large male (Burkhardt & de la Motte 1989). Lesser males are often driven away by a large male. Males of equal size frequently engage in highly ritualized fights to win a harem (de la Motte & Burkhardt 1983). This behaviour was thoroughly studied in the Malay species *Cyrtodiopsis whitei* (Curran), and first observations on two of the Bruneian species, *Teleopsis pallifacies* (Feijen) and *Teleopsis belalongensis*, reveal a similar behaviour: the males display fighting activity on the sleeping sites. Figures 1 to 3 show the allometric data of the Bruneian species in comparison with those of *Cyrtodiopsis whitei*. Here we also find a close concordance.

Tephritidae

On the West ridge of the Belalong study area we encountered a stalk-eyed member of the tephritid family, *Themara hirtipes* (Rondani). In tephritids the female has a normal fly's head configuration; eyestalks are purely a male attribute. As in the diopsids, male eyespan is strongly correlated with the body length (Figure 4). These data are derived from measurements of the population we found on a rotting log on the West ridge slope at an altitude of about 180 m. We measured the visual field of the eyes of males using the pseudopupil method (cf. Burkhardt & de la Motte 1983). The width of the visual field in

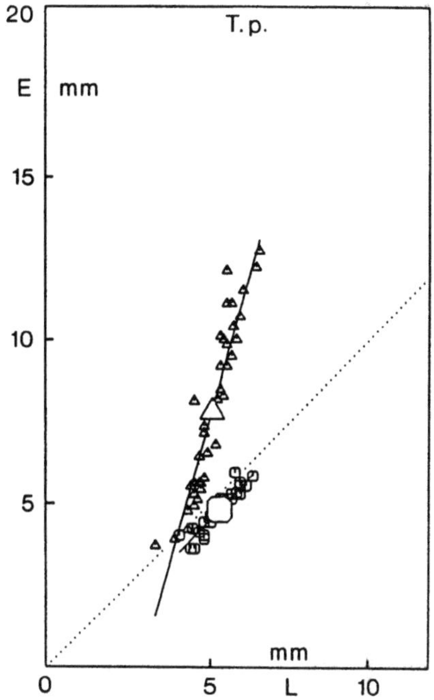

Figure 1. Correlation between eyespan (E) and body length (L) in *Teleopsis pallifacies*, sp.nov. Abscissa: body length in mm, ordinate: eyespan in mm. Triangles: males, squares: females. Large symbols indicate the means. Dotted line: one to one ratio.

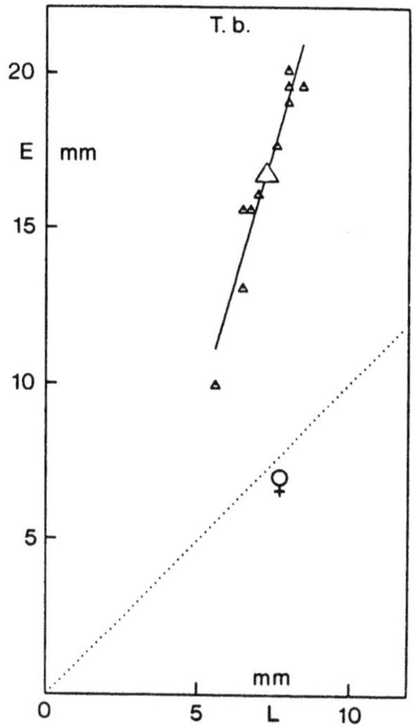

Figure 2. Correlation between eyespan (E) and body length (L) in *Teleopsis belalongensis*, sp. nov. Abscissa: body length in mm, ordinate: eyespan in mm. Values for the males are small triangles. The large triangle indicates the mean value of the 11 males, the female symbol marks the mean value of the densely clustered measurements from 8 females.

tephritid males is about the same as in diopsid males. We observed the population of more than 50 flies for ten days. Usually they arrived at the log in the morning, stayed there during daytime and left for an unknown destination in the late afternoon. About 50 of the flies were captured in small tubes. We took them to the camp to take body measures and to apply marking numbers and released them at the log on the following morning. Some of the marked males returned on ten successive days. The males fought for territories on the log or sometimes on the ground close to the log.

Sizing up the opponent is the initial behaviour in male/male encounters. The smaller of the two usually turns away. Males of about equal size make contact with their heads and eyestalks and try to push each other off the premises. Such contests may last for many minutes and are sometimes repeatedly resumed by the same pair of males. The winner stays in the territory, the looser runs or flies away. Females that pass by are courted. Successful courtship leads to copulation, which lasts 20 minutes on average. Shortly thereafter, the female begins to deposit eggs, searching with her mouthparts and the abdomen for insect-made holes in the log. Normally, after copulation and during egg deposition the male guards the female; this may last for one hour. While mate guarding, even a small male is very pertinacious and successfully fights off much larger intruders. We even observed a small male alternately fighting with two intruders simultaneously. Another female passing the territory will cause most males to turn their interest from the egg laying female to the newcomer.

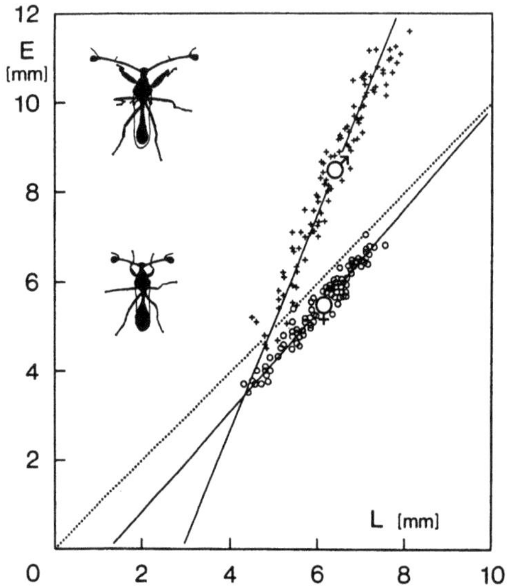

Figure 3. Correlation between eyespan (E) and body length (L) in *Cyrtodiopsis whitei* (specimens from Malay Peninsula). Abscissa: body length in mm, ordinate: eyespan in mm. Crosses: values for males, circles: values for females. The large male and female symbols are positioned at the respective mean values.

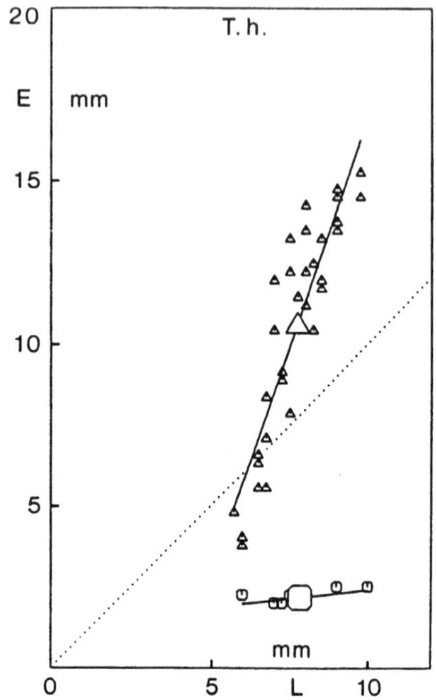

Figure 4. Correlation between eyespan (E) and body length (L) in *Themara hirtipes*. Abscissa: body length in mm. ordinate: eyespan in mm. Males: triangles, females: squares; the larger symbols indicate the mean values.

DISCUSSION

The majority of the newly found diopsid species in the study area belong to the genus *Teleopsis* (see Table 1). As a revision of this genus is in preparation (H. Feijen, personal communication), a full description of these species will be given later.

Eyestalks of flies were often regarded as a hypertely, yet several earlier authors speculated on special visual and various other functions (see Feijen 1989). Recently eyestalks have been mostly discussed in the context of male/male encounters, since head to head encounters between males can be observed in many fly species (see Feijen 1989, Grimaldi & Fenster 1989). Indeed, such male/male encounters have been found in nearly every stalk-eyed species that has been thoroughly investigated.

Thus, at first glance, stalk-eyed flies may strike one as another showcase for a 'caprice of evolution', like horned beetles and the like. But a closer look reveals there is more to it: there are the enhanced visual capacities including a large binocular field which allows accurate size estimation over large distances, there is the unambiguous signal advertising a male's strength, there is the possibility of using a stalked eye to peer around corners, and, finally, the flies can use their eyestalks to push an opponent off a wanted territory (Burkhardt & de la Motte 1987a, Burkhardt & de la Motte 1987b, Burkhardt & de la Motte 1989, de la Motte & Burkhardt 1989).

Table 1. List of stalk eyed flies from the Belalong area, Brunei Darussalam.

Diopsidae

1. *Teleopsis belalongensis*, sp. nov.
2. *Teleopsis pallifacies*, sp. nov. (Feijen, description unpublished)
3. *Teleopsis* sp. nov.
4. *Teleopsis* sp. nov.
5. *Teleopsis* sp. nov., close to *T. discrepans* (Walker)
6. *Megalabops* (= *Teleopsis*) *sexguttata* (Brunetti)
7. *Cyrtodiopsis* sp. nov., close to *C. dalmanni* (Wiedemann)
8. *Cyrtodiopsis* sp. nov., close to *C. quinqueguttata* (Walker)
9. *Eurydiopsis* spec., probably *E. argentifera* (Bigot)
10. *Pseudodiopsis* (= *Sphyracephala*) sp. nov.
11. *Pseudodiopsis* (= *Sphyracephala*) sp. nov., close to *P. detrahens* (Walker)

Tephritidae

12. *Themara hirtipes* (Rondani)

One more undescribed species of the genus *Teleopsis* was found. There are only two specimens (one male and one female) which look very much like *Teleopsis burkhardti* (Feijen sp. nov., unpublished) from the Malay Pensinsula. Resin-embedded samples (males plus females) of each of the listed species 1 to 12 are in the custody of the Brunei Museum.

The question naturally arises whether in a competitive situation larger males do indeed have higher reproductive success than smaller ones. This assumption has been proved in one species using genetically marked pure strains of *Cyrtodiopsis whitei* (Burkhardt *et al.* 1991, Burkhardt *et al.* 1993).

ACKNOWLEDGEMENTS

We are grateful for all the help we received from colleagues at the University of Brunei Darussalam and for their gracious hospitality. We thank the RGS (London) for inviting us to participate in the UBD/RGS Brunei Rain Forest Expedition 1991-92 and for supporting our visit to Brunei in April 1993. We thank Mrs. Eckert and Mr. Schrepfer for taking care of our Diopsids.

LITERATURE CITED

BURKHARDT, D. &. MOTTE, I. DE LA 1983. How Stalk-Eyed Flies Eye Stalk-Eyed Flies: Observations and Measurements of the Eyes of *Cyrtodiopsis whitei* (Diopsidae, Diptera). *Journal of Comparative Physiology* A151:407-421.

BURKHARDT, D. & MOTTE, I. DE LA 1987a. Selective Pressures, Variability, and Sexual Dimorphism in Stalk-Eyed Flies (Diopsidae, Diptera). *Naturwissenschaften* 72:204-206.

BURKHARDT, D. & MOTTE, I. DE LA 1987b. Physiological, Behavioural, and Morphometric Data Elucidate the Evolutive Significance of Stalked Eyes in Diopsidae (Diptera). *Entomologica Generalis* 12:221-233.

BURKHARDT, D. & MOTTE, I. DE LA 1989. Big 'Antlers' are Favoured: Female Choice in Stalk-Eyed Flies (Diptera, Insecta), Field Collected Harems and Laboratory Experiments. *Journal of Comparative Physiology* A152:649- 652.

BURKHARDT, D., MOTTE, I. DE LA & YONG, H. S. 1991. PGM: A Suitable Genetic Marker in Studying Mating Success of *Cyrtodiopsis whitei* (Diopsidae, Diptera). *Naturwissenschaften* 78:76-78.

BURKHARDT, D., MOTTE, I. DE LA & LUNAU, K., in press 1993. Signalling Fitness: Larger Males Sire More Offspring. Studies of the Stalk-Eyed Fly *Cyrtodiopsis whitei* (Diopsidae, Diptera). *Journal of Comparative Physiology* A.

FEIJEN, H. 1989. Diopsidae. Pp. 1-122 in Griffiths, G. C. D. (ed.). *Flies of the Nearctic Region, Vol. IX, Part 12*. E. Schweizerbart'sche Verlagsbuchhandlung, Stuttgart.

GRIMALDI, D. & FENSTER, G. 1989. *Evolution of Extreme Sexual Dimorphisms: Structural and Behavioral Convergence Among Broad-Headed Male Drosophilidae (Diptera)*. American Museum Novitates No. 2939, 25pp, New York.

MOTTE, I. DE LA & BURKHARDT, D. 1983. Portrait of an Asian Stalk-Eyed Fly. *Naturwissenschaften* 70:451-461

SHILLITO, J. F. 1971. Dimorphism in Flies with Stalked Eyes. *Zoological Journal of the Linnean Society* 50:297-305.

STEYSKAL, G. C. 1975. Family Diopsidae. Pp. 32-36 in Delfinado, M. & Hardy, D. E. (eds). *A Catalog of the Diptera of the Oriental Region, Volume III*. Hawaii University Press, Honolulu.

Small mammal diversity in riparian and dipterocarp habitats in Belalong forest, Brunei Darussalam.

JOSEPH K. CHARLES

Universiti Brunei Darussalam, Bandar Seri Begawan 2028, Brunei Darussalam

ABSTRACT. A small mammal study carried out in the forest of Belalong in the Temburong District of Brunei Darussalam, showed that 18 species of small mammals (insectivores, tupaiids and rodents) occurred in the dipterocarp forests as compared to 14 species in the riparian habitats. While some small mammals were exclusive to either the dipterocarp or riparian habitats, others were recorded in both habitats. It is inferred that competition between *Maxomys surifer* and *M. rajah* has displaced the former to the riparian areas. Habitat disturbance of a riparian area appears to have led to a population increase in two murid species, *Maxomys muelleri* and *Rattus exulans*.

KEY WORDS: Oriental, insectivores, tupaiids, rodents, competition, habitat disturbance, Borneo.

INTRODUCTION

Considerable tectonic uplift in geologically recent times in the southern Temburong District of Brunei has given rise to steep slopes and sharp ridges (Dykes, this volume). This has affected the extent of riparian areas in the Belalong valley of Ulu Temburong. The structure and composition of the vegetation and the invertebrate fauna in these areas perhaps determine the presence or absence of certain vertebrate fauna there. Some of the vertebrates may not be found in the adjacent hill dipterocarp forests. A small mammal study was carried out from February 1991 to February 1992 during the Brunei Rainforest Project that was centred in the Kuala Belalong Field Studies Centre (KBFSC) in Belalong (Temburong District, Brunei) to survey the diversity in both riparian and dipterocarp habitats and the movements of some small mammal species in different forest habitats. A wildlife survey carried out in Bukit Tudal, Ulu Temburong, about 25 km south-east of the Kuala Belalong Studies Centre included small mammals but their diversity in relation to different habitats was not investigated (Bennett *et al.* 1987).

METHODS

The study was carried out in two forest types; a variety of survey techniques were employed.

Forest types

Dipterocarp forest: two plots were chosen, as follows:

(a) A 0.9 ha plot at 85 m altitude on the West Ridge slope, close to Sungai (Sg.) Belalong (SM Plot 1).

(b) A 1.0 ha plot at 330 m altitude on the East Ridge (SM Plot 2).

Riparian forest: two plots were chosen, as follows:

(a) An undisturbed 0.22 ha plot (Riparian Plot 1) at 50 m altitude along the Sg. Belalong.

(b) A disturbed 0.22 ha plot at 50 m altitude where the Kuala Belalong Field Studies Centre is situated (Riparian Plot 2).

The distance between Riparian Plots 1 and 2 was about 150 m. Riparian Plot 2 and SM

D. S. Edwards et al. (eds.), Tropical Rainforest Research – Current Issues, 175–182.

Plot 1 were contiguous. It was not possible to use larger study plots in the riparian forest due to to the limited forest fringe available. The Kuala Belalong Field Studies Centre is situated in the valley between the East and West Ridges.

Survey techniques

There are a variety of trapping and marking techniques and their relative merits are discussed by Barnett (1992). The following methods were chosen and modified where necessary, based on previous experience in small mammal research.

Trapping by cage traps

At all sites, standard wire-mesh cage traps for small mammals were set at 10m grid intersection points. The traps were modified to be collapsible and hence permit portability in the forest. The bait used was banana. Previous experience in trapping small mammals in forests in Malaysia and a series of bait acceptance trials in various forests in Brunei (Lamunin, Sg. Akar and Kampung Sg. Dadap) have shown that bananas ('pisang rasthali') were the most acceptable bait under the prevailing conditions (unpublished data).

Unbaited traps were fixed in the open position and left in the respective habitats for one to two weeks. They were then pre-baited for 3 months twice weekly, followed by a 2 to 3 week period of no baiting before trapping began. This method was to permit acceptability of the traps as part of the habitat and to prevent trap shyness in the small mammal species resident in the study plots. To prevent irrecoverable damage to plot vegetation, no trapping session exceeded seven days.

Since the overall study also included investigation of home ranges and population size, marking was necessary. Trapped mammals were first etherised, measured and marked before being released at the point of capture. Previously marked animals were released immediately. Permanent marker pens (Staedtler) and dyes (e.g. Malachite Green, Carbol fuschin) were used to mark the fur; these marks remained discernible for 2 - 3 months. Animals trapped with faint marks were re-marked.

Trapping by pit-fall traps

Buckets buried in the ground with drift fences leading to them were used as pitfall traps to catch very small mammals such as shrews that could not be trapped in the small mammal traps. These traps were located outside the plots to avoid disturbance to the study areas.

Nocturnal surveys

When the weather was favourable (when there was no rain or heavy mist), nocturnal surveys using powerful torches (12 volt Coleman Halogen Spotlights producing approximately 100 watts or 750,000 candle power beam) were conducted for 2 to 3 hours per survey. A distance of 500 m of regular trails was walked at the rate of 150 m per hour from 1900-2200 hrs. The area was carefully scanned for nocturnal mammals. Reflected eyeshine showed the presence of a mammal. Identifications were confirmed using a searchlight and binoculars. This technique was employed in all habitats but much of the time surveys were affected by inclement weather.

RESULTS & DISCUSSION

Though the study covered diversity and home ranges, the results reported here only deal with small mammal diversity, competition between two murids and murid densities in the disturbed and undisturbed riparian areas.

Species richness in the lowland dipterocarp and riparian forests

The term 'small mammal' usually refers to non-flying mammals less than 1 kg in adult weight and in practice includes insectivores, rodents and marsupials. The results presented in this section covers insectivores, tupaiids and rodents and excludes mustelids and viverrids, some of which also occur in both forest types.

Eighteen species of small mammals were recorded in the dipterocarp forest plots in the East and West ridges as compared to 14 in the riparian habitat (Table 1).

Borneo's smallest shrew, Savi's Pygmy Shrew, *Suncus etruscus*, was caught in the pitfall traps in the dipterocarp forest. This is the first record for Brunei. The Bukit Tudal wildlife survey (Bennett *et al.* 1987) showed the presence of the South-east Asian White-toothed Shrew, *Crocidura fuliginosa* rather than *Suncus etruscus* in that area.

Of the 8 species of tupaiids (tree shrews) found in Borneo, one is confined to mountains (Mountain Treeshrew, *Tupaia montana*) while 7 are recorded in the lowland forests. In the

Table 1. Small mammals of the Lowland Dipterocarp Forest (DF) and Riparian (R) areas in Belalong.

	DF	R
Order Insectivora		
Soricidae		
Suncus etruscus, Savi's Pygmy Shrew	*	
Order Scandentia		
Tupaiidae		
Tupaia tana, Large Tree Shrew	*	
Tupaia picta, Painted Tree Shrew	*	*
Tupaia minor, Lesser Tree Shrew	*	
Tupaia gracilis, Slender Tree Shrew	*	
Tupaia dorsalis, Striped Tree Shrew	*	
Order Rodentia		
Muridae		
Rattus exulans, Polynesian Rat		*
Rattus tiomanicus, Malaysian Field Rat		*
Sundamys muelleri, Muller's Rat		*
Maxomys surifer, Red Spiny Rat		*
Maxomys rajah, Brown Spiny Rat	*	*
Maxomys whiteheadi, Whitehead's Rat	*	*
Leopoldamys sabanus, Long-tailed Giant Rat	*	*
Niviventer cremoriventer, Dark-tailed Tree Rat	*	*
Sciuridae		
Callosciurus notatus, Plantain Squirrel	*	*
Callosciurus prevostii, Prevost's Squirrel	*	*
Sundasciurus hippurus, Horse-tailed Squirrel	*	
Sundasciurus lowii, Low's Squirrel	*	*
Exilisciurus exilis, Plain Pygmy Squirrel	*	*
Ratufa affinis, Giant Squirrel	*	*
Rheithrosciurus macrotis, Tufted-ground Squirrel	*	
Hystricidae		
Thecurus crassipinis, Thick-spined Porcupine	*	
TOTAL	18	14

East and West ridges of Belalong, 5 were recorded in the dipterocarp areas (*Tupaia tana, T. picta, T. minor, T. dorsalis* and *T. gracilis*), while only *T. picta* was present in the riparian habitats. *T. glis,* found on Bukit Tudal (Bennett *et al.* 1987), is absent at Belalong.

Eight species of forest rats were trapped, of which only 4, *Maxomys rajah, M. whiteheadi, Leopoldamys sabanus* and *Niviventer cremoriventer,* were found in the dipterocarp forest. These and 4 others, namely *Rattus exulans, R. tiomanicus, Sundamys muelleri* and *Maxomys surifer*, were recorded in the riparian areas. The latter 4 species were resident and exclusive to the riparian zone, and did not occur in the dipterocarp habitat 35 m higher up the slope from the riverine areas. The Red Spiny Rat, *M. surifer,* is a rat of the dipterocarp forests in Malaysia (Harrison 1969, Charles, unpublished data) and Borneo (Payne *et al.* 1985), but at Belalong it was restricted to the riparian areas at very low numbers. The hypothesis that the unusual distribution of this species is the result of competition will be discussed later.

Of the 7 species of squirrels recorded in the forest, 4, namely *Callosciurus notatus, C. prevostii, Sundasciurus hippurus,* and *S. lowii*, were trapped while the other 3, *Exilisciurus exilis, Ratufa affinis* and *Rheithrosciurus macrotis*, were seen. Apart from *C. prevostii* and *R. macrotis*, the other 5 squirrels ranged into the riparian areas during the course of their daily activities. *R. macrotis* is a very rare Bornean endemic.

One porcupine species, *Thecurus crassipinis*, was seen in the forest study plot.

Some tupaiids, murids and sciurids occur in both the dipterocarp and riparian habitats; this could be because the forest small mammals living close to the riparian fringe included parts of it in their activity ranges. Observations of marked animals and preliminary analyses of the home ranges of some species, e.g. *Tupaia picta, Maxomys rajah, M. whiteheadi, Leopoldamys sabanus, Niviventer cremoriventer, Callosciurus prevostii* and *Ratufa affinis* (which are not presented in this paper) have borne out this suggestion.

Some small mammals may be able to live and utilise the resources in both habitats - *Sundasciurus lowii* and *Exilisciurus exilis* seen in the dipterocarp forests were also resident and territorial in Riparian Plot 2.

Some of the other species are probably confined to the riparian habitats owing to competition for some resource. Only a detailed population and resource partition study could throw light on this aspect. *Rattus exulans, Rattus tiomanicus, Sundamys muelleri* and *Maxomys surifer* were resident in the riparian zone though both habitats (Riparian Plot 2 and SM Plot 1) were contiguous but had a 35 m altitudinal difference. The composition and structure of the vegetational communities, food preferences, the availability of food and microhabitats and the need for other environmental resources could determine the occurrence of these species in the riparian habitat alone.

Is there competition between Maxomys surifer and Maxomys rajah?

The Red Spiny Rat, *Maxomys surifer*, is normally a dipterocarp forest rat; at Belalong it is confined to Riparian Plots 1 and 2.

M. rajah, a dipterocarp forest rat, ranges into the riparian areas at Belalong and is probably *M. surifer's* competitor. The hypothesis is that *M. rajah* has probably displaced *M. surifer* in the dipterocarp forest and the latter has been restricted to the riparian habitats, a sub-optimal habitat for its survival.

That some kind of competition has occurred or is occurring between the two species can be inferred from the following observations:

- Though trapping has been going on for one year in the study plots, *M. surifer* has so far not been trapped in the dipterocarp forest but occurs in very low numbers in the contiguous riparian habitat (9 per hectare) where *M. rajah* is still dominant (23 per hectare).

- The lower limit to similarity between competing sympatric species according to Hutchinson (1959) is a 1.3 size ratio and a figure of 2 for the corresponding weight ratio (Hutchinsonian ratios). This 'rule' holds true for many taxonomic groups (Das 1992, Giller

1984).

The Hutchinsonian ratios for both *M. surifer* and *M. rajah* were calculated for mean Head-Body length measurement and Body Weight (Table 2; sample sizes were small for those caught in the riparian areas *M. surifer* = 2; *M. rajah* = 5). The ratios for Head-Body (1.13) and Body-Weight (1.36) were lower than that expected for sympatric congeners to coexist. This implies that the species should not co-exist in the same habitat unless they exploit different resources or adopt different foraging strategies. It has been shown in some anurans that sympatric species overlap broadly in diet but much less in microhabitat usage though the most closely related species overlapped the least (Das 1992).

When the Hutchinsonian ratios for the assemblage of four resident murids in Riparian Plots 1 and 2 (*Rattus exulans, R. tiomanicus, Maxomys surifer, Sundamys muelleri)* are examined (Table 3), the Head-Body ratios generally fit into lower limit of similarity while the Body-Weight ratios are not very clear though they approach the value of 2.0 (values used are not means but the maximum and minimum values for Head-Body and Body-Weight measurements to give the range of the ratios). It implies that these species show some degree of coexistence. To what extent the other 4 dipterocarp forest murids (Table 1) compete with the resident riparian murids during their ranging activities in Riparian Plots 1 and 2 is difficult to know.

Hutchinsonian ratios for the same measurements for the forest murids in SM Plot 1 (West Ridge) (*Maxomys whiteheadi, Niviventer cremoriventer, M. rajah* and *Leopoldamys sabanus*) indicate that there is competition between the first two species but not between the last three (Table 4). Preliminary observations have shown that while *N. cremoriventer* is more arboreal, *M. whiteheadi* is more terrestrial and this spatial separation perhaps reduces competition for food.

Detailed analyses of food (and other resources) and environmental requirement analyses including foraging tactics are required to elucidate the difficult question of whether there is coexistence or competition between *M. surifer* and *M. rajah*. In a resource-abundant environment, two ecologically similar species can coexist indefinitely with complete dietary overlap. Until there is data to show that food or some other resource is limiting in the study area, the question about the role of competition between these two species will remain

Table 2. Hutchinsonian ratios for Head-Body (HB) and Body weight measurements (mean value for one species divided by the corresponding value for the adjacent species).

	HB (mm)	Ratio	Body Weight	Ratio
Maxomys rajah	165.5	-	133.9	-
Maxomys surifer	187.0	1.13	181.5	1.36

Table 3. Hutchinsonian ratios for Head-Body (HB) and Body weight measurements for the resident assemblage in Riparian Plots 1 & 2. (Ratios are based on maximum and minimum values for each measurement.)

	HB ratios	Body Weight ratios
Rattus exulans	-	-
Rattus tiomanicus	1.36 - 1.38	1.73 - 1.92
Maxomys surifer	1.0 - 1.32	1.54 - 2.18
Sundamys muelleri	1.15 - 1.34	1.3 - 2.07

Table 4. Hutchinsonian ratios for Head-Body (HB) and Body weight measurements for the resident assemblage in SM Plot 1, West Ridge. (Ratios are based on maximum and minimum values for each measurement).

	HB ratios	Body weight ratios
Maxomys whiteheadi	-	-
Niviventer cremoriventer	1.03 - 1.07	1.2 - 1.8
Maxomys rajah	1.3 - 1.4	1.8 - 2.2
Leopoldamys sabanus	1.3 - 1.6	2.4 - 2.6

unanswered. The occurrence of *M. surifer* in the Riparian Plots 1 and 2 only might be a local phenomenon since this species occurs in the dipterocarp forest of Bukit Tudal (Bennett *et al.* 1987) and in the Peradayan Forest Reserve about 25 km north of the Belalong forest (personal observation).

Murid diversity and population densities in Riparian Plot 1 and Riparian Plot 2

Riparian Plot 1 was an undisturbed area while Riparian Plot 2 was selectively cleared to build the Field Studies Centre. Some of the riparian vegetation around the buildings has been allowed to regenerate but the habitat has been altered. Many cut tree stumps and logs have been left lying around. This situation allowed investigation of the effect of disturbance on a riparian habitat as seen in the diversity and density of resident murids.

Eight murid species occurred in both the areas (Table 1) - *Rattus exulans, R. tiomanicus, Sundamys muelleri, Maxomys rajah, M. whiteheadi, M. surifer, Leopoldamys sabanus* and *Niviventer cremoriventer.* Disturbance to Riparian Plot 2 has not affected murid diversity between 1991 and 1993.

The 3 murids resident in both riparian areas and not recorded for the dipterocarp forest habitats were *S. muelleri, R. exulans* and *R. tiomanicus. S. muelleri* is a common inhabitant of secondary and beach vegetation habitats and heath forest (kerangas). It is also an urban pest in some parts of Bandar Seri Begawan where it is being replaced by *Rattus rattus* (Charles, unpublished data). *Rattus exulans* and *Rattus tiomanicus* are known to be agricultural pests in Peninsular Malaysia and, though they are not yet regarded as crop pests in Brunei Darussalam, they have the potential to become pest species.

Total removal of the 3 murids (the trap-out technique) was carried out for seven days per period, inundating both riparian areas with cage traps at approximately yearly intervals during March/May 1991, April 1992 and February 1993. Ideally, trapping should have been done during the same months for both riparian areas, but this was not always possible. In 1991, Riparian Plot 1 could only be sampled in May and not in March, as was Riparian Plot 2 (Table 5). No trapping was done in the undisturbed Riparian Plot 1 in Feb 1993. From the number of new animals captured daily and the cumulative number of captures, population size was estimated. In 1991, the densities of *R. exulans* and *R. tiomanicus* in the undisturbed and disturbed areas were similar, *R. exulans* being slightly less common in the disturbed plot. There were more *S. muelleri* in the disturbed area than the undisturbed. A year later in 1992, *S. muelleri* and *R. exulans* showed a clear increase in the disturbed area while their numbers fluctuated slightly in the undisturbed plot. Densities of *R. tiomanicus* did not change. In February 1993, there was a very sharp increase in the *R.exulans* population while *S. muelleri* remained at the level of the previous years. No *R. tiomanicus* were trapped during the 1993 removal. Perhaps the density had fallen below a certain threshold when trapping the animals would be difficult.

The results indicate that habitat disturbance to Riparian Plot 2 has led to a population increase of *S. muelleri* and *R. exulans* when compared to the control situation (Riparian Plot 1). *R. tiomanicus* population was unaffected over a single year but its absence in the 1993

Table 5. Densities of three murid species in riparian areas (numbers per hectare).

Species	Riparian Plot 1 (undisturbed)		Riparian Plot 2 (KBFSC)		
	May 91	Apr 92	Mar 91	Apr 92	Feb 93
Sundamys muelleri	37	30	50	62	55
Rattus exulans	31	37	25	81	250
Rattus tiomanicus	12	15	12	12	-

sampling might be symptomatic of changes in the welfare factors essential to its survival. The high density of *R. exulans* after two years can be attributed to the creation and abundance of new microhabitats by the piling up of logs in various sections of Riparian Plot 2 and the appearance of buildings and new cover. Evidently the area of human disturbance has produced a conducive environment for the survival and reproduction of *S. muelleri* and *R. exulans*. The increase in population size has now produced a problem in the Field Studies Centre - both species have invaded the buildings for foraging and nesting.

CONCLUSIONS

1. Eighteen species of small mammals were recorded in the dipterocarp forests and 14 from the riparian areas of Belalong.
2. The record of *Suncus etruscus*, an insectivore, is the first for Brunei Darussalam.
3. Of 7 tupaiids found in Borneo, 5 have been recorded in the dipterocarp forests of Belalong. Only *T. picta* ranged into the riparian habitat.
4. Four dipterocarp forest murids ranged into the riparian areas where 4 other murids were exclusive to that habitat.
5. Of 7 dipterocarp forest sciurids, 5 ranged into the riparian areas during the the course of their daily activities.
6. Only one hystricid was recorded in the dipterocarp forest.
7. It is inferred that *M. surifer*, a forest rat, competes with *M. rajah* and the former has been displaced to the riparian fringe. This needs to be investigated further.
8. A comparison of the densities of 3 murids (exclusive to the riparian zone) in disturbed and undisturbed riparian areas showed that *S. muelleri* and *R. exulans* have increased their numbers as a result of habitat disturbance while the density of *R. tiomanicus* has remained unchanged.

ACKNOWLEDGEMENTS

I have great pleasure in thanking Samhan bin Nyawa, Ashri bin Hj Ahmad, Dk Mahani bte Pg Momin, Munah bte Hj Yahya and Sarimah bte Abu Bakar for their enthusiastic assistance in fieldwork, and the Standard Chartered Bank (Brunei) for financial support to cover the costs of their visits to the Kuala Belalong Field Studies Centre. This study was supported by a research grant from Universiti Brunei Darussalam.

LITERATURE CITED

BARNETT, A. 1992. *Expedition Field Techniques Small mammals (excluding bats)*. Expedition Advisory Centre, London. 76 pp.

BENNETT. E.L., CALDECOTT, J.O., & DAVISON, G.W.H. 1987. A wildlife survey of Ulu Temburong, Brunei. *Brunei Museum Journal* 6:121-166.

DAS, I. 1992. *Trophic ecology of a community of South Indian anuran amphibians*. Unpublished D. Phil. thesis, University of Oxford, Oxford.

DAVIES, A.G. & PAYNE, J.B. 1982. *A Faunal survey of Sabah*. World Wildlife Fund, Malaysia, Kuala Lumpur.

DYKES, A. P., this volume. Analysis of factors contributing to the stability of steep hillslopes in the tropical rainforest of Temburong, Brunei Darussalam.

GILLER, P.S. 1984. *Community structure and the niche*. Chapman and Hall, London. 175 pp.

HARRISON, J.L. 1969. The abundance and population density of mammals in Malayan lowland forests. *Malayan Nature Journal* 22:174-178.

HUTCHINSON, G. E. 1959. Homage to Santa Rosalia, or why are there so many kinds of animals? *American Naturalist* 93:145-159.

PAYNE, J., & FRANCIS, C. M. 1985. *A Field Guide to the Mammals of Borneo*. The Sabah Society, Kota Kinabalu, Sabah.

Species radiation of predatory empidoids (Insecta, Diptera) in lowland rainforest in northern Papua New Guinea

PATRICK GROOTAERT

Department of Entomology, Royal Belgian Institute of Natural Sciences, Vautierstraat 29, B-1040 Brussels, Belgium

ABSTRACT. Empidoid flies are among the most abundant flies in the lowland rainforests of northern Papua New Guinea. The Empidoidea, as considered here, comprise the families Hybotidae, Empididae, Atelestidae, Microphoridae and Dolichopodidae. Only one representative of the Empididae, a very abundant family in temperate regions, is known from Papua New Guinea. No Atelestidae nor Microphoridae are reported from the area. On the other hand, the Hybotidae and Dolichopodidae have radiated enormously. The stouter hybotids which catch prey with their raptorial hind legs in flight are present but not very common. The small leaf- and especially the soil-dwelling microhybotids are very abundant and very rich in species and they may well be the most important predators of micro-arthropods. The genus *Elaphropeza* (2-3 mm), the vicariant of the genus *Platypalpus* in temperate regions, with its numerous species, can be found running on leaves. The genus *Nanodromia*, known currently only from Papua New Guinea, also has numerous species that are found on leaves or on soil. They possess a piercing proboscis and raptorial front legs. Their small size (about 1 mm) suggests that they feed only on very small arthropods such as mites or collembolans. The dolichopodids occupy two major habitats. The larger, brilliantly green shining Sciapodinae are very commonly seen on leaves of bushes and trees. The other dolichopodids are generally found in humid areas such as the borders of pools, streams or in swamps. The genus *Sympycnus,* with its relatives, is probably the largest genus with more than 200 species still to be described. The males exhibit complicated secondary sexual characters which may reflect a fast radiation in a new environment (cf. species radiation in Hawaiian flies). Their biology and behaviour is unknown. Since some dolichopodid species are known to be predators of mosquito larvae, they warrant special attention.

KEY WORDS: Diptera, Empididoidea, species radiation, *Sympycnus*.

INTRODUCTION

Empidoid flies are among the most abundant flies in the rain forests of lowland Papua New Guinea. An enormous radiation of species is seen in the families Dolichopodidae and Hybotidae. The larger, brilliantly shining Sciapodinae (Dolichopodidae) are very common and they are readily seen sitting on leaves of scrubs and trees. The smaller Hybotidae are also very abundant but less conspicuous and therefore poorly known. Unfortunately we have no idea about their habitat preferences or community structure. At most 10% of the existing fauna has been recorded and named. Even these species are generally only recorded incidentally by non-dipterist explorers. On the whole we estimate that about 90% of the empidoid flies in Papua New Guinea remain undescribed.

MATERIAL AND METHODS

The present study is based on material collected during three short expeditions to the northern coast of Papua New Guinea (Madang province, Bogia district). Samples were taken in mangroves, swamp forests and in secondary and primary rainforest at sea level (Bunapas,

D. S. Edwards et al. (eds.), Tropical Rainforest Research – Current Issues, 183–192.

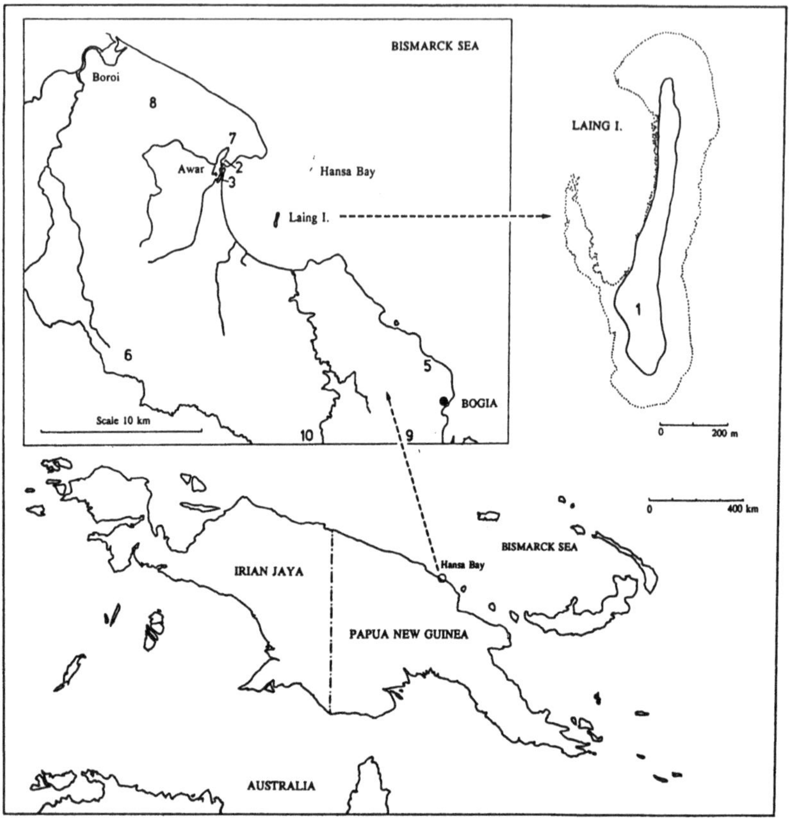

Figure 1. Map of Papua New Guinea with a detail of Hansa Bay and Laing Island

Awar, Boroi bush) and in the hills (alt. 200 m) near Bogia (Apingan, Mangigim) (Figure 1). A large collection of Sympycninae from all over Papua New Guinea was offered to us for study by Dr. Neil Evenhuis of the Bernice P. Bishop Museum in Honolulu.

RESULTS

Current taxonomic knowledge

In recent years many new species of empidoid flies have been reported from Papua New Guinea (De Meyer & Grootaert 1990, Grootaert 1984, 1987, 1992, Grootaert & Meuffels 1984, 1987, 1988, 1989, in press, Grootaert & Van de Velde 1988, McAlpine 1982, Meuffels & Grootaert 1984, 1985, 1986, 1987, 1988). Table I gives a list of genera and the number of species known from Papua New Guinea. The second column contains the number of species mentioned in the catalogue of the Diptera of the Australasian and Oceanic regions (Bickel & Dyte in Evenhuis 1989); the third column contains the number of species published after 1989 and those seen by the author but not yet published. These estimates are cautious and are in fact based only on observations on the northern coast of Papua New Guinea.

The superfamily Empididoidea comprises the families Empididae, Hybotidae, Microphoridae, Atelestidae and Dolichopodidae (Chvala 1983). Apart from a single *Empis* species no other representatives of the family Empididae, and more particularly the Empidinae, are known from Papua New Guinea. The Empidinae are mainly nectar feeders,

Table 1. Empidoidea of Papua New Guinea. The second column gives the number of species cited in the catalogue of the Australasian and Oceanian regions (Evenhuis 1989); the third column gives the number of species published after 1989, together with those seen by Meuffels & Grootaert but unpublished, and the fourth column gives the sum of columns two and three. M: genera with marine or mangrove species.

Taxon	No. of Species (1989)	No. of Species published/ recognised since 1989	Total No. of species	Ecology
Empidoidea				
Empididae				
Empidinae				
Empis	1		1	
Hybotidae				
Hybotinae				
Hybos	1		1	
Syndyas	1		1	
Syneches	6		6	
Tachydromiinae				
Crossopalpus	1	2	3	
Drapetis	2			
Elaphropeza	6	20	26	
Micrempis	1		1	
Platypalpus	4		4	
Tachydromia	1		1	
Chersodromia		5	5	M
Nanodromia		18	18	
Dolichopodidae				
Sciapodinae				
Amblypsilopus	9		9	
Chrysosoma	36	1	37	
Krakatauia	7		7	
Mesorhaga	1	2	3	
Plagiozopelma	4		4	
Neurigoninae				
Neurigona	1		1	
Dolichopodinae				
Lichtwardtia	1		1	
Metaparaclius	1		1	
Paraclius	4	3	7	
Medeterinae				
Medetera	16		16	
Hydrophorinae				
Cymatopus	2	2	4	M
Thinophilus	7		7	
Thinolestris		1	1	M
Diaphorinae				
Asyndetus	7	4	11	
Chrysotus	2	1	3	
Crytophleps	1		1	
Diaphorus	12	5	17	
Mischopyga		2	2	M
Sympycninae				
Chaetogonopteron	1		1	
Phrudoneura	1		1	
Sympycnus	29	100	129	
Teuchophorus	17	2	19	M
Acropsilus		9	9	
Rhaphiinae				
Physopyga		1	1	M
Incertae sedis				
Phacaspis		2	2	M
Total	159	135	294	

a feeding habit that is derived from predatory feeding. However, in almost all species the male hunts prey and offers it to the female before mating. The subfamily is very abundant and rich in species in the temperate zone of the holarctic region and they are also present in Australia and New Zealand. Apparently they failed to colonize successfully even the temperate mountainous areas in Papua New Guinea. The small family Atelestidae is not known from the Australasian region. The Microphoridae are represented by only two species in Australia, none are known from New Guinea.

The families Hybotidae and Dolichopodidae have radiated extensively in Papua New Guinea. The genera *Hybos*, *Syndyas* and *Syneches*, belonging to the ancestral Hybotinae, are rather stout flies with raptorial hind legs. They catch their prey in flight. This subfamily is not very rich in species and is not very common. In contrast, the small leaf- and soil-dwelling microhybotids belonging to the subfamily Tachydromiinae have radiated greatly. These hybotids have the front legs and sometimes the mid legs raptorial instead of the hind legs. They sit or run, preferentially, on horizontal surfaces such as leaves where they search for prey. Prey generally consists of small flies or hymenopterans upon which they jump, hold them firmly with the fore and mid legs, pressing them against the substrate while trying to pierce them with their short but strong proboscis. It is expected that more than 90% of the tachydromiine species still have to be described. They represent probably the most important predators of micro-arthropods.

The genus *Elaphropeza*, which is very rich in species in all tropical regions in the world, is also very diverse in Papua New Guinea. As can be seen in Table I, only six species were known previously from Papua New Guinea, but in the small area under study at least 20 species have now been recognized. *Elaphropeza* is the tropical vicariant of the genus *Platypalpus* in temperate climates. Apart from some bristles on the legs few other male secondary sexual characters (MSSC) have developed. The abdominal pheromone glands show a high morphological variability among the species. The second, third, fourth and/or fifth tergite may be modified and may bear leaf-like bristles and ridges. Underneath the terga are pheromone glands. So instead of visual recognition, chemical communication is highly diversified in *Elaphropeza*.

The new genus *Nanodromia* (Grootaert 1993) is known only from Papua New Guinea and it is also very rich in species. The genus consists of minute flies (1-2 mm). They are very abundant and densities of several tens of individuals per m^2 are commonly observed. They can be seen running on leaves of shrubs and trees or on fallen leaves and on the soil. Their small size suggests that they feed only on small arthropods such as mites, collembolans and small flies. The male secondary sexual characters are also strongly developed in the genus *Nanodromia*. In most species the tarsal segments of the front legs are coloured or annulated. By analogy with the dolichopodids they are probably waved in front of the female before and during mating. The mid legs in males are enlarged and the femora and tibiae bear bristles and spines which are distinct at the species level. These raptorial looking legs are not present in the female. They could be used to handle prey, perhaps to offer prey to the female, but such a habit has never been observed in Tachydromiinae. They may function as a clasping organ to hold the female during mating. One species has the tips of the wings curled up in the male. Apart from the tibial glands no other glands with possible pheromone secreting functions have been observed to date. It is clear that in *Nanodromia* visual communication prevails over chemical communication to permit discrimination among species and sexes. Phylogenetically, *Nanodromia* is closely related to the marine genus *Chersodromia*. A number of species also live on the coast but many are associated with mangrove, swamp forests and drier rainforests.

A number of smaller genera such as *Drapetis*, *Chersodromia*, *Platypalpus* and related new genera occur in Papua New Guinea, but are not as diverse as *Elaphropeza* and *Nanodromia*.

Habitat and niche

The larvae of empidoids and their biology are poorly known. Except for the genus *Medetera*, all others live in the soil and are thought to be predatory like the adults. Dolichopodids are generally found in humid areas near swamps or on the borders of puddles or streams. Although there is a preference for these humid soils, only the larvae of the Hydrophorinae have adaptations for an aquatic existence. The hybotids on the other hand are generally found in drier habitats.

As can be seen in Figure 2, several genera show habitat preferences. On the leaves of herbs, shrubs and trees, the large, brillantly green shining *Chrysosoma*, the somewhat smaller *Sympycnus*, the often yellow to brown coloured *Elaphropeza* or the minute *Nanodromia* are frequently observed. The latter is also active on the soil or on fallen leaves. On tree trunks, in a vertical position, sit *Sciapus*, *Medetera* and *Tachydromia*. The larvae of *Medetera* live in the burrows of scolytid beetles under bark and feed on the beetle larvae, while the adult *Medetera* have been observed feeding on aphids. Nothing is known about the fauna of the canopy but this is likely to be the habitat of the Sciapodinae.

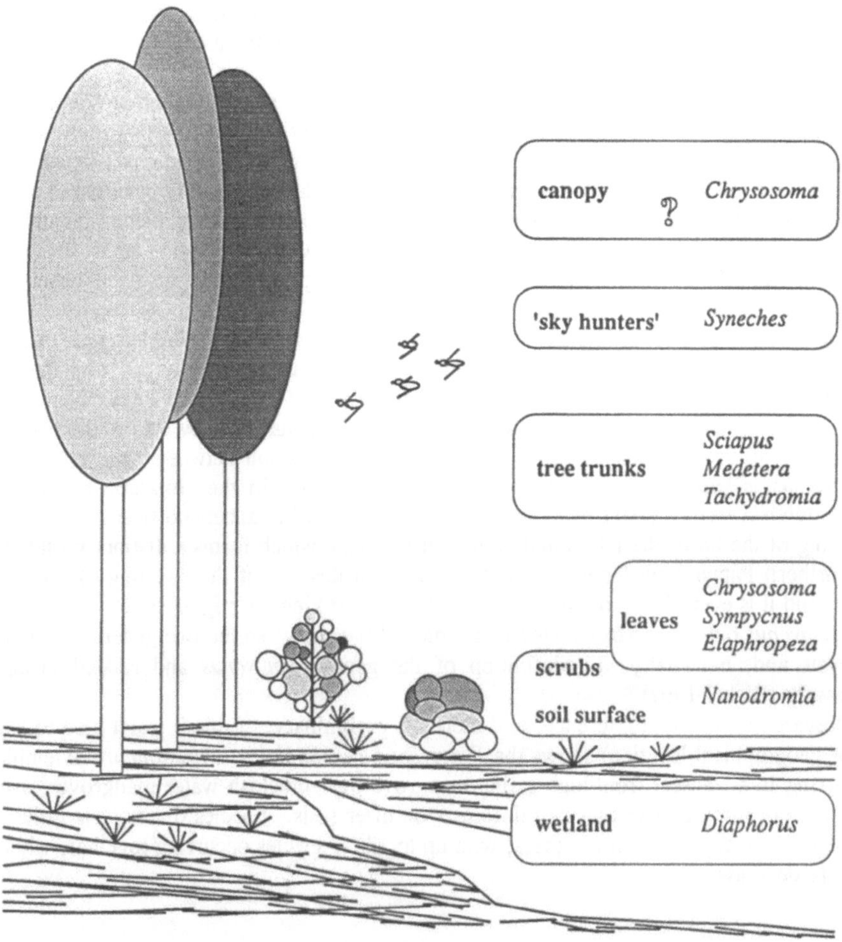

Figure 2. Habitat preference of the major empidoid genera in Papuan rainforest

Species radiation in Sympycnus: a case study

The genus *Sympycnus* is a very large taxon with a worldwide distribution and it is very prolific in the tropics. As can be seen from Table I, the genus is very rich in species. Only a very small portion of the Papuan fauna limited to the northern coast has been described by Meuffels & Grootaert (1987) but at least a hundred species await description (Meuffels & Grootaert, in preparation). The enormous development of male secondary sexual characters (MSSC) is very striking. The antennal arista can bear a terminal thickening comparable to some kind of flag (cf. the related genus *Yumbera* (Bickel 1992)). The eye facets may be enlarged. All pairs of legs may be modified. The fore legs bear peculiar bristles on the femora, tibiae and tarsi (Figure 3). The tarsi may be somewhat compressed (Figure 4) and may be carry various types of bristles. The terminal tarsal segments may be coloured white and bear white or black leaf-like bristles. By analogy with other dolichopodids, the tarsi are probably waved in front of the female during display. The mid legs bear also various hairs and bristles (see Meuffels & Grootaert 1987; e.g. the cork-screw bristle) of which the function is unknown. The hind legs have in many species the two basal tarsal segments shortened (Figure 5). The second segment bears an extension called the clidium (Figure 6). Its morphology is specific but its function is completely unknown. Even the wings may be modified: the tip of the wing may bear a black spot (*S. sticticus*) or the base may be shrunken and bear long bristles (genus *Chaetogonopteron*) or the median vein (vein M) may be abbreviated (*Phrudoneura abbreviatus*).

The male genitalia in dolichopodids may be very complex suggesting the presence of many Lock-Key-Mechanisms (LKM) or even associated with sperm competition. In contrast to the MSSC, the male genitalia in *Sympycnus* are rather simple in structure (Figure 7) but nevertheless specific. The LKM's are apparently poorly developed. They consist of clasping organs, the telomeres, the intromittent organ (aedeagus) and a guiding piece (hypandrium). The cerci are simple in structure and have is no apparent function. Apart from some jumping up and down on leaves, display and mating have not yet been observed in *Sympycnus* species.

Two very closely related genera occur together with *Sympycnus* - *Phrudoneura*, originally described by Meuffels & Grootaert (1987) as a subgenus of *Sympycnus*, and the genus *Teuchophorus*. *Teuchophorus* is also very rich in species in northern Papua New Guinea with no fewer than 16 species described, the same number as are known from the rest of the world (Meuffels & Grootaert 1986). However the distinction between *Teuchophorus* and *Sympycnus* is sometimes unclear in Papua New Guinea. In the Holarctic region, male *Teuchophorus* can be easily recognized by the presence of a lanceolate stigma. This is the swelling of the tip of the R1 (radial vein) and the costa which forms a distinct visual mark. In northern Papua New Guinea, however, different degrees of the swelling of the stigma occur and it is even absent in many species. Species lacking the stigma could well be placed in *Sympycnus* but have to be considered as intermediate between the two genera. A character analysis and, eventually, a redefinition of the genus *Sympycnus* and related groups is necessary (Bickel 1992).

Research on habitat preferences and niches is preliminary. Table 2 (refer also to Figure 1 for geographical location) shows the *Sympycnus-Teuchophorus-Phrudoneura* assemblages occurring in a transect from high saline mangrove over brackish water mangroves, swamp forests to primary and secondary rainforests on drier soils. Species diversity is rather low in the saline mangroves but increases, with up to 12-14 species occurring together, when the salinity decreases.

Figures 3 - 7. *Sympycnus compressipes* Meuffels & Grootaert, male. 3. fore leg with white apical tarsal segments. 4. compressed first tarsal segment of fore leg. 5. hind leg. 6. clidium, appendage on second tarsal segment of hind leg. 7. genitalia (a: aedeagus; c: cerci; h: hypandrium; t: telomeres). Scale bars 0.1 mm.

DISCUSSION

The fauna of Papua New Guinea is interesting for various reasons: (1) its very high biodiversity and the origin of this diversity i. e. the speciation mechanisms responsible for that diversity; (2) the presence of ancestral groups and; (3) the high degree of development of MSSC in many groups and its relationship with the modes of speciation.

Gressitt (1982a) and several other authors (in Gressitt 1982b) state that the insect fauna of Papua New Guinea has a closer relationship with the S.E. Asian fauna than with the Australian fauna and that its origin is probably to be found in the wet tropical rainforests of S.E. Asia rather than in the drier savanna fauna of Australia. This hypothesis seems not unlikely, but since only a very small part of the Papuan empidoid fauna is known and even

Table 2. Habitat preference of species of the *Sympycnus* - complex (*Sympycnus*, *Phrudoneura* and *Teuchophorus*) in Hansa Bay, Bogia district, northern Papua New Guinea (refer to Figure 1 for locations). 1. Laing Island: brackish mangrove; 2. Awar: mangrove with brackish water; 3. Sisimangum: mangrove with low salinity; 4. Awar: puddle with fresh water; 5. Warawaranga: river with fresh water; 6. Bunapas: swamp forest; 7. Awar bush: secondary rainforest; 8. Boroi bush: secondary rain forest; 9. Apingan: primary rainforest in hills (alt. 200 m); 10. Mangigim: secondary rainforest (alt. 50 m).

Species	Locality									
	1	2	3	4	5	6	7	8	9	10
S. teuchophoroides			x					x		
S. campsicnemoides				x			x	x		
S. aechmophorus								x		
S. gigas										
S. sticticus							x			
S. anachrostichus			x				x			
S. tricorniger						x				
S. ictericus					x		x	x		
S. araneipes					x				x	x
S. pallipes							x	x	x	x
S. priapus								x		
S. torrenticola									x	
S. ripicola				x					x	
S. ctenophorus									x	x
S. nigrisquamis								x		
S. alipes									x	
S. miritarsis			x	x			x	x		
S. candidimanus									x	
S. leucotarsus										
S. albiserratus			x					x		
S. compressipes		x					x			
Ph. abbreviatus					x		x			
T. femoratus			x				x			
T. conspicuus					x				x	
T. pusio						x	x	x		
T. denticulatus				x			x	x		
T. costalis			x					x		
T. spatulifer					x				x	
T. quadratus		x	x	x			x	x		
T. digitatus					x			x	x	
T. fuscicornis					x				x	
T. pectinatus		x		x						
T. vanaartseni				x	x				x	
T. parcearmatus					x				x	
T. uncinatus		x	x	x						
T. grandior	x	x	x							
T. laingensis	x									
No. of taxa	2	5	9	8	9	2	12	14	13	3

less is known of that of adjacent areas, we will have to wait before anything definitive can be said on this case.

It is remarkable that the boundaries between certain taxa which are very distinct in other zoogeographical regions fade away in Papua New Guinea. This is seen in *Teuchophorus-Sympycnus* and in the genus *Platypalpus* (Grootaert & Van de Velde 1988). This phenomenon has already been observed by Gressitt (1982) in beetles and by McAlpine (1982) in platystomatid flies. Gressitt suggests that the intermediate forms persisted here because they did not suffer too much during the glacial periods. The presence of these 'missing links' and other ancestral groups (De Meyer & Grootaert 1990) is of course interesting phylogenetically.

Rainforests are well known for their high diversity of plants and animals. However, the numbers of each species per unit of surface are generally rather low. This poses a mate-location problem for the species involved. Two strategies seem to have evolved in the empidoids: a strategy to find the mate directly or the strategy of finding the mate's habitat. In the first strategy there must be long distance communication through pheromones which aggregate the species. In the second type of strategy a species recognizes its correct habitat through structure (vegetation), relative humidity, light intensity or other physical parameters. Adult dolichopodids are very faithful to their habitats and, although they are excelllent flyers, they remain in their habitat even if it is only a few square meters in size. Aggregations of many individuals, even belonging to various species, may thus occur in the same small habitat. Correct partner recognition is important to avoid false matings and over these short distances visual recognition becomes very important.

In both dolichopodids and hybotids there are many species in which the MSSC are well developed. Most of the MSSC play an important role in display and in mate recognition. The high development of MSSC suggests that there exists a sexual selection by the females. This selection could be enhanced by the co-occurrence of several related species in large numbers in the same area and the high development may be due to a reinforcement phenomenon (cf. Butlin 1989). In *Sympycnus* and *Nanodromia* the strategy of early species- and partner-recognition by MSSC is adopted, whereas in *Elaphropeza* chemical communication helps to aggregate members of the same species, and visual MSSC act only over short distances. Many species of dolichopodids which have no chemical, acoustical or other type of long distance communication, probably find their partners because they stay close to their habitats (borders of ponds, rivers) and never stray from them. Visual MSSC are then sufficient in species recognition. *Elaphropeza* has a wider habitat range and members of this genus need long distance communication to find their partners.

ACKNOWLEDGEMENTS

The author acknowledges a grant from the National Fund for Scientific Research (F.K.F.O. project 2.9008.90) to study biodiversity in Papua New Guinea. The author also thanks his colleague Henk Meuffels for many interesting discussions on the subject.

LITERATURE CITED

BICKEL, D.J. 1992. The Australian Sympycninae (Diptera: Dolichopodidae): Introduction and Description of a New Genus, *Yumbera*. *Invertebrate Taxononomy* 6:1005-1017.

BUTLIN, R. 1989. Reinforcement of premating isolation. Pp. 158-179 in Otte & Endler (eds). *Speciation and its consequences*. Sinauer Associates Inc. Massachusetts.

CHVALA, M. 1983. The Empidoidea (Diptera) of Fennoscandia and Denmark. II General Part. The Families Hybotidae, Atelestidae and Microphoridae. *Fauna Entomologica Scandinavica* 12:1-279.

DE MEYER, M. & GROOTAERT, P. 1990 (1991). Pipunculidae (Diptera) from Papua New Guinea. The genus *Cephalosphaera*, *Cephalops* and *Beckerias*. *Zoologica scripta* 19:403-412.

EVENHUIS, N. 1989. *Catalog of the Diptera of the Australasian and Oceanian Regions*. Bishop Museum Press. 1155 pp.

GRESSITT, J.L. 1982a. Ecology and biogeography of New Guinea Coleoptera (Beetles). Pp. 709-734 in Gressitt, J.L. (ed.). *Biogeography and Ecology of New Guinea. Monographiae biologicae* 42.

GRESSITT, J.L. 1982b. Zoogeographical summary. Pp. 897-918 in Gressitt, J.L. (ed.). *Biogeography and Ecology of New Guinea. Monographiae biologicae* 42.

GROOTAERT, P. 1984. Two new species of *Platypalpus* Macquart (Diptera, Empidoidea, Hybotidae) from Papua New Guinea. *Bulletin et Annales de la Société royale belge d'Entomologie* 120:283-290.

GROOTAERT, P. 1987. A new *Tachydromia* from Papua New Guinea (Diptera, Empididoidea, Hybotidae). *Bulletin et Annales de la Société royale belge d'Entomologie* 123:141-144.

GROOTAERT, P. 1992. Description of new species in the genus *Chersodromia* (Diptera, Empidoidea, Hybotidae) from the northern coast of Papua New Guinea. *Bulletin et Annales de la Société royale belge d'Entomologie* 128:235-244.

GROOTAERT, P. & MEUFFELS, H.J.G. 1984. Dolichopodidae (Diptera) from Papua New Guinea. II. Some new species of the genus *Thinophilus* Wahlberg, 1844, from the lowland. *Indo-Malayan Zoology* 2:209-223.

GROOTAERT, P. & MEUFFELS, H.J.G. 1987. Dolichopididae (Diptera) from Papua New Guinea. V. A new species of *Cryptophleps* Lichtwardt, 1898, with notes on other species. *Indo-Malayan Zoology* 4:145-152.

GROOTAERT, P. & MEUFFELS H.J.G. 1988 (1989). Dolichopodidae (Diptera) from Papua New Guinea. VII. Description of *Thinolestris* gen. nov., a new beach dwelling Hydrophorinae. *Indo-Malayan Zoology* 5:31-39.

GROOTAERT, P. & MEUFFELS H.J.G. 1989. Dolichopodidae (Diptera) from Papua New Guinea. IX. Mischopyga gen. nov. and *Physopyga* gen. nov., two new genera from the mangrove in Papua New Guinea. *Indo-Malayan Zoolology* 6:195-206.

GROOTAERT, P. & MEUFFELS, H.J.G. in press. Dolichopodidae (Diptera) from Papua New Guinea. X. Description of new species of the marine genus *Cymatopus* Kertész, and observations on their biology and territorial behaviour. *Invertebrate Taxonomy*.

GROOTAERT, P. & VAN DE VELDE, I. 1988. A new *Platypalpus* species from Papua New Guinea (Diptera Empidoidea Hybotidae). *Bulletin et Annales de la Société royale belge d'Entomologie* 124:223-227.

McALPINE, D.K. 1982. The acalyptrate Diptera with special reference to Platystomatidae. Pp. 659-673 in Gressitt, J.L. (ed.). *Biogeography and ecology of New Guinea. Monographiae biologicae* 42.

MEUFFELS, H.J.G. & GROOTAERT, P. 1984. Dolichopodidae (Diptera) from Papua New Guinea. I: The genus *Cymatopus* Kertész with a discussion on *Abateti* Miller and *Cemocarus* gen. nov. *Indo-Malayan Zoology* 1:141-158.

MEUFFELS, H.J.G. & GROOTAERT, P. 1985. Dolichopodidae (Diptera) from Papua New Guinea. III: New species of the genus *Diaphorus* Meigen, 1824. *Indo-Malayan Zoology* 2:9-28.

MEUFFELS, H.J.G. & GROOTAERT, P. 1986. Dolichopodidae from Papua New Guinea. IV. New species of the genus *Teuchophorus* Loew, 1857. *Indo-Malayan Zoology* 3:217-262.

MEUFFELS, H.J.G. & GROOTAERT, P. 1987. Dolichopididae (Diptera) from Papua New Guinea. VI. Some new species of the genus *Sympycnus*. *Indo-Malayan Zoology* 4:317-397.

MEUFFELS, H.J.G. & GROOTAERT, P. 1988 (1989). Dolichopodidae (Diptera) from Papua New Guinea. VIII. *Phacaspis*, a new genus *incertae sedis* from the mangrove. *Indo-Malayan Zoology* 5:311-319.

The theory and practice of estimating regional species richness from local samples

NICK MAWDSLEY

NERC Centre for Population Biology, Imperial College at Silwood Park, Ascot, Berkshire, & Department of Entomology, The Natural History Museum, Cromwell Road, London, United Kingdom

---

ABSTRACT. Current knowledge of the species richness and spatial distribution of many taxa is insufficient for the basic description of species assemblages, for research into their ecology and evolutionary relationships and perhaps most importantly as the basis of management decisions for their conservation and evaluation. Here I consider the basic problems of sampling diverse insect faunas and how it affects our ability to estimate the species richness of larger areas. This problem is conceptually separated into that of estimating local species richness and species turnover (beta diversity). Using published data for the 64 species of umbellifer (Apiaceae) in Britain it is seen that the error associated with regional species richness estimates is that of estimating species turnover. Moreover widely separated localities tend to greatly overestimate species turnover. But the situation is even worse in the tropics where assemblages contain many more rare species making sampling even more difficult. A number of species richness estimators are compared using data from the canopy fogging of the ant faunas of trees in Brunei. These give a wide range of values and each estimator appears to have its own inherent biases. Essentially the description, recording and estimation of tropical biological diversity requires good quality data which can only be generated by an increase in the amount of standardised field collections being made.

KEY WORDS: species turnover, species spatial distribution, Apiaceae, Dipterocarpaceae, *Dryobalanops*.

INTRODUCTION

The basic question of "How many species are there?" is, and will be for some time, a major question for biologists to answer (May 1988, 1990, 1992a, Stork 1993). A decade ago, Erwin (1982) redefined the problem and suggested that the global total for insects alone may be around 30 million species, highlighting our ignorance of the insect faunas of tropical forests, arguably the most diverse communities on earth. In recent years numerous studies have revealed much of the richness of insects in species and functional significance at the local scale (e.g. Basset & Arthington 1992, Casson & Hodkinson 1991, Erwin 1983, Hammond 1990, Janzen 1988, Stork 1991, Wilson 1987 and many others), but despite this wealth of knowledge we are still in no position to use these local samples to estimate the species richness of larger geographic regions except for just a few relatively well-known taxa such as the larger moths and butterflies (e.g. Holloway 1987).

A variety of approaches have been used to estimate global and regional species totals (see Hammond 1992, May 1992a, Stork 1993). Methods range from collating the views of taxonomists (Gaston 1991), to extrapolating up from local samples using assumptions concerning the host-specificity of tropical insects (Erwin 1982, Thomas 1990), and using the ratio of insect:plant species numbers (Gaston 1992, Hodkinson & Casson 1991). Although the use of several independent methods is useful for overcoming the methodological biases of various estimates, most do not directly increase our knowledge of the distribution of tropical floras and faunas within regions.

D. S. Edwards et al. (eds.), Tropical Rainforest Research – Current Issues, 193–213.

Local samples taken in the tropical system under investigation have the important property that they do provide us with such information. The insights they yield into the diversity of tropical regions are independent of the patterns and assumptions drawn from temperate systems or from disparate taxa. The use of principles gleaned from a small and biased sample of the world's species to predict species numbers of diverse tropical groups is potentially unreliable (Hammond 1992, 1994a). For example, the general trend for species ranges to be narrower at lower latitudes (France 1992, Stevens 1989) suggests that there will be a low species overlap of local faunas in the tropics relative to temperate regions (MacArthur 1969, 1972), and any general theory or estimate of tropical species richness must incorporate these regional-scale patterns. Overall there is a paucity of data on the distribution of species and the degree of faunal overlap between sites in the tropics, despite there being a major need for such data, both for gaining an understanding of biological diversity and for its preservation (Gentry 1992, Pressey *et al.* 1993, Vane-Wright *et al.* 1991).

The challenge to conservationists trying to preserve biological diversity is to make fast, reliable and economic assessments of the relative species richness of different localities and regions world-wide and to assess the distribution of species between such elements (Beattie 1993, Coddington *et al.* 1991, Cranston & Hillman 1992). For the hyperdiverse groups such as arthropods, fungi and microorganisms, statistical methods and empirical data must be used to assess species richness which will need to extrapolate outside the extent of the data, thereby potentially multiplying biases and errors (Colwell & Coddington 1994). Clearly we must have a better understanding of these methods and their application to estimate the species richness of any given area, be it a hectare plot or a continental landmass. Moreover it is in the tropical regions of the world where the bulk of species are found that these methods will be of most use.

Attempts to estimate tropical insect species richness from local samples taken in tropical forests have produced figures spanning the full range of global species richness estimates from 2 million species (Hodkinson & Casson 1991) to 80 million and more (Hammond 1992, Stork 1988). These huge differences arise because we do not have sufficient data, nor do we know how to scale up in this way. It is therefore the intention of this paper to consider some of the basic problems involved. I will show that in theory the estimation of regional richness from local samples is a relatively straightforward process, but that in practice we are faced with some major conceptual and methodological problems. I then go on to analyse the process of scaling up from local to regional scales using data on the distribution of British umbellifers and I compare a variety of methods that have been produced for this purpose using samples of canopy ants fogged from a number of trees at Kuala Belalong in Brunei Darussalam, Borneo. Finally I consider the resources required to estimate the species richness and distribution of a hyperdiverse taxon such as the Coleoptera (beetles) in a region such as the island of Borneo.

SPATIAL SCALE AND SPECIES RICHNESS

Regional species richness describes the species richness of large, subcontinental areas where the large-scale biogeographic processes of speciation and geographic dispersal dominate (Cornell & Lawton 1992, Ricklefs 1987). Within regions there exists spatial heterogeneity in the environment at all scales, ranging from large-scale differences in landscapes, geology and climate, in habitat structure and type, through to the finer-scale patterns of microclimate and resource distribution. Tropical forests, especially, are noted for their many different forest types as well as their high floristic richness (Whitmore 1984). In Amazonia for example, Erwin (1988) considers there to be many more types of floristic assemblage than botanists have yet described (see also Gentry 1988), a feature that he argues is partly behind the great diversity of tropical insect faunas.

It is against this background of floristic and environmental variation that species are adapted and distributed. Within regions, species distributions are highly variable; some will be widespread, but most will have restricted geographical distributions (gamma-rarity) and/or be restricted to certain habitats (beta-rarity) (Cody 1986, Gaston 1990, Rabinowitz *et al.* 1986). The result is that as we move from one locality to another there will be a turnover in the local species composition, and we will accumulate species in our total sample (the familiar species-area relationship; Connor & McCoy (1979), Gleason (1922), Williamson (1988)). Therefore regional richness is related to local richness by the species turnover between localities within a region, so that (Schluter & Ricklefs 1993, Whittaker 1977):

$$\text{Regional species richness} = \text{mean local species richness} \times \text{mean species turnover per locality} \times \text{number of localities} \quad (1)$$

where

$$\text{Regional species turnover} = \text{mean species turnover per locality} \times \text{number of localities} \quad (2)$$

$$= \frac{\text{regional species richness}}{\text{mean local species richness}} \quad (3)$$

(Whittaker's (1977) beta diversity measure)

Theoretically, we simply need to know the mean local richness, the mean species turnover per locality and the number of localities in a region to estimate the regional species richness. In the context of extrapolating from local samples, species turnover per locality can be thought of as the rate at which we accumulate species per additional locality in our total sample (and hence a non-linear function of the number of localities) or as a measure of the average pair-wise similarity (beta diversity) of all the localities within a region. The slope of the *nested* species-area curve is therefore a measure of species turnover. Colwell & Coddington (1994) use the term complementarity to describe the biotic distinctness of two sites - here I use the term species turnover to describe this same concept simply because turnover involves the notion of change independent of scale. Furthermore the concept of species turnover is also well established in the ecological literature in reference to temporal patterns of turnover (MacArthur & Wilson 1967).

Critical to this approach of the measurement of local richness and species turnover is an appreciation of spatial scale, and especially the preservation of the same scale of measurement for studies to be comparative. Local richness is typically defined as the species richness of an homogenous community (Whittaker 1977). But homogeneity is not a feature of the natural world at any scale and the continuum concept of community organisation in space presents a more realistic picture of how species are distributed in response to environmental change (Austin 1985, Auerbach & Shmida 1993), emphasising the individualistic nature of species distributions (Gleason 1939, Hengeveld 1990). The patterns of species turnover will therefore be variable and continuous at all spatial scales, with different patterns and processes dominant at different scales within the spatial heirarchy of ecological systems (Ricklefs 1987). Hence definitions and measures of local richness should ideally consider the scale at which biotic interactions, abiotic fluctuations and disturbances occur (Cornell & Lawton 1992, Lewinsohn 1991). This implies that the scale of measurement should be different for different taxa. For example, small staphylinid beetles perceive and interact with the environment and each other at a much finer scale than do birds, and so the "grain" of the sampling regime used for staphylinids should be smaller than that for birds (Wiens 1989). Selection of appropriate scales must rely on the biological intuition of the investigator or upon initial investigations of species' distribution patterns (Carlile *et al.* 1989, Greig-Smith 1964, Wiens 1989). This is the essential problem in comparing the patterns of diversity between different taxa.

THE EFFECT OF SAMPLING ON LOCAL INVENTORY

It is a huge task to get a full species list for the insect fauna of a locality. The only reasonable option is sampling the flora and fauna in a quantifiable way which allows comparisons to be made between sites and among taxa (Greig-Smith 1964, Krebs 1989, Southwood 1978). However there are two factors that affect the results we gain by sampling, these being that the number of species collected in a sample is related to the area sampled and to the sampling effort used (Gleason 1922, Greig-Smith 1964, Preston 1948, Sanders 1968, Williams 1943).

Sampling and Area

The sampling strategy used to enumerate local sites is crucial because measurement of local richness and species turnover is scale dependant. The relationship between species richness and area illustrates that as we increase our local scale then our estimate of local richness increases; and by definition that for species turnover decreases by an equivalent amount. Thus when estimating regional diversity, our measures of local richness and species turnover must be made at the same scale. Usually this will involve using equivalent sampling methodologies in enumerating local richness and species composition, for instance using the same sized plots for plants or the same trapping methods for animals. It would clearly be invalid to directly compare the species richness and floral turnover between two forest plots of 1 hectare and 5 hectares.

Sampling and Sampling Effort.

Sampling effort has several effects upon enumerating local insect faunas that are crucial to consider. First, there is no single sampling method that will result in a full species list for a locality - different sampling techniques collect different portions of the total fauna at any one site (Southwood 1978). For example, canopy fogging is biased towards flightless arboreal insect taxa, whereas traps depending upon flight activity such as light and malaise traps will be biased towards highly mobile flying groups (Hammond 1990, Mawdsley, in prep.). Generally sampling methodology has more effect on the similarity between samples than site-specific factors (Casson & Hodkinson 1991). In practice it is difficult to combine data from different methods to yield total local richness estimates, requiring careful calibration of methods at sites of known total faunas (Hammond 1994b). Standardisation of a variety of methods to even out the biases of single methods into a standard 'sampling package' used for sampling insect faunas in tropical forests worldwide would maximise the potential of local samples for use in comparative studies and in estimating regional and global species richness (Gadagkar *et al.* 1990, Stork 1994).

A second problem, especially when quantitative data is required, is that many methods for sampling insects such as light and malaise traps are relative methods in that they do not give absolute 'standing crop' estimates of the abundance and species richness in a given area. Patterns of abundance seen in Malaise or light trap samples, for example, are do not directly mirror those of the insect assemblage sampled. Instead, some species are over-represented while others are under-represented, and this is determined by a complex interplay between the nature of the method and species-specific behaviour (Muirhead-Thompson 1987). Knowledge of the bias inherent in the method is crucial here and requires careful interpretation. Moreover the bias associated with any method is likely to vary between habitats and faunas thereby complicating matters further.

Perhaps the most evident point about sampling is that by definition only a certain proportion of the local fauna that is susceptible to being sampled by any given method will be collected (Sanders 1968). A given sample will only collect a certain number of individuals and species alike, and incomplete sampling in this way may be solely responsible for the differences in local richness and the composition between samples collected at different sites (Colwell & Coddington 1994). Methods such as rarefaction have been developed to overcome the problem of incomplete sampling in estimating the local richness of sites for comparative purposes (Sanders 1968, Gage & May 1993, Heck, van Belle & Simberloff 1975). However undersampling proves an even greater problem for estimates of species turnover. A null hypothesis for between-site comparisons would be that between-site differences are entirely due to random sampling from the same local community. This difference, which we can define as "pseudoturnover", will always be part of the measured species turnover and so *must* be estimated (Nilsson & Nilsson 1985). Only when we are dealing with complete inventories will the contribution of pseudoturnover to estimates of species turnover be equal to zero, however complete inventories of local insect faunas are extremely difficult to get (e.g. Hammond 1990). Pseudoturnover may be estimated either by replication of field samples at one locality (which in practice may be difficult because of fine-scale heterogeneity, particularly in tropical forests, and 'pseudoreplication' due to the presence of uncontrolled spatial autocorrelation) or by bootstrapping techniques (see Krebs 1989, Robinson & Tuck 1993).

Finally, the temporal extent of collecting, in a similar way to area, will affect measured local richness. Seasonal patterns of species occurrence vary between taxa and guilds, and species accumulation curves rarely plateau even after many months sampling (Basset & Arthington 1992, Hammond 1990). Moreover, the pattern of species accumulation with time will vary between taxa and guilds. For example, herbivore assemblages, which often have peak abundances correlated with periods of leaf production (Basset & Arthington 1992), can be more effectively sampled in a shorter time frame than taxa and guilds that do not show such temporal concordance. Thus the timing and temporal extent of any sampling programme will affect estimates of local richness and species turnover through temporal turnover.

Tourist species and 'mass effects'.

Local dispersal of individuals from areas where their populations are at high densities ('source populations') to unfavourable areas which could not support a viable population of that species alone ('sink populations') (see Pulliam 1988) will enhance levels of local richness and lower the amount of species turnover between sites. These 'mass effects' of immigration have been documented for desert and tropical forest plant communities (Hubbell & Foster 1986, Shmida & Wilson 1985) and entomologists have used the term 'tourists' to describe immigrant, non-resident species (Moran & Southwood 1982, Stork 1987a). Several studies have classified up to 20-30% of arboreal insect species as tourists (Basset & Arthington 1992, Gaston *et al.* 1993, Stork 1987a), and although this phenomenon distorts our measurements of local richness and species turnover, it is debatable to what extent tourists should be discounted as local community members (Moran & Southwood 1982). For example they may still provide services as pollinators and become food for predators. However, in a qualitative analysis of species diversity and species turnover where species are recorded only in terms of presence and absence, a singleton tourist species would be accorded the same weight as the dominant species in the assemblage. The effect of tourists in qualitative analyses is therefore to cause an overestimation of local species richness but an underestimation of species turnover. Large samples may suffer from this especially because the proportion of species in a sequence of samples that are tourists increases as further samples are added (P.M. Hammond 1994b, and personal communication).

METHODS OF ESTIMATING REGIONAL RICHNESS FROM LOCAL SAMPLES

Estimates of tropical insect richness derived from local samples have used assumptions about the degree of host-specificity of tropical insects (Erwin 1982, Thomas 1990) (i.e. insect:plant species richness ratios (Gaston 1992)) rather than to the spatial patterns of local richness and species turnover, principally because these latter patterns are so poorly known. Generally we can identify four types of methods commonly in use for estimating the species richness of an area from local samples (Table 1), the first three of which represent the data-analytic approach and the latter the sampling-theoretic approach (Bunge & Fitzpatrick 1993, Colwell & Coddington 1994, Hammond 1994b). For a review of the theoretical aspects of various sampling models not listed here see Bunge & Fitzpatrick (1993).

Despite detailing a variety of statistical methods that are directed at the estimation of species richness, Bunge & Fitzpatrick (1993) conclude that the problem is "quite resistant to statistical solution, essentially because no matter how many classes have been observed, there may still be a large number of very small classes" i.e. rare species. They conclude that estimators of coverage, that is the proportion of the total species richness that are represented in the samples, are more reliable and are better understood, and they recommend Chao's method which utilises the concept of coverage. Evidence from studies of single temperate forest plant communities (Palmer 1990, 1991) and computer-simulated communities (Baltanas 1992) using the species-area relationship, the lognormal species-abundance distribution and the jackknife suggests that we should use these methods with caution. Regional species richness was usually underestimated, except for species-area extrapolation applied to plant communities, which provided an overestimate. All the measures considered by these authors were shown to be biased, but the jackknife was the most reliable and the species-area extrapolation the least (Baltanas 1992, Palmer 1990). However, in terms of estimating regional richness the jackknife is likely to be of little value for hyperdiverse taxa that have not been sampled intensively because it cannot give values greater than twice the number of species encountered in samples (Smith & van Belle 1984). The species-area relationship, despite its

TABLE 1: Methods used for estimating the species richness of an area from local samples.

A. Data-Analytic Approach

1. Extrapolation of species accumulation curves (Palmer 1990, Baltanas 1992, Soberon & Llorente 1993).
2. Fitting parametric models
   (a) Relative abundance e.g. log-series, log-normal or negative binomial etc. (Fisher *et al.* 1943, Krebs 1989, Palmer 1990, Coddington *et al.* 1991);
   (b) Fitting parametric models of body size distributions (May 1978)
3. Ratios of known values of species richness (Colwell & Coddington 1994, Hammond 1994):
   (a) Hierarchical
      i. For example, Hemiptera:insect species richness ratio - e.g. Hodkinson & Casson 1991
   (b) Non-hierarchical
      i. For example, insect:plant species richness ratio - e.g. Gaston 1992
      ii. Described:undescribed species richness ratio - e.g. Hodkinson & Casson 1991, Hodkinson 1992, Hodkinson & Hodkinson 1993)

B. Sampling-theoretic Methods

4. Non-parametric statistical methods (Colwell & Coddington 1994)
   (a) The first-order and second-order jackknife (Heltsche & Forrester 1983, Smith & van Belle 1984, Krebs 1989)
   (b) The Bootstrap (Smith & van Belle 1984, Krebs 1989)
   (c) Chao's methods (Chao 1984, 1987, see Bunge & Fitzpatrick 1993, Colwell & Coddington 1994)

shortcomings, is perhaps the most useful of method for estimating the species richness of large areas because area is the most easily measured correlate of species richness and its nature and behaviour is relatively well understood (e.g. Boecklen & Simberloff 1986, Connor & McCoy 1979, Gould 1979, Kelly *et al.* 1989, Martin 1981, Miller & Wiegert 1989, Simberloff 1992, Stevens 1989, Williamson 1988). Why it behaves badly as an estimator of species richness and how to overcome this will be discussed later.

Although the number of species is the quantity of interest in species richness estimation, without any measure of the variance associated with a mean estimate it is difficult to interpret in any common sense way. Variances are as (if not more) important than the actual mean estimate itself. Variances associated with data-analytic measures must be gained by replication, whereas for the sampling-theoretic measures there are a formulae available for the calculation of variance (see Colwell & Coddington 1994).

A final point that relates to the process and dangers of extrapolation in general, and notably to that of the species-area or accumulation curves and integration of distributions of relative abundance for these methods have no intrinsic maximum as do the non-parametirc methods (Colwell & Coddington 1994), is the assumption that the patterns of species accumulation outside the range of the data collected are the same as those within the data. This point has been the focus of much debate concerning high estimates of deep-sea species richness of 10 million species (Grassle & Maciolek 1992, May 1992b, Poore & Wilson 1993). The patterns of local richness and species turnover among a set of local samples need to be representative of the patterns within the whole region (Baltanas 1992). Representativeness is the crucial factor and is extremely difficult to objectively obtain.

ESTIMATING THE REGIONAL RICHNESS OF A KNOWN BIOTA: UMBELLIFERS IN BRITAIN

At present there have been few evaluations made of the efficiency of any of these methods in estimating the species richness of a known regional biota. Miller & Wiegert (1989) fit a negative binomial model to species accumulation curves for the regional flora of the southern Appalachians in the United States, but they do this by randomly sampling the regional pool of species and individuals in a computer-simulation. This acts to erase the spatial heterogeneity present in the region, making it essentially a problem of estimating local species richness without species turnover. Other studies based on single communities (Baltanas 1992, Palmer 1990, 1991) would suggest that these methods would be even less reliable for larger areas where there is greater heterogeneity in species distributions and where a smaller percentage of the regional biota is contained in any one local sample.

The distribution of the British flora has been well documented through the efforts of a great many professional and amateur naturalists (Perring & Walters 1976). In Britain there are 64 species of umbellifer (Apiaceae) whose presence and absence was recorded for the 145 50km x 50km grid squares in Britain. [The choice of this group of plants is arbitrary. It was chosen because it contains a manageable number of species with well-studied biology.] These data were then used to answer the following questions: (1) How well can we estimate this regional richness of 64 species from local samples of grid squares? (2) What sampling regime should we use to maximise the representativeness of our local samples with respect to the regional patterns of local richness and species turnover? The main focus of this analysis was to use the species-area relationship and to partition such estimates into estimated local richness and estimated species turnover. Species-area curves were extrapolated by a power function which showed the best fit to each sample (see Connor & McCoy 1979).

How Many Localities Do We Need To Sample?

Grid squares sampled randomly from the national grid clearly showed a positive correlation between the number of squares sampled and the accuracy of the estimate of regional richness (Greig-Smith 1964, Krebs 1989, Southwood 1978). Relatively good estimates of regional richness were gained when 40 grid squares (28% of the total) were sampled, yielding a mean regional estimate of 76 species (Figure 1 and Table 1). For most tropical regions 28% of the land area inventoried is unrealistically high, and in most countries, National Parks and forest reserves rarely amount to more than 5% of the land area (Hartshorn 1983). Floristic and faunal lists of even these protected areas are scarce. In terms of being able to estimate regional (or national) species richness from such data, it seems likely that large errors would creep in. For the umbellifers in Britain when only 4% (N=5) of grid squares were used, regional richness was overestimated by 250%, giving a mean estimate of 155 species (Figure 2 and Table 1). The bias of these estimates is always positive (overestimation). Given that 4% of the area of a region for which data on floral and faunistic composition is known, how can we improve these estimates of regional richness?

Are Random Local Samples Representative?

The mean local richness and species turnover of umbellifers for all 145 grid squares in Britain is 27.6 and 2.32 respectively. In effect our samples need to be sufficiently representative to yield good estimates of these two quantities. How well do different levels of sampling achieve this? Estimates of mean local richness were close to the population value after 14% of grid squares (N=20) had been sampled (Figure 3), but where only 7% (N=10) or less of grid squares were sampled, the variance of the estimates was high and mean local richness was poorly estimated.

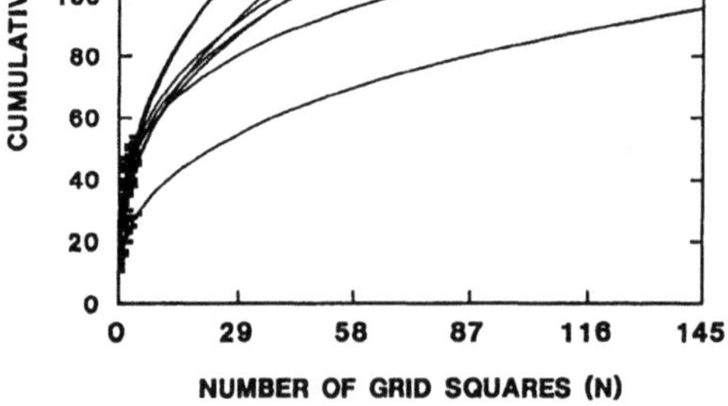

Figure 1. Extrapolated species accumulation curves for umbellifers (Apiaceae) in Britain when 40 grid squares are randomly sampled (equivalent to 28% of the area of Britain). Each curve represents a single sample of 40 grid squares. This is randomly subsampled to generate randomised data of sample sizes of 5, 10, 20 and 40 squares to which the power function if fitted.

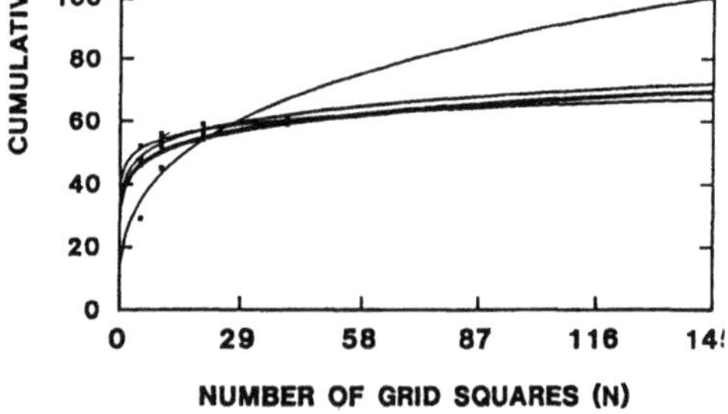

Figure 2. Extrapolated species accumulation curves for umbellifers (Apiaceae) in Britain when only 5 grid squares are randomly sampled, equivalent to 4% of the area of Britain. The same method of randomisation was used as for Figure 1. Note that error and the variance of these estimates are much higher than those in Figure 1.

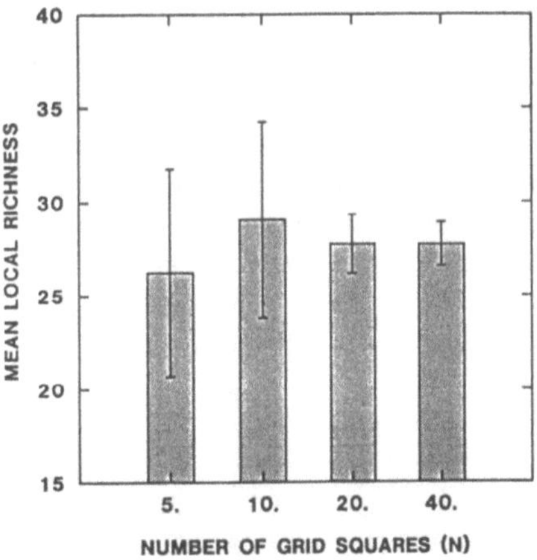

Figure 3. The effect of the number of grid squares sampled randomly on the mean and standard error of mean local richness estimates. These are calculated from five replicates (n=5) for each sample size. The known mean local richness across the UK for umbellifers is 27.6.

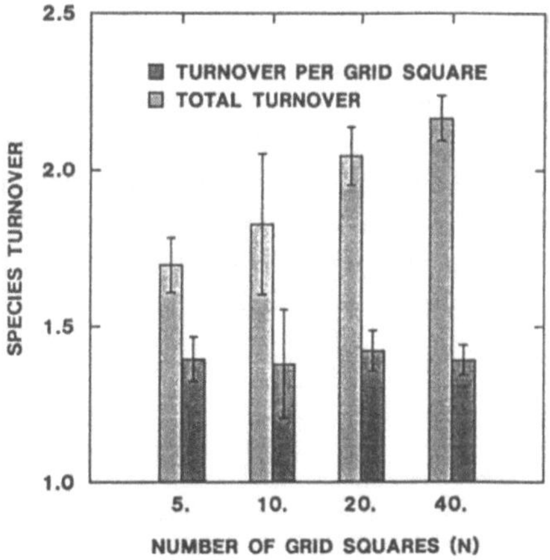

Figure 4. The effect of the number of grid squares sampled randomly on the mean and standard error of mean species turnover estimates. These are calculated from five replicates (n=5) for each sample size. The known mean species turnover per neighbouring grid square is 1.27 i.e. 27% .

Mean species turnover is more difficult to estimate than mean local richness because the total turnover within any sample is highly dependant upon N (Figure 4). This is because of the species-area relationship where larger samples contain more species in total, and this indicates the dependence of extrapolating species-area curves on estimating species turnover. We can compare species turnover from different sample sizes by comparing the amount of species turnover per grid square. For the whole dataset, species richness and species turnover scales with area (N) by $N^{0.12}$, and so by calculating (total sample turnover/$N^{0.12}$) we have a measure of species turnover per unit area. The population value of this `species turnover density' that we want to estimate is $2.32/145^{0.12}=1.27$. Figure 4 shows that our estimates of this quantity are consistently over this by about 40% (note that species turnover has a zero value when equal to 1, i.e. when local richness = regional richness).

Therefore if we sample sufficiently to estimate mean local richness, we have still not sampled enough to accurately estimate mean species turnover. Overestimates of regional richness using species-area curves seem to be because we overestimate species turnover. How then can we improve our estimate of species turnover and therefore regional richness?

The Representativeness of Regional Species Turnover Estimates

The observation that distant sites are usually more different in their environmental attributes than near sites (Bell *et al.* 1993, Williamson 1988) leads to the prediction that species turnover will be correlated with the distance between sites. This relationship has been observed for a variety of taxa along large-scale transects in Britain (Harrison, Ross & Lawton 1992), for continental mammal faunas (Flessa 1980) and for canopy insects in Brunei (Stork 1987b). Its implications for extrapolating species-area curves are that if our sample localities are widely separated then the measured species turnover between them will overestimate the true turnover between localities.

This problem can be clearly observed with umbellifers in Britain. If replicate samples of 5 grid squares (4% of the total) are taken from a contiguous block of adjacent squares (i.e.

localised sampling) rather than randomly from throughout the country (Figure 5), then estimates of regional richness are significantly better (t-test; $t=4.19$, 13 d.f., $p<0.001$) (Figure 2 cf. Figure 6; Figure 7; Table 2). Partitioning the regional estimates into local richness and species turnover (Figures 8 & 9; Table 2) it can be seen that it is species turnover only that is significantly different between random and local sampling regimes. In addition, for localised and random sampling extrapolations combined, the error of regional estimates is proportional to measured species turnover per grid square among all these samples (Figure 10: error of estimate (y) = 0.003(species turnover per grid square (x)) + 1.062; $R^2 = 0.636$; F-ratio = 22.693; $p < 0.001$).

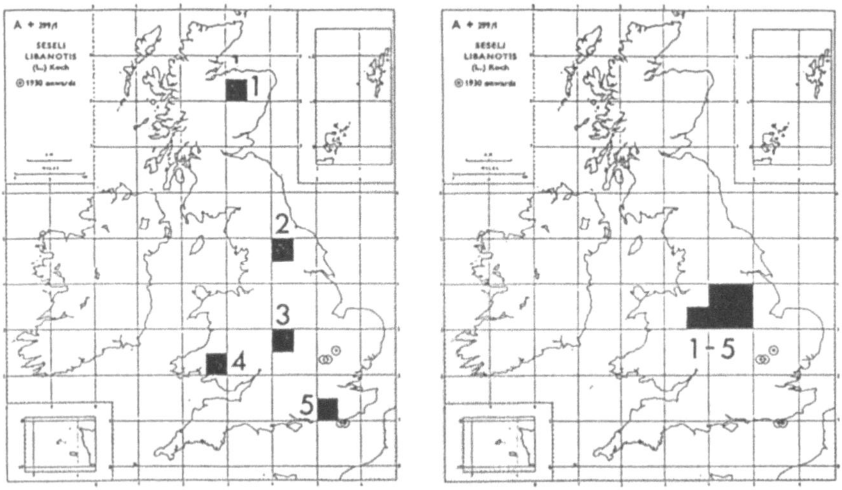

RANDOM LOCALISED

Figure 5. Distribution maps of *Seseli libanotis* (Apiaceae) in Britain showing random and localised sampling distributions.

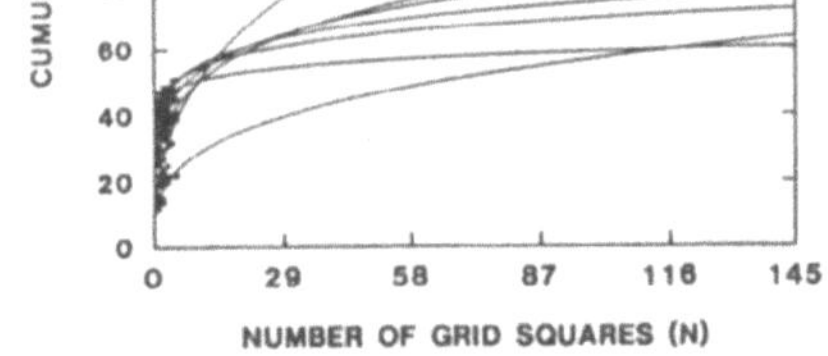

Figure 6. Extrapolated species accumulation curves for umbellifers (Apiaceae) in Britain when 5 grid squares (4%) are sampled from a localised area, as opposed to being sampled at random (see Figure 2 for comparison).

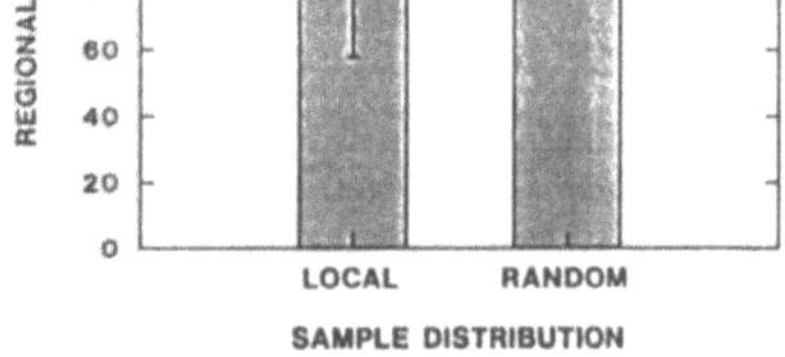

Figure 7. The mean and standard error of regional species estimates of umbellifers (Apiaceae) in Britainwhen samples are taken either locally or randomly from the national grid. These are calculated from seven replications (n=7) for each sample distribution.

TABLE 2. The effect of sample size on the estimation of regional species richness, local species richness and species turnover. Figures are mean estimate ± standard error of the mean for five replicates.

NUMBER OF SQUARES SAMPLED (N)	PROPORTION OF TOTAL N (%)	REGIONAL ESTIMATE (n=5)	LOCAL RICHNESS ESTIMATE (n=5)	REGIONAL SPECIES TURNOVER ESTIMATE + 1 (n=5)
5	4	155 ± 13	26.2 ± 2.5	1.40 ± 0.03
10	7	-	29.0 ± 2.3	1.38 ± 0.08
20	14	89 ± 18	27.7 ± 0.7	1.42 ± 0.03
40	28	76 ± 6	27.7 ± 0.5	1.39 ± 0.02
145	100	64	27.6	1.27

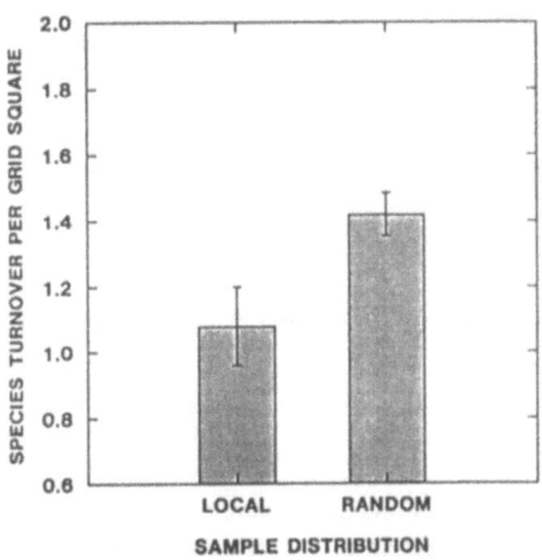

Figure 8. The mean and standard error of "species turnover per grid square" estimates of umbellifers (Apiaceae) in Britain when samples are taken either locally or randomly from the national grid. These are calculated from seven replications (n=7) for each sample distribution.

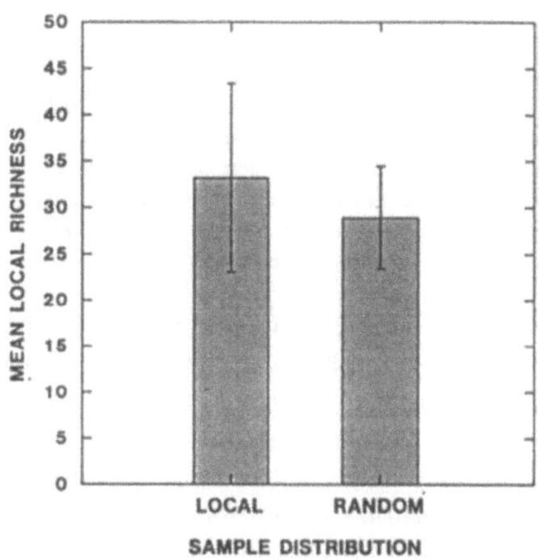

Figure 9. The mean and standard error of mean local richness estimates of umbellifers in Britain when samples are taken either locally or randomly from the national grid. These are calculated from seven replications (n=7) for each sample distribution.

Correcting For Overestimation By Species-Area Extrapolations

A solution to the problem of overestimation by species-area extrapolation from random samples may be found by considering the source and magnitude of the error in the estimates of regional richness when samples are taken randomly. The overestimation of species turnover is dependant upon sample size, mainly in response to the factor of site separation as outlined above. The more sites that are sampled, the closer they are, and the percentage error in the estimate of species turnover shows a strong negative power relationship with the proportion of localities in the sample (Figure 11):

$$\mathrm{ERROR}_{\text{species turnover}} = 4.20(N^{-0.91})$$

This relationship gives us the potential to correct species-area extrapolated estimates of regional richness so long as we are confident that we have sampled sufficiently to get

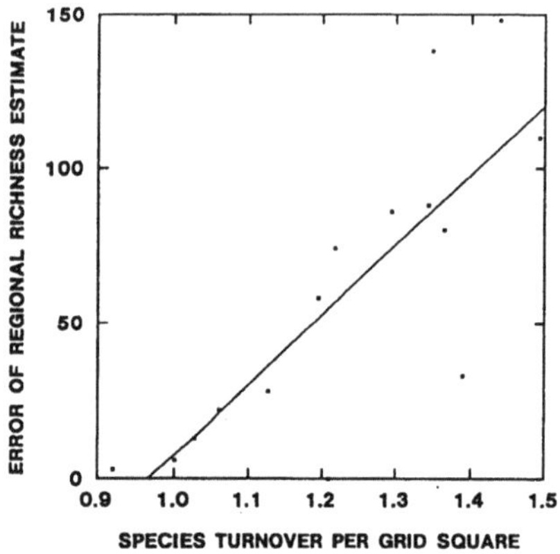

Figure 10. The relationship between the error of the regional richness estimates and species turnover per grid square. Each point represents the error and species turnover of a single extrapolation. Error is measured as the difference between regional estimate and the known regional richness of 64 species. All extrapolations using five grid squares (both random and localised; n=14) have been used.

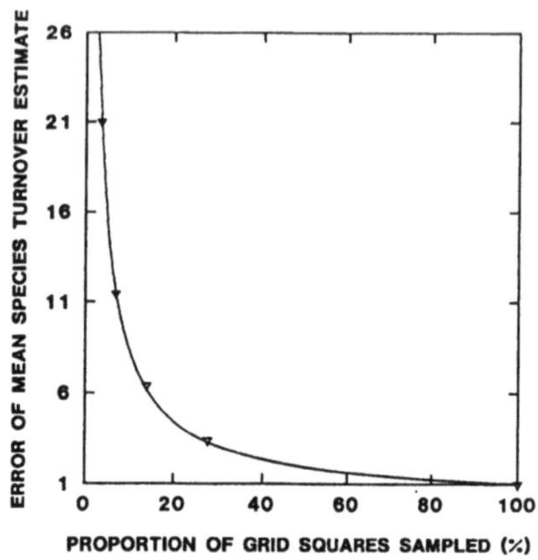

Figure 11. The relationship between the error of the total species turnover estimate and the proportion of grid squares randomly sampled. The points describe the extent to which random sampling overestimates species turnover for sample sizes of 5, 10, 20 and 40 grid squares.

representative samples of local richness and species turnover. Presumably it has some form of inverse relationship with the species-area relationship.

To summarise, the species turnover estimated among local samples of adjacent localities represents a truer measure of the mean species turnover per locality than widely separated sites. This is important because species turnover is the quantity that we need to measure with most accuracy for accurate extrapolation to a regional total. However, random samples do give a better estimate of mean local richness with less variance, but because the effect of species turnover upon the regional estimate is greater, localised sampling performs better (Table 3). An important result is that a given increase in the number of localities sampled does more to reduce the error in species turnover estimation in the early stages of a regional inventory when only a few localities have been sampled (Figure 11). For umbellifers (Apiaceae) in Britain, this analysis suggests that we need species lists of localities representing about 10% of the area of Britain to give sufficient data to extrapolate up to the regional total. Miller & Wiegert (1989) suggest a value of 2-3% for their study of the regional flora of the southern Appalachians. This is likely to be a lower limit because of their random computer-simulated sampling methodology (see above).

TABLE 3: The effect of localised versus random sampling on the estimation of regional species richness, local species richness and species turnover. Figures are mean estimate ± standard error of the mean for five replicates.

	LOCALISED SAMPLING	RANDOM SAMPLING
REGIONAL RICHNESS ESTIMATE	84 ± 10	155 ± 13
LOCAL RICHNESS ESTIMATE	33.2 ± 3.9	26.2 ± 2.0
REGIONAL SPECIES TURNOVER ESTIMATE	1.08 ± 0.04	1.40 ± 0.03

THE TROPICAL PERSPECTIVE OF ESTIMATING REGIONAL RICHNESS.

The distribution and abundance patterns of tropical species are markedly different to those of temperate systems. Tropical systems are more species-rich, have populations at lower densities (Basset & Arthington 1992, Currie & Fritz 1993, Elton 1973, 1975, Watanabe & Ruaysoongnern 1989), and samples of insects and plants from tropical forests usually have a highly inequitable distribution of abundance with many species being relatively rare (e.g. Barlow & Woiwod 1989, Stork 1991, and Figure 12). These features are correlated with inaccuracy of species richness estimates (Baltanas 1992, Mignoti & Meeden 1992), making the goal of producing accurate estimates even more difficult. For tropical regions it is probable that to estimate species richness with any certainty we will have to sample a greater proportion of localities than less diverse regions (Mignoti & Meeden 1992).

A tropical species accumulation curve tends to rise faster than a comparable one from temperate regions and the final asymptote will also be higher. The species accumulation curve for ants fogged from seven trees of the same species (*Dryobalanops beccarii* - Dipterocarpaceae) in Brunei illustrates the problems that tropical biologists face in estimating species richness in that the curve shows no sign of levelling off (Figure 13; Mawdsley unpublished data). For this dataset, the problem of pseudoturnover would be very great if it was used as one of many local faunal lists to estimate regional richness. However we can use these samples to compare the different methods by attempting to estimate the total species richness of canopy ants in the locality at Kuala Belalong. Ideally a variety of real and simulated datasets should be used to explore the performance of these methods. The results presented here are illustrative and limited in their use of a single dataset drawn from a universe of unknown total species richness. Initially it is worth considering what we are extrapolating to for this is far from obvious. The samples are taken from seven trees aligned along a transect 6km long and the tree is taken to be the unit of sampling effort. Each tree is at the large end of the size scale (80-120cm dbh) and trees like this are found at a density of only a few per hectare (e.g. Newberry *et al.* 1992).

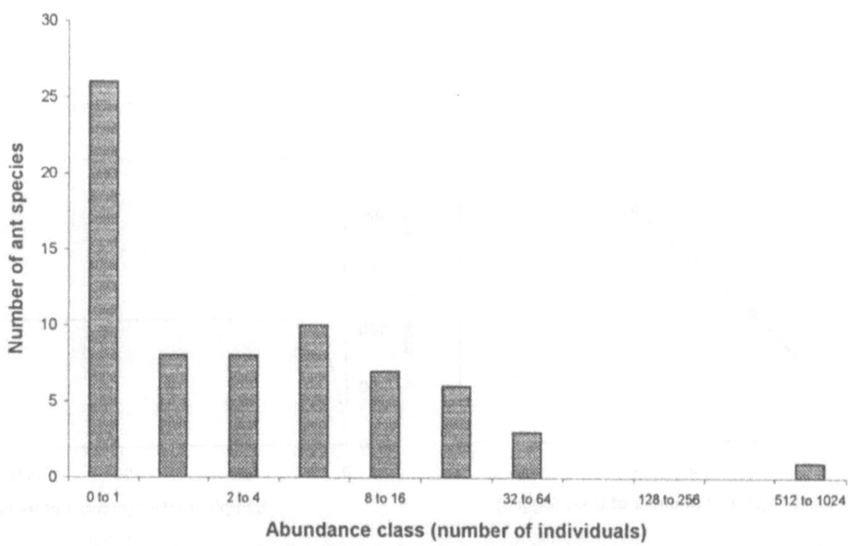

Figure 12. Species-abundance distribution for Brunei canopy ants sampled from seven trees by canopy fogging. Note that these samples have a relatively low proportion of singleton species (abundance class = 1 individual) compared to many other more diverse taxa.

So are the various methods estimating the total species richness of these seven trees only? Or of the area of rain forest in the 6km between them? Or of just the part of this area that is characterised by the seven trees sampled? For example, they are all the same species and thus do not incorporate any species with specific plant-associations. These are important points and applicable to all sampling methods. Their effect will depend both on the taxon considered, the method used and the scale and design of the sampling programme.

Estimates for ant species richness from the samples taken at Kuala Belalong in Brunei are given in Table 4. For the species accumulation curve I have fitted two models (Figure 14, Colwell & Coddington 1994, Miller & Wiegert 1989, Soberon & Llorente 1993):

1. Logarithmic model:

$$S = 93.0 \times \log_{10}(N+1) - 15.6 \qquad R^2 = 0.992$$

where S = the cumulative number of species and N = the sampling effort or number of sites and

2. Negative binomial model:

$$S = S_{max} \times (1 - \exp(-0.15 \times N)) \qquad R^2 = 0.997$$

Log models do not have an asymptote and are considered to be useful for speciose taxa (Soberon & Llorente 1993). In contrast, the negative binomial has an asymptote (S_{max}) which is the estimate of interest. The first-order jackknife is dependant upon the number of singletons in the data while the second-order jackknife and both Chao's methods are functions of the number of singletons and doubletons. Colwell & Coddington (1994) note that Chao #1 provides a lower bound estimate but that it performs well with data containing many rare species as do most tropical insect data. For hyperdiverse tropical insect faunas the non-parametric estimators will certainly underestimate because their estimates are constrained

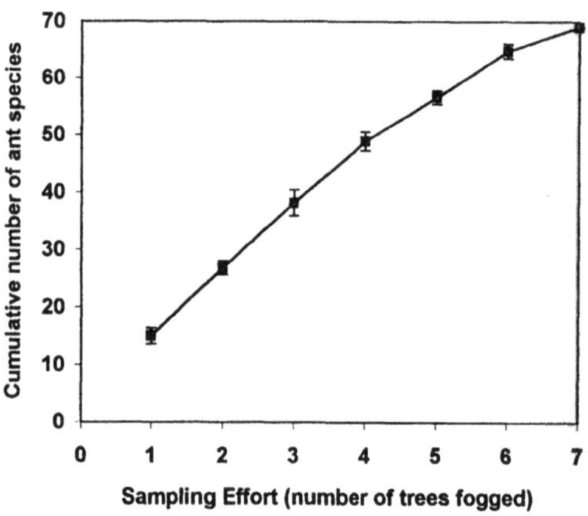

Figure 13. Species-accumulation curve for ants (Formicidae:Hymenoptera) fogged from the canopies of seven *Dryobalanops beccarii* trees at Kuala Belalong FSC, Brunei.

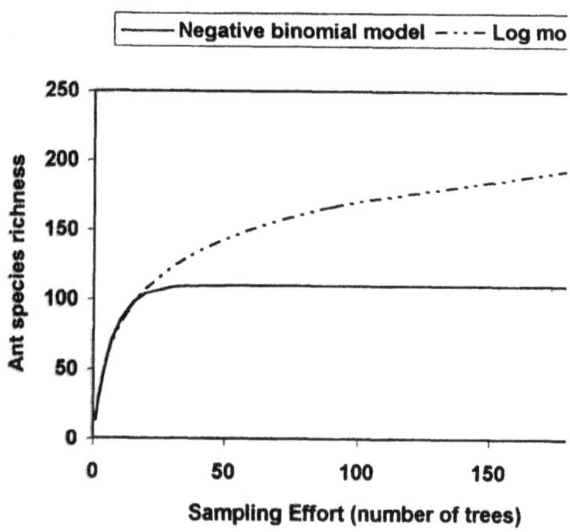

Figure 14. Extrapolated species accumulation curves for the ant data shown in Figure 13. Two models have been used, a log-model and a negative binomial model (see text for details).

TABLE 4: Estimates of species richness using local samples of ants fogged from the canopies of seven *Dryobalanops beccarri* trees in Brunei (see text for discussion).

Estimation Method	Species richness estimate
Observed species richness (7 trees)	*69*
Species accumulation - log model (extrapolation to 30 trees)	123 ± 30
Species accumulation - log model (extrapolation to 5000 trees)	328 ± 32
Species accumulation - log model (extrapolation to 18000 trees)	380 ± 36
Species accumulation - negative binomial model	110
First-order jackknife	105 ± 9
Second-order jackknife	98 ± N/A
Bootstrap	84 ± N/A
Chao's method #1	108 ± 12
Chao's method #2	136 ± 21
2 x 2 contingency tables	102 ± 1
Integration of log-series (30 trees; 6000 individuals)	138
Integration of log-series (5000 trees; 10^6 individuals)	267
Integration of log-series (18,000 trees; 3.5 x 10^6 individuals)	300

to less than twice the observed number of species for the jackknife and bootstrap methods, and to $(S^2_{obs} + 1)/2$ (Colwell & Coddington 1994, Smith & van Belle 1984). Additionally these ignore the percentage of the population that has been sampled (Mignoti & Meeden 1992) unlike the negative binomial in which the probability of gaining another new species in a series of samples is inversely proportional to the number already accumulated (Colwell & Coddington 1994). Two other methods used are 2 x 2 contingency tables which estimate the joint absences between samples (i.e. the number of species not recorded in two samples) given the assumption of random sampling (see Pimm *et al.* 1994) and integration of the area under a log-series abundance distribution fitted to the observed data (Magurran 1988). The assumption of the former that individuals are encountered independently is also an assumption of the rarefaction method and one that clearly is sometimes not tenable (Gage & May 1993).

The estimates are shown in Table 4 and may be divided into: (1) the non-asymptotic estimates (log model of the species accumulation curve; integration of the log-series abundance distribution) where it is necessary to select a value of sampling effort (i.e. trees or individuals sampled) along the x-axis for estimation of species richness, and, (2) the asymptotic or non-parametric measures which give a single estimate independent of any sampling effort parameter. The outstanding pattern in Table 4 is that the asymptotic measures mostly have produced broadly similar estimates of about 100-110 species, whereas the non-asymptotic measures are still rising at this level of species richness (Figure 14). The conclusions drawn from the British umbellifer analysis that random sampling overestimates species turnover per locality means that these non-asymptotic measures are likely to be overestimates. This coupled with the fact that the non-parametric estimates are probably underestimates (see above) suggests that the true answer lies between 100 and 300 species!

Is this a satisfactory estimate? Almost certainly not. The range of estimates in Table 4 is great and although we can generalise about the biases of these methods, the adage that "any method is only as good as the data" must be borne in mind. Generally the estimation of species richness from local samples demands the data to be both representative and of sufficient quality and coverage to be able to estimate with any practical certainty the species richness of a larger region. In the illustrative example of the Bruneian ants it is unlikely that either of these requirements have been met with only seven trees of the same species being sampled.

There is also the problem of what is this estimating? For example, if these samples represents a 6km x 6km area, and there are about 5 large trees per hectare, then the 18,000 trees contained in this area are predicted to be home to roughly 300 to 380 species of ant (integration of log-series and log-model of species accumulation respectively, Table 4), values which seem too large and which have no empirical support (Wilson 1987). Certainly these methods are estimating only the pool of species that are susceptible to fogging (for ants however this is probably a high proportion) and will therefore consider only a proportion of the total fauna. These uncertainties demand increased sampling to gain better estimates.

For taxa other than ants the problem is greater. Although the dominant insect taxa in terms of numbers of individuals and biomass (Erwin 1983, Stork 1991, Wilson 1987), ants are a relatively species-poor group. There is good evidence, both empirical (Wilson 1987) and from consideration of the basic biology of different insect taxa, that the species turnover rates of other taxa such as beetles will be appreciably higher. Species accumulation curves for these groups are likely to be steeper and will continue rising to reach a much higher plateau, lowering the accuracy of our predictions of species richness from a given set of local samples.

SPECIES RICHNESS ESTIMATION: THE SCALE OF THE PROBLEM.

The way forward to estimate regional insect species richness directly as previously discussed requires: (1) better estimates of total local richness because we still do not know how many species in total are found in, for example, one hectare of rain forest and (2) greater understanding of the patterns of species turnover and distribution among different taxa within a variety of regions. More complete inventories necessitate an increase in the amount of *standardised* and therefore *comparable* field collections and records of all poorly known taxa being made, by no means a small undertaking. The value of such information cannot be underestimated and is itself paramount to the goal of sustainable use of the world's natural resources. Only through careful observation of the real world will understanding of its organisation and the magnitude of its diversity become apparent (Hammond 1994a,b).

As a model, how long would it take to make representative collections of beetles in Borneo with the aim of discovering the magnitude of beetle species richness and the distribution of species within Borneo? If we consider, for arguments sake, that 5% is a minimum total area required to be surveyed to estimate regional richness relatively accurately (Figure 6), and a "typical" locality is about 50,000 hectares in size (i.e. 20km x 20 km) (e.g. an ATBI site: Janzen & Hallwachs 1993, Yoon 1993), then we would need to survey about 100 such localities to cover 5% of the area of Borneo.

How long would this take? My PhD has involved putting together such a collection at a single locality in Brunei (11000 specimens representing 3164 species), and the total person-hours needed for this has been about 10,000 hours. To survey 100 localities would therefore take 10^6 hours, roughly 500 person-years. In just 5 years, 100 people (a mix of specialist taxonomists, mostly trained parataxonomists plus a few data analysts) could provide an answer to the regional diversity of Bornean beetles and detailed information on their distribution which would have both great significance both in scientific and management terms. More reliable data might involve sampling perhaps 40,000 specimens per locality, increasing the resources and time required to complete such a programme but not proportionally so. Comparable estimates made for samples of spiders collected in Bolivia (Coddington *et al.* 1991) suggest that to inventory a single 'average' tropical site for a fauna of roughly 800 species might take between 115 and 300 person-days and require a sample of 3000-8000 specimens. 100 sites might therefore take 100 person-years. As anyone who has ever completed faunal surveys knows, the time-consuming element is sorting, especially for difficult and poorly known taxa. Integration of basic research with training of taxonomists and parataxonomists can facilitate much of the preparation and initial sorting of samples taken for biological inventory (see Janzen *et al.* 1993).

The problem of estimating the regional richness in a species-rich group such as Bornean beetles may be simplified by having as a goal the estimation of mean local richness and mean species turnover within the region. The process of scaling up will be best accomplished by the use of ratios and of relative species turnover rates (including comparison of the slopes of nested species area curves) at a variety of scales based on systematically collected samples (Hammond 1994b). Certainly all methods that scale up from local to regional to global faunas must include a comparison of species turnover rates for different taxa at different scales (Hammond 1994a). Explicit consideration of species turnover, long overlooked by animal ecologists, will certainly illuminate the problem of scaling up. Knowledge of the variation of the landscape and environment, and distribution of forest types within a region, is required to determine the localities for sampling to achieve these estimates. For local richness, a wide variety of localities is the requirement, but for species turnover estimates these should be localised. For both, localities need to be representative of the environmental variation within the region.

There is still much to be learnt about the methodology of inventorying and estimating the species richness of regions, however recent initiatives such as the IUBS-SCOPE-UNESCO 'DIVERSITAS' programme (di Castri *et al.* 1992) and 'All Taxa Biodiversity Inventories' ATBI (Janzen & Hallwachs 1993, Yoon 1993) lend weight to the need and importance of such ventures and the need for standard means in which to analyse and interpret such data (Solbrig 1991). Furthermore, since the signing of the Convention on Biological Diversity in June 1992 in Rio de Janeiro (Johnson 1991; note Article 7 of the Convention on Biological Diversity) governments are obliged to make efforts to increase the knowledge of their biodiversity and natural resources as has been done by organisations such as InBio in Costa Rica (GAMEZ 1991). Clearly such a major co-operative enterprise is not beyond the realms of reality - it only needs the will to uncover the true dimensions of the world's biological diversity.

ACKNOWLEDGEMENTS

I am extremely grateful to Peter Hammond for lengthy discussions on the subject of sampling and methodology and for bringing to my attention the paper by Colwell & Coddington (1994). John Lawton and Nigel Stork commented on an earlier version of the manuscript. Special thanks are due to all the members of the RGS/UBD Brunei Rainforest Project who supported and helped with the fieldwork, notably Impin Lasa, Rupert Hanbury-Tennison, Toby Howes, Catriona Prebble, Amanda Simpson, and Mark Welsh. This work forms part of a Natural History Museum funded studentship.

LITERATURE CITED

AUERBACH, M. & SHMIDA, A. 1993. Vegetation change along an altitudinal gradient on Mt. Hermon, Israel-no evidence for discrete communities. *Journal of Ecology* 81:25-33.

AUSTIN, M.P. 1985. Continuum concept, ordination methods and niche theory. *Annual Review of Ecology and Systematics* 16:39-61.

BALTANAS, A. 1992. On the use of some methods for the estimation of species richness. *Oikos* 65:484-492.

BARLOW, H.S. & WOIWOD, I.P. 1989. Moth diversity of a tropical forest in Peninsular Malaysia. *Journal of Tropical Ecology* 5:37-50.

BASSET, Y. & ARTHINGTON, A.H. 1992. The arthropod community associated with an Australian rainforest tree: abundance of component taxa, species richness and guild structure. *Australian Journal of Ecology* 17:89-98.
BEATTIE, A.J. 1993. Rapid biodiversity assessment. *Proceedings of the biodiversity assessment workshop, Macquire University - 1993.* 92 pp.
BELL, G., LECHOWICZ, M. J., APPENZELLER, A., CHANDLER, M., DEBOIS, E., JACKSON, L., MACKENZIE, B., PREZIOSI, R., SCALLENBERG, M., & TINKER, N. 1993. The spatial structure of the physical environment. *Oecologia* 96:114-121.
BOECKLEN, W.J. & SIMBERLOFF, D. 1986. Area-based extinction models in conservation. Pp.247-276 in Elliot, D.K. (ed.). *Dynamics of Extinction.* Wiley, New York.
BUNGE, J., & FITZPATRICK, M. 1993. Estimating the number of species: a review. *Journal of the American Statistical Association* 88:364-373.
CARLILE, D.W., SKALSKI, J.R., BATKER, J.E., THOMAS, J.M. & CULLINAN, V.I. 1989. Determination of ecological scale. *Landscape Ecology* 2:203-213.
CASSON, D. & HODKINSON, I.D. 1991. The Hemiptera (Insecta) communities of tropical rain forest in Sulawesi. *Zoological Journal of the Linnean Society* 102:253-275.
CHAO, A. 1984. Nonparametric estimation of the number of classes in a population. *Scandinavian Journal of Statististics* 11:265-270.
CHAO, A. 1987. Estimating the population size for capture-recapture data with unequal catchability. *Biometrics* 43: 783-791.
CODDINGTON, J.A., GRISWOLD, C.E., DAVILA, D.S., PENARANDA, E. & LAR CHER, S.F. 1991. Designing and testing sampling protocols to estimate biodiversity in tropical ecosystems. Pp. 44-60 in Dudley, E.C. (ed.). *The Unity of Evolutionary Biology: Proceedings of the Fourth International Congress of Systematic and Evolutionary Biology.*Dioscorides Press, Portland, Oregon.
CODY, M.L. 1986. Diversity, rarity, and conservation in Mediterranean-climate regions. Pp.122-152 in Soule, M.E. (ed.). *Conservation Biology: the science of scarcity and diversity.* Sinauer Associates, Sunderland, MA.
COLWELL, R. K., & CODDINGTON, J. A., in press. Estimating the extent of terrestrial biodiversity through extrapolation. *Philosophical Transactions of the Royal Society London B* 345.
CONNOR, E.F. & McCOY, D. (1979). The statistics and biology of the species-area relationship. *American Naturalist* 113:791-833.
CORNELL, H.V. & LAWTON, J.H. 1992. Species interactions, local and regional processes, and limits to the richness of ecological communities. *Journal of Animal Ecology* 61:1-12.
CRANSTON, P., & HILLMAN, T. 1992. Rapid assessment of biodiversity using 'biological diversity technicians'. *Australian Biologist* 5:144-154.
CURRIE, D. J., & FRITZ, J. T. 1993. Global patterns of animal abundance and species energy use. *Oikos* 67:56-68.
DI CASTRI, F., ROBERTSON VERNHES, J. & YOUNES, T. 1992. Inventorying and monitoring biodiversity: a proposal for an international network. *Biology International Special Issue* 27:1-28
ELTON, C. S. 1973. The structure of invertebrate populations inside neotropical rain forest. *Journal of Animal Ecology.*
ELTON, C. S. 1975. Conservation and the low population density of invertebrates inside neotropical rain forest. *Biological Conservation* 7:3-15.
ERWIN, T.L. 1982. Tropical forests: their richness in Coleoptera and other arthropod species. *Coleopterist's Bulletin* 36:74-75.
ERWIN, T.L. 1983. Beetles and other arthropods of the tropical forest canopies at Manaus, Brazil, sampled with insecticide fogging techniques. Pp. 59-75 in Sutton, S.L., Whitmore, T.C. & Chadwick, A.C. (eds). *Tropical Rain Forest: ecology and management.* Oxford, Blackwell.
ERWIN, T.L. 1988. The tropical forest canopy: the heart of biotic diversity. Pp. 123-129 in Wilson, E.O. & Peters, F.M. (eds). *Biodiversity.* National Academy Press, Washington DC.
FRANCE, R. 1992. The North American latitudinal gradient in species richness and geographical range of freshwater crayfish and amphipods. *American Naturalist* 139:342-354.
FISHER, R.A., CORBET, A.S. & WILLIAMS, C.B. 1943. The relation between the number of individuals in a random sample of an animal population. *Journal of Animal Ecology* 12:42-58.
FLESSA, K. W. 1980. Biological effects of plate tectonics and continental drift. *Bioscience* 30:518-523.
GADAGKAR, R., CHANDRASHEKARA, K., & NAIR, P. 1990. Insect species diversity in the tropics: sampling methods and a case study. *Journal of the Bombay Natural History Society* 87:337-353.
GAGE, J. D. & MAY, R. M. 1993. A dip into the seas. *Nature* 365:609-610.
GASTON, K.J. 1990. Patterns in the geographical ranges of species. *Biological Reviews* 65:105-129.
GASTON, K.J. 1991. The magnitude of global insect species richness. *Conservation Biology* 5:283-296.
GASTON, K.J. 1992. Regional numbers of insect and plant species. *Functional Ecology* 6:243-247.
GASTON, K.J., BLACKBURN, T.M., HAMMOND, P.M. & STORK, N.E. 1993. Relationships between abundance and body size: where do tourists fit? *Ecological Entomology* 18:310-314.

GAMEZ, R. 1991. Biodiversity conservation through facilitation of its sustainable: Costa Rica's national biodiversity institute. *TREE* 6:377-378.

GENTRY, A. H. 1988. Changes in plant community diversity and florsitic composition on environmental and geographical gradients. *Annals of the Missouri Botanical Garden* **75**:1-34.

GENTRY, A.H. 1992. Tropical forest biodiversity: distributional patterns and their conservation significance. *Oikos* 63:19-28

GLEASON, H.A. 1922. On the relation between species and area. *Ecology* 3:158-162.

GLEASON, H.A. 1939. The individualistic concept of the plant association. *American Midland Naturalist* 21:92-110.

GOULD, S. J. 1979. An allometric interpretation of species-area curves: the meaning of the coefficient. *American Naturalist* 114:335-343.

GRASSLE, J.F. & MACIOLEK, N.J. 1992. Deep-sea species richness: regional and local diversity estimates from quantitaive bottom samples. *American Naturalist* 139:313-341.

GREIG-SMITH, P. 1964. *Quantitative Plant Ecology*. Second edition. Butterworths, London.

HAMMOND, P.M. 1990. Insect abundance and diversity in the Dumoga-Bone National Park, N. Sulawesi, with special reference to the beetle fauna of lowland rain forest in the Toraut region. Pp. 197-254 in W.J. Knight & J.D. Holloway (eds.). *Insects and the Rain Forests of South East Asia (Wallacea)*. Royal Entomological Society of London.

HAMMOND, P.M. 1992. Species inventory. Pp.17-39 in Groombridge B. (ed.). *Global Biodiversity, Status of the Earth's Living Resources*. Chapman & Hall, London.

HAMMOND, P. M. 1994a. Described and estimated species numbers: an objective assessement of current knowledge. In Allsopp, D., Hawksworth, D.L. & Colwell, R.R. (eds.). *Microbial diversity and ecosystem function*. CABI, Wallingford.

HAMMOND, P. M. 1994b. Practical approaches to the estimation of the extent of biodiversity in speciose groups. *Philosophical Transactions of the Royal Society London B* 345:000-000.

HARRISON, S., ROSS, S. & LAWTON, J.H. 1992. Beta diversity on geographical gradients in Britain. *Journal of Animal Ecology* 61:151-158.

HARTSHORN, G.S. 1983. Wildlands conservation in Central America. In Sutton, S.L., T.C.Whitmore & A.C. Chadwick (eds.). *Tropical Rain Forest: ecology and management.*Blackwell, Oxford.

HECK, K. L. J., VAN BELLE, G., & SIMBERLOFF, D. 1975. Explicit calculation of the rarefaction diversity measurement and the determination of sufficient sample size. *Ecology* 56:1459-1461.

HELTSHE, J., & FORRESTER, N. E. 1983. Estimating species richness using the jackknife proceedure. *Biometrics* 39:1-11.

HENGEVELD, R. 1990. *Dynamic Biogeography*. Cambridge University Press. 249 pp.

HODKINSON, I.D. & CASSON, D. 1991. A lesser prediliction for bugs: Hemiptera (Insecta) diversity in tropical rain forests. *Biological Journal of the Linnean Society* 43:101-109.

HODKINSON, I. D. 1992. Global insect diversity revisited. *Journal of Tropical Ecology* 8:505-508.

HODKINSON, I. D., & HODKINSON, E. 1993. Pondering the imponderable: a probability-based approach to estimating insect diversity from repeat faunal samples. *Ecological Entomology* 18:91-91.

HOLLOWAY, J.D. 1987. Macrolepidoptera diversity in the Indo-Australian tropics: geographic, biotopic and taxonomic variation. *Biological Journal of the Linnean Society* 30:325-341.

HUBBELL, S.P. & FOSTER, R.B. 1986. Commonness and rarity in a neotropical forest: implications for tropical tree conservation. Pp. 205-231 in Soule, M.E. (ed.). *Conservation Biology: the science of scarcity and diversity*. Sinauer Associates, Sunderland, MA.

JANZEN, D.H. 1988. Ecological characterisation of a Costa Rican dry forest caterpillar fauna. *Biotropica* 20:120-135.

JANZEN, D. H., HALLWACHS, W., JIMENEZ, J., & GAMEZ, R. 1993. The role of parataxonomists, inventory managers and taxonomists in Costa Rica's national biological inventory. Pp 223-254 in Reid, V.W. (ed.). *Biodiversity Prospecting*. World Resources Institute, Washington DC.

JANZEN, D. H., & HALLWACHS, W. 1994. All Taxa Biodiversity Inventory ATBI of Terrestrial Ecosystems: a generic protocol for preparing wildland biodiversity for non-damaging use. *Draft report of an NSF Workshop 16-18 April 1993, Philadelphia, Pennsylvania.*

JOHNSON, S.P. 1991. *The Earth Summit: the United Nations Conference on Environment and Development (UNCED)*. International Environmental Law and Policy Series. Graham & Trotman/Martinus Nijhoff. 532 pp.

KELLY, B. J., WILSON, J. B., & MARK, A. F. 1989. Causes of the species-area relation: a study of islands in Lake Manapouri, New Zealand. *Journal of Ecology* 77:1021-1028.

KREBS, C.J. 1989. *Ecological Methodology*. Harper & Row, New York. 654 pp.

LEWINSOHN, T.M. 1991. Insects in flower heads of Asteraceae in southeast Brazil: a tropical case study in species richness. Pp.525-559 in Price, P.W., Lewinsohn, T.M., Fernandes, G.W. & Benson, W.W. (eds.). *Plant-animal interactions: evolutionary ecology in tropical and temperate regions*. Wiley/Interscience, New York.

MACARTHUR, R. H. 1969. Patterns of communities in the tropics. *Biological Journal of the Linnean Society* 1:19-30.

MACARTHUR, R. H. 1972. Patterns in the Distribution of Species. Harper & Row, New York.

MACARTHUR, R. H., & WILSON, E. O. 1967. *The theory of island biogeography*. Princeton University Press, Princeton, N.J. 198pp.

MAGURRAN, A. E. 1988. *Ecological Diversity and Its Measurement* pp. Croom Helm, London.

MARTIN, T. E. 1981. Species-area slopes and coefficients: a caution on their interpretation. *American Naturalist* 118: 823-837.

MAY, R. M. 1978. The dynamics and diversity of insect faunas. Pp. 105-125 in Mound, L. A. & Waloff, N. (eds), *The diversity of insect faunas*. Royal Entomological Society, London.

MAY, R.M. 1988. How many species are there on earth? *Science* 241:1441-1449.

MAY, R.M. 1990. How many species? *Philosophical Transactions of the Royal Society London B* 330:293-304.

MAY, R.M. 1992a. Past efforts and future prospects towards understanding how many species there are. Pp. 71-81 in Solbrig, O.T., van Emden, H.M. & van Oordt, P.G.W.J. (eds). *Biodiversity and Global Change*. IUBS Monograph 8.

MAY, R.M. 1992b. Bottoms up for the oceans. *Nature* 357:278-279.

MIGNOTI, S. A., & MEEDEN, G. 1992. Estimating the total number of distinct species using presence and absence data. *Biometrics* 48:863-875.

MILLER, R. I., & WIEGERT, R. G. 1989. Documenting completenes, species-area relations, and the species-abundance distribution of a regional flora. *Ecology* 70:16-22.

MORAN, V.C. & SOUTHWOOD, T.R.E. 1982. The guild composition of arthropod communities in trees. *Journal of Animal Ecology* 51:289-306.

NEWBERRY, D. M., CAMPBELL, E. J. F., LEE, Y. F., RIDSDALE, C. E., & STILL, M. J. 1992. Primary lowland dipterocarp forest at Danum Valley, Sabah, Malaysia: structure, relative abundance and family composition. *Philosophical Transactions of the Royal Society London B* 335, 341-356.

NILSSON, I.N. & NILSSON, S.G. 1985. Experimental estimates of census efficiency and pseudoturnover on islands: error trend and between-observer variation when recording vascular plants. *Journal of Ecology* 73:65-70.

PALMER, M.V. 1990. The estimation of species richness by extrapolation. *Ecology* 71:1195-1198.

PRESSEY, R. L., HUMPHRIES, C. J., MARGULES, C. R., VANE-WRIGHT, R. I., & WILLIAMS, P. H. 1993 Beyond opportunism: key principles for systematic reserve selection. *TREE* 8:124-128.

PRESTON, F. W. 1948. The commonness, and rarity, of species. *Ecology* 29:254-283.

PULLIAM, H.R. 1988. Sources, sinks, and population regulation. *American Naturalist* 132:652-661.

RABINOWITZ, D., CAIRNS, S. & DILLON, T. 1986. Seven forms of rarity and their frequency in the flora of the British Isles. Pp. 182-205 in Soule, M.E. (ed.). *Conservation Biology: the science of scarcity and rarity*. Sinauer Associates, Sunderland, MA.

RICKLEFS, R.E. 1987. Community diversity: the relative roles of local and regional processes. *Science* 235:161-171.

ROBINSON, G. S., & TUCK, K. R. 1993. Diversity and faunistics of small moths Microlepidoptera in Bornean rainforest. *Ecological Entomology* 18:385-393.

SANDERS, H. L. 1968. Marine benthic diversity: A comparative study. *American Naturalist* 102:243-282.

SCHLUTER, D., & RICKLEFS, R. E. 1993. Species diversity. An introduction to the problem. Pp 1-10 in Ricklefs, R.E. & Schluter, D. (eds). *Species diversity in ecological communities: historical and geographical perspectives*. The University of Chicago Press, Chicago and London.

SHMIDA, A. & WILSON, M.V. 1985. Biological determinants of species diversity. *Journal of Biogeography* 12:1-20.

SIMBERLOFF, D. 1992. Do species-area curves predict extinction in fragmented forest? Pp. 75-89 in Whitmore, T.C. & Sayer, J.A. (eds). *Tropical Deforestation and Species Extinction*. Chapman & Hall.

SMITH, E.P. & VAN BELLE, G. 1984. Nonparametric estimation of species richness. *Biometrics* 40:119-1E29.

SOBERON, J. & LLORENTE, J. 1993. The use of species accumulation functions for the prediction of species richness. *Conservation Biology* 7:480-487.

SOLBRIG, O.T. 1991. From genes to ecosystems: a research agenda for biodiversity. *Report of a IUBS-SCOPE UNESCO workshop, Harvard Forest, Petersham, Ma. USA*. IUBS.

SOUTHWOOD, T.R.E. 1978. *Ecological methods: with particular reference to the study of insect populations*. Second edition. Chapman & Hall.

STEVENS, G.C. 1989. The latitudinal gradient in geographical range: how so many species coexist in the tropics. *American Naturalist* 133:240-256.

STORK, N.E. 1987a. Guild structure of arthropods from Bornean rain forest trees. *Ecological Entomology* 12:69-80.

STORK, N.E. 1987b. Arthropod faunal similarity of Bornean rain forest trees. *Ecological Entomology* 12:219-226.

STORK, N.E. 1988. Insect diversity: facts, fiction and speculation. *Biological Journal of the Linnean Society* 35:321-337.

STORK, N.E. 1991. The composition of the arthropod fauna of Bornean lowland rain forest trees. *Journal of Tropical Ecology* 7:161-180.

STORK, N.E. 1993. How many species are there? *Biodiversity and Conservation* 2:215-232.

STORK, N.E. 1994. Inventories of biodiversity: more than a question of numbers. Pp. 81-100 in Forey, P., Humphries, C.J. & Vane-Wright, R.I. (eds). *Systematics and Conservation Evaluation*. Systematics Association Special Volume 50. Clarendon Press, Oxford.

THOMAS, C.D. 1990. Fewer species. *Nature* 347:247.

VANE-WRIGHT, R.I., HUMPHRIES, C.J. & WILLIAMS, P.H. 1991. What to protect? Systematics and the agony of choice. *Biological Conservation* 55:235-254.

WATANABE, H. & RUAYSOONGNERN, S. 1989. Estimation of arboreal arthropod density in a dry evergreen forest in Northeastern Thailand. *Journal of Tropical Ecology* 5:151-158.

WIENS, J.A. 1989. Spatial scaling in ecology. *Functional Ecology* 3:385-397.

WILSON, E.O. 1987. The arboreal ant fauna of Peruvian Amazon forests: a first assessment. *Biotropica* 19:245-251.

WHITMORE, T.C. 1984. *Tropical Rain Forests of the Far East*. Oxford University Press.

WHITTAKER, R.H. 1977. Evolution of species diversity in land communities. *Evolutionary Biology* 10:1-67.

WILLIAMS, C. B. 1943. Area and number of species. *Nature* 152:264-267.

WILLIAMSON, M. 1988. Relationship of species number to area, distance and other variables. Pp. 91-155 in Pimm, S.L. (ed.). *Analytical Biogeography*.

YOON, C. K. 1993. Counting creatures great and small. *Science* 260:620-622.

Rainforest dynamics: the need for new paradigms

PETER J.GRUBB

Department of Plant Sciences, University of Cambridge, United Kingdom

ABSTRACT. A desire to understand the maintenance of the very high species-richness of tropical lowland rain forest provides the setting for this paper. It is widely accepted that both niche-differentiation and chance effects among ecologically very similar species play an important part in maintaining richness. The purpose of the paper is to provide an improved background of ideas on rainforest dynamics, against which to consider differentiation in the regeneration niche, and against which to plan management and conservation. Paradigms deemed to need review are the following: the all-importance of canopy gaps, the existence of a spectrum of tolerance in which the relative positions of species are not dependent on size of individual, the association of large seed size with shade tolerance, the idea that soil seed banks are composed of species demanding sizeable canopy gaps, and the all-importance of shade as opposed to competition for water or mineral nutrients in limiting regeneration. Recent evidence bearing on these ideas is reviewed briefly, and an alternative set of paradigms is offered.

KEY WORDS: forest dynamics, regeneration niche, gaps, litter, tolerance, seed mass, dispersal, seed banks, root competition.

INTRODUCTION

The setting for this paper is a desire to understand the maintenance of the very high species-richness of tropical lowland rainforest, a topic recently reviewed in the context of conservation by Primack (1992). It is generally accepted that niche-differentiation, particularly differentiation in the regeneration niche *sensu* Grubb (1977), plays a significant part, but it is also accepted that chance effects among ecologically similar species, as argued by Hubbell (1979), are also important. The relative importance of niche-differentiation and chance effects among consistently sparse species is no nearer resolution than it was when discussed by Grubb (1986) and Hubbell & Foster (1986). A third approach to the problem has been the hypothesis that intraspecific density-dependent effects will keep all species sparse and so allow room for many to co-exist (Connell 1971, Janzen 1970). However, in a critical long-term study in Panama it has been established by Hubbell, Condit & Foster (1990) that the density-dependent effects in most species at their site are too weak to make a significant contribution to maintenance of species-richness.

My chief concern in this paper is to try to provide an improved background of ideas on rainforest dynamics, against which to think about differentiation in the regeneration niche and against which to plan management and conservation, including re-establishment of forest on cleared land. The paradigms which have proved valuable in helping us to assemble a mental picture of the rain forest community as a 'working mechanism' (cf. Watt 1947), but which now seem inadequate, have been built up by the influential text-books of Richards (1952) and Whitmore (1975, 1984) and numerous reviews, most notably perhaps those in the volumes edited by Tomlinson & Zimmermann (1978) and Medina, Mooney & Vazquez-Janes (1984), and those by Bazzaz & Pickett (1980), Brokaw (1985), Budowski (1970), Denslow (1987) and van Steenis (1958a, b).

In Table 1 I set out the paradigms to be reviewed. After discussing each of them, I put forward some tentative replacements (Table 2). Other paradigms are in need of replacement,

D. S. Edwards et al. (eds.), Tropical Rainforest Research – Current Issues, 215–233.

Table 1. Paradigms concerning rainforest dynamics that are in need of revision.

1. Canopy-gaps are readily defined and all-important for regeneration.
2. Species can be arranged along a spectrum of shade tolerance; it is implicit that the position of any one species does not change relative to the positions of other species with increased size of individual .
3. Light-demanding plants have small seeds while shade-tolerant plants have large seeds; large seeds are primarily adaptive in tolerating shade.
4. The species represented in the soil seed bank are light-demanders; germination is triggered primarily by a high ratio of red to far-red radiation and/or strong diurnal fluctuation in soil temperature.
5. The light regime is all-important for regeneration; root competition for water and/or mineral nutrients is not important.

Table 2. Paradigms concerning rainforest dynamics proposed in place of those set out in Table 1.

1. Gaps in the litter layer as well as gaps in the canopy are important for regeneration; there is a continuum in conditions from large canopy gaps to microsites in the understorey where the daily total radiation available for plant growth is significantly increased by minute gaps in the canopy.
2. The amount of light needed to sustain a given relative growth rate increases as the size of the individual increases, but the degree of increase in light demand varies widely; therefore there are many cross-overs in shade-tolerance and it is not sufficient to think in terms of a constant spectrum of relative tolerance. The ecological shade tolerance, unlike the physiological shade tolerance, may increase with size of individual, if the importance of absolute growth rate for survival declines after the earliest stages of establishment; such transitions can generate further cross-overs in relative tolerance.
3. There are wide and strongly overlapping ranges of seed size in shade-tolerant and light-demanding plants; minute seeds are typical of species establishing at litter-free sites in shade or in the open on nutrient-poor substrata; common species requiring canopy gaps for establishment have small rather than minute seeds, and their greater size is of significant value in competition during establishment; very large seeds are not primarily adaptive in establishment under shade but in attracting large animals as dispersers.
4. The soil seed bank contains species which display a very wide range of shade-tolerance, and in rainforests not subject regularly to a dry season includes species with minute seeds suited to establish at litter-free sites in deep shade.
5. Periodic competition for water from large trees is a major factor for the growth and survival of seedlings in shade; competition for mineral nutrients is most important for plants in canopy gaps where they are growing faster than in shade; there is commonly an interaction between nutrient supply and degree of shade on growth rate, and the extent of limitation by nutrients varies widely between species.

but I do not have room to consider them in any detail. For example, it is often thought that there is a strong negative relationship between growth rate and longevity of trees. However, Lieberman *et al.* (1985, 1990) have shown that this is not true in the lowland rain forest whichthey studied in Costa Rica; many understorey trees display the combination of slow maximum growth-rate and relatively short life, while many canopy and sub-canopy trees which are shade-tolerant when young but which respond opportunistically to increased light display the opposite combination of fast maximum growth-rate and relatively long life. Received views on the interpretation of 'defence' in leaves of trees with different degrees of shade-tolerance are also in need of revision, but have been discussed recently by Edwards (1989) and Grubb (1992), and will not be treated further here. In at least some rainforests regeneration by sprouting, rather than from seedlings, is much more important than generalized accounts would suggest.

PARADIGMS NEEDING REVISION

All-importance of canopy gaps

Most of the literature on rain forest dynamics is written as though canopy gaps can be defined relatively precisely and are all-important for the regeneration of rain forest trees. In the major reviews it is accepted that the exact size of gap needed varies a great deal with species, but still the emphasis is on 'gap' versus 'non-gap' conditions (Brokaw 1985; Denslow 1987). I think this perspective needs to be revised for two reasons. First, there is a growing appreciation that there is a continuum of conditions in the forest from 'gap' to 'non-gap', and this has important implications for our ideas on the requirements and responses of juvenile plants. Secondly, evidence is accumulating that gaps in the litter, acting on a different scale from gaps in the canopy, are all-important for the establishment of many species.

Popma, Bongers, Martinez-Ramoz & Veneklaas (1988), who used data from the Las Tuxtlas forest in Mexico, were among the first to demonstrate that canopy gaps do not have sharp edges in terms of micrometeorological characteristics. Brown (1993) has published parallel results for the Danum Valley forest in Sabah. The idea of a continuum from fully open to very deeply shaded micro-sites was emphasized by Lieberman, Lieberman & Peralta (1989) with their memorable expression 'forests are not just Swiss cheese'. The latter authors used data from La Selva in Costa Rica for the subcanopy tree *Casearia arborea* (listed as a 'pioneer' by Putz & Appanah 1987) to illustrate the way in which growth can respond continuously to the degree of canopy openness. Others, e.g. Pearcy (1983) who worked in Hawaii, have also shown how slight differences in the light regime in the deeply shaded understorey can have a major impact on the growth rates of saplings.

The importance of litter gaps, as opposed to canopy gaps, was shown by Molofsky &Augspurger (1992) who reported experiments done at Barro Colorado Island in Panama on the impact of litter cover on germination and establishment of species with seeds of widely different sizes. For species with small seeds (less than about 5 mg dry mass), absence of litter was shown to be vital for establishment. Litter-free sites occur not only in canopy gaps but also in deep shade, for example where standing dead trees or small living trees tip up without making an appreciable canopy gap, where trunks sliding down slopes in the shade scrape off the topsoil, where micro-landslips occur, where rocks or large roots protrude, where animals dig or scrape, and where humans create paths along slopes leaving bare banks on the uphill sides (Figure 1). In montane forests and very high-rainfall lowland forests the trunks of tree ferns also provide litter-free sites for establishment.

Kohyama & Grubb (1994) have shown that in a warm temperate rain forest in southern Japan the mean dry mass values of the seeds of the fourteen most abundant shade-tolerant trees span a range of four orders of magnitude (from 0.074 mg for *Rhododendron tashiroi* to 670 mg for *Litsea acuminata*) and have argued that variation in litter depth at micro-sites on the scale of seeds and young seedlings is critically important. Larger-seeded species can cope with accumulated litter, and make use of deeper softer soil, while very small-seeded species cannot become established through litter but can become affixed to a hard substratum and can come to rest on slight irregularities on a steep micro-slope (Figure 2). Although the need for a gap in the litter layer is the prime requirement for the strongly shade-tolerant species with minute seeds, it is possible that most individuals become established at micro-sites which also benefit from tiny canopy gaps - say 1-2% daylight under cloudy conditions as opposed to $<1\%$; critical observations are needed. The responses of these plants to sunflecks also require investigation. I consider the incidence of this type of plant in tropical lowland rain forest later on when discussing the received paradigm for seed size.

Existence of a constant spectrum of tolerance

Trees are conventionally ordered along a linear scale of tolerance from very strongly light-demanding to very strongly shade-tolerant. In temperate deciduous forests of Europe and North America tree species are commonly assigned to five categories of tolerance (Baker 1949, Ellenberg 1963). In primaeval forest all but the most light-demanding trees become

Figure 1. A cross-section through a rainforest showing sites (arrowed) where a litter-free substratum is combined with deep shade. From left to right: trunk of tree-fern, pit and mound of small tipped-up standing-dead tree, heap and pit made by an animal, outcropping rock, major roots above soil surface, micro-landslide, bank on uphill side of path made by humans, topsoil scraped away by sliding fallen trunk.

Figure 2. A comparison of a large-seeded species able to grown down through, or up through,deep litter and penetrate relatively soft soil with its tap root, and a small-seeded species needing a litter-free micro-site for establishment and having a diffuse root system suited to anchor the plant and grow over a hard substratum if necessary.

established in shade so that 'light-demand' generally refers to the phase after germination and early establishment. The mental picture most widely diffused is that the most light-demanding species persist only a short time in deep shade (probably much less than one year on average), but respond to the formation of a sizeable gap above them with rapid growth, while the most shade-tolerant persist for very long periods in deep shade (probably for decades) and achieve only moderate growth rates when a sizeable gap is formed above them. For temperate trees Canham (1989) considered a mechanism which would make this antithesis between persistence in deep shade and growth rate in bright light inevitable. However, in a critical study of six species of canopy-forming tree over six years at La Selva in Costa Rica, Clark & Clark (1992) found no trade-off between the two variables. I suggest that although there is likely to be an antithesis between extreme types one should not expect a precise trade-off as the most important physiological property determining shade-tolerance is likely to be ability to reduce the respiration rate, while the most important in determining ability to respond to increased light are (a) flexibility in the pattern of allocation between organs and (b) flexibility in allocation between biochemical fractions of the cells.

In most of the literature on both temperate and tropical forests no reference is made to tree species changing relative positions along the spectrum of tolerance as their individuals increase in size. However, there is increasing evidence that the relative positions do change. One approach to this problem is to emphasize changes in physiological tolerance that accompany increase in plant size. Givnish (1988) emphasized that the light requirement for survival or a given relative growth rate (increase in dry mass per unit dry mass per unit time) of a tree increases as that tree becomes larger, as a result of the increase in the ratio between the mass of respiring tissues and the mass of photosynthesizing tissues. Dr T. L. Poulson (personal communication, August 1989) has suggested that we may envisage different species having different slopes for the allometric relationship between light requirement for persistence (compensation point) or a given relative growth rate and tree size. In this way species will cross over in their relative light-requirement when successively larger size-classes are considered. It seems that there is not a single case of such a cross-over rigorously proven in the literature. However, in the forest described by Poulson & Platt (1989) Poulson's unpublished records show that as a seedling *Ostrya virginiana* is more light-demanding (for persistence) than *Acer saccharum*, but that it increases in relative shade tolerance to the point where - as a 'subcanopy species' - it ends its life in the shade of the *Acer* (a 'canopy species'). Dr C. D. Pigott (personal communication, April 1989) has found the same for *Carpinus betulus* (subcanopy) in relation to *Tilia cordata* (canopy) in Europe, and Fowells (1965) reports the same for *Lithocarpus densiflora* (subcanopy) in relation to *Sequoia sempervirens* in California. The *Ostrya*, *Carpinus* and *Lithocarpus* appear to be examples of the 'late-tolerant' type of Oldeman (1990), or 'cryptic pioneer' of Hawthorne (1993). It seems very likely that many examples of this kind of cross-over will be found in tropical rain forests, especially among subcanopy trees and treelets of the understorey. Oldeman & van Dijk (1991) give examples from three tropical areas: *Microdesmis puberula* and *Myrianthus liberica* in Cote d'Ivoire, certain *Inga* spp. in French Guyana, and *Piper amalgo* in Central America. Ellison, Denslow, Loiselle & Brenes M (1993) found that for several understorey Melastomataceae in Costa Rica there were significantly more established juveniles in gaps than in the understorey. Possibly related functionally are such Australian species as *Acronychia vestita*, *Euodia xanthoxyloides* and *Melicope broadbentiana* reported by Hyland & Whiffin (1993) to be 'understorey species' but also 'favoured by disturbance'. Herbs of the understorey may be involved as well as woody species, e.g. some of the clonal Zingiberaceae such as *Afromomum* spp. in Africa (W.R. Hawthorne, personal communication, November 1993).

The key to understanding many of these species may lie in assessing ecological tolerance rather than physiological tolerance at different stages of development. While we expect physiological tolerance of shade always to decrease with plant size within a species, ecological tolerance of shade may show an absolute increase. What may matter most for survival through the establishment phase is absolute growth rate; in practice only plants above a certain size may have an appreciable chance of withstanding the hazards of short-lived drought, herbivory or damage by trampling animals or falling debris. Once the plants are established, absolute physiological tolerance of shade may become relatively more important and enable the larger juveniles and the adults to persist in the understorey.

Another approach which has revealed cross-overs in tolerance has been attention to the gap size needed at different stages in growth among species that all mature in the canopy. For example, in temperate deciduous forest in eastern North America *Betula allegheniensis* was found to need gaps of about the size made by single tree-falls to establish but could grow to maturity in such modest-sized gaps, while *Fraxinus americana* could establish in the absence of a canopy gap but needed very large gaps, such as are caused by cyclones, to reach its full potential during onward growth (Merrens & Peart 1992). Tropical parallels certainly seem to exist. In northern Queensland an example of a species needing formation of a canopy gap at least the size of a single treefall for establishment but able to mature in a such a gap would be *Polyscias murrayi*, which has a narrow crown and rapid height growth, while an example of a species able to establish in the absence of a sizeable canopy gap but needing a large gap to reach full potential would be *Flindersia brayeleyana* which has a wide crown at maturity (A.K. Irvine, personal communication, April 1993).

Large seed size as an adaptation to shade

The idea that shade-tolerant species of plant have large seeds and light-demanders small seeds goes back at least as far as the classic text of Salisbury (1942). In fact, in respect of plants in northern Europe, he proved the idea only for herbs, and he did not present a comparable analysis for shrubs or trees. Indeed, in the temperate deciduous forests of Europe the largest-seeded trees are not the most shade-tolerant (Grubb 1977). Nevertheless it is widely accepted that in tropical rain forests shade-tolerators have large seeds and strongly light-demanding species have small seeds (Clark & Clark 1992, Foster 1986, Foster & Janson 1985, Ng 1978, Swaine & Whitmore 1988). Dissent from this view has been expressed by Ellison *et al.* (1993), Kelly & Purvis (1993) and Putz & Appanah (1987).

In this account the mass values quoted refer to the true seed plus fibrous endocarp when one is present. For species with an aril, wing or pappus on the seed this was removed before the mass was determined.

The line between 'shade-tolerators' and 'light-demanders' is inevitably arbitrary because at any one stage of development there is a continuum of tolerance. In the following account I treat as 'light demanders' species with the characterstics of 'pioneers' *sensu* Swaine & Whitmore (1988), plus the most light-demanding of the non-pioneers, i.e. species which routinely become established in shade but cannot persist there for long (probably less than twelve months) and depend on the early creation of a sizeable gap for further growth. Species considered here as 'shade-tolerators' persist longer (but for very varying times) as seedlings in deep shade, say $<2\%$ daylight under cloudy conditions, and achieve a wide range of growth rates when released in a canopy-gap. Species which we suspect to be 'late-tolerants' *sensu* Oldeman are omitted from the analysis at this stage. In theory, the shade-tolerators could be subdivided on their degree of shade-requirement. This point leads to a further problem. It is virtually certain that some species can regenerate under a wider range of shade than others; for example, in the Malay Peninsula the treelet or shrub *Ficus grossularoides* germinates freely in the shade (Raich & Gong 1990), and can be numerous as shaded juveniles in high forest (D.J. Metcalfe & author, unpublished), but it appears to establish at open micro-sites as well and can certainly be common in secondary forest (Sim, Tan & Turner 1992). Apart from the problems of definition, we suffer in most parts of the tropics from a lack of critical quantitative data on degrees of tolerance. In practice, we are often dependent on the accumulated experience of those with an extensive knowledge of local natural history.

It was mentioned above that in a Japanese warm temperate rain forest Kohyama & Grubb (1994) found that the commonest shade-tolerant species ranged in mean seed dry mass over four orders of magnitude. In this case shade-tolerance has been proven not only by extensive quantitative stand analysis but also by obervations on mapped plants over about 15 years. In the same forest the commonest strongly light-demanding species cover a parallel but less extensive range (from c. 1 mg for *Aralia elata* to c. 730 mg for *Aleurites cordata*; P.J. Grubb & T. Kohyama, unpublished).

Recent work in Singapore (Metcalfe & Grubb, submitted) has revealed a very wide range in mean seed dry mass in shade-tolerant trees and shrubs in coastal hill dipterocarp forest. Species which appear to establish in deep shade on litter-free sites, notably on the steep banks uphill of the paths through the Bukit Timah Nature Reserve, and which certainly persist and grow at micro-sites with about 1% of daylight under cloudy conditions, have values for seed dry mass as low as 50 mg in the case of the 20-m-tall tree *Pternandra echinata* (Melastomataceae) and 30 mg for the 3-m-tall treelet or shrub *Urophyllum hirsutum* (Rubiaceae). Kiew (1988) drew attention to the minute seeds of several genera of autotrophic shade-tolerant herbs in Malaysian rain forests, and Metcalfe & Grubb (submitted) have obtained values of 5-15 mg for species of Gesneriaceae, Melastomataceae, Pentaphragmataceae and Rubiaceae of the type to which she referred. Although tiny-seeded shade-tolerant herbs are rare in temperate deciduous forest, a few such species are known (Struik 1965).

Are the very wide and overlapping spectra of seed mass values of shade-tolerators and light-demanders to be found in other areas too? That is certainly what was found by Foster & Janos (1985) for 202 species of trees, shrubs and vines at Manu in Peru; 'pioneers' had wet seed mass values ranging from 7 mg to 58.6 g, and 'mature forest' species 1.2 mg to 54.7 g, and although the mean values for the two groups were highly significantly different they were similar in absolute terms (1.39 and 2.19 g respectively). In the last year my wife and I have determined mean seed dry mass for numerous species of tropical lowland rain forest plants in northern Queensland (Figure 3). There is a general trend for the maximum, median and mean values of mean seed mass to be lower in species which form smaller individuals at maturity, especially if the strange herbaceous cycad *Bowenia spectabilis* is excluded (tall and medium-sized trees > small trees and shrubs > herbs). Within each of these three growth-forms there is wide overlap between the ranges of seed size for shade-tolerators and light-demanders. However, there is no shade-tolerant tree with seeds as small as those of the *Pternandra* spp. of the Malay Peninsula, and there are very few shade-tolerant small trees and shrubs with seeds of < 100 mg, and no herbs in this range. The most likely explanation is the fact that the Australian forests have a much more strongly seasonal pattern of rainfall so that plants growing slowly from minute seeds in deep shade are at risk of desiccation virtually every year. Quite likely the even drier periods during the Pleistocene (cf. Hopkins *et al.* 1993, Kershaw 1988) should be seen as equally responsible for the complexion of the present flora. The thesis that dry spells, on whatever time-scale, are responsible is supported by a consideration of the habitats of the shade-tolerant species with minute seeds (< 100 mg) known in northern Australia. The shrub *Cyrtandra baileyana* of the Gesneriaceae (mean seed dry mass 10 mg) is virtually confined to the edges of streams and to deeply incised ravines. The one *Pternandra* species in Australia, *P. coerulescens*, is confined to swampy forest (Hyland & Whiffin 1993); in Singapore and Western Malaysia (where mean seed dry mass values of 25-30 mg have been recorded, D. J. Metcalfe unpublished) this species occurs in swampy forest but is not confined to it. One other species should be mentioned: *Saurauia andreana*, a small tree in the Actinidiaceae; it combines minute seeds (57 mg) and ability to flower and fruit in deep shade, but is not shown in Figure 3 because it is suspected of being a 'late-tolerant'. It is most often found in the same situations as the *Cyrtandra*.

Thus the picture of seed size that we now have is quite different from that commonly canvassed. Where prolonged effectively moist periods occur, shade-tolerant species may

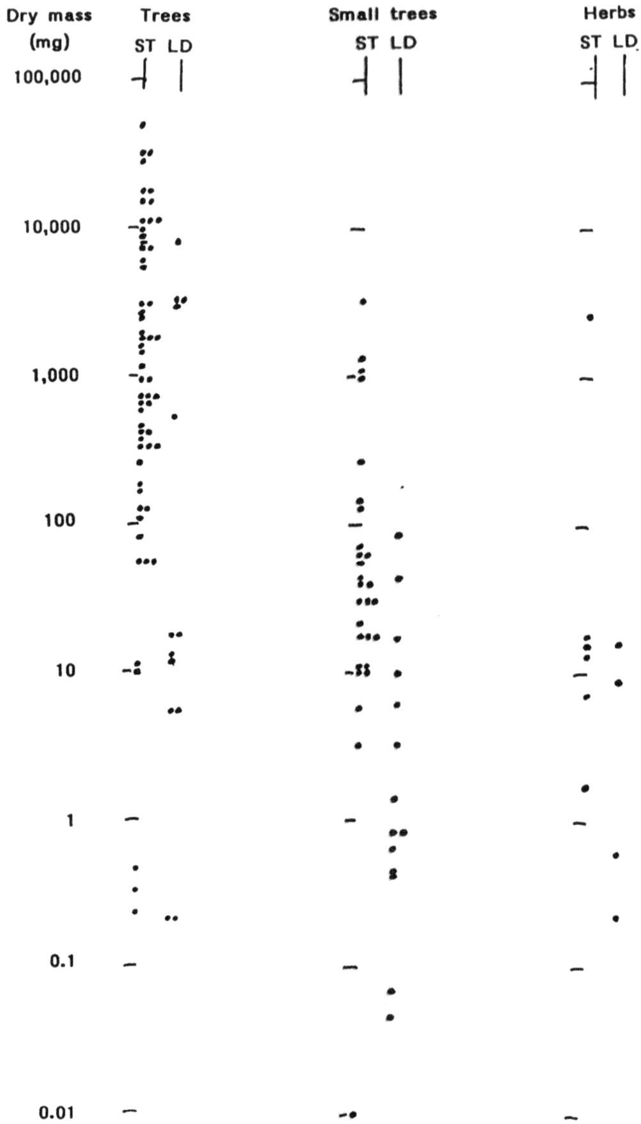

Figure 3. The distributions of mean seed dry mass values for species of trees (tall and medium height at maturity), small trees and shrubs (commonly maturing at <9 m tall), and herbs in rain forests of northern Queensland (unpublished data of P.J. & E.A.A. Grubb): ST, shade-tolerant; LD, light-demanding (see text for definitions of these two groups).

display approximately the same range of mean seed mass as the light-demanders. The smaller-seeded species need litter-free sites for establishment, as emphasized above, and such sites are produced in the shade in a variety of ways summarized in Figure 1. We consider this issue still further below in connection with the 'seed bank' in the soil.

Turning from shade-tolerant species to light-demanders, it is a striking fact that the common light-demanding species of secondary forests in the Malay Peninusula and the island of Borneo, species of such genera as *Commersonia*, *Glochidion*, *Macaranga*, *Mallotus* and *Trema*, undoubtedly small-seeded relative to, say, dipterocarps or laurels of primary forest, actually have much larger seeds than the tiny-seeded species we have just discussed as being markedly shade-tolerant during establishment such as species of *Pternandra* and *Urophyllum* (mean dry mass values in the range 2-20 mg vs 20-50 mg). So the question of interest becomes not 'Why are the common weedy trees small-seeded ?' but 'Why are the common weedy trees as large-seeded as they are?'. In fact, there are indeed tiny-seeded weedy trees in the Malay Peninsula and Borneo. Several are characteristic of strongly infertile soils: species of *Adinandra*, *Fagraea*, *Melastoma* and *Ploiarium* (cf. Sim *et al.* 1992). They have mean seed dry mass values in the range 30-1100 mg (Metcalfe & Grubb, submitted). The only *Macaranga* species on such soils (*M. heynei*) has one of the smallest *Macaranga* seeds we have studied (3.2 mg vs a range of 3-30 mg). The same trend is found in northern Queensland where the dominant weedy trees of roadsides on relatively fertile basalt-derived soils have seeds with dry mass values in the range 5-20 mg (species of *Alphitonia*, *Mallotus*, *Omalanthus* and *Polyscias*) while four of the six weedy small trees and shrubs with tiny seeds shown in Figure 2 (range 49-800 mg) are all characteristic of much less fertile soils on granite or metamorphic rocks (species of *Abrophyllum*, *Geniostoma*, *Melastoma*, *Rhodomyrtus*). In the Blue Mountains of Jamaica the characteristic colonizers of nutrient-poor stony micro-sites on landslips studied by Dalling (1992) all have very small seeds (notably species of *Clethra*, *Cyrilla*, *Lyonia* and *Vaccinium*), while the colonizers of micro-sites with more fine earth, and dominants of secondary forest on non-truncated profiles have larger seeds. I suggest that the seeds of weedy species on more fertile soils are as large as they because that gives them a necessary advantage in competition during establishment. We are lacking comparative studies on relative growth rate for these species but for other plant groups it has been found to be strongly negatively correlated with seed mass; nevertheless seedlings of larger-seeded species may be larger than those from smaller-seeded species for prolonged periods after germination (Maranon & Grubb 1993). By hypothesis competition among colonizers is less critical on soils which supply smaller amounts of nutrients. I have recently argued that where plants invade soil free of roots the impact of competition among invaders will be greater on soil that is richer in nutrients, but that where plants invade soil already fully occupied by roots the impact of root competition on colonizers will be greater on soil that is inherently nutrient-poor (Grubb 1994). Certainly the same trend to smaller seeds in plants of open sites on more nutrient-poor substrata is seen in communities quite different from those of the wet tropical lowlands - notably the annual grasslands of the Mediterrranean Basin and the commuities of annuals on sand dunes of northern Europe (Maranon & Grubb 1993). Qualitatively the same argument may apply to the tiny-seeded species suited to establish under damp conditions in deep shade on litter-free soil; they have to tolerate strong competition for light from established trees but the impact of competition for resources from fellow colonizers may be appreciably less than for species which are suited to establish in high irradiance.

Not all light-demanding tropical rainforest trees with very small seeds are associated with nutrient-poor soil. *Nauclea officinalis* (dry mass 30 mg) and *Anthocephalus chinensis* = *Neolamarkia cadamba* (wet mass 50 mg, Foster & Janos 1985) are associated in the least disturbed areas with rivers, and tiny-seededness may perhaps be related to establishment at very wet micro-sites.

So far my discussion has centred on small and minute seeds. Two points have to be made now about especially large seeds. First, we have to consider what is the factor to which particularly large seeds in general are adapted. Secondly, we have to treat the large-seeded gap-demanding species.

Foster (1986) attempted a comprehensive discussion of the adaptive value of large seeds for tropical rainforest trees, and considered large size in relation to temperature, moisture, 'light intensity', soil factors, competition, predation, pathogens and mycorrhiza. There is no convincing evidence that very large seeds (say, 10-100 g dry mass) allow trees to establish at micro-sites which are more deeply shaded or drier than large seeds (say, 1-10 g). Physiological properties of the shoot and root are more important in that context than the difference between large and very large seeds. I suggest that the primary adaptive value of very large seeds is not to be found in connection with any of the physical factors traditionally considered by plant ecologists but in attracting relatively large animals as dispersers: mammals generally, but in the special case of northern Queensland and Papua New Guinea cassowaries - the huge flightless rainforest birds (*Casuarius* spp.; see Stocker & Irvine 1983). I suggest that, other things being equal (e.g. nutritive value and taste), larger seeds, or seeds with a larger mass of flesh around them, will be more attractive to larger dispersers whose livelihood arguably depends on intake per unit time (Martin 1985) - a larger fruit or seed may be taken in as quickly as a smaller one. There is a related issue: the provision of an adequately nutritious food to support the disperser. In the context of dispersal by squirrels in northern temperate deciduous forest attention has been paid to the relative amounts of energy supplied by various large seeds among species of *Carya*, *Juglans* and *Quercus* (Stapanian 1986). Perhaps equally important is the supply of protein.

A distinction should be made between species in which the flesh around the seed is the attractant to dispersers, and those in which the seed itself is the attractant. In the Lauraceae of northern Queensland my wife and I have found a negative relationship between the nitrogen concentration of the embryo (probably proportional to protein concentration) and the mean dry mass of the embryo (Figure 4, circles). The relatively large seeds (say, mean dry mass 5-10 g) believed to be dispersed mainly by cassowaries and the musky rat kangaroo (*Hypsiprymnodon moschatus*) rather than by flying birds have embyos which are relatively unnutritious in protein terms, but this is irrelevant to the animals which consume the flesh. However, in the largest-seeded species, which are eagerly sought by white-tailed rats

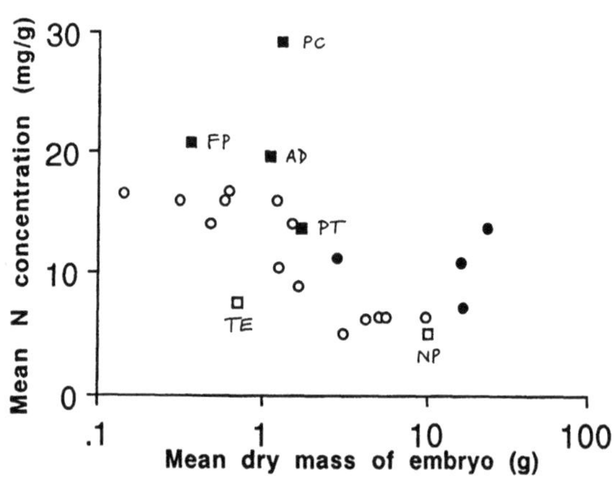

Figure 4. The mean concentration of nitrogen in the embryo as a function of mean embryo dry mass in a range of Lauraceae (circles), and in a subjective selection of other tree species in northern Queensland (squares). Open symbols for species dispersed only by birds, and filled symbols for species dispersed by mammals as well as birds or by mammals alone . The birds use only the flesh, while the mammals mostly use the embryo. AD, *Athertonia diversifolia*; FP, *Fontainea picrosperma*; NP, *Niemeyera prunifera*; PC, *Pouteria castanosperma*; PT, *Prunus turneriana*; TE, *Triunia erythrocarpa*. The Lauraceae (in order of increasing mean embryo dry mass) are *Neolitsea dealbata*, *Cryptocarya grandis*, *Beilschmiedia brunnea*, *Litsea leefeana*, *Cinnamomum laubatii*, *Beilschmiedia recurva*, *Endiandra bessaphila*, *Cryptocarya angulata*, *Endiandra monothyra*, *Cryptocarya pleurosperma*, *Endiandra montana*, *Beilschmiedia tooram*, *Endiandra sankeyana*, *E. xanthocarpa*, *Cryptocarya oblata*, *Endiandra insignis*, *E. palmerstonii* and *Beilschmiedia bancroftii*. Unpublished data of P.J. & E.A.A. Grubb and G.D. Jones.

(*Uromys caudimac-ulatus*) and scattered by them (Hyland & Whiffin 1993, Wellesley-Whitehouse 1983), the nitrogen concentration in the embryos is higher than would be expected by extrapolation from the smaller-seeded species (Figure 4, filled circles); clearly this feature could be regarded as adaptive. Perhaps it is not by chance that nuts in general are found by humans to provide not only plenty of energy but an effective source of protein!

Three of the most widespread very large-seeded Lauraceae in northern Queensland (mean seed dry mass 18-43 g) have relatively thick and hard endocarps (c. 28-44% of seed dry mass) probably broken into only by the white-tailed rat (A.K. Irvine, personal communication, April 1993) - rather as some of the largest seeds of the Leguminosae in the neotropics can be broken into only by certain ground-dwelling rodents (Forget 1990, Hallwachs 1986) and some northern temperate seeds can be broken into only by certain squirrels, *Sciurus* spp. (cf. Stapanian 1986). Another of the laurels shown in Figure 4, *Cryptocarya pleurosperma*, with a somewhat smaller seed (c. 11 g), has an even thicker endocarp (78% of dry mass) and is commonly taken by *Uromys*. It has a nitrogen concentration on the upper edge of the cloud of points for the Lauraceae in Figure 4. Several non-Lauraceous species in northern Queensland with smaller seeds (1.3-11 g) have very thick endocarps which, again, are probably bitten through only by the white-tailed rat, e.g. *Athertonia diversifolia* in the Proteaceae (85% of dry mass in endocarp), *Fontainea picrosperma* in the Euphorbiaceae (62%), and *Pouteria castanosperma* (73%) in the Sapotaceae. Such species (filled squares in Figure 4) are found to have a nitrogen concentration in the embryo appreciably higher than expected for their mean embryo dry mass on the trend line for the Lauraceae, and high relative to other members oftheir families with seeds for which the flesh is the attractant (*Niemeyera prunifera*, Sapotaceae; *Triunia erythrocarpa*, Proteaceae; unfilled squares in Figure 4). It seems as though the seeds represent a package of especially nutritious food which only very few dispersers can utilize, and which will be prized items in the diet of the animal concerned. Certainly very high proportions of all the seeds of the species mentioned as attractive to the white-tailed rat are consumed by that species (G.N. Harrington & A.K. Irvine, personal communication, April 1993). The seeds of some species with a fairly thick and hard endocarp, and a relatively high nitrogen concentration in the embryo, e.g. *Prunus turneriana* (3 g, 40% in endocarp; Figure 4) are taken less by the white-tailed rat than by the musky rat kangaroo (A. Dennis, personal communication, April 1993).

We are still in the earliest stages of understanding the evolution and present-day dispersal of these large seeds (see also Howe 1986), but I suggest that there is enough evidence to make us think in terms of large seed size and high protein content as being significantly adaptive in attracting dispersers.

It often seems to be taken for granted that we should expect strongly light-demanding species to produce large numbers of small seeds - how else will they produce enough to ensure their arrival (or presence in the seed bank) when a canopy gap is formed? But in fact it is quite clearly possible for a plant with a large seed (say mean seed dry mass 730-7800 mg) to be an effective colonizer of canopy-gaps. *Aleurites cordata* at the site studied by Grubb (1992) and Kohyama & Grubb (1993) in southern Japan, and *Aleurites moluccana* in northern Australia (as well as much of Malesia) are striking examples; they behave essentially as 'pioneers' in the sense of Swaine & Whitmore (1988). The seeds of *A. moluccana* can persist and remain viable a long time in the soil, probably much longer than two years (Hopkins & Graham 1987). Both these species have large seeds containing large embryos (embryo and endosperm make up 3.1 g out of total 7.8 g in *A. moluccana*, which has a hard thick testa). Other notable light-demanding species have large seeds but small embryos: various *Elaeocarpus* species such as *E. angustifolius* in Australia (97% of dry seed mass as endocarp), and *Spondias* species in C. America and S.E. Asia (cf. Ng 1991, Roth 1987). In tropical Queensland the *Aleurites* and the *Elaeocarpus* are very much sought after by the white-tailed rat, and appear to be scatter-hoarded by them. In South America the agouti (*Dasyprocta punctata*) utilizes and disperses *Spondias* (Roth 1987). It is not easy to see why large-seeded gap-demanders have not evolved in more families. It is certainly striking that most of the largest-seeded trees of the temperate forests of the northern hemisphere (*Carya*, *Castanea*, *Corylus*, *Juglans* and *Quercus*) are strongly light-demanding (Ellenberg 1963; Fowells 1965), not 'pioneers' sensu Swaine & Whitmore (1988) but light-demanding 'non-pioneers'. Platt & Hermann (1986) give a valuable account of dispersal, establishment and onward growth for species of *Carya* and *Quercus* at a site in the south-eastern USA.

To summarize the revised perspective on seed size, very wide and overlapping ranges of seed mass values are found for shade-tolerant and strongly light-demanding species. Tiny-seeded shade-tolerant species need reliably moist and litter-free sites for establishment. Large-seeded strongly light-demanding species are effectively dispersed into actual and potential gaps by large animals, and may be enabled to persist in the soil seed bank by their thick hard coats. The smallest-seeded light-demanders are mostly characteristic of nutrient-poor soils, where (it is hypothesized) competition among colonizers is less critical. The possession of especially large seeds, among both shade-tolerant and strongly light-demanding species, is seen as being primarily adaptive in attracting dispersers rather than in enabling the seedlings to be especially resistant to drought or shade. There is preliminary evidence that the large seeds are especially valuable to potential dispersers through having high concentrations of protein as well as providing a large amount of energy per unit picked up. Soil seed banks are composed of species which need canopy gaps.

The definition of soil seed bank species is difficult because of our lack of precise information on the longevities of seeds of different species in the soil (Vazquez-Yanes & Orozco-Segovia 1993). A theoretical qualification might be that a significant proportion of the seeds can remain viable in the soil for more than one year. In practice, one goes on the density of viable seeds inthe soil relative to the probable current seed rain. Where viable seeds are numerous in the soil,and fruiting adults are rare or absent in the vicinity, persistence in the soil appears very probable.

It is commonly believed that in all kinds of forest the soil seed bank is composed of species which need canopy gaps for establishment, and this idea is linked with the idea that germination is triggered mainly by changes in the red/far red ratio of the radiation incident on the soil, and by increased daily amplitude of soil-surface temperature (cf. Vazquez-Yanes & Orozco-Segovia 1984). I have myself contributed to the perpetuation of this idea (Grubb 1988). However, while working in Queensland, I have found that the most abundant species in the seed bank may display a very wide range of shade-tolerance. In a forest on a strongly leached basalt-derived soil I have found that in terms of ecological tolerance the seed bank species range from those found naturally on edges, e.g. *Alphitonia petriei*, through those found commonly in the centres of single-treefall gaps as well as on edges, e.g. *Polyscias murrayi*, to those found typically in small gaps or at the edges of gaps and occasionally in extremely small gaps as well as in the centres of gaps, e.g. *Dendrocnide moroides* and *Solanum dallachyi*. In terms of physiological tolerance I have found that the first two are hardly able to grow under a shade screen letting through about 2% of daylight (Osunkoya 1992 found much the same) while the latter two can grow appreciably under those conditions and show a marked positive response to the addition of mineral nutrients in such deep shade.

Most interestingly Kennedy (1991) has found that the most abundant species in the soil seed bank of the Danum Valley forest in Sabah include a number with minute seeds, closely related to species known from the work of D.J. Metcalfe to be shade-tolerant, e.g. *Pternandra azurea*, species of the scandent shrub *Dissochaeta* (also Melastomataceae), and species of *Urophyllum*, notably *U. glabrum*. Other species, e.g. *Callicarpa* spp. (Verbenaceae) and *Uncaria* spp. (Rubiaceae) seem likely to be much more light-demanding.

The picture is thus of a wide spectrum of shade-tolerance among the soil seed bank species, as in Queensland, but in this case extending to very shade-tolerant and very tiny-seeded; *Urophyllum glabrum* probably has a mean seed dry mass in the range 20-40 mg, judging by the four species of that genus studied in Singapore). D.J. Metcalfe (unpublished) has similarly found shade-tolerant tiny-seeded species in the soil seed bank in Singapore. Very significantly Raich & Gong (1990) showed for two species in this group (*Pternandra coerulescens* and *Urophyllum glabrum*) that they can show high percentage germination under a forest canopy (27-67%) and more there than in a large clearing.

Much more work is needed to establish the full and true position, but it looks as though the seed bank species in tropical rain forests without frequent risks of marked desiccation cover about as wide a range of shade-tolerance as the non-seed-bank species - possibly wider if virtually all the most light-demanding species are present in the seed bank. Among the species which combine strong shade-tolerance as juveniles and possession of minute seeds there is evidence of a wide range in requirements for onward growth, some being able to flower and fruit in deep shade (e.g. species of *Urophyllum*) while others flower only when they have reached a considerable size and have direct sunlight on at least a large part of their crown (e.g. *Pternandra echinata*). These species whose start in life is geared to litter-gaps thus parallel the species with large seeds which germinate and establish in shade without the need for a litter-gap, and which display the whole spectrum of canopy-gap requirement for persistence and onward growth.

A lesser revision of the paradigm is needed in respect of those soil seed bank species which do indeed need a sizeable canopy gap for establishment and onward growth. There is increasing evidence that bare soil open to unfiltered radiation and subject to large daily fluctuations in temperature is not necessary for germnation. Physical disturbance of the upper soil, even where the litter layer is replaced, and even under a forest canopy, can lead to the germination of numerous seeds (Dalling 1992, Hopkins & Graham 1984, Kennedy & Swaine 1992). Also Raich & Gong (1990), in their trials with fresh seed in western Malaysia, showed that very few 'gap species' conformed strictly to the supposed characteristic of a 'pioneer' proposed by Swaine & Whitmore (1988): no germination under forest shade and abundant germination in unfiltered daylight. One 'gap species', *Trema tomentosa*, did conform, showing negligible germination in repeated trials under a forest canopy but abundant germination in a large clearing, but others formed a continuum to the 'non-pioneer species', as illustrated by the sequence *Melastoma malabathricum* (3-6% germination in different trials under a forest canopy), *Commersonia bartramia* (0-14%) and *Porterandia anisophylla* (9-31%). The shade regime is all-important for regeneration.

Compared with the innumerable references to shade effects on plants in tropical rainforest, very little has been published on the effects of root competition. In the 1930s, in a number of papers thoroughly summarized by Richards (1952), C. Coster drew attention to the potential role of root competition, but there have been few attempts to follow up the idea with appropriate experiments. There are several significant questions that might be asked about the impact of shortage of water and/or mineral nutrients, induced by the roots of established trees.

We start by asking whether species found to establish only in sizeable canopy gaps are excluded from the understorey to a significant extent by root competition as well as by shade. In temperate forests this idea has received support from the finding that the gap-demanding species *Galeopsis tetrahit* and *Rubus idaeus* invaded plots freed of root competition by trenching under *Fagus sylvatica* and *Picea abies* respectively (Slavikova 1958; V.G. Karpov 1962, cited by Walter & Breckle 1985). In temperate deciduous forest the mass of fine roots per unit volume of topsoil can drop appreciably in canopy gaps (Wilczynski & Pickett 1993), and the same was established for lowland rain forest in Costa Rica by Sanford (1989). In the same forest the water content of the topsoil is greater in canopy gaps (Vitousek &

Denslow 1986). Also in a Panamanan moist forest with a pronounced dry season the water potential of the topsoil was found not to drop as much in gaps as in the understorey (Becker, Rabenold, Idol & Smith 1988). It seems that the only experimental study showing desiccation rather than lack of light to be the apparent cause of exclusion from the understorey is that of Mulkey, Smith & Wright (1991) for the giant herb *Pleiostachya pruinosa* (Marantaceae) in the Panamanian forest studied by Becker *et al.* (1988); survival of plants in the understorey over 3.5 years was markedly and significantly greater in an irrigated plot than in a control plot.

The reduction in rooting density in canopy-gaps may lead to significantly greater availability of nutrients for colonizing plants but that will depend on the relationship between the mean concentration of a nutrient ion in the soil solution and rooting density. Tilman & Wedin (1991) have illustrated this relationship for a number of temperate grasses; the mean nitrate concentration in the soil solution fell very markedly with a modest rise in root mass per unit volume and then only a little more over a wide range of increasing root mass values. If the availability of nutrients is increased significantly in gaps, then this is likely to have a marked positive effect on growth rate and survival of seedlings of gap-demanders. As explained below, there is commonly a marked positive interaction between irradiance and the supply of limiting nutrients on growth, and at low irradiance many species are not responsive to nutrients. Such species of small gaps as *Dendrocnide moroides*, mentioned above as responsive to nutrients in about 2% daylight, seem likely to be kept in gaps by root competition as well as by shade.

Of special interest is the preferential occurrence of certain 'pioneers' within canopy gaps on the mounds of soil on tipped-up root systems (Putz 1983); it has been suggested that the effective supply of limiting mineral nutrients may be greater in root-free subsoil than in topsoil densely filled with roots (Grubb 1992) and the idea has been justified in a laboratory study of simulated grassland gaps (Reed & Grubb submitted) but a critical test is needed for rainforest tip-ups.

A further set of questions concerns the impact of root competition for species which are markedly shade-tolerant at the stage of establishment. First, does it have any significant effect on the growth or survival of seedlings and saplings in deep shade? Secondly, does it affect rates of onward growth in sizeable canopy gaps? Thirdly, do different species of tree display different combinations of sensivity to shade, and sensitivity to nutrient supply or water supply, such that different species will be favoured by different combinations of conditions in different parts of a single gap? The last question represents a special case of the more general question asked by Orians (1982).

A distinction should be made between the effects of water shortage and those of nutrient shortage. Very few studies have been made on the combined effects of different degrees of shade and different degrees of water shortage, but Papavasilliou (1991) found for four European shrub species no interaction; the proportional deleterious effect of drought on growth was the same at different irradiances. The same has been found by Burslem (in press) for some tropical rain forest trees. In contrast, addition of nutrients to plants in pots or solution culture commonly fails to increase the growth rate in deep shade, even where it has a marked positive effect in bright light (references summarized by Peace & Grubb 1982; see also Latham 1992). Turner (1991) found no interaction between irradiance and nutrient-addition effects on the growth of potted plants of the 'pioneers' *Melastoma malabathricum* and *Trema tomentosa*, grown in the open and in the shade of *Eugenia grandis*; an interaction might have been found at lower irradiances. Turner, Brown & Newton (1993) found no increase in growth of potted *Shorea macroptera* in the shade of the same Eugenia, but Sundralingam (1983) did find an increase in growth of *Dryobalanops* spp. fertilized in relatively bright light. Failure to respond to nutrient addition in a pot experiment does not mean that a species would not have its growth inhibited when deprived of nutrients in the same soil by roots of established plants. However, on the basis of the majority of the evidence one might expect competition for water from the roots of tall trees always to be liable to cause a reduction in growth and survival of seedlings, while for most tree species competition for nutrients might have significant effects only in canopy gaps - the size of gap needed for an effect depending on the species. I have written 'liable to cause a reduction' because what actually happens will depend on the sensitivity of seedlings to water status. Burslem (in press) has found a striking ability of seedlings of a number of tree species in Singapore to maintain growth rates even though watered infrequently in deep shade - a result which could be interpreted to show that there has been selection for an ablity to resist periodic desiccation even in plants of 'aseasonal' rain forest.

What has been found out in experiments on root-trenching and on addition of water and/or nutrients to plants in different degrees of shade in rainforests? Connell (1970) reported no effect of root-trenching on survival for spontaneously established seedlings of one species of strongly shade-tolerant Sapotaceae in two degrees of shade (under a mother tree and away from it) in northern Queensland, but Fox (1973) reported slight increases in growth rate of naturally developed dipterocarp seedlings in Sabah. In both cases there were only pseudo-replicates in the sense of Hurlbert (1984). More recently Denslow, Newell & Ellison (1991) found no effect of root-trenching on either survival or growth of seedlings of two shade-tolerant *Inga* spp. sown into the understorey at La Selva in Costa Rica. Working in the strongly seasonal moist forest of Panama, Fisher, Howe & Wright (1991) found that irrigation in the dry season tended to increase survival of first-year seedlings of the nutmeg *Virola surinamensis* in the understorey (19/21 vs 16/24), but the difference was not statistically significant. Turner, Brown & Newton (1993) found no effect of adding NPK to dipterocarps in the understorey in western Malaysia and Sabah. More provocatively Denslow, Schultz, Vitousek & Strain (1990) found no effect of adding a complete fertilizer to cuttings of various shrubs (species of *Miconia* and *Piper*) in gaps in a Costa Rican rainforest. Denslow *et al.* did not establish the primary limiting nutrient for the test plants in soil densely occupied by roots, and it might well have been nitrogen (cf. Burslem in press); it is notable that there was no increase in nitrogen concentration in the test plants or in the soil - perhaps because all the added nitrogen was taken up by competitors.

Clearly there is a need for further, more extensive and more elaborate experiments covering a range of species, soils and climatic regimes. In the meantime we are left with the suggestive observational studies of Gentry & Emmons (1987) which showed that there are markedly fewer species of herb and shrub per unit area in lowland rain forests with a marked seasonality in rainfall or marked nutrient-deficiency in the soil. This result would be explained most economically by an increased impact of root competition in a given degree of shade at sites where water and/or nutrients are in short supply. The experimental work on the effects of seasonal drought on *Pleiostachya* in Panama, mentioned above, supports the explanation, but we are lacking experimental data on the sensitivity of herbs to the relevant range of nutrient-availabilities.

CONCLUDING REMARKS

In this review I have summarized the evidence that there is a need to revise at least five commonly quoted paradigms regarding rainforest dynamics, and I have set out suggested replacement paradigms (Table 2). Critical studies are needed to test their generality.

There is general agreement between the points made in this review and those made in certain papers which had not reached me when I was writing my lecture for the symposium, most notably those by Clark & Clark (1992), Ellison *et al.* (1993) and Kelly & Purvis (1993). I thoroughly endorse the plea by Clark & Clark (1992) that attention be directed to studies which embrace all life history stages, and that comparisons be based on operational criteria. I also support the view of Ellison *et al.* (1993) that in the recent past insufficient effort has been put into studies of life-forms other than trees that have the potential to reach the top of the canopy. We are on the verge of obtaining a much better balanced understanding of rainforest dynamics, which will enable us to interpret more soundly evolution of life histories, the significance of morphological and physiological differences between species, and the maintenance of species-richness, while also providing a firmer basis for the planning of conservation and replanting schemes.

ACKNOWLEDGEMENTS

This paper was written at the Tropical Forest Research Centre of the CSIRO Division of Wildlife and Ecology at Atherton in northern Queensland. I thank the staff there for making facilities freely available to me and for many kindnesses; Tony Irvine, in particular, introduced me to the local rainforests and collected a number of important seeds. I am indebted to Ian Turner for unstinted hospitality in Singapore in recent years, and for valuable comments on early drafts of this paper, to David Burslem for important suggestions, and also to Daniel Metcalfe for permission to quote unpublished results. The Royal Society of London provided a grant to cover the expenses of the work done by my wife and myself in Australia.

LITERATURE CITED

BAKER, F.S. 1949. A revised tolerance table. *Journal of Forestry* 47:179-181.

BAZZAZ, F.A. & PICKETT, S.T.A. 1980. Physiological ecology of tropical succession: a comparative review. *Annual Review of Ecology and Systematics* 11:287-310.

BECKER, P., RABENOLD, P.E., IDOL, J.R. & SMITH, A.P. 1988. Water potential gradients for gaps and slopes in a Panamanian tropical moist forest's dry season. *Journal of Tropical Ecology* 4:173-184.

BROKAW, N.V.L. 1985. Treefalls, regrowth, and community structure in tropical forests. Pp. 53-69 in Pickett, S.T.A. & White, P.S. (eds). *The ecology of natural disturbance and patch dynamics*. Academic Press, Orlando.

BROWN, N.D. 1993. The implications of climate and gap microclimate for seedling growth conditions in a Bornean lowland rain forest. *Journal of Tropical Ecology* 9:153-168.

BUDOWSKI, G. 1970. The distinction between old secondary and climax species in tropical central American lowland forests. *Tropical Ecology* 11:44-48.

BURSLEM, D.F.R.P., in press. Differential responses to nutrients, shade and drought among tree seedlings of lowland tropical rain forest in Singapore. *Proceedings of the Workshop on Tropical Tree Seedling Ecology*. University of Aberdeen, September 1992.

CANHAM, C.D. 1989. Different responses to gaps among shade-tolerant tree species. *Ecology* 70:548-550.

CLARK, D.A. & CLARK, D.B. 1992. Life history diversity of canopy and emergent trees in a tropical rain forest. *Ecological Monographs* 62:315-344.

CONNELL, J.H. 1971. On the role of natural enemies in preventing competitive exclusion in some marine animals and in rain forest trees. Pp. 298-312 in den Boer, P.J. & Gradwell, G.R. (eds). *Dynamics of populations*. Centre for Agricultural Publishing and Documentation, Wagenginen.

DALLING, J.W. 1992. *Regeneration on landslides in the Blue Mountains, Jamaica*. PhD dissertation, University of Cambridge.

DENSLOW, J.S. 1987. Tropical rainfrest gaps and tree species diversity. *Annual Review of Ecology and Systematics* 18:431-451.

DENSLOW, J.S., NEWELL, E. & ELLISON, A.M. 1991. The effect of understorey palms and cyclanths on the growth and survival of *Inga* seedlings. *Biotropica* 23: 225-234.

DENSLOW, J.S., SCHULTZ, J.C., VITOUSEK, P.M. & STRAIN, B.R. 1990. Growth responses of tropical shrubs to treefall gap environments. *Ecology* 71:165-179.

EDWARDS, P.J. 1989. Insect herbivory and plant defence theory. Pp. 275-297 in Grubb P.J. & Whittaker, J.B. (eds). *Toward a more exact ecology*. Blackwell Scientific Publications, Oxford.

ELLENBERG, H. 1963. *Die Vegetation Mitteleuropas mit den Alpen*. Ulmer, Stuttgart.

ELLISON, A.M., DENSLOW, J.S., LOISELLE, B.A. & BRENES M., D. 1993. Seed and seedling ecology of neotropical Melastomataceae. *Ecology* 74:1733-1749.

FISHER, B.L., HOWE, H.F. & WRIGHT, S.J. 1991. Survival and growth of *Virola surinamensis* yearlings: water augmentation in gap and understory. *Oecologia* 86:292-297.

FORGET, P.-M. 1990. Seed-dispersal of *Vouacapoua americana* (Caesalpiniaceae) by cavimorph rodents in French Guiana. *Journal of Tropical Ecology* 6:459-468.

FOSTER, S.A. 1986. On the adaptive value of large seeds for tropical moist forest trees: review and synthesis. *Botanical Review* 52:260-299.

FOSTER, S.A. & JANSON, C.H. 1985. The relationship between seed size and establishment conditions in tropical woody plants. *Ecology* 66:773-780.

FOX, J.E.D. 1973. Dipterocarp seedling behaviour in Sabah. *Malaysian Forester* 36:205-214.

FOWELLS, H.A. 1965. Sylvics of forest trees of the United States. *Agriculture Handbook* 271, US Department of Agriculture, Washington.

GENTRY, A.H. & EMMONS, L.H. 1987. Geographical variation in fertility, phenology, and composition of the understory of neotropical forests. *Biotropica* 19:216-227.

GIVNISH, T.J. 1988. Adaptation to sun and shade: a whole-plant perspective. Pp. 63-91 in J.R. Evans, S. von Caemmerer & W.W. Adams (eds). *Ecology of photosynthesis in Sun and Shade*. CSIRO, Melbourne.

GRUBB, P.J. 1977. The maintenance of species richness in plant communities: the importance of the regeneration niche. *Biological Reviews* 52:107-145.

GRUBB, P.J. 1986. Problems posed by sparse and patchly distributed species-rich plant communities. Pp. 207-225 in Diamond, J. & Case, T.J. (eds). *Community ecology*. Harper & Row, New York.

GRUBB, P.J. 1988. The uncoupling of disturbance and recruitment, two kinds of seed bank, and persistence of plant populations at the regional and local scales. *Annales Zoologici Fennici* 25:23-36.

GRUBB, P.J. 1992. A positive distrust in simplicity - lessons from plant defences and from competition among plants and among animals. *Journal of Ecology* 80:585-610.

GRUBB, P.J., in press. A paradox of root competition: a possible resolution and implications. **Phytocoenologia.**

HALLWACHS, W. 1986. Agoutis (*Dasyprocta punctata*), the inheritors of guapinol (*Hymenaea courbaril*: Leguminosae). Pp. 285-304 in Estrada, A. & Fleming, T.H. (eds). *Frugivores and seed dispersal*. Junk, Dordrecht.

HAWTHORNE, W.D. 1993. *Forest regeneration after logging in Bia South G.P.R., Ghana*. Overseas Development Agency Forestry Series No. 4. N.R.I., Chatham.

HOPKINS, M.S., ASH, J., GRAHAM, A.W., HEAD, J. & HEWETT, R.K. 1993. Charcoal evidence of the spatial extent of the Eucalyptus woodland expansions and rainforest contractions in North Queensland during the late Pleistocene. *Journal of Biogeography* 20.

HOPKINS, M.S. & GRAHAM, A.W. 1984. The role of soil seed banks in regeneration in canopy gaps in Australian tropical lowland rain forest - preliminary field experiments. *Malaysian Forester* 47:146-158.

HOPKINS, M.S. & GRAHAM, A.W. 1987. The viability of seeds of rainforest species after experimental soil burials under tropical wet lowland forest in north-east Australia. *Australian Journal of Ecology* 12:97-108.

HOWE, H.F. 1986. Seed dispersal by fruit-eating birds and mammals. Pp. 123-189 in Murray, D.R. (ed.). *Seed dispersal*. Academic Press, Sydney.

HUBBELL, S.P. 1979. Tree dispersion, abundance and diversity in a tropical dry forest. *Science* 203:1299-1309.

HUBBELL, S.P., CONDIT, R. & FOSTER, R.B. 1990. Presence and absence of density-dependence in a neotropical tree community. *Philosophical Transactions of the Royal Society of London* B330:269-281.

HUBBELL, S.P. & FOSTER. R.B. 1986. Biology, chance and history in the structure of tropical rain forest tree communities. Pp. 314-329 in Diamond, J. & Case. T.J. (eds). *Community ecology*. Harper & Row, New York.

HURLBERT, S.H. 1984. Pseudoreplication and the design of ecological field experiments. *Ecological Monographs* 54:187-211.

HYLAND, B. & WHIFFIN, T.R. 1993. *Key to the tropical forest trees of Australia*. CSIRO, Melbourne.

JANZEN, D.H. 1970. Herbivores and the number of tree species in tropical forests. *American Naturalist* 104:501-528.

KELLY, C.K. & PURVIS, A. 1993. Seed size and establishment conditions in tropical trees. On the use of taxonomic relatedness in determining ecological patterns. *Oecologia* 94:356-360.

KENNEDY, D.N. 1991. *The role of colonizing species in the regeneration of dipterocarp rain forest*. Ph.D. thesis, University of Aberdeen.

KENNEDY, D.N. & SWAINE, M.D. 1992. Germination and growth of colonizing species in artificial gaps of different sizes in dipterocarp rain forest. *Philosophical Transactions of the Royal Society* B335:357-367.

KERSHAW, A.P. 1988. Australasia. Pp. 127-306 in Huntley, B. & Webb, T. (eds). *Vegetation history*. Kluwer, Dordrecht.

KIEW, R. 1988. Herbaceous flowering plants. Pp. 56-76 in Cranbrook, The Earl of (ed.). *Key environments: Malaysia*. Pergamon, Oxford.

KOHYAMA. T. & GRUBB, P.J., in press, 1994. Below- and above-ground allometries of shade-tolerant seedlings in a Japanese warm-temperate rain forest. *Functional Ecology* 8.

LATHAM, R.E. 1992. Co-occuring tree species change rank in seedling performance with resources varied experimentally. *Ecology* 73:2129-2144.

LIEBERMAN, D., LIEBERMAN, M., HARTSHORN, G. & PERALTA, R. 1985. Growth rates and age-size relationships of lowland tropical wet forest trees in Costa Rica. *Journal of Tropical Ecology* 1:97-109.

LIEBERMAN, D., HARTSHORN, G.S., LIEBERMAN, M. & PERALTA, R. 1990. Forest dynamics at La Selva Biological Station 1969-85. Pp. 509-521 in Gentry, A.H. (eds). *Four neotropical rainforests*. Yale University Press, New Haven.

LIEBERMAN, M., LIEBERMAN, D. & PERALTA, R. 1989. Forests are not just Swiss cheese: canopy stereogeometry of non-gaps in tropical forests. *Ecology* 70:550-552

MARTIN, T.E. 1985. Resource selection by tropical frugivorous birds: integrating multiple functions. *Oecologia* 66:563-573.

MEDINA, E., MOONEY, H.A. & VAZQUEZ-JANES, C. (eds) 1984. *Physiological ecology of plants of the wet tropics*. Junk, The Hague.

MERRENS, E.J. & PEART, D.R. 1992. Effects of hurricane damage on individual growth and stand structure in hardwood forest in New Hampshire, USA. *Journal of Ecology* 80:787-795.

METCALFE, D.J. & GRUBB, P.J. Seed weights and light requirements for regeneration in south-east Asian rain forest. Submitted to *New Phytologist*, February 1994.

MOLOFSKY, J. & AUGSPURGER, C.K. 1992. The effect of litter on early seedling establishment in a tropical forest. *Ecology* 73:68-77.

MULKEY, S.S., SMITH, A.P. & WRIGHT, S.J. 1991. Comparative life history and physiology of two understorey tropical herbs. *Oecologia* 88:263-273.

NG, F.S.P. 1978. Strategies of establishment in Malayan forest trees. Pp. 129-162 in Tomlinson, P.B. & Zimmermann, M.H. (eds). *Tropical trees as living systems*. Cambridge University Press.

NG, F.S.P. 1991. *Manual of forest fruits, seeds and seedlings, Volume 1*. Forest Research Institute, Kepong, Malaysia.

OLDEMAN, R.A.A. 1990. *Forests: elements of silvology*. Springer, Heidelberg,

OLDEMAN, R.A.A. & DIJK, J. VAN 1991. Diagnosis of the temperament of tropical rain forest trees. Pp. 21-65 in Gomez-Pompa, A., Whitmore, T.C. & Hadley, M. (eds). *Rain forest regeneration and management*. UNESCO, Paris.

ORIANS, G.H. 1982. The influence of tree-falls in tropical forests on tree species richness. *Tropical Ecology* 23:255-279.

OSUNKOYA, O.O. 1992. *Ecology of seed survival and seedling establishment in Australian tropical forest*. Ph.D. thesis, Australian National University, Canberra.

PAPAVASILLIOU, S. 1991. *Ecophysiological studies on chalkland shrubs*. M.Phil. thesis. University of Cambridge.

PEACE, W.J.H. & GRUBB, P.J. 1982. Interaction of light and mineral nutrient supply in the growth of *Impatiens parviflora*. *New Phytologist* 90:127-150.

PEARCY, R.W. 1983. The light environment and growth of C_3 and C_4 tree species in the understorey of Hawaiian forest. *Oecologia* 58:19-25.

PLATT, W.J. & HERMANN, S.M. 1986. Relationships between dispersal syndrome and characteristics of populations of trees in a subtropical forest. Pp. 309-321 in Estrada, A. & Fleming, T.H. (eds). *Frugivores and seed dispersal*. Junk, Dordrecht.

POPMA, J., BONGERS, F., MARTINEZ-RAMOS, M. & VENENKLAAS, E. 1988. Pioneer species distribution in treefall gaps in neotropical rain forest; a gap definition and its consequences. *Journal of Tropical Ecology* 4:77-88.

POULSON, T.L. & PLATT, W.J. 1989. Gap light regimes influence canopy tree diversity. *Ecology* 70:553-555.

PRIMACK, R.B. 1992. Tropical community dynamics and conservation biology. *BioScience* 42: 818-821.

PUTZ, F.E. 1983. Treefall pits and mounds, buried seeds, and the importance of soil disturbance to pioneer trees on Barro Colorado Island, Panama. *Ecology* 64:1069-1074.

PUTZ, F.E. & APPANAH, S. 1987. Buried seeds, newly dispersed seeds, and the dynamics of a lowland forest in Malaysia. *Biotropica* 19:326-333.

RAICH, J.W. & GONG, W.K. 1990. Effects of canopy openings on tree seed germination in a Malaysian dipterocarp forest. *Journal of Tropical Ecology* 6:203-217.

REED, R.A. & GRUBB, P.J. Submitted. Nutrient supply available to species colonizing gaps in chalk grassland. Submitted to *Journal of Vegetation Science*, March 1994.

RICHARDS, P.W. 1952. *The tropical rain forest*. Cambridge University Press.

ROTH, I. 1987. *Stratification of a tropical forest as seen in dispersal types*. Junk, Dordrecht.

SALISBURY, E.J. 1942. *The reproductive capacity of plants*. Bell, London.

SANFORD, R.L. 1989. Fine root biomass under a tropical forest light gap opening in Costa Rica. *Journal of Tropical Ecology* 5:251-256.

SIM, J.W.S., TAN, H.T.W. & TURNER, I.M. 1992. *Adinandra* belukar: an anthropogenic heath forest in Singapore. *Vegetatio* 102:125-137.

SLAVIKOVA, J. 1958. Einfluss der Buche (*Fagus silvatica* L.) als Edifikator auf die Entwicklung der Krautschicht in Buchenphytozonosen. *Preslia* 30:19-42.

STAPANIAN, M.A. 1986. Seed dispersal by birds and squirrels in the deciduous forests of the United States. Pp.225-236 in Estrada, A. & Fleming, T.H. (eds). *Frugivores and fruit dispersal*. Junk, Dordrecht.

STEENIS, C.G.G.J. van 1958a. *Basic principles of rain forest sociology*. Pp. 159-162 in Proceedings of the Kandy symposium on the study of tropical vegetation. UNESCO, Paris.

STEENIS, C.G.G.J. van 1958b. *Rejuvenation as a factor for judging the status of vegetation types: the biologcal nomad theory*. Pp. 212-218 in same.

STOCKER, G.C. & IRVINE, A.K. 1983. Seed dispersal by cassowaries (*Casuarius casuarius*) in north Queensland's rainforest. *Biotropica* 15:170-176.

STRUIK, G.J. 1965. Growth patterns of some native annual and perennial herbs in southern Wisconsin. *Ecology* 46:401-420.

SUNDRALINGHAM, P. 1983. Response of potted seedlings of *Dryobalanops aromatica* and *Dryobalanops oblongifolia* to commercial fertilizers. *Malaysian Forester* 46:86-92.

SWAINE, M.D. & WHITMORE, T.C. 1988. On the definition of ecological species groups in tropical rain forests. *Vegetatio* 75:81-86.

TILMAN, D. & WEEDIN, D. 1991. Plant traits and resource reduction for five grasses growing on a nitrogen gradient. *Ecology* 72:685-700.

TOMLINSON, P.B. & ZIMMERMANN, M.H. (eds) 1978. *Tropical trees as living systems*. Cambridge University Press.

TURNER, I.M. 1991. The effects of shade and fertilizer addition on the seedlings of two tropical woody pioneer species. *Tropical Ecology* 32:24-29.

TURNER, I.M., BROWN, N.D. & NEWTON, A.C. 1993. The effect of fertilizer application on dipterocarp seedling growth and mycorrhizal infection. *Forest Ecology and Management* 57:329-337.

VAZQUEZ-YANES, C. & OROZCO-SEGOVIA, A. 1984. Ecophysiology of seed germination in the tropical humid forest of the world. Pp. 37-50 in Medina, E., Mooney, H.A. & Vazquez-Yanes, C. (eds). *Physiological ecology of plants in the wet tropics*. Junk, The Hague.

VAZQUEZ-YANES, C. & OROZCO-SEGOVIA, A. 1993. Patterns of seed longevity and germination in the tropical rainforest. *Annual Reviews of Ecology and Systematics* 24:69-87.

VITOUSEK, P.M. & DENSLOW, J.S. 1986. Nitrogen and phosphorus availability in treefall gaps of lowland tropical rainforest. *Journal of Ecology* 74:1167-1178.

WALTER. H. & BRECKLE, S.-W. 1985. *Ecological systems of the biosphere. I. Ecological principles in global perspective*. Springer, Berlin.

WATT, A.S. 1947. Pattern and process in the plant community. *Journal of Ecology* 35:1-22.

WELLESLEY-WHITEHOUSE, H. 1983. White-tailed rat. Pp. 371-373 in Strahan, R. (ed.). *Complete book of Australian mammals*. Angus & Robertson, Sydney.

WHITMORE, T.C. 1975. *Tropical rain forests of the Far East*. Clarendon Press, Oxford.

WHITMORE, T.C. 1984. *Tropical rain forests of the Far East, 2nd edn*. Clarendon Press, Oxford.

WILCZYNSKI, C.J. & PICKETT, S.T.A. 1993. Fine root biomass within experimental canopy gaps: evidence for a below-ground gap. *Journal of Vegetation Science* 4:571-57.

The interaction of rainfall, tree diameter and sulphuric-acid treatments on resin yield of Palosapis (*Anisoptera thurifera* (Blanco) Blume ssp. *thurifera*)

ARSENIO B. ELLA

Forest Products Research and Development Institute, Department of Science and Technology College, Laguna 4031, Philippines

ABSTRACT. Thirty-two trees of palosapis (*Anisoptera thurifera* (Blanco) Blume ssp. *thurifera*), naturally growing in the logging concession of the Eastern Plywood Corporation in Llavac, Real, Quezon, Luzon Island, Philippines were tapped (boxed) for resin or oil yield. Four diameter at breast height (dbh) classes: 30-39, 40-49, 50-59 and 60 cm and over were selected. Eight trees were designated for each of four levels of aqueous sulphuric acid treatment at 0%, 40%, 50% and 60%. Resin yield per tree was determined and the influence of rainfall, diameter classes and sulphuric acid concentration was studied. Months with higher rainfall in most cases gave greater resin yield. The effect of different diameter classes on resin yield was not statistically significant except in the case of 60 cm and over which gave the highest yield. The four levels of acid concentration had no significant effect on the resin yield.

KEY WORDS: resin yield, *Anisoptera*, Dipterocarpaceae.

INTRODUCTION

Tapping or boxing of dipterocarp species for resin provides a livelihood for people in parts of the Philippines. In Quezon province for instance, many eke out their living by tapping dipterocarp species while waiting for the harvest of their farm crop. Some of those whose farm lands are already depleted of soil nutrients become full-time tappers. Boxing of this species thus provides a means of generating employment.

One of the most common resin producing dipterocarp species in the Philippines is *Anisoptera thurifera* (Blanco) Blume ssp. *thurifera* locally known as palosapis. The resin or 'oil of palosapis' is very similar to 'balau' from apitong (*Dipterocarpus grandiflorus* Blanco). It is obtained in the same manner and is used for the same purposes, e.g., in the manufacture of varnish which makes a brilliant, tough and durable coating (Brown 1921). Clover (1906), as cited by Brown (1921), said that the oil from palosapis resembles that from apitong, but dries much faster. In addition, the 'oil of palosapis' is light colored, apparently homogenous in composition and so viscous that it can scarcely be poured.

The traditional method of obtaining the resin is by boxing and refreshing by burning at frequent intervals to expose the cut (West & Brown 1921). This method is the same as that used in the production of dammar (Tongacan 1973). Either by tapping or by accidental injury (Tanchico *et al.* 1934), resin is exuded from the resin ducts or gum canals found in the sapwood. This practice however, is still crude and often results in the entry of wood-destroying organisms, exposure to wind and rain and/or subsequent death of the tree.

Reports of studies conducted by Parameswaran & Zamuco (1976) based on the miscroscopy of the bark of four species of the so-called 'Philippine mahogany' and by Whitmore (1962) on the bark morphology of Malaysian Dipterocarpaceae make no mention of the presence of resin ducts in the bark. Metcalfe & Chalk (1957), however, reported that in the young stems of dipterocarps, resin canals are always present in the pith, especially in

D. S. Edwards et al. (eds.), Tropical Rainforest Research – Current Issues, 235–240.

Figure 1. Initial cut is made but limited only up to the sapwood-heartwood boundary. The initial height of the box is about 25 mm and is V-shaped, in order to facilitate resin flow. The width of boxing should be 1/3 the girth of the tree.

Figure 2. Sulphuric acid solution is applied to the freshly tapped palosapis tree using a one-pint squeeze sprayer.

the perimedullary region, and that they also occur in the phloem and cortex.

Few studies of the boxing of dipterocarp species have been conducted. Toñgacan and Ordinario (1976) studied the boxing of apitong for balau production and the results revealed that monthly rainfall was found to have an inverse influence on monthly balau yield. Within the limits of sulphuric-acid concentrations used in the study, acid treatments were found to have no effect on balau yield. Nor were there significant effects reported for either interactions between rainfall and diameter, or rainfall and acid concentrations on resin yield.

The objective of this study was to determine the effects of rainfall, tree diameter and sulfuric acid treatments in resin yield of boxed palosapis trees.

Figure 3. Photograph showing the correct way to securely attach a polyethylene sheet before sealing with plastic roofing cement. This prevents seepage of water into the tacked plastic receptacle below.

MATERIALS AND METHODS

The tapping experiment

The tapping or boxing experiment was conducted in the logging concession of Eastern Plywood Corporation in Barangay Llavac, Real, Quezon, Luzon Island, Philippines for a duration of 12 months. Thirty-two healthy palosapis trees belonging to four diameter classes i.e. 30-39, 40-49, 50-59 and ≥60 cm (8 trees each) were included in the study. Four levels of sulphuric acid concentration vis.: 0%, 40%, 50% and 60% (by volume) were applied using 2 trees each. The experimental design of this boxing experiment is a 4 x 4 factorial in completely randomized design with only two replications owing to the scarcity of suitable, vigorous, healthy trees.

The tapping procedure

Palosapis bark was first cleaned or scraped off, using a sharp bolo or wood chisel. The depth of tapping conformed to the thickness of sapwood as determined by incremental boring; the initial height of the box was about 25 mm and its form V-shaped to facilitate resin flow; the width was 1/3 of the girth of the tree. Subsequent rechippings proceeded vertically upward at a width of about 20 mm to make sure that a minimum amount of the tissues destroyed by acid in the previous tapping was removed.

Sulphuric acid solution was immediately applied evenly in a fine mist after cutting the streaks, using a US standard plastic squeezer of one pint capacity. Plastic bags were used as receptacles. Each tree boxed was properly labelled. The collection of resin was conducted weekly. Rechipping was done after each collection and sulphuric acid solution was applied. The amount of resin gathered was weighed and recorded.

The effect of variation in rainfall on resin yield in relation to different diameter classes and sulfuric-acid treatment was determined. Monthly rainfall data was obtained from the Philippine Atmospheric Geophysical and Astronomical Services Station (PAGASA), in Llavac, Real, Quezon, Luzon Island. The experimental site falls under the second climatic type, having no dry season but with a very pronounced maximum rain period from November to January.

Figure 4. Photograph showing a tapped palosapis tree (40-49 cm dbh) all set for next resin collection, rechipping, and sulfuric acid application.

Table 1. Annual resin yield from different diameter classes. r values = correlation coefficients; p values = probability. *total yield of 2 trees; **total yield of 8 trees.

Diameter class (cm)	Annual resin yield (kg/yr)* Acid concentration (% v.v)				Total Yield (kg/yr)	Average yield (kg/tree/yr)
	0	40	50	60		
30-39	7.51	5.49	8.22	8.74	29.96	3.74
40-49	5.05	9.77	6.75	4.92	26.49	3.31
50-59	7.61	7.19	4.26	7.92	26.98	3.37
60 & up.	10.59	14.68	13.31	8.69	47.27	5.91
r values	0.59	0.74	0.30	0.14	0.58	0.58
p values	0.4098	0.2697	0.7005	0.8561	0.4227	0.4218
Total yield (kg/yr)**	30.76	37.13	32.54	30.27		
Average yield (kg/tree/yr)	3.845	4.64125	4.0675	3.78375		

RESULTS AND DISCUSSION

Effect of diameter of the tree on resin yield

The results are shown in Table 1. Statistical analysis revealed that diameter of the tree has no significant effect on the resin yield (Table 2). This lack of significance, however, is

Table 2. Analysis of variance of total resin yield. ns = not significant at 5% level.

SOURCE OF VARIATION	d.f	SS	MS	F
Treatment (T)	15	63.224		
Diameter	3	35.779	11.933	2.9027ns
Acid concentration	3	3.157	1.052	0.2558ns
Diameter x Acid concentration	9	24.276	2.697	0.6560ns
Residual	16	65.783	4.111	
TOTAL	31	129.007		

Table 3. Monthly resin yield per tree. r values = correlation coefficient; p values = probability.

Monthly rainfall in mm (2nd climatic type)		Monthly resin yield (kg) Acid concentration (% v/v)				Total monthly resin yield (kg)
		0	40	50	60	
1978						
February	161.4	1.68	2.23	2.8	1.19	7.9
March	77.4	2.66	2.97	2.67	2.1	10.4
April	282.7	2.37	2.19	2.15	2.09	8.8
May	249.2	2.14	2.91	1.93	1.92	8.9
June	323.6	1.68	2.64	2.05	2.74	9.1
July	218.8	1.89	2.85	2.31	2.37	9.42
August	563.9	1.64	2.98	2.39	1.79	8.8
September	366.5	1.97	2.83	1.52	2.29	8.61
October	1397.4	4.14	4.87	2.94	2.86	14.81
November	637.8	2.66	4.04	3.61	3.29	13.6
December	687.4	2.85	3.2	2.97	3.61	12.63
1979						
January	148.4	2.65	3.83	4.16	3.75	14.4
r values		0.72	0.74	0.16	0.33	0.58
p values		0.0082	0.0060	0.6247	0.2917	0.0490

probably the result of low population (n) as comparison among the 4 diameter classes showed trees with diameter of 60 cm and up do have higher resin yield per tree than the 3 other diameter classes (Table 1). This might be due to the fact that larger trees have larger bark surface area and wider tapping cut exposure that have direct positive effect on the resin production. Table 1 also shows that the annual resin yield per tree is correlated to tree diameter ($r = 0.58$; $p = 0.4218$). A similar degree of correlation was shown at 0% acid treatment ($r = 0.59$; $p = 0.4098$) and slightly higher at 40% acid treatment ($r = 0.74$; $p = 0.2697$). However, at higher acid treatments (50% and 60%) poor correlation was observed $r = 0.3$; $p = 0.7005$ and $r = 0.14$; $p = 0.8561$, respectively). This is consistent with the yield decrease at 50% acid treatment in 40-49 cm, 50-59 cm diameter classes and at 60% acid treatment in the 60 cm and greater diameter class. It can be seen that trees with larger diameter need higher acid concentration to lower their yield.

Effect of rainfall on resin yield

The monthly resin yield is highly correlated with monthly rainfall at 0% and 40% v/v acid treatment with r values of 0.72 and 0.74 respectively (Table 3). However, such correlation was not observed at higher acid concentration treatments (50% and 60%) in which lower r values were obtained (0.16 and 0.33 respectively).

Effect of sulphuric acid on resin yield

The four levels of acid concentration had no significant effect on the resin yield (Table 2). Hence, there is no need to compare the means of the four levels of acid studied. Although not statistically significant (Table 2) acid concentration did appear to interact with dbh (i.e. only at low acid concentrations was there a high r of resin yield on dbh, Table 1).

CONCLUSIONS

Monthly rainfall was found to have a direct influence on monthly resin yield. Specifically, months with higher rainfall in most cases gave more resin yield. Acid concentration had no significant effect on the resin yield.

Further experimental work using similar variables and tapping techniques but involving a larger number of replicates should be done in a region of the country with pronounced wet and dry seasons and also in a region with no dry season to serve as a basis of comparison of resin yields, and the influence of climatic conditions.

LITERATURE CITED

BROWN, W.H. 1921. Philippine resins. *Minor products of Philippine forests*. Vol. 2:52-64.

CLOVER, A.M. 1906. Philippines wood oils. *Philippine Journal of Science* 1:181-202.

KRAMER, F.J. & KOZLOWSKI, T.T. 1960. *Physiology of trees*. McGraw-Hill, New York.

METCALFE, C.R. & CHALK, C. 1957. *Anatomy of the dicotyledons*. Clarendon Press, Oxford. Vol. 1:212-220.

PARAMESWARAN, N. & ZAMUCO, G.I., JR. 1976. *Microscopy of barks of the so- called "Philippine Mahogany" species*. Unpublished report, FPRDI, College, Laguna.

TANCHICO, S.S., WEST, A.P. & FORTANOZA, J. 1934. Philippine Panau (Dipterocarp) resin. *Philippine Journal of Science* 54:75-81.

TOÑGACAN, A.L. 1973. *FPRDI Technical Note No. 136*. FPRDI, College, Laguna.

TOÑGACAN, A.L. & ORDINARIO, F.F. 1976. *The interaction of tree diameter rainfall and sulfuric acid on "Balau" yield of apitong (Dipterocarpus grandiflorus Blume)*. Unpublished report, FPRDI, College, Laguna.

WEST, A.P. & BROWN, W.H. 1921. Philippine resins, gums, seed oils and essential oils. *Minor Forest Products of Philippine Forests*. Volume 2:5-225.

WHITMORE, T.C. 1962a. Studies in systematic bark morphology. I. Bark morphology in Dipterocarpaceae. *The New Phytologist* 62:191-207.

Preliminary studies of altitudinal zonation and nitrogen supply on Bukit Belalong, Brunei Darussalam

C.A. PENDRY *and* J. PROCTOR

Department of Biological and Molecular Sciences, University of Stirling, Stirling FK9 4LA, United Kingdom

ABSTRACT. Forest structure is briefly described from plots at 200 m, 500 m and 850 m on Bukit Belalong (913 m). Evergreen lowland rain forest is found at 200 m and 500 m with tall lower montane rain forest at 850 m. Automatic weather stations at the base and the summit suggest that the summit receives 20% more rainfall and mean temperatures there are 3.9 °C lower. A cloud cap does not usually form and mean daily totals of solar radiation are similar at the base and summit. Total small litterfall was measured for one year and the following values (t ha^{-1} yr^{-1}) were recorded: 10.56 (200 m), 10.48 (500 m), and 8.26 (850 m). Leaf litterfall values (t ha^{-1} yr^{-1}) were 7.88 (200 m), 7.89 (500 m) and 6.02 (850 m). Nitrogen concentrations (mg g^{-1}) in total litterfall were 14.5 (200 m), 12.7 (500 m), 12.7 (850 m) and in leaf litterfall 11.4 (200 m), 10.6 (500 m) and 11.3 (850 m). Total litterfall and leaf litterfall were significantly lower at the highest altitude ($p < 0.05$), but differences in nitrogen concentrations were not significant.The following soil nitrification rates (kg ha^{-1} 21 d^{-1}) were measured: 4.65 (200 m), 2.60 (500 m) and 3.72 (850 m), with no significant differences between altitudes. A bioassay using rice (*Oryza sativa* L.) was carried out near the base and summit of Bukit Belalong using soils from both altitudes. There was a dramatic response to altitude, a lesser one to soil type and only a slight response to nitrogen fertilisation. It is concluded that nitrogen is not more limiting in the montane forest compared with the lowland forest and that forest differences with altitude are most likely to be related to changes in temperature.

KEY WORDS: Litterfall, nitrogen, nitrification, bioassay, montane tropical rain forest.

INTRODUCTION

Floristic and physiognomic changes with altitude are commonly observed features of tropical mountains (Whitmore 1984), but the causes are still unclear. They include: lower temperatures (Grubb 1977), high winds (Sugden 1986), low soil pH (Grubb & Tanner 1976), nutrient limitation due to reduced decomposition and mineralisation rates (Edwards & Grubb 1977, Heaney & Proctor 1989), low radiation totals (Grubb 1977), periodic water shortage (Whitmore 1984) and soil toxicity (Bruijnzeel *et al.* 1993). These factors may operate singly or in combination and the reasons for the replacement of lowland forest by montane forest certainly vary from site to site. Attention has recently focussed on the stunting of upper montane forest (Bruijnzeel *et al.* 1993, Veneklaas 1992) which has been associated with the presence of a cloud cap (Grubb 1977, Whitmore 1984). The cloud causes reduced solar radiation and higher soil moisture which reduces the nitrogen supply. Nitrogen has been considered to be the nutrient most likely to limit montane forests (Grubb 1989, Heaney & Proctor 1989), and in this paper we examine this hypothesis for Bukit Belalong.

This paper examines the forests of Bukit (Mount) Belalong which has relatively tall lower montane forest at its summit and develops a cloud cap infrequently. We try to explain the limited altitudinal changes in the forests on Bukit Belalong and contrast it with other mountains where there is stunted forest and we present some of our data on climate and nitrogen supply. We report litterfall nitrogen concentrations and soil nitrification at three altitudes and part of a bioassay experiment using hill rice *(Oryza sativa* L.). The studies on Bukit Belalong are still continuing and further data about the site will be reported elsewhere.

D. S. Edwards et al. (eds.), Tropical Rainforest Research – Current Issues, 241–257.

STUDY SITE

Bukit Belalong (913 m) is in the Temburong District of Brunei, north western Borneo (Figure 1). The mountain forms part of the Batu Apoi Forest Reserve (4° 30' north, 115° 10' east), and has never been logged. The nearest settlement, Kampong Batang Duri, is 8 km downstream from the Field Studies Centre (FSC) on the Temburong river, and though shifting cultivation is practised along the river it appears that no land has been cleared upstream from Kuala Belalong. Human impact on the forest is limited to hunting and the infrequent gathering of forest products.

Evergreen lowland rain forest with emergent trees up to 60 m tall is found up to c. 750 m. Above this altitude the forest is lower montane rain forest (*sensu* Whitmore 1984) in which the trees are smaller but reach heights of 33 m. Dipterocarps are present at all altitudes.

Climate

Two automatic weather stations (AWS) were used: one at the summit of Bukit Belalong, from 7 May 1991 to 24 April 1992; and another on a small cleared hill (60 m) near Kuala Belalong, from 18 June 1991 to 23 April 1992. Rainfall was also measured in a standard rain gauge at the FSC from 1 February 1991 to 31 May 1992. The rain gauge was sited within 3 m of the nearest building, and recorded 6% more rain than the properly sited AWS at Kuala Belalong. The mean annual rainfall (1973 to 1992) at Semabat Agricultural Station, 20 km to the north of Kuala Belalong was 4086 mm. There is no dry season, though February and March and June to October tend to be the drier months. Since the Semabat values are similar to those at Kuala Belalong for the period of measurement we estimate the mean annual rainfall at Kuala Belalong to be c. 4100 mm. Whilst the AWS was at the summit rainfall there was 21% higher than at Semabat so the mean annual rainfall is estimated at c. 5000 mm.

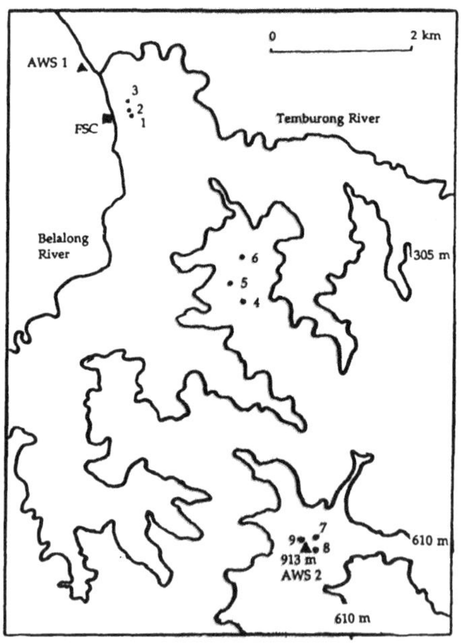

Figure 1. The locations of study plots, automatic weather stations (AWS) and the Field Studies Centre, Batu Apoi Forest Reserve, Brunei Darussalam.

The highest daily rainfalls were for 24 September 1991 when 181.5 mm rain fell at Bukit Belalong and 133 mm at Kuala Belalong. The longest dry periods were 7 days at Kuala Belalong and 12 days at Bukit Belalong. During February and March 1992 Semabat received 251.8 mm of rain which is 45% of the mean rainfall for those months measured between 1973 and 1992. This is the second lowest total recorded for these months (237.2 mm fell in the 'El Niño' year 1983). The rainfall totals for February and March 1992 were 224.0 mm at Kuala Belalong and 327.0 mm at Bukit Belalong. Rainfall data for the entire period of the study are summarised in Figure 2.

Mean daily temperature (July 1991-March 1992) was 25.7°C at Kuala Belalong and 21.8°C at Bukit Belalong. The lapse rate is therefore estimated at 0.44°C per 100 m, which is typical for a dry adiabatic tropical lapse rate (Bruijnzeel *et al.* 1993). The ranges of mean monthly maxima are 4.5°C at Kuala Belalong and 5.5°C at Bukit Belalong, whilst the corresponding ranges of mean monthly minima are 2.4°C and 3.2°C. The mean daily solar radiation for this period was 15.63 MJ m^{-2} at Kuala Belalong and 15.71 MJ m^{-2} at Bukit Belalong, but it should be noted that there is greater variability at the summit. Temperature data are summarised in Figure 3 and solar radiation data in Figure 4.

Geology

The geology of Brunei is dominated by older Tertiary sandstones and shales. Narrow ridges with steep sides and deep valleys are characteristic of the Belalong area and are a consequence of Miocene folding and lifting of the Setap shales. Large remnants of sandstone have been reported from further south in Temburong (Ashton 1964), but on Bukit Belalong these are small and localised, so the mountain is essentially continuous shale from bottom to top. Erosional processes are exceeded by rates of tectonic uplift so slope failures are frequent and are the principal determinant of the area's geomorphology (A. Dykes, personal communication).

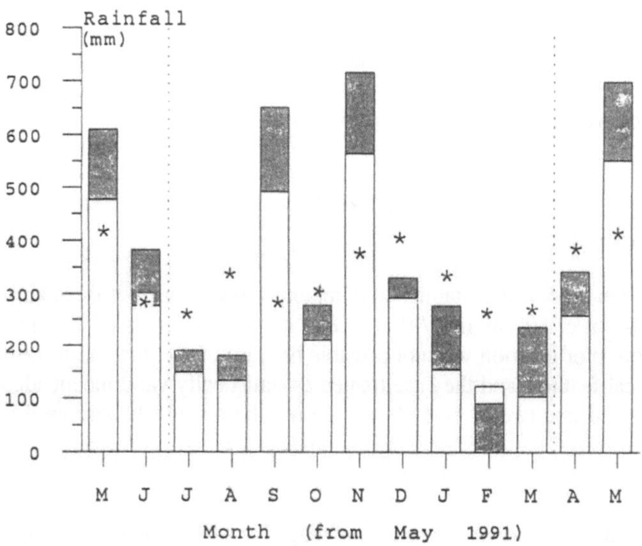

Figure 2. Monthly rainfall at Kuala Belalong and Bukit Belalong, May 1991 to May 1992. The unshaded portion of the column represents rainfall at Kuala Belalong and the shaded portion the extra rainfall at Bukit Belalong (in February 1992 more rain fell at Kuala Belalong than Bukit Belalong). The values for May and June 1991 and April and May 1991 are estimated from a rain gauge at Kuala Belalong Field Studies Centre. * represents the mean monthly rainfall (1973-1992) at Semabat Agricultural Station.

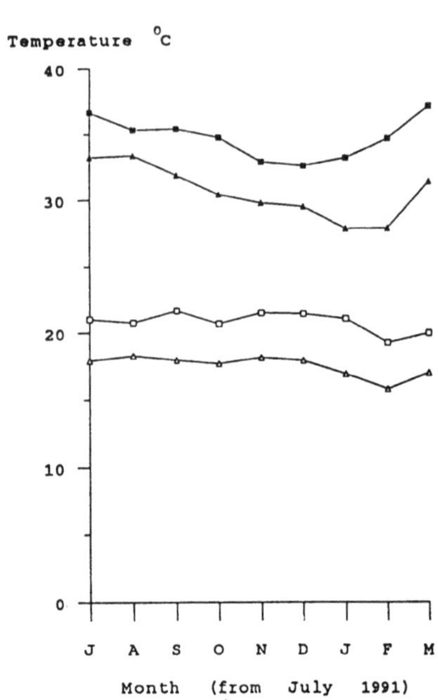

Figure 3. The maximum (closed symbols) and minimum (open symbols) monthly temperatures at Kuala Belalong, 60 m (▬) and Bukit Belalong, 913 m (▲) between July 1991 and March 1992.

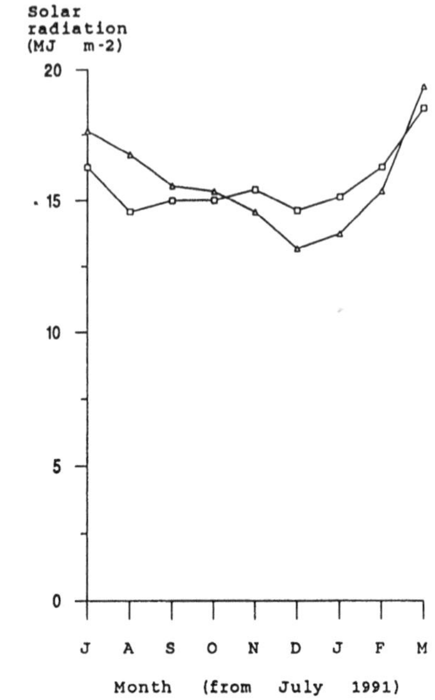

Figure 4. Mean daily totals of solar radiation (MJ m^{-2}) at Kuala Belalong, 60 m (▬) and Bukit Belalong, 913 m (▲) between July 1991 and March 1992.

Soils from low altitudes have been classified as orthic acrisols (FAO/UNESCO, S.M. Ross, personal communication). The soils lack distinct horizons, are acid (mean pH_{H2O} was 3.9 at 200 m and 3.8 at 880 m) base-poor and highly weathered. Soil pits were dug beside each plot and although the soils from higher altitudes had more organic matter in the upper horizon the profiles were similar to those at lower altitudes.

METHODS

Plot selection

Three 50 m x 50 m plots were established in mature forest at each of three altitudes: plots 1-3 at c. 200 m, 4-6 at c. 500 m and 7-9 at c. 850 m. Some details of the plots are given in Table 1. Random plot-location was not possible because of the difficult terrain. None of the plots had atypical features and they are treated as statistically valid independent samples. The plots were divided into twenty five 10 m x 10 m subplots which were used as a sample grid.

Forest description

All trees (≥ 10 cm dbh) were enumerated and tagged at 140 cm above the ground. Girths were measured 10 cm below the tag except where buttresses, prop roots or damage distorted the trunk, and on these trees girth was measured at 30 cm above the protrusion. A second nail then marked 10 cm above the point of measurement. Trees with multiple stems were tagged on the main trunk, and the points of measurement on the other trunks (≥ 10 cm dbh

Table 1. Mean altitude, aspect and slope of nine 50 m x 50 m plots on Bukit Belalong.

Plot number	Altitude	Aspect	Slope
1	200	210	23
2	200	240	25
3	210	280	26
4	480	80	16
5	520	130	21
6	510	150	30
7	830	40	33
8	860	130	35
9	880	30	30

only) marked with nails but not tagged. The position of each tree was measured to the nearest 50 cm.

A 60 m x 7.5 m strip of representative mature forest was selected for a profile diagram from plots 1, 5 and 9. The heights of all the trees in these plots were measured and the size of each crown was estimated by measuring the height to the emergence of the lowest branch and the furthest horizontal projections of the crown in the plane of the diagram.

Small litterfall

Small litterfall was collected from 8 May 1991 to 8 May 1992 in plots 1-6 and 7 May 1991 to 7 May 1992 in plots 7-9. The litter traps were baskets of perforated metal sheeting, with a horizontal area of 40 cm x 40 cm, mounted on poles 1.2 m above the ground. There were thirteen litter traps placed in a restricted random design in each plot.

It was difficult to collect litterfall at regular intervals because of the remoteness of the sites, but collections were usually at 14-20 day intervals. Two collections (9-10 December 1991 and 5-6 April 1992) were after 12 days and two other collections were after 36 days (26-27 June 1991) and 39 days (17-18 January 1992).

The litter traps were emptied into cotton bags and the litter was air dried within two days of collection. Alternate collections (eleven out of nineteen) were sorted into four fractions, following the recommendations of Proctor (1983): leaves (including stipules), small wood (twigs with diameter <2 cm, and bark fragments smaller than 2 cm along their longest dimension), reproductive parts and trash (any plant fragments with a longest dimension smaller than 2 mm and faeces, insect bodies and resin). The sorted samples were weighed and combined to give bulked samples of the litterfall fractions for each collection from each plot. The bulked samples were ground and a 20 g sub-sample dried at 105°C before chemical analysis. When the total mass of the bulked sample was less than 20 g the entire sample was oven dried. Unsorted collections were oven dried at 105°C and weighed.

0.4 g of each subsample was digested in 4.4 ml of a sulphuric acid/hydrogen peroxide mixture (Moore & Chapman 1986). Samples were digested at 100°C for 20 min, followed by 2 h at 370°C. Nitrogen was determined as ammonium on a Tecator FIAstar 5000 flow injection analysis system.

It was not practical to analyse all the samples in duplicate so the precision of the method was tested by analysing five replicates of eight samples. The results were always within 10% of the mean and usually within 5%. Analysis of reference material supplied by E.V.J. Tanner gave results within 3% of those found by investigators in other laboratories.

Nitrification rates

In April 1992 one random position was located in each of ten of the undisturbed sub-plots (not used for litterfall collection) in each of the nine plots. At each position the litter layer was removed and a 5-cm deep soil sample collected. The sample was thoroughly sorted to remove stones, roots and large fragments of organic matter and crushed by hand till no particles were bigger than 10 mm. The soil was mixed thoroughly and a c. 200-g sub-sample placed in a plastic bag and replaced at the collection site and covered with litter. The top of the bag was folded down to allow gaseous exchange, but to prevent the entry of water. The soil was left to incubate for 21 d before collection and extraction. A second sub-sample was returned to the FSC for immediate extraction and for determination of the moisture content (after oven drying at 105°C).

20 g soil (wet weight) was extracted in 100 ml 1.5 M potassium chloride containing 3 mg l^{-1} mercuric chloride to inhibit bacterial activity. The samples were shaken and left to equilibrate for 4 h before filtration and bottling. They were then frozen until sent to the UK where they were stored in the dark at 2°C until analysis. Soil nitrate/nitrite nitrogen was determined as nitrite after reduction by a copperised cadmium column on the FIA system.

In each of the sub-plots used for the collection of soil samples a second random point was located for the collection of a soil bulk density sample. At each point the litter layer was removed and a 10 cm x 10 cm open-ended stainless steel box was inserted to a depth of 5 cm and the soil excavated from within it. The samples were oven dried at 105°C, weighed and then the stones (>5 mm) and roots were sorted and weighed separately.

Bioassay

The species used in the bioassay was a variety of hill rice (*Oryza sativa* L.) obtained from local farmers. Local tree species might have been preferable to rice for the test plant but this was not practical and in any case species from a nutrient-poor environment may not necessarily be responsive to fertilisation (Grubb 1989).

A total of about 6 kg of soil was collected to a depth of 5 cm from six random positions located around each of plots 3 (bottom soil - BS) and 9 (summit soil - SS). The soil was sorted to remove stones, roots and larger fragments of organic matter (>1 cm along any dimension), crushed by hand to give aggregates no larger than 10 mm, and thoroughly mixed. The soils were stored in sealed plastic bags for up to 5 days before the experiment was set up.

The two sites for the experiment were the summit of Bukit Belalong and an open space at the FSC. At both sites the plants were grown on a table with a screen of green plastic netting to exclude flying insects and with metal sheeting on the legs to exclude rats and mice. It is accepted that the netting will have reduced the photosynthetically active radiation available to the plants. At each site forty-eight 14 cm diameter pots were each filled with 500 cm^3 of BS and a further forty-eight pots filled with a similar quantity of SS and the pots placed on saucers. Seven pregerminated rice seedlings were placed in each pot and after emergence the plants were thinned to five per pot. Nitrogen was added to six replicates of each soil as 100 mg sodium nitrate in 10 ml of water (from the nearest stream) and the six control pots were watered with 10 ml of stream water only. The fertilisation rate was adapted from Healey (1989) but reduced by 50% because of the shorter growing period and was equivalent to 38 kg nitrogen ha^{-1}.

The nitrogen treatment was part of a larger experiment consisting of eight nutrient treatments and the pots were laid out in six blocks containing one replicate of each treatment. The pots were randomised within each block and rerandomised after 7 days at the summit site and after 8 days and 11 days at the FSC. The plants were watered as required, using water from the nearest stream. After 14 days the plants were harvested, the soil was washed from their roots and they were oven dried and the shoots and roots were weighed separately.

Table 2. Basal area, mean tree (≥10 cm dbh) height, maximum tree height, tree density and percentage of trees in the Dipterocarpaceae and Myrtaceae in plots at three altitudes on Bukit Belalong.

	Plot 1	Plot 5	Plot 9
Altitude (m)	200	520	880
Basal area (m^2 ha^{-1})	58.8	43.5	32.0
Mean tree height (m)	22.9	22.2	17.8
Maximum tree height (m)	60	45	33
Tree density (ha^{-1})	470	610	670
Dipterocarpaceae (%)	16.0	14.5	9.7
Myrtaceae (%)	1.6	4.4	16.1

Statistical analyses

The data were analysed by ANOVA on the Minitab statistical package. Normality was tested by the Shapiro-Wilk test and non-normal data were log transformed.

RESULTS

Forest description

Table 2 summarises data from plots 1, 5 and 9. The maximum tree height declines with altitude and is reduced from 60 m in plot 1 to 33 m in plot 9, though the mean tree height declines only slightly because of the preponderance of understorey trees. The number of trees per hectare increases by almost half between plots 1 and 9. The profile diagrams (Figures 5-7) give an overall impression of the forest structure at the three altitudes and show that emergents are an important feature of the lower plots, with the canopy becoming more even at the higher altitudes. The proportion of Myrtaceae increases and that of Dipterocarpaceae decreases with altitude.

On the basis of the reduction in stature and frequency of tall (>2 m) buttresses (10-11% trees at low altitudes, 5% of trees at highest altitude), increase in bryophytes and disappearance of large emergents, the forest at 850 m can be considered Lower Montane Rain Forest according to Whitmore's physiognomic classification (1984), whilst the forests at 200 m and 500 m are Evergreen Lowland Rain Forest.

Litterfall

The values for small litterfall declined from a mean of 10.56 t ha^{-1} yr^{-1} in the low plots to a mean of 8.26 t ha^{-1} yr^{-1} in the top plots (Table 3). All fractions except flowers and fruits decreased with altitude and ANOVA shows that the reductions in total litterfall and leaf litterfall in the top plots were significant ($p<0.05$). There was a marked peak in leaf litterfall production in April-May 1992 (Figure 8) which followed the low rainfall during January-March. The peak was highest in the mid plots and least in the top plots.

Nitrogen concentrations varied in no obvious pattern during the year and the results are summarised as annual means in Table 4. An ANOVA showed no significant differences in nitrogen concentrations between altitudes for leaf and small wood litter, but significantly lower concentrations were found in the reproductive and trash fractions in the mid and top plots compared with the low plots ($p<0.05$). The mean concentration values in Table 4 were multiplied by the values for annual litterfall to give estimates of the annual rate of addition of nitrogen to the forest floor via each litter fraction at each altitude (Table 5). The inputs followed a similar temporal pattern to those of small litterfall mass.

Figure 5. Profile diagram (60 m x 7.5 m) of Plot 1 (200 m), Bukit Belalong. Trees under 6 m excluded.

Figure 6. Profile diagram (60 m x 7.5 m) of Plot 5 (520 m), Bukit Belalong. Trees under 6 m excluded.

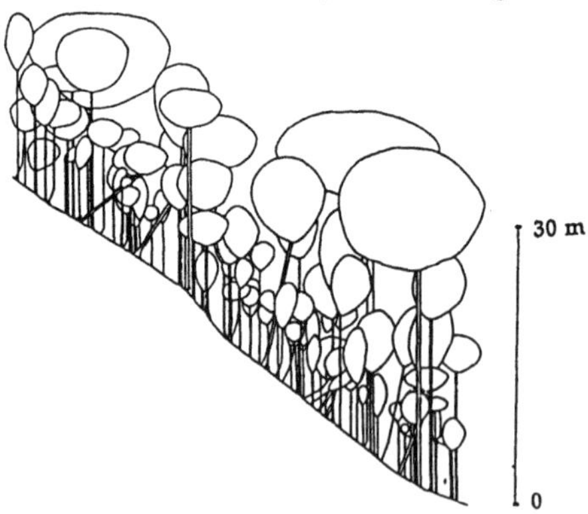

Figure 7. Profile diagram (60 m x 7.5 m) of Plot 9 (880 m), Bukit Belalong. Trees under 6 m excluded.

Table 3. Estimated litterfall (t ha^{-1} yr^{-1}) from thirty-nine 40 cm x 40 cm litter traps at each of three altitudes on Bukit Belalong. Values are means ± 95% confidence limits.

Litter fraction	Altitude		
	200 m	500 m	850 m
Leaves	7.88 ±0.44	7.89 ±0.46	6.02 ±0.41
Wood (<2cm diameter)	1.76 ±0.57	1.60 ±0.40	1.32 ±0.24
Flowers & fruits	0.47 ±0.12	0.53 ±0.14	0.53 ±0.22
Trash	0.44 ±0.06	0.46 ±0.08	0.40 ±0.10
Total	10.55 ±0.80	10.48 ±0.69	8.27 ±0.63

Table 4. Mean nitrogen concentrations (mg g^{-1}) in small litterfall from three altitudes on Bukit Belalong. Values are the means of eleven samples ±95% confidence limits.

Litter fraction	Altitude		
	200 m	500 m	850 m
Leaves	11.4 ±0.41	10.6 ±0.21	11.3 ±0.28
Wood (<2cm diameter)	7.9 ±0.50	7.2 ±0.40	7.6 ±0.41
Flowers & fruits	17.8 ±1.68	16.0 ±1.21	13.7 ±1.17
Trash	20.8 ±0.82	17.2 ±1.03	18.1 ±0.69
Total	14.5	12.7	12.7

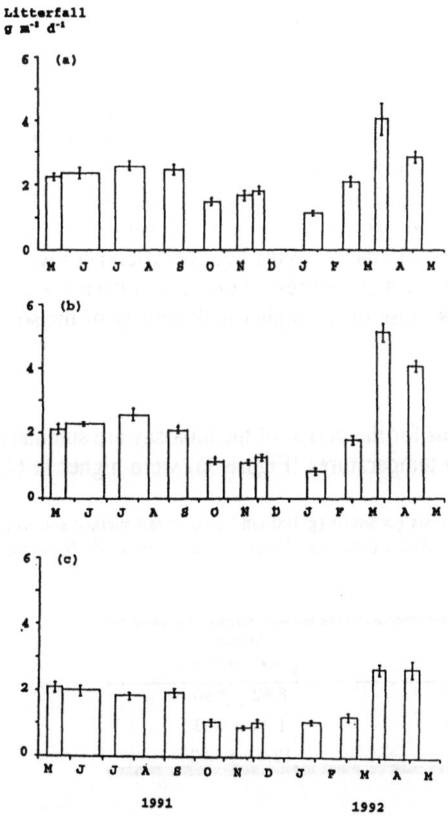

Figure 8. The rate of leaf litterfall (g m^{-2} d^{-1}) from May 1991 to May 1992 at three altitudes on Bukit Belalong. (a) 200 m, (b) 500 m, (c) 850 m. Leaves were sorted from eleven collections out of a total of nineteen and the columns represent the means of the sorted collections and the error bars are the 95% confidence limits.

Table 5. Estimated rate of addition of nitrogen (kg ha^{-1} yr^{-1}) to the forest floor from small litterfall fractions for three altitudes on Bukit Belalong. Values are the means for three plots at each altitude with the ranges in parentheses.

Litter fraction	Altitude		
	200 m	500 m	850 m
Leaves	90.0 (81.4-95.5)	83.1 (73.5-94.9)	67.8 (63.2-72.3)
Wood (<2cm diameter)	14.4 (12.3-17.2)	11.1 (9.4-12.3)	10.1 (8.0-12.0)
Flowers & fruits	8.5 (7.1-9.5)	8.4 (6.2-12.2)	7.7 (5.4-10.1)
Trash	9.2 (6.8-10.6)	7.8 (6.5-9.5)	7.3 (5.6-9.4)
Total	122.1 (114.2-127.8)	110.4 (102.6-122.2)	92.9 (90.4-95.2)

The quantity of nitrogen in the litterfall declined with altitude for all fractions, and the total annual input in the top plots was 76% of the annual total in the bottom plots. The differences in nitrogen input between the top plots and the low plots were significant for total (ANOVA $p<0.05$) and leaf ($p<0.01$) litterfall but differences between the top and middle plots were not significant.

Nitrification rates

The soils are more organic in the middle and top plots and hence bulk density lower than in the low plots (Table 6). The difference in bulk density (defined as mass of soil in unit volume, excluding stones and roots) between soils at these two altitudes can be attributed to the greater stoniness of the top plots.

Pigs and termites caused considerable damage to the samples left to incubate in the field particularly in the top plots where only nine out of thirty bags were undisturbed. Table 7 summarises the results from the soil analyses. The initial concentrations of extractable nitrate showed no significant differences between altitudes (one way ANOVA) and the results from the top plots showed the greatest variability. Nitrification rates were between 11.6 μg N g^{-1} 21 d^{-1} in the mid plots and 16.6 μg N g^{-1} 21 d^{-1} in the top plots but the differences were not significant (one way ANOVA). When the rates were expressed on an area basis the lowest rate was still found in the middle plots but because of the higher bulk density of the soils in the low plots the rate was highest there.

Bioassay

Climatic data from the two weather stations for the period of the bioassay are summarised in Figure 9. Maximum and minimum daily temperatures (Figure 9a) were higher at 60 m

Table 6. Bulk density (g cm^{-3}) and concentration of stones (>5mm) (g 100 cm^{-3} soil) in ten surface soil samples (10 cm x 10 cm area and 5 cm deep) at three altitudes on Bukit Belalong. Values are the means ±95% confidence limits of thirty samples.

Altitude (m)	Bulk density (g cm^{-3})	Stones (g 100 cm^{-3} soil)
200	0.619 ±0.06	8.62 ±4.96
500	0.448 ±0.06	1.28 ±1.13
850	0.398 ±0.05	12.82 ±5.82

Table 7. Concentrations of nitrate/nitrite nitrogen and nitrification rates in soil (0-5 cm) at three altitudes on Bukit Belalong. Values are the means ±95% confidence limits where applicable. n=23 at 200 m, 17 at 500 m and 9 at 850 m.

Altitude (m)	NO_2/NO_3-N (μg N g^{-1})	Nitrification rate (μg g^{-1} 21 d^{-1})	(kg ha^{-1} 21 d^{-1})
200	5.2 ±1.2	15.0 ±4.6	4.65
500	6.4 ±2.5	11.6 ±5.4	2.60
850	4.8 ±2.0	16.6 ±10.7	3.72

and the mean difference was 3.13°C for maxima and 1.95°C for minima. Solar radiation (Figure 9b) was generally higher at 913 m and the mean values were 14.47 MJ m^{-2} at 60 m and 17.03 MJ m^{-2} at 913 m. The Penman transpiration estimate (Figure 9c) followed a similar pattern to solar radiation and the total values were 51.5 mm at 60 m and 58.2 mm at 913 m.

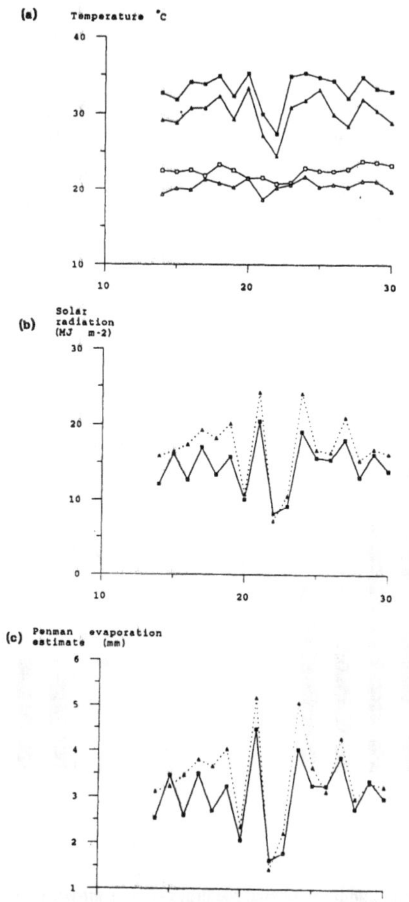

Figure 9. (a) Maximum (closed symbols) and minimum (open symbols) daily temperature (°C), (b) daily total solar radiation (MJ m^{-2}) and (c) Penman evaporation (mm) at Kuala Belalong, 60 m (▬) and Bukit Belalong, 913 m (▲) between 12 and 30 August 1991.

The results of the bioassay experiment are summarised in Figure 10. These results are reported fully in Wilkie (1992). There were substantial differences in the rice plants grown at the two altitudes. The shoot height (Figure 10a) was significantly reduced ($p<0.001$, three way ANOVA) for plants grown in all the treatments at 913 m compared with those at 61 m. Shoot mass (Figure 10b) showed a different response. In BS the plants grown at 913 m had increased shoot mass compared with those at 61 m whilst in SS the opposite trend was found.

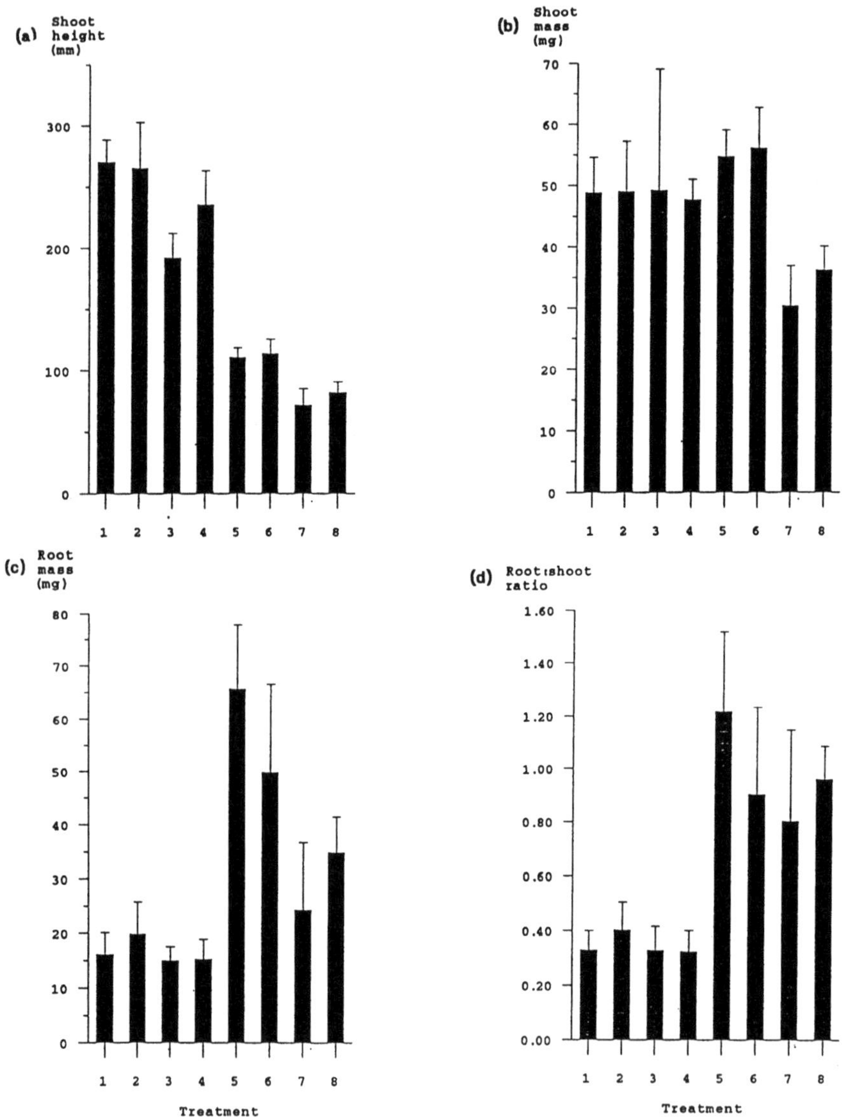

Figure 10. The response to nitrogen addition of rice seedlings (*Oryza sativa*) grown in two soils (Base soil, BS and Summit soil SS) at two altitudes on Bukit Belalong. (a) shoot height, (b) shoot mass, (c) root mass, (d) ratio of root mass : total mass of plant. Treatments: (1) BS, 61 m. (2) BS +N, 61 m. (3) SS, 61 m. (4) SS +N, 61 m. (5) BS, 913 m. (6) BS +N, 913 m. (7) SS, 913 m. (8) SS +N, 913 m. The columns represent the mean values for six replicates and the error bars are the 95% confidence limits.

The differences between the two altitudes were significant ($p<0.01$) only for the plants grown in SS without the addition of nitrogen (Treatments 3 and 7). Root mass (Figure 10c) was significantly increased ($p<0.001$) in plants grown in BS at 913 m compared with 61 m. The increases were not so marked for plants grown in SS and were significant ($p<0.01$) only for the fertilised plants (Treatments 4 and 8). There were large differences in the root:shoot ratios (Figure 10d) with higher values at 913 m ($p<0.001$, three-way ANOVA).

The addition of nitrogen had some effects, but these were slight compared with the effects of site microclimate and were only significant for shoot height in SS ($p<0.05$). Plants responded differently to fertilisation in the two soil types. Without addition of nitrogen shoot height was significantly smaller in SS compared with BS at both 61 m ($p<0.001$) and 913 m ($p<0.05$) but there were no significant differences between fertilised soils. At 61 m differences in shoot mass between control and fertilised treatments were negligible, but at 913 m fertilised plants had greater shoot mass in BS than SS ($p<0.05$) whilst there was no significant difference between shoot mass on unfertilised soils. There were also negligible differences in root mass between soils and treatments at 61 m but plants responded to fertilisation differently at 913 m. In the control pots root mass was significantly higher ($p<0.001$) in BS compared with SS but differences were no longer significant after fertilisation.

DISCUSSION

Climate

Gunung Silam, in Sabah, is similar in height (884 m) and latitude (5° N) to Bukit Belalong, but has very stunted forest at its summit and its climate also shows some significant differences. The most up-to-date account of the climate on Gunung Silam is contained in Bruinzeel *et al.* (1993). Mean temperatures on Bukit Belalong are c. 2°C lower than at similar altitudes on Gunung Silam and almost twice as much rain falls on Bukit Belalong as on Gunung Silam (c. 5000 mm vs c. 2700 mm). A cloud cap usually forms on Gunung Silam early each afternoon and consequently daily solar radiation totals at the summit are reduced by 34% compared with the base. A cloud cap does occasionally form on Bukit Belalong, there is no reduction in the mean solar radiation because on clear days radiation is higher at the summit than at the base. The cloud cap also reduces evaporation and leads to much wetter conditions and the stunted forest is characterised by thick growths of epiphytic and forest floor bryophytes. Occult precipitation is significant on Gunung Silam, but may also be important on Bukit Belalong. On several rain free evenings in March it was observed that fog in the early evening deposited moisture. The deposition was heavy enough to leave the forest floor damp for several hours in the morning. On colder, clear nights dew formation may also have been an important source of unrecorded precipitation. Epiphytic bryophytes are more abundant near the summit (though not to the same degree as on Gunung Silam) and terrestrial herbs are also considerably more abundant there, and this has been related to a moister environment (Poulsen, this volume).

Litterfall

Table 8 gives data from litterfall studies in other South East Asian primary forests. It can be seen that the values for the forests up to 500 m on Bukit Belalong are at the upper end of the range for lowland forests (exceeded only by Alluvial forest at Gunung Mulu) whilst the lower montane forest has a higher litterfall than any other montane forest from the region, for which litterfall was measured for a complete year (Proctor *et al.* (1983b) sampled litterfall on Gunung Mulu for only 16 weeks). The Belalong forests thus appear to be highly productive, but the results must be treated with caution since the study lasted for only one

Table 8. Leaf and total small litter production (t ha^{-1} yr^{-1}) and litter nitrogen concentrations (mg N g^{-1}) in South East Asian primary rain forests. 1; This study, 2; Yamada (1976), 3; Bruijnzeel (1984), 4; Lim (1978), 5; Edwards (1982), 6; Proctor *et al* (1989), 7; Proctor *et al.* (1983a), 8; Proctor *et al.* (1983b).

Location	Forest type	Altitude	Total litter		Leaf litter	
		(m)	(t ha^{-1} yr^{-1})	(mg N g^{-1})	(t ha^{-1} yr^{-1})	(mg N g^{-1})
Brunei, Belalong[1].	Lowland	200	10.7	14.5	7.7	11.4
	Lowland	500	10.4	12.7	7.5	10.6
	Lower montane	850	8.3	12.7	5.7	11.3
Java, Cibodas[2].	Lower montane	1550	6.0	-	4.5	-
Java, Pringombo[3].	Lower montane	650	7.4	-	5.8	11.2
Malaya, Pasoh[4].	Lowland	100	8.9	11.0	6.4	12.0
New Guinea, Kerigomna[5].	Lower montane	2500	7.6	12.0	6.1	13.0
Sabah, Silam[6].	Lowland	280	6.5	13.0	3.9	12.6
	Lowland	330	7.4	10.4	4.5	10.8
	Lowland	480	5.2	11.5	3.4	11.2
	Lower montane	610	5.6	10.7	4.1	10.5
	Lower montane	790	5.5	10.8	3.7	10.5
	Lower montane (stunted)	870	4.8	8.6	3.3	8.3
Sarawak, Mulu[7].	Lowland (alluvial)	50	11.5	9.7	6.6	9.0
	Lowland	200	8.8	9.2	5.4	9.5
	Heath	200	9.2	6.0	5.6	5.7
	Forest over limestone	300	12.0	12.0	7.3	12.0
Sarawak, Mulu[8].	Lowland	500	10.9	10.0	6.7	10.0
	Lower montane	1310	11.0	7.8	5.7	8.1
	Upper montane	1860	3.6	7.9	2.3	7.4

year and longer studies have shown up to a two-fold variation in litterfall from year to year (Blasco & Tassy, 1975). After the dry period at the beginning of 1992 there was an increase in leaf litterfall at all altitudes, but especially in the mid plots where the peak value was 2.35 times the mean annual rate compared with 1.85 times higher in the low plots and 1.65 times higher in the top plots. The mid plots are sited close to a ridge top and may be more vulnerable to drought than the top plots despite the steeper slopes and shallow, stony soils of the latter (Table 6).

Litter production, but not its nitrogen concentration, is significantly lower in the top plots. The absence of changes in nitrogen concentrations with altitude on Bukit Belalong contrasts with studies on Gunung Mulu, Gunung Silam, Volcán Barva and in the Luquillo Mountains reviewed by Heaney & Proctor (1989). On all these mountains decline in forest stature was correlated with reduced litterfall nitrogen concentration and it was concluded that nitrogen is the element most likely to restrict ecosystem processes on wet tropical mountains. The data for Bukit Belalong thus reinforce these conclusions since it has a relatively large stature forest with an apparently adequate supply of nitrogen at the highest altitude. The high nitrogen concentrations in litterfall from the forest at 850 m on Bukit Belalong are striking when compared with the low concentrations from the stunted forest at 870 m on Gunung Silam (maximum height of emergent tree 16 m) and they resemble nitrogen concentrations in the tall montane New Guinean forest at 2500 m studied by Edwards & Grubb (1977) where the largest trees were 37 m tall.

Compared with other forests in South East Asia (Table 8) the litterfall from the forests at all altitudes on Bukit Belalong has relatively high nitrogen concentrations. The total small litterfall at 200 m on Bukit Belalong has the highest concentration of nitrogen in Table 8 and those at 500 m and 850 m are exceeded only in the forest at 280 m in Sabah (Proctor *et al.* 1989). Leaf litterfall nitrogen concentrations are also relatively high though lower than several values in Table 8.

Nitrification rates

Marrs *et al.* (1988) found considerable differences in nitrate availablity and production along an altitudinal transect on Volcán Barva, Costa Rica. In that study soil nitrate/nitrite-nitrogen concentrations decreased from 11.6 to 1.5 μg N g^{-1} soil between 100 m and 2600 m, and rates of nitrification also declined with increasing altitude.

On Bukit Belalong however, there was no significant difference in nitrate supply with altitude which is not unexpected since the climatic differences on Bukit Belalong are much less pronounced than on Volcán Barva. Soil moisture content partially regulates the rate of nitrification (Robertson 1989), with nitrate concentrations decreasing as a soil becomes wetter (Chandler 1985). An increase in soil moisture due to elevated humidity within a cloud cap (leading to the accumulation of mor humus or peat) has been recorded on mountains with very stunted forests (Whitmore 1984, Bruijnzeel *et al.* 1993), and the absence of a daily cloud cap on Bukit Belalong is reflected in the similarity between nitrate concentrations at all three altitudes.

The results for Bukit Belalong indicate low rates of nitrification compared with the Volcán Barva data (3.0 to 21.8 μg N g^{-1} soil 14 d^{-1}), but the results are within the ranges reported by Vitousek & Matson (1988) for undisturbed forest soils (6 to 25 μg N g^{-1} soil 10 d^{-1}), and are similar to rates measured by Chandler (1985) in soils from two Malaysian lowland forests (6.3 and 14.8 μg N g^{-1} soil 16 d^{-1}) and one 'sub-montane' forest (5.0 μg N g^{-1} soil 16 d^{-1}).

Bioassay

Nitrogen limitation of bioassay plants in soils from stunted Jamaican montane forests has been shown by Healey (1989). Significant differences were found between leaf and root mass of the grass *Melinis multiflora* Beauv. grown in lowland and montane soils, and the addition of nitrogen and phosphorus significantly increased growth of plants in soils from both upper montane and lower montane forests. This was in contrast to the results of Tanner (1977) whose bioassay on the same soils showed no such limitation. Healey attributed the conflicting results to flaws in Tanner's experimental design in which the test plants were grown at a single lowland site but not at a montane site with a climate similar to the forests from which the soils were collected. The bioassay in this study was replicated at two altitudes, but provided only slight evidence that rice plants are more nitrogen limited in montane soil compared with lowland soil, and whilst it would have been desirable to have extended the length of the experiment to ensure exhaustion of the seedling reserves, and possibly to have used a higher rate of nitrogen addition, the results are revealing. The large differences in partitioning between root and shoot at the two altitudes (Figure 10d) may be at least partly explained by the higher insolation and greater potential evapotranspiration at 913 m (Figures 9b, 9c). The long term data (Figure 4), however indicate no significant differences in solar radiation between altitudes. We are left with temperature changes as themost likely cause of the relatively slight differences between the forests at different altitudes on Bukit Belalong.

CONCLUSION

Litterfall nitrogen concentrations, soil nitrate concentrations and rates of nitrification do not indicate any reduction in nitrogen supply in the lower montane forest on Bukit Belalong compared with surrounding lowland forests, and the bioassay experiment gave little evidence of nitrogen limitation at higher altitude. The marked dry season peak of leaf litterfall in the mid plots compared with the top plots suggests that the montane forest is not more susceptible to drought than the lowland forest, and despite the high rainfall at the summit there are no indications that waterlogging might adversely affect growth. In the absence of a cloud cap and substantial differences in solar radiation, altitudinal effects on Bukit Belalong seem more likely to be related to differences in temperature than differences in other factors including nitrogen.

ACKNOWLEDGEMENTS

The authors would like to thank Universiti Brunei Darussalam and the Royal Geographical Society for the opportunity to participate in the 1991-1992 Brunei Rainforest Project, and the Leverhulme Trust and the Royal Geographical Society for financial support. We particularly thank Mr Peter Wilkie for his assistance with the bioassay and Ramlah anak Sapong for her help throughout the field work.

LITERATURE CITED

ASHTON, P.S. 1964. Ecological studies in the mixed dipterocarp forests of Brunei State. *Oxford Forestry Memoirs* 25.

BLASCO, F. & TASSY, B. 1975. Etude d'un ecosysteme forestier montagnard du Sud de l'Inde. *Bulletin d'Ecologie* 6:525-539.

BRUIJNZEEL, L.A. 984. Elemental concentration of litterfall in a lower montane rain forest in Central Java, Indonesia. *Malayan Nature Journal* 37:199-208.

BRUIJNZEEL, L.A., WATERLOO, M.J., PROCTOR, J., KUITERS, A.T., & KOTTERINK, B. 1993. Hydrological observations in montane rain forests on Gunung Silam, Sabah, Malaysia, with special reference to the *Massenerhebung* effect. *Journal of Ecology* 81:145-168.

CHANDLER, G. 1985. Mineralization and nitrification in three Malaysian forest soils. *Soil Biology and Biochemistry* 17:347-353.

EDWARDS, P.J. 1982. Studies of mineral cycling in a montane rain forest in New Guinea V. rates of cycling in throughfall and litterfall. *Journal of Ecology* 70:807-827.

EDWARDS, P.J. & GRUBB, P.J. 1977. Studies of mineral cycling in a montane rain forest in New Guinea. I. The distribution of organic matter in the soil and vegetation. *Journal of Ecology* 65:943-969.

FAO/UNESCO. 1974. *Soil map of the world* 1: 5 000 000. I. Legend. UNESCO, Paris.

GRUBB, P.J. 1977. Control of forest growth and distribution on wet tropical mountains: with special reference to mineral nutrition. *Annual Review of Ecology and Systematics* 8:83-107.

GRUBB, P.J. & TANNER, P.J. 1976. The montane forests and soils of Jamaica: a reassessment. *Journal of the Arnold Arboretum* 57:313-368.

HEALEY, J.R. 1989. A bioassay study of soils in the Blue Mountains of Jamaica. Pp. 273-283 in Proctor, J. (ed.). *Mineral Nutrients in Tropical Forest and Savanna Ecosystems*. Blackwell Scientific Publications, Oxford.

HEANEY, A. & PROCTOR, J. 1989. Chemical elements in litterfall in forests on Volcan Barva, Costa Rica. Pp. 255-271 in Proctor, J. (ed). *Mineral Nutrients in Tropical Forest and Savanna Ecosystems*. Blackwell Scientific Publications, Oxford.

LIM, M.T. 1978. Litterfall and mineral nutrient content of litter in Pasoh Forest Reserve. *Malayan Nature Journal* 30:375-380.

MARRS, R.H., PROCTOR, J., HEANEY, A. & MOUNTFORD, M.D. 1988. Changes in soil nitrogen mineralization and nitrification along an altitudinal transect in tropical rain forest in Costa Rica. *Journal of Ecology* 76:466-482.

MOORE, P.D. & CHAPMAN, S.B. 1986. *Methods in Plant Ecology*. (2nd edition). Blackwell Scientific Publications, Oxford.

PROCTOR, J. 1983. Tropical forest litterfall I. Problems of data comparison. Pp. 267-273 in Sutton, S.L., Whitmore, T.C. & Chadwick, A.C. (eds). *Tropical rain forest: ecology and management*. Blackwell Scientific Publications, Oxford.

PROCTOR, J., ANDERSON, J.M., FOGDEN, S.C.L. & VALLACK, H.W. 1983a. Ecological studies in four contrasting lowland rain forests in Gunung Mulu National Park, Sarawak. II. Litterfall, litter standing crop and preliminary observations on herbivory. *Journal of Ecology* 71:261-283.

PROCTOR, J., ANDERSON, J.M. & VALLACK, H.W. 1983b. Comparative studies on forests, soils and litterfall at four altitudes on Gunung Mulu, Sarawak. *The Malaysian Forester* 46:60-76.

PROCTOR, J., LEE, Y.F., LANGLEY, A.M., MUNRO, C.M. & NELSON, T. 1988. Ecological studies on Gunung Silam, a small ultrabasic mountain in Sabah, Malaysia. I. Environment, forest structure and floristics. *Journal of Ecology* 76:320-340.

PROCTOR, J., PHILLIPS,C., DUFF,G.K., HEANEY,A. & ROBERTSON,F.M. 1989. Ecological studies on Gunung Silam, a small ultrabasic mountain in Sabah, Malaysia. II. Some forest processes. *Journal of Ecology* 77:317-331.

ROBERTSON, G.P. 1989. Nitrification and denitrification in tropical forests. Pp. 55-69 in Proctor, J. (ed.). *Mineral Nutrients in Tropical Forest and Savanna Ecosystems*. Blackwell Scientific Publications, Oxford.

SUGDEN, A.M. 1986. The montane vegetation and flora of Margarita island, Venezuala. *Journal of the Arnold Arboretum* 67:187-232.

TANNER,E.V.J. 1977. Four montane forests of Jamaica: a quantitative characterization of the floristics, the soils and the foliar mineral levels and a discussion of the interrelations. *Journal of Ecology* 65:883-918.

VENEKLAAS, E.J. 1992. Litterfall and nutrient fluxes in two montane tropical rain forests, Colombia. *Journal of Tropical Ecology* 7:319-336.

VITOUSEK, P.M. & SANFORD, R.L.Jr. 1986. Nutrient cycling in moist tropical forest. *Annual Review of Ecology and Systematics* 17:137-167.

VITOUSEK, P.M. & MATSON, P.A. 1988. Nitrogen transformations in a range of tropical soils. *Soil Biology and Biochemistry* 20:361-367

WHITMORE, T.C. 1984. *Tropical Rain Forests of the Far East*, (2nd edition). Clarendon Press, Oxford.

WILKIE, P.W. 1992. *A Bioassay to Investigate the Causes of Altitudinal Zonation on Bukit Belalong, Brunei*. Unpublished BSc thesis, University of Stirling, Stirling, U.K.

YAMADA, I. 1976. Forest ecological studies of the montane forest of Mt. Pangrango, West Java. III. Litterfall of the tropical montane forest near Cibodas. *South east Asian Studies* 14:193-229.

Soil conditions, erosion and nutrient loss on steep slopes under mixed dipterocarp forest in Brunei Darussalam

S.M. ROSS[1] *and* A. DYKES[2]

[1]*Department of Geography, University of Bristol, United Kingdom*
[2]*Department of Geography, King's College, London, United Kingdom*

ABSTRACT. Soil physical and chemical characteristics are reported under virgin, mixed dipterocarp forest growing on steep slopes in the Temburong District of Brunei Darussalam. Hydrological, sedimentological, soil organic matter and nutrient losses from experimental erosion plots are presented for both wet and dry season conditions. Total annual runoff losses from bare soil surfaces under an intact rainforest canopy exceed 107 litres per hectare, while total annual sediment losses exceed 46 tonnes per hectare. These values are approximately 6 and 20 times higher respectively than those measured from the undisturbed forest floor, comprising leaf litter layer and root mat. The combined root mat and litter layers are more important in protecting the soil surface against sediment loss than they are in reducing overland flow. These high values indicate how serious runoff and soil erosion can be on such steep slopes with very low permeability clay textured soils. Losses of soil nitrogen, phosphorus and potassium in eroded sediments are serious, particularly from the bare soil treatment, where 388 kg N ha^{-1}, 13 kg P ha^{-1} and 196 kg K ha^{-1} were lost in twelve months.

KEY WORDS: orthic acrisols, soil chemistry, sediment loss, throughfall, sediment yields, Borneo.

INTRODUCTION

Brunei Darussalam is one of the few countries in SE Asia and in the tropics generally where there has been no widespread felling of forests either for logging or for shifting cultivation. Around 81% of the land remains forested, with as much as 59% under primary forest (Anderson & Marsden 1988, Collins *et al.* 1991). Under these circumstances, field experimentation and discussions of geomorphological and soil effects of large scale deforestation are inappropriate for the study region in the Temburong District of Brunei which is characterised by dense dipterocarp forest clothing steep river valleys. Instead we report here the results of a series of field experiments designed to elucidate detail about soil erosion processes under the forest canopy.

Evidence from previous soil erosion and soil process studies under tropical rain forest have indicated that the forest canopy, the shrubs, herbs and the leaf litter layer at the soil surface all act to intercept raindrops and to dissipate their energy (e.g. Wiersum 1984). Where raindrop energy is dissipated directly at the mineral soil surface, soil particles are detached and moved downslope. Whether or not soil particles are eroded from the soil surface during rainstorms is determined by three main factors: (i) rainfall intensity, (ii) soil surface conditions (iii) soil hydrological conditions. Physical conditions of the surface soil determine how much water infiltrates into the soil profile and how much water flows away over the soil surface as overland flow or runoff. Soil texture and soil structure, which determine soil bulk density, antecedent soil moisture conditions, as well as the presence of leaf litter at the soil surface, are the main controlling factors at the forest floor.

In our experimental plots on the West Ridge of the Belalong valley, our aim was to study the processes of soil hydrology and erosion under the intact dipterocarp canopy. In particular, our objective was to examine the roles of the root mat and leaf litter at the soil surface in partitioning incoming throughfall between overland flow over the soil surface and

D. S. Edwards et al. (eds.), Tropical Rainforest Research – Current Issues, 259–270.

infiltration into the soil profile. A subsequent aim was to quantify sediment, organic matter and nutrient losses in erosion from the three treatments.

In this paper we firstly report details of the soil conditions in the Belalong valley under mixed dipterocarp forest. Secondly, we describe field experiments designed to aid our understanding of soil surface hydrology and erosion processes. Finally we compare our results to those reported elsewhere in rain forests of SE Asia.

SOIL PROFILE CHARACTERISTICS

The steep slopes and ridges of the Temburong District of Brunei are covered by relatively thin, silty clayey soils, developed over weathered and permeable shales. The entire region shows much evidence of small and large scale landslides, manifest in the forest by treefall gaps. These breaks in the canopy allow direct rainfall-fed soil erosion. Soil survey work by Ashton (1964) and his soil physical and chemical analysis indicate that the soils of the region are relatively nutrient poor, of varying depth and rather impervious, due to their high clay content. Ashton (1964) identified four soil types in the Belalong region:

(i) Ridge soils, developed on sandstone, described as yellow clay latosols.
(ii) Upper slope shale lithosols - shallow and stony.
(iii) Lower Hillslope clay lithosols - two types:
  a. steep slopes, often just below ridges, little litter, often signs of surface erosion.
  b. more gentle slopes, deeper soils, often sticky and humid, with leaf litter at the soil surface.

Our descriptions of midslope soils of the western and eastern valley slopes are given in Tables 1a and 2a respectively. Of the 12 soil profiles excavated and described in detail from various slope locations up to 913 m, all show a lack of organic matter beneath the immediate soil surface and a relatively friable topsoil containing numerous weathered soft shale clasts. Both examples reported here would appear to represent Ashton's gentler slope clay lithosol category, with sticky, moist surface conditions after rainfall, a complete cover of leaf litter at the soil surface and weathered shale fragments throughout the profile. Soil physical and chemical characteristics for the two example profiles are given in Tables 1b and 2b respectively. The downprofile particle size data indicate relatively sandy textured topsoils,

Table 1a. Soil profile description, Hydrology Pit No. 6, West Ridge.

General Description: Altitude 110m above the R. Belalong. 30° slope; aspect 65°. Tree canopy 60-70 m, with light sapling understorey. 100% superficial cover of leaf litter, averaging 2-3 cm depth.

Soil Profile:	Two horizons identified.
0-4/7cm	Very fine crumb (granular) structure with tiny fragments of leaf litter. Silty clay texture. Colour: between 5YR 3/3 and 5YR 3/2. Visibly porous and friable with numerous small pores <1mm. Root mat at the soil surface with numerous fine roots up to 1mm diameter. No true mottles. A very few tiny stones <2mm diameter on the soil surface. Diffuse and regular boundary to:
>7cm	Medium angular blocky structure about 5cm diameter. Silty clay texture. Colour 10YR 5/8 (yellowish brown). Few visible pores with no cracks. Frequent medium sized roots (<4mm) to 17cm, but very infrequent below. 3 large roots (5cm diameter) at 10cm depth. Stones frequent below 35cm, including large (10-15cm diameter) weathered shale fragments, which make up 40% of the profile below 35cm depth.

Table 1b. Soil profile characteristics, Hydrology Pit No. 6, West Ridge.

Depth (cm)	%LOI	Particle Size (%)			Bulk Density (g cm^{-1})
		sand	silt	clay	
0-4	35.4	53.9	28.3	17.8	0.703
5-10	18.1	27.2	42.5	30.3	1.180
>25	12.2	6.5	39.4	54.0	1.193

Soil Total Nutrients (mg g^{-1})

Depth (cm)	N	P	K	Na	Ca	Mg
0-4	4.65	0.36	9.73	1.86	0.15	0.08
5-10	2.27	0.26	11.2	2.49	0.17	0.23
>25	0.50	0.20	13.57	2.90	0.05	0.03

Table 2a. Soil profile description, Rattan Plot No. D10, East Ridge.

General Description: Altitude 180m. 27° slope, aspect 275°. Tree canopy 60-70 m, with very dense throughstorey. Understorey of saplings, palms and rattans to a height of 1.5 m. 100% leaf litter cover at the soil surface, averaging 1-2 cm depth. 20-30% woody litter cover, no bare soil.

Soil Profile:	Three horizons identified:
0-2.5cm	Very superficial organic horizon, with numerous roots up to 1 mm diameter. Much very finely fragmented litter mixed into the soil. Very friable silty clay; matrix colour 10YR 4/4 (dark yellowish brown). No stones. Occasional larger roots up to 2cm diameter. Diffuse, regular boundary to:
2.5-35cm	Medium angular blocky structure, peds 5cm. Cracks between peds up to 1mm wide. Ped faces not visibly porous. Silty clay loam, matrix colour 10YR 6/6 (brownish yellow). Frequent roots up to 8mm diameter. No stones. Very diffuse and irregular boundary to:
>35cm	Poor angular blocky structure of medium size; ped up to 5cm across. Fairly consolidated silty clay matrix plastered around numerous shale fragments up to 20cm across. Stones make up 70-80% of the horizon. No crack between peds or between matrix and stones. Not porous. Orange/brown bands (2-3cm wide) of deposited iron oxides. Matrix colour 10YR 6/6 (brownish yellow); pale weathered shale 10YR 8/4 (very pale brown); deposited iron oxides 7.5 YR 6/8 (reddish yellow). Infrequent roots to a depth of 50cm.

changing to very clayey textures below 5 cm depth. All loss on ignition percentages are higher than the visible organic matter content would suggest. This may be interpreted as weight loss due to interstitial water in weathered secondary clays, or to volatilisation of carbonates or salts. The very low calcium and magnesium concentrations, however, would suggest that the soils are not high in these compounds. As expected, our soil chemical analysis indicates higher total soil nitrogen and phosphorus in the surface 0-5 cm where the soil organic matter contents are higher. All soils are poor in both calcium and magnesium, with much higher potassium content, reflecting intense weathering and the presence of potassium in secondary clay minerals.

According to the FAO-Unesco (1974) world soil classification, the Belalong soils are classed as orthic acrisols. This means that they are relatively poor in bases and have an argillic (a clay-enriched) B-horizon. Determinations of cation exchange capacity and percentage base saturation are in progress and will be used to assess how typical are the Belalong soils of this class.

Table 2b. Soil profile analytical characteristics, Rattan Plot No. D10, East Ridge.

Depth (cm)	%LOI	Particle Size (%)			Bulk Density (g cm^{-1})
		sand	silt	clay	
0-2.5	28.0	59.9	23.5	16.6	na
2.5-10	9.8	14.4	43.9	41.7	na
>35	6.58	25.5	32.3	42.2	na

Soil Total Nutrients (mg g^{-1})

Depth (cm)	N	P	K	Na	Ca	Mg
0-2.5	4.75	0.51	12.66	2.80	0.20	0.80
2.5-10	0.98	0.31	12.80	3.48	0.14	1.01
>35	0.03	0.24	15.68	4.22	0.13	1.10

FIELD EXPERIMENTS ON SEDIMENT LOSS

Materials and Methods

(a) Field Site and Methods

Soil hydrology and erosion studies were carried out 0.5 km from the Kuala Belalong Field Studies Centre (KBFSC) in the Temburong District of Brunei (Figure 1). The plots were located on a North Easterly facing slope, described elsewhere as the West Ridge.

Hourly rainfall measurements were monitored during the entire study period at Kuala Belalong (Dykes, this volume), 1.5 km from the forest study site. Throughfall measurements were made daily using 19 simple funnel collectors, 25 cm in diameter, which were randomly relocated every week in the soil hydrology study area of 1500 m^2, during three and a half months between early November 1991 and mid February, 1992, thus including both wet and dry season conditions. In order to estimate throughfall from measured rainfall, fifty two throughfall collection dates were used to calculate an empirical relationship between rainfall and throughfall. To measure runoff generation and sediment losses under the forest canopy, three treatment plots were constructed at a height of 110 m above the R. Belalong, on the west slope of valley, above the KBFSC. In the three treatments, we aimed to create three different soil surface conditions which would allow us to elucidate the roles of leaf litter and the root mat in surface soil hydrology and sediment loss processes. The treatments were: (T1) intact forest floor, retaining the surface root mat and fallen leaf litter, (T2) soil with surface root mat intact and all leaf litter removed, and (T3) completely cleared and bare soil surface. These Gerlach-type plots were 5 m long by 2.5 m wide, bounded by 20 cm timber edges, with plot slope angles of 30°. Remoteness from easy access, laboratory analysis and supplies limited our ability to replicate the experiment. Results and discussion acknowledge the lack of replication and small size of runoff plots. Runoff collection was measured daily or after major storms at the base of each plot and sediments were collected at the same times using very fine mesh bags inserted into the runoff collectors. Runoff volumes reported in this paper are currently uncorrected for direct precipitation into the collection vessels. After collection, the sediments were sieved into 5 size fractions: <0.25 mm, 0.25-1 mm, 1-4.75 mm, 4.75-11.2 mm and >11.2 mm. All fractions were then oven dried at 105°C, weighed and transported to the UK for chemical analysis.

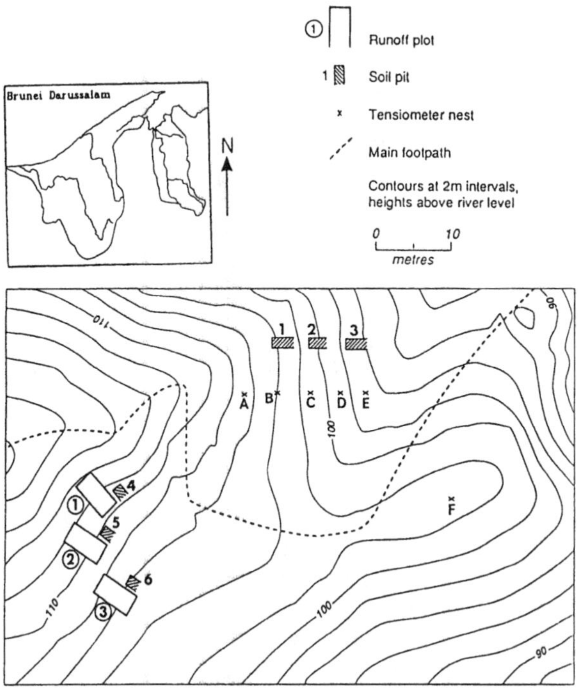

Figure 1. Map showing field study site and field plot locations in the Belalong valley, Temburong District, Brunei Darussalam.

(b) Laboratory Analysis.

Sediment samples were hand sorted into organic and mineral fractions, then weighed to determine proportional organic matter contents. Loss on ignition determinations were also carried out on the sediment samples, using a 24 hour ignition period at 450°C. Total N, P and K in eroded sediments were determined using a microdigestion in concentrated sulphuric acid with hydrogen peroxide (Allen 1989), followed by automated colorimetric determinations for total nitrogen as ammonium (Crooke & Simpson 1971) and phosphorus (Murphy & Riley 1962), and using atomic emission spectrophotometry for the determination of potassium.

RESULTS

Rainfall, Interception and Throughfall Characteristics

Monthly rainfall totals for Kuala Belalong (Figure 2) indicate that September to December marked a distinct wet season, despite rather low rainfall in October, while January to April marked a distinct dry season. Rainfall comparisons with measured throughfall amounts showed a good correlation of r = 0.989. Interception and percentage throughfall measurements for tropical rainforests have shown great variability in quantity and distribution of 'net rainfall'. Bruijnzeel (1990) for example comments on more than 70 studies from natural rain forest sites whose results differ greatly and cites several climatic, vegetative and procedural reasons for this. Calculated on the basis of the throughfall study period only,

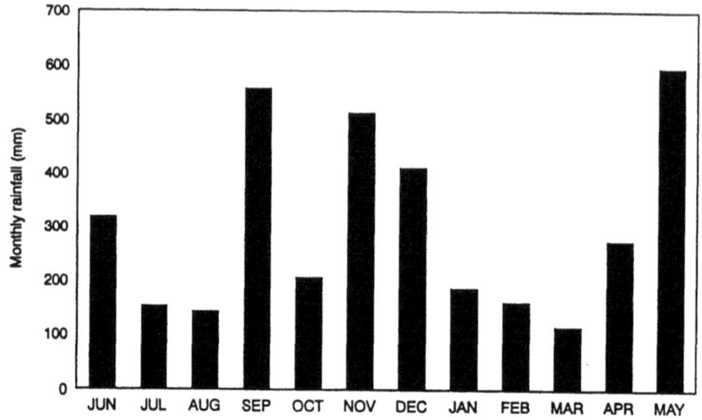

Figure 2. Monthly rainfall totals recorded at the automatic rainstation at Kuala Belalong, June 1991 - May 1992.

interception accounted for 20% of rainfall, 1% was assumed to reach the forest floor in stem flow, and the remaining 79% of rainfall was collected as throughfall. The throughfall patterns will be discussed in detail elsewhere. Calculated over the 12 month field project period, a range of 3-93% of rainfall was intercepted by the canopy of mixed dipterocarp forest at Belalong. This wide range of values can be accounted for by variations in the rainfall amounts and intensities, combined with differences in canopy structure and foliage throughout the year. We shall consider only throughfall, or net rainfall, as pertinent to our runoff and erosion studies since the forest canopy above the erosion plots remained completely intact throughout the study period.

Runoff Generation

The relationship between throughfall amounts and runoff volumes for the three plots are given in Table 3. All three treatments show extremely good r^2 values. Very similar throughflow/runoff patterns were observed for T2 and T3, both of which showed runoff rates around six times higher than the intact root mat and litter soil surface. Total annual runoff quantities, calculated on the basis of six months of data: three months representing the 'wet' season (November - January) and three months representing the 'dry' season (February - April), are given for the three treatments in Table 3. Calculated as a percentage of throughfall, runoff percentages were 0.26-4.97% for T1, 0.3-7.1% for T2 and 0.75-6.98% for T3. These low values agree well with those calculated for runoff generation elsewhere in Borneo. In undisturbed forest in Sabah, Sinun *et al.* (1992) reported runoff percentages

Table 3. Relationship between actual throughfall and runoff volumes and between runoff volumes and sediment yield for three surface soil treatments under Mixed Dipterocarp Forest (based on seven months of measurements: October 1991 - May 1992). Th = Throughfall; Rv = Runoff volume; Sy = Sediment yield.

Treatment	Actual Throughfall vs Runoff Volume	r^2	Runoff Volume vs Sediment Yield	r^2
Intact forest floor (root mat + litter) (T1)	Rv = -5.76 + 1.01 Th	84.0	Sy = 17.5 + 0.484 Rv	39.3
Root mat only (T2)	Rv = -38.3 + 5.97 Th	95.3	Sy = 11.1 + 1.79 Rv	71.1
Bare soil (T3)	Rv = -24.5 + 6.00 Th	94.6	Sy = -14.3 + 3.37 Rv	62.5

of 2-3%; while Douglas *et al.* (1992), working in the same area, quotes runoff percentages of 5%. Only on logging tracks were higher runoff percentages of 55% reported. Root mat and bare soil plots showed significantly lower runoff percentages in the very dry season months of March and April. Seasonal patterns in runoff volumes follow the seasonal rainfall patterns (Figure 2) very closely. Greatest runoff volumes were generated in the peak wet season months of November and December 1991, with increasing runoff volumes again apparent in May 1992 (Figure 3). In all months, T1 showed only small quantities of runoff, reaching a maximum at 31 litres/m^2 for the month of November 1991, compared to 135 and 151 litres/m^2 for T2 and T3 respectively.

Sediment loss

Total annual mineral and organic sediment losses, calculated on the basis of six months of data; three months representing the 'wet' season (November - January) and three months representing the 'dry' season (February - April), are given for the three treatments in Table 4. Combining amounts of mineral and organic material, a calculated total of more than 46 tonnes per hectare of sediments were eroded from bare soil under the forest canopy. This amounted to 18.6 times more than from the intact forest floor. Regression equations for sediment yield on runoff volumes are presented in Table 3. Both the removal of leaf litter (T2) and the removal of the combined root mat and litter (T3) cause incremental increases in sediment loss. This indicates that both leaf litter and root mat play an important part in sediment retention. The strength of relationship between runoff volume and sediment loss over the study period is very good for T2 and T3, showing r^2 values of 71.1% and 62.5% respectively. The relationship between runoff and sediment loss from the undisturbed forest floor (T1) is poorer at $r^2 = 39.3$. This can be explained by the same argument given for variability in canopy interception of rainfall - namely that the variability is due to different quantities and intensities of net rainfall coupled with different litterfall rates and root growth dynamics throughout the year. Seasonal patterns of sediment loss follow the seasonal runoff patterns (Figure 3) very closely. Greatest sediment losses, like runoff, occur in the peak of the wet season (Figure 4). Increases in sediment losses in May 1992 are not as great as the increases in runoff volumes. There are several possible explanations for this observation: (i) that much of the easily detachable surface soil particles had already been eroded, (ii) that

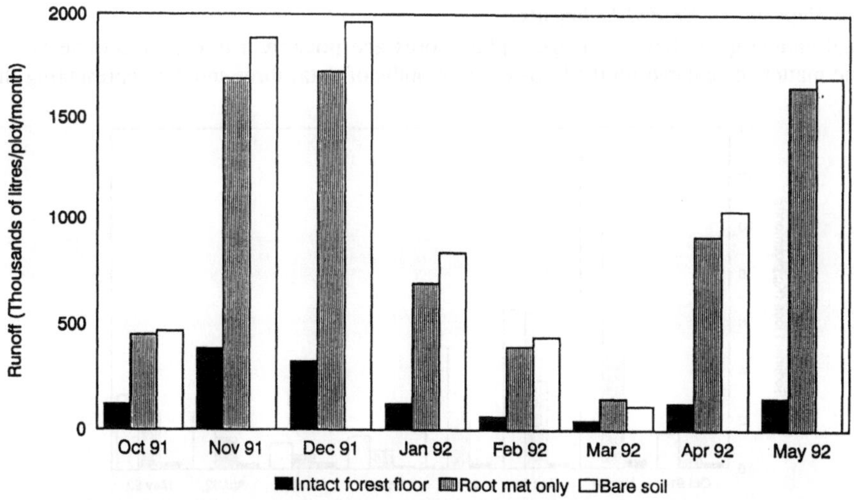

Figure 3. Monthly runoff totals for treatment plots T1 (intact forest floor), T2 (root mat only) and T3 (bare soil) (all values in thousands of litres per plot).

Table 4. Total runoff and eroded mineral sediments and organic litter under Mixed Dipterocarp Forest soil surface treatments (calculated in kg ha^{-1} y^{-1} from six months of field data).

Treatment	Runoff (l ha^{-1}) x 10^6	Mineral Sediment Loss (kg ha^{-1} y^{-1})	Organic Litter (kg ha^{-1} y^{-1})
Intact Forest Floor (T1) (Root mat + Litter)	0.913	1198	1308
Root mat only (T2)	4.715	23070	868
Bare soil (T3)	5.245	45689	936

previous wet season rainsplash followed by dehydration during the dry season caused surface crusting which enhances runoff but reduces sediment loss, or (iii) that active root growth at the beginning of the 1992 growing season as the soil is wetting up after the dry conditions of February, March and April, stabilised the surface soil. The last suggestion seems untenable since T3 (bare soil surface) also showed little sediment loss and it was not protected by a root mat.

Leaf Litter Removal

Very clear differences were observed in the amounts of leaf litter eroded from treatment plots. The actual quantities of leaf litter compared to mineral sediments lost for one example collection date in each month of the study period are illustrated in Figure 5. The organic matter data are calculated for hand-sorted litter > 11.2 cm in size. Substantial quantities of leaf litter fell onto treatment plots T2 and T3, and were eroded before they could be cleared by hand. These litter losses undoubtedly made significant contributions to nutrient losses described below. Proportionally larger quantities of leaf litter were found in eroded sediments in March and April 1992, at the end of the dry season, after the main period of leaf fall. During these two months, the material eroded from T1 consisted almost entirely of leaf litter. This is an interesting result since it illustrates that significant loss of soil organic matter and nutrients bound up in leaf litter can be eroded from steep forested hillslopes even without any canopy removal.

Loss of Nutrients in Eroded Sediments

Total annual quantities of nitrogen, phosphorus and potassium in eroded sediments and organic matter, calculated on the basis of six months of data; three months representing the

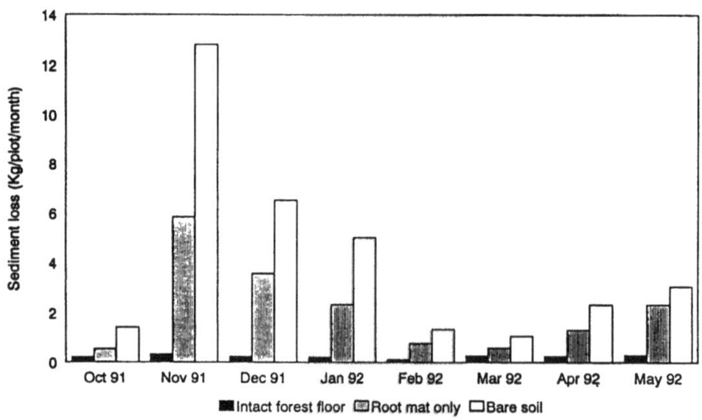

Figure 4. Monthly sediment loss totals for treatment plots T1 (intact forest floor), T2 (root mat only) and T3 (bare soil) (all values in kg per plot).

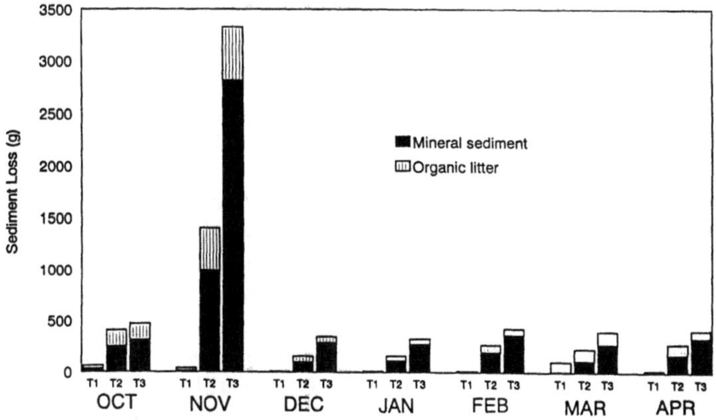

Figure 5. Quantities of mineral sediment and organic litter eroded from three treatment plots on three example collection dates selected from each month of the study period. T1 = Intact forest floor, T2 = root mat only, T3 = bare soil (all values in g per plot).

Table 5. Total Nitrogen, Phosphorus and Potassium losses in eroded sediments under Mixed Dipterocarp Forest soil surface treatments. (Calculated in kg ha^{-1} y^{-1} from six months of field data).

Treatment	Nutrient loss (kg ha^{-1} y^{-1})		
	N	P	K
Intact Forest Floor (T1) (Root mat + Litter)	29.02	1.454	20.066
Root mat only (T2)	131.57	8.545	106.400
Bare soil (T3)	388.324	13.077	195.79

'wet' season (November-January) and three months representing the 'dry' season (February-April), are given for the three treatments in Table 5. The amounts of all three nutrients lost from a cleared bare soil surface are very high. For nitrogen, an extremely high 388 kg N ha^{-1} y^{-1} was lost, amounting to 13 times that lost from the intact forest floor. Nine times more phosphorus and ten times more potassium were lost from the bare soil plot than from the intact forest floor.

DISCUSSION

The complexity and dynamism of physical, chemical and biological activity in the forest floor of tropical rain forests is just being realised. Coelho Netto's (1987) description of the rain forest litter layer also describes our conditions exactly: "...a surface layer, representing the current fall of leaves which show little evidence of decay, and a second layer comprising all partially decomposed material above the mineral soil. Fine roots with variable mat thickness overlie the mineral soil, being easily peeled back from the soil like a carpet."

Leaf litter and roots at the soil surface play several roles in the physical and chemical processes occurring on the forest floor. Firstly, they provide both absorption and channelling of incoming throughfall, which can result in higher infiltration rates and lower quantities of runoff. Secondly, litter prevents rainsplash at the mineral soil surface. Litter and leaf fragments can, however, become entrained within the surface mineral soil and within transported slope sediments. Thirdly, organic litter is the main source of recyclable nutrients for feeding roots. Thus, if litter is removed from the site by erosion, it may result in significant nutrient depletion.

The exceedingly steep 30° slopes of the Belalong valley provided an unusually extreme testbed for our experiments on the role of soil surface leaf litter and roots in protecting the soil under mixed dipterocarp rain forest from erosion.

Since several studies confirm the theory that the kinetic energy of rain forest throughfall exceeds that of above-canopy rainfall (e.g. Brandt 1988, Vis 1986), our experiments were designed to examine the role of protective litter layers in dissipating this energy. Other authors have alluded to the role of leaf litter as a dissipator of rainsplash energy and have shown how sediment losses can be enhanced when the litter layer is removed (eg. Ross *et al.* 1990). Their Brazilian site, however, had no significant root mat at the soil surface. Since runoff from T2 and T3 plots at Belalong are very similar, but sediment losses from T3 are significantly higher than from T2, it appears to be the litter rather than the root mat which provides the main protection for the mineral soil and prevents rainsplash and particle detachment.

Working in Atlantic rain forest on silty clay latosols near Rio de Janeiro, Coelho Netto (1987) identified three routes of water movement at the soil surface: (i) over-litter flow, (ii) through-litter flow and (iii) within the root litter carpet. All three routes are likely to occur during the heaviest rainstorms at Belalong. Although we did not specifically measure the flows associated with the litter layers, they must have provided significant transporting power since as much as 130 g (oven dry weight) of leaf litter per m^2 was calculated to have been removed by erosion processes in a year from the intact forest floor. While no other quantification of litter transport is available for tropical rain forest conditions Van Zon (1980), reported very significant leaf transport from hillslopes under temperate beech, pine and spruce forests in Luxembourg. It is interesting to compare the quantities of litter lost in erosion to rates of litterfall and litter decay measured elsewhere for similar dipterocarp forest in the Danum valley in Sabah. Burghouts *et al.* (1992) report litterfall and litter disappearance (kL) rates for primary forest ranging from 3 - 11 t ha^{-1} y^{-1}. The 1.3 t ha^{-1} y^{-1} litter erosion rates recorded in this study indicate that erosion could account for loss of as much as 40% of the organic matter on the forest floors of some dipterocarp forests with lower litterfall rates. Young feeding roots actively grow through and into decaying leaf litter at the soil surface since it is the main location of nutrient release and recycling in tropical forests. The roots anchor the litter to the surface in the vicinity of the mother tree, providing an easily accessible supply of nutrients. Under these conditions, the scale of litter transport from the intact forest floor even on very steep slopes at Belalong was something of a surprise. Malaysian dipterocarp litterfall amounts are around the lowest of any tropical forests (Bray & Gorham 1964). The loss of nutrients in eroded litter undoubtedly contributes to the already low fertility of soils on these steep slopes.

The sediment yields recorded for intact forest in this study are of the same order of magnitude as values reported elsewhere in SE Asia, despite different methods of measurement and calculation. Table 6 lists a range of different sediment loss determinations for tropical rain forest in SE Asia. The higher sediment yields recorded for bare soil under the forest canopy at Belalong are more similar to rates reported by Morgan *et al.* (1982) for secondary forest, also on steep, 25°, slopes in Sarawak, at 35-462 kg ha^{-1} y^{-1}. Morgan *et al.* (1982) describe both primary and secondary forests as having "thin" soil litter layers, so the litter layer itself may not be the main influencing factor here. Soil organic matter plays an important role in stabilising the soil structure. When the organic matter is removed, aggregate disruption makes particles more easily detached by raindrop splash. Wiersum (1984), summarising the work of many studies on clean weeded forests and forests whose

Table 6. Sediment yield in a range of tropical rain forest locations in SE Asia, using three estimation techniques: (a) Catchment Measurements, (b) Erosion Plot Measurements, (c) Numerical Simulation Modelling. All values in t $ha^{-1}y^{-1}$.

Study Location	Sediment Yield (t ha^{-1} y^{-1})	Author
(a) Catchment Studies		
Sg. Tenom, Cameron Highlands	0.34-0.35	Shallow (1956)
Sg. Gombak, Selangor	0.97	Douglas (1975)
Ulu Segama, Sabah	3.12	Douglas *et al.* (1992)
(b) Erosion Plot Studies		
Semongok, Sarawak	0.035-0.0247	Morgan *et. al.* (1982)
Ulu Segama	0.16-0.29	Sinun *et. al.* (1992)
Belalong, Brunei		This study
Intact forest Floor	2.42	
Root mat only	24.05	
Bare Soil	461.2	
(c) Numerical Simulation Studies		
Semongok, Sarawak	0.277	Morgan *et. al.* (1982)
Kalimantan	0.0085	Besler (1987)

litter had been removed or burnt, showed that sediment yields could be increased by around 20-200 times if the treatments were applied repeatedly.

Our very simple experiments at Belalong have indicated that even very minimal disturbance of the forest system, by altering the forest floor conditions, can have a deleterious effect on soil quality. These relatively small scale soil surface mass movements contribute to both slope form and sediment loading to streams. The steep slopes of this part of Brunei Darussalam show frequent small landslides, controlled in part by shallow soils and by reduced soil strength under wet conditions. The presence of both leaf litter and particularly the root mat have an effect on soil strength, through altering surface soil moisture conditions and by welding soil particles and aggregates together. Further work on the Belalong sites is elucidating how surface soil conditions affect infiltration capacity and influence mass movements at all scales.

SUMMARY OF CONCLUSIONS

1. The combined root mat and litter layers are more important in protecting the soil surface against sediment loss than they are in reducing overland flow.
2. Around 20 times more sediment is eroded from bare soil under a dipterocarp canopy than from the intact forest floor.
3. The mass:mass ratio of organic matter to mineral sediments lost from the bare soil surface was 1:49 compared to that of the intact forest floor at 1:0.9.
4. For low litterfall dipterocarp forests, erosion of leaf litter, particularly on steep slopes, may cause up to 40% depletion of incoming soil organic matter.
5. Almost 400 kg N ha^{-1} y^{-1} and 200 kg K ha^{-1} y^{-1} can be lost from the bare soil surface under the forest canopy.

ACKNOWLEDGEMENTS

The authors gratefully acknowledge the logistical support of The Royal Geographical Society and the University of Brunei Darussalam, particularly Dr David Edwards, Catriona Prebble,

Samhan bin Nyawa, Dr Kamariah Abu Salim, Amanda Simpson and John Wills. The financial support of the Greencard Trust, Lloyds Bank, Royal Brunei Airlines and Brunei Shell is also gratefully acknowledged. We are particularly indebted to numerous Iban and British colleagues who assisted both in field construction and sampling, particularly Kuni Anggat. We also wish to acknowledge the laboratory assistance in the UK of Aysen Brown and Bob Iles. Figure 1 was drawn by the Cartographic Unit of Kings College, Geography Department, London.

LITERATURE CITED

ALLEN, S.E. 1989. *Chemical Analysis of Ecological Materials*. Blackwell Scientific Publications, Oxford, UK.

ANDERSON, J.A.R. & MARSDEN, D. 1988. *Brunei Forest Resources and Strategic Planning Study*. Unpublished report to the Government of Brunei (quoted by Collins *et al.* 1991).

ASHTON, P.S 1964. *Ecological Studies in the Mixed Dipterocarp Forests of Brunei State*. Oxford Forestry Memoirs No.25. Clarendon Press, Oxford, UK.

BESLER, H. 1987. Slope properties, slope processes and soil erosion risk in the tropical rain forest of Kalimantan Timur (Indonesian Borneo). *Earth Surface Processes and Landforms* 12:195-204.

BRANDT, J. 1988. The transformation of rainfall energy by a tropical rain forest canopy in relation to soil erosion. *Journal of Biogeography* 15:41-48.

BRAY, J.R. & GORHAM, E. 1964. Litter production in forests of the world. *Advances in Ecological Research* 2:101-157.

BRUIJNZEEL, L.A. 1990. *Hydrology of Moist Tropical Forests and Effects of Conversion: A State of Knowledge Review*. Netherlands IHP Committee for UNESCO and IAHS.

BURGHOUTS, T., ERNSTING, G., KORTHALS, G. & DE VRIES, T. 1992. Litterfall, leaf litter decomposition and litter invertebrates in primary and selectively logged dipterocarp forest in Sabah, Malaysia. *Philosophical Transactions of the Royal Society, London, B* 335:407-416.

COELHO NETTO, A.L. 1987. Overland flow production in a tropical rainforest catchment: the role of litter. *Catena* 14:213-231.

COLLINS, N.M., SAYER, J.A. & WHITMORE, T.C. (eds) 1991. *The Conservation Atlas of Tropical Forests: Asia and the Pacific*. MacMillan and the World Conservation Union.

CROOKE, W.M. & SIMPSON, W.E. 1971. Determination of ammonium in kjeldahl digests of crops by an automated procedure. *Journal of the Science of Food and Agriculture* 22:9-10.

DOUGLAS, I. 1975. The impact of urbanisation on river systems. *Proceedings of the IGU, Regional Conference, 8th New Zealand Geographical Conference* pp. 307-317.

DOUGLAS, I., SPENCER, T., GREER, T., BIDIN, K., SINUN, W., & WAI WENG, W. 1992. The impact of selective commercial logging on stream hydrology, chemistry and sediment loads in the Ulu Segama rain forest, Sabah, Malaysia. *Philosophical Transactions of the Royal Society, London, B* 335:397-406.

FAO-UNESCO 1974. *Soil Map of the World, 1:5000000*. (18 sheets).

MORGAN, R.C.P., HATCH, T., SULAIMAN, W. & HARUN, W. 1982. A simple procedure for assessing soil erosion risk: a case study for Malaysia. *Zeitschrift fur Geomorphologie, Suppl.-Bd.* 44:69-68.

MURPHY, J. & RILEY, J.P. 1962. A modified single solution method for the determination of phosphate in natural waters. *Analytica Chemica Acta* 27:31-36.

ROSS, S.M., THORNES, J.B. & NORTCLIFF, S. 1990. Oil hydrology, nutrient and erosional response to the clearance of terra firme forest, Maraca Island, Roraima, northern Brazil. *The Geographical Journal* 156:267-282.

SHALLOW, P.G.D. 1956. *River flow in the Cameron Highlands*. Hydroelectric Technical Monograph 3, Central Electricity Board, Kuala Lumpur, Malaysia.

SINUN, W., WAI MENG, W., DOUGLAS, I. & SPENCER, T. 1992. Throughfall, stemflow, overland flow and throughflow in the Ulu Segama rain forest, Malaysia. *Philosophical Transactions, Royal Society London, B* 335:389-395.

VAN ZON, H.J.M. 1980. The transport of leaves and sediment over a forest floor. A case study in the Grand Duchy of Luxembourg. *Catena* 7:97-110.

VIS, M. 1986. Interception, dropsize distributions and rainfall kinetic energy in four Columbian forest ecosystems. *Earth Surface Processes and Landforms* 11:591-603.

WIERSUM, K.F. 1984. Surface erosion under various tropical agroforestry systems. In O'Loughlin, C.L. & Pearce, A.J. (eds). *Proceedings of a Symposium on Effects of Forest Land Use on Erosion and Slope Stability*. IUFRO, Vienna, and East-West Centre, Honolulu, Hawaii.

Construction of a low-cost runoff and sediment monitoring plot

ALEXANDER K. SAYOK[1] *and* NALAU BINGEDING[2]

[1]*Department of Forestry, PNG University of Technology, Lae, Papua New Guinea*
[2]*PNG Forest Products, Bulolo, Papua New Guinea*

ABSTRACT. A low-cost runoff and sediment monitoring technique which uses rectangular plots and collection devices is described. The technique is presently being tested near Hobu Community School, approximately 15 km NE of the Papua New Guinea University of Technology, Lae. Sediment and runoff are being monitored on four plots of *Leucaena leucocephala* seedlings, with two of these plots interplanted with sweet potato (*Ipomea batatas*). The plots are inexpensive to construct and maintain, as well as being easy to replicate and control.

KEY WORDS: soil loss, Universal Soil Loss Equation, Papua New Guinea.

INTRODUCTION

Instream Monitoring Systems

Traditionally, runoff and sediment from a watershed area or tract of land are sampled directly from the stream in that particular watershed. However, watershed runoff discharge and sediment transport is a complex process due to diversified topographic configuration, surface roughness, and drainage systems, in addition to changes in land-use. As such, instream runoff and sediment transport rates are highly variable both in time and space (Beschta 1985). A soil particle may be trapped or deposited in a micro-depression at a site with changes in slope, just beside any debris, or at the bottom of a stream for days, weeks, or even years before it is monitored at the stream gauging station (Sayok *et al.* 1993). Thus, traditional instream sediment monitoring method may not be a sensitive indicator of changes in watershed conditions (Jackson *et al.* 1985).

Alternative Monitoring Systems

Among the alternatives to instream sampling for direct watershed monitoring are erosion net studies and erosion conditions assessments (Clark 1974), retention basin studies (Lusby 1979), erosion transect studies (Blaney & Warrington 1983), simulated rainfall studies (Meyer 1988), and laboratory erosion studies (Mutchler *et al.* 1988).

Each of the systems above has advantages as well as disadvantages. Both erosion net studies and erosion transects are amenable to replication and control, and are easy to maintain. In this they are similar to the modified Universal Soil Loss Equation (USLE) plot technique described here. However, they only provide soil-loss data, not runoff data. Erosion condition assessments which provide relative ratings are best suited only to inventory-type studies. Small retention basin studies are also similar in concept to the technique described here, but are less amenable to the design principles of control and replication. With respect to simulated rainfall studies using plots large enough for soil-loss assessments which can be carefully designed and controlled, Jackson *et al.* (1985) stated that: "They are labor-intensive as well as expensive. Furthermore, data obtained may not be representative of natural runoff and erosion rates because simulated storms generally have little resemblance to natural storms, and temporal variations in infiltration and soil erodibility are not usually sampled."

D. S. Edwards et al. (eds.), Tropical Rainforest Research – Current Issues, 271–275.

Small laboratory plots are best for studying basic erosion phases that are difficult to study in detail on larger ones (Mutchler *et al.* 1988, Meyer 1988). On the other hand, their operation inevitably requires rainfall simulators and disturbed soils, therefore, it inherits all the problems faced by the simulated rainfall studies above.

USLE Plot System

Erosion research plots large enough to represent the complete process of rill and inter-rill erosion and equipped with retention boxes have been successfully used to measure long-term runoff and erosion rates. A good example is the 0.01 ac plot of 72.6 ft (22.1 m) long which was a standard runoff unit used to develop the Universal Soil Loss Equation (USLE) by the Agricultural Research Service to estimate water erosion on agricultural land under different cultivation practices (Wischmeier & Smith 1978). This plot measures erosion from the top of a slope where runoff begins. It is sufficiently wide to minimize edge or border effects and large enough for development of downslope rills. Today, the system has been applied in the United States to estimate soil losses or sediments for construction sites (Wischmeier & Meyer 1973), irregular slopes (Foster & Wischmeier 1974), forest land (Dissmeyer 1973) and a surface-mine spoil bank (McKenzie & Studlick 1979). It has also been used in Canada, Peru, Ivory Coast, and Nigeria (Soil Conservation Society of America 1976). The system has been modified by various researchers to particular requirements (Jackson *et al.* 1985, Lal 1981). Jackson *et al.* (1985) found the modified system to be extremely accurate when used on grazed and ungrazed plots in Denver, Colorado, as well as being easily replicated and controlled. In addition, it is cheap to install and maintain. Chang *et al.* (1992) stated that the system could be a useful and powerful tool for estimating annual, as well as single, storm soil losses in forested areas. The system in Papua New Guinea, described below, is also a modified USLE system. The objective is to determine whether the technology is applicable to Papua New Guinea where rainfall is over 4 000 mm per year.

THE PAPUA NEW GUINEA EROSION MONITORING SYSTEM

Plot Construction

Four plots were established on an abandoned food garden (sugar cane and sweet potato) near Hobu Community School compound. These plots are constructed from low cost and readily available materials. The cost of material per plot including collecting boxes is approximately US$130.00. The list below summarises the materials required for plot installation. The time required to construct each plot is approximately three man days.

Material List for Each Runoff-Sediment Plot :

- 4 sheets 4 mm exterior plywood (for construction of plot borders)
- 3 sheets 8 mm exterior plywood (for storage boxes)
- 25 30 cm sticks (diameter, 0.50 cm)
- ½ kg 2.5 cm flat-head galvanized nails
- ½ kg 3.75 cm flat head galvanized nails
- 4 3 m PVC rain gutters
- 10 cm black plastic sheet (cut into 4) gutter end caps
- 4 gutter connecting sleeves
- 1 gutter corner
- 1 downspout
- 1 40 kg bag cement
- 3 50 kg bag sand and small pebble mix

Approximate cost of materials = US$130.00

The size of each plot is 0.015 ha (15 m long and 10 m wide - Figures 1 and 2). Its side

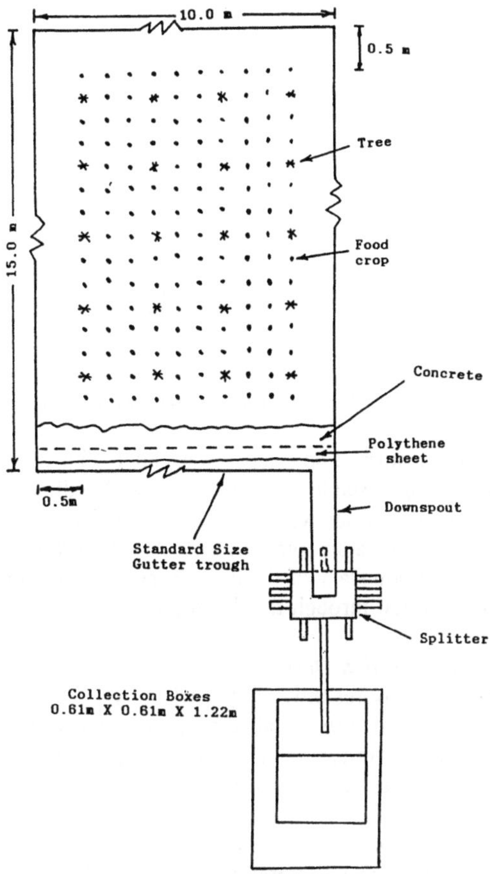

Figure 1. Plan view of runoff and soil erosion plot along with collection devices.

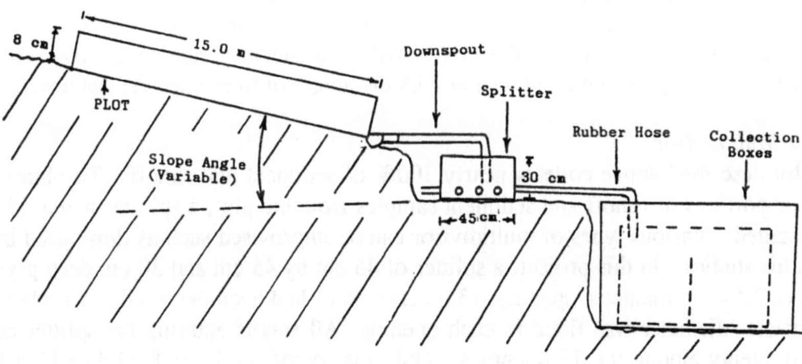

Figure 2. Cross-section view of runoff and erosion plot along with collection devices.

and upper boundaries are made from 15 cm plywood strips set about 8 cm into the soil and 7 cm sticking above ground. These strips are stabilized with sticks, 17 cm long, obtained from the surrounding bush. At the lower boundary, a standard PVC rain gutter was placed with its upper edge flush with ground level. To ensure that runoff and sediment will move through the gutter, a slight drop was maintained. A length of black plastic sheet (60 cm

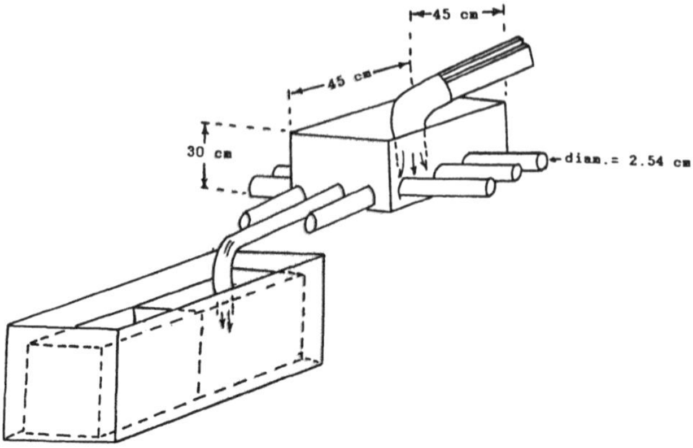

Figure 3. Cross-section view of splitter and collection boxes.

wide) was laid on the disturbed soil above the gutter with its lower edge overhanging the edge of the gutter giving a stable flow of runoff and sediment into the gutter. Cement, mixed at an appropriate consistency, was laid on the disturbed soil above the gutter. When set, this cement mix creates a smooth surface. A PVC drain (downspout) was attached at one corner of the lower boundary of the plot. Runoff and sediment that drop from the plot into the gutter are delivered through the downspout to a collection box below.

Runoff and Sediment Collection Devices

Each collection box measures 0.61 m by 0.61 m by 1.22 m, large enough to store runoff for a 50-year 48-hour storm, was installed at the lower end of the downspout. Inside this box, a small storage box with two compartments was placed for convenience in measuring smaller runoffs. Both boxes are made from exterior grade plywood.

Automatic Sampling Equipment and Continuous Monitoring

To augment or replace manual sampling, a stage-height recorder and automatic sediment samplers can be installed. However, due to budget constraints and high incidences of vandalism in some parts of Papua New Guinea, neither of these two pieces of equipment was installed. So, only runoff depth for a particular rain event or period (a week) was calculated based on the amount collected in the box. Sediment delivery rate (mass over time) is determined by dividing the sediment weight with the length of time since the last observation.

Splitter or Multidivisor

The plot described above collects nearly 100% of sediment and runoff. To obtain only aliquots or portions of runoff and sediment samples from the plot, a splitter or multidivisor can be installed. Various types of multidivisor can be improvised such as those used by Lal (1981) in his studies. In this project, a splitter of 45 cm by 45 cm and 30 cm deep plywood with twelve 2.54 cm diameter openings (3 on each side) had been designed. A rubber hose of appropriate diameter was fitted to each opening. All runoff entering the splitter can be distributed evenly among the 12 openings. Only that runoff (1/12 or 8.33%) which flows through the opening at the lower side of the splitter will flow via the hose into the collection box (Figure 3).

CONCLUSION

The runoff and sediment monitoring plots described above are quick and inexpensive to construct as well as easy to replicate and control. Sample plots were installed in June 1992 near Hobu Community School, approximately 15 km NE of the Papua New Guinea University of Technology. The plots were set up in four plots of *Leucaena leucocephala* seedlings, two of which were interplanted with sweet potato (*Ipomea batatas*). The plots have been monitored since 1992. Soil loss details will be available after one year's monitoring.

ACKNOWLEDGEMENTS

The authors would like to thank the Research Committee, PNG University of Technology, Lae, for funding this project. Presentation of this paper was made possible through a grant awarded by the International Tropical Timber Organization (ITTO) under its Fellowship/Small Grant Scheme.

LITERATURE CITED

BESCHTA, R. L. 1985. Conceptual Model of Sediment Transport in Streams. In C. R. Thorne (ed.). *Proceeding of the International Workshop on Problems of Sediment Transport in Gravel-Bed Rivers.* Colorado State University, Colorado.

BLANEY, D. G. & WARRINGTON, G.E. 1983. *Estimating Soil Erosion using an Erosion Bridge.* US Department of Agriculture Forest Service. WSDG-TP-00008. 5 pp.

CHANG, M., SAYOK, A.K. & WATTERSTON, K.G. 1992. *Comparison of Three Proposed C-Values of the Soil Loss Equation for Various Forested Conditions.* Proceedings of Conference XXIII. International Erosion Control Association. Feb 18-21. Reno, Nevada.

CLARK, R. 1974. *Erosion Conditions Classification System.* US Department of the Interior, Bureau of Land Management Technical Note 346. BLM Service Center, Denver, Colorado. 47 pp.

DISSMEYER, G. E. 1973. Evaluating the Impact of Individual Forest Management Practices as on Suspended Sediment, a Presentation at the National Meeting of Soil Conservation. Society of America, Hot Springs, Arkansas.

FOSTER, G. R. & WISCHMEIER, W.H. 1974. Evaluating Irregular Slopes for Soil Loss Prediction. *Transactions of the American Society of Agricultural Engineers.* 17:305-309.

JACKSON, W. L., KNOOP, K., SZALONA, J.J. & HUDSON, S. 1985. *A Runoff and Soil-loss Monitoring Technique using Paired Plots.* US Department of the Interior, Bureau of Land Management. Technical Note 368. Denver, Colorado. 20 pp.

LAL, R. 1981. Deforestation of Tropical Rainforest and Hydrologic Problems. In Lal, R. & Russell, E.W. (eds). *Tropical Agricultural Hydrology.* John Wiley and Sons Ltd.

LUSBY, G. C. 1979. *Effects of Grazing on Runoff and Sediment Yield from Desert Rangeland at Badger Wash in Western Colorado, 1953-1973.* US Geological Survey Water Supply Paper 1532-I. 134 pp.

MCKENZIE, G. D. & STUDLICK, J.R.J. 1979. Erodibility of Surface-mine Spoil Banks in Southeastern Ohio: An Approximation. *Journal of Soil and Soil Conservation* 34:187-190.

MEYER, L. D. 1988. Rainfall Simulators for Soil Conservation Research. Pp. 75-95 in Lal, R. (ed.). *Soil Erosion Research Methods.* Soil and Water Conservation Society. Ankeny, Iowa.

MUTCHLER, C. K., MURPHREE, C.E. & MACGREGOR, K.C. 1988. Laboratory and Field Plots for Soil Erosion Studies. Pp. 9-36 in Lal, R. (ed.). *Soil Erosion Research Methods.* Soil and Water Conservation Society. Ankeny, Iowa.

SAYOK, A. K., CHANG, M. & WATTERSTON, K.G. 1993. Forest Clearcutting and Site Preparation on a Saline Soil in East Texas: Impact on Soil Losses. Paper presented during IAMAP/IAHS'93 Meeting. July 11-23, 1993. Yokohama, Japan.

SOIL CONSERVATION SOCIETY OF AMERICA 1976. *Soil Erosion: Prediction and Control.* Special Publication No. 21. 393 pp.

WISCHMEIER, W. H. & MEYER, L.D. 1973. *Soil Erodibility on Construction Areas.* Special Report 135. Highway Research Board. National Academy of Sciences, Washington, D.C. Pp. 20-29.

WISCHMEIER, W. H. & SMITH, D.D. 1978. *Predicting Rainfall Erosion Losses: A Guide to Conservation Planning.* US Department of Agriculture, Handbook 537. 58 pp.

The growth performance of four dipterocarp species under different terrain and shade conditions

LOK ENG HAI, HASHIM MD. NOOR *and* FAUZIDAH AHMAD

Forest Research Institute of Malaysia (FRIM), Kepong, 52109 Kuala Lumpur, Malaysia

ABSTRACT. Four dipterocarp species, *Shorea curtisii*, *S. leprosula*, *S. parvifolia* and *Dryobalanops aromatica*, were planted on level terrain and on hill sites, with and without the shade of nurse trees (*Gmelina arborea*). Twelve years of growth indicate that trees on level terrain attain greater heights than those on hill tops. Shade did not significantly influence the performance of the dipterocarps. Interactions of the three factors, terrain, shade and species, had no influence on height and diameter growth, but interactions of terrain-shade and terrain-species have a significant affect on survival. Overall performance in terms of diameter, height and survival of the dipterocarps can be arranged in the order *S. leprosula* > *D. aromatica* > *S. parvifolia* > *S. curtisii*.

KEY WORDS: *Shorea, Dryobalanops*, plantations, tropical forests, Malaysia.

INTRODUCTION

Forest plantations in the tropics play an important role in ensuring a continuous supply of timber from desired species. Future large-scale forest plantations could supplement the depleted natural rainforest resources which have been reduced in area due to extensive and damaging logging activities. In forest plantations the better quality species can be selected, thus maximising growing space, growth rates and allowing easier manipulation of growing conditions.

In Malaysia, research into establishing forest plantations was started in FRIM in the early 50's and 70's with the establishment of both exotic and indigeneous species. Based on their growth data, some commercially important timber species belonging to the Dipterocarpaceae family have been identified, including species of *Dryobalanops, Shorea* and *Hopea*. The family Dipterocarpaceae comprises about 515 species with 16 genera (Jacobs 1981) and occurs naturally in the forests of Guyana and Venezuela in South America, as well as Africa and the Indo-Malayan region (Jacobs 1981, Symington 1941). In Peninsular Malaysia, there are about 150 species of dipterocarps occurring in lowland forests; the number of species decreases as altitude increases (Ashton 1964, Symington 1941).

This paper highlights some of the silvicultural requirements of four dipterocarp species grown in cleared secondary lowland forest.

MATERIALS AND METHODS

Location

The experiment was located in Block P, Compartment 10 of Sungai Buloh Forest Reserve, Selangor, Malaysia (3° 18' N, 101°34'E). The sites were on a south-facing slope which has annual rainfall of 2613 mm. The lower and upper terrain sites are at an altitude of 50 and 100 m above sea level respectively. The soil is a sandy clay of the Rengam series. The planting sites were initially clear-felled and burnt in March 1977 but planting was delayed until November 1977, when suitable planting stock was available; consequently, the prepared

D. S. Edwards et al. (eds.), Tropical Rainforest Research – Current Issues, 277–281.

sites were colonised by perennial weeds such as *Musa acuminata*, *Macaranga gigantea*, *Mallotus paniculatus*, *Trema* spp. and *Mikania micrantha*. These weeds were cut down immedidately prior to planting the dipterocarps.

Experimental design

Planting was based on a square spacing pattern of 2.4 x 2.4 m for the four species of dipterocarps: *S. curtisii, S. leprosula, S. parvifolia* and *Dyrobalanops aromatica*. The experiment was laid out using a split-split plot design, with terrain (lower and upper slope) as the main plot, shade (with and without shade tree intercropping) as sub-plots and species as the sub-sub plot. The design was replicated three times. Fast growing seedlings of *Gmelina arborea* were planted between the dipterocarp species. These shade seedlings were much taller than the dipterocarps when planted in November 1977 and provided some shade with their sparse crowns.

The initial height of the dipterocarp seedlings was recorded immediately after completion of planting. Survival and height were monitored quarterly during the first year of establishment and yearly thereafter. Diameter measurement at breast height (dbh) was recorded when the trees reached the minimum girth of 3.0 cm. Weeding and climber cutting was performed regularly.

The growth performance data at age 12 years were analysed using the SAS software Package. Means of the treatments were compared using Duncan's Multiple Range Test.

RESULTS AND DISCUSSION

Tables 1a and 1b give the summary results of the average diameter, height and survival rate at both sites and shade conditions. *S. leprosula* gave the greatest increase in growth at both sites and shade conditions. The height increments (final-initial heights) of the dipterocarps were observed to be significantly affected by both terrain and species factors at $P = 0.05$ and $P = 0.01$ respectively but not by shade. There was no differential effect of terrain-species, shade-species and terrain-shade-species upon the height increment (Table 2).

Table 1a. Average diameter, height and survival of four Dipterocarp species, 12 years old - level terrain conditions (lower slope).

Species	Diameter (cm)		Height (m)		Survival (%)	
	A	B	A	B	A	B
S. curtisii	13.9	12.7	11.2	12.1	14.6	28.8
S. leprosula	14.8	15.9	12.9	14.0	62.1	69.6
S. parvifolia	14.7	13.6	10.9	10.6	26.5	39.8
D. aromatica	13.1	13.6	13.1	13.4	57.6	62.9

Table 1b. Average diameter, height and survival of four Dipterocarp species, 12 years old - hill terrain conditions (upper slope).

Species	Diameter (cm)		Height (m)		Survival (%)	
	A	B	A	B	A	B
S. curtisii	10.3	13.5	8.7	8.8	35.5	21.0
S. leprosula	13.1	14.9	10.2	9.7	41.1	31.0
S. parvifolia	10.4	10.8	8.1	7.3	24.4	21.1
D. aromatica	10.1	10.8	9.3	8.8	38.9	41.1

Table 2. Anova result for height increment on four dipterocarps. N.S. = Not Significant; * = Significant at P = 0.05; ** = Significant at P = 0.01.

Source of Variation	df	SS	MS	F value	Remarks
Replicate	2	0.90	0.45	0.18	N.S.
Terrain (T)	1	171.00	171.01	69.39	*
Shade (SH)	1	1.40	1.40	0.65	N.S.
Species (SP)	3	53.08	17.69	6.51	**
T x SH	1	8.67	8.67	4.04	N.S.
T x SP	3	3.47	1.16	0.43	N.S.
SH x SP	3	2.78	0.93	0.34	N.S.
T x SH x SP	3	4.94	1.65	0.61	N.S.

Table 3. Anova result for diameter increment on four dipterocarps. N.S. = Not Significant; * = Significant at P = 0.05.

Source of Variation	df	SS	MS	F value	Remarks
Replicate	2	0.59	0.29	0.03	N.S.
Terrain (T)	1	97.36	97.36	9.46	N.S.
Shade (SH)	1	0.25	0.25	0.09	N.S.
Species (SP)	3	73.55	24.52	4.53	*
T x SH	1	1.09	1.09	0.40	N.S.
T x SP	3	10.33	3.44	0.80	N.S.
SH x SP	3	12.95	4.32	0.80	N.S.
T x SH x SP	3	1.29	0.43	0.80	N.S.

Table 4. Anova result for survival percentage on four dipterocarps. N.S. = Not Significant; * = Significant at P = 0.05; ** = Significant at P = 0.01; *** = Significant at P = 0.001.

Source of Variation	df	SS	MS	F value	Remarks
Replicate	2	482.44	241.22	2.41	N.S.
Terrain (T)	1	2073.36	2073.36	7.96	N.S.
Shade (SH)	1	0.26	0.26	0.00	N.S.
Species (SP)	3	7553.70	2517.90	25.14	***
T x SH	1	840.59	840.59	9.13	*
T x SP	3	1932.11	644.04	6.43	**
SH x SP	3	155.46	51.82	0.52	N.S.
T x SH x SP	3	455.03	148.34	1.48	N.S.

Concerning diameter growth, the species had a significant affect at P = 0.05, but terrain and shade had no significant influence. There were also no differential interactions among the three factors on the diameter of the trees (Table 3). Survival was significantly influenced by the species of dipterocarps. However, this was not affected by terrain and shade factors alone. Nevertheless, the interaction of terrain-shade and terrain-species had a significant affect on survival of the trees (Table 4). Shade and open plantings made no significant

Table 5. Duncan new multiple range tests on height, diameter increment and survival percentage of four dipterocarps. Note: Values in a column with the same letter are not significantly different at P = 0.05.

Species	Height increment (m)	Diameter	Survival %
S. leprosula	11.7[a]	14.7[a]	50.8[a]
S. parvifolia	9.3[b]	12.4[b]	26.9[b]
D. aromatica	11.2[a]	11.9[b]	50.0[a]
S. curtisii	9.5[b]	11.4[b]	23.9[b]

difference to height or diameter increment. This may have been due to poor growth of the *Gmelina arborea* shade trees in the plots, which provided little shade for the dipterocarps.

Of the species tested, *S. leprosula* and *D. aromatica* had the greatest height increment, while *S. parvifolia* and *S. curtisii* the least (Table 5). *S. leprosula* had the greatest diameter increment, a value which was significantly different from the diameter increments of the other three species. The outstanding performance of *S. leprosula*, which is considered as a late secondary species (Koyama, 1981) was expected since it is a light-demanding fast growing species and is commonly found in young regenerating forests (Symington 1941). The species has a very high photosythetic rate (Kayoma 1981, Kira 1982). In fact, Watson (1935) has reported that *S. leprosula* was the most vigorous of the light hardwood dipterocarp species. Watson (1935) recorded an average height of 10.7 m and diameter of 9.7 cm at an age of seven years for this species, which was better that the mean height recorded in the present study. He also listed *D. aromatica* among the fast-growing trees. According to Walton (1938), the growth of *S. parvifolia* was slower that *S. leprosula* but the former was more shade tolerant. Although *S. leprosula* surpassed *S. parvifolia* during the early growth, the latter is likely to overtake the former after about 25 years (Symington 1941). Both *S. leprosula* and *S. parvifolia* occur naturally on the gentle slopes and in the valleys.

CONCLUSION AND RECOMMENDATIONS

The present study indicated that dipterocarp trees growing on the lower slope gave better growth and survival than trees growing on the upper hill slope. However, only height increment was significantly affected. Shade had no significant effect on the growth of the dipterocarps. There was a significant variation in the performance of the dipterocarp species. Overall, performance in terms of diameter, height growth and survival of the dipterocarps can be arranged in the order of *S. leprosula* > *D. aromatica* > *S. parvifolia* > *S. curtisii*. Interactions among the investigated factors showed no significant effect on the height and diameter of the dipterocarps, but interactions of terrain-shade and terrain-species did influence survival. However, further trials are needed to test the various site effects and silvicultural treatments upon the performance of *S. leprosula* and *D. aromatica* under plantation conditions.

ACKNOWLEDGEMENTS

We wish to thank FRIM for sponsorship and help in carrying out the projects. We are also grateful to our Research Assistant, Mr. M. Markandan, and the staff of the Plantation Silviculture Unit for their help in plot maintenance and data collection. We would also like to record our appreciation to Mr. Chew Teng Kok, who initiated the experimental trials, and Dr. Zakaria Ibrahim for comments on the first draft of this manuscript.

LITERATURE CITED

ASHTON, P.S. 1964. *Ecological studies in the mixed dipterocarp forests of Brunei State.* Clarendon Press, Oxford.

BURGESS, P.F. 1975. *Silviculture in the hill forests of the Malay Peninsula.* Forest Reserach Institute, Research Pamphlet 66. Kepong, Malaysia.

JACOBS, M. 1981. Dipterocarpaceae: The taxonomic and distribution framework. *The Malayan Forester* 44:168-189.

KAMIS AWANG, MOHD BASRI HAMZAH, TAJUDDIN BAHAROM & SAHIR OTHMAN 1981. Effects of some environmental factors on the growth of three hill forest dipterocarps. *BIOTROP Special Publication* 13:9-21.

KAYOMA, H. 1981. Photosynthetic rates in lowland rainforest trees of Peninsular Malaysia. *Japanese Journal of Ecology* 31:361-369.

KIRA, T. 1982. Patterns of productivity in tropical trees and forests. *Proceedings of the workshop on Ecological basis for rational resource utilization in the humid tropics of South East Asia.* UPM Serdang.

LONDON, F.H. 1948. Planting experiments with Kapur. *The Malayan Forester* 11:146-153.

SYMINGTON, C.F. 1941. Forester's manual of dipterocarps. *Malayan Forest Record* 16.

VINCENT, A.J. 1961. *A note on the growth of three meranti (Light Heavy Wood Shorea) hill forest species in naturally and artificially regenerated forest of Malaya.* Forest Research Institute - Forest Department Federation of Malaya Research Pamphlet.

WALTON, A.B. 1938. Some notes on nursery and planting technique. *The Malayan Forester* 7:122.

WATSON, J.G. 1935. Plantation experiments at Kepong. *The Malayan Forester* 4:110-119.

WHITTY, R.H. 1932. The use of nursery transplants in planting. *The Malayan Forester* 1. 228 pp.

Variation in reproductive success of gynodioecious figs (*Ficus* spp., Moraceae) and their pollinators (Hymenoptera: Agaonidae) in New Guinea

GEORGE D. WEIBLEN

Cape Tribulation Tropical Research Station, P.M.B. 5, Cape Tribulation via Mossman, 4873, Queensland, Australia

ABSTRACT. Variability in ovary utilization by figs and their obligate wasp pollinators was measured in middle and lower montane rainforest in New Guinea. Examination of fruit crops from individuals of nine gynodioecious tree species revealed unprecedented levels of variation in seed set and wasp predation. Approximately one-third of all fig flowers were preyed on by wasps. In female figs, the proportion of flowers containing seeds ranged 18 - 97% among the species sampled. In hermaphroditic figs, the proportion of seeds killed by wasps ranged from 12 - 87%. Contrary to expectation, results suggest that fig and wasp reproductive success is positively related in gynodioecious species. The ecological significance of fig reproductive output to tropical rain forests is discussed.

KEY WORDS: *Ficus*, Agaonidae, gynodioecy, pollination, seed predation, New Guinea.

INTRODUCTION

The pollination of figs (*Ficus* spp., Moraceae) by species-specific wasps (Hymenoptera: Agaonidae) is arguably the most widely known example of an obligatory mutualism between plant and pollinator (cf. Bronstein 1992). Evidence of extreme host-specificity of agaonids to figs (est. 800 species worldwide) suggests that figs and their exclusive wasp pollinators have evolved in parallel (Wiebes 1979). The persistence of the mutualism in some tropical rain forests has been identified as critical to community-level conservation (McKey 1987, Terborgh 1986). Figs are food for a range of frugivorous animals during periods of scarcity, and this role as possible 'keystone' species stems from the peculiar nature of fig pollination.

The fig/pollinator interaction is thoroughly described in Bronstein (1992). In monoecious species, flowers are arranged inside a closed receptacle (syconium) which is accessible only through a bract-lined opening (ostiole). This arrangement of flowers is technically an inflorescence, although ecologically it is a fruit. In a young fig, hundreds of female flowers line the internal wall of the syconium. Pollen-carrying female wasps enter syconia through the ostiole near the time of female flower receptivity. Inside the fig cavity, female wasps deposit eggs in fig ovaries and simultaneously pollinate the flowers. Developing seeds are consumed by wasp larvae. As the fig ripens, adult female wasps hatch into the fig cavity after being fertilized by wingless males, which die shortly after chewing a hole through the ostiole. Females collect pollen from male flowers usually clustered around the ostiole while leaving the fig. Pollen-carrying females then find receptive figs in which to oviposit, thus completing their life cycle and at the same time achieving pollination.

Wasp survival is ensured by the availability of figs at differing stages of development. Flowering asynchrony at the population-level is therefore common in figs (Janzen 1979a), although there are exceptions (Kjellberg *et al.* 1987). Population asynchrony can thus provide ripe figs for frugivores at times when food is otherwise scarce (Terborgh 1986). Maintenance of this keystone species interaction obviously depends on continued reproductive

D. S. Edwards et al. (eds.), Tropical Rainforest Research – Current Issues, 283–289.

success of the fig and its pollinator. However, natural history suggests that a fundamental conflict of interest imposes reciprocal constraints on the success of each partner. Seeds containing wasp larvae are destroyed and consequently, wasps are both pollinators and seed predators. Viable seed production in figs is therefore constrained by the percentage of ovaries used by the wasps. The conflict of interest is determined by a single floral trait, the length of the style (Bronstein 1992, Janzen 1979a). The ovaries of short-styled flowers are accessible to wasp ovipositors while those of long-styled flowers are not. Wasps oviposit in both types but successfully leave offspring in only the short-styled flowers. Janzen (1979b) was first to ask: what proportion of fig ovaries are donated to wasps in exchange for seeds? Bronstein (1992) further defined the tradeoff between seeds and wasps based on the premise that they alternately use fig ovaries. The expectation that wasp and seed maturation are negatively related can be tested by measuring reproductive success (i.e. numbers of offspring) using fig ovaries as 'common currency'.

This approach has not yet been applied to gynodioecious figs, which comprise over 300 species in the Old World tropics (Berg 1987). Gynodioecious populations consist of trees whose figs contain both male and female flowers (hermaphrodites) and trees with only female flowers (females). Short- and long-styled flowers occur together in monoecious figs, but gynodioecious figs are completely heterostylous. Female trees have long-styled flowers and hermaphroditic trees apparently have only short-styled flowers. Accordingly, females set seed and hermaphrodites produce mostly wasps (Verkerke 1987). Hermaphrodites producing few or no seeds are functionally male.

The conflict of interest for functionally dioecious species is different than for their monoecious relatives. Note that in dioecious species, wasps pollinating female figs leave no offspring. There should therefore be strong evolutionary pressure for wasps to exploit hermaphroditic trees while avoiding lethal female trees (Kjellberg *et al.* 1987). I tested this possibility by measuring variation in seed set and wasp maturation in pairs of hermaphroditic and female trees. This comparative approach, using fig ovaries as 'currency', allowed me to investigate fig and wasp success for nine gynodioecious species in New Guinea.

METHODS

Fruiting trees were sampled in disturbed and undisturbed forests around Wau, Papua New Guinea during October and November of 1992. Sampling localities included the Wau Ecology Institute arboretum, the slopes of Mount Kaindi (7° 24' S, 146° 44' E), and the Kuper Range Wildlife Conservation Area (8° 77' S, 146° 80' E). Figs were collected in lower montane (1100 - 1300 m altitude) and mid-montane forests (1300 - 1800 m altitude).

The climate in this area is highly humid and quite aseasonal, with rainfall distributed somewhat evenly throughout the year. Lower and mid-montane forests on Mount Kaindi receive approximately 2000 and 3000 mm of rain per year, respectively. Mean annual temperature is 21°C at 1200 m and falls to around 16°C at 1800 m altitude. Average daily temperature during the year is nearly constant.

Natural vegetation in the study area is mixed *Lithocarpus-Castanopsis* rain forest in the mid-montane and mixed rain forest dominated by *Araucaria* in the lower montane zone (Gressitt & Nadkarni 1978). However, anthropogenic grassland has replaced much of the lower montane forest due to extensive clearing and burning. Fig trees were especially common in regrowth forests and abandoned gardens, where they comprised up to 25% of the standing trees with diameters exceeding 10 cm (personal observation).

Pairs of hermaphroditic and female trees representing nine gynodioecious species were sampled. I collected figs when ripe over an 8-week period. Ten figs from each of 18 trees were stored in 70% ethanol. Figs were later sectioned into quarters along the longitudinal axis and one quarter was chosen at random for dissection under 15x magnification. There were three visually distinguishable classes of fig seeds. (1) Mature seeds were white-to-yellow with hard seed coats. (2) Seeds killed by wasps had conspicuous exit holes in the seed coats, were dark brown in color, or contained dead wasps. (3) Abortive ovaries were considerably smaller than (1) and (2). I counted totals for each class and calculated percentages for each based on the total number of ovaries in the sample.

Table 1 lists collection numbers for fig species employed in the study. Species identification required considerable effort due to the unresolved status of *Ficus* taxonomy in New Guinea. Many of the 138 fig species recorded for New Guinea are known from only one or a few specimens. Nearly all species selected for this study were common around Wau and well-represented in the PNG National Herbarium collection. Plant specimens were identified using the diagnostic keys of Corner & Henty in Johns (1988) and checked against the reference collection at the National Herbarium, where vouchers are deposited. Pollinator species (also listed in Table 1) were later checked against Wiebes (1982).

RESULTS

An average of 65 percent of ovaries were occupied by seeds (33%) or wasps (32%), while the remaining 35 percent were aborted (Table 2). The mean percentage of fig ovaries producing both seeds and wasps ranged from 21±7 (X±SD) in *F. trachypison* to 96±3 in *F. complexa*. Seed set averaged 66±26 percent in female figs and 7±7 percent in hermaphroditic figs (Figure 1). Wasp larvae exploited an average of 58±24 percent of hermaphroditic fig ovaries. The percentage of wasps in hermaphroditic figs and of seeds in female figs showed wide variation among the sampled species. Female and hermaphroditic trees in *F. complexa* produced the highest percentage of seeds and wasps on average (97±2 and 87±5 percent) and *F. trachypison* matured the lowest percentage of seeds and wasps (18±5 and 12±6 percent).

Table 1: Gynodioecious figs (*Ficus spp.*) and their wasp pollinators (family Agaonidae) collected at Wau, Morobe Province, Papua New Guinea.

Fig	Collection N° Female	Collection N° Bisexual	Pollinator
Ficus complexa Corner	074	103	*Ceratosolen gressitti* Wiebes
F. erythrosperma Miq.	049	072	undescribed
F. mollior F.v. M. ex Benth.	112	071	*C. medlerianus* Wiebes
F. nodosa Teysm. et Binn.	002	065	*C. nexilis* Wiebes
F. robusta Corner	068	067	undescribed
F. septica Burm.f.	085	004	*C. bisulcatus* Mayr
F. trachypison K. Schum var. *pallida* Corner	005	108	undescribed
F. virgata Reinw. ex Bl. var. *sessilis* (Bur.) Corner	076	101	*Liphorrhopalum sessilis* Hill
F. wassa Roxb.	003	089	*Blastophaga wassae* Wiebes

Table 2: Mean (±SD) fig diameter in millimetres and total female flowers per fig for nine species. Total flowers per fig was calculated by multiplying the number of flowers per sample by a factor of 4. Percent ovary use was calculated by pooling data from hermaphroditic and female trees of the same species (N = 20).

Ficus species	diameter		flowers per fig		% ovary use	
complexa	20	(2)	668	(87)	96	(3)
erythrosperma	10	(1)	449	(40)	72	(6)
mollior	19	(1)	828	(119)	36	(4)
nodosa	34	(1)	2648	(231)	63	(8)
robusta	46	(1)	3515	(282)	83	(5)
septica	18	(1)	576	(55)	51	(6)
trachypison	15	(1)	616	(44)	21	(7)
virgata	12	(1)	503	(59)	86	(8)
wassa	12	(2)	513	(138)	79	(9)
Average (N= 180)	21	(11)	1146	(1072)	65	(24)

The mean percentage of aborted fig ovaries was remarkably similar among female and hermaphroditic figs (34±26 and 35±23 percent), despite a highly variable percentage among species. Vacant ovaries ranged 2-82 percent for females and 5-75 percent for hermaphrodites.

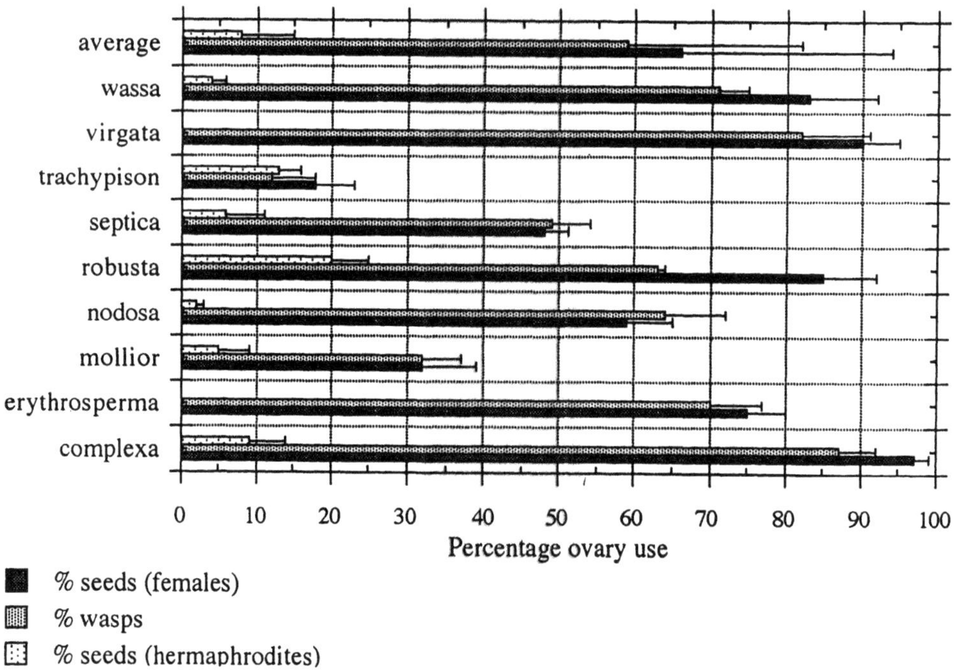

Figure 1: Levels of seed set and wasp predation for nine gynodioecious figs in New Guinea. Average seed set for female and hermaphroditic trees for each species is listed separately (N = 10). Error bars indicate 95% confidence intervals.

Average percent seed and wasp maturation varied significantly among sampled species (one-way ANOVAs, $F = 200.2$, $P < 0.001$ and $F = 353.4$, $P < 0.001$). There was a strong positive correlation between percent seed production in female figs and percent wasp predation in hermaphroditic figs ($r = 0.882$, $P < 0.001$, $N = 90$). Fig diameter was obviously correlated with numbers of female flowers per sample ($r = 0.674$, $P < 0.001$, $n = 161$), but interestingly, female figs were on average significantly larger than hermaphroditic figs (mean difference = 0.773 mm, ANOVA, $F = 33.7$, $P < 0.001$).

DISCUSSION

Although wasp pollinators had access to every ovary within hermaphroditic figs, nearly 35 percent remained unoccupied on average. In some hermaphroditic figs, a fraction of the unused ovaries produced seed, suggesting that wasps pollinate slightly more flowers than they occupy. There are several possible reasons why wasps may not successfully exploit all available ovaries. (1) Wasps may fail to pollinate or oviposit in all flowers (i.e. too few 'foundresses'). (2) Limitation of plant resources may limit the number of wasps a fig can support (Bronstein 1988a). (3) Predators and parasites of fig wasps may limit maturation success. Bronstein (1988b) suggested that factors extrinsic to the mutualism, such as ant predation, can impose constraints on wasp and fig reproductive success. There are presently not enough data available to conclude exactly which factors determine the relative success of fig and wasp.

The seemingly logical tradeoff between seed and wasp production draws no support from the findings for figs in New Guinea. I found that wasp and seed maturation was positively correlated among nine species sampled. Surprisingly, this result for gynodioecious species is consistent with the findings of Bronstein (1986) for monoecious *F. pertusa*. We based our expectation that wasp and fig success should be negatively related on the premise that the partners alternately exploit fig ovaries. In the case of gynodioecy, however, it is false to assume that wasp overexploitation in hermaphroditic figs is disadvantageous. Increased wasp production enhances male fig reproductive success by increasing the amount of pollen donated to female figs. I found that nearly 90% of the ovaries were exploited in some hermaphroditic figs. Hermaphrodites therefore appear to donate the bulk of their ovaries to wasps (and lose female fitness) in exchange for increased male fitness.

I found that hermaphrodites occasionally set a few seeds and that females often contain the remains of aborted anthers clustered around the ostiole. The presence of relictual 'maleness' and 'femaleness' in both sexes conforms to the taxonomists' belief that gynodioecious figs evolved from monoecious ancestors (Wiebes 1979). A simple change in style length could facilitate the transition to dioecy, which is thought to have occurred at least twice (Berg 1987). The evolution of dioecy would have replaced the potential for a fig/pollinator tradeoff in ovary utilization with a different conflict of interest.

Conflict between gynodioecious figs and their pollinators results from the separation of seed and wasp production among two sexes. Wasps leave no offspring in female figs, but have access to every ovary in hermaphroditic figs. Thus, the seed-producing (female) trees are evolutionarily lethal to their pollinators. It seems logical to suppose that strong selection against pollinators of female trees would favor wasps with a means of discriminating between the sexes. Individual wasps visiting only hermaphroditic figs would achieve higher reproductive success. I found that ovary utilization among the sexes were remarkably similar across nine species (Figure 1). Surprisingly, there were more seeds on average in female figs than wasps in hermaphrodites. These results suggest that wasps are efficient pollinators of both sexes, although pollination efficiency for females appears slightly higher than for hermaphrodites. Considering the obvious disadvantage to individual pollinators of female figs, what explains the apparent stability of gynodioecious fig pollination?

The persistence of over 300 gynodioecious species and their pollinators is possibly explained by flowering phenology at the population level. Kjellberg *et al.* (1987) showed for gynodioecious *F. carica* that the majority of wasps emerged from hermaphroditic figs when the female figs were receptive. Asynchronous flowering among the sexes can thus eliminate selection against wasps which pollinate females. Kjellberg *et al.* (1987) suggest that sexes flower at different times in seasonal environments, where the relative costs of wasp and seed production vary throughout the year. However, this explanation does not account for dioecious figs in aseasonal environments. The New Guinea highlands climate, for example, may not offer the necessary environmental cues for flowering synchrony. Answers to this question for figs in aseasonal climates will remain a matter of speculation until new information is available. Only a single study at present has detailed the wasp associations of a fig species in New Guinea (Godfray 1988) and there are no published data on flowering phenology. Fruit production in gynodioecious *F. variegata* in nearby tropical North Queensland, however, appears to be highly seasonal (H. Spencer, unpublished data).

Fig reproductive output is of critical importance to conservation in some tropical forests where figs are a keystone resource for frugivorous animals (Terborgh 1986). The observation that female figs are on average slightly larger and qualitatively sweeter than hermaphroditic figs is consistent with a hypothesis that seed-bearing female figs are more attractive to animal dispersers (H. Spencer, personal communication). Frugivores in New Guinea include megabats, marsupials, parrots, fruit doves, and birds of paradise. The production of female figs as food for these animals of high conservation value appears to depend on the successful exploitation of hermaphroditic trees by pollinating wasps. Whether this food resource is available continuously or patchily during the year has yet to be determined.

In conclusion, this study suggests that for gynodioecious species: (1) wasps do not successfully exploit all available fig ovaries; (2) figs do not show an evolved means of lowering wasp success in overexploited syconia, and (3) fig and wasp maturation success are positively correlated. These results contribute to a collection of data which has failed to find evidence for a simple trade-off between figs and their pollinators. The persistence of the fig-pollinator interaction in New Guinea rain forests currently under threat could decide the fate of a host of other important species.

ACKNOWLEDGMENTS

The Institute of Papua New Guinea Studies, Wau Ecology Institute and PNG National Herbarium made the study possible. Many thanks to Harry Sakulas and Robert Kiapranis. I am grateful for criticism and discussion provided by Yves Bassett and Robert Bino. I am also indebted to 'papa bilong graun' who allowed me access to their lands and Deren and Roy who climbed the tall trees. The study was supported by a post-graduate fellowship from the Thomas Watson Foundation.

LITERATURE CITED

BERG, C. C. 1987. Classification and distribution of *Ficus*. *Experientia* 45:605-611.

BRONSTEIN, J. L. 1986. *Coevolution and constraints in a neotropical fig-pollinator wasp mutualism*. Ph.D. Dissertation, University of Michigan, Ann Arbor.

BRONSTEIN, J. L. 1988a. Limits to fruit production in a monoecious fig: consequences of an obligate mutualism. *Ecology* 69:207-214.

BRONSTEIN, J. L. 1988b. Predators of fig wasps. *Biotropica* 20: 215-219.

BRONSTEIN, J. L. 1992. Seed predators as mutualists: ecology and evolution of the fig/pollinator interaction. Pp. 1-47 in Bernays, E. A. (ed.). *Insect-Plant Interactions* Vol. 4. CRC Press, Boca Raton, Florida.

GODFRAY, H. C. J. 1988. Virginity in haplodiploid populations: a study on fig wasps. *Ecological Entomology* 13:283-291.

GRESSITT, J.L. & NADKARNI, N. 1978. *Guide to Mt. Kaindi: Background to Montane New Guinea Ecology*. Wau Ecology Institute Handbook No. 5. Wau, Papua New Guinea, 135 pp.

JANZEN, D. H. 1979a. How to be a fig. *Annual Review of Ecology and Systematics* 10:13-51.

JANZEN, D. H. 1979b. How many babies do figs pay for babies? *Biotropica* 11:48-50.

JOHNS, R.J. 1988. *The Flowering Plants of Papuasia: Hamamellidae*. P.N.G. University of Technology, Lae, Papua New Guinea, 139 pp.

KJELLBERG, F., GOUYON, P. H., IBRAHIM, M., RAYMOND, M., & VALDEYRON, G. 1987. The stability of the symbiosis between dioecious figs and their pollinators: a study of *Ficus carica* L. and *Blastophaga psenes* L. *Evolution* 41:693-701.

McKEY, D. 1987. Population biology of figs: applications for conservation. *Experientia* 45:661-673.

VERKERKE, W. 1987. Structure and function of the fig. *Experientia* 45:612-621.

WIEBES, J. T. 1982. Fig wasps (Hymenoptera). Pp. 735-755 in Gressitt, J. L. (ed.). *Biogeography and Ecology of New Guinea*. Vol. 2. W. Junk Publishers, The Hague.

WIEBES, J. T. 1979. Co-evolution of figs and their insect pollinators. *Annual Review of Ecology and Systematics* 10:1-12.

Interactions between lycaenid butterflies and ants in Peninsular Malaysia

KONRAD FIEDLER

Zoologisches Institut II, Universität Würzburg, Am Hubland, D-97074 Würzburg, Federal Republic of Germany

ABSTRACT. Interactions between lycaenid butterflies and ants were studied in West Malaysia. The majority of species observed so far are facultative myrmecophiles that interact unspecifically with a broad range of ant taxa. A few cases of obligate myrmecophily have been detected, including the polyphagous herbivore *Rapala dieneces* and two aphytophagous members of the genus *Logania* (Miletinae). Myrmecoxeny, i.e. lack of ant-association, is rare among Oriental lycaenids, but was confirmed for two species. Together with data from available literature, these findings suggest that in tropical rainforests the proportion of obligate myrmecophiles is not much higher than in the northern temperate zone. Therefore, myrmecophily has probably contributed little to the speciation events themselves, but its role in maintaining species diversity through resource partitioning and niche segregation remains to be discovered.

KEY WORDS: Ants, butterflies, evolution, Lycaenidae, Malaysia, mutualism, myrmecophily, rainforest, species diversity, zoogeography.

INTRODUCTION

Interspecific, and in particular mutualistic, interactions between organisms of very different kinds are extremely widespread and highly important in tropical ecosystems. Such symbiotic interactions are thought to play a significant, but still insufficiently understood role in the evolution and maintenance of species diversity (e.g. contributions in Huxley & Cutler 1991). Ants are well-known for their involvement in various interactions with plants, Homoptera and a large guild of myrmecophilous arthropods (Hölldobler & Wilson 1990). An interesting paradigm for ecological and evolutionary studies on myrmecophily is provided by the butterfly family Lycaenidae. Many lycaenid species are myrmecophilous, i.e. they live in association with ants. In general, myrmecophily occurs in the later larval stages, but in certain taxa interactions with ants also involve the pupae and/or adult butterflies. The larvae and pupae of lycaenids possess a variety of epidermal glands the secretions of which manipulate ant behaviour and thereby enable the caterpillars to enter into persistent associations with ants (Cottrell 1984, Fiedler 1991a, Malicky 1969, Pierce 1989).

It has been suggested by various authors that myrmecophily has made an important contribution to the evolution and diversification of lycaenid butterflies (e.g. Atsatt 1981). Specifically, Pierce (1984) described two possible ways in which ants could influence the evolution of diversity in the Lycaenidae: through increase of 'oviposition mistakes' in relation to ant-dependent host plant selection, and through local fragmentation of populations that require both specific host plants and host ants. Both mechanisms only apply to obligatorily myrmecophilous lycaenids, i.e. butterfly species that depend on specific host ants for survival (see Fiedler 1991a). A zoogeographic survey by Pierce (1987) provided supporting evidence that cases of very close, obligate myrmecophily are particularly common on the southern hemisphere. Her sample mostly covered subtropical faunal regions (India, Australia, South Africa) with limited representation of tropical rainforests.

It therefore remains to be tested whether myrmecophily has played an equally important role in the evolution of lycaenids in tropical rainforests. This biome supports by far the

D. S. Edwards et al. (eds.), Tropical Rainforest Research – Current Issues, 291–296.

largest number of species. From West Malaysia, for example, 395 Lycaenidae species are recorded (excluding the Riodinidae; Eliot 1992), and Seki *et al.* (1991) list 379 species for the island of Borneo alone. If the scenario of Pierce (1984, 1987) is true for the centre of the Oriental faunal region, then one should postulate that the majority of Malaysian lycaenid butterflies are myrmecophilous. Certainly a greater proportion of obligate myrmecophiles is to be expected in comparison with, for example, the northern temperate Holarctic region. In addition, many species should either be local endemics or should occur only in scattered, fragmented populations.

A complete analysis of this question is still out of reach, because life-history data are as yet available for only a fraction of Malaysian lycaenid butterflies. During the past decade, however, intensive efforts have been made to collect larger sets of pertinent biological data in Peninsular Malaysia. In this paper I will firstly outline some of the most recent findings, then discuss these observations in terms of the importance of myrmecophily and species diversity. Questions of endemism and population structure will not be addressed in this paper.

My observations were conducted in the vicinity of the Ulu Gombak Field Studies Centre of the Universiti Malaya (20 km north of Kuala Lumpur), in an advanced secondary dipterocarp forest at 200-400 m elevation.

OBSERVATIONS

Facultative myrmecophiles

Lycaenidae species whose larvae are unspecifically visited by various ant genera are termed facultative myrmecophiles. Such species are not fully dependent on ant-association for their survival, although tending ants can provide substantial fitness benefits (Pierce & Easteal 1986). Moreover, ant-associations are not necessarily universal or permanent in facultative myrmecophiles. A fraction of larvae can usually be found without attendant ants. Among facultative myrmecophiles, the intensity of ant-interactions varies from very loose and casual associations to almost persistent tending. Facultative interactions with ants are the dominant type of myrmecophily among Holarctic Lycaenidae (Fiedler 1991b).

During my fieldwork in West Malaysia I documented a number of facultative, non-specific ant-associations. These included *Actyolepis puspa* (Horsfield 1828), 7 species of *Jamides*, *Prosotas dubiosa* (Semper 1879), *Surendra florimel* Doherty 1889, *Arhopala buddha* Bethune-Baker 1903 and *A. metamuta* (Hewitson 1863), *Eooxylides tharis* (Geyer 1837), *Drupadia ravindra* (Horsfield 1828), *Hypolycaena othona* (Hewitson 1865), *Rapala iarbus* (Fabricius 1787) and *R. pheretima* (Hewitson 1863) (cf. Fiedler 1992a, b & in press a, b). Within the genus *Jamides*, there is no evidence that myrmecophily has played any role in speciation. In all *Jamides* species with documented life-histories, the caterpillars possess a full set of myrmecophily organs and are facultatively tended by an overlapping range of ant taxa (especially *Crematogaster*, *Rhoptromyrmex*, *Technomyrmex*, and *Camponotus*). No preferences of particular *Jamides* species for certain ants could be observed during oviposition or in the larval stage.

Weak and casual ant-associations occur in *Prosotas dubiosa* and *Hypolycaena othona*. Contacts with ants observed in the field were usually short, although in captivity more or less stable tending behaviour could be induced in ants such as *Crematogaster* or *Meranoplus mucronatus*.

Obligate myrmecophiles

Only a few species with an obligate symbiosis with ants have yet been found in West Malaysia. Prominent and well-documented examples include *Anthene emolus* (Godart 1824) and *Hypolycaena erylus* (Godart 1824) both of which are exclusively associated with the weaver ant, *Oecophylla smaragdina* (Jacobson 1912, Fiedler & Maschwitz 1989a). In both species, oviposition occurs on a variety of plants, but always in the presence of the host ant. The larvae are constantly tended by weaver ants throughout their development, and the ants carry the caterpillars to feeding places without harming them. A new case of obligate myrmecophily was observed in *Rapala dieneces* (Hewitson 1878). Caterpillars of this species feed on a variety of host plants (young trees and shrubs in the Fabaceae, Mimosaceae, Caesalpiniaceae, Annonaceae, Myrtaceae, Sapindaceae and further families), where they are invariably visited by 10-40 worker ants of an *Iridomyrmex* species (Dolichoderinae). If disturbed, these ants are highly aggressive and might be able to ward off potential enemies of the caterpillars (in spite of this, tachinid flies have repeatedly been bred from field-collected larvae). A first instar larva was once observed being carried by its host ants as if it was ant brood, and larvae as well as pupae can be found in the galleries of *Iridomyrmex* ants (A. G. Orr, personal communication). Another new obligate myrmecophile record with a similar life-history is the polyphagous *Drupadia theda* (C. & R. Felder 1862) whose larvae are specifically associated with a black *Crematogaster* species.

Myrmecoxenous species

Lycaenid larvae that do not possess a dorsal nectary organ have been termed myrmecoxenes by Kitching & Luke (1985). Myrmecoxenous caterpillars are hardly ever found in stable associations with ants. Hence, such larvae are not protected by ants against enemies. Whilst myrmecoxeny is associated with .a number of temperate zone lycaenids, very few confirmed records of tropical myrmecoxenes are available. In Malaysia, I have observed *Cheritra freja* (Fabricius 1793) and two *Curetis* species (*santana* (Moore 1858), *bulis* (Westwood 1851)) whose legume-feeding larvae are not tended by ants. In addition, the Rhamnaceae-feeding *Caleta roxus* (Godart 1824) has turned out to be myrmecoxenous; the larvae have neither a dorsal nectary organ nor eversible tentacle organs, and they are never tended by ants. Even in captivity, they were constantly ignored by *Iridomyrmex* and only very briefly antennated by *Rhoptromyrmex wroughtonii* (Fiedler in press a). For systematic reasons, myrmecoxeny in *Ch. freja* and *C. roxus* must be seen as a secondary character. In closely related genera like *Drupadia* or *Discolampa*, caterpillars are myrmecophilous.

The selective forces that have repeatedly promoted the loss of myrmecophily in a number of lycaenid lineages are still poorly understood. This phenomenon of secondary myrmecoxeny is particularly unexpected in tropical rainforests where ants are virtually omnipresent in large numbers.

Carnivorous species

In the almost exclusively paleotropical subfamily Miletinae, the caterpillars feed on sap-sucking Homoptera (aphids, coccids, or membracids). Some species even drink honeydew or feed on ant brood (Cottrell 1984). Certain Miletinae larvae are not tended by ants, but are largely ignored, even when feeding in trophobiotic associations of homopterans with ants (e.g. some *Allotinus* species, *Spalgis*, *Taraka*). All known Miletinae larvae lack a nectary organ and hence are morphologically myrmecoxenous. Some Miletinae taxa, however, are true myrmecophiles and may even be intensively tended.

Two novel examples of obligate myrmecophily in the subfamily Miletinae have been discovered. In the Oriental genus *Logania*, the caterpillars as well as the pupae live in permanent association with specific host ants (Fiedler 1993). The ants constantly harvest secretions from specialized epidermal organs. These organs occur on the prothorax and the seventh abdominal segment and consist of pairs of lateral hemispherical bulges equipped with dense clusters of pore cupola organs. In *Logania malayica* Distant 1884, young larvae are occasionally fed by the host ant, *Rhoptromyrmex wroughtonii*, with trophallactic regurgitations. Older larvae drink extrafloral plant nectar as well as honeydew from aphids and membracids. They also prey upon these homopterans. The adult butterflies of *L. malayica* spend their whole life, including feeding, mating and oviposition, in association with these host ants. Adults drink extrafloral nectar and homopteran honeydew, but never visit flowers. Like the caterpillars, the butterflies are never attacked by the ants.

A closely related species, *Logania marmorata* (Moore 1884), has a very similar life cycle, but utilizes a different, unrelated host ant (*Hypoclinea thoracicus* [Dolichoderinae]). Associations with specific ants have also been found in certain *Allotinus* species (with *Anoplolepis* or *Myrmicaria*) and in *Miletus* (with *Hypoclinea*; see Fiedler & Maschwitz 1989b; Maschwitz *et al.* 1985, 1988). Therefore, specializations towards particular host ants may have contributed to the radiation of Miletinae butterflies in South East Asia.

DISCUSSION

The observations summarized here indicate that in oriental rainforests facultative myrmecophily is prevalent among the Lycaenidae butterflies. This is consistent with available life-history data extracted from the literature, covering 112 of the 395 Lycaenidae species of Peninsular Malaysia (28.4%; see Eliot 1992, Fiedler 1991a). Of these 112 species, 63 (56.3%) are facultative myrmecophiles, 15 (13.4%) are myrmecoxenous and only 21 (18.8%) are obligatorily associated with specific ants. For 10 additional species the degree of myrmecophily is unknown, but they are probably facultative myrmecophiles as are their congeners. Three species are so poorly known that not even a tentative scoring of their relationship with ants is possible. In all, 75-85% of the species with recorded life-histories are associated with ants, but less than one fifth are specifically dependent on ants.

If only the most species-rich subfamily Lycaeninae (sensu Scott & Wright 1990, as adopted by Eliot 1992) is considered, the pattern becomes even more obvious. Of 93 species for which information is available, 61 (65.6%) are facultative myrmecophiles, 8 (8.6%) are myrmecoxenous, and 11 (11.8%) are obligatorily myrmecophilous. Again 10 additional species with insufficiently well-recorded life-histories are most likely to be facultative myrmecophiles, while the status of 3 species must be left open. These figures are very similar to the results of an analysis of the Palaearctic fauna (Fiedler 1991b), where the proportion of all myrmecophiles was about 80%, and obligate myrmecophiles accounted for roughly 10% of the entire Lycaenidae fauna.

There are two alternative explanations for this remarkable similarity in the proportions of myrmecophiles when comparing the Oriental and Palaearctic regions:

1) The present knowledge of Oriental lycaenids is too poor and strongly biased towards species with facultative ant-associations, possibly because of a still scanty representation of canopy species in the available data base. If so, one must then postulate that further intensive fieldwork will elucidate a considerable, and proportionally higher, number of additional obligate myrmecophilous systems.

2) The pattern is real, indicating quite a low proportion of obligate myrmecophiles in tropical rainforests in contrast to the semi-arid biotas of South Africa and Australia.

At present, it is impossible to definitively decide between these alternatives. Three lines of evidence, however, support the second view.

a) Obligate myrmecophiles are particularly attractive and apparent to students of natural history, hence such interactions are more likely to be reported, at least anecdotally, whereas even in well-surveyed areas like Europe or North America numerous facultative ant-associations have gone undetected for a long time.

b) Obligate and specific interactions with ants are much more likely to evolve in relation to abundant and dominant ant species that form large, long-lived colonies (Hölldobler & Wilson 1990). In the extremely complex and dynamic ant mosaics of tropical rainforests, only a few ant species, such as *Oecophylla smaragdina*, are sufficiently dominant or occur predictably to foster the establishment of stable, specific symbioses.

c) A comparative survey of myrmecophily data from other tropical rainforest areas (Africa, South America) revealed that in these regions, as well, the proportions of obligate myrmecophiles among lycaenid butterflies appear to be quite low (Fiedler 1991a).

Symbioses with ants might therefore have been less important in the radiation of tropical rainforest lycaenid butterflies than believed by some authors. Nevertheless, specific myrmecophily systems have repeatedly evolved in certain taxonomic subgroups. In the subfamily Miletinae there is increasing evidence that specific ant-associations are an important facet of their life-history patterns, and investigations into the role of ant-associations for resource partitioning and niche segregation among tropical lycaenids have just started. If myrmecophily has played only a limited role in speciation and radiation (i.e. in the origin of species diversity), this does not necessarily imply it has a minor role in the maintenance of species richness.

As a next step, a more complete inventory of individual life-histories is required, especially in the highly diverse faunas of tropical rainforests. In parallel, detailed case studies should focus on the mechanisms that allow coexistence of large species numbers with similar myrmecophily systems. In combination with other studies on interspecific interactions (e.g. ant-plant symbioses), investigations into lycaenid myrmecophily may then aid in the understanding of the general significance of mutualistic interactions in terms of the evolution and maintenance of tropical biodiversity.

ACKNOWLEDGEMENTS

My sincere thanks go to Dr. Azarae H. Idris and Prof. Dr. Yong Hoi Sen (Universiti Malaya, Kuala Lumpur) for their hospitality and help during my visits to the Ulu Gombak Field Studies Centre. Prof. Dr. Ulrich Maschwitz (University of Frankfurt) first introduced me to the biology of tropical lycaenids, and Lt. Col. John N. Eliot (Taunton, UK) graciously provided his unsurpassed expertise in the systematics of oriental lycaenids. Thanks also to Jutta Klein for her steady assistance during my studies. My fieldwork was generously supported by a post-doctoral fellowship from the Leibniz Award of the Deutsche Forschungsgemeinschaft (DFG) to Prof. Dr. Bert Hölldobler (University of Würzburg) as well as through DFG grant Fi 547/1-1.

LITERATURE CITED

ATSATT, P. R. 1981a. Lycaenid butterflies and ants: selection for enemy-free space. *American Naturalist* 118:638-654.

COTTRELL, C. B. 1984. Aphytophagy in butterflies: its relationship to myrmecophily. *Zoological Journal of the Linnean Society* 79:1-57.

ELIOT, J. N. 1992. *The butterflies of the Malay Peninsula*. Fourth edition, Malayan Nature Society, Kuala Lumpur. x + 595 pp., 69 pls.

FIEDLER, K. 1991a. Systematic, evolutionary, and ecological implications of myrmecophily within the Lycaenidae (Insecta: Lepidoptera: Papilionoidea). *Bonner zoologische Monographien* 31:1-210.

FIEDLER, K. 1991b. European and Northwest African Lycaenidae (Lepidoptera) and their associations with ants. *Journal of Research on the Lepidoptera* 28:239-257.

FIEDLER, K. 1992a. Notes on the biology of *Hypolycaena othona* (Lepidoptera: Lycaenidae) in West Malaysia. *Nachrichten des entomologischen Vereins Apollo, Frankfurt, N. F.* 13:65-92.

FIEDLER, K. 1992b. The life-history of *Surendra florimel* Doherty 1889 (Lepidoptera: Lycaenidae) in West

Malaysia. *Nachrichten des entomologischen Vereins Apollo, Frankfurt, N. F.* 13:107-135.

FIEDLER, K. 1993. The remarkable biology of two Malaysian lycaenid butterflies. *Nature Malaysiana* 18:35-43.

FIEDLER, K. in press a. The life-history of *Caleta roxus* (Lepidoptera: Lycaenidae). *Nachrichten des entomologischen Vereins Apollo, Frankfurt, N. F.* 14.

FIEDLER, K. in press b. Observations on the biology of *Eooxylides tharis* (Lepidoptera: Lycaenidae). *Nachrichten des entomologischen Vereins Apollo, Frankfurt, N. F.* 14.

FIEDLER, K. & MASCHWITZ, U. 1989a. The symbiosis between the weaver ant, *Oecophylla smaragdina*, and *Anthene emolus*, an obligate myrmecophilous lycaenid butterfly. *Journal of Natural History* 23:833-846.

FIEDLER, K. & MASCHWITZ, U. 1989b. Adult myrmecophily: the role of the ant *Anoplolepis longipes* in the feeding and oviposition behaviour of *Allotinus unicolor* (Lepidoptera: Lycaenidae). *Tyô to Ga* 40:241-251.

HÖLLDOBLER, B. & WILSON, E. O. 1990. *The ants.* Harvard University Press, Cambridge (Massachusetts). xii + 732 pp.

HUXLEY, C. R. & CUTLER, D. F. (eds) 1991. *Ant-plant interactions.* Oxford University Press, Oxford. xviii + 601 pp.

JACOBSON, E. 1912. Symbiose zwischen der Raupe von *Hypolycaena erylus* Godart und *Oecophylla smaragdina* Fab. *Tijdschrift voor Entomologie* 55:9-14.

KITCHING, R. L. & LUKE, B. 1985. The myrmecophilous organs of the larvae of some British Lycaenidae (Lepidoptera): a comparative study. *Journal of Natural History* 19:259-276.

MALICKY, H. 1969. Versuch einer Analyse der ökologischen Beziehungen zwischen Lycaeniden (Lepidoptera) und Formiciden (Hymenoptera). *Tijdschrift voor Entomologie* 112:213-298.

MASCHWITZ, U., DUMPERT, K. & SEBASTIAN, P. 1985. Morphological and behavioural adaptations of homopterophagous blues (Lepidoptera: Lycaenidae). *Entomologia Generalis* 11:85-90.

MASCHWITZ, U., NÄSSIG, W. A., DUMPERT, K. & FIEDLER, K. 1988. Larval carnivory and myrmecoxeny, and imaginal myrmecophily in Miletine lycaenids (Lepidoptera: Lycaenidae) on the Malay Peninsula. *Tyô to Ga* 39:167-181.

PIERCE, N. E. 1984. Amplified species diversity: a case study of an Australian lycaenid butterfly and its attendant ants. Symposium of the *Royal Entomological Society of London* 11 (The biology of butterflies):197-200.

PIERCE, N. E. 1987. The evolution and biogeography of associations between lycaenid butterflies and ants. *Oxford Surveys in Evolutionary Biology* 4:89-116.

PIERCE, N. E. 1989. Butterfly-ant mutualisms. Pp 299-324 in P. J. Grubb & J. Whittaker (eds). *Towards a more exact ecology.* Oxford: Blackwell Science Publishers.

PIERCE, N. E. & EASTEAL, S. 1986. The selective advantage of attendant ants for the larvae of a lycaenid butterfly, *Glaucopsyche lygdamus. Journal of Animal Ecology* 55:451-462.

SCOTT, J. A. & WRIGHT, D. M. 1990. Butterfly phylogeny and fossils. Pp 152-208 in O. Kudrna (ed). *Butterflies of Europe. Volume 2: Introduction to lepidopterology.* Wiesbaden: Aula-Verlag.

SEKI, Y., TAKANAMI, Y. & OTSUKA, K. 1991. *Butterflies of Borneo, Volume 2, no. 1: Lycaenidae.* Tobishima Corp., Tokyo, x + 113 pp. (English section), 72 pls.

A quantitative survey of the termite assemblage and its consumption of food in lowland mixed dipterocarp forest of Brunei Darussalam

DAVID T. JONES[1]

Department of Biology, Universiti Brunei Darussalam, Bandar Seri Begawan 2028, Brunei Darussalam

ABSTRACT. Preliminary results of sampling the termite assemblage in the lowland mixed dipterocarp forest of Batu Apoi Forest Reserve in Temburong District, Brunei Darussalam, are presented. Numbers of epigeal and arboreal nests in a one hectare plot were recorded and the populations of termites in soil and dead wood were measured. Termites were initially separated into four taxonomic groups: the family Rhinotermitidae, and the Termitidae subfamilies Termitinae, Macrotermitinae and Nasutitermitinae. The percentage of total density represented by each taxon was 9.2%, 28.8%, 33.9% and 28.1%, respectively. Density above-ground (513 m^{-2}), in dead wood (320 m^{-2}) and in soil (616 m^{-2}) suggests that the total termite population exceeds 1500 m^{-2}. Generalized rates of consumption by each taxon were used to estimate that total consumption of organic matter by termites was 35.2 g (dry weight) m^{-2} yr^{-1}, which represents approximately 3% of the annual litter production. *Odontotermes* spp. (Macrotermitinae) were responsible for consuming 67% of the total amount of organic matter taken by termites. Comparisons were made with data from Mulu National Park, Sarawak, and Pasoh Forest Reserve, Peninsular Malaysia. The combined data suggest that the density of Macrotermitinae (fungus-growers) increases with decreasing annual rainfall.

KEY WORDS: termites, density, rain forest, organic matter, consumption, Macrotermitinae, Borneo.

INTRODUCTION

Termites are a major invertebrate group in the tropics and can have a dramatic influence on the environment (see the reviews of Collins 1989, Wood & Sands 1978). Stork (this volume) recognizes termites as one of several key animal groups in tropical forests and emphasizes their contribution to decomposition processes and forest dynamics. In many tropical forests termites are present at very high densities and are the dominant arthropod detritivores. For example, in the lowland rain forests of Malaysia, termite populations can exceed 38 million individuals per hectare (Abe & Matsumoto 1979) and can represent 60% of individuals among the soil macro-invertebrate assemblage and 42% of its biomass (Collins 1979a). The ecological impact of termites is derived from this vast abundance.

Since approximately 63% of the net primary production in tropical forests enters directly into the decomposer system (Kukihara & Kikkawa 1986), termite feeding habits (reviewed by Wood 1978) play a critical role in the release of nutrients and carbon via the comminution and mineralization of dead plant material. In addition, termites modify the physical structure of the environment by translocating organic matter away from feeding sites and by building nests and galleries. Their role in soil genesis, altering physical and chemical properties of the soil, and the effect of their subterranean tunnels on water infiltration has been reviewed by Anderson (1988) and Lobry de Bruyn & Conacher (1990).

Given the cryptic habits of most termite species plus their heterogenous spatial distribution, appropriate sampling techniques are required to ensure their numbers are not

[1]Present address: Department of Entomology, Natural History Museum, Cromwell Road, London SW7 5BD, United Kingdom.

D. S. Edwards et al. (eds.), Tropical Rainforest Research – Current Issues, 297–305.

severely underestimated, as occurred in the survey by Stork & Brendell (1993). To date, only two studies have attempted to quantify the abundance and distribution of the termite fauna in Southeast Asian rain forests; Abe & Matsumoto (1979) worked in Pasoh Forest Reserve, Peninsular Malaysia, while Collins (1983) worked in Mulu National Park, Sarawak, east Malaysia.

This paper reports some general findings based on field work designed to assess the density and diversity of termites in the lowland rain forest of Brunei Darussalam, and makes estimates of their consumption of organic matter. Termites were separated initially into four taxa (the family Rhinotermitidae, and three Termitidae subfamilies i.e. Termitinae, Macrotermitinae and Nasutitermitinae) to allow direct comparisons with the results from Mulu and Pasoh.

MATERIALS AND METHODS

Study site

Field work was conducted in the Batu Apoi Forest Reserve, Temburong District, based at the Kuala Belalong Field Studies Centre. The terrain is extremely rugged, with steep sided, v-shaped valleys separated by sharp, narrow ridges. Hillslopes of 35° are common and occasionally exceed 55°. Geomorphological processes in the Belalong valley have been described by Dykes (this volume). The relatively thin soils, classified as orthic acrisols, are mainly silty clays derived from the eroded shale bedrock (Ross 1994). The vegetation was described as mixed dipterocarp forest by Ashton (1964). The Kuala Belalong area receives between 4000 - 4500 mm rainfall per annum (Dykes, this volume), with 4135 mm being recorded in the first year of measurement (Poulsen, this volume).

Termite studies were concentrated in a one hectare research plot (100 m x 100 m) on which a 5 m x 5 m grid had been marked using aluminium poles and string. The plot is situated on the north east facing slope of a spur leading from the West Ridge of the Belalong valley (115°10'W, 4°32'N). The ground surface is steeply sloping and the middle of the plot lies at an altitude of approximately 250 m. The location and topography of the plot is shown in Poulsen (this volume, referred to as Plot A in Figure 1). The plot is designated as a permanent research plot and efforts were made to minimise disturbance. Therefore, all soil pits and 'dead wood quadrats' (described below) were located outside the hectare plot within 40 m of its perimeter.

Density of epigeal and arboreal termite nests

The plot was censused systematically for epigeal mounds and arboreal nests. When searching, special attention was paid to the base of trees, buttresses, and the ground beside fallen trunks. All vegetation and the surface of tree trunks were inspected for arboreal nests to a height of 3 m. Mounds and nests were opened with minimal disturbance to the colony and termites collected for identification. Termites in the forest canopy were excluded from this survey.

The estimates of mean nest population size employed by Collins (1983) were used to produce figures for the total number of termites in epigeal and arboreal nests. As the relationship between nest volume and population size was unknown it was assumed that all nests contained the quoted mean population. The total arboreal and epigeal populations were combined to give the 'above-ground' population size within the hectare plot and the above-ground density of termites per m^2.

Density of termites in dead wood

Ten 'dead wood quadrats', each 5 m x 5 m in area, were randomly located. Within each quadrat all items of dead wood with a diameter greater than 1 cm were examined. Earlier qualitative examination revealed that twigs with a diameter less than 1 cm were rarely occupied by termites. The diameter and length of each item of dead wood was recorded before being opened with a machete to search for termites. The number of termites in each item was measured using a method of visual estimation developed by Collins (1983). As the item of dead wood was successively broken open and the termites revealed, the number of termites seen was counted at an appropriate order of magnitude (i.e. 10, 10^2 or 10^3). Specimens from each item were retained for identification. The total population of each taxon within the dead wood quadrats was converted to an abundance in dead wood per m^2 of forest floor.

Density of subterranean termites

Ten 'soil quadrats', each 0.25 m^2 (50 cm x 50 cm) in surface area were randomly located. Loose leaf litter was removed and the soil in each quadrat was then excavated to a depth of 30 cm. Soil and leaf litter was hand-sorted in trays on site and all termites preserved for identification and enumeration. Soil was removed as rapidly as was practicable in order to prevent termites escaping from the disturbed area.

Another four randomly positioned soil pits were dug in order to search for subterranean termite nests. Three of these additional pits were 3 m x 3 m in surface area and the fourth was 1 m x 3 m. In each case the soil was dug out to a depth of 30 cm to search for nests.

RESULTS

The density of termites

A total of 76 epigeal and arboreal nests were found in the hectare plot (Table 1), of which 46 (60% of nests) were mounds of *Dicuspiditermes nemorosus*. The above-ground density of individual termites totalled 513 m^{-2} (Table 1), of which 89% was represented by three species: *D. nemorosus* (41% of above-ground density), *Hospitalitermes medioflavus* (35%) and *Longipeditermes longipes* (12%).

A total of 204 items of dead wood were examined for the presence of termites. Of these, 31% contained termites, with some items having more than one species. The total density in dead wood was 320 m^{-2} (Table 2) of which the Rhinotermitidae, which feed mainly on sound or rotten wood, constituted the largest proportion at 36%. This sampling method possibly gives figures on the low side as isolated populations in larger logs and branches may have been overlooked.

The density of subterranean termites was 616 m^{-2} (Table 3). This figure was derived from two sources, firstly 592 m^{-2} averaged from the ten 0.25 m^2 soil pits, and secondly, an additional 24 m^{-2} due to eight subterranean nests found during the hectare plot census (listed in Table 1). The Macrotermitinae constituted 78% of soil termites, and all of these were *Odontotermes* spp. Very few individual termites were found in the leaf litter samples and these were included with the termites from the soil pit from above which the leaf litter was collected.

The use of a greater number of soil pits would have been preferable (see Baroni-Urbani *et al.* 1978) but limitations of time and man-power prevented this. Wood *et al.* (1977) reported that 78% - 92% of termites were collected by hand-sorting soil. It is believed that due to careful hand-sorting an equally high rate of extraction was achieved in this survey. Undoubtedly, some nests which did not protrude far above the soil surface will have been overlooked in the census of the hectare plot. In Pasoh, 25% of *Homallotermes foraminifer* nests went unrecorded during an initial census of a 200 m^2 area and were only found when the entire area was dug up (Abe & Matsumoto 1979).

Table 1. Termite nest densities and population densities estimated from the census of the hectare plot at Batu Apoi. Above-ground and subterranean nests are listed separately.

	Above-ground nest density (No. ha^{-1})	Population per nest (x 10^4)	Density (No. m^{-2})
Rhinotermitidae			
Parrhinotermes spp.	2	3*1	6
Termitinae			
Microcerotermes spp.	2	3*1	6
Termes spp.	6	3*2	18
Homallotermes spp.	3	1.26*3	4
Dicuspiditermes nemorosus	46	4.6*3	212
Subtotal	59		246
Nasutitermitinae			
Bulbitermes spp.	5	4.24*3	21
Longopeditermes longipes	8	7.87*3	63
Hospitalitermes medioflavus	3	60*4	180
Subulitermes spp.	1	3*1	3
Subtotal	17		267
Grand Total	76		513
	Subterranean nest density (No. ha^{-1})	**Population per nest (x 10^4)**	**Abundance (No. m^{-2})**
Termitinae			
Prohamitermes spp.	5	3*1	15
Labritermes spp.	1	3*1	3
Procapritermes spp.	1	3*2	3
Subtotal	7		21
Nasutitermitinae			
Subulitermes spp.	1	3*1	3
Grand Total	8		24

*1 Conservative estimate of nest population.

*2 from Collins (1983)

*3 from Abe & Matsumoto (1979)

*4 from Collins (1979b)

Very few termites were found below a depth of 20 cm. This is consistent with the findings of Abe & Matsumoto (1979) who observed at Pasoh that most termites occurred in the top 15 cm of the soil profile and were very scarce below 30 cm. During the excavation of the ten 0.25 m^2 pits and the four larger pits (total surface area 32.5 m^2) accumulations of termites were found but no hard subterranean nest structures (made of hardened soil or faecal material) were unearthed.

Table 2: Estimate of density of termites in dead wood (ten 5 m x 5 m quadrats sampled) in the forest at Batu Apoi. Number of woody items examined = 204; Number of items attacked by termites = 64 (31%).

	Density of termites in dead wood (No. m^{-2} of forest floor ± S.E.M.)
Rhinotermitidae	116 ± 52
Termitinae	98 ± 59
Macrotermitinae	11 ± 7
Nasutitermitinae	95 ± 45
Total	320

Table 3. Estimate of the density of termites in soils (from hand-sorting 10 soil pits 50 cm x 50 cm x 30 cm depth, and the number of subterranean nests {see Table 1}) at Batu Apoi.

	Density of termites (No. m^{-2} ± S.E.M.)		
	Soil pits	Soil nests	Total
Rhinotermitidae	11 ± 10	0	11
Termitinae	58 ± 23	21	79
Macrotermitinae	481 ± 295	0	481
Nasutitermitinae	42 ± 40	3	45
Total abundance	592	24	616

The total density of 1449 m^{-2} (Table 4) is an underestimate because termites infesting living trees and those inhabiting the canopy were ignored. For example, *Coptotermes* spp. often nest inside mature trees (Collins 1984) and colony populations can exceed 5 x 10^5 (Hadlington 1987). Therefore, the total density of termites in the forest at Batu Apoi will probably exceed 1500 m^{-2}. Of the four taxonomic groups none were recorded from either all ten 'dead wood quadrats' or from all ten 0.25 m^2 soil pits, reflecting the highly heterogenous distribution of termites.

Consumption of organic matter

Table 4 gives the total density and biomass of the four taxa (combining the data from Tables 1-3) and estimations of their consumption of organic matter. Termites are estimated to consume 35.2 g (d.w.) m^{-2} yr^{-1} of organic matter, of which 67% is accounted for by Macrotermitinae. As all the Macrotermitinae collected were *Odontotermes* spp. a rate of consumption specific to this genus was used to estimate the consumption

Table 4. Estimates of total density and biomass of termites and their consumption of organic matter in the lowland mixed dipterocarp forest of Batu Apoi. *[1] from Collins (1983); *[2] from Abe (1979); *[3] mean (alcohol w.w.) of 300 *Odontotermes* individuals collected from soil pits at Batu Apoi (this study); *[4] rate for Macrotermitinae is mean of two figures for *Odontotermes* spp. given in Collins (1983).

	Total density (No. m^{-2})	biomass of individual (mg w.w.)	Total biomass (g m^{-2})	Rate of consumption (mg d.w. g^{-1} day^{-1})	Total consumption (g d.w. m^{-2} yr^{-1})
Rhinotermitidae	133	3.0 *[1]	0.399	20 *[1]	2.91
Termitinae	417	1.674 *[2]	0.681	30 *[1]	7.46
Macrotermitinae	492	2.855 *[3]	1.405	46.2 *[4]	23.69
Nasutitermitinae	407	0.758 *[2]	0.309	10 *[1]	1.13
All termites	1449		2.794		35.19

by Macrotermitinae (see Table 4). The biomass (alcohol wet weight) of *Odontotermes* was measured as the mean of 300 individuals from the soil pits (after specimens had been blotted on filter paper to remove surface liquid). All other values of biomass and rates of consumption are from Collins (1983).

The annual small litter (leaves, twigs, fruits, flowers and trash) production at Batu Apoi is 10.67 t ha^{-1} yr^{-1} (C.A. Pendry, personal communicaion) but the production of large woody litter (> 2 cm diameter) is not known. Assuming that the ratio of woody litter to small litter at Batu Apoi is the same as that at Mulu (Proctor *et al.* 1983) then the total annual litter production would be 11.56 t ha^{-1}. In this case termites would be responsible for consuming 3.0% of the annual litter produced in the lowland forests of Batu Apoi.

DISCUSSION

Comparison of termite density between sites

Table 5 compares the density at Batu Apoi with that of Mulu and Pasoh. The Nasutitermitinae, Termitinae and Rhinotermitidae are all less abundant in Batu Apoi than at Mulu but overall termite density is similar due to the greater number of Macrotermitinae at Batu Apoi. The estimates for all taxa enumerated at Pasoh are higher than at the other two sites and the overall total would have been even larger had the Rhinotermitidae been estimated.

A total of 17 epigeal and arboreal nests were recorded from a one hectare plot at Mulu (Collins 1983). The resulting above-ground density of 249 m^{-2} was only half that recorded at Batu Apoi and much of this difference is due to the relatively low density of *Dicuspiditermes nemorosus* mounds at Mulu (2 ha^{-1} compared with 46 ha^{-1} at Batu Apoi). However, striking differences are seen when Batu Apoi is compared with Pasoh. A hectare plot at Pasoh had 249 epigeal and arboreal nests with a total above-ground density of 870 - 1520 m^{-2} (Abe & Matsumoto 1979). Of these nests, 131 were mounds of *D. nemorosus*, while 85 and 20 were mounds of *Homallotermes foraminifer* and *Macrotermes carbonarius* respectively. Only three nests of *Homallotermes* spp. were found at Batu Apoi and none at Mulu. Nests of *Macrotermes* spp. were not found in the plot at either Batu Apoi or Mulu.

At Mulu the density of termites in dead wood was 153 m^{-2} (Collins 1983), only half that of Batu Apoi. However, while the minimum diameter for items examined at Batu Apoi was 1 cm, it was only 2 cm at Mulu. Of the 46 items from Batu Apoi with a diameter in the range 1 - 2 cm only three contained termites.

Table 5. Comparisons of annual rainfall, the density of termites and their consumption of organic matter in the forests of Batu Apoi (Brunei Darussalam), Mulu (Sarawak) and Pasoh (Peninsular Malaysia).

	Mulu		Batu Apoi		Pasoh	
Source of data	Collins (1983)		This study		Abe & Matsumoto (1979)	
Rainfall (mm yr^{-1})	5107		4000 - 4500		2000	
	Density (No. m^{-2})	Consumption (g m^{-2} yr^{-1})	Density (No. m^{-2})	Consumption (g m^{-2} yr^{-1})	Density (No. m^{-2})	Consumption (g m^{-2} yr^{-1})
Rhinotermitidae	293	6.42	133	2.91	Not estimated	
Termitinae	612	11.21	417	7.46	1280-1730	23.76-31.32
Macrotermitinae	5	0.70	492	23.69	934-984	129.43-138.41
Nasutitermitinae	616	1.70	407	1.13	943-1093	2.23-3.47
Total	1526	20.03	1449	35.19	3157-3807	155.42-173.20

Excluding these 46 items has a negligible effect upon abundance since the populations in the three occupied twigs were very small. The higher annual rainfall at Mulu compared with Batu Apoi (see Table 5) may increase the initial rate of fungal and microbial attack on freshly fallen wood and thus help to make a greater proportion of the dead wood more palatable to termites and therefore more susceptible to invasion. The Rhinotermitidae are usually associated with wood which is already in a state of decay due to the presence of fungi (Sands 1969). The density of termites in dead wood at Pasoh was not estimated.

The density of subterranean termites at Batu Apoi is only about half that of Mulu (1125 m^{-2}; Collins 1979a) and only one third that of Pasoh (1898 m^{-2}; Abe & Matsumoto 1979). The absence of hard subterranean nests in 32.5 m^2 of soil pits at Batu Apoi indicates that their density may be low compared with Mulu and Pasoh. For example, Collins (1983) searched a 50 m^2 area at Mulu to a soil depth of 15 cm and recorded 15 nests. Abe & Matsumoto (1979) reported 4.5 nests of *Procapritermes* sp. in a soil pit of only 16 m^2 at Pasoh.

Subterranean nests may be relatively more vulnerable in steep areas experiencing soil instability, erosion and a higher level of water runoff. Ashton's (1964) description of the hillside soils in the Belalong valley states that "evidence of surface erosion and profile truncation was rarely absent". If physical conditions at Batu Apoi are less favourable for soil-nesting termites then epigeal mounds may also be exposed to similar destructive forces. Very few of the *D. nemorosus* mounds at Batu Apoi are free standing but instead are supported on the base of tree trunks or around the basal stems of palms and saplings, which may reflect the need for anchorage on steeper and more unstable slopes.

Comparison of consumption by termites between sites

The estimates in Table 5 indicate that consumption of organic matter by termites at Pasoh is approximately 8 times higher than at Mulu and about 4.5 times that at Batu Apoi. At Pasoh and Batu Apoi the Macrotermitinae are the dominant taxa in terms of consumption, being responsible for 80% - 83% and 67% of the total consumption at these respective sites. At Mulu the Macrotermitinae are less important and consume only 3.5% of the organic matter taken by termites. In terms of their overall consumption, termites consume an estimated 2.1% of the total annual litter production at Mulu (Collins 1983), 3.0% at Batu Apoi, and 14.7% - 16.3% at Pasoh (Matsumoto & Abe 1979). The high total consumption at Pasoh relative to the other two sites (Table 5) is a result of the greater overall density of termites, but more importantly, it is due to the dominance of the Macrotermitinae which have relatively higher weight-specific rates of consumption (Collins 1989). However, as Collins (1983) states, the use of generalized rates of consumption to estimate total consumption by termites is only approximate and figures should be regarded as minimum values.

The Macrotermitinae cultivate mutualistic fungi on faecal combs in their nests, a feeding habit unique amongst termites. This enables them to feed on fresh litter while other groups of termites prefer litter which has already been partially decayed. The mixed dipterocarp forests of Mulu and Pasoh lie at opposite ends of a climatic spectrum (see rainfall figures in Table 5). Collins (1980, 1983) speculates that as a result the Macrotermitinae are more prevalent in drier forests such as Pasoh where they dominate the decomposition process but in the much wetter forests of Mulu, where the activity of free-living fungi and microbes is far greater, they lose their ecological advantage and are virtually absent. The intermediate level of rainfall at Batu Apoi may have given rise to the relatively high density of *Odontotermes* compared with Mulu, but it is still too wet for the mound-building *Macrotermes* spp. favoured at Pasoh. In the Western Ghats of India, Ferry (1992) reported that the Macrotermitinae are the dominant Isoptera taxa in drier forests but become less common at sites with increasing rainfall. Wood & Sands (1978) suggested that the Macrotermitinae's share of the organic matter consumed by termites in West African savannas was lower in those areas with higher climatic humidity.

The experiments of Matsumoto & Abe (1979) showed that litter removal at Pasoh was dominated by invertebrate feeding activities, in particular the termite species *Macrotermes carbonarius* and *Longipeditermes longipes* which both feed on leaf litter. In contrast, at Mulu, termites consume less and play a smaller role in the initial decomposition process (Anderson *et al.* 1983, Collins 1983). With its higher density of *Odontotermes* (which have been observed feeding on leaf litter at Batu Apoi; unpublished data, D.T. Jones) and *L. longipes*, termites at Batu Apoi will have a greater role in the consumption and comminution of fresh litter when compared with Mulu but less than at Pasoh.

Although termites may only consume a small proportion of the annual litter production in South east Asian ever-wet rain forests such as Batu Apoi and Mulu, their impact on the forest ecosystem is still highly significant through indirect means. Termites introduce microorganisms into their feeding and nest sites (Collins 1988, Sands 1969) while comminution of litter increases surface area thereby encouraging additional microbial and fungal activity and thus enhancing overall rates of decomposition. The feeding and nesting habits of several species can also affect forest dynamics by killing mature trees and saplings. For example, Tho (1982) found that *Microcerotermes dubius* kills a range of tree species and is involved in canopy gap formation in the primary forests of Peninsular Malaysia. Similar behaviour by *M. dubius* has been observed in the forests of Brunei Darussalam (unpublished data, D.T. Jones).

ACKNOWLEDGEMENTS

This work was funded by a research grant from the Universiti Brunei Darussalam (grant No. UBD/T6/RG.02). The author is grateful for assistance in the field from the Venturers and staff of Raleigh International. Thanks are given to Bert Orr, Paul Eggleton and an anonymous referee for their helpful comments on the manuscript.

LITERATURE CITED

ABE, T. 1979. Studies on the distribution and ecological role of termites in a lowland rain forest of West Malaysia. 2. Food and feeding habits of termites in Pasoh Forest Reserve. *Japanese Journal of Ecology* 29:121-135.

ABE, T. & MATSUMOTO, T. 1979. Studies on the distribution and ecological role of termites in a lowland rain forest of West Malaysia. 3. Distribution and abundance of termites in Pasoh Forest Reserve. *Japanese Journal of Ecology* 29:337-351.

ASHTON, P.S. 1964. Ecological studies in the mixed dipterocarp forests of Brunei State. *Oxford Forestry Memoirs* No. 25. Clarendon Press, Oxford.

ANDERSON, J.M. 1988. Invertebrate-mediated transport processes in soils. *Agriculture, Ecosystems and Environment* 24:5-19.

ANDERSON, J.M., PROCTOR, J. & VALLACK, H.W. 1983. Ecological studies in four contrasting lowland rain forests in Gunung Mulu National Park, Sarawak. 3. Decomposition processes and nutrient losses from leaf litter. *Journal of Ecology* 71:503-527.

BARONI-URBANI, C., JOSENS, G. & PEAKIN, G.J. 1978. Empirical data and demographic parameters. Pp. 5-44 in Brian, M.V. (ed.). *Production ecology of ants and termites*. IBP Vol. 13. Cambridge University Press, Cambridge.

COLLINS, N.M. 1979a. A comparison of the soil macrofauna of three lowland forest types in Sarawak. *Sarawak Museum Journal* 27:267-281.

COLLINS, N.M. 1979b. Observations on the foraging activity of *Hospitalitermes umbrinus* (Haviland), (Isoptera: Termitidae) in Gunung Mulu National Park, Sarawak. *Ecological Entomology* 4:231-238.

COLLINS, N.M. 1980. The effect of logging on termite (Isoptera) diversity and decomposition processes in lowland dipterocarp forests. Pp. 113-121 in Furtado, J.I. (ed.). *Tropical ecology and development.* International Society of Tropical Ecology, Kuala Lumpur.

COLLINS, N.M. 1983. Termite populations and their role in litter removal in Malaysian rain forests. Pp. 311-325 in Sutton, S.L., Whitmore, T.C. & Chadwick A.C. (eds). *Tropical rain forest: Ecology and Management.* Blackwell, Oxford.

COLLINS, N.M. 1988. Termites. Pp. 196-211 in Earl of Cranbrook (ed.). *Key environments - Malaysia.* Pergamon, Oxford.

COLLINS, N.M. 1989. Termites. Pp. 455-471 in Lieth & Werger (eds). *Tropical rain forest ecosystems: Biogeography and ecological studies.* Elsevier, Oxford.

DYKES, A. (this volume). Analysis of factors contributing to the stability of steep hillslopes in the tropical rainforest of Temburong, Brunei Darussalam.

FERRY, B. 1992. Distribution of the important litter decomposing termites (Isoptera) in the Western Ghats forests of Karnataka (India). *Pedobiologia* 36:193-211.

HADLINGTON, P. W. 1987. *Australian termites.* New South Wales University Press, Kensington, NSW.

KUKIHARA, Y. & KIKKAWA, J. 1986. Trophic relations of decomposers. Pp. 127-160 in Kikkawa, J. & Anderson, D.J. (eds). *Community ecology: Pattern and process.* Blackwell, Oxford.

LOBRY DE BRUYN, L.A. & CONACHER, A.J. 1990. The role of termites and ants in soil modification: A review. *Australian Journal of Soil Research* 28:55-93.

MATSUMOTO, T. & ABE, T. 1979. The role of termites in an equatorial rain forest ecosystem of West Malaysia. 2. Leaf litter consumption on the forest floor. *Oecologia* 38:261-274.

POULSEN, A.D. (this volume). The herbaceous ground flora of the Batu Apoi Forest Reserve, Brunei Darussalam.

PROCTOR, J., ANDERSON, J.M., FOGDEN, S.C.L. & VALLACK, H.W. 1983. Ecological studies in four contrasting lowland rain forests in Gunung Mulu National Park, Sarawak. 2. Litterfall, litter standing crop and preliminary observations on herbivory. *Journal of Ecology* 71:261-283.

ROSS, S.M. 1994. Soils and hydrology. Chapter 4 in Cranbrook, Earl of, & Edwards, D.S. *Belalong, A tropical rainforest.* Sun Tree Publishing, Singapore.

SANDS, W.A. 1969. The association of termites and fungi. Pp. 495-524 in Krishna K. & Weesner, F.M. (eds). *Biology of termites.* Vol. 1. Academic Press, New York & London.

STORK, N. (this volume). Tropical forest dynamics: the faunal components.

STORK, N.E. & BRENDELL, M.J.D. 1993. Arthropod diversity studies in lowland rainforest of Seram, Indonesia. Pp. 115-130 in Edwards, I.H., Macdonald, A.A. & Proctor, J. (eds). *The natural history of Seram.* Intercept, Andover.

THO, Y.P. 1982. Gap formation by the termite *Microcerotermes dubius* in lowland forests of Peninsular Malaysia. *Malaysian Forester* 45:184-192.

WOOD, T.G. 1978. Food and feeding habits of termites. Pp. 245-292 in Brian, M.V. (ed). *Production ecology of ants and termites.* IBP Vol. 13. Cambridge University Press, Cambridge.

WOOD, T.G., JOHNSON, R.A., OHIAGU, C.E., COLLINS, N.M. & LONGHURST, G. 1977. *Ecology and importance of termites in crops and pastures in northern Nigeria.* Project report. Centre for Overseas Pest Research, London.

WOOD, T.G. & SANDS, W.A. 1978. The role of termites in ecosystems. In Brian, M.V. (ed.). *Production ecology of ants and termites.* IBP Vol. 13. Cambridge University Press, Cambridge.

Resource use in a freshwater fish community of a tropical rainforest stream in northern Borneo

SATISH C. CHOY[1], SALWANA A. LATIF *and* YONG N. YUNG

Universiti Brunei Darussalam, Bandar Seri Begawan 2028, Brunei Darussalam

ABSTRACT. Partitioning of key environmental resources (food, space and time) was studied in a freshwater fish community in the Belalong River, Brunei Darussalam, northern Borneo. Of a total of twenty-five species recorded, eight were confined to the lentic (slow-flowing) zone and seven to the lotic (fast-flowing) zone of the stream. An additional ten species were found in both these zones. In the lentic zone, fish species partitioned the habitat vertically into surface, pelagic, demersal, benthic and substratum niche types. In the lotic zone this stratification was not observed as the vertical component was compressed. Here, the abundance of all but two species of fish was low. Ecological indices (prey diversity, prey dominance, niche breadth and niche overlap) were used to analyse the data on food items ingested by each species. Although there was some overlap in food niche, the fish species were ecologically separated by differences in morphological features related to foraging tactics, feeding behaviour and space utilisation. Species that were close trophically either occupied different micro-habitats or foraged at different times. Results indicate that differences in the utilisation of space, time and food play an important role in the co-existence of species in the freshwater communities.

KEY WORDS: fish, lentic, lotic, resource partitioning, space, food, time, Brunei Darussalam.

INTRODUCTION

Although tropical communities are considered to be complex and highly structured, with the presence of many specialised species (Goulding 1980, Lowe-McConnell 1975), extreme specialisation need not be necessary for the existence of high species diversity (Connell 1978; Sale 1982). Differential use of environmental resources by closely-related sympatric species is thought to be a mechanism for their co-existence within a community (Pianka 1981, Schoener 1986). Fishes, like other animals, partition resources along habitat, time and food axes (Helfman 1978, Lowe-McConnell 1987, Wooton 1992). Although habitat and food resource use and partitioning in tropical freshwater fishes have been extensively studied (Angermeier & Karr 1983, Balon & Stewart 1983, Bishop 1973, Inger & Chin 1962, Moyle & Senanayake 1984, Watson & Balon 1984a, 1984b), little consideration has been simultaneously given to all three resource dimensions: space, food and time.

The aim of the present study was to determine patterns of resource partitioning in a tropical freshwater fish community and see if such partitioning could explain the co-existence of sympatric species.

MATERIAL AND METHODS

Fieldwork was carried out in a slow-flowing (lentic) zone and a fast-flowing (lotic) zone of the Belalong River within the Batu Apoi Forest Reserve, close to the Kuala Belalong Field Studies Centre (KBFSC) in Temburong district, Brunei Darussalam. The Belalong River is

[1]Present address: DPI Water Resources Technical Centre, 1345 Ipswich Road, Rocklea, Qld 4106, Australia.

D. S. Edwards et al. (eds.), Tropical Rainforest Research – Current Issues, 307–314.

about 5-30 m wide and 0.3-5 m deep and runs through primary mixed dipterocarp forest. Details of its hydrology and water quality are given in Dykes (this volume) and Choy (this volume), respectively.

In the study area the deep, slow-flowing channel (lentic zone) was 3-7 m wide, 1.0-3.0 m deep and flowed at about 0.1 m s^{-1}. The shallow, fast-flowing (lotic zone) was 5-15 m wide, 0.2-1.0 m deep and flowed at 0.5-1.0 m s^{-1}. The fish species composition in each of these zones is different, but with some overlap (Choy, this volume).

Observations on microhabitat utilisation were made visually using mask and snorkel. Investigations relating to time utilisation were based on visual observations and on diurnal variation in fish catches, using hand nets and gill nets. To study food utilisation, the stomach contents and certain morphological features of the fish were analysed. The morphological attributes measured were biomass, total length, standard length, head length, head width, gape width, cleft length, eye diameter, gut weight, gut length and gut volume. The morphology of the head, in terms of the orientation of the mouth, its protrusibility and the presence or absence of barbels was also examined. These attributes are considered to be those most likely to reflect habitat utilisation, diet and foraging tactics of the fish (Watson & Balon 1984a). Description of the morphological features, their measurement and ecological significance can be found in Gatz (1979a, 1979b) and Watson & Balon (1984a). For stomach content analysis, all fish caught were either dissected while fresh or preserved in 10% formalin. Stomach contents were examined under the microscope and the food items were identified to the lowest possible taxon. Gut volume was calculated based on its cylindrical dimension and the volume of each food item was calculated from its proportion in the gut multiplied by the total volume of the gut.

To measure the prey diversity in each species, the Shannon-Wiener diversity index was used:

$$H' = -\sum p_i \ln p_i$$

where p_i is the volume or frequency of occurrence of a food item (i^{th}) in the stomach.

Prey evenness within a species was calculated using: $E = H'/H_{max}$, where $H_{max} = \ln N$, N being the total number of food items consumed by the species. The degree of dominance of a particular food item in each species was calculated using the Berger-Parker Index: $\underline{d} = d_{max}/N$, where N is the total volume of all food items and d_{max} is the volume of the most abundant food item. Since $\underline{d}$ is obtained as a fraction, it has been expressed in its reciprocal form. Niche breadths of each sympatric species were estimated using the Brillouin Index:

$$HB = \frac{\ln N! - \sum \ln(n_i)}{N}$$

where n_i is the volume of each food item and N is the total volume of all food items in each species. For $N > 10$, the Sterling's Slacker approximation ($\ln N! = n \ln N - N$) was used. Niche overlap between the fish species was calculated using Pianka's Index:

$$O_{jk} = \frac{\sum p_{ij} p_{ik}}{\sqrt{\sum (p_{ij})^2 (p_{ik})^2}}$$

where p_{ij} and p_{ik} are volumes of the ith food item used by j^{th} and k^{th} species. The value of this index varies between 0 (no overlap) and 1 (total overlap). For justification, strengths and shortcomings of the above-mentioned indices see Magurran (1988).

RESULTS

Species composition and microhabitat utilisation

A total of twenty five species of fish were recorded from the Belalong River near KBFSC (Table 1). Of these, eight species were confined to the lentic (slow-flowing, deep) zones and seven to the (fast-flowing, shallow) lotic zones. An additional ten species were found in both these zones. Most of these latter were daytime lentic dwellers which made nocturnal foraging excursions into the lotic zone. For details on distribution and ecological zonation of these fish see Choy (this volume).

The majority of the species in the lentic zone belong to the family Cyprinidae while those in the lotic zone belong to the family Balitoridae, particularly the subfamily Gastromyzontinae. Fish of the lentic zone are mainly surface, pelagic and demersal dwellers, only one is strictly benthic (Table 1). Juveniles of many pelagic species were observed near the banks and in flood pools while those of benthic species were found in flood pools and in the accumulated plant litter on the beds of streams.

Fishes in the lotic zone are mainly demersal and benthic species, with the exclusive members of this community (the gastromyzontins) belonging to the latter group (Table 1).

Table 1. Fish species composition (numbers caught) and microhabitat utilisation. L, lentic zone; Lo, lotic zone; Dl, diurnal; N, nocturnal; S, surface; P, pelagic; D, demersal; B, benthic.

Species	Composition		Microhabitat	Activity	
	L	Lo		Dl	N
Nematabramis steindachneri	21	0	S	21	4
Rasbora argyrotaenia	17	0	S	17	0
Barbodes collingwoodi	11	0	P, D	5	6
Hampala bimaculata	8	2	D	3	7
Tor douronensis	6	2	P, D	2	6
Lobocheilus bo	5	2	D	2	5
Osteochilus spilurus	5	0	P, D	3	2
Puntius binotatus	4	0	P, D	0	4
Rasbora sumatrana	3	0	S, P	1	0
Glyptothorax major	2	1	D, B	1	2
Rasbora sp.	1	0	S, P	1	0
Pseudogobiopsis sp.	1	0	B	1	0
Anguilla marmorata	1	1	D, B	0	2
Mystus baramensis	1	1	D, B	0	2
Clarias teijsmanni	1	1	D, B	0	2
Pangio mariarum	1	1	D, B	1	1
Macrognathus maculatus	1	1	D, B	1	1
Silurichthys hasselti	1	1	D, B	1	1
Paracrossochilus acerus	0	12	B	12	0
Gastromyzon borneensis	0	8	B	8	0
Neogastromyzon nieuwenhuisi	0	3	B	3	0
Gastromyzon monticola	0	3	B	3	0
Gastromyzon lepidogaster	0	2	B	2	0
Gastromyzon ridens	0	1	B	1	0
Protomyzon whiteheadi	0	1	B	1	0

Time utilisation: activity patterns

In the lentic zone five species (*Nematabramis steindachneri*, *Rasbora argyrotaenia*, *R. sumatrana*, *Rasbora* sp. and *Pseudogobiopsis* sp.) were mainly diurnally active while eight species were mainly nocturnally active (Table 1). The remaining five species seemed to be active at both times. The seven species confined only to the lotic zone are all diurnal. The ten migratory species are bottom-dwelling and are either nocturnal or active during both times. However, their presence in the lotic zone was mainly during nocturnal hours.

Fish morphology in relation to micro-habitat and diet

Fishes confined to the lotic zone (the gastromyzontins and *Paracrossocheilus acerus*) are generally smaller and more dorso-ventrally flattened (low CI and IVF values) than those of the lentic zone (predominantly cyprinids) (Table 2). Most of the former are benthic and have fins which are modified for clinging on to the substrate. These and other benthic species (such as *Lobocheilus bo*, *Mystus baramensis* and *Leiocassis micropogon*) have their eyes placed more dorsally than demersal, pelagic or surface species. Their mouths are ventrally placed, enabling them to feed on items found on the substrate. Surface and pelagic species are more laterally compressed (deep-bodied) than demersal or benthic ones. Surface dwellers such as *Nematabramis steindachneri* and *Rasbora argyrotaenia* are small and their mouths are placed more dorsally than species of the other strata. This allows them to feed on items (such as insects) falling on the surface of the water.

The relative gut lengths of the herbivores are greater than those of the carnivores (Table 2). In general, the gut of most species is relatively long, reflecting their omnivorous diets (see next section).

Diet

Items identified in the stomachs of the fishes were plant material, insect parts, fish bones and scales, nematodes and sediment (sand and small pebbles). Unidentifiable material was clumped and greenish, probably comprising partly digested food. Many specimens had either empty stomachs or had in them only unidentifiable material; these have been excluded from the data (Table 3). Similarly, nematodes which were ubiquitous and are thought to be part of the stomach fauna have not been included.

Most species had a predominance of either plant or insect material. Three species (*Nematabramis steindachneri*, *Puntius binotatus* and *Mystus baramensis*) did not have any plant material in their stomachs. Insects contained in the stomachs of the surface dwellers (*N. steindachneri* and *R. argyrotaenia*) belonged mainly to the terrestrial orders Diptera, Hemiptera, Homoptera, Hymenoptera, Coleoptera, Orthoptera and Lepidoptera whereas those contained in the stomachs of bottom dwellers (*M. baramensis*, *Leiocassis micropogon*, *P. acerus* and *N. nieuwenhuisi*) belonged to the aquatic larvae of the orders Ephemeroptera, Trichoptera, Odonata and Plectoptera. Pelagic species had a combination of all these orders in their stomachs. Plant material in the stomachs of the lotic dwellers (*P. acerus* and the gastromyzontins) was mainly algal whereas those in the stomachs of the lentic dwellers (*Barbodes collingwoodi*, *Hampala bimaculata*, *Tor douronensis* and *Osteochilus spilurus*) were remains of vascular plants. Only *H. bimaculata* and *B. collingwoodi* had remains of fish in their stomachs. No crustaceans or molluscs were identified in the stomach of any fish.

Large lentic-dwelling omnivores such as *H. bimaculata*, *O. spilurus*, *T. douronensis* and *L. bo*, had greater prey diversity and prey evenness than the food specialists and their smaller lotic counterparts (*P.acerus* and *N. nieuwenhuisi*). As expected, the niche breadths of the food generalists were greater than those of the food specialists. Insects were the dominant prey for all but two (*O. spilurus* and *L. bo*) lentic species whereas they were dominant in

Table 2. Morphological characteristics of the seventeen predominant fish species. The last four species are found exclusively in the lotic zone. The wet weights (Wt) and standard lengths (SL) are mean values. RGL, relative gut length (GL/SL); RED, relative eye diameter (ED/SL); CI, compression index (greatest body depth/greatest body width); IVF, index of ventral flattening; M, position of mouth; EP, position of eyes; B, barbels; L, lateral; D, dorsal; L/D dorso-lateral; T, terminal to oblique; V, ventral; A, absent; P, present.

Species	Wt (g)	SL (cm)	RGL	RED	CI	IVF	M	EP	B
Nematabramis steindachneri	3.0	5.7	11.48	0.059	2.33	0.70	D	L	P
Rasbora argyrotaenia	3.7	5.8	11.04	0.072	2.00	0.67	D,T	L	A
R. sumatrana	3.9	6.1	11.13	0.070	1.98	0.64	D,T	L	A
Rasbora sp.	2.9	4.6	10.96	0.069	2.03	0.66	D,T	L	A
Barbodes collingwoodi	29.0	10.7	11.51	0.071	2.24	0.52	T	L	P
Hampala bimaculata	71.2	15.2	11.55	0.054	2.01	0.48	T	L	P
Tor douronensis	38.8	15.0	6.35	0.057	1.71	0.51	T	L	P
Lobocheilus bo	45.5	12.7	7.69	0.066	1.75	0.51	T	L,D	P
Osteochilus spilurus	35.1	11.2	8.65	0.063	2.03	0.51	T	L	P
Puntius binotatus	33.5	15.5	9.03	0.039	2.20	0.51	T	L	P
Mystus baramensis	48.1	14.4	11.81	0.035	1.20	0.55	T	L,D	P
Macrognathus maculatus	40.9	24.0	3.75	0.025	1.69	0.57	T	L	A
Leiocassis micropogon	33.9	13.3	9.61	0.032	1.18	0.52	T	L,D	P
Paracrossochilus acerus	1.9	4.3	5.63	0.041	1.24	0.52	V	L,D	P
Gastromyzon borneensis	3.3	4.0	12.19	0.030	0.71	0.31	V	D	P
Neogastromyzon nieuwenhuisi	0.9	4.6	10.75	0.054	0.93	0.37	V	D	P
Gastromyzon monticola	1.6	3.2	12.38	0.029	0.70	0.30	V	D	P

only one (*N. nieuwenhuisi*) lotic species. A cladogram of niche overlap, based on diet is shown in Figure 1. Although there were some large niche overlaps amongst lentic species, most of them had either different micro-habitats or foraged at different times (see Table 1).

Resource partitioning

Based on resource use by each fish species, a model of resource partitioning was constructed (Figure 2). Diet for species which had empty stomachs was obtained from Inger & Chin (1962) and Bishop (1973). As a result of resource partitioning, very few species occupied the same niche. Those that did were either the migratory carnivores or the small gastromyzontin herbivores. The abundance of these was low and the former could be expected to be in different areas at any time while the latter had an abundance of food and space resource.

DISCUSSION

Fish assemblages in tropical rainforest streams are highly structured (Moyle & Senanayake 1984, Watson & Balon 1984a, 1984b). However, the degree of ecological segregation can differ amongst streams (Moyle & Senanayake 1984). In the Belalong River, it is clear that the degree of ecological segregation is very marked when all three niche dimensions (space, time and diet) are considered. Longitudinal zonation alone accounts for 60% segregation: the overlap between the lotic and lentic zone is only 40%. Those species confined to the lotic zone are morphologically well adapted to living under fast-flowing conditions while species in the lentic zone are not. Within each of these zones, there is further habitat partitioning, particularly in the

Table 3. Stomach contents of some freshwater fish in the Belalong River. V, mean percentage volume; F, percentage frequency of occurrence.

Species	Plants		Insects		Fish		Unident.	
	V	F	V	F	V	F	V	F
Nematabramis steindachneri	0	0	79.3	100	0	0	20.7	37.9
Rasbora argyrotaenia	1.5	8.3	98.0	100	0	0	0.9	16.7
Barbodes collingwoodi	7.1	33.3	45.9	77.8	1.0	11.1	45.9	100
Hampala bimaculata	15.4	61.5	16.5	38.5	1.6	5.5	68.0	67.0
Tor douronensis	29.3	16.7	66.7	50.0	0	0	3.8	50.0
Lobocheilus bo	24.5	69.2	3.6	15.4	0	0	71.8	84.6
Osteochilus spilurus	52.6	83.3	9.4	25.0	0	0	37.9	51.3
Puntius binotatus	0	0	80.2	100	0	0	19.8	100
Mystus baramensis	0	0	87.5	100	0	0	12.5	100
Leiocassis micropogon	1.3	25.0	90.6	100	0	0	8.0	75.0
Paracrossochilus acerus	59.6	30.8	13.2	23.1	0	0	26.3	23.1
Gastromyzon borneensis	42.9	33.3	0	0	0	0	57.1	25.0
Neogastromyzon nieuwenhuisi	9.8	25.0	81.3	50.0	0	0	8.9	75.0
Gastromyzon monticola	10.0	33.3	0	0	0	0	90.0	100

lentic zone. Fish species diversity is correlated with habitat complexity (Gorman & Karr 1978), with most concentrated into deep pools (Angermeir & Karr 1983).

The diversity of body shape and morphological specialisation associated with habitat and feeding in the Belalong River is typical of tropical streams elsewhere (Bishop 1973, Gatz 1979, Moyle & Senanayake 1984, Watson & Balon 1984a). It has been suggested that morphological diversity reflected ecological diversity, implying co-evolution to reduce competition for limited resources (Gatz 1979).

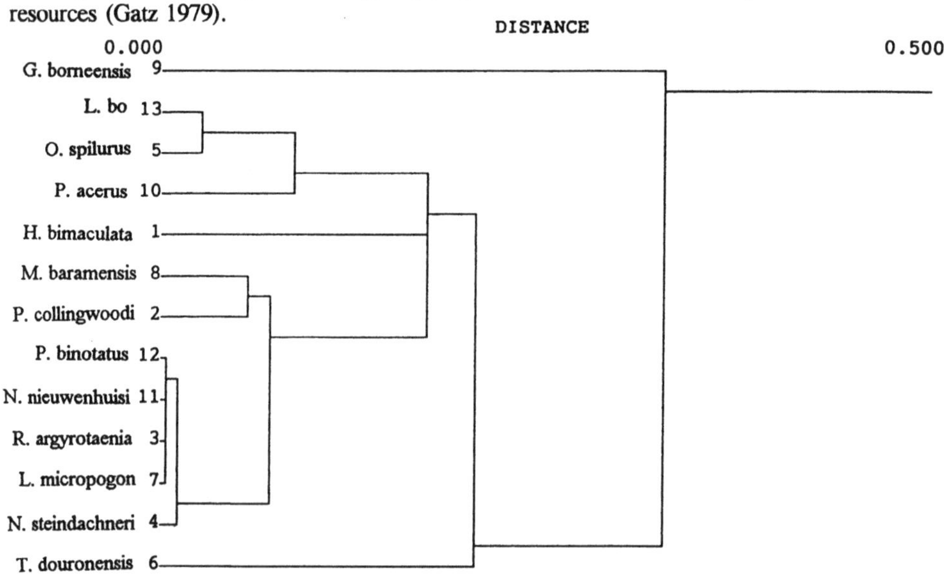

Figure 1. Cladogram of the thirteen predominant fish species, based on diet similarity. Scale is euclidean distance obtained using single linkage method. For full names of species see Table 1.

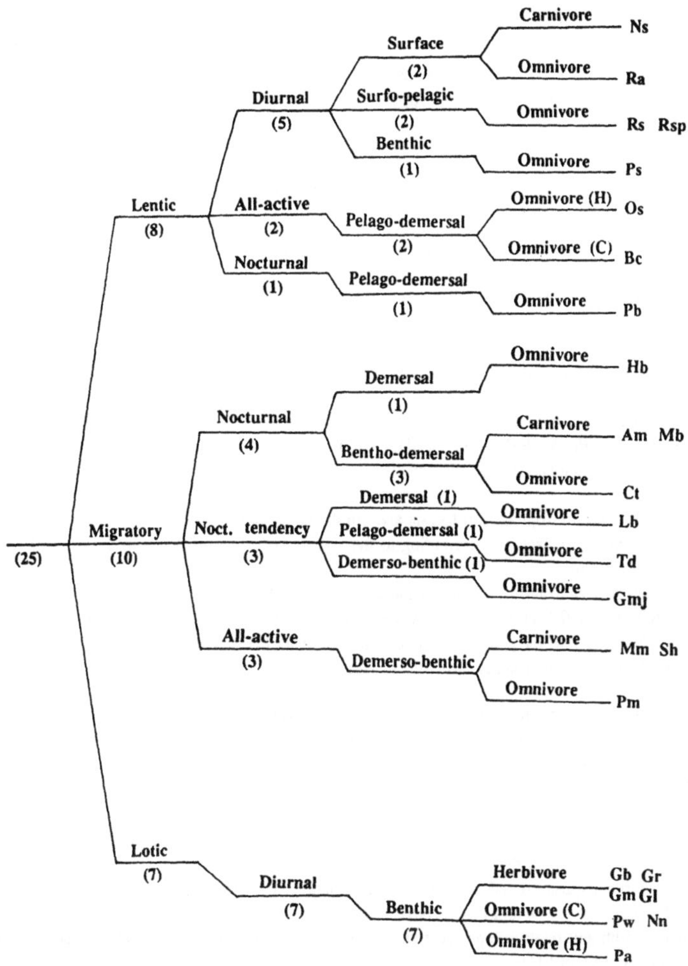

Figure 2. A model of ecological segregation based on partitioning of space, time and diet. Lentic and lotic refer to species resident in this zone; migratory refers to species that move between these two zones; diurnal and nocturnal refer to times of activity; all-active, active at all times; omnivore (H) and omnivore (C) refer to omnivorous diets but with herbivorous and carnivorous tendencies, respectively; numbers in brackets indicate the number of species in that category; acronyms are of species listed in Table 1.

Most studies of the feeding habits of tropical stream fish have found a large degree of dietary overlap and a heavy reliance on allochthonous food, mainly terrestrial arthropods and detritus of terrestrial origin (Lowe-McConnell 1975). However, Moyle & Senanayake (1984) found that overlap was low and that there was heavy reliance on autochthonous material, mainly algae and aquatic invertebrates. Methodology and time of study have been suggested as possible reasons for the different findings (Moyle & Senanayake 1984). The present study suggests that reliance on autochthonous or allochthonous material depended on whether the fish inhabited lotic or lentic zones and, in the case of the latter, whether a particular species was surface or bottom dwelling. The degree of dietary overlap between species can be either high or low depending on factors such as food abundance, seasonality and the ability of species to utilise different diets. Furthermore, this study suggests that even though dietary overlap may be high, times of foraging activity may be different, resulting in a decrease in the level of competition.

This study indicates that differences in space, time and food utilisation play an important role in the co-existence of freshwater fish species. It also supports the suggestion that fish community structure is determined by the range of available resources and the associated specialisation of the co-existing species (Watson & Balon 1984a).

LITERATURE CITED

ANGERMEIER, P.L. & KARR, J.K. 1983. Fish communities along environmental gradients in a system of tropical streams. *Environmental Biology of Fishes* 9:117-135.

BALON, E.K. & STEWART, D.J. 1983. Fish assemblages in a river with unusual gradient (Luongo - Zaire System, Africa), reflections on river zonation, and description of another new species. *Environmental Biology of Fishes* 9:225-252.

BISHOP, J.E. 1973. Limnology of a small Malaysian river, Sungai Gombak. *Monographiae Biologicae* 22, Junk, The Hague.

CHOY, S.C. & CHIN, P.K. 1994. Freshwater fishes from the headwaters of the Belalong-Temburong river system, Brunei Darussalam, Borneo. *Raffles Bulletin of Zoology* 42 (4):757-774.

CHOY, S.C., this volume. Distributional ecology of freshwater fishes in the Belalong-Temburong river system, Brunei Darussalam.

CONNELL, J.H. 1978. Diversity in tropical rain forests and coral reefs. *Science* 199:1302-1310.

DYKES, A., this volume. Analysis of factors contributing to the stability of steep hillslopes in the tropical rainforest of Temburong, Brunei Darussalam.

GATZ, A.J. 1979. Community organization in fishes as indicated by morphological features. *Ecology* 60:711-718.

GORMAN, D.T. & KARR, J.R. 1978. Habitat structure and stream fish communities. *Ecology* 59:507-515.

GOULDING, M. 1980. *The fishes and the forest.* University of California Press, Berkeley.

HELFMAN, G.S. 1978. Patterns of community structure in fishes: summary and overview. *Environmental Biology of Fishes* 3:128-148.

INGER, R.F. & CHIN, P.K. 1962. The freshwater fishes of Northern Borneo. *Fieldiana: Zoology* 45:1-268.

LOWE-McCONNELL, R.H. 1975. *Fish communities in tropical freshwaters*. Longman, London.

LOWE-McCONNELL, R.H. 1987. *Ecological studies in tropical fish communities*. Cambridge University Press, Cambridge.

MAGURRAN, A.E. 1988. *Ecological diversity and its measurement*. Croom Helm, London.

MOYLE, P.B & SENANAYAKE, F.R. 1984. Resource partitioning among the fishes of rainforest streams in Sri Lanka. *Journal of Zoology, London* 202:195-223.

PIANKA, E.R. 1981. Resource acquisition and allocation among animals. Pp. 300-314 in Townsend, C.R. & Calow, P. (eds). *Physiological ecology: an evolutionary approach to resource use*. Blackwell Scientific, Oxford.

SALE, P.F. 1982. Stock-recruitment relationships and regional coexistence in a lottery competitive system: a simulation study. *American Naturalist* 120:139-159.

SCHOENER, T.W. 1986. Resource partitioning. Pp. 91-126 in Kikawa, J. & Anderson, D.J. (eds). *Community ecology: patterns and processes*. Blackwell Scientific, Melbourne.

WATSON, D.J. & BALON, E.K. 1984a. Ecomorphological analyses of fish taxocenes in rainforest streams of northern Borneo. *Journal of Fish Biology* 25:371-384.

WATSON, D.J. & BALON, E.K. 1984b. Structure and production of fish taxocenes in tropical rainforest streams of northern Borneo. *Canadian Journal of Zoology* 62:927-940.

WOOTON, R.J. 1992. *Fish ecology*. Thompson, New York.

Spatio-temporal resource utilization by a Bornean rainforest herpetofauna: Preliminary results

INDRANEIL DAS[1]

Department of Biology, Universiti Brunei Darussalam, Bandar Seri Begawan 2028, Brunei Darussalam

ABSTRACT. Field work was conducted on the community organization of the herpetofauna at Batu Apoi Forest Reserve, a lowland mixed dipterocarp forest in Brunei Darussalam. The present communication is a preliminary analysis of the patterns of utilisation of microhabitat and diel-time resources by the more abundant species of amphibians and reptiles at the study site. A total of fifty-two species of amphibians and forty-six species of reptiles were recorded, both faunas composed of a species-poor riparian and a species-rich non-riparian assemblage. Several species recorded at the Reserve are frequently human commensals, and therefore indicators of disturbed habitats. Human activities tend to create new niches or expand existing ones that are rapidly filled by these weed species associated with man. Analysis of the microhabitat and diel-time reveal that the non-riparian assemblage has more space and time-specialists than the riparian one. It is concluded that high diversity is maintained in the more climatically-stable forested zones by species specializing in the use of both microhabitat and diel-time.

KEY WORDS: Brunei Darussalam, amphibians, reptiles, rainforest, diversity, community structure, microhabitat, diel-time, resource use, human commensals.

INTRODUCTION

The organization of communities and assemblages of tropical vertebrates continue to be poorly-understood, although a plethora of recent studies in the past two decades have shown that various factors, including competition (Pianka 1986) and predation (Lowe-McConnell 1975) play important roles in shaping natural communities. Holt (1984) suggested that both competition and predation may complement each other in structuring communities.

I commenced long-term field studies on the herpetological community at Batu Apoi, Brunei Darussalam, Borneo, in January, 1991. The two goals of this research were to prepare an inventory of the amphibians and reptiles of the Batu Apoi Forest Reserve, soon to be gazetted a National Park, and to examine the organization of the herpetological assemblages. This paper presents preliminary results of investigations of the organisation of amphibian and reptile species in time and space at Batu Apoi.

STUDY AREA AND METHODS

Field work was conducted at Batu Apoi, a lowland rainforest site in Temburong District, Brunei Darussalam, in north-western Borneo. A general description of the site, including topography and vegetation, has been given by Edwards (1992). In January, 1991, a *ca.* 6 hectare area on the West Ridge of the Belalong Valley, in the vicinity of the Kuala Belalong Field Studies Centre, was selected for intensive field work. This area included both riparian and non-riparian habitats and covered an altitudinal range from ca. 60-80 m (the riparian

[1]Present address: Centre for Herpetology, Madras Crocodile Bank Trust, Post bag 4, Mamallapuram 603104, Tamil Nadu, India

D. S. Edwards et al. (eds.), Tropical Rainforest Research – Current Issues, 315–323.

zones, where the rivers Sg. Belalong and Sg. Temburong are > 10 m in width) to 80-110 m (the maximum height of the non-riparian zone that was sampled, which included streams < 2 m in width). Of the species listed in Tables 1 and 2, only two (*Megophrys baluensis* and *Ptychozoon horsfieldii*) were taken outside this altitudinal range, both from the East Ridge of the Belalong Valley at an altitude of ca. 440 m. Four transects were opened, three along forest streams, one through a mixed dipterocarp forest. Over one thousand four hundred hours were spent sampling the fauna. The effective observation area was *ca.* 2 m off either side of the transect line and about 3.0 m off the ground, and observations were thus over an area of 206,500 m^2, the sampling volume being 619,500 m^3 of the study area. These figures refer to the total area and volume sampled, the same transects being repeatedly sampled during all months of the year and during all times of the day. In addition to hand collecting along the transects, pit-fall traps in association with drift fences were employed, as were blow-pipes and rat-glue on tree-trunks, the last two techniques with limited success. The collection has been deposited at the Department of Biology, Universiti Brunei Darussalam, Bandar Seri Begawan, with a voucher series being deposited at the Zoological Reference Collection, National University of Singapore, Singapore. All visual and many acoustic observations were noted. Data recorded for each individual specimen observed and/or collected include time and date of observation, vertical position, microhabitat and behaviour. General locations of collection were noted and the data subsequently entered into a Geographical Information System (GIS) database, maintained by the Universiti Brunei Darussalam (Wills, this volume). To prepare a more complete inventory of the species of amphibians and reptiles of the Reserve, species that were collected by colleagues have also been included. Resource data matrices (m x n) were created for microhabitat and diel-time resources separately for the riparian and non-riparian assemblages. The microhabitat categories were based on subjective evaluations (e.g., saplings 0-30 cm above the forest floor, under fallen logs, tree-holes), the diel-time categories were one-hour partitions of the 24-hour diel cycle. Only the relatively more abundant species (n = 5 and above) were utilised for studies on the use of time and space.

Niche overlaps were calculated using a symmetrical version (*fide* Pianka 1986) of MacArthur and Levins' (1967) equation:

$$O_{jk} = \frac{\sum_{i=1}^{n} p_{ij}\, p_{ik}}{\sqrt{\sum p_{ij}2\; p_{ik}2}}$$

where, p_{ij} and p_{ik} are the proportions of the *i*th resource state used by the *j*th and *k*th species. Cluster analyses on the overlap data were conducted using SYSTAT, Version 5.03 (Wilkinson 1990).

RESULTS

Species diversity

Tables 1 and 2 list the identified amphibian and reptile species recorded from the Batu Apoi Forest Reserve. A total of 446 individual animals were collected and/or observed, not including a large series of eggs and larvae that were also collected. Table 3 shows the number of species represented by numbers of individuals.

The inventory of the Reserve's herpetofauna is far from complete: Reading (1992) conducted an eight-week extensive survey of the frogs of Batu Apoi in April-May, 1992 and reported 28 species. His work, combined with my more intensive long-term sampling from a smaller area (mostly the West Ridge) have resulted in a better knowledge of the amphibian

Table 1. Checklist of amphibians verified at the Batu Apoi Forest Reserve.

GYMNOPHIONA

FAMILY: ICHTHYOPHIIDAE

1. Yellow-black caecilian *Caudacaecilia nigroflava* (Taylor, 1960)

ANURA

FAMILY: PELOBATIDAE

2. Large-eyed litter frog *Leptobrachium montanum* Fischer, 1885
3. Slender litter frog *Leptolalax gracilis* (Günther, 1872)
4. Bornean horned toad *Megophrys nasuta* (Schlegel, 1858)
5. Kinabalu horned toad *Megophrys baluensis* (Boulenger, 1899)
6. Cricket litter frog *Nesobia mjobergi* (Smith, 1925)

FAMILY: BUFONIDAE

7. White-eared stream toad *Ansonia albomaculata* Inger, 1960
8. Slender-footed stream toad *Ansonia leptopus* (Günther, 1872)
9. Long-fingered stream toad *Ansonia longidigita* Inger, 1960
10. Spiny stream toad *Ansonia spinulifer* (Mocquard, 1890)
11. Malayan giant toad *Bufo asper* Gravenhorst, 1829
12. Giant river toad *Bufo juxtasper* Inger, 1964
13. Double-ridged toad *Bufo divergens* Peters, 1871
14. Brown tree toad *Pedostibes hosei* (Boulenger, 1892)
15. Green tree toad *Pedostibes rugosus* Inger, 1958
16. Short-legged toadlet *Pelophryne brevipes* (Peters, 1867)

FAMILY: MICROHYLIDAE

17. Saffron-bellied frog *Chaperina fusca* Mocquard, 1892
18. Intermediate sticky frog *Kalophrynus intermedius* Inger, 1966
19. Red-sided sticky frog *Kalophrynus pleurostigma* Tschudi, 1838
20. Bornean tree-hole frog *Metaphrynella sundana* (Peters, 1867)
21. Rufous froglet *Microhyla borneensis* Parker, 1926
22. Stony froglet *Microhyla petrigena* Inger & Frogner, 1979

FAMILY: RANIDAE

23. Blotch-thighed stream frog *Meristogenys poecilus* (Inger & Gritis, 1983)
24. Dwarf mountain frog *Micrixalus baluensis* (Boulenger, 1896)
25. Seep frog *Occidozyga baluensis* (Boulenger, 1896)
26. Yellow-bellied puddle frog *Occidozyga laevis* (Günther, 1858)
27. Blyth's frog *Rana blythi* (Boulenger, 1920)
28. White-lipped frog *Rana chalconota* (Schlegel, 1837)
29. Paddyfield frog *Rana erythraea* (Schlegel, 1837)
30. Poisonous rock frog *Rana hosei* Boulenger, 1891
31. Ridged river frog *Rana ibanorum* Inger, 1964
32. Kuhl's creek frog *Rana kuhli* Duméril & Bibron, 1841
33. Smooth guardian frog *Rana palavanensis* Boulenger, 1894
34. Spotted stream frog *Rana signata* (Günther, 1872)
35. Rock skipper *Staurois latopalmatus* (Boulenger, 1887)
36. Black-spotted rock frog *Staurois natator* (Günther, 1858)
37. Green-spotted rock frog *Staurois tuberilinguis* Boulenger, 1918

FAMILY: RHACOPHORIDAE

38. Painted tree frog *Nyctixalus pictus* (Peters, 1871)
39. Obscure bush frog *Philautus tectus* Dring, 1987
40. Collett's tree frog *Polypedates colletti* (Boulenger, 1890)
41. Five-lined tree frog *Polypedates leucomystax* (Gravenhorst, 1829)
42. Dark-eared tree frog *Polypedates macrotis* (Boulenger, 1894)
43. File-eared tree frog *Polypedates otilophus* (Boulenger, 1893)
44. Jade tree frog *Rhacophorus dulitensis* Boulenger, 1892
45. Mossy tree frog *Rhacophorus everetti* Boulenger, 1894
46. Harrisson's tree frog *Rhacophorus harrissoni* Inger & Haile, 1959
47. Emerald tree frog *Rhacophorus kajau* Dring, 1983
48. Wallace's flying frog *Rhacophorus nigropalmatus* Boulenger, 1895
49. Harlequin tree frog *Rhacophorus pardalis* Günther, 1858

fauna than that currently exists for the reptiles. The snake and lizard lists appended identify perhaps 70% and 90%, respectively, of the total fauna of the Reserve.

Inger and Voris (1992) reported that the maximum number of amphibian species collected from a single site on Borneo (Nanga Tekalit, in Sarawak, East Malaysia) was 57. The 52 amphibians and 46 reptiles represent 36% and 15.1%, respectively, of the total Bornean amphibian fauna (currently 145 amphibian and 305 reptile species; Das, unpublished). Thus the amphibian fauna of Batu Apoi must be described as relatively diverse for Borneo.

Table 2. Checklist of reptiles verified at the Batu Apoi Forest Reserve.

SAURIA

FAMILY: GEKKONIDAE
1. Four-clawed gecko *Gehyra mutilatus* (Wiegman, 1835)
2. Barking gecko *Gekko smithii* Gray, 1841
3. Kinabalu bent-toed gecko *Gonydactylus baluensis* (Mocquard, 1890)
4. Peters' bent-toed gecko *Gonydactylus consobrinus* (Peters, 1871)
5. Inger's bent-toed gecko *Gonydactylus ingeri* (Hikida, 1990)
6. Grooved bent-toed gecko *Gonydactylus pubisulcus* (Inger, 1958)
7. Asian house gecko *Hemidactylus frenatus* Duméril and Bibron, 1836
8. Dwarf gecko *Hemiphyllodactylus typus* Bleeker, 1860
9. Horsfield's flying gecko *Ptychozoon horsfieldii* (Gray, 1827)

FAMILY: AGAMIDAE
10. Green crested lizard *Bronchocela cristatella* (Kuhl, 1820)
11. Crested flying lizard *Draco cristatellus* Günther, 1872
12. Black-bearded flying lizard *Draco melanopogon* Boulenger, 1887
13. Six-lined flying lizard *Draco quinquefasciatus* Hardwicke & Gray, 1827
14. Bornean angle-headed lizard *Gonocephalus borneensis* (Schlegel, 1848)

FAMILY: SCINCIDAE
15. Common tree skink *Apterygodon vittatum* Edeling, 1864
16. Three-keeled ground skink *Mabuya rudis* Boulenger, 1887
17. Blue-bellied litter skink *Sphenomorphus cyanolaemus* Inger & Hosmer, 1965
18. Beccari's water skink *Tropidophorus beccarii* (Peters, 1871)
19. Brooke's water skink *Tropidophorus brookei* (Gray, 1945)

FAMILY: VARANIDAE
20. Water monitor lizard *Varanus salvator* (Laurenti, 1768)

SERPENTES

FAMILY: BOIDAE
21. Reticulated python *Python reticulata* (Schneider, 1801)

FAMILY: COLUBRIDAE
22. Speckle-headed whip snake *Ahaetulla fasciolatus* (Fischer, 1885)
23. Oriental whip snake *Ahaetulla prasinus* (Boie, 1827)
24. White-fronted water snake *Amphiesma flavifrons* (Mocquard, 1887)
25. Blunt-headed tree snake *Aplopeltura boa* (Boie, 1828)
26. White-spotted catsnake *Boiga drapiezii* (Boie, 1827)
27. Twin-barred flying snake *Chrysopelea pelias* (Linnaeus, 1758)
28. Striped bronzeback tree snake *Dendrelaphis caudolineatus* Gray, 1834
29. Spotted water snake *Enhydris punctata* (Gray, 1849)
30. Yellow-striped water snake *Hydrablabes periops* (Günther, 1872)
31. Orange-bellied snake *Liopeltis baliodeirus* (Boie, 1827)
32. Three-coloured ringnecked snake *Liopeltis tricolor* (Schlegel, 1837)
33. Malayan banded wolf snake *Lycodon subcinctus* Boie, 1827
34. Annulated kukri snake *Oligodon annulifer* (Boulenger, 1893)
35. Sinister water snake *Opisthotropis typica* (Mocquard, 1890)
36. Painted mock viper *Psammodynastes pictus* Günther, 1858
37. White-collared reed snake *Pseudorabdion albonuchalis* (Günther, 1896)
38. Dwarf reed snake *Pseudorabdion longiceps* (Mocquard, 1892)

FAMILY: ELAPIDAE
39. Blue coral snake *Maticora bivirgatus* (Boie, 1827)
40. Malayan striped coral snake *Maticora intestinalis* (Laurenti, 1768)
41. King cobra *Ophiophagus hannah* (Cantor, 1836)

FAMILY: VIPERIDAE
42. Bornean flat-nosed pit viper *Trimeresurus borneensis* (Peters, 1871)
43. Wagler's pit viper *Tropidolaemus wagleri* (Boie, 1827)

TESTUDINES

FAMILY: BATAGURIDAE
44. Malayan flat-shelled turtle *Notochelys platynota* (Gray, 1834)
45. Spiny turtle *Heosemys spinosa* (Gray, 1831)

FAMILY: TRIONYCHIDAE
46. Malayan softshell turtle *Dogania subplana* (Geoffroy Saint-Hillaire, 1809)

Included in the fauna from this primarily lowland site is *Microhyla petrigena*, a species considered to be restricted to montane zones (> 800 m) by Inger & Stuebing (1992). Correlates of species diversity are poorly-known, and, besides historical components, may include litter production (Inger 1980) and annual moisture regime (Duellman 1978). Scott

Table 3. Number of specimens and of species of amphibians and reptiles recorded at Batu Apoi. Figures in parentheses are percentages of total species count (N).

No. of specimens	No. of amphibian species	No. of reptile species
1	10 (22.7)	23 (57.5)
2	7 (15.9)	6 (15.0)
3	3 (6.8)	2 (5.0)
4	0 (0)	0 (0)
5	4 (9.1)	3 (7.5)
6	3 (6.8)	1 (2.5)
7	1 (2.3)	1 (2.5)
8	2 (4.5)	1 (2.5)
9	2 (4.5)	1 (2.5)
10	0 (0)	1 (2.5)
11	1 (2.3)	1 (2.5)
12	0 (0)	0 (0)
13	1 (2.3)	0 (0)
14	1 (2.3)	0 (0)
15	2 (4.5)	0 (0)
16	2 (4.5)	0 (0)
17	2 (4.5)	0 (0)
18	1 (2.3)	0 (0)
19	0 (0)	0 (0)
20	1 (2.3)	0 (0)
21	0 (0)	1 (2.8)
22	0 (0)	0 (0)
23	0 (0)	0 (0)
24	1 (2.3)	0 (0)
	N = 44 (100)	N = 40 (100)

(1976) found a negative correlation between the length of the dry season and species richness of rainforest litter herpetofauna.

Diel-temporal resource use

Fifty-two species of amphibians were recorded from the study site (Table 1 lists only the verified species). The amphibian fauna can be divided into three components: aquatic or aquatic-margin dwelling (23 species), terrestrial and/or fossorial (14 species) and arboreal (15 species). All species of amphibians, at least as adults, at the site are nocturnal, except *Rana palavanensis*, which was collected both by day and at night (but see below). Forty-six species of reptiles were recorded from Batu Apoi (Table 2). Three turtles were recorded, one (*Dogania subplana*) from the main river, one from the forest streams (*Notochelys platynota*) and one from the dipterocarp forest floor (*Heosemys spinosa*). No tortoises have been recorded (only one species of the family Testudinidae, *Manouria emys*, occurs on Borneo, *fide* Iverson 1992). Freshwater turtles in general are scarce in the study area, possibly a result of hunting in the past by Iban tribesmen, and on account of their rarity, no further mention of turtles will be made in the subsequent discussion on community structure.

The squamates include 12 riparian lizards (two, *Sphenomorphus cyanolaemus* and *Gonydactylus consobrinus*, also occurring on the non-riparian sites), 11 riparian snakes (two, *Trimeresurus borneensis* and *Tropidolaemus wagleri*, also occurring on the non-riparian sites); 13 non-riparian lizards, 14 non-riparian snakes. The lizard fauna is composed of a greater number of diurnal species (12) than nocturnal ones (9); the snake fauna has equal number of diurnal and nocturnal species (13). Classified by substrates used, 15 lizards are arboreal, while six are terrestrial or with limited scansorial capabilities. Among snakes, eight are arboreal, ten largely terrestrial/fossorial and four primarily aquatic.

All nocturnal lizards (nine species) are arboreal. Among the snakes, the diurnal species have six terrestrial/fossorial and four arboreal members, while the nocturnal/crepuscular ones have five terrestrial/fossorial and five arboreal forms. Aquatic snakes at Batu Apoi include both diurnal (two species) and nocturnal (two species) forms.

Classifying species according to activity patterns is beset with at least one problem: some species show no clear-cut pattern. The pit vipers, *Tropidolaemus wagleri* and *Trimeresurus borneensis* were found abroad both by day and at night, but have been considered nocturnal/crepuscular here, based on data available from other sites. Rand and Myers (1990) mentioned that ambush hunters at a New World tropical forest site show similar diel-activity patterns. A possible explanation for this observation is that the sedentary foraging mode is energetically less demanding, and predators practicing it can afford (and probably need) to forage for longer periods, according to the optimum foraging theory. Juveniles of the aquatic *Rana kuhlii* were also taken both by day and at night, the day time activity of these size-classes possibly due to the need to feed continuously in order to grow out of the more vulnerable smaller body size. Excluding these problematic species, a single frog appear to show bimodal peaks of diel-activity: *Rana palavanensis*, the adults of which were found on the leaf litter of the forest floor during the afternoons and during the late evenings.

Microhabitat use

Two assemblages of amphibians and reptiles are readily differentiated, with little overlap in terms of species composition, the assemblages being spatially separated along a topographical gradient: the riparian and the non-riparian assemblages. The first is composed of 18 medium to large amphibians, including four bufonids, nine ranids, three rhacophorids, one microhylid and one ichthyophid. The non-riparian assemblage is much more diverse (38 species) but its members are smaller in mean body size, and include six pelobatids, six bufonids, five microhylids, nine ranids, twelve rhacophorids and one ichthyophid.

The riparian assemblage of reptiles is composed of 23 species, including one trionychid, five gekkonids, two agamids, three scincids, one varanid, one boid, seven colubrids, one elapid and two viperids. The non-riparian assemblage is slightly more speciose (27 species), and is composed of two batagurids, five gekkonids, three agamids, three scincids, ten colubrids, two elapids and two viperids.

Three amphibians (*Rana blythi, Staurois natator* and *Rhacophorus kajau*) and four reptiles (*Gonydactylus consobrinus, Sphenomorphus cyanolaemus, Trimeresurus borneensis* and *Tropidolaemus wagleri*) were found in both riparian as well as non-riparian habitats, the taxa thus shared between the two zones being 5.8% for amphibians and 8.7% for reptiles. *Rana blythi* is a large ranid that utilises forest streams, but may also be found along large rivers, while *Rhacophorus kajau*, usually a forest species, has been recorded breeding in the riparian zone. *Staurois natator* has been collected from riparian habitats as well as from non-riparian ones, far from any permanent waterbodies. Both *Gonydactylus consobrinus* and *Sphenomorphus cyanolaemus* are associated with large trees, the first utilising tree-trunks, the second a tree-buttress specialist. Both are comparatively rare in the riparian zones. *Trimeresurus borneensis* and *Tropidolaemus wagleri* have been found in both the riparian and non-riparian zones, microhabitats associated with this species being low saplings as well as from the litter, and on account of their comparative rarity, no generalisation of habitat preference will be made. The non-riparian amphibian (38) and reptile (27) faunas were thus found to be more diverse than the riparian ones (18 and 23, respectively), a pattern similar to that reported from the Western Ghats, south-western India by Inger *et al.* (1987). Another similar trend observed is the scarcity of animals under logs: a single *Sphenomorphus cyanolaemus* was found in such a microhabitat.

Structure of space and time guilds

Cluster analysis provides a visual representation of guild structure based on the niche overlap data. A general observation of the figures produced using cluster analyses (which will be presented in a later paper on resource use when the studies currently in progress are complete) of the niche overlap data on diel-time and microhabitat use separately for the riparian and non-riparian assemblages is that the relatively more speciose non-riparian assemblage has a larger cluster with larger guilds than the depauperate riparian one. A few congeneric species that are phylogenetically (and therefore morphologically) close fall far out ecologically, including the species-pairs *Rana kuhlii - R. palavanensis, Ansonia albomaculata - A. spinulifer* and *Leptolalax gracilis - Leptobrachium montanum.* Inexplicably, the above pattern is thus similar to that observed by Inger & Colwell (1977) at a seasonal locality in Thailand, but at variance with the observations of Inger (1969) at a non-seasonal site in Borneo. Frogs and lizards are part of the same assemblage, if microhabitat use is considered. These two groups also utilised similar trophic resources (studies on their foraging ecology are now in progress), although to recognise functional guilds (*sensu* Root 1967), a broader taxonomic range of organisms needs to be studied. For instance, the diurnal, arboreal, insectivorous lizard *Gonocephalus borneensis* may be a member of a forest guild along with understorey birds. The cluster showing the microhabitat-association of riparian species shows a tightly constrained group of two true frogs (*Rana chalconota* and *R. signata*) and a toad (*Bufo juxtasper*), all of which utilise the banks of rivers. The rest of the riparian species are loosely associated, the species clearly set apart from its sympatrics in its use of space being the agamid *Bronchocela cristatella*, which perches on low vegetation and on tree trunks along river banks. In its use of temporal resources, the agamid is once again far apart from the rest of the members of its assemblage, being diurnal, similar to the monitor lizard *Varanus salvator*. However, the latter species utilises river-edges and river banks, and is thus spatially separated. *Rana kuhlii* shows another remarkable pattern as it falls out from the rest of the non-riparian species in both the use of time and space. Adults of the species are nocturnal, newly-metamorphosed juveniles being taken both by day and at night. The juveniles of *Occidozyga baluensis* show this pattern too, but to a lesser degree. Relatively fewer species-pairs in the non-riparian habitats show high overlaps (O_{jk} >0.06) compared to riparian ones for both diel-time (13.9% versus 14.3%) and microhabitat (0% versus 17.9%) resources. This indicates the presence of many more time and space specialists in the non-riparian zones.

New niches or better opportunities?

Several species normally associated with humans have been recorded from within the boundaries of the Reserve several months after the establishment of the Kuala Belalong Field Studies Centre at Batu Apoi and are thus considered human commensals. These include the frogs *Rana erythraea* and *Polypedates leucomystax* and the lizard *Hemidactylus frenatus*. These co-exist with man and are therefore indicators of disturbed habitats. The gecko was transported with the logs used in the construction of the Field Studies Centre (Earl of Cranbrook, personal communication 1992), but the frogs were recorded approximately one year later. Human activities tend to create new niches that are rapidly occupied by these weed species which appear to follow man, but may also persist in low densities at the edges (such as riparian zones) of undisturbed forests. All three newcomers have been recorded only from the riparian areas. Any effects of these edge species on the local herpetofauna are at present unknown.

DISCUSSION

Diversity and inventory methods

Preparing inventories of rainforest herpetofaunas is an extremely labour-intensive activity, and rapid assessment methods are likely to include only the more abundant species. Over

45 percent of the amphibians and 77 percent of the reptiles are represented by three or less specimens (Table 3), although collections spanned more than a year, and included both dry and wet periods. In the case of at least the tree frogs, the low numbers obtained may be attributed to their use of arboreal (and therefore difficult to sample) habitats. The same explanation may hold for the eight arboreal snakes that are represented in the collection by a single specimen. However, the scarcity of many other groups of amphibians and reptiles, including species from 'accessible' habitats, such as true toads, caecilians, litter skinks and terrestrial and aquatic snakes, may not be a sampling bias, and appears to represent their true rarity in the Reserve. Therefore, rapid assessment methods to identify areas of high biological diversity for conservation purposes may be unsuitable for amphibians and reptiles in rainforest habitats, where species diversity is apparently offset by low densities of many individual species. Short-term surveys thus would tend to be biased towards the more abundant, *r*-selected species, such as weed species of the forest edges, rather than the true rainforest endemics.

Patterns of spatio-temporal resource use

The interactions of the habitat and species is evident from the faunal list: the non-riparian zone has six species of pelobatids (litter frogs) that live on the leaf litter, the riparian zone has none; the non-riparian zone also has twelve species of rhacophorids (tree frogs), only one of which co-occurs in the riparian zone. Another family of anurans well represented in the first zone, with five species (all ant and small beetle specialists), is the Microhylidae, which has a single species in the riparian areas. The non-riparian zone's greater species diversity, relative to that of the riparian zone, is attributed to its greater habitat diversity, complexity (including the thick accumulated layer of leaf litter and mature dipterocarp trees over 30 m in height) as well as less stable environmental factors in the riparian areas (such as the river, whose level shows considerable variation on a temporal scale).

The presence of more time- and space-specialists at the non-riparian sites is in concordance with Margalef's (1963) prediction that as relationships between species become more specialised, stenophagy and other precise relationships increase, leading to reduced niche overlap in rainforest communities. Since the non-riparian assemblage is much more species-diverse than the riparian one, with species showing relatively less resource sharing habits, specialization in the use of time and space resources appears to be a factor allowing high species diversity in the more climatically stable environment of the non-riparian zones.

ACKNOWLEDGEMENTS

A large number of people contributed to this research, from arranging logistics to keeping an eye open for specimens. A partial list includes: Webber Booth, Joe Charles, The Earl of Cranbrook, Satish Choy, David Edwards, David Jones, Chris Kofron, Ruth Levy, Hanafi Hj Manaf, Samhan Nyawa, Bert Orr, Catriona Prebble, Kamariah Abu Salim and John Wills. Robert F. Inger and David Jones read earlier drafts of the manuscript. The project was supported by a grant (UBD/T6/RG 03) from Universiti Brunei Darussalam.

LITERATURE CITED

DUELLMAN, W. W. 1978. The biology of an equatorial herpetofauna in Amazonian Ecuador. *Miscellaneous Publications of the Museum of Natural History, University of Kansas* 65:1-352.

EDWARDS, D.S. 1992. The riches of Batu Apoi. *Muhibbah* Jan/Feb 1992:24-26.

HOLT, R. D. 1984. Spatial heterogeneity, indirect interactions, and the coexistence of prey species. *American Naturalist* 124:377-406.

INGER, R. F. 1969. Organization of communities of frogs along small rain forest streams in Sarawak. *Journal of Animal Ecology* 38:123-148.

INGER, R. F. 1980. Densities of floor-dwelling frogs and lizards in lowland forests in southeast Asia and Central America. *American Naturalist* 115:761-770.

INGER, R. F. & COLWELL, R. K. 1977. Organization of contiguous communities of amphibians and reptiles in Thailand. *Ecological Monographs* 47:229-253.

INGER, R. F., SHAFFER, H. B., KOSHY, M. & BAKDE, R. 1987. Ecological structure of a herpetological assemblage in South India. *Amphibia-Reptilia* 8:189-202.

INGER, R. F. & STUEBING, R. B. 1992. The montane amphibian fauna of northwestern Borneo. *Malayan Nauture Journal* 46:41-51.

INGER, R. F. & VORIS, H. K. 1992. Variation in communities of amphibians in Bornean forests. Pp. 494-495 in Ismail, G., Mohamed, M. & Omar S. (eds). *Forest biology and conservation in Borneo*. Proceedings of the International Conference on Forest Biology and Conservation in Borneo. July 30-August 3, 1990, Kota Kinabalu. Yayasan Sabah, Kota Kinabalu.

IVERSON, J. B. 1992. *A revised checklist with distributional maps of the turtles of the world*. Privately printed, Richmond.

LOWE-McCONNELL, R. H. 1975. *Fish communities in tropical freshwaters: their distribution, ecology and evolution*. Longman, New York.

MACARTHUR, R. H. & LEVINS, R. 1967. The limiting similarity, convergence, and divergence of coexisting species. *American Naturalist* 101:377-385.

MARGALEF, R. 1963. On certain unifying principles in ecology. *American Naturalist* 97:357-374.

PIANKA, E. R. 1986. *Ecology and natural history of desert lizards*. Princeton University Press, Princeton, New Jersey.

RAND, A. S. & MYERS, C. W. 1990. The herpetofauna of Barro Colorado Island, Panama: An ecological summary. Pp. 386-409 in Gentry, A.H. (ed.). *Four Neotropical rainforests*. Yale University Press, New Haven & London.

READING, C. J. 1992. *Frogs found within the Batu Apoi Forest Reserve in Brunei*. Unpublished Final Report, UBD-RGS Rainforest Project, 1991-1992.

ROOT, R. 1967. The niche exploitation pattern of the blue-gray gnat-catcher. *Ecological Monographs* 37:317-350.

SCOTT, N. J. 1976. The abundance and diversity of the herpetofaunas of tropical forest litter. *Biotropica* 8:41-58.

WILKINSON, L. 1990. *SYSTAT: The system for statistics*. SYSTAT, Inc., Evanston, Illinois.

WILLS. J. T. (this volume). The creation of a Geographical Information System for the Kuala Belalong Field Studies Centre.

The altitudinal distribution of forest tree families on the western slopes of Mount Banahao, Luzon, Philippines

EUSTAQUIO G. ARAGONES

Forest Products Research and Development Institute, Department of Science and Technology, College, Laguna 4031, Philippines

ABSTRACT. The vertical distribution of forest tree species represented by families was sampled in 0.42 ha quadrats at different altitudes between 750 m and 2100 m on the western slopes of Mount Banahao, Luzon, Philippines. Species and family relative occurrence at different elevational sampling sites was assessed. Some plant groups such as the Meliaceae, Lauraceae and Myrtaceae displayed wider distribution vertically, being represented from 750 m to 1500 m. Other groups were confined to the summit zone, e. g. Podocarpaceae, Taxaceae, Ericaceae and Winteraceae. Families typical of lowland rainforest, such as Anacardiaceae, Euphorbiacae, Clusiacaeae, Myristicaceae, and the Nyctaginaceae did not extend beyond 750 m. The distribution pattern observed is discussed in relation to plant adaptation, dispersal and the effect of elevation.

KEY WORDS: ecological response, megatherm, microtherm, elevation effects.

INTRODUCTION

On tropical mountains both woody and herbaceous plants at all altitudes, according to Van Steenis (1961, 1962), are derived from autochthonous (locally derived) and allochthonous sources (from plants that are not native to the area). Autochthonous families or genera such as the Annonaceae, Bombacaceae, Dipterocarpaceae, Meliaceae, Rhizophoraceae, Marantaceae and *Ficus* mostly find their uppermost limit in the colline subzone below 1000 m. Normally, they fade rapidly away between 1000 m and 2000 m and only a few are represented at higher elevations. Conifers, tree ferns and some angiosperm families familiar to temperate botanists may exist in the tropics only on mountains. Some families, however, e. g. Labiatae, Oleaceae, Scrophulariaceae and Theaceae have wide altitudinal distribution in the tropics and wide latitudinal distribution over the globe. Certain species of oaks and chestnuts, *Ajuga*, *Daphne*, *Podocarpus*, *Rhododendron*, and *Vaccinium* occur even at sea level in Malesia. It is accepted that these genera or families have wide ecology and are able to tolerate a wide range of temperature. For these, Van Steenis (1961) used the term megatherm for heat-adapted and microtherm for cold-adapted species.

The effect of elevation influences various ecological factors, the most important of which are temperature, soil, autecology of the species, dispersal biology and man's activities.

This paper presents and discusses the composition of forest tree species and families in six 0.42 ha quadrats positioned along an altitudinal gradient on the western slopes of Mt. Banahao, Luzon, Philippines.

METHODS

The Study Area

The study was carried out on the western slopes of Mt. Banahao, Luzon (14.5^0 N, 121.5^0

D. S. Edwards et al. (eds.), Tropical Rainforest Research – Current Issues, 325–330.

E) from the low-lying Barangay (Bgy) Kinabuhayan, Dolores, Quezon, some 550 m above sea level, to the summit at 2100 m.

From Bgy Santa Lucia to Bgy Kinabuhayan, the steep slopes and the general ruggedness of Mt. Banahao are evident. Huge basaltic and andesitic rock outcrops are prominent within upland farms and in the forest itself, to an elevation of 750 m.

The climate map of Quezon province, Luzon (1975 Soil Survey Report) indicates a Type II climate for Dolores Quezon, that is with a pronounced maximum rain period from August to December and without a dry season. Annual rainfall averages 4000 mm. The mountain, especially at higher altitudes, may have different patterns of rainfall and temperature from the areas lying below owing to its vegetation and relief.

Sampling Procedure

Natural forest communities encountered at different altitudes (750, 950, 1200, 1500, 1800 and 2100 m) were sampled by a modified quadrat method as employed by Whittaker & Niering (1975). Two contiguous quadrats each 30 m x 70 m were laid upslope of each other at each of the six altitudes. A systematic sampling interval of 200 m elevation, as used by Khan (1969) in Mt. Makiling, Luzon, was adopted up to 1200 m elevation, beyond which it was adjusted to 300 m in succeeding sites when slopes became much steeper and vertical cliffs were encountered more often.

In each of the two quadrats at each altitude, all trees greater than 10 cm diameter at breast height (DBH) were censused. Height and diameter were measured with a Haga altimeter and diameter tape respectively and herbarium specimens collected from each species tallied.

RESULTS AND DISCUSSION

The vertical range of species distribution based on their observed occurrence at each altitudinal sampling site is presented in Tables 1 and 2.

It is evident from the data that certain plant groups such as the Myrtaceae, Lauraceae, Fagaceae and Meliaceae have a wide vertical distribution (Figure 1). The Meliaceae were represented from 750 m to 1500 m altitude while the Fagaceae, which were not encountered until 950 m, peaked at 1500 m elevation then declined at 1800 m. The Lauraceae were present in plots from 750 m to 1500 m. The Myrtaceae were present in plots from 750 m to the summit (2100 m). Members of the Melastomataceae, Theaceae and Symplocaceae on the other hand, appeared to be high elevation families as they were observed not lower than 950 m altitude and extended vertically to higher sites close to the summit. Other families such as the Myristicaceae, Nyctaginaceae, Staphyleaceae, and Rutaceae were observed only at 750 m.

Such patterns of distribution, particularly along gradients of elevation, seem to indicate major differences in the adaptation of different plant groups to varying niche and habitat types as influenced by relief and soil conditions (Whittaker 1965, Whittaker *et al.* 1968). The existence of what Van Steenis (1934, 1961, 1962) called microtherm taxa (e. g. *Drimys, Rhododendron*, *Podocarpus* and *Taxus*) at the summit zone of Mt. Banahao's western slopes (1800 - 2100 m) may well be considered an ecological response to the much lower temperature. Van Steenis (1934, 1961, 1962) further stressed that these groups of temperate origin may only be present at high altitudes in the tropics.

The absence of high altitude microclimatological data for most Philippine mountains makes it difficult to determine exact differences in climate, particularly of temperature, from low to high elevations. Nonetheless, if an adiabatic lapse rate of 0.67^0 C is assumed per 100 m rise in elevation, the summit of Mt. Banahao (2100 m) would have a temperature of about 13 - 14^0 C, compared to a mean of about 26 - 27^0 C at sea level. This approximates the

Table 1. Species observed at two or more altitudinal sampling sites on the western slopes of M. Banahao, Luzon.

Altitude (m)	Species	Family
550-750	*Pipturus arborescens*	Urticaceae
	Ficus nota	Moraceae
550-950	*Mallotus multiglandulosus*	Euphorbiaceae
	Ficus septica var. *septica*	Moraceae
750-950	*Parashorea malaanonan*	Dipterocarpaceae
	Ficus gul	Moraceae
	Aglaia llanosiana	Meliaceae
	Ficus minahassae	Moraceae
	Elaeocarpus calomala	Elaeocarpaceae
	Saurauia latibractea	Actinidiaceae
	Cinnamomum mercadoi	Lauraceae
	Dracontomelon edule	Anacardiaceae
	Syzygium polycephaloides	Myrtaceae
	Palaquium philippense	Sapotaceae
	Ficus ampelas var. *ampelas*	Moraceae
	Neonauclea media	Rubiaceae
750-1200	*Euphoria didyma*	Sapindaceae
	Pedicellia fuscescens	Sapindaceae
	Cryptocarya ampla	Lauraceae
	Acmena acuminatissima	Myrtaceae
	Acalypha stipulacea	Euphorbiaceae
	Palaquium merrillii	Sapotaceae
	Palaquium foxworthii	Sapotaceae
750-1500	*Aglaia luzoniensis*	Meliaceae
	Syzygium nitidum	Myrtaceae
950-1200	*Ficus balete*	Moraceae
	Dysoxylum altissimum	Meliaceae
	Aglaia diffusa	Meliaceae
	Palaquium luzoniensis	Sapotaceae
	Syzygium simile	Myrtaceae
	Eusideroxylon zwageri	Lauraceae
950-1500	*Cleistocalyx operculata*	Myrtaceae
950-1800	*Lithocarpus* spp.	Fagaceae
950-2100	*Symplocos* spp.	Symplocaceae
1200-1500	*Syzygium mainitense*	Myrtaceae
	Memecylon spp.	Melastomataceae
	Syzygium decipiens	Myrtaceae
1500-2100	*Eurya* sp.	Theaceae

record of Brown (1919) from the northeastern slopes of Mt. Banahao, and that of Mohr & Van Baren (1954) in the mountains of Indonesia.

The higher number of species occurring at low and medium altitudes (Tables 1 and 2) could attest to a preponderance of generalist species in the tropics which have wide ecological tolerance. In contrast, the most distinctive specialist group in this study is confined to the summit zone (Table 3), where temperature is much lower, making it difficult for the natural megatherms to adapt (Cockburn 1978, Corner 1978, Whitmore 1975). Along this line of reasoning, the altitudinal distribution observed of families (Figure 1) and even that of genera

Table 2. Species observed at only one altitudinal sampling site.

Altitude (m)	Species	Family
750	*Goniothalamus amuyon*	Annonaceae
	Neotrewia cumingii	Euphorbiaceae
	Artocarpus ovatus	Moraceae
	Pisonia umbelliferae	Nyctaginaceae
	Pterospermum niveum	Sterculiaceae
	Diplodiscus paniculatus	Ulmaceae
	Octomeles sumatrana	Datiscaceae
	Pagiantha plumerilifolia	Flacourtiaceae
	Ficus botryocarpa var. *botryocarpa*	Moraceae
	Mallotus philippinensis	Euphorbiaceae
	Myristica philippinensis	Myristicaceae
	Grewia multiflora	Tiliaceae
	Ficus pubinervis	Moraceae
	Alstonia scholaris	Apocyanaceae
	Garcinia binucao	Clusiaceae
	Homalanthus sp.	Euphorbiaceae
	Aglaia iloilo	Meliaceae
	Diospyros philippense	Ebenaceae
	Toona calantas	Meliaceae
	Chisocheton pentandrus	Meliaceae
	Dysoxylon arborescens	Meliaceae
	Laportea luzonensis	Urticaceae
	Melicope triphylla	Rutaceae
	Ficus congesta	Moraceae
	Allantus integrifolia	Simaroubaceae
	Litsea perrotetii	Lauraceae
	Pterocymbium macrocater	Sterculiaceae
	Palaquium sp.	Sapotaceae
	Cratoxylum sumatranum ssp. *sumatranum*	Hypericaceae
	Ervatamia sp.	Apocyanaceae
	Canarium asperum	Burseraceae
	Lepisanthes schizolepis	Sapindaceae
	Ahernia glandulosa	Flacourtiaceae
	Phaleria capitata	Thymelaeaceae
	Casearia fuligunosa	Flacourtiaceae
	Terminalia foetidissima	Combretaceae
	Epicahris cumingiana	Meliaceae
	Myristica elliptica var. *simiarum*	Myristicaceae
	Bischofia javanica	Bischofiaceae
950	*Nephelium* sp.	Sapindaceae
	Aphanamixis perrottetiana	Meliaceae
	Leucosyke capitellata	Urticaceae
	Duabanga moluccana	Sonneratiaceae
	Celtis luzonica	Ulmaceae
	Cratoxylum arboreum	Hypericaceae
	Trichadenia macrantha	Flacourtiaceae
	Cratoxylum formosum ssp. *formosum*	Hypericaeae
	Diospyros copelandii	Ebenaceae
	Ficus variegata var. *variegata*	Moraceae
1200	*Aglaia longipes*	Meliaceae
	Ardisia sp.	Myrsinaceae
1500	*Alseodaphne longipes*	Lauraceae
	Cryptocarya sp.	Lauraceae
	Lansium dubium	Meliaceae
	Neolitsea sp.	Lauraceae
	Syzygium decipiens	Myrtaceae

Table 3. Species observed exclusively at the summit zone (1800 - 2100 m).

Species	Family
Dacrycarpus cumingianus	Podocarpaceae
Podocarpus rotundus	Podocarpaceae
Taxus sumatrana	Taxaceae
Clethra sp.	Clethraceae
Rhododendron sp.	Ericaceae
Drimys piperita	Winteraceae

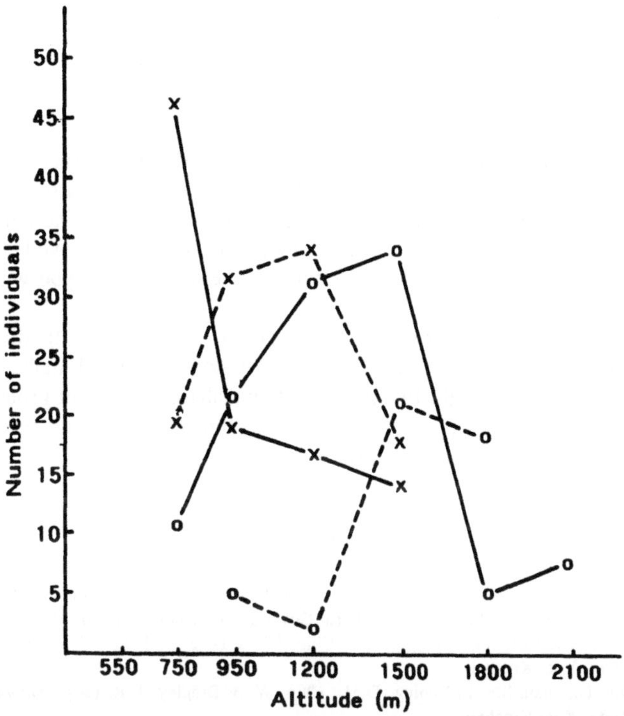

Figure 1. Altitudinal distribution of Meliaceae, Lauraceae, Myrtaceae and Fagaceae on the western slopes of Mt. Banahao, Luzon. x———x Meliaceae; x-------x Lauraceae; o———o Myrtaceae; o-------o Fagaceae.

and species could be purely ecological in context and the effect of elevation obscured (Van Steenis 1934, 1962). Even so, the elevation effect may still fall within the bounds of ecological preferences as the latter takes into consideration and gives emphasis to the significance of dispersal or dissemination of propagules (Hengeveld 1985).

Altitudinal Distribution in relation to elevation effect

To illustrate the elevation effect in the sense used by Van Steenis (1934, 1961, 1962), the vertical range of distribution of the four most important families of trees in the western slopes of Mt. Banahao (Figure 1) is further assessed.

The Myrtaceae had a peak density at 1500 m altitude and extended up to the summit zone, where only one species, *Leptospermum flavescens*, was represented. Members of the Meliaceae were abundant at 750 m altitude but abruptly declined beyond this elevation and faded out beyond 1500 m. The Lauraceae was abundant at 950 to 1200 m, less abundant at 1500 m and finally disappeared beyond there. The Fagaceae was absent from plots below 950 m, was abundant at 1500 m, diminished at 1800 m and was finally absent at the summit plot (2100 m).

According to Van Steenis (1934, 1961, 1962) and Backhuys (1969) the mountain flora could have a permanent zone of establishment above a certain level and on both sides above and below this, upper and lower zones of temporary establishment could occur. Both authors stressed the significance of downward dispersal of propagules from the permanent zone above and also beyond it for successful development of upper and lower temporary zones. For example, the Fagaceae could well have 1500 m altitude as their permanent zone of establishment. However, as Hengeveld (1985) warns, drawing such conclusions from relatively few samples and a small segment of the mountain could be hazardous.

SUMMARY AND CONCLUSIONS

Some so-called microtherm families, e.g. Winteraceae, Ericaceae and Taxaceae, are confined to the summit zone of Mt. Banahao's western slopes where their occurrence is probably influenced greatly by the low temperatures to which they are adapted. The generalist plant groups like the Meliaceae, Myrtaceae, Lauraceae, Fagaceae and others proved to be of wider distribution vertically as they adapt more to varying niche and habitat type with change in altitude.

The elevation effect as conceived by Van Steenis (1934) may apply to the different forest communities and plant groups observed on the western slopes of Mt. Banahao, Luzon. In mountains, especially of great height like that of Mt. Banahao, each plant group apparently displays permanent and temporary zones of establishment depending on source and availability of propagules and manner of dispersal.

LITERATURE CITED

BACKHUYS, W. 1969. Der elevations-effect bei siningen alpenplanzen der scherreiz. *Blumea* 16:273-320.

BROWN, W.H. 1919. Vegetation of Philippine Mountains. Government Printing Bureau, Manila.

COCKBURN, R.F. 1978. The flora. In Luping, D.M., Chin, W. & Dingley, E.R. (eds). *Kinabalu, Summit of Borneo*. Sabah Society, Kota Kinabalu.

CORNER, E.J.H. 1978. The plant life. In Luping, D.M., Chin, W. & Dingley, E.R. (eds). *Kinabalu, Summit of Borneo*. Sabah Society, Kota Kinabalu.

HENGEVELD, R. 1985. On the explanation of the elevation effect by a dynamic interpretation of species distribution along altitudinal gradients. *Blumea* 30:353-361.

KHAN, M. 1969. *Some physical and chemical properties of the forested soils of Mt. Makiling and its vegetation*. Ph. D. thesis, University of the Philippines, Los Banos College, Laguna.

MOHR, E.C.J. & VAN BAREN, F.A. 1954. *Tropical soils*. Interscience Publishers, The Hague.

VAN STEENIS, C.G.G.J. 1934. On the origin of the Malaysian mountain flora. Part 2. Altitudinal zones, general considerations and renewed statement of the problem. *Bulletin du Jardin Botanique Buitenzorg* 111:289-518.

VAN STEENIS, C.G.G.J. 1961. An attempt towards an explanation of the effect of mountain mass elevation. *Proceedings of the Royal Academy of Sciences of the Netherlands* C64:435-442.

VAN STEENIS, C.G.G.J. 1962. The mountain flora of the Malaysian tropics. *Endeavour* 21:183-193.

WHITMORE, T.C. 1975. *Tropical rain forests of the far east*. Clarendon Press, Oxford.

WHITTAKER, R.H. 1965. Dominance and diversity in land plant communities. *Science* 147:250-260.

WHITTAKER, R.H. 1972. *Ordination and classification of communities*. Junk, The Hague.

WHITTAKER R.H. & NIERING, W.A. 1975. Vegetation of the Santa Catalina Mountains, Arizona. *Journal of Ecology* 56:523-544.

WHITTAKER, W.H., BUOL, S.W. & NIERING, W.A. 1968. A soil and vegetation pattern in the Santa Catalina Mountains, Arizona. *Soil Science* 105:440-450.

Interspecific colony dispersion and niche relations of three large tropical rain forest ant species

RUTH LEVY

Department of Zoology, University of Oxford, South Parks Road, Oxford OX1 3PS, United Kingdom.

ABSTRACT. Three large tropical rain forest ant species (*Camponotus gigas, Diacamma intricatum* and *Odontoponera transversa*) in primary lowland dipterocarp forest, Brunei Darussalam, were not found to be significantly negatively associated spatially or in occurrences at baits. Food overlap was found to be high, and time overlap quite low, but not in the species of high food overlap. No evidence of resource partitioning was found, although each species had food preferences. Positive associations are likely to be microhabitat mediated.

KEY WORDS: resource partitioning, *Camponotus, Diacamma, Odontoponera*, Borneo, Brunei Darussalam.

INTRODUCTION

Since the pioneering work of Lotka (1932) and Volterra (1926) on predictive models of coexistence of competing organisms, and competitive exclusion experiments by Gause (1934), competition and resource partitioning have frequently been regarded as major community structuring forces. These concepts have been investigated in three major ways: workers have searched for 'substantial differences' in the niches of supposedly competing species (Schoener 1986); empirical data on niche distributions have been compared with null models in which niches are distributed at random (May 1984); and theoretical studies have been used to predict exactly how much overlap is permissible for species to coexist (Macarthur & Levins 1967).

Most workers have adhered to Hutchinson's definition of the niche as a 'multi-dimensional hypervolume' of space within the limits of tolerance of factors that affect an organism's survival (Hutchinson 1957); Pianka (1981) summarises the definitions. Today, the niche is commonly defined as a 'utilisation distribution' (Arthur 1987, Giller 1984, Schoener 1986), or frequency histogram, of food size classes, for example.

The relationship between niche overlap and competition between species has been quite confused in the past (Schoener 1974), authors recently concluding that the magnitude of overlap is a bad indicator of the level of competition (Keddy 1987, Schoener 1986).

The sorts of resources which have been found to be significantly partitioned in communities include habitat or micro-habitat, food type or size, and time, in that order (Pianka 1969, Schoener, 1986). This is because altering spatial utilisation has the least cost/benefit ratio of all resource types. Not feeding at any time costs most of all. Schoener (1986) notes that organisms most likely to be partitioned by time include predators and ectotherms. This is less likely as species number increases. Those with a relatively high prey size/body size ratio are often partitioned by prey type.

Workers usually choose to investigate niche relations in species of the same 'guild' (Root 1967). These are groups of species, the members of which are similar in their trophic roles, and which interact weakly with other guilds (Pianka 1983).

However, despite the interest in resource partitioning as a mechanism enabling coexistence

D. S. Edwards et al. (eds.), Tropical Rainforest Research – Current Issues, 331–340.

in communities, Arthur (1987) reminds us that there are at least five methods of enabling stable coexistence in communities, and only one of these is resource partitioning.

Lynch *et al.* (1980) examined ecological segregation in a guild of surface foraging, generalist ants in north American temperate forests; they found that the less aggressive species overlapped less in their resources with the most aggressive species, and that the weakest competitors had broader niches, an assumption of the competitive hierarchy model of community structure (Keddy 1989). However, Culver (1974) found no relationship between niche width and aggressiveness. Lynch (1981) found that habitat was a major factor contributing to community diversity. Differential uses of food sizes and type have been found to be an effect of differential desiccation resistance, body size, nest site and activity, and to promote coexistence in a grassland ant community (Hansen 1978). Briese (1982) found that food type and time overlaps fluctuated greatly in response to environmental changes in semi-arid Australia, and that there was little evidence for exploitative competition. Marsh (1985) found the same patterns in Namibia. Davidson (1977) found that differences in foraging strategy allowed coexistence of ant species in north American deserts.

In this study, the niche relations and interspecific dispersion of a guild of three large, generalist, solitary foraging, ground nesting ants, *Camponotus gigas* (subfamily Formicinae), *Diacamma intricatum* and *Odontoponera transversa* (both Ponerinae) in a hectare plot in Bruneian primary dipterocarp rain forest, were examined. These species were common at all primary lowland dipterocarp rainforest sites, at all times of year, at baits, in pitfall traps and in litter samples. The following questions were posed:

1) Is there any evidence of negative interspecific interactions spatially or at baits?
2) Do this guild of species' niches overlap substantially?
3) Are resource overlap and interspecific interactions related?

We might expect that low niche overlap would promote coexistence between the species, and that any negative interactions between them might be evidenced by relative nest distributions and associations at baits: if avoiding each other, nearest neighbour nests of different species would be situated further away from each other than expected randomly, and the species would forage less together.

This study was also intended as a descriptive one, as we have very little idea of what tropical rain forest ant species do, and especially what they eat (Naomi Pierce, personal communication).

The resources used in the analysis were those of food, space and time (Schoener 1974), namely, food type, food size, relative occurrence on the trees and the ground (space), and numbers of entrances to and exits from the nest entrance during 24 hours (time).

METHODS

Fieldwork

The fieldwork took place in June-December 1992 in plot AAU1, on the West Ridge of the Kuala Belalong Field Centre Reserve, Temburong, Brunei Darussalam. Nest entrance distributions of the three species of ant were mapped by following sardine baited foragers to nest entrances in the hectare, and noting locations of colony entrances on a map of the plot. If multiple nest entrances were found (only in the case of *C. gigas*), the geometric centre of these was noted as the colony location. The whole 100 x 100 m was mapped for *C. gigas* (45 colonies), a 60 x 60 m area for *D. intricatum* (62 colonies) and a 40 x 40 m area for *O. transversa* (367 colonies). It was not feasible to map the whole hectare for all species, particularly *O. transversa*. As the frequencies of all three species were found to be even over the hectare at baits, I considered it valid to implement the randomisation procedure described below on a section of the whole hectare's ant colonies. Interspecific nearest neighbour nest distances of 45 pairs of *Camponotus-Diacamma, Camponotus-Odontoponera* and *Diacamma-Odontoponera* were recorded.

Food samples were collected as follows:- food items carried by returning foragers were collected at the nest entrance of the three species (at least six colonies of each species were sampled), during the day and after dark, and preserved for later examination in 80% alcohol. Samples were measured and identified in the laboratory.

Use of space was assessed by placing protein and carbohydrate baits at 100 points at two levels in the hectare plot, on the litter and at eye level on tree trunks, and recording occurrences of these species at the baits, day and night. This was repeated three times. Each bait was visited only once in the day or in the night, in order to keep each observations independent over time. Diel cycles were recorded by observing ant activity at the nest entrance throughout four 24 hour cycles. All entrances and exits were recorded for four different colonies of each species.

ANALYSES

Interspecific Nest Entrance Dispersions

It was not possible to assess the interspecific association of the three species in the same way as other authors i.e. by using a G-test of contingency, as numbers of colonies of each species per unit area were so vastly different, and mapping an entire hectare of *O. transversa*, around 3000 nest entrances, was not feasible. I therefore decided to approach the problem in a different way.

The median values of observed interspecific nest entrance distances of 45 nest pairs for each species combination were compared with 1000 randomised replicates of interspecific nest distances. For each replicate, the larger (bottom) map (*C. gigas* or *D. intricatum*) was replicated ninefold, and the smaller (top) map (*D. intricatum* or *O. transversa*) was effectively moved around relative to it, always keeping at least a small fraction of the smaller map inside the centre of the nine larger map replicates. Nearest neighbour nest distances for the pair of species were then calculated. The ninefold replication which was used as the overlaid map was liable to go over the edge of the larger map when moved, and drastically alter inter-nest distances. It ensured that edge effects of the larger map were not a problem (Ryti & Case 1992), as the distance from the edge of the smaller to the edge of the larger map replicates was always at least 20 m. Edge effects caused by proximity of the larger map species to the edge of the smaller map were resolved by taking distances of nests only where the colony of the larger species fell inside a prescribed boundary area (Figure 1). Edge effects of this sort are liable to create spuriously high values for inter-nest distances. The width of this boundary was 2 m for *O. transversa* and 7.5 m for *D. intricatum.* Colonies of the larger map species within the internal boundary of the smaller map species were selected and interspecific nearest neighbour distances calculated. This method also kept the colony distributions intact during replication. This was necessary as nest distributions could not be assumed to be random. This was repeated for each replicate (see Figure 2).

The frequency distributions of 1000 randomised replicate medians were used to find directly the probability of the observed median result happening by chance. The number of randomised medians smaller or larger than the observed value were counted, and a p-value was calculated directly from this number. A negative association of nests (overdispersion) is shown when most randomised replicates are smaller than the observed value, and a positive association (clumping) when most randomised values are larger. The 0.05% level of significance (50 replicates) was used.

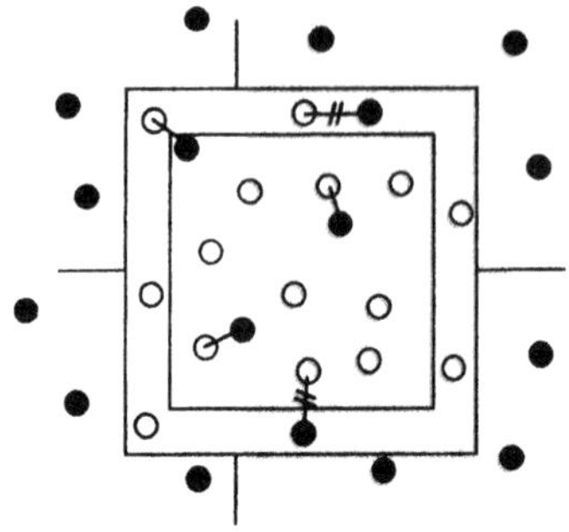

Figure 1. Close up view of nearest neighbour relations in randomised replicates of two species nest distributions. The species of smaller area (open circles) is overlaid on the nine larger species replicates (closed circles), e.g. *D. intricatum* overlaid onto *C. gigas*, or *O. transversa* on to *D. intricatum*. Valid internest distances are shown by lines joining the nests concerned. Invalid distances are shown by scored lines. This occurs if the larger species occurs outside the internal boundary area set for *D. intricatum* and *O. transversa*.

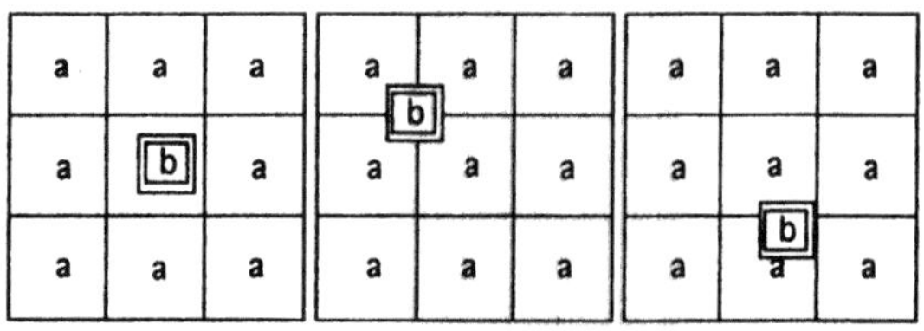

Figure 2. Successive replications of randomised nearest neighbour distance calculations for two ant species (a and b). Nearest neighbour distances between colonies of species a (larger ninefold replicated map) and b (smaller map) are calculated for colonies of species a occurring within the internal boundaries of the map of b.

Species Association at Baits

In order to assess interspecific association between the three species of interest at baits, contingency tests of association (Ludwig & Reynolds 1988) were performed on presence/absence data for the three two species interactions.

For each two species interaction, the following contingency table was constructed:

		species 1: present	species 1: absent	
species 2	present	a	b	a + b
	absent	c	d	c + d
		a + c	b + d	a + b + c + d

where: expected values (e.g. $E_{(a)}$ for cell a)

$$E_{(a)} = \frac{(a+b)(a+c)}{a+b+c+d} \qquad \text{(Equation 1)}$$

No correction for small sample size was necessary (Ludwig & Reynolds 1988), as numbers were large.

Niche Breadths and Overlaps

It was assumed that resources were limited in availability. Niche width and resource overlaps for food size, food type, use of vertical foraging space and use of diel time were calculated. The Levins index of niche breadth was used (Krebs, 1989):

$$B_i = \frac{1}{\sum_{a=1}^{n} p_{ai}^2} \qquad \text{(Equation 2)}$$

where B_i = Levins niche breadth of species i ; p_{ai} = proportion of use of resource class a by species i and n = number of resource classes.

As the index is maximum when all resource classes are used in equal proportion, the maximum possible niche breadths for each resource for each species were calculated. From this, an overall mean percentage niche breadth for each species was found.

The percentage overlap was calculated for each two species interaction (Krebs 1989):

$$p_{ij} = \left[\sum_{a=1}^{n} (minimum\ p_{ai}, p_{aj})\right] 100 \qquad \text{(Equation 3)}$$

where p_{ij} = percentage overlap of species i and j; p_{ai} = proportion of use of resource class a by species i and n = number of resource classes.

Food size niche parameters were calculated from the probability distribution function generated from the mean and standard deviation of the logged food size data, in order to facilitate smoothing the distribution. Food type overlap was calculated from the numbers of prey in each of the classes for each species. Time overlap was calculated for entrances and exits, and reduced to two classes of activity in the day and night (6 am - 6 pm and vice versa). Space utilisation (foraging height) was calculated from the relative numbers of occurrences of each species at tree and ground baits, day and night.

Overall percentage overlap for each pair of species was calculated in the following way: as spatial utilisation data were dependent on day and night (time), overall percentage overlap in time and space was calculated as a single parameter for each species. The percentage overlap of each of four classes, ground bait/day, ground bait/night, tree bait/day and tree bait/night were calculated and added together to give this parameter (May 1975). This percentage was then multiplied by those for food type and food size overlap to give an overall multiplicative percentage overlap. No inter-dependence of other resource axes were found for any of the species.

RESULTS

Interspecific Nest Entrance Dispersions

Table 1 shows the results of 1000 replicated randomised tests on the medians of the interspecific nest entrance distances, and the kinds of interactions going on. There are no significant negative interactions, but the nests of *D. intricatum* and *O. transversa* are significantly positively associated.

Species Associations at Baits

Table 2 shows the probabilities of observed occurrences of species with each other, and the kind of interactions, positive or negative, occurring.

Table 1. Observed median inter-nest distances for three species pairs, numbers of randomised inter-nest median distances smaller than the observed median from 1000 replicates, probability and type of interaction happening between measured nests.

species pair	observed median	number of randomised replicates	p	association of observed nests
C. gigas/D. intricatum	4.10	857	0.143	negative
C. gigas/O. transversa	1.00	350	0.350	none
D.intricatum/O.transversa	0.65	4	0.004	positive

Table 2. Results of test of interspecific association of three species with each other.

Species	χ_1^2	p	interaction
C. gigas/ D. intricatum	0.29	NS	none
C. gigas/ O. transversa	22.10	<0.001	+
D. intricatum/ O. transversa	0.67	NS	none

C. gigas and *O. transversa* are significantly positively associated at baits.

Niche Breadths and Overlaps

Figure 3 shows the percentage utilisation of the three commonest food resources, ants, termites and fungus by the three species. Each species differs in its favoured food resource.

Other food types utilised by the species included: arthropod parts (usually Orthoptera legs), droppings (mostly lizard), bees, termites, spiders, Orthoptera, insect pupae and larvae (occasionally of each other's species), seeds and seed parts (frequently dipterocarp seeds). I also observed these species respond to fresh vertebrate carrion.

Figure 4 shows the probability distribution of log food size classes consumed by the three species. Resource widths and overlaps were calculated from this distribution. Figures 5-7 show the combined diel schedules for each species.

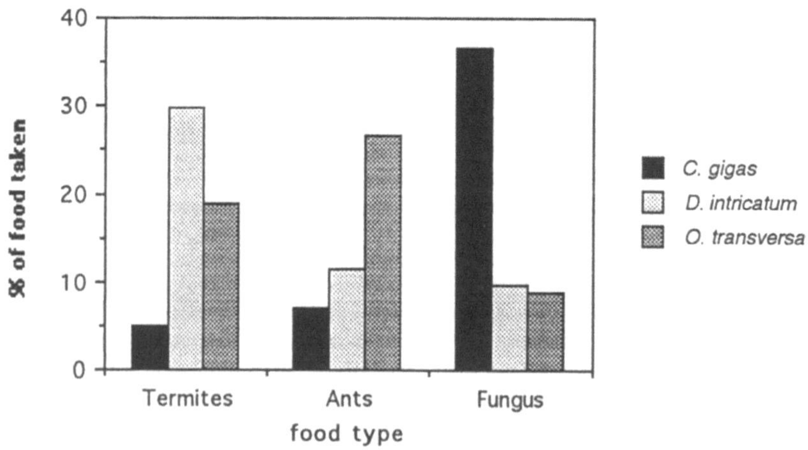

Figure 3. Percentage food type utilisation of the three commonest food types by the three species.

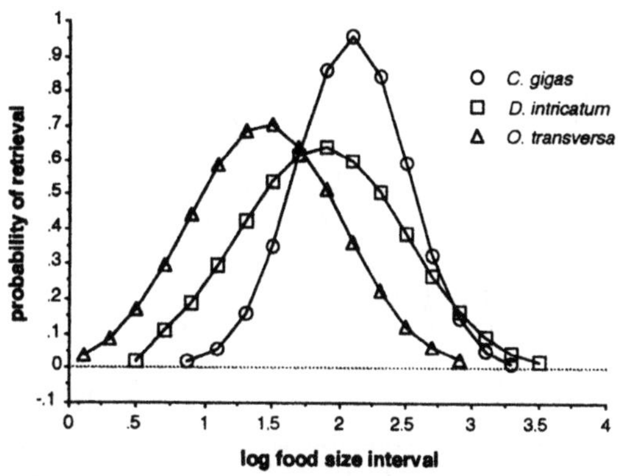

Figure 4. Probability distribution of food size classes collected by the three ant species, constructed from the means and Standard Deviations of the raw data.

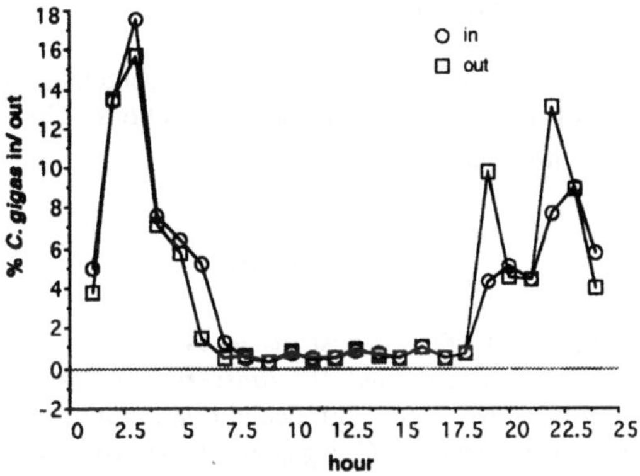

Figure 5. *Camponotus gigas* diel schedule. This ant is primarily nocturnal.

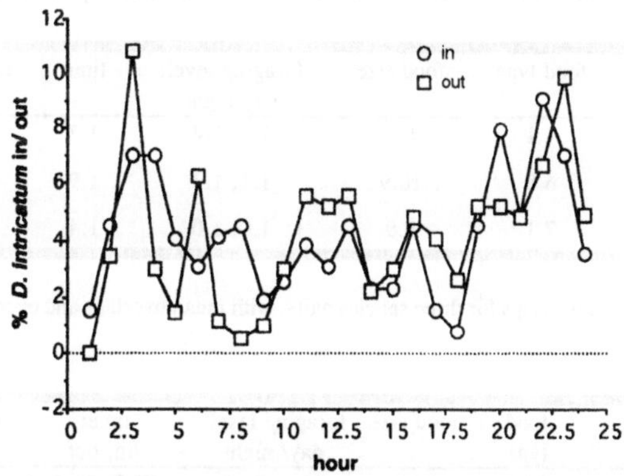

Figure 6. *Diacamma intricatum* diel schedule. This ant is nocturnal and diurnal.

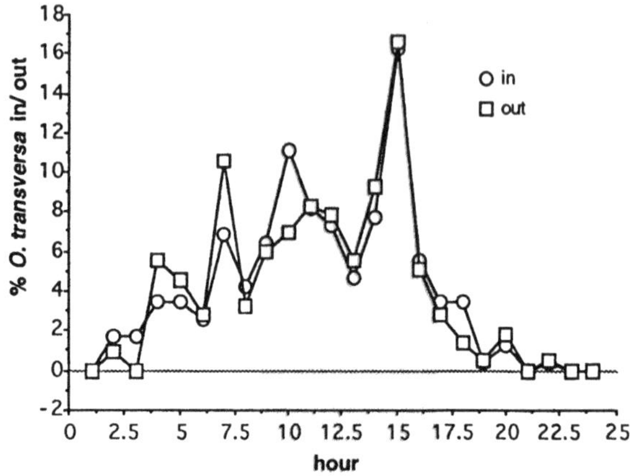

Figure 7. *Odontoponera transversa* diel schedule. This ant is primarily diurnal.

Table 3 shows Levins niche breadths for each species: *C. gigas* has the narrowest niche, and *D. intricatum* the widest.

Table 4 shows percentage resource overlaps, mean percentage overlaps and overall multiplicative overlaps for the three two species interactions. Greatest niche overlap is between *C. gigas* and *D. intricatum.*

DISCUSSION

I found that observed nearest neighbour nest distances were overwhelmingly larger than expected in the case of *D. intricatum* and *O. transversa*, indicating positive association, clumping, of their colonies. No other significant relationships were found. The only significant relationship at baits, between *C. gigas* and *O. transversa*, was also positive: this may be because resource overlap is lowest for *C. gigas* and *O. transversa*, and highest for *C. gigas* and *D. intricatum.* These two species' nest distributions tended non-significantly

Table 3. Levins niche breadths of three species, and mean percentage of total possible niche breadth.

Species	food type	food size	foraging level: day/night	time	mean % of total niche breadth
C. gigas	4.1	6.8	1.1, 1.4	1.2	39.6
D. intricatum	6.8	10.9	1.1, 1.2	1.9	52.0
O. transversa	7.1	9.9	1.0, 1.0	1.4	44.2

Table 4. Percentage resource overlaps for three species pairs, with mean overlaps and overall multiplicative overlaps for each species pair.

species pair	food type	food size	foraging level: day/ night	time: in, out	mean % overlap	overall % overlap
C. gigas/ D. intricatum	54.7	76.2	99.4, 91.8	72.7, 69.3	77.4	29.48
C. gigas/ O. transversa	44.5	50.28	96.8, 84.4	23.4, 24.3	53.9	5.33
D. intricatum/ O.	70.4	71.5	97.5, 92.6	50.7, 55.0	73.0	26.12

towards avoidance. The high resource overlap between these species might be viewed as surprising as they are in different subfamilies. The evidence for any pattern of associationcorrelating with niche overlaps in these species is slight. Positive association relationships may reflect habitat preferences in foraging area, or responses to competitive pressure from other aggressive species.

The resource creating most difference in species overall resource overlap in this study is time. Time is known to be an important factor allowing ecological segregation in ant communities (Hölldobler & Wilson 1990), often depending on physiological tolerance limits of species involved. Time segregation is likely to reduce interference competition between species, as opposed to food type or size, which reduces exploitative competition (Putman 1994). *C. gigas*' primarily nocturnal habit reflects that of many other *Camponotus* species (personal observations).

All three species are relatively large, compared with other ants in the community, but *C. gigas* is the largest and *O. transversa* the smallest. Mean food size collected is distributed as expected from rank of body size.

Although negative interspecific competitive interactions have usually been found between ant species of higher abundance and social organisation, these three species apparently form an ecologically distinct guild of their own, that of large, ground dwelling, generalist solitary foragers. It is not clear whether such a group of species would be expected to compete interspecifically, but as Savolainen & Vepsäläinen (1988) note, within the broad hierarchy of the Finnish taiga ant community, each tier is further subdivided into further hierarchical tiers which compete amongst themselves.

The old problem that species differ in their resource utilisation whether they are under competitive pressure or not (Giller 1984), occurs in this study. A null model would be needed to test whether the realised niches of these species are random or not. An improvement would be to have heeded May (1975) and to have calculated multidimensional resource overlap directly, by creating a single resource spectrum of all combinations of food type, food size, time and space variables. It would then be possible to avoid the problems of estimating multidimensional overlap indices indirectly, by adding or multiplying unidirectional measures, as I have done. However, in the complicated environment of a tropical rain forest, this is easier said than done. It might be possible with the use of specially designed baits, administered at prescribed times and places (e.g. Lynch *et al.* 1980), but this might also lose important qualitative information.

Putman (1994) discusses the fundamental problems with studies like the present one. He notes that it must be shown first that resources are in fact in limited supply. This is notoriously difficult to establish in field studies, and is usually assumed, as I have done. Another problem is non-independence of resource axes. I found, on analysing data collected for this study, dependence only of space and time utilisation, and combined these into a single parameter. Resource type may also segregate species in different ways: differential utilisation of certain resources may influence potential for interference competition (e.g.space and time), or for exploitation competition (e.g. food type and size): models of resource partitioning are invariably assembled to examine exploitative competition. Diffuse competition from the rest of the community may also alter the effect individual species have on each other. Putman concludes that characterisation of niche parameters is useful primarily as a descriptive tool of an organism's role in a community (its original function), and in fact has little predictive power.

All three species are frequently found in the community: they are distributed quite evenly over the hectare and at all other sites I visited. They are obviously able to coexist adequately with this magnitude of resource overlap. The next step in understanding the dynamics of the relationships between these species would be to perform resource and colony density manipulation experiments, and to look at resource utilisation responses and fitness of the species as a result.

LITERATURE CITED

ARTHUR, W. 1987. *The Niche in Competition and Evolution*. John Wiley & Sons, Chichester.

BRIESE, D. T. 1982. Partitioning of resources amongst seed-harvesters in an ant community in semi-arid Australia. *Australian Journal of Ecology* 7:299-307.

CULVER, D. C. 1974. Species packing in Caribbean and north temperate ant communities. *Ecology* 55:974-988.

DAVIDSON, D. W. 1977. Species diversity and community organisation in desert seed-eating ants. *Ecology* 58:711-724.

GAUSE, G. 1934. *The Struggle for Existence*. Haefner, New York.

GILLER, P. S. 1984. *Community Structure and the Niche*. Outline studies in ecology. Chapman and Hall, London.

HANSEN, S. R. 1978. Resource utilisation and coexistence of three species of *Pogonomyrmex* ants in an Upper Sonoran grassland community. *Oecologia (Berlin)* 35:109-117.

HÖLLDOBLER, B. & WILSON, E.O. 1990. *The Ants*. Belknap of Harvard University Press, Cambridge, Mass.

HUTCHINSON, G. E. 1957. Concluding Remarks. *Cold Spring Harbor Symposiun on Quantitative Biology* 22:415-427.

KEDDY, P. A. 1987. *Competition*. Population and Community Biology Series. Chapman and Hall, London.

KREBS, C. S. 1989. *Ecological Methodology*. Harper, New York.

LOTKA, A. J. 1932. The growth of mixed populations: two species competing for a common food supply. *Journal of the Washington Academy of Science* 22:461-469.

LUDWIG, J. A. & REYNOLDS, J.F. 1988. *Statistical Ecology*. Wiley, New York.

LYNCH, J. F. 1981. Seasonal, successional and vertical segregation in a Maryland ant community. *Oikos* 37:183-198.

LYNCH, J. F., BALINSKY, E.C. & VAIL, S.G. 1980. Foraging patterns in three sympatric forest ant species. *Ecological Entomology* 5:353-371.

MACARTHUR, R. H. & LEVINS, R. 1967. The limiting similarity, convergence and divergence of coexisting species. *American Naturalist* 101:377-385.

MARSH, A. C. 1985. Forager abundance and dietary relationships in a Namib desert ant community. *S.-Afr. Tydskr. Dierk* 20:197-203.

MAY, R. M. 1975. Some notes on estimating the competition matrix, alpha. *Ecology* 56:737-741.

MAY, R. M. 1984. Real and apparent patterns in community structure. Pp. 3-18 in Strong, D.R., Simberloff, D., Abele, L.G. & Thistle A.B. (eds). *Ecological Communities: Conceptual Issues and the Evidence*. Princeton University Press, Princeton, New Jersey..

PIANKA, E. R. 1969. Sympatry of desert lizards (*Ctenotus*) in Western Australia. *Ecology* 50:1012-1030.

PIANKA, E. R. 1981. Competition and Niche Theory. Pp. 167-196 in May, R.M. (ed.). *Theoretical Ecology*. Blackwell Scientific, Oxford.

PIANKA, E. R. 1983. *Evolutionary Ecology*. Harper and Row, New York.

PUTMAN, R. J. 1994. *Community Ecology*. Chapman and Hall, London.

ROOT, R. B. 1967. The niche exploitation pattern of the blue-gray gnatcatcher. *Ecological Monographs* 37:317-350.

RYTI, T. R. & CASE, T.J. 1992. The role of neighbourhood competition in the spacing and diversity of ant communities. *The American Naturalist* 139:355-374.

SAVOLAINEN, R. & VEPSÄLÄINEN, K. 1988. A competition hierarchy among boreal ants: impact on resource partitioning and community structure. *Oikos* 51:135-155.

SCHOENER, T. W. 1974. Resource partitioning in ecological communities. *Science* 185:27-39.

SCHOENER, T. W. 1986. Resource Partitioning. Pp. 91-126 in Kikkawa, J. & Anderson, D.J. (eds). *Community Ecology: Pattern and Process*. Blackwell Scientific, Victoria, Australia.

VOLTERRA, V. 1926. Variations and fluctuations of the number of individuals in animal species living together. Reprinted in: *Animal Ecology*, Chapman, R N. (ed.). McGraw-Hill, New York.

Immediate effects on wildlife of selective logging in a hill dipterocarp forest in Sarawak: mammals

ZAINUDDIN DAHABAN[1], M. NORDIN[1] *and* ELIZABETH L. BENNETT[2].

[1]*Department of Zoology, Universiti Kebangsaan Malaysia, 43650 Bangi, Selangor Darul Ehsan, Malaysia*
[2]*Wildlife Conservation International, 7 Jalan Ridgeway, 93200 Kuching, Sarawak, Malaysia*

ABSTRACT. The effects of selective timber extraction on wildlife communities were studied in detail in a mixed hill dipterocarp forest of the Kapit Division, Sarawak. The main study area was systematically surveyed for one year before it was logged and for another year immediately after logging. This paper highlights the changes in diversity and density of mammals in the area following logging, with special emphasis on the population of primates and squirrels. Comparisons are made to the results of similar studies carried out in Peninsular Malaysia and Sabah. As a result of logging, some species decline in number or become locally extinct, some remain relatively stable, others increase, and a few new species come into the logged area. A system of totally protected areas and large undisturbed areas within logging concessions are needed to ensure continued survival of the fauna.

KEY WORDS: Borneo, density, diversity, hill dipterocarp forest, logging, primates, squirrels.

INTRODUCTION

In earlier studies of this type, Johns (1983, 1989) studied the effects on wildlife of logging hill dipterocarp forests in Peninsular Malaysia and in Sabah. These areas were on good soils, with high initial densities of wildlife, and the virtual absence of hunting. On the other hand, Sarawak has notoriously poor soils and exceptionally high rainfall, low natural densities of wildlife (Bennett 1992), and a high level of hunting which almost invariably accompanies logging (Caldecott 1986). These were likely to make the impact of logging on wildlife in Sarawak more severe.

With 60% of the Sarawak's land area licensed for logging (Kavanagh *et al.* 1989) and 4.1 million hectares of production forest logged by 1990 (Mok *et al.* 1991), there was an urgent need to determine its effect on wildlife populations. In this paper, we present preliminary data collected by ZD for two years on the responses of wildlife to logging. At around the same time, the effects of hunting and shifting cultivation on wildlife of Sarawak were studied by Bennett (Bennett 1992, Bennett & Dahaban 1992).

The main objective of this work was to document the effects on wildlife of logging a hill dipterocarp forest in Sarawak. It is essential to know the true impacts of logging on wildlife and the environment so that effective management programs can be formulated to benefit both wildlife and humans.

STUDY AREA

The study was carried out in a hill mixed dipterocarp forest logging concession near Kapit. The 100 ha main study area (MSA) at 01°39'N, 113°24'E, comprised steep terrain between 250 m and 380 m altitude, and the soils were red-yellow podzols. The forest vegetation was dominated by trees of the Dipterocarpaceae, especially *Shorea* species. For a full description

D. S. Edwards et al. (eds.), Tropical Rainforest Research – Current Issues, 341–346.

of this habitat type, see Whitmore (1984, 1992).

Rainfall data were gathered by the logging company at their camp about 65 km north of the MSA since September 1988. There was no distinct seasonal pattern. Mean annual rainfall was 5550 mm while monthly totals ranged from 195 mm to 731 mm. Ambient temperature ranged from 22°C to 32°C.

The site was chosen mainly because of a highly organised infrastructure within the logging company and its predictable cutting rate. This meant that an area due to be logged halfway through the study period could be accurately delimited.

Data collection started in December 1989 when the MSA was primary forest, and about 7 km from the nearest access road or any disturbance. From December 1990 to March 1991, the area was selectively logged. Data collection continued until the end of November 1991, and a further survey was done in April 1992, a year after the felling had finished.

In addition, surveys were conducted in two other parts of the same concession area. One, the two-year old logged forest study area, had been logged in December 1988, and was surveyed in April and November 1991. The other, the four-year old logged forest study area, had been logged in January 1987, and was surveyed in September and November 1991.

In all three areas, logging was done using similar methods. As elsewhere throughout Sarawak's hill forests, tractor logging was employed. Trees were cut using chainsaws, pulled out by bulldozer to a loading site, loaded onto a truck and transported by road to a log pond. Timber extraction rates depended on the available volume. In the MSA, 54% of 4039 trees of at least 30 cm gbh tagged before logging were removed or destroyed. This treatment was likely to have been similar in all three sites. All three were part of the same forest block, the maximum distance between them was only 35 km, and the terrain and slope were similar in all. Thus the flora and fauna prior to logging were likely to have been similar at all sites.

Prior to logging, the whole area was remote and about 40 km from the nearest human settlement; hunting would have been negligible. Once access roads had been built and during the logging process, some hunting occurred. It ceased in the MSA two months after logging finished because a bridge along the access road collapsed. Even during logging, hunting in the MSA was lighter than elsewhere because ZD's presence acted as a deterrent. In general, most hunting was from vehicles along the road. Any animal seen was usually shot at, including primates, sun bear (*Helarctos malayanus*), clouded leopard (*Neofelis nebulosa*), and all species of ungulate. Hunting at night using vehicles and spotlights occurred in all areas with road access, and was especially heavy in weeks preceding major local festivals.

METHODS

The diversity of mammals in the study area was monitored throughout the study period. Any species of animal seen at any time in all three sites was noted.

The basic method to assess animal abundance used was the line transect method (Brockelman & Ali 1987, Burnham *et al.* 1980, Davies & Payne 1982, Marsh & Wilson 1981). At each study site, one survey trail two km long was cut. For the post logging surveys, new trails were cut; old logging roads were not used for surveys since they give a biased sample of the forest and its wildlife.

Survey sessions were divided by months. In each month, about 10 mornings, equivalent to 20 km of survey trail were walked. Data from different months were pooled in the analysis, resulting in a single data block each for the MSA before logging, the MSA after logging, the two-year old logged forest, and the four-year old logged forest.

Trapping of small mammals was not feasible due to time constraints. Hence, no data was available on the effect of logging on small mammals.

RESULTS

Diversity of primates and squirrels

The total number of species of all diurnal primates and squirrels seen in the MSA declined from 13 before logging to 12 after logging. Those seen in the two-year old logged forest and four-year old logged forest were 11 and 12 species, respectively. Analysis at species level showed the composition changed greatly. Some species seen before logging disappeared after logging and vice versa. Table 1 summarises the changes in species occurrence of primates and squirrels at all study sites.

Diversity of other mammals

Sixteen other species of mammals were recorded in the MSA before logging and 14 species after. Among those that were seen before but not after logging were clouded leopard, marbled cat (*Felis marmorata*), sun bear, thick-spined porcupine (*Thecurus crassispinis*), and oriental small-clawed otter (*Aonyx cinerea*). Conversely, the lesser treeshrew (*Tupaia minor*), slender treeshrew (*T. gracilis*), and collared mongoose (*Herpestes semitorquatus*) appeared after logging.

The diversity of other mammals observed in the two-year old and four-year old logged forests was extremely low. Only six species were seen in the two-year old logged forest, namely the common treeshrew (*Tupaia glis*), lesser treeshrew, sambar deer (*Cervus unicolor*), barking deer (*Muntiacus muntjac*), lesser mousedeer (*Tragulus javanicus*), and bearded pig (*Sus barbatus*). In the four-year old logged forest, the five species seen were common treeshrew, striped treeshrew (*Tupaia dorsalis*), barking deer, bearded pig, and malay weasel (*Mustela nudipes*).

Density of primates and squirrels

The immediate effect of logging on the primates (Table 2) proved to be severe. Bornean gibbons (*Hylobates muelleri*) in the MSA decreased more than 70% in group density after

Table 1. Presence and absence of primates and squirrels in main study area (MSA) before logging and after logging, in two-year old logged forest, and in four-year old logged forest. P=present, A=absent.

Species	MSA before logging	MSA after logging	Two-year old logged forest	Four-year old logged forest
Hylobates muelleri	P	P	P	P
Presbytis frontata	P	P	P	P
Presbytis rubicunda	P	P	P	P
Macaca fascicularis	P	A	A	P
Macaca nemestrina	A	P	A	P
Ratufa affinis	P	P	P	P
Callosciurus prevostii	P	P	P	P
Callosciurus notatus	A	P	P	P
Rheithrosciurus macrotis	P	P	P	P
Sundasciurus hippurus	P	A	A	P
Sundasciurus tenuis	P	P	P	P
Sundasciurus lowi	P	P	P	A
Lariscus hosei	P	A	A	A
Nannosciurus melatonis	P	P	P	A
Exilisciurus exilis	P	P	P	P

Table 2. Density of primates (groups per sq km) in main study area (MSA) before logging and after logging, in two-year old logged forest, and in four-year old logged forest. P = present but not seen during transect surveys.

Species	MSA before logging	MSA after logging	Two-year old logged forest	Four-year old logged forest
Hylobates muelleri	4.73	1.36	2.08	2.00
Presbytis rubicunda	0.38	0.21	1.25	0.05
Presbytis frontata	0.87	0.57	P	1.50

Table 3. Density of squirrels, treeshrews, and barking deer (individuals per sq km) in main study area (MSA) before logging and after logging, in two-year old logged forest, and in four-year old logged forest. P = present but not seen during transect surveys.

Species	MSA before logging	MSA after logging	Two-year old logged forest	Four-year old logged forest
Ratufa affinis	5.18	1.48	3.75	1.00
Other squirrels	15.82	24.08	103.50	19.00
All treeshrews	10.28	4.92	10.04	38.45
Muntiacus muntjac	2.91	0.79	10.22	P

logging. The group density of maroon langurs (*Presbytis rubicunda*) in the MSA dropped about 45% after logging while that of the white-fronted langurs (*P. frontata*) decreased by 35%.

Giant squirrels (*Ratufa affinis*) in the MSA showed a reduction of 71% in individual density after logging. On the other hand, the smaller squirrels in the MSA increased by 52% in combined individual density after logging. Table 3 shows the squirrel densities at the three study sites.

Density of other mammals

The combined treeshrews' individual density in the MSA decreased by 52% after logging. The individual density of barking deer in the MSA declined by 73% after logging (Table 3).

DISCUSSION

Some species never seen in the MSA before logging appeared after logging. Those that appeared are all abundant in secondary forests and gardens; for example the pig-tailed macaque (*Macaca nemestrina*), plaintain squirrel (*Callosciurus notatus*), and shrews (*Tupaia* spp.). Those animals which disappeared after logging were all endangered or totally protected species or both.

The logged-over forest supported less than half the number of gibbons than did primary forest. White-fronted and maroon langurs declined significantly. Giant squirrels were greatly affected after cutting. On the other hand, smaller squirrels increased in number greatly. The number of treeshrews and barking deer was less than half immediately after logging.

The purpose of surveying the two logged forest areas was to determine the abundance of animals there. This was to see if colonisation processes could take place from the old logged forest to the recently logged forest. From the current study, only the two langurs, the smaller squirrels, the treeshrews, and the ungulates survived well after logging. It is difficult to draw any firm conclusions on the fate of animals that vanished after logging and were not seen in the two older logged forest study sites.

In previous studies of similar nature in Malaysia (e.g. Johns 1983, 1989, 1992, Johns & Skorupa 1987), wildlife survival in general was apparently greater than that observed in the present study. One of the contributing factors could have been that the original densities of wildlife were lower in the current study area than in Johns' study sites (Tekam, Peninsular Malaysia and Ulu Segama, Sabah). Lower natural densities might mean that animals are more vulnerable to disturbance.

All of the studies, however, show that as a result of logging, some species decline in number or become locally extinct, some remain relatively stable, others increase and a few new species come into the area. The trend is for edge or coloniser species to replace primary forest species. Thus, even though the overall species diversity is similar with disturbance, the change is not necessarily beneficial from the conservation point-of-view. The species which are most threatened in Sarawak are those which depend on tall forests. Animals coming into disturbed sites such as plaintain squirrels are common in gardens throughout South-east Asia and are not likely to become extinct. As disturbance increases, they take over from species which depend on undisturbed forest and which cannot survive elsewhere.

The aim of conservation is not to maintain or increase the total number of species in a small area, but within a country or region. It is well known that localised diversity can often be increased by disturbing the habitat to create more edges (Murphy 1989). However, this does not increase biodiversity in the whole of Sarawak. In the long run, it decreases it.

Loss of animals from a forest is not just a conservation problem for their own sake, or even for the sake of maintaining a protein supply for rural people. Animals are an integral part of a tropical forest. Without them, the forest itself is likely to suffer a loss of diversity because of the animals' role as pollinators, dispersers, and as grazers stimulating plant diversity.

It is clear that the long-term effects of logging on wildlife are not yet fully understood. At the same time, the resilience of tropical forests to logging is still largely unknown. At this stage, an effective system of totally protected areas is desperately needed, to act as reservoirs during logging and to allow ample opportunity for animals to recolonise the disturbed areas. In addition, large primary forest areas should be retained within logging concession for the same purpose.

ACKNOWLEDGEMENTS

Studies such as this are rarely successful without a great deal of help from many people, too many to mention here. We would like to thank the then Sarawak State Secretary, Tan Sri Datuk Haji Bujang Nor for permission to conduct the study. Members of the Sarawak Forest Department were invaluable in their support in so many ways, especially Datuk Leo Chai, Abang Haji Kassim Abang Morshidi, Philip Ngau Jalong, Ngui Siew Kong, Francis Gombek, and Haji Othman Ismawi. The study could not have taken place without the constant help from B. Ross Ibbotson, David Manjah, Francis Goh, Patrick Braganza and other staff of Batang Baleh Forest Enterprises Sdn. Bhd. Others of enormous help include Dr Mikaail Kavanagh, Cynthia Low, Dr Mary Pearl, Martha Schwartz, and Abdullah Che Sahamat. The work was funded by Wildlife Conservation International of the New York Zoological Society and World Wide Fund for Nature Malaysia.

LITERATURE CITED

BENNETT, E.L. 1992. *A wildlife survey of Sarawak*. Wildlife Conservation International of the New York Zoological Society and WWF Malaysia.

BENNETT, E.L. & DAHABAN, Z. 1992. *Responses of wildlife to different types of disturbance in Sarawak, and implications for forest management.* Paper presented at the International Workshop on Ecology, Conservation and Management of South-east Asian Rainforest. Kuching, Sarawak, Oct 12-14, 1992. Proceedings in press.

BROCKELMAN, W.Y. & ALI, R. 1987. Methods of surveying and sampling forest primates populations. Pp. 23-62 in Marsh, C.W. & Mittermeier, R.A. (eds). *Primate Conservation in the Tropical Rain Forest.* Alan R. Liss Inc., New York.

BURNHAM, K.P., ANDERSON, D.R., & LAAKE, J.L. 1980. Estimation of density from line transect sampling of biological populations. *Wildlife Monographs* 72.

CALDECOTT, J.O. 1986. *Hunting and wildlife management in Sarawak.* World Wildlife Fund Malaysia and National Parks and Wildlife Office, Kuching.

DAVIES, A.G. & PAYNE, J.B. 1982. *A faunal survey of Sabah.* WWF Malaysia, Kuala Lumpur.

JOHNS, A.D. 1983. *Ecological effects of selective logging in a West Malaysian rain-forest.* D. Phil. dissertation, University of Cambridge.

JOHNS, A.D. 1989. *Timber, the environment and wildlife in Malaysian rain forests.* Unpublished report, Institute of South-east Asian Biology, University of Aberdeen.

JOHNS, A.D. 1992. Species conservation in managed tropical forests. Pp. 15-53 in Whitmore, T.C. & Sayer, J.A. (eds.). *Tropical Deforestation and Species Extinction.* Chapman and Hall, London.

JOHNS, A.D. & Skorupa, J.P. 1987. Responses of rain forest primates to habitat disturbance: a review. *International Journal of Primatology* 8: 157-191.

KAVANAGH, M., ABDULLAH ABDUL RAHIM & HAILS, C.J. 1989. *Rainforest Conservation in Sarawak, An International Policy for WWF.* WWF Malaysia and WWF International, Kuala Lumpur and Gland.

MARSH, C.W. & WILSON, W.L. 1981. *A survey of primates in peninsular Malaysian forest.* Final report to Universiti Kebangsaan Malaysia, Kuala Lumpur.

MOK SIAN TUAN, ALIAS ABDUL JALIL & DAWEND JIWAN, 1991. *A WWF Strategy for Tropical Forest in Sarawak.* WWF Malaysia and WWF International, Kuala Lumpur and Gland.

MURPHY, D.D. 1989. Conservation and confusion: wrong species, wrong scale, wrong conclusions. *Conservation Biology* 3: 82-84.

WHITMORE, T.C. 1984. *Tropical Rain Forests of the Far East, 2nd Edition.* Clarendon, Oxford.

WHITMORE, T.C. 1992. *An Introduction to Tropical Rainforests.* Clarendon Press, Oxford.

Growth performance of a naturally regenerated stand of Kapur (*Dryobalanops aromatica* Gaertn. f.) 37 years after a heavy low thinning treatment

ABD. RAHMAN, K.[1], SHAHRULZAMAN, I.[1] *and* WEINLAND, G.[2]

[1]*Forest Research Institute of Malaysia (FRIM), Kepong 52109, Kuala Lumpur*
[2]*Technical Advisor, German Technical Cooperation (GTZ), Malaysian-German Forestry Research Project, Forest Research Institute, Malaysia*

ABSTRACT. Growth increment and forest structure within a naturally regenerated Kapur Forest after 37 years following a heavy low thinning in Compartment 15, Kanching Forest Reserve, Malaysia, is described. Two sample plots (SP 85 and SP 86), each consisting of a thinned subplot (A) and an unthinned subplot (B), were analysed. Increments in diameter, basal area and volume of the dominant trees in the treated stands were generally higher than in the untreated stands. The mean annual volume increment for the four subplots was between 7.1 m^3 and 9.7 m^3. Considering the abundance of regeneration in all plots, research on natural regeneration systems for *Dryobalanops aromatica* is recommended.

KEY WORDS: Dipterocarpaceae, low thinning, increment, Malaysia.

INTRODUCTION

Kapur forests are named after the dominant species, *Dryobalanops aromatica* Gaertn.f., or Kapur as in the vernacular. In Peninsular Malaysia, the Kapur forests are confined to two large blocks at the east coast (in the south of Trengganu to north of Pahang and south of Pahang to south, southeast and east of Johor) and one in two small pockets at the west coast in Bukit Lagong F.R and Kanching F.R. (see Figure 1). The Kapur stands at the west coast have obviously been introduced sometime in the past, when camphor was traded between the east coast and the west coast of the Peninsula (Wyatt-Smith 1963).

Kapur is regarded as the most suitable species for forest management. The reasons are: (i) satisfactory and regular seed supply, (ii) high germination rates, (iii) easy nursery handling, (iv) good survival rates of planted seedlings even under adverse conditions. Barnard (1954) placed Kapur at the top of his silvicultural list of trees for plantations. Kapur seedlings and saplings are vigorous in growth. The young plant requires shade and responds later readily to canopy openings. It recovers quickly from mechanical injuries (Lee 1967, Foxworthy 1932).

In its natural habitat, Kapur tends to dominate the stands, so much so that a forest type has been named after the species. The stands in the Kanching Forest Reserve are generally pure stands. In such natural stands Kapur is strongly represented in all size classes, a feature which is not a characteristic of other emergent species (Wyatt-Smith 1963). Lim (1992) reported that litter production of a 70 year old Kapur stand in Compartment 15 of Kanching F.R. was 12.2 t/ha/year, while litter accumulation on the forest floor was 9.5 t/ha, giving a turnover rate of 1.3 which is fairly typical of many tropical forests. The high production of litter suggests that the biological productivity of the tree is high (Lim 1992).

Kapur was probably the most widely studied dipterocarp species in Peninsular Malaysia before the 1960's. During that period, a lot of effort was placed on research to develop establishment and management procedures for Kapur. Apart from the Kapur plantations

D. S. Edwards et al. (eds.), Tropical Rainforest Research – Current Issues, 347–355.

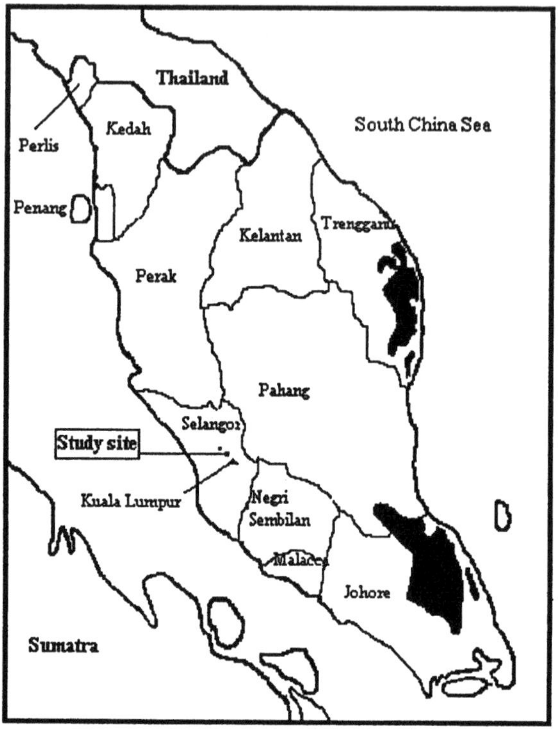

Figure 1. Location of Study Site and Distribution of Kapur Forest (shaded, after Wyatt-Smith 1963) in Peninsular Malaysia.

within the FRIM area which received some early thinning, only three thinning trials on a probing scale had been set up. These were in Panti Forest Reserve, Mersing Forest Reserve and Kanching Forest Reserve. This paper presents results of the thinning trial in Kanching Forest Reserve.

MATERIAL AND METHODS

The plots are located in Compartment 15, Kanching Forest Reserve, Ulu Selangor (see Figure 1) about 20 kilometres north of Kuala Lumpur. The topography is very steep to moderately steep with an elevation of approximately 150 m asl. SP 85 is facing south, while SP 86 varies from a western to a southern aspect. Substrate parent material is granite. Soil pits were dug in the vicinity of the plots when the trial was established. Soil texture is of a clay-loam with a higher content of sand on the upper slope.

At the time of establishment of the trial, the Kapur trees formed a more or less even-aged pole-size forest with individuals of other species such as Meranti tembaga (*Shorea leprosula*), Pulai (*Alstonia* spp.), Terentang daun besar (*Campnospermum* spp.) and bamboo. There were dense patches of Kapur regeneration with heights to about 4 m. Saplings of other species were also present. The forest was well stocked, containing some 5-10 per cent forked stems (Vincent 1963). The forest, in which both sample plots are located, has been treated since 1910, the treatment following the normal system of improvement fellings used in Selangor. This consisted of two improvement fellings. The first included a release of the crowns of the superior trees from competition of inferior trees, and a cutting of bertam palms (*Eugeissona triste*) and creepers if they were smothering or threatening to smother the crop trees. The treatment was repeated in the second improvement felling after about two years. In 1914 all trees interfering with the selected crop trees were felled and all bertam palms cut back. From 1914 until 1934, mature Kapur trees were removed from the stands and the final felling carried out in 1934. The majority of the mature Kapur trees and unclassified trees were removed between 1919 and 1923.

The trial consists of the two FRIM Sample Plots, SP 85 and SP 86. Each sample plot is divided into two subplots, a thinned (subplot A) and an unthinned (subplot B). The size of each subplot of SP 85 is 0.36 ha (40 m x 90 m) and of SP 86 is 0.40 ha (40 m x 100 m).

The goal of the trial was to determine the effect of a thinning treatment on volume increment. The crop age was unknown when the trial was established in 1954, but was assumed to be about 33 years (± 5 years). The mean annual increment is calculated based on the crop age of 70 years in 1991 assuming that all regeneration was established in 1921.

In both sample plots, subplot B was left as an unthinned control. In subplot A, about half of the trees, including dominant stems and defective trees, were felled. Volume before thinning, removal and remaining volume, both for stand total and the tree average, are presented in Table 1. All data are converted from imperial to metric units.

As can be seen from the ratio v_t/v_r, the thinning method applied was in both cases a low thinning. For low thinning the average size of the remaining trees shifts to a higher value, because mostly trees smaller than the average size are removed. This shift in average size can be expressed by the ratio of the average size of thinned trees (v_t) to the average size of the remaining trees (v_r). If this ratio is equal to unity ($v_t/v_r = 1.0$), trees were removed systematically favouring no size class. If the ratio is between 0.5 and 1.0, a low thinning was applied with the removal of mostly smaller trees. A ratio of $v_t/V_r \geq 1.0$ indicates that a crown thinning was applied, favouring selected trees by removing competing dominant and co-dominant trees.

Both treatments were distinct low thinnings. How the removal might have appeared is visualised in Figure 2. Figure 3 shows the stem-diameter distributions for the thinned plots. The majority of the trees have been removed from the left arm of the distribution, which is typical for a low thinning.

Unfortunately, no individual measurement data were recorded, only the averages and totals, including averages and totals of stand, averages and totals of population of the 25 largest and 50 largest trees per acre. As a result, no comparison of individual tree growth was possible. With two replicates only, no reasonable statistical comparisons could be made.

Table 1. Stand volume and average tree volume before and after thinning at commencement of the trial in 1954. v_r = average tree volume of the trees remaining; v_t = average tree volume of the felled trees.

Plots	N/ha	d_g	G/ha	V/ha	V/tree
85A					
Before thinning	181	40.4	23.20	260	1.4365
Removal	86	36.9	9.20	100	1.1765
After thinning	95	43.3	14.00	160	1.6842
86A					
Before thinning	220	36.8	23.40	240	1.0909
Removal	111	32.0	8.94	90	0.8108
After thinning	109	41.1	14.46	150	1.3762

The ratios for the thinning plots were: SP 85A : $v_t/v_r = 1.1765 : 1.6842 = 0.6989$
SP 86A : $v_t/v_r = 0.8108 : 1.3762 = 0.5892$

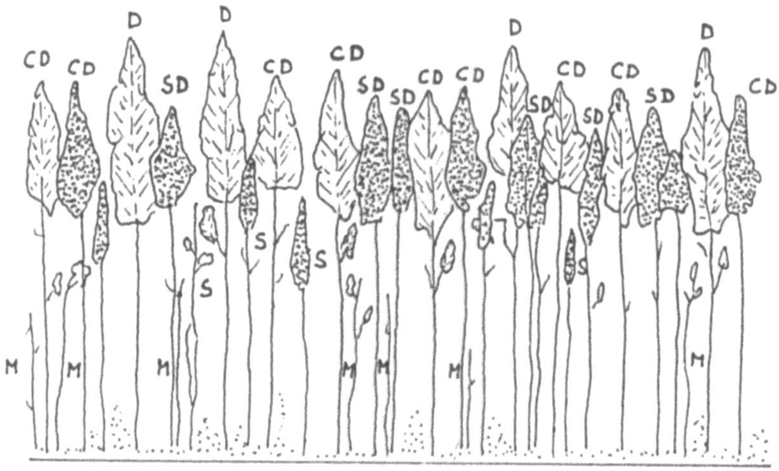

Figure 2. Visualisation of how the heavy low thinning in the Kapur thinning trial could have looked (drawing and description of tree classes modified from Evans 1982). Legend: D - Dominant trees, tallest and most vigorous trees, most of the crown free; CD - Upper canopy trees, below the crown level of the dominants; SD - Not in the upper canopy, leaders still have free access to overhead light; S - Branches do not reach up into the branches of the upper canopy, or leading shoot under some part of the crown of another tree; M - Dying or dead.

The height curves were fitted by the method of least squares using a quadratic function:

$$h = b_0 + b_1 * d + b_2 * d^2$$

where: h = dependent variable, height, $b_{0,1,2}$ = regression coefficients, d = independent variable, diameter. From the individual height curves the average height of the dominant stems was calculated. The height-diameter scatter discloses peculiarities of the canopy structure.

The comparability of the plots depends on the achieved top height at a certain age, which is regarded as an index for site quality. This top height, which is defined as the average height of the 100 largest trees per hectare, is usually taken as a basis for comparison. In our case, the average height of the 62 largest trees per hectare was taken instead, because individual tree heights from which the average height of the 100 largest could have been calculated were not available. However, the average height of the 62 trees/ha, which is here referred to as dominant height, can be regarded as a close approximation to the top height.

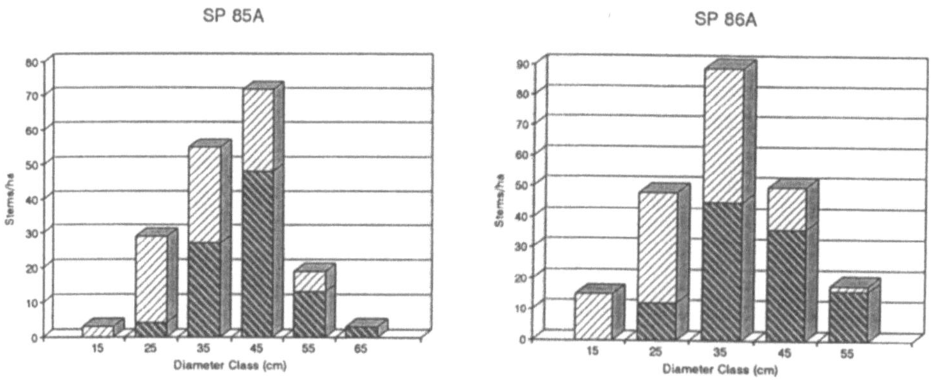

Figure 3. Stem-diameter distributions for the thinned plots, SP85A and SP 86A at stand age of 30 years (1954) before and after thinning.

Table 2. Dominant stand height at stand age of 33 years and 70 years.

Subplot	Height (m)		Increment (m)
	1954	1991	
85A	36.5	51.7	15.2
85B	35.5	49.6	14.1
86A	33.1	45.7	13.6
86B	33.4	50.0	16.6

Closer inspection of the height values indicate that the subplots in each sample plot obviously belong to different sites. Table 2 contains the average height of the dominant trees at commencement of the trial and at stand age of 70 years.

When considering the achieved dominant height at stand age of 70 years, it would be justified to compare the results of subplot 85A with those of subplot 86B and the results of subplot 86A with those of subplot 85B. The results are arranged accordingly.

The average diameter is calculated as the diameter of the mean basal area. Since no double-entry volume table was available for *Dryobalanops aromatica*, stand volume was determined approximately by: V = G x Bole Length x 0.6, where the term 0.6 is a reduction factor for stem taper. The height curves were calculated with STATGRAPHICS software.

RESULTS AND DISCUSSION

Stand and stock tables

Table 3 contains the stand and stock tables of the four experimental thinning plots. Except for stem number (N), it was not possible to retrieve information on the mortality during the period 1954-1991.

In 1954 the dominant heights ranged from 33.1 m to 36.5 m with the highest values for the subplots of SP85. In 1991, however, the largest dominant heights were achieved in the unthinned subplot of sample plot 85 (51.7 m) and in the thinned subplot of sample plot 86 (50.0 m). The average stand height was highest for the unthinned subplot of sample plot 85 (50.2 m). The average stand height of the remaining three subplots differed little (44.8 m - 46.3 m). The difference in clean bole height among the subplots was also small, in 1954 as well as in 1991. At the last measurement clean bole height ranged from 24.0 m to 28.5 m.

Larger differences were found for average stand diameter. The thinned stand of SP 85A achieved the highest average diameter (64.9 cm), the unthinned stand of SP 85B the smallest diameter (52.7 cm). Overall, the stand diameter reacted as we might expect following thinning: in both groups the stand diameter of the thinned plots is distinctly higher than the diameter of the unthinned plots. In sample plot 85 the highest stand basal area and stand volume was achieved in the thinned stand. In sample plot 86 the unthinned stand had higher basal area and volume than the thinned stand. Even in sample plot 85 the superiority of the thinned plot in stand basal area and stand volume over the unthinned plot is not pronounced. Unlike stand averages the stand totals such as stand basal area and stand volume depend on the number of stems per unit area as well. Thinning effects might not become evident until the average stand increment is inspected. In sample plot 85 the increment for all stand averages and stand totals is higher in the thinned plots than in the unthinned plots. In sample plot 86, however, the thinned plot shows a higher value only for the average diameter (D_g). Since the average diameter in the unthinned plot is considerably smaller than the average

Table 3: Stand and stock tables for four experimental thinning plots of *Dryobalanops aromatica*

Year/period	N	H_{dom} (m)	H_g (m)	H_{bole} (m)	D_g (cm)	G (m^2)	V* (m^3)
Group 1							
			Plot SP 85A - thinned				
1954 stocking	181	36.5	34.5	18.4	40.4	23.20	260
1954 removal	86		33.2	17.5	36.9	9.20	100
1954 remaining	95	36.5	35.0	18.7	43.3	14.00	160
1954-1991 mortality	-	-	-	-	-	-	-
1991 stocking	95	51.7	50.2	28.5	64.9	31.43	540
1954-1991 increment					21.6	17.43	380
			Plot SP 86B - unthinned				
1954 stocking	190	33.4	31.0	17.2	39.0	22.70	270
1954-1991 mortality	15	-	-	-	-	-	-
1991 stocking	175	50.0	46.9	27.5	54.8	41.37	680
1954-1991 increment	-	-	-	-	15.8	18.67	410
Group 2							
			Plot SP 86A - thinned				
1954 stocking	220	33.1	30.6	17.1	36.8	23.40	240
1954 removal	111	-	29.8	16.6	32.0	8.94	90
1954 remaining	109	33.1	31.4	17.3	41.1	14.46	150
1954-1991 mortality	-	-	-	-	-	-	-
1991 stocking	109	45.7	44.0	24.0	58.4	29.20	420
1954-1991 increment	-	-	-	-	17.3	14.74	270
			Plot SP 85B - unthinned				
1954 stocking	165	35.5	34.1	19.1	39.3	20.04	230
1954-1991 mortality	30	-	-	-	-	-	-
1991 stocking	135	49.6	47.0	28.2	52.7	29.48	500
1954-1991 increment	-	-	-	-	13.4	9.44	270

* No double-entry volume tables were available for *Dryobalanops aromatica*. The approximative volume is determined by: G x H_{bole} x 0.6, where the term 0.6 is assumed as a reduction factor for stem taper. H_{dom}: Average height of the dominant trees. H_g: Average stand height. H_{bole}: Average clean bole height. D_g: Diameter of the basal area mean stem. G: Stand basal area. V: Stand volume.

diameter in the thinned plot, the larger stand basal area of the unthinned plot is obviously brought about by the distinctly higher stand density (stem numbers). The main crop in all plots achieved harvestable size.

Dominant trees

In both groups, the thinned plot had a distinctly higher diameter, basal area and volume increment than the unthinned plot (Table 4). Periodic height increments, however, were, as expected, not different. The dominant height, in general, is not affected by the treatments.

The increment differences between unthinned and thinned plots are more pronounced, when the increment percentages are compared. The increment percent (i_p%) is the ratio of the increment value of a growth period to the growth value at the beginning of the period. The use of i_p% is recommended, especially when the initial growth values of the plots being compared are different.

The i_p% for basal area were 119% and 113% in the thinned plots and 82% and 77% in the unthinned plots. The i_p% for the volume were 233% and 200% in the thinned plots and

Table 4. Standing stock and periodic increment of the group of dominant trees (62 largest trees/ha). i_p: periodic increment; i_p%: increment percent; *The volume was determined approximately by: G/ha x Bole Height x 0.6; **Average height in 1954 was calculated based on 28 trees per hectare for SP 85 and 25 trees per hectare for SP 86. For 1991, average height for both sample plots were based on the 62 largest trees/ha.

Plot	Year	Height** (m)	Diameter (cm)	Basal Area (m^2)	Volume* (m^3)
85A	1954	36.5	47.1	10.8	120
	1991	51.7	69.7	23.66	400
	i_p	15.2	22.6	12.86	280
	i_p%	-	-	119	233
86B	1954	33.4	50.0	12.17	130
	1991	50.0	67.4	22.12	360
	i_p	16.6	17.4	9.95	230
	i_p%	-	-	82	177
86A	1954	33.1	44.5	9.62	100
	1991	45.7	64.9	20.51	300
	i_p	12.6	20.4	10.89	200
	i_p%	-	-	113	200
85B	1954	35.5	47.1	10.44	120
	1991	49.6	61.7	18.53	310
	i_p	14.1	14.6	8.09	210
	i_p%	-	-	77	175

177% and 175% in the unthinned plots. We may assume that there was a stimulating effect on the increment of the dominant trees following thinning. However, the thinning effect could not be confirmed as being statistically significant owing to lack of replicates.

Mean annual volume increment

The total stocking of all Kapur trees at the age of 70 including the thinning volume, which accrued in 33 years, is given in Table 5. The volume from mortality for the unthinned plots, which has to be known for an exact calculation of the mean annual volume increment, could not be included because no such information was available. However an approximate, albeit too low, mean annual increment can be derived.

The approximate mean annual volume increment (mai) at a stand age of 70 years is between 7.1 m^3 and 9.7 m^3. These values are similar to those of broad-leaved species stands of the same age on good sites in Europe (Anonymous 1990).

The mai figures of Group 1 are larger than those of Group 2 and in Group 1 the mai of the unthinned plot is larger than that of the thinned plot. Positive increment effects following a thinning intervention may not be visible in the mean annual increment. On the contrary,

Table 5. Approximate mean annual volume increment for the total stand at the age of 70 years. *Volume is determined by: G/ha x Bole height x 0.6. **mai does not contain mortality (self-thinning) volume.

Plot	Age (years)	Volume/ha * (m^3)	Mean annual volume increment (m^3)
Group 1			
SP 85A	70	640	9.1
SP 86B	70	≈ 680	≈ 9.7 **
Group 2			
SP 86A	70	510	7.3
SP 85B	70	≈ 500	≈ 7.1 **

if growing stock is removed beyond a certain limit, there will be no positive effect on the mai. The margins are small within which the mai can be raised through thinnings (Assmann 1970).

Structure

The height-diameter curves (Figure 4) cover a height range of about 35 m to 55 m. The height range of 35 m to 40 m is represented by only a small number of trees. Most of the trees are in the height range of 40 m to 50 m, which corresponds to the visual impression of a distinctly mono-layered upper canopy in all the four plots.

The understorey of *Dryobalanops aromatica* (trees >5 cm diameter) ranges in height from about 4 m to 32 m (Table 6).

However, most of the understorey trees are in the range 10 m to 15 m, which gives the impression of distinctly two-layered stands. The density in this height class ranges from 70 stems/ha to 141 stems/ha. Altogether, the number of stems/ha of understorey *Dryobalanops aromatica* ranges from 116 to 261. The trees smaller than 5 cm diameter have not been counted. However, all plots contain plentiful seedlings and saplings. Naturally regenerating stands of *Dryobalanops aromatica* should not pose major problems as far as recruitment is concerned.

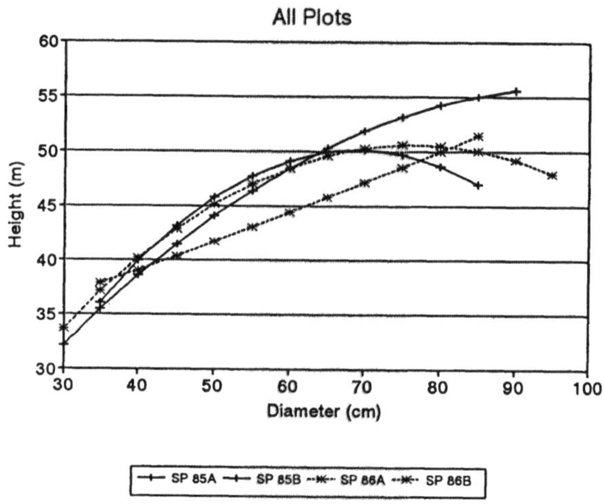

Figure 4. Height-diameter curves (1991) for Sample Plot 85 and 86.

Table 6. Stem-height distribution of understorey *Dryobalanops aromatica* trees in Sample Plot 85 and 86.

Height-Class	Stems/ha			
	SP 85A	SP 85B	SP 86A	SP 86B
0 - 5 m			2	
5 - 10 m	22	14	25	49
10 - 15 m	91	110	74	141
15 - 20 m	50	16	10	57
20 - 25 m	10	12	5	5
25 - 30 m	3	9		7
30 - 35 m	1	2		2
Total	177	163	116	261

CONCLUSIONS

The thinning trial in the Kanching Forest Reserve showed that *Dryobalanops aromatica* stands seem to react favourably to heavy low thinnings. Even better results could be expected from crown thinnings. Considering the ease with which the species establishes abundant seedlings under the canopy of the parent trees, research on regeneration systems for *Dryobalanops aromatica* stands would be a profitable exercise.

ACKNOWLEDGEMENTS

We wish to thank all officers and staff of the Forest Research Institute (FRI) and Forestry Department of Selangor who initiated the study, established, monitored and maintained the Sample Plots. Thanks are also due to Dr. Wan Razali and Dr. Appanah who read through the final draft and made valuable comments on the manuscript. The assistance of En. Ahmad Zuhaidi, who provided his field crew for control measurements, is gratefully acknowledged. Many thanks are extended to En. Norazmi Yaakub and En. Muzaid Hj. Salleh and other field staff for the assistance rendered in the data capturing and handling. Lastly we thank the Director-General of the Forest Research Institute of Malaysia for allowing the authors to present this paper.

LITERATURE CITED

ANONYMOUS 1990. *Hilfstafeln für die Forsteinrichtung*. Bayerisches Staatsministerium für Ernährung, Landwirtschaft und Forsten.

ASSMANN, E. 1970. *The Principles of Forest Yield Study*. Pergamon Press, Oxford.

BARNARD, R.C. 1954. *A manual of Malayan silviculture for inland forests*. Malayan Forest Research Pamphlet No. 14.

EVANS, 1982. *Plantation in the tropics*. Clarendon Press Oxford. 472 pp.

FOXWORTHY, F.W. 1932. Dipterocarpaceae of the Malay Peninsula. *Malayan Forest Records* No. 10:105-109.

LEE, P.C. 1967. *Ecological studies on Dryobalanops aromatica Gaertn. F.* Unpublished Ph.D. thesis. University of Malaya. Kuala Lumpur.

LIM, M.T 1992. Ecological studies on *Dryobalanops aromatica. Proceedings of the National IRPA (Intensification of Research Priorities Areas) Seminar (Agricultural Sector). 6-11 January 1992.* Kuala Lumpur. Volume II, p.431.

VINCENT, A.J. 1961. A survey of the Kapur silvicultural treatment research plots in naturally and artificially regenerated forest, Malaya. *Research Pamphlet No. 36.* FRIM, Kepong. 96 pp.

WYATT-SMITH, J. 1963. Manual of Malayan silviculture for inland forests. *Malayan Forest Records* 23.

A Comparative Study of the Ant Fauna in a Primary and Secondary Forest in Sabah, Malaysia

ARTHUR Y.C. CHUNG & MARYATI MOHAMED

Biology Department, Science & Natural Resources Faculty, Universiti Kebangsaan Malaysia, Sabah Campus, Locked Bag No. 62, 88996 Kota Kinabalu, Sabah

ABSTRACT. The diversity and ecology of ants in a primary and secondary selectively logged forest was compared over a 12 month period at Danum Valley, Sabah, Malaysia. Species richness and diversity were higher in the primary forest. A total of 192 species from seven subfamilies were recorded. The Myrmicinae was the most common subfamily in species and abundance, for both habitats. Rainfall had some negative effects on subterranean fauna in primary and secondary forest while high temperature diminished the number of arboreal species collected in the primary forest.

KEY WORDS: ant diversity, ecology, Myrmicinae, Formicinae, Ponerinae.

INTRODUCTION

Ants are extremely abundant, species-rich, and ecologically important in tropical lowland rain forests (Atkin & Proctor 1988, Burghouts *et al.* 1992, Levings 1983). Measurements suggest that about one third of the entire animal biomass of the Amazonian *terra firme* rain forest is composed of ants and termites, with each hectare of soil containing in excess of eight million ants and one million termites (Fittkau & Klinge 1973). These insects, along with bees and wasps, account for more than 75 percent of the total insect biomass in that forest. The diversity of ants is substantial, far exceeding that of other social insects and reflecting the manner in which ant species have evolved to saturate a wide range of feeding niches in soil and vegetation (Holldobler & Wilson 1990). In a lowland rain forest in Papua New Guinea, Wilson (1959) collected 172 species of ants belonging to 59 genera in an area of about one square mile (2.6 km^2). Olson (1991) recorded 41 genera and 122 species from three primary forest sites in Costa Rica, while Levings (1983) collected 49 genera and 127 species in a tropical moist forest in Panama. Bolton (in Room 1971) recorded 219 species in 63 genera in a square mile of a cocoa plantation and forest in Ghana.

Tropical rain forests are the richest ecosystems that the world has ever known (Marshall 1992). However, it has been estimated that 11 million hectares of mature tropical forests, of which 7.5 million hectares are rain forests, are converted each year to other uses. The great majority is changed to non-forest uses, only 600,000 hectares becoming timber plantations (Gomez-Pompa & Burley, in Marshall 1992), resulting in a loss of global biodiversity. Ants, one of the most diverse and abundant group of organisms in the rainforest, have great potential as indicator species for studying the effects of deforestation. The aims of this research were to describe the ant fauna in the rain forest, to study the effects of logging and to look at the effects of biotic and abiotic factors on the diversity and ecology of ants.

Study Area and the Environment

A 12-month study was conducted in Danum Valley Conservation Area, Sabah, Malaysia (4^0 58' N and 117^0 48' E), which comprises 43,800 hectares of tropical lowland evergreen

D. S. Edwards et al. (eds.), Tropical Rainforest Research – Current Issues, 357–366.

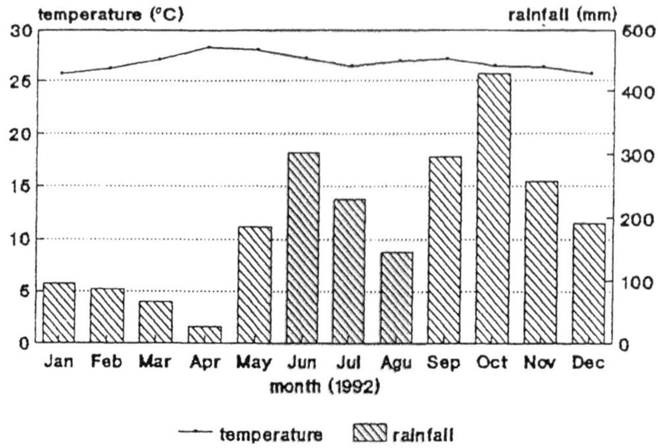

Figure 1. Mean temperature and monthly rainfall in Danum Valley Field Centre (1992).

dipterocarp rain forest. This area is situated in eastern Sabah at the upper reaches of the Segama River, approximately 70 km from the coast.

The total rainfall recorded at Danum Valley Field Centre (DVFC) for 1992 was 2314 mm, which was lower than the annual rainfall average of 2822 mm (based on 5 years of records). It was dry at the beginning of the year until April and two rainfall peaks in June (302.7 mm) and October (428.1 mm) were observed. This pattern (Figure 1) is influenced by the edge effects of two monsoons; the wetter northeast monsoon from November to March and the drier but more consistent southwestern monsoon in June and July (Marsh & Greer 1992). Mean annual temperature at DVFC in 1992 was 26.7°C. The highest mean monthly temperature recorded was 28.3°C in April, and the lowest was 25.6°C in December.

Two permanent 100 x 150 m plots were used to study the ant fauna; one each in a primary and secondary forest. The altitude of the plots was 180-230 m. In both plots, three line transects (150 m in length and 50 m apart) were established to collect ants. The primary forest plot was located at 0.6 km from DVFC, and was composed primarily of dipterocarps with a high canopy at about 40 m. The secondary forest plot was located within a three year old selectively logged forest, 2.0 km from DVFC.

MATERIAL AND METHODS

Sampling was done once a month along the transects in each plot. Three methods were used to collect ants at the arboreal, terrestrial and subterranean levels.

Manual Sampling

A sweep net, an aspirator and forceps were used to collect arboreal ants. The maximum height that could be reached using this method was about 3 m above the ground. Sampling was carried out in the morning (0830-1130 hrs), by collecting ants found on vegetation along the transects.

Baited Pitfall Sampling

Plastic cups with a diameter of 7 cm and a depth of 12 cm were planted in the soil, every 15 m along the transects. Eleven pitfall traps were set up in each transect, resulting in a total of 33 for one plot. Tuna flakes in oil were used as bait. The bait was wrapped in a piece of cloth which was hung across the cup with a small stick. The traps were 1/8-filled with water and liquid detergent was added to reduce water tension. They were left overnight in the forest and emptied the following day. This method was used for collecting nocturnal and terrestrial ants.

Winkler's Sampling

Leaf litter and soil to about 5 cm depth were collected and sifted from eleven 0.25 m^2 quadrats chosen at random along one transect. The sifted litter and soil were transferred to debris bags in the field. When the bags were brought back to DVFC, the contents were transferred to several flat mesh bags which were hung inside an outer cloth sack (Winkler's bag). Soil organisms worked their way out of the litter and dropped into a container which was placed at the base of the Winkler's bag. Specimens were collected and sorted the next day.

Sorting and Identification of Ants

Most of the ants were identified to generic level using Bolton (1990) and Murphy (1973). Specimens are deposited with the Entomology Section, Universiti Kebangsaan Malaysia, Sabah Campus.

Statistical Analysis

For the analyses of data, the Wilcoxon paired-sample test, which is a non parametric analogue to the paired sample t-test, was used as the data was not normally distributed.

Shannon's Index (H') was used to evaluate species diversity. This provides the average degree of uncertainty in predicting what species an individual chosen at random from a sample will belong to. Hutcheson 't' test was applied to test for the difference between two diversity indices in each month (Zar 1984).

Hill's Modified Index (E5) was used to calculate the degree of evenness of ant species distribution. The value approaches zero as a single species becomes more and more dominant in a community. This index was recommended as it is the least ambiguous and easiest to interpret. It does not require an estimate of the number of species in the community, a feature which is affected by sample size (Ludwig & Reynolds 1988).

Multiple correlation was used to detect relationships among variables throughout the year. The variables were number of species and individuals collected with different methods, total species, abundance, diversity, evenness, rainfall and temperature.

RESULTS AND DISCUSSION

Species Richness

A total of 192 ant species from more than 50 genera and seven subfamilies were recorded (see Appendix). However, some taxa were identified only to subfamily level. The percentage of species and abundance recorded for each subfamily in the primary and secondary plot are shown in Figure 2(i) and (ii). Myrmicinae was the most abundant and diverse subfamily in both plots, followed by Formicinae and Ponerinae. Percentage of species and abundance of the myrmicine ants were relatively higher in the secondary plot. The abundance of formicine and ponerine ants were higher in the primary plot. Dorylinae, Pseudomyrmecinae and Leptanillinae were represented by very few species and individuals in both plots.

Throughout the year, species richness was higher in the primary plot (Wilcoxon test, $P<0.01$). 164 (86%) species (out of the total of 192) were recorded in the primary plot, while 155 (81%) species were found in the secondary plot. 19.3% of the total species were found exclusively in the primary plot and 14.6% were recorded exclusively in the secondary plot. The monthly overlap of species was 33-44%. The annual overlap of species was 66%.

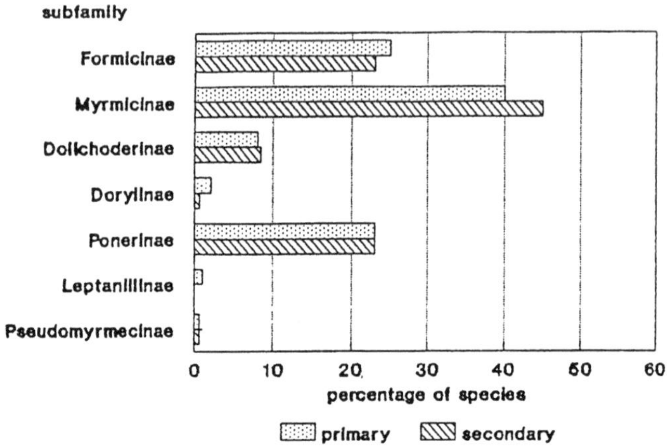

Figure 2 (i). Percentage of species in each subfamily, in primary and secondary plots.

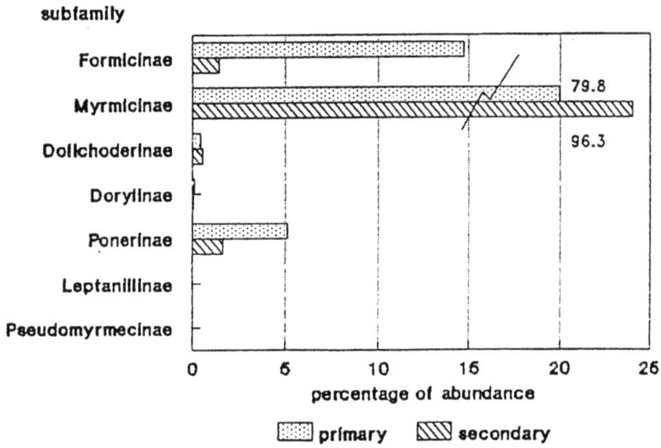

Figure 2 (ii). Percentage of ant abundance in each subfamily, in primary and secondary plots.

Diversity

The Shannon's diversity indices were significantly higher in the primary plot with the ratio of 10:2 for the 12 months' sampling (Hutcheson 't' test, $P<0.05$). The average value of Shannon's index (formula based on ln) was 1.83 (SD=0.48), compared to 1.47 (SD=0.37) in the secondary plot. The higher value in diversity index suggests that either one or both species richness and evenness were higher in the primary plot because diversity indices incorporate species richness and evenness into a single value.

Evenness

The average value of the evenness index (E5) was fairly low; 0.40 (SD=0.08) in the primary plot and 0.39 (SD=0.08) in the secondary plot. This suggests that dominance of certain ant species occurred in both plots and different ant-species had very different colony sizes. It also showed that higher diversity in the primary plot was due to species richness and the low value of evenness had indirectly affected the diversity values in both plots.

Table 1. Mean total number of species and individuals (±SD) collected each month with different methods in primary and secondary plots throughout the year.

	Pitfall	Manual	Winkler's
Species			
primary plot	21.5±3.5	20.8±3.1	33.9±3.8
secondary plot	20.7±2.6	22.6±2.6	23.8±3.4
Individuals			
primary plot	5498±2682	131±34	337±147
secondary plot	6818±3445	156±29	218±67

Table 2. Ant species that achieved more than 5% in the total abundance in respective plots throughout the year.

Species	code	% of abundance
Primary plot		
Pheidole sp.	DV83	49.0
Euprenolepis sp.	DV121	8.3
Pheidologiton sp.	DV45	7.8
Pheidole sp.	DV14	6.5
Secondary plot		
Pheidole sp.	DV83	66.0
Pheidologiton sp.	DV45	8.6

Abundance

Although more ants were collected from the secondary plot, the difference was not significant. Abundance in both plots varied and fluctuated throughout the year. The inconsistency in ant abundance may be the result of the methods used to collect ants. Baited pitfall sampling collected the most ants (Table 1), thus it was this method that primarily affected the total abundance of ants collected. At times, thousands of ants from one species were found in one pitfall trap because the trap was placed within the foraging trail of the species. Catches of any ant species depends on the size and abundance of its foraging population, the dispersion of colonies, levels of activity, the size of individual ants, and the thoroughness with which foragers cover their territory (Southwood 1966). The abundances in both plots were dominated by a myrmicine species, *Pheidole* sp. (DV83), large colonies of which were collected, mainly by pitfall sampling. Table 2 shows that four ant species from the primary plot and only two from the secondary plot achieved more than 5% of the total abundance in the respective plots.

Frequency of Occurrence

The number of individuals collected in an area is not a good indication of how common a species is, because of differences in foraging behaviour of different ant species (e.g, solitary or in colonies). The frequency of occurrence seems to give a more comprehensive picture. In this context, 'common' refers to the chances of being encountered. Figure 3 shows the frequency of occurrence of ant species at 12 sampling times. High percentages of species that occurred once only throughout the year in both plots suggests that a large number of ant species migrate from one place to another for food and shelter. 9.1% of the ant species in the primary plot and 8.4% in the secondary plot achieved 100% occurrence (Table 3).

Table 3. Ant species that were present throughout the year in primary and secondary plot.

Primary plot		Secondary plot	
Species	code	Species	code
Odontoponera sp.	DV1	*Odontoponera* sp.	DV1
Camponotus gigas	DV3	*Camponotus gigas*	DV3
Diacamma rugosum	DV10	*Diacamma rugosum*	DV10
Pheidole sp.	DV14	*Pheidole* sp.	DV14
Pheidole sp.	DV83	*Pheidole* sp.	DV83
Ponera sp.	DV23	*Ponera* sp.	DV23
Hypoponera sp.	DV87	*Hypoponera* sp.	DV87
Odontomachus rixosus	DV11	*Polyrhachis* sp.	DV7
Pheidole sp.	DV20	*Crematogaster* sp.	DV18
Pheidole sp.	DV118	*Polyrhachis ypsilon*	DV32
Tapinoma sp.	DV21	*Pheidologiton* sp.	DV45
Colobopsis sp.	DV30	*Strumigenys* sp.	DV48
Polyrhachis bihamata	DV31	*Pheidole* sp.	DV113
Discothyrea sp.	DV74		
Euprenolepis sp.	DV121		

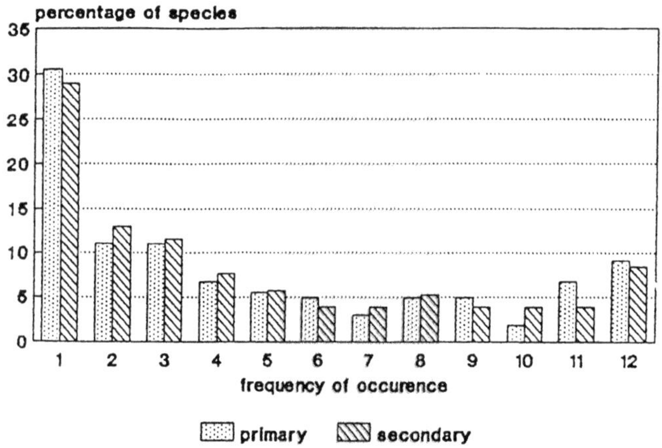

Figure 3. Frequency of occurrence in primary and secondary plots throughout the year.

Comparison of Species and Individuals Collected with Different Methods in Primary and Secondary Forest

Table 1 shows the mean total number of species and individuals collected each month. Number of species (Wilcoxon test, $P < 0.001$) and number of individuals ($P < 0.05$) collected with the Winkler's method were significantly lower in the secondary plot. Korthals (1990), in his research in Danum Valley, reported that the soil in the secondary forest contained more sand, less clay and seemed to be more compact than the primary forest. Conductivity, percentage of carbon, nitrogen, cation exchange capacity, and total and soluble phosphorus were lower in the secondary forest. The soil in the primary forest is less disturbed and has a higher nutrient status, thus making it more suitable for the soil fauna.

The number of individuals collected by manual sampling were higher (Wilcoxon test, $P < 0.05$) in the secondary plot. Manual sampling collected mainly arboreal ants. In general,

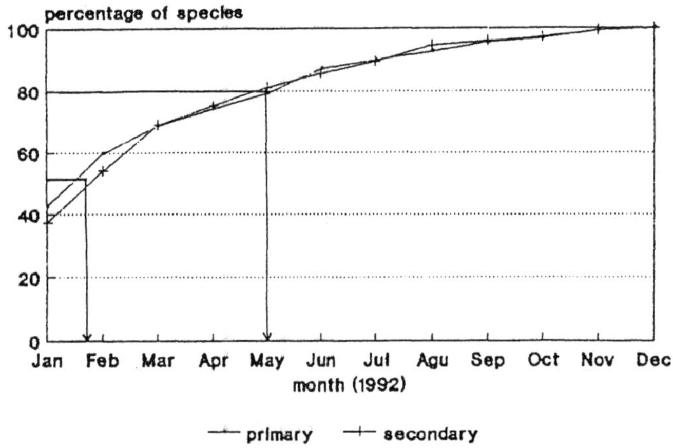

Figure 4. Cumulative percentage of species in both plots throughout the year.

trees in the secondary forest are lower with their canopy not as dense as those in the primary forest. Consequently, more light reaches the lower strata of the forest, resulting in an undergrowth which is relatively dense in certain parts of the plot. Climbers and shrubs like *Eupatorium, Jacquemontia, Callicarpa* and *Merremia* are abundant in such vegetation. Many of these plants produce extrafloral nectaries and food bodies which are a source of food for the ants. In return, the ants protect the plant from herbivore attack. Some plants, such as *Macaranga hypoleuca* and *Korthalsia furtadoana*, also provide shelter for ants. Ants stay in the stem and branch cavities of *M. hypoleuca* and in the ochreas (apical leaf sheath) of *K. furtadoana*. In primary forest, ground vegetation is scarce and most of the arboreal ants are found in the canopy (Fittkau & Klinge 1973) which was beyond the limit of the manual sampling.

Figure 4 estimates that to obtain 50% of the ant species recorded in this study, sampling using the three different methods should be carried out twice; to obtain 80% of the species, sampling should be carried out five times (sampling once a month).

Effects of Rainfall and Temperature

The number of species collected with the Winkler's method in the secondary plot and the number of individuals in the primary plot, collected by the same method, showed negative correlation with monthly rainfall ($P<0.05$). In the rain forest, 80.7% of the annual rainfall in 1990 reached the forest floor as throughfall and 1.9% as stemflow, giving an interception loss of 17.4% (Sinun *et al.* 1992). This percentage of throughfall affected the soil fauna. Rainfall reduced population density in the primary forest but did not affect the population density in the secondary forest. However, it caused a reduction in the number of species observed in the secondary forest, suggesting that while some species were disappearing, the remaining species were thriving after rainfall.

Ant species collected by manual sampling in the primary plot correlated negatively with high monthly mean temperature ($P<0.001$) and high daily temperature ($P<0.05$) (temperature taken at 0800hrs of the sampling day) whereas no correlation was detected in the secondary plot. This suggests that arboreal ant species in the primary forest have a lower tolerance towards high temperatures, compared with those of the secondary forest. Possibly temperature affected only the arboreal ants because microhabitats of arboreal ants are more exposed to changes of temperature, when compared to terrestrial and subterranean species.

Terrestrial and subterranean microhabitats are often surrounded by soil and leaf litter where the humidity is constantly high. Torres (1984), in his research on upland tropical forest, grassland and agricultural land in Puerto Rico, reported that forest ants had a lower tolerance to high temperatures than ants from non-forested areas.

ACKNOWLEDGEMENTS

We would like to thank Danum Valley Management Committee for permission to work at DVFC. We are grateful to the Danum researchers, especially Dr. J. Willot, Dr. P.M. Costa and R. Nussbaum, for comments on an earlier version of this manuscript. This study was funded by a M.Sc. Scholarship from Sabah Shell and partly from IRPA grant 04-07-03-07, and supported by the Royal Society, UK.

LITERATURE CITED

ATKIN, L. & PROCTOR, J. 1988. Invertebrates in the litter and soil on Volcan Barva, Costa Rica. *Journal of Tropical Ecology* 4:307-310.

BOLTON, B. 1990. Key to the ant genera of Oriental region. Pp. 46-53 in Holldobler, B. & Wilson E.O. (eds). *The ants*. Belknap Press of Harvard University Press, Cambridge, Massachusetts.

BURGHOUTS, T., KORTHALS, G. & VRIES, T.D. 1992. Litterfall, leaf litter decomposition and litter invertebrates in primary and selectively logged dipterocarp forest in Sabah, Malaysia. Pp. 407-416 in Marshall, A.G. & Swaine, M.D. (eds). *Tropical rain forest: disturbance and recovery*. The Royal Society, London.

FITTKAU, E.J. & KLINGE, H. 1973. On biomass and trophic structure of the Central Amazonian rain forest ecosystem. *Biotropica* 5:2-14.

HOLLDOBLER, B. & WILSON, E.O. 1990. *The ants*. Belknap Press of Harvard University Press, Cambridge, Massachusetts. 732 pp.

KORTHALS, G.W. 1990. *A description of the litter dwelling arthropod fauna in a primary and a secondary rain forest in North-East Borneo*. Ph.D. Thesis, Free University, Amsterdam.

LEVINGS, S.C. 1983. Seasonal, annual and among site variation in the ground ant community of a deciduous tropical forest: some causes of patchy species distributions. *Ecological Monographs* 53:435-455.

LUDWIG, J.A. & REYNOLDS, J.F. 1988. *Statistical ecology*. John Wiley & sons, Inc. New York. 337 pp.

MARSH, C.W. & GREER, A.G. 1992. Forest land-use in Sabah, Malaysia: an introduction to Danum Valley. Pp. 331-340 in Marshall, A.G. & Swaine, M.D. (eds). *Tropical rain forest: disturbance and recovery*. The Royal Society, London.

MARSHALL, A.G. 1992. The Royal Society's South-east Asia rain forest research programme: an introduction. Pp. 327-330 in Marshall, A.G. & Swaine, M.D. (eds). *Tropical rain forest: disturbance and recovery*. The Royal Society, London.

MURPHY, D.H. 1973. Preliminary key to genera of Malaysian ants, based on workers. *Bukit Timah Survey: Supplement 1*.

OLSON, D.M. 1991. A comparison of the efficacy of litter and pitfall traps for sampling leaf litter ants (Hymenoptera: Formicidae) in a tropical wet forest, Costa Rica. *Biotropica* 23:166-172.

ROOM, P.M. 1971. The relative distributions of ants species in Ghana's cocoa farms. *Journal of Animal Ecology* 40:735-751.

SINUN, W., WONG, W.M., DOUGLAS, I. & SPENCER, T. 1992. Throughfall, stemflow, overland flow and throughflow in the Ulu Segama rain forest, Sabah, Malaysia. Pp. 389-396 in Marshall, A.G. & Swaine, M.D. (eds). *Tropical rain forest: disturbance and recovery*. The Royal Society, London.

SOUTHWOOD, T.R.E. 1966. *Ecological methods, with particular reference to the study of insect populations*. Chapman & Hall, London. 542 pp.

WILSON, E.O. 1959. Some ecological characteristics of ants in New Guinea rain forests. *Ecology* 40:437-447.

ZAR, J.H. 1984. *Biostatistical analysis*. Prentice Hall, New Jersey. 718 pp.

Appendix. Ant species recorded in Danum Valley Conservation Area (1972). Species not in italics are identified to subfamily. sf (Subfamily) - Myrmicinae (M.), Formicinae (F), Ponerinae (P), Dolicherodinae (D), Dorylinae (DO), Pseudomyrmicinae (PS), Leptanillinae (L). Size in mm. Method - Manual (M), Pitfall (P), Winkler (W).

No.	Genus/Species	Code	s.f.	Size	Method	No.	Genus/Species	Code	s.f.	Size	Method
1	*Acanthomyrmex* A	DV 71	M	3	M,W	57	*Eurhopalothrix* A	DV 99	M	3	W
2	*Acanthomyrmex* B	DV 164	M	3	W	58	*Eurhopalothrix* B	DV 97	M	2	W
3	*Acropyga* A	DV 12	F	4	M	59	*Eurhopalothrix* C	DV 143	M	3	W
4	*Aenictus* A	DV 16	DO	4	M	60	Formicine A	DV 104	F	4	M
5	*Aenictus* B	DV 66	DO	4	M	61	Formicine B	DV 108	F	5	M
6	*Aenictus* C	DV 151	DO	3	M	62	Formicine C	DV 181	F	4	M
7	*Anochetus* A	DV 17	P	4	W	63	Formicine D	DV 187	F	3	P
8	*Anochetus* B	DV 95	P	2	W	64	Formicine E	DV 127	F	13	M
9	*Anochetus* C	DV 186	P	6	W	65	Formicine F	DV 169	F	1	W
10	*Bothroponera* A	DV 165	P	7	W	66	*Gnamptogenys* A	DV 57	P	6	M
11	*Calyptomyrmex* A	DV 50	M	3	W	67	*Gnamptogenys* B	DV 98	P	5	W
12	*Calyptomyrmex* B	DV 93	M	2	W	68	*Gnamptogenys* C	DV 100	P	3	W
13	*Camponotus* A	DV 70	F	5	M,P	69	*Gnamptogenys* D	DV 148	P	5	M
14	*Camponotus* B	DV 79	F	3	W	70	*Gnamptogenys* E	DV 177	P	2	W
15	*Camponotus* C	DV 107	F	12	M	71	*Hypoponera* A	DV 26	P	7	W
16	*Camponotus gigas*	DV 3	F	25	M,P	72	*Hypoponera* B	DV 73	P	2	W
17	*Camponotus variegatus*	DV 22	F	7	P,M	73	*Hypoponera* C	DV 75	P	1	W
18	*Cataulacus* A	DV 115	M	4	M	74	*Hypoponera* D	DV 87	P	3	W
19	*Ceraphachys* A	DV 135	P	2	W	75	*Hypoponera* D	DV 77	P	2	W
20	*Ceraphachys* B	DV 136	P	2	W	76	*Hypoponera* E	DV 176	P	6	W
21	*Colobopsis* A	DV 30	F	5	M	77	*Labidogenys* A	DV 49	M	2	W
22	*Colobopsis* B	DV 59	F	6	M	78	*Labidogenys* B	DV 142	M	2	W
23	*Colobopsis* C	DV 43	F	6	M	79	*Leptanilla* A	DV 91	L	2	W
24	*Colobopsis* D	DV 179	F	7	M	80	*Leptanilla* B?	DV 140	L	4	W
25	*Crematogaster* A	DV 5	M	4	M	81	*Leptogenys* A	DV 2	P	8	M,P,W
26	*Crematogaster* B	DV 18	M	2	M,P,W	82	*Leptogenys* B	DV 144	P	6	P
27	*Crematogaster* C	DV 34	M	4	M	83	*Leptogenys* C	DV 6	P	9	M,P,W
28	*Crematogaster* D	DV 62	M	3	M	84	*Leptogenys* D	DV 25	P	3	M
29	*Crematogaster difformis*	DV 28	M	5	M	85	*Liometopum* A?	DV 114	D	5	M
30	*Crematogaster* E	DV 102	M	2	M	86	*Liomyrmex* A?	DV 124	M	3	P
31	*Crematogaster* F	DV 184	M	2	M	87	*Mesoponera* A	DV 53	P	6	M
32	*Crematogaster* G	DV 105	M	2	M	88	*Mesoponera* B	DV 90	P	7	W
33	*Crematogaster* H	DV 156	M	2	W	89	*Monomorium* A	DV 68	M	1	M
34	*Crematogaster* I	DV 192	M	3	P	90	*Myrmicaria* A	DV 9	M	6	M,P
35	*Crematogaster inflata*	DV 33	M	5	M	91	*Myrmicaria* B	DV 110	M	5	M
36	*Diacamma* A	DV 162	P	9	M	92	*Myrmicaria* C	DV 155	M	5	M
37	*Diacamma rugosum*	DV 10	P	12	M,P	93	Myrmicine A	DV 188	M	2	W
38	*Discothyrea* A	DV 74	P	2	W	94	Myrmicine B	DV 191	M	1	W
39	*Discothyrea* B	DV 133	P	3	W	95	Myrmicine C	DV 63	M	4	M
40	Dolichoderine A	DV 55	D	2	P	96	Myrmicine D	DV 85	M	4	M,W
41	Dolichoderine B	DV 123	D	5	P	97	Myrmicine E	DV 86	M	3	W
42	Dolichoderine C	DV 152	D	5	M	98	Myrmicine F	DV 92	M	5	W
43	*Dolichoderus* A	DV 4	D	5	M	99	Myrmicine H	DV 96	M	1	W
44	*Dolichoderus* B	DV 38	D	4	M	100	Myrmicine I	DV 111	M	4	M
45	*Dolichoderus* C	DV 67	D	5	M	101	Myrmicine J	DV 119	M	2	P
46	*Dolichoderus* D	DV 167	D	3	M	102	Myrmicine K	DV 125	M	2	P
47	*Dolichoderus* E	DV 58	D	5	M	103	Myrmicine L	DV 131	M	5	W
48	*Dolichoderus* F	DV 126	D	8	P,M	104	Myrmicine M	DV 132	M	3	W
49	*Dolichoderus* G	DV 44	D	6	M	105	Myrmicine N	DV 145	M	3	P
50	*Dolichoderus* H	DV 64	D	6	M	106	Myrmicine O	DV 149	M	1	M
51	*Dolichoderus* I	DV 190	D	3	M	107	Myrmicine P	DV 157	M	3	W
52	*Dolichoderus thoracicus*	DV 60	D	4	M,P	108	Myrmicine Q	DV 158	M	1	W
53	*Echinopla* A	DV 35	F	6	M	109	Myrmicine R	DV 159	M	3	W
54	*Epitritus* A	DV 94	M	1	W	110	Myrmicine S	DV 160		3	W
55	*Euprenolepis* A	DV 121	F	5	P	111	Myrmicine T	DV 163		2	W
56	*Euprenolepis* B	DV 13	F	3	M,P	112	Myrmicine U	DV 171		3	P

Appendix (continued). Ant species recorded in Danum Valley Conservation area (1972). Species not in italics are identified to subfamily. s.f. (subfamily) - Myrmicinae (M), Formicinae (F), Ponerinae (P), Dorylinae (DO), Pseudomyrmicinae (PS), Leptanillinae (L). Size in mm. Method - Manual (M), Pitfall (P), Winkler (W).

No.	Genus/Species	Code	s.f.	Size	Method	No.	Genus/Species	Code	s.f.	Size	Method
113	Myrmicine V	DV 180	M	2	W	153	*Polyrhachis* L	DV 154	F	8	M
114	*Myrmoteras* A	DV 27	F	4	W	154	*Polyrhachis* M	DV 54	F	7	M
115	*Myrmoteras* B	DV 134	F	4	W	155	*Polyrhachis* N	DV 168	F	7	M
116	*Mystrium* A	DV 84	M	4	W	156	*Polyrhachis* O	DV 174	F	10	M
117	*Odontomachus rixosus*	DV 11	P	12	M,P,W	157	*Polyrhachis* P	DV 175	F	7	M
118	*Odontoponera transversa*	DV 1	P	10	M,P,W	158	*Polyrhachis* Q	DV 182	F	5	M
119	*Oligomyrmex* A?	DV 41	M	2	M	159	*Polyrhachis rastellata*	DV 37	F	5	M
120	*Pachycondyla* A	DV 47	P	11	W	160	*Polyrhachis ypsilon*	DV 32	F	12	M
121	*Pachycondyla* B	DV 52	P	8	M,W	161	*Ponera* A	DV 23	P	3	P,W
122	*Pachycondyla* C	DV 128	P	5	M	162	Ponerine A	DV 138	P	2	W
123	*Paratrechina* A	DV 15	F	3	M,P,W	163	Ponerine B	DV 172	P	6	M
124	*Paratrechina* B	DV 103	F	2	M	164	Ponerine C	DV 183	P	2	W
125	*Pheidole* A	DV 14	M	2	M,P,W	165	Ponerine D	DV 189	P	4	M
126	*Pheidole* B	DV 42	M	4	M	166	*Proatta* A	DV 141	M	2	W
127	*Pheidole* C	DV 69	M	1	M,P	167	*Proceratium* A	DV 80	P	4	W
128	*Pheidole* D	DV 83	M	3	M,P,W	168	*Proceratium* B	DV 101	P	3	W
129	*Pheidole* E	DV 113	M	2	M	169	*Proceratium* C	DV 161	P	2	W
130	*Pheidole* F	DV 118	M	1	P	170	*Proceratium* D	DV 185	P	4	W
131	*Pheidole* G	DV 116	M	3	P	171	*Pseudolasius* A?	DV 139	F	3	W
132	*Pheidole* H	DV 122	M	1	P	172	*Smithistruma* A?	DV 56	M	2	P
133	*Pheidole* I	DV 20	M	3	M,P	173	*Strumigenys* A	DV 48	M	1	W
134	*Pheidole* J	DV 129	M	3	M	174	*Strumigenys* B	DV 76	M	1	W
135	*Pheidole* K	DV 178	M	1	W	175	*Strumigenys* C	DV 170	M	2	W
136	*Pheidologiton* A	DV 45	M	2	M,P	176	*Strumigenys* D	DV 24	M	3	W
137	*Pheidologiton* B	DV 88	M	2	W	177	*Strumigenys* E	DV 82	M	3	W
138	*Pheidologiton* C	DV 146	M	3	P	178	*Tapinoma* A	DV 166	F	2	M
139	*Polyrhachis* A	DV 7	F	6	M	179	*Tapinoma* B	DV 21	F	4	M,P,W
140	*Polyrhachis armata*	DV 39	F	7	M,P	180	*Tapinoma* C	DV 29	F	1	M
141	*Polyrhachis* B	DV 8	F	4	M	181	*Tetramorium* A	DV 51	M	2	W
142	*Polyrhachis bihamata*	DV 31	F	11	M	182	*Tetramorium* B	DV 72	M	4	M,P
143	*Polyrhachis* C	DV 36	F	6	M	183	*Tetramorium* C	DV 81	M	3	W
144	*Polyrhachis calybea*	DV 173	F	9	M	184	*Tetramorium* D	DV 89	M	1	W
145	*Polyrhachis* D	DV 40	F	5	M	185	*Tetramorium* E	DV 112	M	4	M
146	*Polyrhachis* E	DV 46	F	6	M	186	*Tetramorium* F	DV 117	M	2	P
147	*Polyrhachis* F	DV 61	F	5	M	187	*Tetramorium* G	DV 137	M	3	W
148	*Polyrhachis* G	DV 105	F	5	M	188	*Tetramorium* H	DV 150	M	2	M
149	*Polyrhachis* H	DV 109	F	7	M	189	*Tetramorium* I	DV 78	M	2	W
150	*Polyrhachis* I	DV 120	F	9	M	190	*Tetraponera* A	DV 147	PS	7	M
151	*Polyrhachis* J	DV 130	F	4	M	191	*Trigonogaster* A	DV 19	M	2	P
152	*Polyrhachis* K	DV 153	F	7	M	192	*Vollenhovia* A?	DV 65	M	4	M

Comparative morphology and physiology of gap and understorey seedling leaves from a mixed dipterocarp forest

WEBBER E. BOOTH

Biology Department, Universiti Brunei Darussalam, Bandar Seri Begawan 2028, Brunei Darussalam

ABSTRACT. Leaf morphology and rates of stomatal conductance, net photosynthesis and dark respiration were determined for seedlings associated with a gap-understorey situation in the mixed dipterocarp rainforest of the Batu Apoi Forest Reserve in the Temburong District of Brunei Darussalam. The species investigated included four emergent tree species, ***Shorea falciferoides*** Foxw., ***Shorea leprosula*** Miq., ***Parashorea macrophylla*** Wyatt-Sm. ex Ashton, and ***Dryobalanops lanceolata*** Burck., two canopy species *Aporosa lunata* (Miq.) Kurz and *Drypetes* sp. and two pioneer species, *Macaranga trachyphylla* Airy Shaw and *Musa* sp. The pioneer species were confined to the gap while all other species were present in both the gap and understorey. Comparative morphological studies indicated that leaves of understorey seedlings are thinner, have a lower stomatal density and higher specific leaf area than leaves of the same species growing in the gap. Stomatal conductances for leaves of gap seedlings fall into two categories; high rates for the pioneers and more conservative rates for the non-pioneers. In the understorey low photosynthetically active photon flux density values are associated with relatively high stomatal conductance. Rates of light saturated net photosynthesis were significantly higher for pioneer species than non-pioneers. Based on apparent quantum yields, understorey seedlings tended to be photosynthetically more efficient than gap plants; also, light compensation points and rates of dark respiration were lower for understorey plants than for those of the same species growing in the gap. These features may be important adaptations which allow seedlings to establish in the very low light of the understorey.

KEY WORDS: tropical rainforest, seedlings, gap, understorey, leaf morphology, photosynthesis, respiration, Borneo.

INTRODUCTION

The death of an individual tree or group produces a gap in the canopy into which other trees grow. These in turn reach maturity and perhaps senescence, then die (Whitmore 1975). As a result of such natural disturbance the tropical rainforest is in a state of dynamic equilibrium being a mosaic of patches in different stages of growth (Chazdon & Fetcher 1984, Martinez-Ramos *et al.* 1988, Whitmore 1989). Based on the understanding that the successional status of a species is determined by the ability of its seedlings to exploit the available photosynthetically active photon flux density (PPFD) and the validity of the observation that in Asian forests the seedlings of canopy dominants and emergent trees often establish in the understorey then later are dependent on the formation of a gap for further development (Whitmore 1975, 1978), then it must follow that these seedlings are able to initially exploit the very low PPFD of the understorey and subsequently the high PPFD of the gap. Tropical rainforest seedlings therefore require a high plasticity in terms of their physiology and morphology as they encounter contrasting environments (Canham 1989, Osunkoya & Ash 1991, Popma & Bongers 1991). The capacity to tolerate different light regimes involves not only adjustments of the photosynthetic apparatus but also of the leaf structure (Bjorkman 1981, Loach 1970, Riddoch *et al.* 1991). In contrast to the canopy and emergent species, short-lived pioneer plants establish only after the gap is formed and therefore need only an enhanced capacity to utilize the high PPFD of the gap (Bazzaz & Pickett 1980, Martinez-Ramos 1988, Whitmore 1989). They can therefore be expected to

D. S. Edwards et al. (eds.), Tropical Rainforest Research – Current Issues, 367–375.

have a distinctive physiology.

Differences in structure and physiology of leaves of seedlings from gap and understorey situations as suggested above, are widely accepted as being applicable in the tropical rainforest situation but few supporting *in situ* data exist. It was therefore the objective of this study to gather such data from a gap-understorey situation within the lowland mixed dipterocarp rainforest in the Temburong District of Brunei Darussalam for a selection of emergent, canopy and pioneer seedlings.

MATERIALS AND METHODS

Site

The study site was located beside the Belalong river approximately 2 km upstream from the Kuala Belalong Field Studies Centre in the Batu Apoi Forest Reserve. It consisted of a gap approximately 20 x 100 m with the longitudinal axis oriented E-W and extending across the river so providing exposure to high irradiance from about 0900 to 1500 hours. For information on the geology and geomorphology of this area refer to Ross & Dykes (this volume).

Species

The species studied included four emergent trees, *Shorea falciferoides* Foxw., *Shorea leprosula* Miq., *Parashorea macrophylla* Wyatt-Sm. ex Ashton and *Dryobalanops lanceolata* Burck; two canopy trees, *Aporosa lunata* (Miq.) Kurz and *Drypetes* sp. and two pioneer species, *Macaranga trachyphylla* Airey Shaw, a tree, and *Musa* sp., a tall, robust herb. The two pioneers were confined to the gap while the six other species were present both in the gap and understorey. Only seedlings 0.5 to 2.0 m tall were included in this study.

Leaf morphology

Recently expanded leaves from five seedlings for each species represented in the gap and understorey were selected for stomatal density and leaf thickness determinations. Total leaf thickness and the thicknesses of the adaxial epidermis and cuticle, palisade mesophyll, spongy mesophyll and abaxial epidermis and cuticle were determined from examination of hand-cut transverse sections from the mid-lamina region of the leaves using a calibrated ocular micrometer at 400x.

Stomatal densities were determined from examination of nail-varnish impressions of ad- and abaxial leaf surfaces using a calibrated ocular grid at 400x. Specific leaf area (SLA) determinations were based on leaf area/weight measurements for recently expanded leaves from ten seedlings for each species in each situation. Areas were calculated from leaf tracings on even density paper and weights from specimens dried to constant weight at 80^{0}C in a forced draught oven.

Gas exchange

Measurements of rates of carbon dioxide and water vapour exchange were made using a LI-COR LI-6200 transportable photosynthetic system. The IRGA component was calibrated against gas of known carbon dioxide content and the RH sensor against a salt-cup and subsequently factory calibrated LI-1600 steady state porometer. Recently expanded leaves from three seedlings for each species present in the gap and understorey were tagged and gas exchange rates were determined for them at one to two hourly intervals. Because of functional limitations of the LI-6200 system, measurements at ambient RH above 95% were avoided. Only two sets of diurnal records are presented in the results, one for a comparison of the eight gap species measured on 24 August 1991, and the other a gap-understorey comparison using *Shorea falciferoides* on 29 July 1991. These, together with results from other occasions from July 1991 through to February 1992, have allowed comparison of rates of light saturated carbon assimilation for gap plants and apparent quantum yields and light compensation points to be determined for both gap and understorey plants. Values influenced by obvious signs of water stress are not included in these comparisons. Dark respiration rates were determined from measurements conducted on darkened leaves from five seedlings for each species in each situation except for *Musa* sp. which, because of the form of its leaf, was not able to be satisfactorily darkened.

RESULTS

Leaf morphology

As shown in Table 1, leaf morphologies of gap and understorey seedlings differed in a number of respects. Total leaf thickness was significantly greater for gap plants than understorey plants. All leaf tissues contributed to this 22-35% difference. However, the greatest contribution was from the spongy mesophyll, except in the case of *Shorea falciferoides* and *Parashorea macrophylla*, where the palisade mesophyll contributed most. The gap species *Macaranga trachyphylla* and *Musa* sp. contrasted in their leaf thickness the former having a thin leaf and the latter a thick leaf. *Musa* sp. had, within its spongy mesophyll, distinctive substomatal, tunnel-like cavities. Stomatal densities were substantially higher (15-52%) for gap plants than those of the understorey. The gap plant, *Musa* sp. was distinctive in that it was amphistomatal, with a very low density of large raised stomata on the adaxial leaf surface (17 ± 2 mm^{-2}) and moderate density on the abaxial surface. Specific leaf areas for the understorey plants were 31-68% higher than those of the gap. The pioneer species, *Macaranga trachyphylla*, had a typically low SLA, whereas *Musa* sp. had a high SLA, greater than any of the understorey seedlings.

Gas exchange

Diurnal patterns of stomatal conductance (Figure 1a) and net photosynthesis (Figure 1b) were recorded for all gap species on 24 August 1991, during a period of relatively dry weather. Rapid changes in both PPFD and RH at 0900 hours make detailed comparisons between species impossible at this time although the pioneers are clearly set apart with very high rates of stomatal conductance and high net photosynthesis. Rates at 1000 hours and 1200 hours are associated with saturating PPFD, low RH and relatively high temperatures. The two pioneer species maintained high rates of gaseous exchange at 1000 hours while for all other species there was little difference between them except in the case of *Drypetes* sp. which displayed relatively low rates of net photosynthesis, a feature maintained at the 1200 hours sampling. Net photosynthesis and stomatal conductance rates for *Shorea leprosula* were high relative to the other non-pioneer plants at the midday sampling. At 1400 hours, with a substantial drop in PPFD, there was a reduction in rates of a gas exchange for the pioneers but little change for the other species. All species showed a marked reduction in net photosynthesis and stomatal conductance with substantially reduced PPFD at 1600 hours. High RH precluded further, late afternoon readings.

Figure 2 shows the diurnal patterns of net photosynthesis and stomatal conductance for *Shorea falciferoides* both in the gap (Figure 2a) and the understorey (Figure 2b) as measured on 29 July, 1991, during a dry period. In the morning the PPFD and, as a result, the rates of net photosynthesis in the understorey were extremely low compared to those of the gap. Stomatal conductance values, however, did not differ between the two environments to the same extent; in the understorey low PPFD values were associated with relatively high

Table 1. Properties of gap (G) and understorey (U) seedling leaves. Mean ± SD; n = 5-10.

	Shorea falciferoides		*S. leprosula*		*Parashorea macrophylla*		*Dryobalanops lanceolata*		*Aporosa lunata*		*Drypetes* sp.		*Macaranga trachyphylla*	*Musa* sp.
	G	U	G	U	G	U	G	U	G	U	G	U	G	G
Total leaf thickness μm	133±12	98±3	108±11	83±9	124±4	102±7	243±15	179±24	255±23	206±24	261±25	208±11	82±13	241±23
Adaxial epidermis + cuticle μm	17±2	15±2	20±2	19±2	16±1	14±1	26±1	25±2	27±2	26±2	23±1	19±2	11±1	53±9
Palisade mesophyll μm	46±6	29±3	42±6	33±4	49±2	36±4	48±9	33±5	87±11	77±6	67±21	57±8	35±9	50±9
Spongy mesophyll μm	54±5	42±5	38±8	20±5	47±3	41±7	153±10	107±28	129±12	91±20	155±40	118±16	27±6	122±18
Abaxial epidermis + cuticle μm	11±1	10±1	11±1	10±0	11±1	9±1	14±2	14±2	11±1	10±2	15±2	13±1	8±1	21±3
Abaxial stomatal density mm^{-2}	441±44	384±39	404±50	265±34	525±58	390±52	263±10	189±32	438±38	295±28	181±29	130±24	174±23	111± 10
Specific leaf area $cm^2\ g^{-1}$	140±19	197±27	164±13	215±18	125± 8	196±16	119±10	164±18	103±13	150±20	109± 8	183±23	109±8	290±15

Table 2. Gas exchange characteristics for gap (G) and understorey (U) seedling leaves. Mean ± SD.

		Shorea falciferoides	*S. leprosula*	*Parashorea macrophylla*	*Dryobalanops lanceolata*	*Aporosa lunata*	*Drypetes sp.*	*Macaranga trachyphylla*	*Musa sp.*
Net photosynthetic rate μ mol $CO_2\ m^{-2}\ s^{-1}$ PPFD >1,000μmol $m^{-2}\ s^{-1}$	G	7.82±2.08 n = 21	7.73±2.21 n = 13	5.68±1.20 n = 8	6.22±1.32 n = 8	7.29±1.64 n = 8	5.02±1.41 n = 10	12.84±1.86 n = 12	11.82±1.66 n = 14
Apparent quantum yield mol CO_2 mol photons^{-1} PPFD 20-100μmol $m^{-2}\ s^{-1}$	G	0.029±.005 n = 21	0.026±.006 n = 20	0.031±.004 n = 5	0.026±.003 n = 11	0.035±.008 n = 6	0.024±.004 n = 8	0.040±.008 n = 10	0.038±.011 n = 8
	U	0.038±.013 n = 18	0.040±.018 n = 24	0.044±.016 n = 13	0.039±.013 n = 17	0.036±.005 n = 13	0.044±.019 n = 12	NA	NA
Light compensation μmol photons $m^{-2}\ s^{-1}$ n = 3	G	11.9±1.2	18.6±1.1	17.4±1.3	16.7±1.6	15.1±0.2	15.7±1.6	16.9±2.0	18.1±0.6
	U	8.1±1.5	9.2±2.1	13.1±1.6	9.3±1.6	8.6±0.8	8.0±1.5	NA	NA
Dark respiration μmol $CO_2\ m^{-2}\ s^{-1}$ n = 5	G	-0.93±.18	-0.99±.32	-1.26±.47	-.63±.12	-0.98±.33	-0.69±.15	-1.35±.32	-0.93±.26
	U	-0.23±.12	-0.40±.20	-0.33±.18	-0.21±.11	-0.33±.07	-0.25±.08	NA	NA

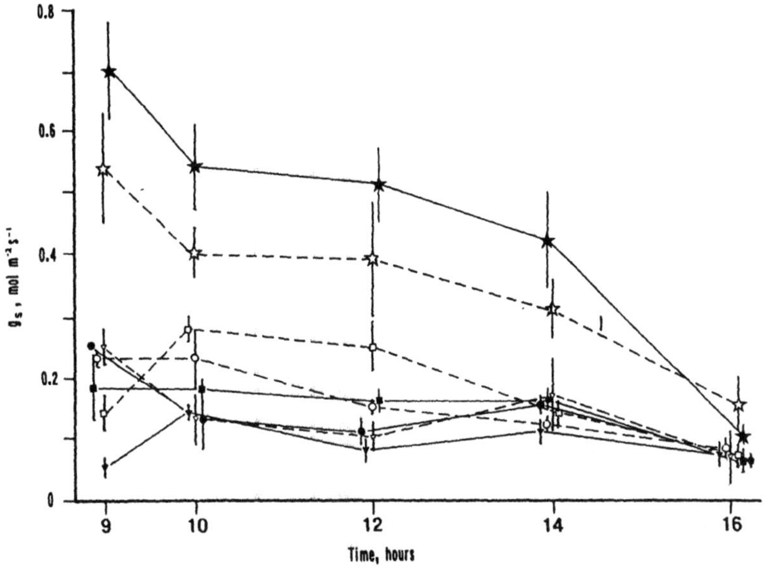

Figure 1a. Diurnal patterns of stomatal conductance for seedlings of the gap on 24 August 1991. Points mean values ± SD, n = 3. *Musa* sp. (closed stars), *Macaranga trachyphylla* (open stars), *Shorea falciferoides* (closed squares), *Shorea leprosula* (open squares), *Parashorea macrophylla* (closed circles), *Aporosa lunata* (open circles), *Dryobalanops lanceolata* (closed triangles), *Drypetes* sp. (open triangles).

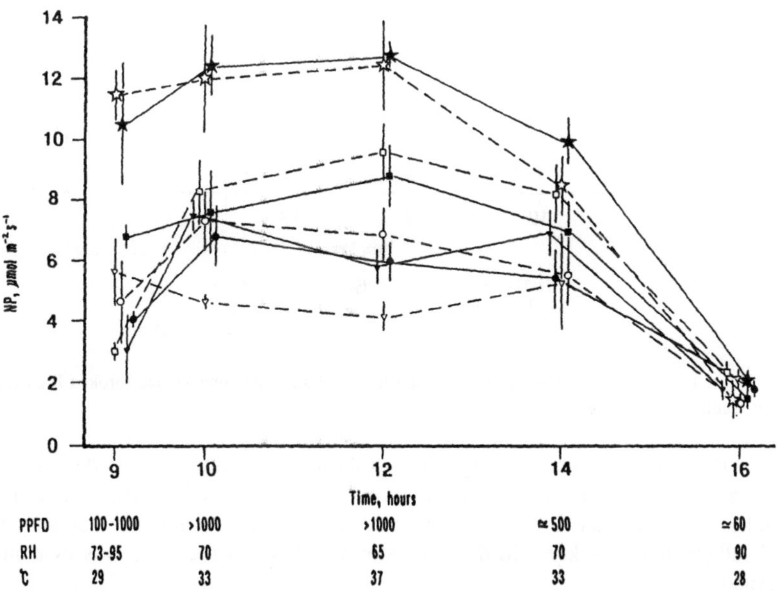

Figure 1b. Diurnal patterns of net photosynthesis for seedlings of the gap on 24 August 1991. Symbols as above. Points mean values ± SD, n = 3.

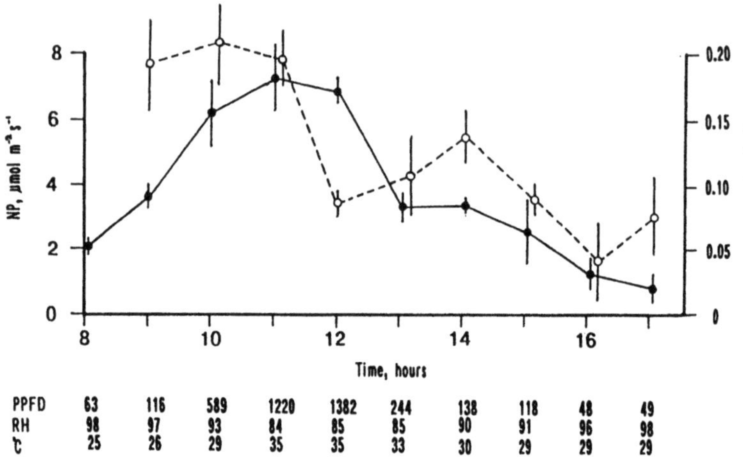

Figure 2a. Diurnal pattern of net photosynthesis (solid line) and stomatal conductance (broken line) for *Shorea falciferoides* seedlings in the gap on 29 July 1991. Points mean values ± SD, n = 3.

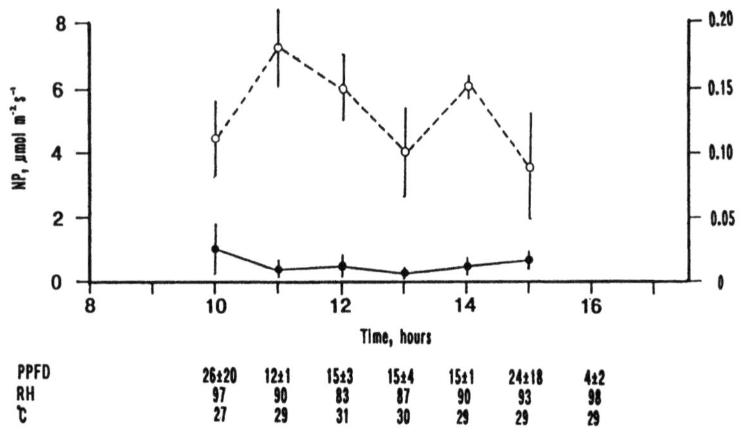

Figure 2b. Diurnal pattern of net photosynthesis (solid line) and stomatal conductance (broken line) for *Shorea falciferoides* seedlings in the understorey on 29 July 1991. Points mean values ± SD, n = 3.

stomatal conductances, similar in magnitude to those of the gap. At midday, in the gap, there was a dramatic reduction in stomatal conductance and subsequently a drop in net photosynthesis, a sequence of events caused by water stress, a feature that may also be observed, albeit more weakly, in the understorey. A slight recovery was evident in the mid-afternoon.

Table 2, line 1, based on a large number of measurements of light saturated net photosynthesis for all gap species, indicates that the pioneers have higher rates than the non-pioneers. Although not differing significantly from each other, mean values for the non-pioneers suggest slightly higher rates may be associated with *Shorea leprosula* and *Shorea falciferoides* and a lower rate with *Drypetes* sp., a pattern also evident in the diurnal curve of Figure 1.

In an attempt to determine if there were differences in photosynthetic efficiency between gap and understorey seedlings apparent quantum yields were calculated for plants in both situations Table 2 line 2. It was assumed that photosynthetic rate plotted against PPFD values between 20 and 100 μmol m^{-2} s^{-1} would fall within the initial slope of the P vs. I curve for all species in each situation. The results, although not statistically significantly different, suggest that for each species the apparent quantum yields for the understorey plants are greater than those for the gap plants.

Table 2, line 3, presents light compensation point PPFD values for all species in the gap and understorey. These values are based on three measurements where there was little or no change in the CO_2 level within the chamber of the photosynthetic system over a period of three minutes. For each species the gap value was significantly higher than that associated with the understorey seedlings.

A comparison of rates of dark respiration between gap and understorey seedling leaves (Table 2, line 4) indicate that, for each species, understorey plants have substantially lower rates of respiration than gap plants.

DISCUSSION

The results of the comparison of the morphological features of gap and understorey seedling leaves suggest typical sun/shade associated patterns as described by Boardman (1977). The leaves of the plants grown in the understorey are thinner, have a higher SLA and a lower stomatal density than those of the gap seedlings. Oberbauer & Strain (1985), in their investigations into the effects of different light regimes on seedling growth reported a similar set of associated features. Langenheim *et al.* (1984) showed differences in SLA (converted from SLW) close to 50% higher for understorey seedlings compared to those of the gap for all except one of the Australian and Amazonian rainforest tree species studied. Pearcy (1987) showed the same trend with a 19% difference in SLA between understorey and gap plants for *Argyrodendron peralatum*.

Stomatal conductance and net photosynthesis for the range of seedlings in the gap indicate that the pioneer species have higher rates for both of these parameters than any of the canopy or emergent species, a feature also reported by Oberbauer & Strain (1984) from their comparative studies in the Costa Rica rainforest. *Musa* sp. and *Macaranga trachyphylla* net photosynthetic rates are close to the 12.6-15.8 μmol m^{-2} s^{-1} given for the typical pioneer trees, *Mallotus*, *Macaranga* and *Glochidion* by Bazzaz & Pickett (1980), but are considerably lower than the 27.7 μmol m^{-2} s^{-1} reported for the pioneer *Ochroma lagopus* of Costa Rica by Oberbauer & Strain (1984). The light saturated net photosynthetic rates of between 5 and 10 μmol m^{-2} s^{-1} for the seedlings of canopy and emergent tree species in this study are of similar magnitude to those reported by Kwesiga *et al.* (1986), Langenheim *et al.* (1984) and Pearcy (1987) and match the shade type upper canopy category of Koyama (1981) based on his studies of lowland rainforest trees in Peninsular Malaysia. They range from values that Bazzaz & Pickett (1980), in their review, associate with early successional trees (8.9 μmol m^{-2} s^{-1}) to close to those of canopy trees (4.4 μmol m^{-2} s^{-1}).

The diurnal patterns of stomatal conductance and net photosynthesis for the seedlings of the gap (Figure 1) are based only on a few points spread over a limited part of the day and therefore do not include such extensive detailed information as presented in Roberts *et al.* (1990); they do, however, give a broad outline of responses which are similar in pattern and magnitude to those presented by Pearcy (1987) and discussed by Tenhunen *et al.* (1987).

The stomatal conductances for the pioneers, although not unique for tropical trees (see Roberts *et al.* 1990), are high (they exceeded 0.5 mol m^{-2} s^{-1} in the early part of the day) and such rates may in part be associated with temperature control for these large leaved species.

The diurnal patterns of gas exchange for *Shorea falciferoides* seedlings in the gap and understorey (Figures 2a & b) indicate that the midmorning stomatal conductance rates do not differ greatly between the two environments in spite of big differences in PPFD and associated net photosynthesis. This result contrasts with those obtained by Pearcy (1987) and Roberts *et al.* (1990) who recorded substantially lower stomatal conductance values for plants of the same species at low PPFD. In fact, in this present study, all species surveyed displayed similar patterns of relatively higher stomatal conductance/PPFD ratios at lower irradiances compared to high, both for plants in the gap and in the understorey. The midday stomatal closure and resultant depression of stomatal conductance and net photosynthesis for the gap seedlings in Figure 2a illustrate dramatically the effects of water stress that can occur in wet tropical rain forest during dry periods. Schulze *et al.* (1987) discuss direct stomatal response to soil drought, the probable sequence of events involved in this present situation.

The photosynthetic response as a function of PPFD for leaves of understorey seedlings showed lower compensation points and higher apparent quantum yields (based on mean values only) than those of the gap, patterns consistent with maximizing use of the limited light resources of the understorey. The apparent quantum yield differences were, however, not statistically significant, and in this respect there is agreement with the observations of Ramos & Grace (1990) and Sims & Pearcy (1989) that there is surprising insensitivity to the light regime. Apparent quantum yield values obtained were of similar magnitude to those of Oberbauer & Strain (1985) and Ramos & Grace (1990) but low when compared with some of those reported by Kwesiga *et al.* (1986), Langenheim *et al.* (1984) and Pearcy (1987).

The light compensation points for understorey seedlings were always lower than those for gap seedlings, an adaptive feature associated with growth in a low light environment (Bjorkman 1981). The gap associated values which ranged from 12 to 18 μmol m^{-2} s^{-1} and understorey values of 8 to 13 μmol m^{-2} s^{-1} are similar to those reported for many tropical forest seedlings grown under such contrasting conditions of PPFD (Kwesiga *et al.* 1986, Langenheim *et al.* 1984, Oberbauer & Strain 1985, Ramos & Grace 1990).

Gap seedlings displayed relatively high rates of dark respiration which, as suggested by Ramos & Grace (1990), is probably a consequence of their enhanced growth rates. These respiration rates which ranged from -0.63 to -1.35 μmol CO_2 m^{-2} s^{-1} are of similar magnitude to those reported by Kwesiga *et al.* (1986) and Ramos & Grace (1990) but higher than the -0.14 to -0.48 μmol CO_2 m^{-2} s^{-1} values of Langenheim *et al.* (1984), Oberbauer & Strain (1985) and Pearcy (1987). The low levels of understorey dark respiration which ranged from -0.21 to -0.33 μmol CO_2 m^{-2} s^{-1} correspond closely to most published records for tropical rainforest seedlings.

In conclusion, the sets of morphological and physiological data obtained for seedlings grown in the contrasting gap and understorey environments conform to those expected of plants from respectively high and low light regimes. The understorey plant strategy is one of maintenance at very low PPFD, with maximized light harvest and minimized energy expenditure, while that of the gap plant is rapid growth which is made possible by relatively high rates of photosynthesis and, inevitably, high respiration. Pioneer plants have an enhanced capacity to use the high PPFD of the gap.

ACKNOWLEDGEMENTS

This study was supported by a research grant from Universiti Brunei Darussalam. I would like to thank the members of staff of KBFSC and the Biology Department for their support during this study.

LITERATURE CITED

BAZZAZ, F.A. & PICKETT, S.T.A. 1980. Physiological ecology of tropical succession: a comparative review. *Annual Review of Ecology and Systematics* II:287-310.

BJORKMAN, O. 1981. Responses to different quantum flux densities. Pp. 57-108 in Lange, O.L., Nobel, P.S., Osmond, C.B. & Ziegler, H. (eds). *Encyclopedia of plant physiology, new series. Physiological plant ecology I, Volume 12A*. Springer Verlag, Berlin. 625 pp.

BOARDMAN, N.K. 1977. Comparative photosynthesis of sun and shade plants. *Annual Review of Plant Physiology* 28:355-377.

CANHAM, C.D. 1989. Different responses to gaps among shade-tolerant tree species. *Ecology* 70:548-550.

CHAZDON, R.L. & FETCHER, N. 1984. Photosynthetic light environments in a lowland tropical rain forest in Costa Rica. *Journal of Ecology* 72:553-564.

KOYAMA, H. 1981. Photosynthetic rates in lowland rain forest trees of Peninsular Malaysia. *Japanese Journal of Ecology* 31:361-369.

KWESIGA, F.R., GRACE, J., & SANDFORD, A.P. 1986. Some photosynthetic characteristics of tropical timber trees as affected by the light regime during growth. *Annals of Botany* 58:23-32.

LANGENHEIM, J.H., OSMOND, C.B., BROOKS, A. & FERRAR, P.J. 1984. Photosynthetic responses to light in seedlings of selected Amazonian and Australian rainforest tree species. *Oecologia (Berlin)* 63:215-224.

LOACH, K. 1970. Shade tolerance in tree seedlings II: Growth analysis of plants raised under artificial shade. *New Phytologist* 69:273-286.

MARTINEZ-RAMOS, M., SARUKHAN, J. & PINERO, D. 1988. The demography of tropical trees in the context of forest gap dynamics: the case of *Astrocaryum mexicanum* at Los Tuxtlas tropical rain forest. Pp. 293-313 in Davy, D.J., Hutchings, M.J. & Watkinson, A.R. (eds). *Plant population ecology*. Blackwell, Oxford.

OBERBAUER, S.F. & STRAIN, B.R. 1984. Photosynthesis and successional status of Costa Rican rain forest trees. *Photosynthetic Research* 5:227-232.

OBERBAUER, S.F. & STRAIN, B.R. 1985. Effects of light regime on the growth and physiology of *Pentaclethra macroloba* (Mimosaceae) in Costa Rica. *Journal of Tropical Ecology* 1:303-320.

OSUNKOYA, O.O & ASH, J.E. 1991. Acclimation to a change in light regime in seedlings of six Australian rainforest tree species. *Australian Journal of Botany* 39:591-605.

PEARCY, R.W. 1987. Photosynthetic gas exchange responses of Australian tropical forest trees in canopy, gap and understorey micro-environments. *Functional Ecology* 1:169-178.

POPMA, J. & BONGERS, F. 1991. Acclimation of seedlings of three Mexican tropical rain forest tree species to a change in light availability. *Journal of Tropical Ecology* 7:85-97.

RAMOS, J. & GRACE, J. 1990. The effects of shade on the gas exchange of seedlings of four tropical trees from Mexico. *Functional Ecology* 4:667-677.

RIDDOCH, I., LEHTO, T. & GRACE, J. 1991. Photosynthesis of tropical tree seedlings in relation to light and nutrient supply. *New Phytologist* 119:137-147.

ROBERTS, J., CABRAL, O.M.R. & DE AGUIAR, L.F. 1990. Stomatal and boundary-layers conductances in an Amazonian terra firme rain forest. *Journal of Applied Ecology* 27:336-353.

ROSS, S.M, & DYKES, A. (this volume). Soil conditions, erosion and nutrient loss on steep slopes under mixed dipterocarp forest in Brunei.

SCHULZE, E.D., TURNER, N.C., GOLLAN, T. & SHACKEL, K.A. 1987. Stomatal responses to air humidity and to soil drought. Pp. 311-321 in Zeiger, E., Farquhar, G.D. & Cowan, I.R. (eds). *Stomatal Function*. Stanford University Press.

TENHUNEN, J.D., PEARCY, R.W. & LANGE, O.L. 1987. Diurnal variations in leaf conductance and gas exchange in natural environments. Pp. 323-351 in Zeiger, E., Farquhar, G.D. & Cowan, I.R. (eds). *Stomatal Function*. Stanford University Press.

WHITMORE, T.C. 1975. *Tropical Rain Forests of the Far East*. Clarendon Press, Oxford, 352 pp.

WHITMORE, T.C. 1978. Gaps in the forest canopy. Pp. 639-655 in Tomlinson, P.B. & Zimmermann, M.H. (eds). *Tropical Trees as Living Systems*. Cambridge University Press, Cambridge.

WHITMORE, T.C. 1989. Canopy gaps and the two major groups of forest trees. *Ecology* 70:536-538.

Distributional ecology of freshwater fishes in tropical rainforest streams of Borneo

SATISH C. CHOY[1]

Biology Department, Universiti Brunei Darussalam, Bandar Seri Begawan 2028, Brunei Darussalam

---

ABSTRACT. A survey was conducted to study the distribution of freshwater fishes in the middle and headwater streams of the Belalong-Temburong watershed, Brunei Darussalam, northern Borneo. The watershed is covered by a mixed dipterocarp primary rainforest. The number of species found at any one locality ranged from zero to twenty five; the number of localities occupied by any one species ranged from one to twenty six. Distribution and abundance of fish were not related to water quality, which was good at all the sampled sites. Drainage area, elevation, gradient and stream order were all significantly correlated with the number of species collected at a site. Species richness was highest in large watersheds, at low elevation, low gradient and in streams of high order. In most streams, species addition occurred with increasing drainage area, increasing stream order, decreasing altitude and decreasing gradient, but species replacement was frequent as well, particularly when waterfalls were encountered. Species belonging to the family Balitoridae dominated channels above waterfalls, whereas those belonging to the family Cyprinidae were predominant in streams below waterfalls. In deep channels, different fish species occupied different zones of the vertical stratum. Shallowing of channels corresponded to the compression of this stratum, resulting in lower species richness.

KEY WORDS: species richness, stream order, altitude, gradient, drainage area, Brunei Darussalam.

INTRODUCTION

Biogeographic distributional patterns of freshwater fishes of Southeast Asia have been discussed by Burridge (1992), Cranbrook (1981), Inger & Chin (1962), Kottelat (1989), Rainboth (1991) and Zakaria-Ismail (1991). There seems to be general agreement that the fish fauna within Borneo is divided into several distinct biogeographic zones (Inger & Chin 1962, Rainboth 1991) but the demarcations are not clear.

On a smaller scale, some quantitative information on the distribution and abundance of the freshwater fish fauna within watersheds of Borneo have been provided by Inger (1955), Inger & Chin (1962), Lee *et al.* (1990), Samat (1990a, b), Taylor (1989) and Watson & Balon (1984a, b). However, only Samat (1990a) and Watson & Balon (1984b) quantitatively analysed for correlates between physico-chemical characteristics of streams and species diversity.

This paper is based on some quantitative analyses between fish species richness and physical attributes such as drainage area, stream order, elevation and gradient within the Belalong-Temburong watershed. The objective of the study was to see if any significant correlation could be found and, if so, how it may be used to estimate or predict species richness in other watersheds of the region.

[1]Present Address: DPI Water Resources Technical Centre, 1345 Ipswich Rd., Rocklea, Qld. 4106, Australia

D. S. Edwards et al. (eds.), Tropical Rainforest Research – Current Issues, 377–386.

MATERIALS AND METHODS

Sampling

Fish were caught from the two main rivers, Sungai Belalong and Sg. Temburong as well as from most of their tributaries (Figure 1). The watershed is covered by a mixed dipterocarp primary rainforest. A list of the sampling stations and their physiographic features are given in Table 1. Fishing techniques utilized were gill netting, cast netting, seining, and hand netting. For the present study, fishing effort was standardised to 1 hour per sampling station. Fish collected from the different micro-habitats were enumerated and recorded. They were later fixed in 10% formalin and then preserved in 70% ethanol. The specimens are deposited in the UBD/RGS Reference Collection at the Department of Biology, Universiti Brunei Darussalam and in the Reference Collection of the Fisheries Department of Sabah, Likas, Kota Kinabalu, Sabah under catalogue numbers F/1-F/146. Water samples collected from different localities were analysed by the Chemistry Department of Universiti Brunei Darussalam and the Soils Laboratory of Kilanas Agricultural Station.

Analysis

Stream order, altitude, gradient and drainage areas were determined from 1:50,000 topographic maps. For stream order, the Horton-Strahler stream classification was used. First order streams are headwater streams with no tributaries. Stream order increases by one each time two streams of the same order flow together. Gradient at a sampling locality was determined by dividing the altitudinal difference between points 0.5 km downstream and 0.5 km upstream of the sampling site and then converted to a percentage. Jaccard's coefficient of similarity (Southwood 1978) was used to access species similarity in the different micro-habitats. Statistical procedures follow Schefler (1980) and Daniel (1990).

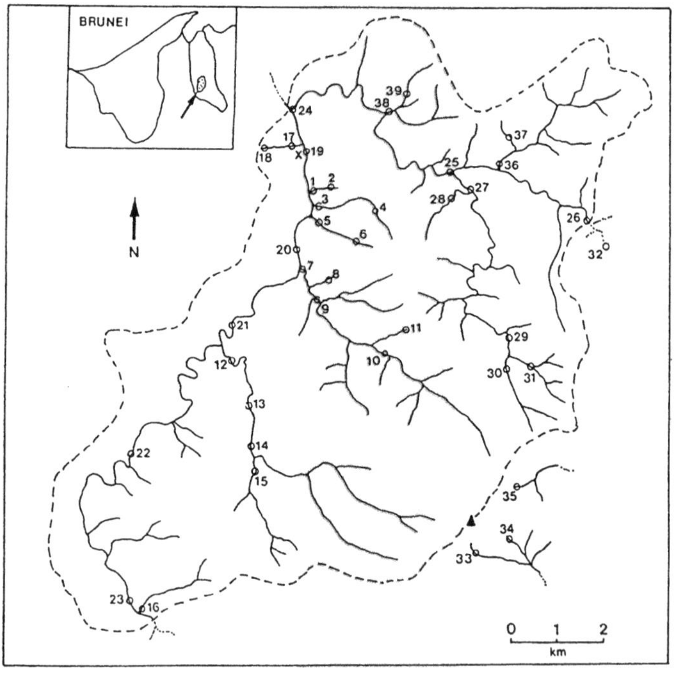

Figure 1. Locality map of part of the Belalong-Temburong watershed showing the UBD-RGS Rainforest Project boundary (----) and sampling stations 1-39, as listed in Table 1. Station 40 (not shown) is about 10 km southeast of Bukit (Mt.) Belalong (▲).

Table 1. Sampling station numbers, localities, stream order, altitude, gradient and the number of fish species for each locality. Stream order, altitude, gradient and drainage areas were determined from 1 : 50,000 topographic maps. a.w., above waterfall areas; b.w., below waterfall areas; KBFSC, Kuala Belalong Field Studies Centre; S, sungai (river).

Station	Locality	Stream Order	Altitude (m)	Gradient (%)	Number of fish species
1.	S. Anak Esu, b.w.	1	60	41	5
2.	S. Anak Esu, a.w.	1	152	21	0
3.	S. Esu, b.w.	1	60	15.5	6
4.	S. Esu, a.w.	1	183	12.4	2
5.	S. Engkabang, b.w.	1	60	7.8	7
6.	S. Engkabang, a.w.	1	152	15.5	2
7.	Lower S. Sitam, b.w.	3	67	4.3	8
8.	Tributary of S. Sitam, a.w.	1	183	30.5	0
9.	Middle S. Sitam, b.w.	3	92	3	9
10.	Upper S. Sitam, below Pondok Busiri, b.w.	2	131	3.3	9
11.	Bathing stream, below Pondok Busiri. a.w.	1	335	30.5	0
12.	Lower S. Engkiang, b.w.	3	76	2.5	8
13.	Middle S. Engkiang, b.w.	3	107	2.7	8
14.	Upper S. Engkiang, b.w.	2	128	3.8	9
15.	Upper right trib. S. Engkiang, a.w.	1	134	5	3
16.	S. Split, b.w.	2	152	6	7
17.	Bathing stream, near KBFSC, b.w.	2	60	10	6
18.	Bathing stream, near KBFSC, a.w.	1	122	20	3
19.	S. Belalong, near KBFSC, b.w.	4	55	0.7	25
20.	S. Belalong, btw. S. Engkabang & S. Sitam, b.w.	4	67	0.7	21
21.	S. Belalong, near S. Engkiang, b.w.	4	70	0.5	13
22.	S. Belalong, btw. S. Engkiang & S. Split, b.w.	3	100	0.8	10
23.	S. Belalong, near S. Split, b.w.	3	137	0.6	6
24.	Confluence of S. Belalong & S. Temburong, b.w.	5	49	0.7	13
25.	S. Temburong, near S. Babi, b.w.	4	61	0.9	24
26.	S. Temburong, near S. T-Machang, b.w.	4	79	0.6	12
27.	Lower S. Babi, b.w.	3	92	5	15
28.	Trib. Lower S. Babi, a.w.	1	152	12.4	0
29.	Upper S. Babi, b.w.	3	214	6	10
30.	Upper S. Babi, above runs, b.w.	2	275	10	11
31.	Trib. Upper S. Babi, a.w.	2	335	15	3
32.	Lower S. T-Machang, b.w.	3	100	1.5	19
33.	Trib. S. T-Machang, below Bt. Belalong, a.w.	1	610	15.5	1
34.	Trib. Upper S. T-Machang, a.w.	1	579	15	1
35.	Trib. Upper S. T-Machang, a.w.	1	457	10	3
36.	S. Inging, b.w.	3	70	1.9	11
37.	Trib. S. Inging, a.w.	1	140	31	0
38.	S. Apan, b.w.	2	61	10	12
39.	S. Apan, a.w.	1	137	15	4
40.	Stream below Bukit Pagon	1	1400	?	4

RESULTS

Forty four species of fish, belonging to ten families and thirty genera, from a total of forty sites were collected. The numbers of species collected at each sample site are shown in Table 1 and a checklist of species with the sites of their occurrence are given in Table 2. Species belonging to the families Balitoridae and Cyprinidae dominated the catch, with fifteen and thirteen species, respectively. Most members of the latter family were found in the lower, slow-flowing channels of streams while those of the former family were mainly found

Table 2. Species checklist and sites of occurrence.

Family Anguillidae
Anguilla marmorata Quoy & Gaimard, 1924 - Stations 19, 20, 25, 27, 30 & 32.
Family Bagridae
Mystus baramensis (Regan, 1906) - Stations 5, 19, & 25.
Mystus sp. - Stations 24 & 25.
Leiocassis micropogon (Bleeker, 1852) - Stations 19 & 24.
Family Balitoridae
Gastromyzon borneensis (Gunther, 1874) - Stations 15, 19-21, 25, 26, 29 & 30.
Gastromyzon fasciatus Inger & Chin, 1961 - Stations 19 & 35.
Gastromyzon monticola (Vaillant, 1902) - Stations 7, 10, 14, 19, 20, 25, 27 & 32.
Gastromyzon lepidogaster (Roberts, 1982) - Stations 3, 4, 9, 10, 15, 18-20, 25, 29, 30, 32, 35 & 38
Gastromyzon punctulatus Inger & Chin, 1962 - Stations 6, 19, 25 & 27.
Gastromyzon ridens (Roberts, 1982) - Stations 19, 24 & 25.
Glaniopsis gossei - Station 40.
Glaniopsis hanitschi Boulenger, 1899 - Stations 31, 33 & 34.
Nemacheilus sp. - Station 31
Neogastromyzon nieuwenhuisi Popta, 1905 - Stations 3-6, 9, 10, 12-15, 17-27, 35, 38 & 39.
Neogastromyzon sp. - Station 40.
Protomyzon whiteheadi (Vaillant, 1893) - Stations 16, 18-20, 24-26, 29-32 & 39.
Protomyzon griswoldi (Hora & Jayaram, 1951) - Station 32.
Protomyzon sp. - Station 39.
Sundareonectes sabanus Chin, 1990 - Station 40
Family Clariidae
Clarias leiacanthas (Bleeker, 1851) - Stations 19, 25 & 32.
Clarias teijsmanni Bleeker, 1857 - Station 19.
Family Cobitidae
Pangio mariarum (Inger & Chin, 1962) - Stations 19-21, 27, 32 & 36.
Family Cyprinidae
Barbodes collingwoodi (Gunther, 1868) - Stations 19-22, 27, 29, 30, 32 & 36.
Chela sp. - Station 23
Hampala bimaculata (Popta, 1905) - Stations 7, 9, 10, 12-14, 16, 19-27, 29, 30, 32, 36 & 38
Lobocheilus bo (Popta, 1904) - Stations 7, 9, 10, 12-14, 16, 19-22, 25-27, 32 & 38.
Nematabramis steindachneri (Popta, 1905) - Stations 1, 3, 5, 7, 9-10, 12-14, 16-17, 19-27, 29, 30, 32, 36 & 38.
Osteochilus spilurus (Bleeker, 1851) - Stations 19, 24 & 36.
Paracrossochilus acerus Inger & Chin, 1962 - Stations 5, 10, 12-14, 19-22, 25-27, 29, 30, 32, 36 & 38
Paracrossochilus vittatus (Boulenger, 1894) - Station 36.
Puntius binotatus (Valenciennes, 1842) - Stations 19 & 24
Rasbora argyrotaenia (Bleeker, 1850) - Stations 1, 3, 5, 7, 9-10, 12-14, 16-17, 19-27, 29, 30, 32, 36 & 38.
Rasbora sumatrana (Bleeker, 1852) - Stations 1, 3, 5, 7, 9, 12-14, 16, 19-26, 29, 30, 32 & 36.
Rasbora sp. - Stations 19-21, 25, 29 & 38.
Tor douronensis (Valenciennes, 1842) - Stations 1, 3, 5, 7, 9, 10, 12, 13, 16, 19-27, 29, 32 & 36.
Family Gobiidae
Parawaous sp. - Stations 1, 7, 9, 19, 20, 24, 25, 27, 32 & 38.
Pseudogobiopsis cf. *oligactus* (Bleeker, 1875) - Stations 1, 19 & 38.
Family Mastacembelidae
Macrognathus maculatus (Cuvier, 1831) - Stations 19-27, 32, 36-38.
Mastacembelus unicolor Cuvier, 1831 - Stations 24 & 36.
Family Siluridae
Silurichthys hasselti Bleeker, 1858 - Stations 17, 19 & 25.
Ompak sp. - Station 19.
Kryptopterus sp. - Station 32.
Silurus sp. - Station 32.
Family Sisorodae
Glyptothorax major (Boulenger, 1894) - Stations 17, 19, 24, 25 & 27.

in the upper, faster flowing channels. No Cyprinidae was found above any waterfall greater than 5m high.

The number of fish species collected at a site ranged from 0 to 25, with the main channels of Sungai Belalong and Sg. Temburong having the highest species richness and diversity (Table 1). The number of localities where a species was found ranged from 1 to 26 (Table 3).

Table 3. Number of occurrences of fish species in different stream orders, total number of individuals per species (N) and its percentage frequency (%).

Stream Order	1	2	3	4	5	Total	N	%
Number of stations	17	7	10	5	1	40		
Species								
Anguilla marmorata	0	2	2	3	0	7	8	0.8
Mystus baramensis	1	0	0	2	0	3	6	0.6
Mystus sp.	0	0	0	1	0	1	1	0.1
Leiocassis micropogon	0	0	0	1	1	2	18	1.8
Clarias leiacanthus	0	0	1	1	1	3	3	0.3
Clarias teijsmanni	0	0	1	0	0	1	1	0.1
Pangio mariarum	0	0	3	3	0	6	8	0.8
Sundareonectes sabanus	1	0	0	0	0	1	2	0.2
Nemacheilus sp.	0	1	0	0	0	1	4	0.4
Chela sp.	0	1	0	0	0	1	1	0.1
Hampala bimaculata	0	5	10	5	1	21	65	6.6
Lobocheilus bo	0	4	7	5	0	16	58	5.9
Nematabramis steindachneri	5	6	9	5	1	26	153	15.5
Osteochilus spilurus	0	0	1	1	1	3	7	0.7
Paracrossochilus acerus	1	4	7	5	0	17	102	10.4
Paracrossochilus vittatus	0	0	1	0	0	1	4	0.4
Puntius binotatus	0	0	0	1	1	2	5	0.5
Bardodes collingwoodi	0	1	5	3	0	9	71	7.2
Rasbora argyrotaenia	3	6	10	5	1	25	148	15.0
Rasbora sumatrana	3	3	9	5	1	21	32	3.2
Rasbora sp.	0	1	0	4	0	5	8	0.8
Tor douronensis	3	2	10	5	1	21	51	5.1
Parawaous sp.	1	1	4	3	1	10	15	1.5
Pseudogobiopsis sp.	1	1	1	0	0	3	3	0.3
Gastromyzon borneensis	1	1	1	5	0	8	33	3.3
Gastromyzon fasciatus	1	0	0	1	0	2	3	0.3
Gastromyzon monticola	0	2	3	3	0	8	22	2.2
Gastromyzon lepidogaster	5	3	3	3	0	14	18	1.8
Gastromyzon punctulatus	1	0	1	2	0	4	7	0.7
Gastromyzon ridens	0	0	0	2	0	2	3	0.3
Glaniopsis hanitschi	3	1	0	0	0	4	7	0.7
Glaniopsis gossei	1	0	0	0	0	1	1	0.1
Neogastromyzon nieuwenhuisi	8	4	6	5	1	26	54	5.4
Neogastromyzon sp.	1	0	0	0	0	1	2	0.2
Protomyzon whiteheadi	2	3	2	4	1	12	16	1.6
Protomyzon griswoldi	0	0	1	0	0	1	1	0.1
Protomyzon sp.	1	0	0	0	0	1	2	0.2
Macrognathus maculatus	1	1	5	5	1	13	18	1.8
Mastacembelus unicolor	0	0	1	0	1	2	4	0.4
Krytopterus sp.	0	0	1	0	0	1	1	0.1
Ompak sp.	0	0	1	0	0	1	1	0.1
Silurichthys hasselti	0	1	0	2	0	3	7	0.7
Silurus sp.	0	0	1	0	0	1	1	0.1
Glyptothorax major	0	1	1	2	1	5	9	0.9
Total no. species in each stream order	20	22	27	30	15	44		
Max. no. species in each stream order	7	12	19	25	13			

The most widespread species, collected from over twenty sites, were *Nematabramis steindachneri, Rasbora argyrotaenia, R. sumatrana, Tor douronensis, Neogastromyzon nieuwenhiusi* and *Hampala bimaculata.* Of these, only *N. nieuwenhuisi* is a rheophilic species. Overall, *Nematabramis steindachneri, Rasbora argyrotaenia* and *Paracrossochilus acerus* were most numerous, with over a hundred specimens of each being caught. Eight species (*Clarias teijsmanni, Glaniopsis gossei, Protomyzon griswoldi, Chela* sp., *Mystus* sp., *Krytopterus* sp., *Ompak* sp. and *Silurus* sp.) seem rare (Table 3).

Water quality of the streams was good and unpolluted, with low nutrient and high dissolved oxygen content (Table 4). Smaller streams generally had lower conductivity and total suspended solids (TSS). Following rain, anion, cation and total suspended solids increased markedly. There was no correlation between species richness or diversity and water quality ($P>0.10$).

Only seven species were collected at any site above 400 m. They were *Gastromyzon fasciatus, G. lepidogaster, Glaniopsis hanitschi, G. gossei, Neogastromyzon nieuwenhiusi, Neogastromyzon* sp. and *Sundareonectes sabanus*, all belonging to the torrent loach family Balitoridae. With a single exception, no species was collected at a site with a gradient exceeding 20%. The exception (Station 1, gradient 41%) is an artifact of the method of calculating gradient: all species were collected in an area just before a series of steep slopes with waterfalls; the slopes were included in the calculation of the gradient, according to the standardised methodology. No fish was found beyond these slopes, even though the gradient eventually decreased.

The mean number of fish species per site increased from 2.4 ($\pm$ 0.53 S.E.) for first order streams through 8.1 ($\pm$ 1.08 S.E.) for second order streams, 10.4 ($\pm$ 1.17 S.E.) for third order streams and 19.0 ($\pm$ 2.46 S.E.) for fourth order streams, but decreased to 13 ($\pm$ 0 S.E.) for the fifth order stream.

Table 4. Physico-chemical characteristics of some of the streams in the study area. For stations 19 & 25 the upper limits were obtained following rain. S, sungai (river).

	S. Sitam (Stn 7)	Trib. S. Sitam (above Stn 9)	S. Belalong (Stn 19)	S. Temburong (Stn 25)
Width (m)	5	0.4	15	20
Depth (m)	0.3	0.04	1.5	1.1
Current (ms^{-1})	0.75	0.17	0.65-3.7	2.0
Temperature (°C)	23-25	24-25	24-25	24-25
Conductivity (μScm^{-1})	36	15	24-35	26
pH	6.8	6.8	5.9-6.9	5.4
Diss. Oxygen (mgl^{-1})	8	7.6	7.6	7.7
CO_2 (ppm)	3	3	3	3
Total Hardness ($mgl^{-1}CaCO_3$)	14	14	8	8
Ca Hardness (mgl^{-1})	4	4	4	4
NO_3-N (mgl^{-1})	<0.2	0.83	0.12-0.42	0.07-0.21
PO_4-P (mgl^{-1})	<0.2	0.04	0.04	0.04
Calcium (mgl^{-1})	-	0.49	0.78-1.50	0.27-0.31
Magnesium (mgl^{-1})	-	0.79	0.91-1.30	0.02-0.88
Sodium (mgl^{-1})	-	0.99	1.32-1.90	1.49
Potassium (mgl^{-1})	-	0.21	0.20-0.72	0.15-0.22
Chloride (mgl^{-1})	-	0.23	0.23-0.85	0.15-0.31
SO_4 (mgl^{-1})	-	1.85	3.30-5.54	4.31
TSS (mgl^{-1})	-	7.1	10-250	10-320

Table 5. Regression equations for log transformed data.

Log (no. species)= 5.137 - 0.672 log (elevation), n=35, r=0.65
Log (no. species)= 2.580 - 0.464 log (gradient), n=35, r=0.70
Log (no. species)= 1.067 + 1.199 log (stream order), n=26, r=0.82
Log (no. species)= 1.990 + 0.297 log (drainage area),n=12, r=0.95

Table 6. a) Regression equations and correlation, r for log transformed data. All are significant at $P<0.01$ (2-tailed). b) Kendall's rank correlation coefficient (Kendall's tau) between number of species and physical attributes of sites. All Kendall's tau are significant at $P<0.01$ (2-tailed). Highest number of species, largest drainage area, lowest stream order, lowest elevation and lowest gradient were ranked as 1. c) Partial Kendall's tau with elevation effect controlled (Daniel 1990). All are significant at $P<0.01$ (2-tailed).

	Elevation	Gradient	Drainage area	Stream order
a) No. species	0.65 n=35	0.70 n=34	0.95 n=12	0.82 n=35
b) No. species	0.45 n=40	0.42 n=40	0.83 n=12	0.74 n=40
c) No. species	n.a.	0.31	0.61	0.52

The number of fish species at a site was generally higher at high drainage area, low elevation, high stream order and low gradient in decreasing order of relationship ($P<0.01$, 2-tailed; Tables 5 & 6). Partial Kendall's tau (Daniel 1990) was also calculated to control the influence of elevation on Kendall's tau values for gradient, drainage area and stream order with number of species (Table 6). Again, the relationships were significant.

A cladogram of ecological zones based on Jaccard's coefficient of similarity of species is given in Figure 2. Two distinct ichthyological groupings were identified in the study area; those species which are confined almost exclusively to areas below waterfalls (which act as dispersal barriers) and those which are found mainly above. The latter is confined to twelve species, belonging to the torrent loach family Balitoridae. These species are *Neogastromyzon nieuwenhuisi*, *Gastromyzon borneensis, G. fasciatus*, *G. lepidogaster*, *G. punctulatus*, *Protomyzon whiteheadi*, *Protmyzon* sp., *Nemacheilus* sp., *Neogastromyzon* sp., *Glaniopsis hanitschi, G. gossei* and *Sundareonectes sabanus*; the last six being found exclusively above waterfalls.

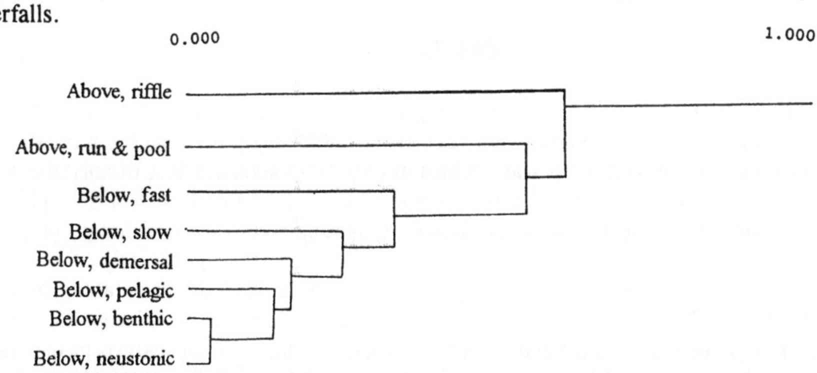

Figure 2. Cladogram of ecological zones based on species similarity. Scale is euclidean distance obtained using single linkage method. Above, streams above first encountered major waterfall (> 3m high), acting as barrier; Below, streams below any first major waterfall; fast (velocity > 0.5 ms^{-1}) and slow (velocity < 0.5 ms^{-1}) refer to flowing condition; neustonic (near surface), pelagic (mid-water), demersal (near stream bed) and benthic (on stream bed) refer to the zones in the water column.

Table 7. Habitat classification of the freshwater fish in the Belalong-Temburong watershed.

1. BELOW WATERFALLS:
   a) Pools & slow-flowing channels
      i) Neustonic: *Nematabramis steindachneri* & *Rasbora argyrotaenia*.
      ii) Pelagic: *Hampala bimaculata*, *Lobocheilus bo*, *Osteochilus spilurus*, *Barbodes collingwoodi*, *Puntius binotatus*, *Rasbora sumatrana*, *Rasbora* sp., *Tor douronensis*, *Krytopterus* sp. & *Chela* sp.
      iii) Demersal: *Anguilla marmorata*, *Clarias leiacanthus*, *Clarias teijsmanni*, *Glyptothorax major*, *Macrognathus maculatus*, *Mastacembelus unicolor*, *Mystus baramensis*, *Silurichthys hasselti*, *Silurus* sp., *Leiocassis micropogon*, *Pangio mariarum*, *Lobocheilus bo*, *Hampala bimaculata*, *Tor douronensis*, *Osteochilus spilurus*, *Barbodes collingwoodi*, *Puntius binotatus*, *Ompak* sp., *Krytopterus* sp. & *Mystus* sp.
      iv) Benthic: *Pangio mariarum*, *Parawaous* sp. & *Pseudogobiopsis* sp.
   b) Runs & riffles (fast flowing channels)
      i) Demersal: *Lobocheilus bo*, *Hampala bimaculata*, *Glyptothorax major*, *Silurichthys hasselti*, *Pangio mariarum*, *Mystus baramensis*, *Mystus* sp., *Leiocassis micropogon*, *Clarias leiacanthus*, *C. teijsmanni*, *Anguilla marmorata*, *Tor douronensis*, *Macrognathus maculatus*, *Mastascembelus unicolor*, *Silurus* sp., *Ompak* sp. & *Krytopterus* sp.
      ii) Benthic: *Gastromyzon borneensis*, *G. lepidogaster*, *G. monticola*, *G. punctulatus*, *G. ridens*, *G. fasciatus*, *Neogastromyzon nieuwenhuisi*, *Paracrossochilus acerus*, *P. vittatus*, *Protomyzon whiteheadi*, *P. griswoldi*, *Parawaous* sp. & *Pseudogobiopsis* sp.

2. ABOVE WATERFALLS: all benthic species
   a) Pools: *Neogastromyzon nieuwenhuisi*, *Neogastromyzon* sp., *Nemacheilus* sp., *Glaniopsis gossei*, *Sundareonectes sabanus*, *Gastromyzon borneensis* & *G. lepidogaster*.
   b) Runs & riffles: *Glaniopsis hanitschi*, *Gl. gossei*, *Nemacheilus* sp., *Neogastromyzon nieuwenhuisi*, *Gastromyzon borneensis*, *G. fasciatus*, *G. punctulatus*, *Protomyzon whiteheadi* & *Protomyzon* sp.

In pools and slow flowing areas below waterfalls, *Nematabramis steindachneri* and *Rasbora argyrotaenia* were most abundant numerically while in faster flowing zones, *Paracrossochilus acerus* and *Gastromyzon* spp. were predominant (Table 7). In the deep, slow-flowing channels, neustonic or surface species were *Nematabramis steindachneri* and *Rasbora argyrotaenia*, pelagic species were other cyprinids such as *Hampala bimaculata*, *Lobocheilus bo*, *Osteochilus spilurus*, *Tor douronensis*, *Puntius binotatus* and *Barbodes collingwoodi*, demersal species were the catfishes and the benthic species were the gobies and loaches. There was little vertical stratification in the shallow fast-flowing zones, due to the compression of this stratum.

DISCUSSION

The water quality of the streams in the Belalong-Temburong watershed, above the confluence (Kuala Belalong) is good and unpolluted. Lee *et al.* (1990) suggested that the most suitable aquatic environment for fish life is water which has pH of between 6.5-9.0, dissolved oxygen content of greater than 5 mgl^{-1} and total suspended solids (TSS) of less than 80 mgl^{-1}. All these conditions are being met in the Belalong-Temburong river system although they may not be fulfilled temporarily after heavy rain, when TSS increases significantly.

Increase in species richness with increasing stream order and drainage area has been demonstrated by Mahon (1984) for the rivers in the temperate region. Except for Bishop (1973), Livingstone *et al.* (1982) and Watson & Balon (1984a, b), such analysis has not been done for tropical streams. Livingstone *et al.* (1982) and Watson & Balon (1984a, b) found species richness to be positively correlated to drainage area.

Fausch *et al.* (1984) used relationships between stream order and number of fish species to develop indices of biotic integrity for use in comparing streams within a biogeographic region. Stream order is an indicator of stream size (width, depth and discharge). Gradient and elevation and water velocity decrease with increase in stream order and generally substrates change from rocky to silty (Kuehne 1970). The present study suggests that species richness increases with increase in stream order. Reduced species richness in low order streams results from steep gradient, rapid changes in flow, unstable substratum, small habitat volume, limited habitat complexity and low autochthonous production (Vannote *et al.* 1980). Because stream order is generally easy to determine from topographic maps, and because it reflects other physical attributes which could influence absence or presence of aquatic species, it could form a basis for comparing streams in icthyological studies (Kuehne 1970). If these relationships are valid, then stream order could be useful in predicting species numbers or presence of particular species in different watersheds within a given region.

In this study, however, drainage area was more strongly correlated than stream order. Analysis of published data from Southeast Asia (Bishop 1973, Furtado & Mori 1982, Inger 1955, Lee *et al.* 1990, Roberts 1989, Samat 1990a, b, Taylor 1989) also indicated this, suggesting that drainage area is a better physical attribute to use than stream order, at least for Southeast Asia. Estimating drainage area from topographic maps is more accurate than determining stream order. In tropical rainforest areas, where rainfall is high and intermittent many streams are ephemeral and attributing stream order to a particular stream may not be accurate. Furthermore, most of the smaller streams are not shown on topographic maps, which makes assessment difficult and unreliable.

The lower correlation of stream order, elevation and gradient to species richness is likely to be the result of geographic heterogeneity of the areas under study. Streams having low and high topographic relief within a small area result in extremes of gradients and low species richness.

The segregation of fish above and below any major waterfall is expected. Waterfalls and cascades form effective barriers to migration and dispersal (Schlosser 1982). All the species found above waterfalls belong to the family Balitoridae and exhibit morphological and behavioural specialisations to live in torrent streams. In above-waterfall streams, several of these species occur in slow flowing pools whereas in below-waterfall streams none are found such areas. Above waterfalls, there is an absence of other fish but in pools below any major waterfall there is an abundance of competitors and predators such as the cyprinids, bagrids, clarids and silurids. Thus, predation by and/or competition from other species prevent the balitorids from colonising the slow-flowing areas of below-waterfall streams.

Many published studies have shown increased community diversity in stream fishes from upstream to downstream areas (Bishop 1973, Gorman & Karr 1978, Sheldon 1968) with most emphasizing species addition. Bishop (1973) also noted replacement of one species by another. In the present study, increased species richness from upstream to downstream was initially by species replacement, particularly when a major waterfall was encountered, and then for below-waterfall streams, by species addition. Increase in species diversity has been attributed to increase in available habitat and decrease in environmental fluctuations (Lowe-McConnell 1975). Downstream, there is an increase in the availability of deep channels. Increased species diversity in these channels was a result of different species occupying the different zones of the vertical stratum, suggesting that fish are prone to using the stream environment in a 3-dimensional way (Gorman & Karr 1978, Sheldon 1968).

ACKNOWLEDGEMENTS

I wish to express my gratitude to the guides and RGS employees from the Batang Duri longhouse who helped with logistics and sampling. Special thanks to Dr. D.S. Edwards, other Biology Dept. colleagues and the RGS personnel for support and encouragement and to Mr. Samhan Nyawa for his company on many of the rather difficult sampling trips. This research was funded by grant number UBD/T/RG.12 from Universiti Brunei Darussalam.

LITERATURE CITED

BISHOP, J.E. 1973. Limnology of a small Malayan river, Sungei Gombak. *Monographiae Biologicae* 22. Junk, The Hague. 485 pp.

BURRIDGE, M.G. 1992. Systematics of the *Acanthophthalmus kuhlii* complex (Teleostei: Cobitidae), with the new description of a new species from Sarawak and Brunei. *Copeia* 1:172-186.

CRANBROOK, EARL OF. 1981. The vertebrate faunas. In Whitmore, T.C. (ed.). *Wallace's Line and plate tectonics*. Clarendon Press, Oxford.

DANIEL, W.W. 1990. *Applied non-parametric statistics*. PWS Kent. Boston, USA. 635 pp.

FAUSCH, K.D., KARR, J.R. & YANK, P.R. 1984. Regional application of an index of biotic integrity based on stream fish communities. *Transactions of the American Fisheries Society* 113: 39-55.

FURTADO, J.I & MORI, S. 1982. *Tasek Bera*. Junk, The Hague. 413 pp.

GORMAN, O.T. & KARR, J.R. 1978. Habitat structure and stream fish communities. *Ecology* 59: 507-515.

INGER, R.F. 1955. Ecological notes on the freshwater fishes of a coastal drainage of North Borneo. *Fieldiana: Zoology* 37:47-90.

INGER, R.F. & CHIN, P.K. 1962. The freshwater fishes of Northern Borneo. *Fieldiana: Zoology* 45:1-268.

KOTTELAT, M. 1989. Zoogeography of the fishes from Indo-Chinese inland waters with an annotated checklist. *Bulletin of the Zoological Museum, University of Amsterdam* 12(1):1-54.

KUEHNE, R.A. 1970. Application of the Horton stream classification to evaluate faunal studies. Pp. 367-370 in Weist, W.G. Jr. & Greeson, P.E. (eds). *Hydrobiology - bioresources of shallow water environments*. American Water Resources Association. Urbana, Illinois.

LEE, Y.H., LEE, N., GHAFFER, M.A. & SAMAT, A. 1990. *Water quality and fish fauna distribution in the rivers of the Tawau Hills Park, Sabah, East Malaysia*. Pp. 132-133, in Extended abstracts of papers presented at the International Conference on Forest Biology and Conservation in Borneo. Kota Kinabalu, Sabah, 30 July-3 August, 1990.

LIVINGSTONE, D.A., ROWLAND, M. & BAILEY, P.E. 1982. On the size of African riverine fish faunas. *American Zoologist* 22:361-369.

LOWE-McCONNELL, R.H. 1975. *Fish communities in tropical freshwater: their distribution, ecology and evolution*. Longman, London. 337 pp.

MAHON, R. 1984. Divergent substructure in fish taxocenes of north temperate streams. *Canadian Journal of Fisheries and Aquatic Sciences* 41:330-350.

RAINBOTH, W.J. 1991. Cyprinids of Southeast Asia. In Winfield, I.J & Nelson J.S. (eds). *Cyprinid fishes - systematics, biology and exploitation*. Chapman & Hall, London.

ROBERTS, T.R. 1989. The freshwater fishes of western Borneo (Kalimantan Barat, Indonesia). *Memoirs of the California Academy of Science* 14:1-210.

SAMAT, A. 1990a. Taburan dan populasi ikan air tawar di beperapa altitud di Taman Kinabalu, Sabah, Malaysia. *Pertanika (Sabah)* 13(3):341-348.

SAMAT, A. 1990b. *A study of freshwater fish ecology of streams in logged and unlogged areas of Danum Valley*. Technical Report, Danum Valley Committee/Royal Society. 5 pp.

SCHEFLER, W.C. 1980. *Statistics for biological sciences*. 2nd Edition. Addison-Wesley, California, USA. 230 pp.

SCHLOSSER, I.J. 1982. Fish community structure and function along two habitat gradients in a headwater stream. *Ecological Monographs* 52:395-414.

SHELDON, A.L. 1968. Species diversity and longitudinal succession in stream fishes. *Ecology* 49:193-198.

SOUTHWOOD, T.R.E. 1978. *Ecological methods with particular reference to the study of insect populations*. Chapman & Hall, London. 524 pp.

TAYLOR, R.S. 1989. A survey of the fish fauna of the Tabin Wildlife Reserve, Sabah. *Sabah Museum Monographs* 3:87-99.

VANNOTE, R.L., MINSHALL, G.W., CUMMINS, K.W., SEDELL, J.R. & CUSHING, C.E. 1980. The river continuum concept. *Canadian Journal of Fisheries and Aquatic Sciences* 37:130-138.

WATSON D.J. & BALON, E.K. 1984a. Structure and production of fish communities in tropical rainforest streams of northern Borneo. *Canadian Journal of Zoology* 62:927-940.

WATSON D.J. & BALON, E.K. 1984b. Ecomorphological analysis of fish taxocenes in rainforest streams of northern Borneo. *Journal of Fish Biology* 25:371-384.

ZAKARIA-ISMAIL, M. 1991. *Zoogeography and biodiversity of the freshwater fishes of South-east Asia*. Paper presented at the Regional Seminar on Ecology and Conservation of Southeast Asian Marine and Freshwater Environments including Wetlands. Institute of Advanced Studies, University of Malaya, Malaysia, November 1991.

Analysis of factors contributing to the stability of steep hillslopes in the tropical rainforest of Temburong, Brunei Darussalam

A.P. DYKES

Department of Geography, King's College, London, United Kingdom

ABSTRACT. Tectonic uplift and associated channel incision in southern Temburong District, Brunei Darussalam, and adjacent parts of Sarawak, have given rise to a deeply dissected topography which is characterised by steep slopes and sharp ridges. These steep hillslopes are prone to both shallow and deep-seated failure although all recent examples (not yet revegetated) fall into the former category. The Infinite Slope Model for shallow planar failures is used to examine the stability of characteristic slopes for the range of hydrological conditions observed in the field. Matric suction in unsaturated soil significantly enhances the ability of these slopes to resist failure, but typical wet-season rain events can cause failures due to positive pore water pressures beneath perched water tables in the soil. Seasonality of the climate is shown to be an important factor in the stability regime of the hillslopes.

KEY WORDS: hydrology, hillslope stability, climate, topography, soils, tectonics, Borneo.

Reality is a dynamic system, variable in time and space, integrating all pathways, and ranging between the extremes. (C.A. Troendle 1985, p.350)

INTRODUCTION

This paper is intended to give a broad introduction to the physical characteristics and processes of the tropical rainforest of Brunei Darussalam's Temburong District. It will describe the main features of the climate, hydrology and geomorphology of the region, before focussing on the apparent contemporary dominance of shallow mass movement over other hillslope processes in order to illustrate how seasonal climatic patterns can directly affect geomorphic processes beneath a primary rainforest vegetation cover. This region of rainforest was the focus of the Brunei Rainforest Project 1991-92, an international research effort organised jointly by the Universiti Brunei Darussalam and the Royal Geographical Society (U.K.).

The research described in this paper was conducted within a designated 50 km^2 area based on the new Field Studies Centre (FSC) south of Kuala Belalong, in the Temburong District of Brunei. Brunei Darussalam is an oil-rich microstate of 5,765 km^2 situated on the northwestern coast of Borneo. 59% of the country is covered with primary rainforest, with a further 22% also under forest cover (Collins *et al.* 1991). Its population of 268,000 (1991 census) is largely concentrated along the coastal strip. Temburong District is physically separated from the remainder of Brunei by a narrow region of Sarawak. Most of Temburong's 1,288 km^2 is covered with primary rainforest, the southernmost 488 km^2 belonging to the Batu Apoi Forest Reserve. Kuala Belalong lies on the northwestern boundary of this reserve, and the designated research area extended southwestwards into the reserve. All field data was collected in the lower Belalong valley and in the valley of the Engkiang river, a major tributary of the Belalong river (Figure 1).

D. S. Edwards et al. (eds.), Tropical Rainforest Research – Current Issues, 387–409.

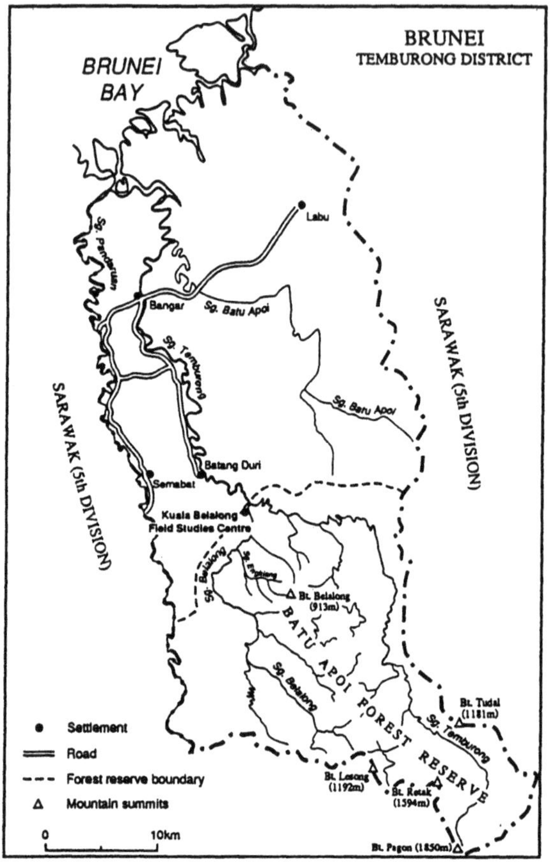

Figure 1. Map of Temburong District, Brunei Darussalam, showing locations referred to throughout the text.

THE RESEARCH ENVIRONMENT

Climate

Brunei experiences a tropical climate which is influenced by the monsoon seasons of South East Asia. These are caused by the seasonal movement of the pressure systems in tropical latitudes. The Inter Tropical Convergence Zone (ITCZ), which defines the Equatorial low pressure belt, moves south across the Equator during November and, from December to March, Brunei and northern Borneo experience northeasterly winds (the 'North East Monsoon'), which curve into the ITCZ over Indonesia. The ITCZ moves north across the Equator during May, and the 'South West Monsoon' sets in over Brunei from June to September as the southeast trade winds curve across the Equator to become southwesterly winds over northern Borneo (Figure 2). Since the dominant direction of the monsoon winds is roughly parallel with the northwest coast of Borneo during both seasons, the effects of the monsoons in terms of rainfall seasonality in Brunei are greatly ameliorated. Long-term means of rainfall data (e.g. Figure 3), show peaks in May and November which coincide with the overhead passage of the ITCZ, and in some areas the start of the North East Monsoon appears to bring slightly enhanced rainfall.

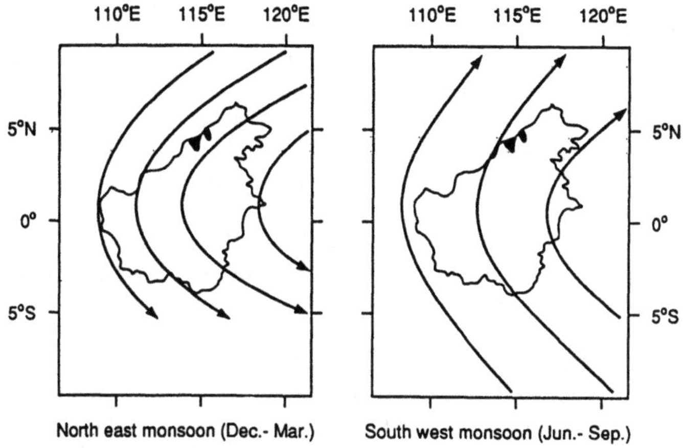

Figure 2. Monsoon wind patterns over Brunei and the rest of Borneo.

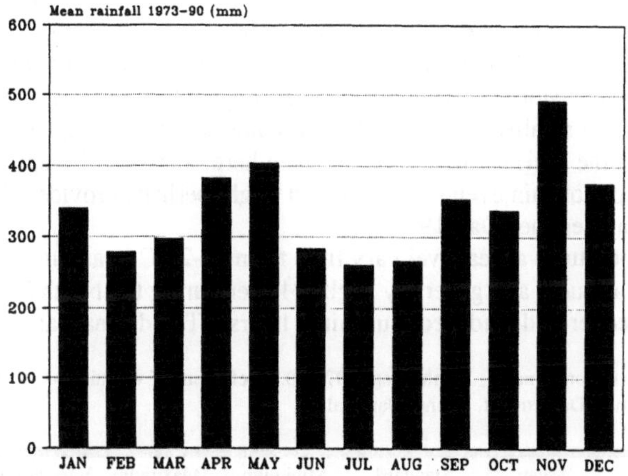

Figure 3. Mean monthly rainfall at Semabat Agricultural Station (see Figure 1), 1973-90 inclusive.

Much of Brunei's rain falls in short duration, high intensity and localised events, mostly due to small-scale convection cells which are accentuated in areas of high relief, although longer periods of steadier, less intense rainfall can occur in the interior due to the rising and cooling of moist monsoon air masses over the mountains. The combination of monsoon, convection and orographic effects results in a rainfall regime noted for its inconsistency and variability, both spatially and temporally, although broad seasonal trends are apparent in the long term. The mean annual rainfall exceeds 2300 mm throughout Brunei (Grant 1984); in Temburong, the annual mean of around 3600 mm at Bangar increases inland to over 4000 mm at Semabat, and the area around Kuala Belalong should expect 4000-4500 mm of rain annually. Table 1 shows the monthly and annual variability of the rainfall recorded at Semabat, which is just 9 km west-northwest of Kuala Belalong, whilst Table 2 illustrates the orographic effect on rainfall totals using data recorded by automatic weather stations (AWS) located at Kuala Belalong (800 metres north of the FSC) and on the summit of Bukit Belalong, just 7.6 km apart horizontally but at altitudes 853 metres different, between July 1991 and March 1992 inclusive.

Dry periods lasting three or four months are irregular but not infrequent in this part of Borneo. The total of 377 mm of rain recorded between 1 January 1992 and 31 March 1992 at Kuala Belalong is comparable with dry spells recorded at Semabat and other nearby stations in Sarawak at other times. At Nanga Medamit, 32 km south-west of Kuala Belalong, 1962.5 mm of rain fell during the 12 months June 1984-May 1985 inclusive (c.f. annual mean 4040.7 mm; data supplied by the Malaysian Meteorological Service). Such dry periods usually include the months of February and March, or July and August, but appear to be local rather than regional in extent.

The incidence of extremely wet periods is equally variable. At Semabat, 1559.1 mm fell in Apr/May 1984 and 1602.1 mm in Nov/Dec 1987, whilst at Ukong (35 km west of Kuala Belalong), 1063.2 mm fell in January 1963 (data supplied by the Malaysian Meteorological Service). It is likely that most of this monthly total resulted from a very small number of extreme rain events, or even a single exceptional storm. At Kuala Belalong, an event of 116.5 mm of rain falling in 75 minutes was recorded at the Field Studies Centre on 19 September 1991. Five days later, 139.5 mm fell throughout an 11-hour period. This latter storm also produced 181.5 mm of rain in 10 hours at Bukit Belalong, of which 67 mm fell in one hour, and 100.5 mm within two hours. Figure 4 illustrates the daily rainfall pattern recorded by the weather station at Kuala Belalong throughout September 1991, and shows how one or two large rainfall events can dominate the monthly total. The hourly rainfall intensity profile of the large storm of 24 September 1991 is characteristic of tropical thunderstorms generally, although five-minute intensities of over 168 mm/hr have been recorded during smaller storms. The storm of 24 September 1991 is shown in Figure 5 along with those of the smaller events on the preceding and following days, as recorded at Bukit Belalong. On the basis of the examples given here, analysis of magnitude-frequency-intensity relationships for rain events in the region might perhaps provide a better indication of the overall rainfall regime (Day 1979).

Daily mean temperatures at sea level vary little from 25-26^{0}C in any month, falling to 20-23^{0}C at 915 metres altitude, and generally slightly lower during the North East Monsoon due to increased cloud cover and reduced sunshine hours. The diurnal range is around 10^{0}C

Table 1: Rainfall summary for Semabat, Temburong; 1973-90. Coefficient of variation = (St.Dev./Mean) x 100. Source: Kilanas Agriculture Department, Brunei Darussalam.

Month	Mean (mm)	Standard Deviation	Minimum (mm)	Maximum (mm)	Variation Coefficient (%)
JAN	354.8	169.6	29.7	637.5	47.8
FEB	275.0	179.6	89.8	655.7	65.3
MAR	305.4	146.5	135.9	724.0	48.0
APR	384.7	158.5	101.3	734.3	41.2
MAY	409.4	184.8	216.9	824.8	45.1
JUN	289.3	196.5	104.9	890.8	67.9
JUL	269.2	113.9	117.2	482.6	42.3
AUG	273.5	200.9	21.3	721.2	73.5
SEP	351.1	198.8	73.1	783.8	56.6
OCT	338.7	109.7	231.0	691.2	32.4
NOV	492.1	151.3	289.9	757.2	30.7
DEC	376.1	185.5	69.9	752.2	49.3
ANNUAL	4119	779	2845	5328	18.9

Table 2: Monthly rainfall and rainless days in the Kuala Belalong research area, 1991-92.

	Kuala Belalong Altitude: 60 metres		Bukit Belalong Altitude: 913 metres	
Month	Rainfall (mm)	R/less days	Rainfall (mm)	R/less days
JUL '91	148.0	17	190.0	14
AUG '91	131.5	14	180.5	15
SEP '91	492.5	4	651.5	3
OCT '91	210.5	9	278.5	5
NOV '91	564.0	3	717.0	1
DEC '91	292.0	6	330.0	5
JAN '92	153.0	14	277.5	12
FEB '92	121.5	17	91.5	13
MAR '92	102.5	18	235.5	15
TOTAL	2215.5	102	2952.0	83

throughout the year. Mean onshore wind speeds tend to be low, but instantaneous speeds associated with vigorous convection cells can reach 25-30 m/s for short periods during storms in Brunei generally (Grant 1984). An hourly average wind speed of 4 m/s was attained only once between May 1991 and April 1992, on the exposed summit of Bukit Belalong.

Relative humidity in the open varies daily between 95-100% during the night to around 75% in the early afternoon (Grant 1984); relative humidities at Kuala Belalong exceeded 99% every night, and fell as low as 57% between 1pm and 2pm on 19 June 1991, although the low-70s are the usual lower values. On Bukit Belalong, relative humidities tended to stay lower during the night, but rarely dropped below the low-80s during the day, due to the higher vapour pressures in the cooler air at the higher altitude. Grant (1984) also reports a mean annual evaporation of 1831 mm at Brunei Airport for the period 1978-81, measured using an evaporation pan. This appears significantly greater than estimates for the mountainous interior: the mean annual evaporation measured at Ukong was 1382 mm for the period 1963-77, the monthly mean ranging from 104.1 mm in February to 128.5 mm in May (data from the Malaysian Department of Irrigation & Drainage). A water balance calculation of mean annual evapotranspiration over a 5-year period for mature dipterocarp forest at Mendolong, Sabah, came to 1540 mm (Malmer 1992), with a mean annual rainfall of 3352 mm for the same period, whilst Bruijnzeel (1983) reported a mean annual evaporation of 1345 mm for a forested drainage basin in central Java (mean annual rainfall 4770 mm). On the basis of these figures, southern Temburong might expect 1350-1450 mm of evapotranspiration losses annually.

Geology

The underlying geology of the research area is generally very simple, although the regional geological history is extremely complex (see James 1984, Wilford 1961). The whole of the Batu Apoi Forest Reserve is underlain by the Temburong Formation, sedimentary rocks of Middle Oligocene to Early Miocene age (around 16-30 million years old) of shale lithology, described by Wilford (1961) as a sandy, dark grey to black rock with abundant plant remains, occasionally encompassing thin beds of hard, fine to medium grained quartzitic sandstone, and thin limestone beds (the latter extremely rare). The formation is strongly folded, though the intensity and amplitude of folding decreases northwestwards. In the lower Belalong valley the shale beds dip eastwards at 40-60^0 and occasionally steeper (Wilford 1961, Map Sheet 2).

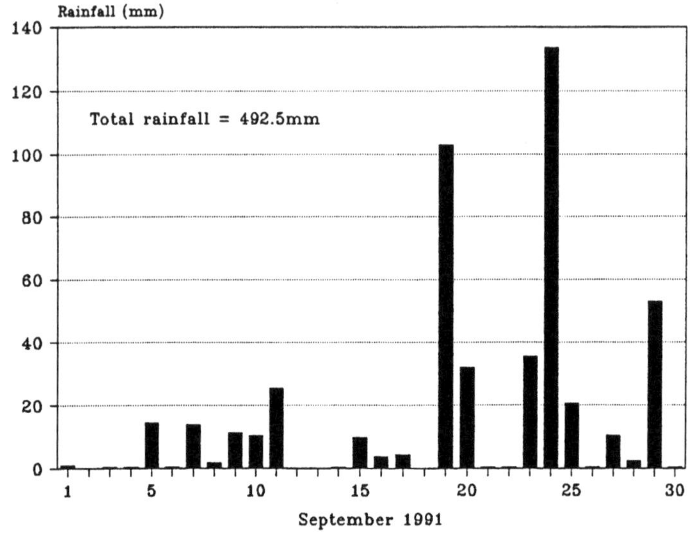

Figure 4. Daily rainfall recorded in September 1991 by the automatic weather station at Kuala Belalong.

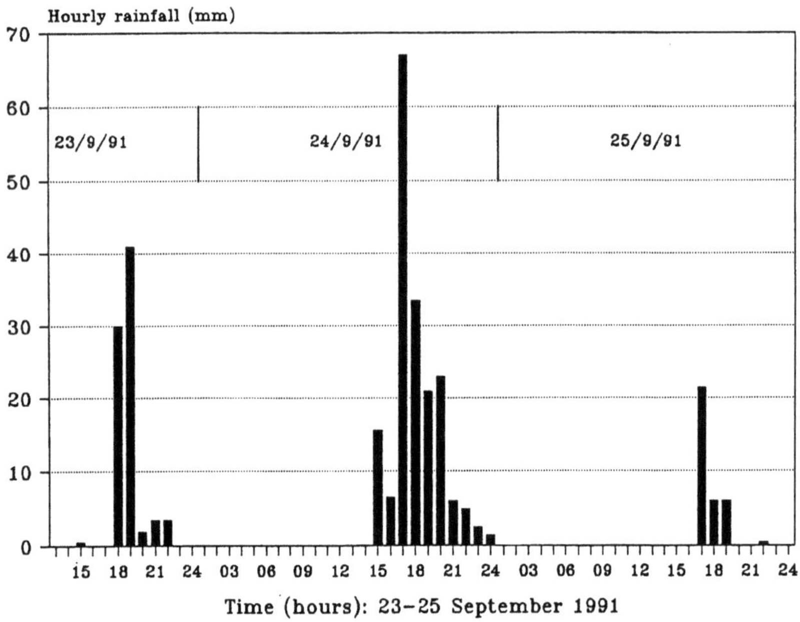

Figure 5. Hourly rainfall intensity profiles of the storms of 23-25 September 1991 at Bukit Belalong.

The area has experienced considerable tectonic uplift in geologically recent times, although estimates of uplift rates vary. Wilford provides a detailed denudation chronology of the entire region of Sarawak within which Brunei is enclosed, though it lacks measured dates for any of the geomorphological features he uses to justify his account. He suggests a rate of uplift of 1.1 mm/year in southeast Temburong, decreasing towards the northwest (giving an estimate of 0.75 mm/year at Kuala Belalong). Recent palaeomagnetic dating of cave sediments in the nearby Gunung Mulu National Park, approx. 40 km southwest of

Temburong's southern border, suggests a rate of uplift there of 0.19 (±0.02) mm/year for at least the last 2 million years (A. Farrant & P. L. Smart, personal communication), which is probably the best available upper estimate for southern Temburong since the Mulu anticline has uplifted to a much greater extent than regions surrounding it.

Soils

Intense chemical weathering of the shale bedrock has given the area a relatively thin cover of residual clay soil. The depth of undisturbed soil cover is thought to only rarely exceed a couple of metres or so, although considerable depths of soil and unsorted weathered shale fragments are likely to occur in some restricted localities marked by the deposition zones of major landslides. The soils of the research area were first described by Ashton (1964), who summarised his profile observations by concluding: "The catena at Belalong from high ridge to river side showed ... a sequence from friable relatively dry soils, with roots continuing deep below the surface, to wet sticky impervious soils with shallow roots" (p.26). Soil profiles examined in detail during 1991/92 have been characterised by Ross & Dykes (this volume), and physical properties of the soils which affect the hillslope hydrology will be discussed later in this paper.

Vegetation

Ashton (1964) gave the first detailed account of the vegetation of this region, describing it as 'Lowland Mixed Dipterocarp Forest'. The results of detailed studies and surveys from the 1991-92 expedition (this volume) will provide a more extensive coverage of species assemblages and physiographic characteristics to complement Ashton's earlier work. The implications of the vegetation cover for any analysis of slope stability may be considerable, given especially the dominance of the giant dipterocarp trees over other species in terms of physical size and species assemblages; these implications and assumptions made in respect of them will be discussed later.

Topography and Drainage

The topography of the lower Belalong Valley and its tributaries is characteristic of the whole of southern Temburong. Channel gradients are steep: the mean slope of the Belalong river for the first few kilometres upriver from Kuala Belalong exceeds 0.5%, and the Engkiang river gradient increases to over 2.4% within 3 km upstream from its confluence with the Belalong river. The channels are cut into the bedrock of the valley floors, and in many places the larger channels contain large quantities of coarse bedload material. Steep slopes rise from the sides of the channels to form deep, V-shaped valleys. There are no floodplains. Minor tributaries with channel gradients often exceeding 40° (84%) descend the main valley sides within their own sub-valleys.

Main valley hillslopes occasionally exceed 45° or even 60°, whilst slope angles in the tributary valleys are usually in excess of 45°, although the slopes are much shorter. The major catchments and the minor sub-valleys are separated by narrow, often knife-edged ridges, and almost all hillslopes bear the scars of major landslide events. Permanent drainage density in the region appears to be fairly high, but cannot be reliably determined from maps since the maps are often produced from air photographs which do not show ground surface features such as drainage channels because of the dense vegetation cover (Morgan 1976). There is a much higher density of storm-activated channels, which should be included in any estimate of drainage density because "... most landforming processes are more active ... when water is flowing in the intermittent first-order channels than during base flow" (Morgan 1976, p.323). Morgan (1976) estimated the "effective" drainage density (i.e. including storm-activated channels) of the dissected uplands of West Malaysia to be around 8-12 km/km^2, subject to the assumptions he applied to his map data.

The total relief of the research area is about 863 metres, from Kuala Belalong itself (the river junction) at 50 metres above sea level, rising via the main ridge east of the Belalong river (c.500 m) to the summit of Bukit Belalong at 913 m a.s.l.; the ridge on the west side of the Belalong river averages 300 m. Beyond the research area, the Belalong catchment rises to 1192 m at Bukit Lesong, at the southern edge of Temburong, and the Temburong river itself rises on the slopes of Bukit Pagon (1850 m).

HYDROLOGICAL PROCESSES

The rainfall regime experienced in this region of northern Borneo is highly variable in space, time and magnitude. The frequency distribution of rainfall events determines the hydrological regime of the hillslope soil profiles, and this in turn controls the development of the slope forms by denudation processes and the flood response of the channels. The nature of channel discharge regimes can, depending on the fluvial characteristics of the channel, directly affect slope form by undercutting of the lower parts of the hillslopes, and the overall slope form greatly influences the hillslope hydrology. The problem of analysing any single component of the drainage basin system stems from the interrelationships identified above, and the integration of hillslope and channel processes within a single model to simulate drainage basin evolution is now being attempted by a number of workers (e.g. Willgoose *et al.* 1991). This paper will describe the general hydrological characteristics of the Temburong rainforest, before analysing the linkages between hillslope hydrology and slope form with respect to shallow mass movement.

Water enters the catchment as rainfall, and that water which is not transferred back to the atmosphere by evaporation or transpiration, leaves the catchment in the form of channel flow. Some of the intermediate components of the tropical rainforest hydrological cycle can not yet be confidently quantified (Bruijnzeel 1990), especially the forest canopy factors of interception, throughfall and stemflow, which determine the amount of rain which reaches the ground. Bruijnzeel's (1990) review of published interception and throughfall data concludes that whilst stemflow usually accounts for no more than 0.5-2% of rainfall, average interception for lowland rainforest is 13% (range 4.5-22). At Kuala Belalong, measured canopy interception was exactly 18% of rainfall if 1% stemflow is assumed.

The rainfall patterns experienced around Kuala Belalong have been described already, but as yet the likely evaporation and evapotranspiration rates can only be estimated from data collected elsewhere. River discharges fluctuate widely and rapidly in response to rainfall, as the stage hydrograph of the Belalong river (catchment 150 km^2) shows (Figure 6). The lowest observed flow was measured near Kuala Belalong towards the end of the extremely dry period in early 1992. The discharge of 1.17 cumecs probably represents the groundwater base flow from the entire catchment (i.e. a catchment yield of approx. 6700 litres/hectare/day). Flash floods with estimated peak discharges in excess of 120 cumecs occurred three times within a 14-month period ending 14 June 1992 (equivalent to a catchment-wide yield of 29,000 litres/hectare/hour). The stage hydrographs shown in Figure 6 illustrate the typical wet season condition, with discharge rarely falling below about 10 cumecs (calculated for a stage of 2 metres). Neither the location, spatial extent or the intensity distributions of the storms which produced these flood events are known, but rainfall on Bukit Belalong which contributed to these flood peaks is shown for illustration.

What is clear from the discharge pattern in Figure 6 is that rain falling on any part of the catchment is producing extremely rapid flood responses. Overland flow is not a rare occurrence in tropical forests (e.g. Bonell *et al.* 1983), but it is usually of limited spatial extent due to the high infiltration capacities of undisturbed forest soils (Bonell & Gilmour

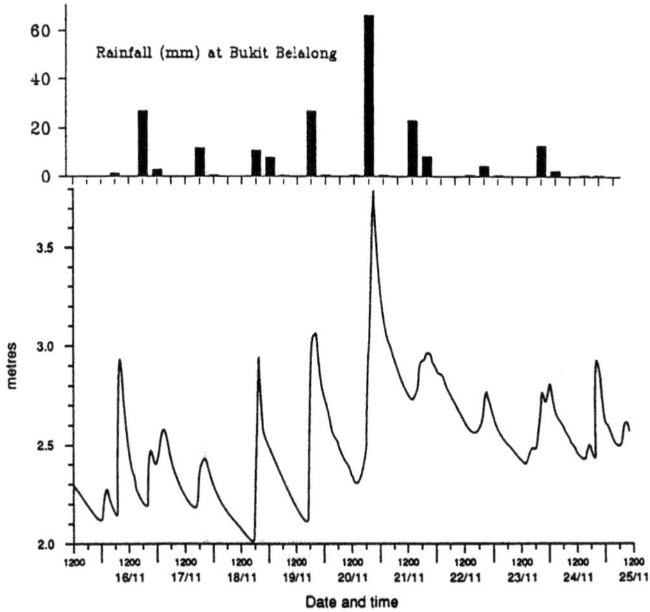

Figure 6. Stage hydrograph for the Belalong river at the FSC for a 10-day period in November 1991.

1978) and results from the build-up of saturated wedges in the soil due to rapid subsurface flow from upslope, which reach the surface in hollows or adjacent to channels (e.g. Moore *et al.* 1986). Overland flow was observed on undisturbed ground in the Kuala Belalong area in the form of storm-activated channels originating from hollows where saturated wedges would be expected to develop. Measured saturated infiltration rates (subject to allowance for problems as discussed by Youngs 1991) were extremely high at the soil surface (average 656.2 mm/hr, n=6, range 503-878) and at surfaces cut into the B-horizon (wet season average 69 mm/hr, n=9, range 6-267), thus under the range of 60-minute rainfall intensities recorded at Kuala Belalong between June 1991 and April 1992 (up to 67 mm/hr), almost all the rainwater which reaches the ground should enter the soil. Subsurface throughflow therefore appears to be the dominant contributor to the storm hydrographs. Considering the hillslope profile in a schematic manner, and leaving aside the mathematical treatments of soil-water relationships (the interested reader is referred to Hillel 1971, for these; or to Kirkby 1985, for their incorporation within hillslope hydrological analyses), it is assumed here that all of the infiltrating rainwater moves vertically down through the soil profile as far as the bedrock interface, then downslope along the base of the soil profile parallel to the surface with negligible losses to the bedrock. Unsaturated flow between storms will supply enough water to the saturated lower parts of the hillslopes to maintain discharge of water into the channels between rainfall events (Weyman 1970), whilst downslope flow during storms is likely to be saturated, creating perched water tables above the bedrock within the soil profile if the average subsurface flow velocity is not too great (Pearce & O'Loughlin 1985). The extent of the saturated zone at the lower extent of the soil profile varies within and between storms, hence the area of slope which contributes to the discharge hydrograph of the channel, and the channel stormflow itself, similarly vary (see review by Troendle 1985).

That saturated wedges rapidly develop in hollows as well as adjacent to channels during heavy rain is demonstrated by soil tension measurements from mercury manometer tensiometers near Kuala Belalong. Figure 7 illustrates the rapid build-up of saturated conditions at a depth of 90 cm in the soil at tensiometer nest D, near the bottom of a small

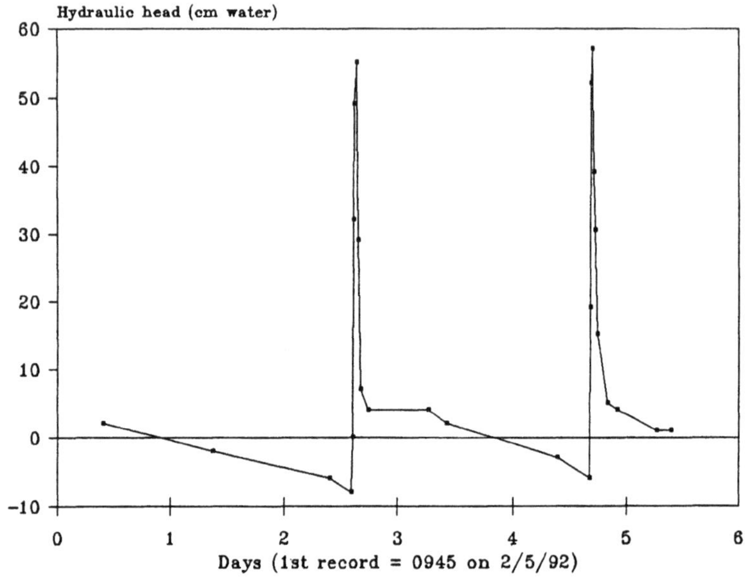

Figure 7. Changing soil water conditions over six days in early May 1992, measured at tensiometer D3 (90 cm deep in the soil). A positive hydraulic head shows the height of a water table above the tensiometer cup; negative values are tensions.

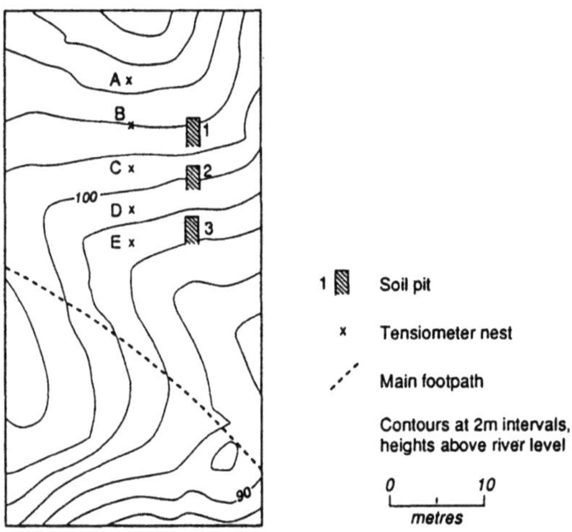

Figure 8. Map showing arrangement of tensiometers and soil pits on the research hillslope, with tensiometer nest E nearest the bottom of the small topographic hollow.

topographic hollow (see Figure 8), during two storms in May 1992 (86.8 mm and 47.0 mm of rain respectively), and the very fast saturated subsurface drainage of the perched groundwater once the input of rainwater has eased. During both of these storms, a perched water table had developed within 20 minutes of the onset of rain, but most of this water had drained within about 90 minutes at this site, and within 5-6 hours at the deepest tensiometer nearest the slope foot. In neither of these events did the saturated wedge reach the deepest tensiometer furthest up the slope. (Similarly large moisture fluxes are experienced during rainfall under dry season conditions, although there is rarely sufficient rainfall at these times of year to cause saturation.) 'Translatory flow', i.e. displacement of downslope soil water into the channels by new rainwater from above (e.g. Bruijnzeel 1983), may account for the initial rapid channel responses to rainfall, but the rapid drainage of saturated zones as illustrated above suggests that direct throughflow from a given storm dominates the hydrograph, with significant contributions from the numerous storm-activated channels which originate in saturated hollows on the hillslopes.

The rate and nature of water movement through the hillslope profile is normally dependent on the moisture status of the soils, which is controlled jointly by the physical characteristics of the soils and by the climatic regime. The soils of southern Temburong have very high clay contents (Ross & Dykes, this volume), which enables them to retain more moisture at a given tension than coarser-textured soils due to adsorption of water onto the clay minerals as well as capillary resistance to unsaturated flow in the micropores (Hillel 1971). The relationship between moisture content and soil tension is usually unique to a particular soil: the moisture retention curve for the soils at Kuala Belalong shows an initially rapid decline of soil moisture with low applied tensions, unusual for clay soils generally but indicative of the occurrence of macropores within these soils, and is fully described by the empirical equation:

$$\Theta = 83.7 - 3.87 \ln(\Phi) \qquad (1)$$

for $\Phi > 0$, where Φ is the tension (bars) and Θ is the percentage saturation of the soil: the mean saturated moisture content for the soil (i.e. $\Phi=0$) was 105.10% (n=12, s.d.=5.99) by volume pore space, based on a mean porosity of 49.53% (n=12, s.d.=6.08). Most of the measured change in the soil moisture content occurred over the first bar (100 kPa) of applied tension, which is slightly more than the highest tensions measured in the field. The seasonal variation of soil tensions encountered here therefore covers most of the range of moisture contents possible in these soils; this seasonal variation not only affects the channel response to rainfall at different times, but also the probability of slope failures occurring at any given time.

HILLSLOPE PROCESSES AND SLOPE FORM

Introduction

Hillslopes are initially produced by channel incision into an uplifted land mass, their length and slope angle being largely determined by the available relief and drainage density (Strahler 1950). Thus an area with a high drainage density and high relief will be characterised by short but steep slopes, and since steep slopes supply large amounts of material to the channels, then the channels must be steep in order to remove the eroded material. The actual removal of material from a hillslope will either be weathering-limited or transport-limited; in humid tropical regions the latter is dominant, i.e. rates of weathering and soil production are greater than (or, on a geological timescale, equal to) the available long-term transport capacity provided by water or gravity. Rapid mass movement is the most significant slope process in steep, forested tropical landscapes (Chorley *et al.* 1985), with water and gravity being the causative agents. Small-scale erosion of surface soil by overland flow, splash erosion and creep is largely inhibited by the vegetation, especially the dense networks of shallow and surface roots (Bonell *et al.* 1983); overland flow generally only occurs in undisturbed forests on saturated channel margins and stream head hollows (Chorley *et al.* 1985). Preliminary data from a hillslope plot experiment near Kuala Belalong clearly illustrates the degree of protection from splash and runoff erosion afforded to the soil by the surface litter layer (Ross & Dykes, this volume).

Hillslopes evolve by erosion, either retreating parallel to themselves or declining from the top downwards, with the dominant erosion mechanism being largely determined by the climate. The concept of threshold slopes may enable the process history of a given landscape to be interpreted using the observable range of characteristic slope angles (i.e. the inclination of the steep, straight part of any given profile: Chorley *et al.* 1985). Hillslopes which initially exceed their calculated maximum stable angles will decline by mass movement processes, until they are lowered sufficiently to be stable with respect to further movement and so enable small scale denudation processes of creep and wash, etc., to control the subsequent evolution of the hillslopes. Parallel retreat by mass movement may also occur if the threshold slope angle for stability is exceeded and shallow failures dominate. This situation arises where channel incision is rapid, and hillslope erosion occurs at a similar rate: a dynamic equilibrium situation in which shallow failures maintain steep slopes. In this situation, steep and straight slope forms will tend to dominate the observed slope features. In cases where the hillslope response is not rapid enough to maintain an equilibrium form with the channel, convex slope forms may occur, indicating that the hillslopes have not yet adjusted to very rapid incision. "Differences in hillslope form should not be the result of the disparity between uplift and denudation, but probably are the result of the difference between the two components of denudation, rates of channel incision and hillslope erosion" (Schumm 1963, p.H8).

Shallow Mass Failure: the Infinite Slope Model

The simplest form of failure is that of a layer of soil or weathered material sliding downslope along a surface which is parallel to the surface profile of the slope. The analysis of hillslope stability with respect to this type of failure is also simple, involving a single linear equation, but is direct in its application. Although this infinite slope model was developed to analyse shallow landslides in the London Clay (Skempton 1964, Skempton & Delory 1957), it has been widely used for interpreting observed limiting or threshold inclination of slopes where landslides are thought to be responsible for slope decline (Chandler 1982), and has been adapted and used in analyses of stream head extension through shallow failure (e.g. Dietrich *et al.* 1992, Montgomery & Dietrich 1989). However, "the question of whether conventional methods of stability analyses are equally applicable to very steep slopes as to relatively gentle ones is not one which has received much attention in the soil mechanics literature" (Wesley 1977, p.127, in reference to massive, shallow surface failures on steep slopes in Indonesia).

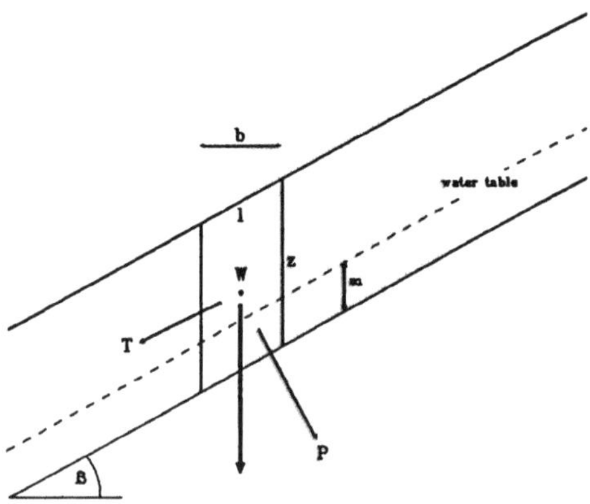

Figure 9. The Infinite Slope Model. See text for explanation.

The infinite slope analysis is illustrated in Figure 9, and is used in this paper to calculate the 'Factor of Safety' of a hillslope under given input conditions. This is generally defined as the ratio of the forces which promote downslope movement to those which resist movement, i.e. the tangential (T) and normal (P) components of the vertical force due to gravity acting on the mass W, which act on the base of a given slice of the hillslope profile. A factor of safety of 1.0 indicates the critical condition at which the slope reaches the point of instability; below this value, the slope will fail since the downslope force exceeds the strength of the slope material. Since the single slice is assumed to represent conditions at every slice and therefore the hillslope as a whole, lateral forces acting on it from the slices on either side are assumed to be equal and opposite in effect, and can therefore be removed from the equation. The factor of safety is thus:

$$F = \frac{c' + (yz\cos^2\beta - u)\tan\varnothing'}{yz\sin\beta\cos\beta} \qquad (2)$$

here the geometric terms are ß, the slope angle, and z, the depth of the slip plane; the soil shear strength terms are c', the cohesion and Ø', the angle of internal resistance (friction angle) of the soil; and the hydrologic terms are y, the bulk unit weight of the soil (dependent on the moisture content) and u, the pore water pressure, where:

$$u = y_w m \cos^2\beta \qquad (3)$$

for positive pressures, with y_w = unit weight of water, and m = height of water table above the slip plane. Negative pore pressures (soil tensions) can be incorporated directly in the "u" term of the equation for tensions of up to about 20 kPa (Professor E.N. Bromhead, personal communication), although Fredlund (1987) has analysed the effect of suction on the effective shear strength of a soil and has detailed more rigorous methods for incorporating suction effects into stability analyses.

One of the major problems with stability analysis such as this is the extent to which soil strength parameters obtained in laboratory tests reflect the true conditions in the field (e.g. Bromhead 1978). Strength tests on samples from Temburong have not yet been completed, and so this problem will not be addressed here.

THE SOUTHERN TEMBURONG LANDSCAPE

Analysis of Hillslope Stability

Given that soil shear strength data are not yet available from Kuala Belalong, a modified version of the infinite slope model has been run for assumed soil strength parameters, based on Wesley's (1977) Indonesian results and on a simple back-analysis of a recent landslide. That the landscape in this region is inherently unstable has already been indicated; the purpose of this paper is to evaluate the relative importance of environmental factors on stability conditions in this region.

The analyses described below assume a friction angle of 42.5°, slightly higher than Wesley's (1977) results, and a soil cohesion of 10.5 kPa. The latter was obtained by conducting a back-analysis of a hypothetical slope of 45° which had failed (i.e. F=1.0), a soil depth of 2.0 metres to the shear plane (as observed at the upper edge of a shallow failure in the Engkiang valley thought to have occurred on 24 September 1991), the friction angle

42.5° and worst-case groundwater conditions with the water table at the ground surface and therefore maximum possible pore water pressures within the slope. This is likely to have been the case during the rainfall of 24 September 1991. The parameter values assigned above are those which were used as the basis for the analyses which follow, with the soil moisture characteristic for these soils (equation (1)) being explicitly incorporated within the stability model, and saturation ($\Phi=0$) assumed when $\Theta=105\%$.

A theoretical evaluation of the relative importance of each of the variables incorporated in the stability model can be made by means of a sensitivity analysis of the model (Pearce & O'Loughlin 1985). This procedure involves the calculation of a base factor of safety using the medians of realistic ranges of values for all variables, and its subsequent recalculation as the value of each variable in turn is changed whilst the others are held constant. The chosen base values are those given in the previous paragraph, and the results have been plotted as relative percentages (Figure 10). The analysis shows that the factor of safety is most sensitive to changes in ß and Ø' and is also quite sensitive to the cohesion. It is relatively insensitive to higher unit weights or greater depths than those used as the basis for this analysis, although they do become progressively more influential with respect to increasing stability as they become unrealistically low in value. The influence of pore water pressures is presented by the height of the water table above the shear plane, m, which increases the factor of safety as the water table, and pore water pressures, fall from their maximum at the point of failure. The importance of the soil shear strength parameters, and the caution with which this and the following discussion based on assumed soil parameters should be treated, is therefore demonstrated.

In attempting to account for ultimate stable slope angles under natural conditions, Chowdhury (1978) raises the potential problem for evolutionary studies of the possible decrease of soil shear strength parameters due to weathering or geological factors, but demonstrates that "... long-term stability is controlled by slope inclination, pore water pressure and effective shear strength parameters" (p. 91). In this study, the emphasis is on

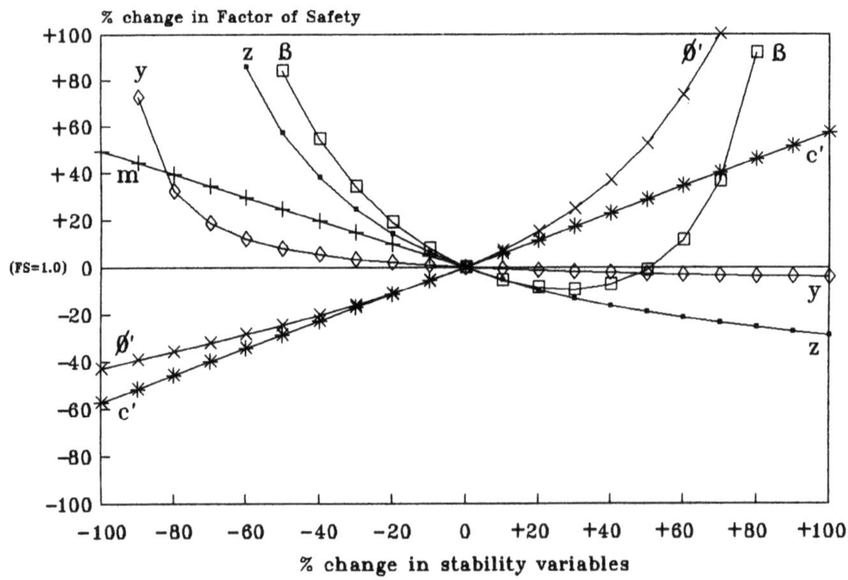

Figure 10. Sensitivity analysis of variation in factor of safety for infinite slope stability analysis.

accounting for the occurrence of landslides under the present climatic and tectonic regime, and the initial model results in the sensitivity analysis presented above agree with Chowdhury's comment on the dominant controls on stability. In the Temburong setting, soil shear strength variation is assumed to be negligible in the fully weathered residual soil profile, and the inclination of any given slope is assumed to be constant over time (due to parallel retreat) as long as the equilibrium status of the slope/channel unit under analysis does not change. The soil profiles on the steep slopes under discussion may not, however, attain the fully weathered state: this may result in the Factor of Safety being underestimated slightly, which is not serious as it errs on the side of caution (the slope may not fail even if predicted to do so). Shear strength and slope angles provide static controls on the stability analysis of a hillslope in the context of this study, whereas the hydrological regime determines the actual stability at any given moment in time.

The dynamic hydrological processes illustrated earlier can now be incorporated into the discussion of the stability regime. The theoretical effects of soil tension and of water table depth (which produces positive pore water pressure according to equation (3), above) on the calculated factor of safety are plotted for different soil profile depths in Figure 11 and for different slope angles in Figure 12. The large increase of stability for very small rises in tension (-1.0 m head = 0.1 bars (10 kPa) tension), and the rapid decrease of stability as pore water pressures increase show that the hydrological inputs are of significant importance. The seasonal pattern of soil moisture/tension changes (Figure 13) shows that any given slope would be extremely unlikely to fail during dry season conditions due to the high tensions experienced within the soil profile, and that even during a very wet season rapid drainage of the soils result in dominantly unsaturated conditions except during and immediately following heavy rain (Figure 7). As Figure 11 shows, even very low tensions can significantly raise the factor of safety of a hillslope, especially on steeper slopes, thus maintaining stable conditions between potentially critical rainfall events.

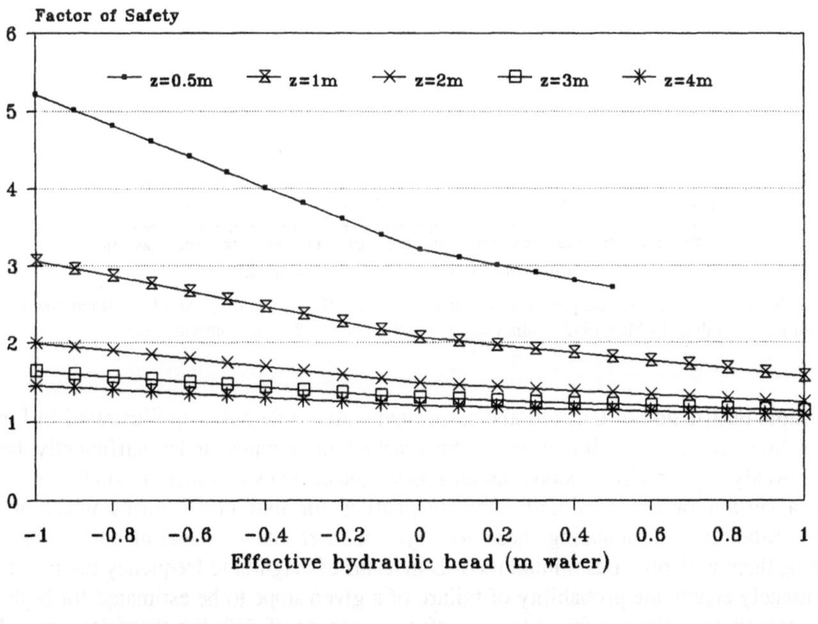

Figure 11. Variation in calculated slope stability with changing soil hydrological conditions, showing how these are affected by the soil profile depth.

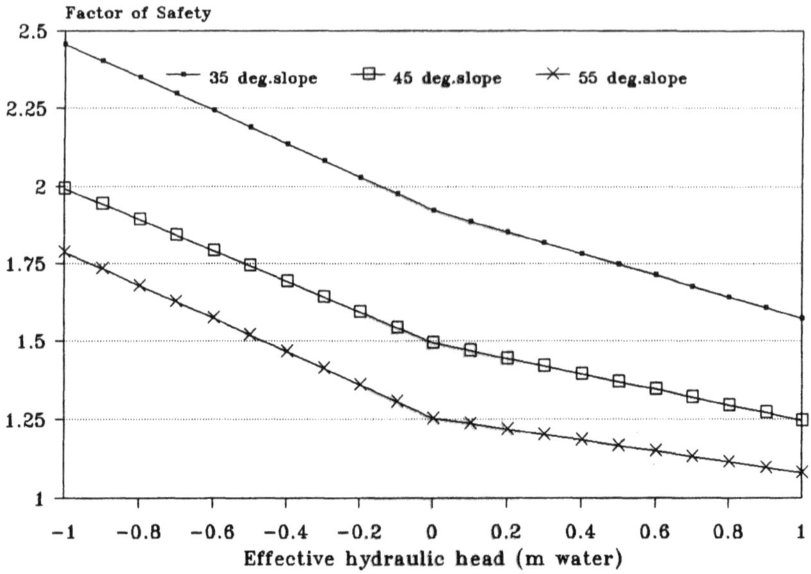

Figure 12. Variation in calculated slope stability with changing hydrological conditions on slopes of different inclinations.

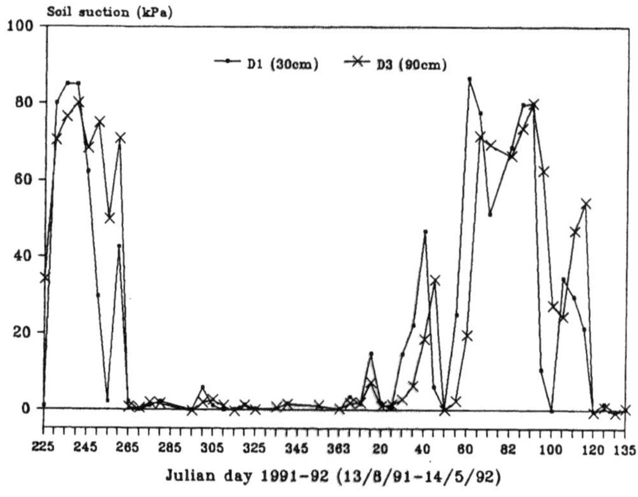

Figure 13. Seasonal soil suction changes measured at tensiometers D1 (30 cm deep) and D3 (90 cm deep) over a nine month period ending 14 May 1992. Julian day 1 = 1st January; 32 = 1st February, etc.

The rapid destruction of tensions and associated water table build-up illustrated in Figure 7 shows how unstable conditions could be reached in minutes under sufficiently heavy rainfall. Analysing for a given slope, under known antecedent soil moisture conditions, there will be a certain range of rainfall intensity profiles for individual storms which would generate instability. Calculating these for 'typical' wet and dry season conditions, and comparing them with previous rainfall records in terms of magnitude-frequency relationships, will ultimately enable the probability of failure of a given slope to be estimated for both wet and dry season conditions, subject to a satisfactory means of defining these seasons. This part of the research is still under development.

There are problems with the analysis of the steep lower valley slopes which have not yet been addressed. The first is specific to this geomorphic situation, and is the model requirement of an infinitely long, planar hillslope with a shallow slip plane parallel to the surface, and the assumption that the depth of water table above the slip plane is also constant along the entire slope length. The development of saturated wedges in the lower parts of a hillslope, adjacent to a channel or hollow, will result in significantly higher positive pore water pressures in the lower slope region, but there may be no saturated conditions upslope. In addition, active channel erosion and undercutting of slope foot locations may remove the lowest part of the soil profile, leaving an unsupported bottom 'slice' and reducing the overall stability of the slope by unloading of the toe. Under these conditions a factor of safety can be calculated using the method of analysis developed by Morgenstern & Price (1965) for slip surfaces, ground surfaces and water table surfaces of any shape (Bromhead 1988).

The second problem is that of quantifying the effect of vegetation on the stability of the steepest slopes with respect to shallow failure. "Vegetation that may be growing on a slope has traditionally been considered to have an indirect or minor effect on stability, and it is usually neglected in stability analyses. This assumption is not always correct, and for certain forested slopes with relatively thin soil mantles, it has been shown to be significantly in error" (Greenway 1987, p. 187).

Greenway's (1987) review considers vegetation effects as being either hydrological or mechanical, and discusses in particular the impact of root strength, distribution and morphology on the overall shear strength of the soil mass. There does not yet appear to be any published data concerning the physical or spatial properties of root distributions for rainforest tree species, nor on the actual spatial size distribution of trees within natural forest. In Temburong, it appears that few trees of significant size occurred on the steepest slopes, and that throughout the forest few roots penetrated more than a few centimetres into the soil, except perhaps directly beneath some of the largest trees. There was no visible evidence of roots penetrating the bedrock beneath a soil mantle, although Ashton (1964) reported that this did occur on ridgetop sites. In terms of the slopes under analysis, it is assumed that no trees of significant size occur, so the surcharge due to the weight of the trees can be ignored, and it is also assumed that the reinforcing effect of those roots which do reach the depth of the bedrock interface is negligible.

The hydrological effects of vegetation are less difficult to estimate, but accurate quantification remains a considerable problem for hydrological and stability research (Bruijnzeel 1990). The hydrological factors are either adverse to stability, in that vegetation promotes soil infiltration capacities by increasing the soil surface roughness and organic content and by causing dessication cracks to appear in the soil following moisture uptake by roots, or beneficial to stability, by reducing the amount of rainfall which reaches the soil by as much as 24% (Vis 1986) and then by removing the water which does enter the soil and transpiring it to the atmosphere. On the steep hillslopes in Temburong, the role of slope inclination in *reducing* infiltration is an unknown but possibly significant factor which may actually *increase* the effective stability of the steepest slopes under wet conditions. Soil moisture losses through evapotranspiration are of great importance in maintaining stable conditions; in this environment plants are likely to transpire at their maximum potential rate almost all the time, since soil moisture deficits are not expected to occur in any month on the basis of rainfall and mean evaporation data discussed earlier.

The relationship between slope stability and the depth of the soil profile above the shear plane is worthy of further comment. Figure 10 implies that if all other parameters are held constant, there is a threshold profile depth with respect to stability, in that any increase in depth beyond that threshold has relatively little influence on the stability of the hillslope. In terms of the varying hydrological conditions within the profile, it is obvious that a given water table depth affects a far smaller proportion of a deep profile than of a shallow one.

The major implication is in analysing the evolution of the hillslope in terms of material production by weathering, and material loss by erosion or solution. For a soil profile of any depth to occur, weathering rates must exceed rates of removal of the weathered products. As the threshold depth for stability is approached, the probability of failure of the slope increases, until failure occurs and erosion by mass movement instantaneously exceeds long-term production. Thus one might expect to find a long-term equilibrium between weathering and erosion, an equilibrium which cannot be tested for because the timescale over which it applies is a function of the prevalent weathering rates, but which is concordant with the earlier discussion of slope development by parallel retreat.

Given that the initial parameter values used in the calculations above have been assumed, largely on the basis of a back-analysis of one small (approx. 500 m^3) landslide, the high sensitivity of the model to the shear strength of the soil means that any application of results obtained so far on the basis of assumed parameter values must be regarded as somewhat tenuous. However, the preceding paragraph provides an approach for analysing the actual contemporary status of the Temburong landscape in evolutionary terms. Retaining the assumed parameters used above, for a 45^0 slope there appears to be a threshold soil profile depth of about two metres, above which the slope may fail if no other parameter value changes. The threshold profile depth with respect to failure will be smaller for steeper slopes and greater for less steep slopes. The return period of rainfall events required to cause a hillslope of given profile depth to fail, will also need to be calculated and accounted for in probabilistic terms. If the topographic lowering of the landscape by shallow mass movements is in equilibrium with channel incision, most soil profiles at any given slope angle would be expected to be at, or close to, the calculated threshold depth for that slope. Stability analyses using measured values of soil shear strength will ultimately be performed to predict the expected hillslope inclination-profile depth relationship for the range of slopes encountered in the field. These results will be taken to the landscape to see how, or even if, they account for the observed hillslope forms.

Hillslope Characteristics in southern Temburong

Hillslope profiles measured in the Belalong and Engkiang valleys in Temburong display many distinctive features. Twenty-two profiles were measured, and four of these are shown in Figure 14. The profiles were measured by taking slope angles over lengths of 5 to 15 metres, along a compass bearing at right angles to the line of the valley floor (whenever possible). The lines of the profiles were chosen in the field in an attempt to avoid distortions due to tributary valleys. The main effect of the generally high drainage density is to limit the areal extent of any given valley side hillslope, so that the sides of the main valleys comprise short steep footslopes with long, undulating, knife-edge ridges, or relatively steep tributary valleys leading to very steep stream-head slopes.

The profiles in Figure 14 show many of the features encountered in the field. The lower slopes of the Engkiang valley are the steepest slopes generally, with the mean slope angle exceeding 48^0 over a 100+ metre long segment on profile 4E. The hillslopes on the west side of the valley are less steep, sloping with the dip of the bedrock as in the main Belalong valley. These lower slopes are either straight or slightly convex, although the lower part of profile 4W displays a markedly concave section due to an old large, deep-seated landslide. Many of the lower slopes display more recent signs of shallow failures. The upper slopes in the Engkiang valley are generally less steep, with segments averaging around 30^0 but always with small sections considerably greater than 45^0, usually marking the headwalls of large, deep-seated movements. The slopes of the tributary valleys appeared to be generally steeper than many of the main valley slopes. These slopes often exceeded 50^0 as measured over several tens of metres.

In the main Belalong valley, the eastern slopes are again the steepest, with mean

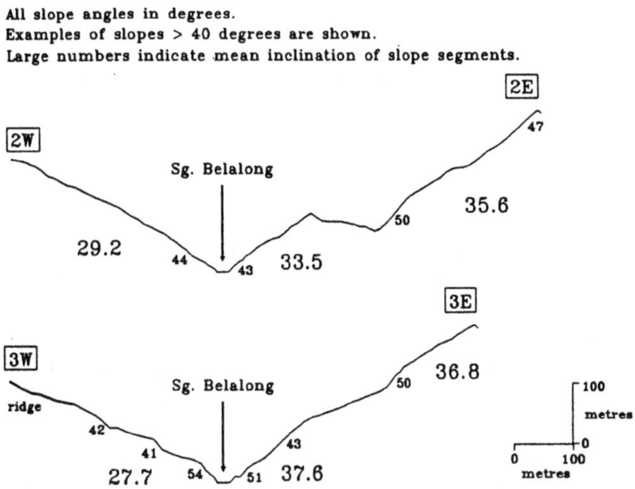

Figure 14a. Valley side profiles of the Belalong valley just south of the FSC, looking downstream.

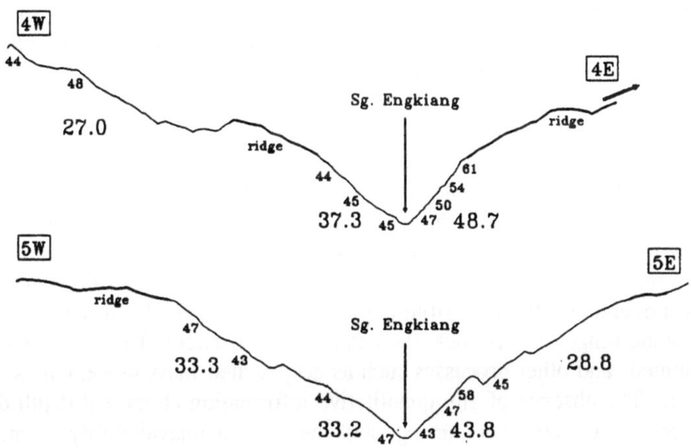

Figure 14b. Valley side profiles of the lower Engkiang valley, within 2.5 km of the Belalong river. Diagram scale and details as for Figure 14a.

inclinations exceeding 35^0 in some places. The western slopes rarely average 30^0, but, as in the Engkiang valley, all the slopes contain very steep sections (up to 60^0) which are either adjacent to the river channel or at the top of depressions caused by deep-seated mass movements. Profile 2E includes the slope directly above a stream-head position and clearly shows the steep character of stream-head locations in this environment.

The observed spatial distribution of major deep-seated landslides on the measured profiles did not follow any significant pattern, although they were relatively uncommon on the steep, straight slopes adjacent to the main rivers, where shallow failures predominated. At least 31 major deep-seated landslides (e.g. 20-400 metres wide, lengths usually unknown) were directly encountered whilst measuring the profiles, together with a further 10 smaller failures and a number of slopes in the Engkiang valley where shallow slips had occurred relatively recently (e.g. within the previous couple of years).

Shallow failures seem to be confined to the steep, straight slopes adjacent to the channels, with several recent (1991) examples being noted along the Engkiang, Belalong and Temburong rivers. They can also be found on the steep upper slopes adjacent to the ridges, as a late 1992 example near the Field Studies Centre shows. The scar of an enormous hillside collapse above the confluence of the Engkiang and Belalong rivers was also found to be retreating by shallow failure, part of it having done so in 1990. In addition, it is likely that such failures occur in stream-head hollows (Deitrich *et al.* 1992, Montgomery & Dietrich 1989), although no obvious examples of this have been encountered so far.

A digital elevation model (DEM) produced from the Geographical Information System (GIS) database for the research area (Wills, this volume), was analysed to produce a map of slope angles for the entire area. This approach to landscape stability evaluation is well established, but is dependent on a good digitised topographic dataset being available (e.g. Carrara *et al.* 1991). Despite inaccuracies inherited from the original aerial survey of the region, from which the DEM is derived, the obtained map of slope angles does show that the steepest slopes are generally of limited spatial extent, and are concentrated either near the valley floors, which is not unexpected given the regional context of rapid uplift and incision which prevails here, or adjacent to the crests of the interfluve ridges. Francis (1987) notes that "... in areas apparently dominated by steep slopes, low-angles slopes are always present and often cover a large proportion of catchment areas" (p. 608). The most frequently occurring slope angle class within the DEM is $25\text{-}29^0$, with almost as many slopes of $30\text{-}34^0$. In other words, the research area is *actually* dominated by steep slopes.

Provisional Interpretation of Observed Slope Forms

A very crude application of the stability analysis results to the suite of measured slope profiles would imply that soil profiles on the lower slopes of the Engkiang valley will be thinner than on the upper slopes of the main valley or in the Belalong valley, and that even less soil could remain stable on the steepest slopes of the minor tributary valleys, if the valley slopes are in equilibrium with the contemporary channel positions with respect to shallow mass movement. If soil profile depths generally differ from the critical predicted depths over a wide range of hillslopes, then equilibrium between the hillslopes and channels cannot be assumed, and other processes such as deep-seated mass movement would need to be investigated. The absence of any quantitative information about soil depth distributions on the hillslopes in southern Temburong, and the current unavailability of measured soil shear strength parameters, precludes a more detailed application of the analysis at this stage.

The observed steep and straight or slightly convex lower valley slopes with generally thin soil profiles, and the concentration of shallow failures along these slopes, suggests that in the Engkiang valley at least, the lower slopes are probably in equilibrium with the rapidly incising channel, and that shallow slope failures are encouraged by progressive steepening of the slope by channel incision. The apparent convexity of the Engkiang valley sides as a whole suggests that incision by the Engkiang river is currently proceeding more rapidly than the valley as a whole can adjust to. In the Belalong valley, where the main valley slope forms are generally straight but where soil profiles are thought to be considerably less deep than their predicted thresholds using the infinite slope analysis, the slopes may be in equilibrium with the channel due to deep-seated failures, with shallow landslides occurring where lateral erosion and undercutting has produced localised steepening of slopes at the channel edges.

When the work currently in progress reaches the stage of allowing more concrete theories to be formulated concerning the equilibrium status of the hillslopes in southern Temburong, a wider view of the evolution of the entire drainage basin will be possible for this regional setting of tectonic uplift and subsequent landscape denudation beneath a primary rainforest cover.

CONCLUSIONS

The physical characteristics of the landscape of southern Temburong have been described and, to some extent, explained in terms of the interrelated hydrological and geomorphic processes active within the drainage basin system. The tropical climate allows a complete rainforest cover to flourish, but its seasonality of rainfall can directly influence the temporal occurrence of major geomorphic events such as landslides.

Analysis of the stability of a hypothetically characteristic 45^0 slope, and further stability calculations using the modified infinite slope model, demonstrate the critical importance of hydrological conditions on the instantaneous stability of the hillslopes, and show that matric suction in unsaturated soil (soil tension) is probably the most important factor in accounting for the maintained stability of these hillslopes where the soil profile is close to a critical threshold depth for the angle of slope. High tensions experienced by the soils during the dry seasons will prevent failure conditions from being attained during many storms whose magnitudes could cause failure under wet season conditions.

No firm conclusions can yet be drawn concerning the equilibrium status of any hillslope-channel units in the research area, but it appears that parallel slope retreat by means of shallow failure is maintaining steep, straight slope forms in the lower slopes of the Engkiang valley, and that these slopes are in equilibrium with the Engkiang river channel. This situation is thought to prevail along most of the tributary valleys in the Belalong and Temburong catchments.

Many of the geomorphic processes acting on this region are not yet known in any quantitative sense, and even some of the basic hydrological parameters can not be utilised with any great degree of confidence. The slope stability work introduced in this paper would be greatly enhanced by data on spatial variations in soil profile depth, spatial distribution of tree sizes with respect to topographic and geomorphological features, and some quantitative data on root size and depth distributions through the soil profiles.

Finally, more general research questions raised by this work include the effect of slope angle on soil surface processes and the spatial variability of infiltration generally, rates of weathering and denudation by solution, and the links between mean rates of sediment production and transportation in the rivers and how they relate to seasonal hydrological conditions.

ACKNOWLEDGEMENTS

This research was undertaken as part of the Brunei Rainforest Project 1991-92 organised by Universiti Brunei Darussalam and the Royal Geographical Society. The author would like to acknowledge the support of the Project's corporate patrons (Royal Brunei Airlines, The Baring Foundation, Diawa-DICAM, Greencard Trust, Hongkong Bank, Morgan Grenfell and Nomura-NIMCO) and Brunei Shell Petroleum Sdn Bhd, for their commitment to the Kuala Belalong Field Studies Centre. In particular, RGS sponsorship of this research is gratefully acknowledged, together with the logistical support provided in Brunei by C. Prebble, A. Simpson and J.T. Wills (RGS), and Dr D.S. Edwards, Dr Kamariah Abu Salim and Samhan bin Nyawa (Biology Dept., UBD).

The advice and encouragement given throughout by Professor John Thornes (King's College London) has been invaluable, both in the field and subsequently.

At Kuala Belalong, assistance was provided by many people throughout the 17 months of fieldwork. In particular, special thanks go to Kuni Anggat, whose dedication and effort towards the fieldwork for this research during the latter 10 months was of immense value. Thanks also go to Gawing Anggat, Jali Amba, Simpurai Anggat, Inggul Timbang and Samsudin Panya, who gave considerable help at various times, with further assistance from Impin Lasah, Nyua Anggat, Ramlah Sapong, Isah Laju, Mujah Ngambat, Osman Linggi, and others. Additional help in the field came from Prof J.B. Thornes, J.T. Wills, Dr S.M. Ross, M. Mulligan, Y. Willcox, D. Buck, A. Simpson, C.A. Pendry, M. Welsh, and members of 2nd Battalion, King Edward's Own Gurkha Rifles. The helicopter was provided by Royal Brunei Armed Forces.

Additional practical and logistical support in Brunei was provided at UBD by all members of the Biology Dept., and by Dr. Goh Kim Chuan and Dulima Jali (Geography Dept.). The Dept. of Civil Aviation provided the Semabat rainfall data.

In the U.K., the practical and logistical help of all the technical/support staff in the Geography Dept. at King's College London, and Ms C. Prebble and Mrs V. Simonds at the RGS are acknowledged, as are NERC (Scientific Services) and Mr. M. Walker (IoH) for the preparation and loan of two automatic weather stations to the Project. Thanks also go to the many people with whom relevant discussions took place; also to Halcrow Engineering Ltd. for providing the Malaysian meteorological data, and to Mr. I.D. Bloomfield for assistance with the computer programming.

LITERATURE CITED

ASHTON, P.S. 1964. Ecological studies in the Mixed Dipterocarp forests of Brunei state. *Oxford Forestry Memoirs No. 25.* Clarendon Press, Oxford.

BONELL, M. & GILMOUR, D.A. 1978. The development of overland flow in a tropical rainforest catchment. *Journal of Hydrology* 39:365-382.

BONELL, M., GILMOUR, D.A. & CASSELLS, D.S. 1983. Runoff generation in tropical rainforests of northeast Queensland, Australia, and the implications for land use management. In *Hydrology of Humid Tropical Regions with Particular Reference to the Hydrological Effects of Agriculture and Forestry Practice.* (Proceedings of the Hamburg Symposium, August 1983). IAHS Publ. No. 140.

BROMHEAD, E.N. 1978. Large landslides in London Clay at Herne Bay, Kent. *Quarterly Journal of Engineering Geology* 11:291-304.

BROMHEAD, E.N. 1988. *Slope Stability Analysis Package: User's Manual.* Unpublished handbook.

BRUIJNZEEL, L.A. 1983. Evaluation of runoff sources in a forested basin in a wet monsoonal environment: a combined hydrological and hydrochemical approach. In *Hydrology of Humid Tropical Regions with Particular Reference to the Hydrological Effects of Agriculture and Forestry Practice.* (Proceedings of the Hamburg Symposium, August 1983). IAHS Publ. No. 140.

BRUIJNZEEL, L.A. 1990. *Hydrology of moist tropical forests and effects of conversion: a state of knowledge review.* International Hydrological Programme (UNESCO), Paris. 224 pp.

CARRARA, A., CARDINALI, M., DETTI, R., GUZZETTI, F., PASQUI, V. & REICHENBACH, P. 1991. GIS techniques and statistical models in evaluating landslide hazard. *Earth Surface Processes and Landforms* 16:427-445.

CHANDLER, R.J., 1982. Lias Clay slope sections and their implication for the prediction of limiting or threshold slope angles. *Earth Surface Processes and Landforms* 7:427-438.

CHORLEY, R.J., SCHUMM, S.A. & SUGDEN, D.E. 1985. *Geomorphology.* Methuen, London. 605 pp.

CHOWDHURY, R.N. 1978. *Slope Analysis (Developments in geotechnical engineering; 22).* Elsevier, Amsterdam. 423 pp.

COLLINS, N.M., SAYER, J.A., & WHITMORE, T.C. 1991. *The Conservation Atlas of Tropical Forests: Asia and the Pacific.* Macmillan, London.

CROZIER, M.J. 1986. *Landslides: causes, consequences and environment.* Croom Helm, London. 252 pp.

DAY, M.J. 1979. *Landslides in the Gunong Mulu National Park, Sarawak, E. Malaysia.* Unpublished report to the Royal Geographical Society (Mulu Expedition Library).

DIETRICH, W.E., WILSON, C.J., MONTGOMERY, D.R., McKEAN, J. & BAUER, R. 1992. Erosion thresholds and land surface morphology. *Geology* 20:675-679.

FRANCIS, S.C. 1987. Slope Development Through the Threshold Slope Concept. Pp. 601-624 in Anderson, M.G. & Richards, K.S. (eds). *Slope Stability*. Wiley, Chichester.

FREDLUND, D.G. 1987. Slope Stability Analysis Incorporating the Effect of Soil Suction. Pp. 113-144 in Anderson, M.G. & Richards, K.S. (eds). *Slope Stability*. Wiley, Chichester.

GRANT, C.K. 1984. Geographical Overview. Pp. 11-33 in James, D.M.D. (ed.). *The Geology and Hydrocarbon Resources of Negara Brunei Darussalam*. Muzium Brunei, Bandar Seri Begawan.

GREENWAY, D.R. 1987. Vegetation and Slope Stability. Pp. 187-230 in Anderson, M.G. & Richards, K.S. (eds). *Slope Stability*. Wiley, Chichester.

HILLEL, D. 1971. *Soil and Water: Physical Principles and Processes*. Academic Press, Inc., New York. 288 pp.

JAMES, D.M.D. 1984. Regional Geological Setting. Pp. 34-42 in James, D.M.D. (ed.). *The Geology and Hydrocarbon Resources of Negara Brunei Darussalam*. Muzium Brunei, Bandar Seri Begawan.

KIRKBY, M.J. 1985. Hillslope hydrology. Pp. 37-75 in Anderson, M.G. & Burt, T.P. (eds). *Hydrological Forecasting*. Wiley, Chichester.

MALMER, A. 1992. Water-yield changes after clear-felling tropical rainforest and establishment of forest plantation in Sabah, Malaysia. *Journal of Hydrology* 134:77-94.

MONTGOMERY, D.R. & DIETRICH, W.E. 1989. Source areas, drainage density, and channel initiation. *Water Resources Research* 25:1907-1918.

MOORE, I.D., MACKAY, S.M., WALLBRINK, P.J., BURCH, G.J. & O'LOUGHLIN, E.M. 1986. Hydrological characteristics and modelling of a small forested catchment in southeastern New South Wales. Pre-logging conditions. *Journal of Hydrology* 83:307-335.

MORGAN, R.P.C. 1976. The Role of Climate in the Denudation System: a Case Study from West Malaysia. Pp. 317-343 in Derbyshire, E. (ed.). *Geomorphology and Climate*. Wiley, London.

MORGENSTERN, N.R. & PRICE, V.E. 1965. The analysis of the stability of general slip surfaces. *Geotechnique* 15:79-93.

PEARCE, A.J. & O'LOUGHLIN, C.L. 1985. Slope Stability and Land Use. *Water Resources Monograph 11*. American Geophysical Union.

ROSS, S.M. & DYKES, A.P. (this volume). Soil conditions, erosion and nutrient loss on steep slopes under mixed dipterocarp forest in Brunei Darussalam.

SCHUMM, S.A. 1963. The disparity between present rates of denudation and orogeny. *US Geological Survey Professional Paper 454-H*.

SKEMPTON, A.W. 1964. Long-term stability of clay slopes. *Geotechnique* 14:77-102.

SKEMPTON, A.W. & DELORY, F.A. 1957. Stability of natural slopes in London Clay. Pp. 378-381 in *Proc. 4th Int. Conf. of Soil Mechanics & Foundation Engineering* Vol. 2. London.

STRAHLER, A.M. 1950. Equilibrium theory of erosional slopes approached by frequency distribution analysis. *American Journal of Science* 248:673-696, 800-814.

TROENDLE, C.A. 1985. Variable source area models. Pp. 347-403 in Anderson, M.G. & Burt, T.P. (eds). *Hydrological Forecasting*. Wiley, Chichester.

VIS, M. 1986. Interception, drop size distributions and rainfall kinetic energy in four Colombian forest ecosystems. *Earth Surface Processes and Landforms* 11:591-603.

WESLEY, L.D. 1977. Shear strength properties of halloysite and allophane clays in Java, Indonesia. *Geotechnique* 27:125-136.

WEYMAN, D.R. 1970. Throughflow on hillslopes and its relation to stream hydrographs. *International Association of Scientific Hydrology. XV Annee, No.3, September*:25-33.

WILFORD, G.E. 1961. The Geology and Mineral Resources of Brunei and adjacent parts of Sarawak with descriptions of Seria and Miri Oilfields. *Memoir 10, Geological Survey Department, British Territories in Borneo*. 319 pp.

WILLGOOSE, G., BRAS, R.L. & RODRIGUEZ-ITURBE, I. 1991. Results from a new model of river basin evolution. *Earth Surface Processes and Landforms* 16:237-254.

WILLS, J.T. (this volume). The creation of a Geographical Information System for the Kuala Belalong Field Studies Centre.

YOUNGS, E.G. 1991. Infiltration measurements - a review. *Hydrological Processes* 5:309-320.

The effects of timber harvesting and forest conversion on peat swamp forest dynamics and environment in West Kalimantan

HERUJONO HADISUPARTO

Department of Forestry, Faculty of Agriculture, Tanjungpura University, Pontianak, West Kalimantan, Indonesia

ABSTRACT. The peat swamp forests of Kalimantan, formerly covering 6.3 million hectares, are found primarily along the western and southern parts of Indonesian Borneo and are separated by the Sunda shelf from those of Sumatra's east coast. Studies conducted on production forests in West Kalimantan indicate that the presence of natural regeneration stages favour continued sustainability. Silvicultural practices ensure sustainability when the residual trees left after logging are adequate and there is protection of the resource from unplanned relogging and forest conversion.

KEY WORDS: peat swamp forest, sustainability, silviculture, forest population dynamics.

INTRODUCTION

The remaining tropical rain forests in the world are largely concentrated in three regions, namely America (Amazon basin), Africa (Zaire and Congo) and Indo-Malaya (primarily in South East Asia). Among the three regions, the Indo-Malayan rain forest contains the greatest richness in species (Van Steenis 1971, Whitmore 1985). The Indo-Malayan rain forest region includes different forest formations such as dry-land rain forest, wet-land rain forest and monsoon forest.

Peat swamp forest is a very distinct forest formation of wet-land rain forest, which is generally found on flat terrain, along with fresh water swamp forest and mangrove forest. According to Jacobs (1988) peat swamp forest is a successional stage of fresh water swamp forest which is characterised by having soils with a low pH, lack of minerals, and very slow natural decomposition of the litter.

In Southeast Asia, extensive peat swamp forests are found in western and southern parts of Borneo, primarily in West Kalimantan (West Indonesian Borneo) in the down-stream area of large drainage basins. Such forests are also found in eastern Sumatra and Malaya. Anderson (1963) indicated that peat swamp forests in Sarawak and Brunei occupy nearly the entire coastline and the largest, Maludam Peninsula peat swamp, stretches inland up to 64 km. Two of the most important timber species in these forests are *Shorea albida*, a Bornean endemic with a range extending from north-west Kalimantan to Brunei and *Gonystylus bancanus*, one of the most valuable endemic timber species of West Kalimantan. This latter species is endangered as a result of logging and forest conversion.

This paper describes the effects of timber harvesting and conversion on peat swamp forest dynamics and environment with special reference to West Kalimantan. Field data included here was obtained from ecological studies conducted primarily by the staff and students of the Department of Forestry - Faculty of Agriculture at the Tanjungpura University in Pontianak.

D. S. Edwards et al. (eds.), Tropical Rainforest Research – Current Issues, 411–415.

EFFECTS OF LOGGING ON PEAT SWAMP FOREST DYNAMICS

Total forest land area in West Kalimantan according to the Recognised Forest Land Use Survey (Menteri Pertanian 1982) comprises 9,204,425 hectares (ha) or 62.5% of the total provincial area. It consists of protection forest (2,047,125 ha) parks and forest reserve (1,339,880 ha) and production forest in all types of formation totalling 5,817,420 ha. This includes an assigned convertible production forest area of 1,505,670 ha. Peat swamp forest area in the province makes up 1,878,200 ha including a total of 857,460 ha which has been designated as protected (Tim RSTRP-Kalbar 1991).

Timber harvesting of tropical rainforest in Indonesia following Foreign Investment Law was probably initiated in the peat swamp forest. In West Kalimantan, semi-mechanical logging using small railroads was introduced by several foreign loggers coming from Sarawak (Malaysia) in the late sixties. The timber trade boom of the early eighties has caused significant changes in forest structure and composition. The application of the Indonesian selective Cutting System (TPI/TPTI) has not been successfully practiced. In order to attain forest sustainability damage to residual growing trees and regeneration stages should be avoided by preventing unplanned re-logging and uncontrolled forest conversion.

Ecological studies on peat swamp forests were conducted in several selected locations in West Kalimantan. Study locations included: Mandor nature reserve (unlogged forest); logged-over forest of Sukadana; logged-over forest of Air Hitam near Pontianak; production forest of Nanga Tayap (Ketapang). These four sites represent different facets with respect to forest population dynamics, as summarized in the following paragraphs.

Unlogged peat swamp forest (Mandor Nature Reserve)

This peat swamp forest is located on the fringe of kerangas (heath) forest within the nature reserve. The peat swamp forest was found to be more diverse than the kerangas forest (Table 1). These two adjacent habitats had a relatively high species similarity of poles (SS index = 0.64); on the other hand both habitats showed a lower species similarity of trees (SS index = 0.30). This indicates that disturbance had occurred in the nature reserve and that the vegetation is still in the process of succession towards a climax forest (Hadisuparto & Syafruddin 1990). There were 86 species of poles and trees in the peat swamp forest, whereas in the kerangas 75 species were found. species with the highest importance value indexes in the peat swamp forest were *Dryobalanops beccarii*, *Xylopia malayana*, *Calopyllum* sp. and *Gonystylus bancanus*, whereas in the kerangas forest the highest values were for *Tristania* sp., *Dryobalanops beccarii*, *Shorea pachyphylla* and *Calophyllum* sp. Forest gaps, in terms of canopy openings, was found to be 10.5% lower in peat swamp forest compared to Kerangas. The natural forest community of peat swamp forest in Mandor nature reserve consisted of 14.7% seedlings, 54.7% saplings, 20.0% poles, and 10.7% trees. This suggests that a seasonal natural regeneration has occurred which may result in a normal forest stand in the future.

Table 1. Species diversity and similarity indexes between peat swamp and kerangas forests at Mandor Nature Reserve.

	Diversity Index		Similarity Index	
Forest type	Pole	Tree	Pole	Tree
Kerangas	1.33	0.67	0.64	0.30
Peat swamp	1.42	0.73		

Table 2. Species number and similarity indexes between unlogged and logged-over forest of Sukadana. From Hadisuparto *et al.* 1990.

Regeneration stage	Number of species		Similarity Index (%)
	Unlogged	Logged-over	
Seedling	61	40	49.9
Sapling	59	52	46.8
Pole	53	35	33.7
Tree	73	32	31.3

Logged-over peat swamp forest of Sukadana

This forest is growing on an alluvial plain and overlies white sand deposits (Leighton, 1990). It is located between a low-land mixed dipterocarp forest and brackish water swamp and mangrove forests. This peat swamp forest was designated as production forest and was logged more than 10 years ago. The study compares the residual growing stand with the unlogged forest in the area.

In the unlogged forest seedlings and saplings were dominated by *Diospyros* sp. while trees were dominated by *Dipterocarpus* sp., whereas in the logged-over forest *Gluta renghas*, *Diospyros* sp. and *Litsea* sp. were dominant. Ramin (*Gonystylus bancanus*) was not found at any stage of its growth in the logged-over forest. However, the species similarity indexes between unlogged and logged-over forests indicate that naturally regenerating seedling and saplings of many species grew commonly in both unlogged and logged-over forest. In other words, the forest stand could be restored gradually if no further disturbance or interference of the successional stages occurred.

Logged-over forest of Air Hitam

The study conducted here included the analysis of vegetation following timber harvesting. Sample plots were located in 5-year and 10-year logged-over forests. The results indicate that 5 and 10 years after logging, recovery is occurring, although some changes in forest structure and composition might have occurred. Table 3 suggests that the residual stand was sufficient to provide enough stock of each regeneration stage from seedling to tree. Dominant species present in both logged-over forests (with the highest importance value indexes) were: *Gonystylus bancanus*, *Dryobalanops* sp., *Litsea* sp., *Gluta renghas* and *Cratoxylon arborescens*. Non-dipterocarp species dominate the forest community in this peat swamp formation.

Nanga Tayap production forest

The effects of timber harvesting in this peat swamp forest formation was studied by means of vegetation analyses before and after logging. This 'Ramin Forest' was assessed through

Table 3. Forest dynamics on logged-over peat swamp forest of Air Hitam.

Regeneration stage	5-year logged- over		10-year logged-over	
	No. species	Stems/ha	No. species	Stems/ha
Seedling	24	28375	26	38938
Sapling	24	2790	23	3600
Pole	24	830	26	915
Tree	25	182	25	188

Table 4. Population dynamics of Ramin (*Gonystylus bancanus*) on the peat swamp production forest of Nanga Tayap.

Regeneration stage	Stems/ha Before logging	Stems/ha After logging
Seedling	6450	5600
Sapling	656	488
Pole	26	18

consideration of its dominant species (Table 4). Ramin (*Gonystylus bancanus*) is the most valuable timber in the peat swamp forest which may have been seriously affected by logging practices in West Kalimantan. Natural regeneration of this species has been considered more successful than artificial planting. There are still adequate seedlings and saplings in the forest following the logging, although poles are relatively rare, indicating heavy damage during logging. Silvicultural practices aimed at maintaining natural regeneration are needed to ensure future standing stock of this valuable species.

EFFECTS OF FOREST CONVERSION ON ENVIRONMENT

Peat swamp forest in Kalimantan (Indonesian Borneo) formerly covered 6.3 million ha. Like fresh water swamp forest, the peat swamp forest has been rapidly cleared for agriculture and other land uses. The acceleration of peat swamp forest conversion in Kalimantan is primarily linked to the geographical distribution of the forests which are mostly along rivers and on coastal plains, hence they are more readily exploited for timber. Increased transportation facilities in such areas has promoted settlement and other land uses and has exacerbated the situation.

In West Kalimantan, the peat swamp forest comprises 1,878,200 ha or about 12.8% of the province. The forest extends from the north to south to follow the coast behind the brackish waters. The ecological role of peat lands, especially in conserving water and controlling flooding in down-stream areas, was given due attention and through a Presidential Decree, peat swamp forest with a peat depth of 3 or more meters was assigned as protected areas. In accordance with the Presidential Decree, 45.7 % of the peat land area in West Kalimantan was designated as protected forest, which can only be cultivated with a special release by the government. However, assignment to protected status is not easy and is not instantly applicable because it requires a comprehensive study on peat depth all over the forests.

The remaining peat swamp forests have been continuously threatened by local migrants who wrongly consider that peat lands in their forested state are less useful (economically) than cleared land. After the forests are cleared however, it is found that the peat cannot be cultivated directly and as a result, people would rather continue exploiting the residual trees of the remaining forests. A very degraded forest is finally cleared by burning (in the dry season) and drainage channels made to allow cultivation. Deeper peats have recently become a commercial target as energy sources and such use will eventually destroy the environment completely if there is no integrated central planning followed by proper implementation.

Population growth and the need for land in Indonesia, both for agriculture and settlement, in some parts through transmigration projects, has to be carefully planned. Conversion of peat swamp forests should be based on firm ecological principles. Peat swamp forest is known to be a renewable resource but once it is cleared, the peat will eventually degrade and the site cannot return to the original state. It becomes non-renewable as a natural forest habitat.

CONCLUSION

The peat swamp forest of West Kalimantan is one of the largest natural ecosystems in the Indonesian rainforest. The forests extend along the coastal plain behind the brackish water and along the rivers inland within the upstream section of Kapuas river basin.

The effects of timber harvesting in this distinctive forest formation have changed significantly the forest structure and composition. However, studies show that forest sustainability may be attained if sufficient residual growing trees are left standing and if enough natural regeneration occurs, provided that unplanned re-logging and uncontrolled conversion of the forest can be avoided. Ramin (*Gonystylus bancanus*), the dominant species of the peat swamp forest in this area, has possibly become an endangered species, with natural regeneration considered more successful than artificial planting for this species. The remaining peat swamp forests have been threatened by increasing population pressure, mainly by migrants.

Forest conversion has been accelerated since most people consider that peat lands in their forested state are less useful than if they are cleared. Conversion of peat swamp forest to other uses should be based on ecological as well as economic principles.

ACKNOWLEDGEMENTS

I would like acknowledge the work of our staff and students for the preparation and field data collection; in particular:- Mr. Abdurrani Muin, Mr. Syafruddin Said, Ms. Augustin Lumangkun, Mr. Dirhamsyah, Mr. Kambodja, Ms. Ina Kartini and Mr. Agus Purnomo.

LITERATURE CITED

ANDERSON, J.A.R. 1963. The flora of peat swamp forests of Sarawak and Brunei including a Catalogue of all recorded species of flowering plants, ferns and fern allies. *Gardens' Bulletin, Singapore* 20:131-228.

HADISUPARTO, H. & SYAFRUDDIN S. 1990. *Species diversity of two adjacent minor habitats in Mandor nature reserve of West Kalimantan*. Paper presented at the International Conference on Forest Biology and Conservation in Borneo. Kota Kinabalu, Sabah, Malaysia. July 30 - August 3, 1990.

HADISUPARTO, H., ABDURRANI M., LUMANAGKUN, A. & DIRHAMSYAH. 1990. *Evaluasi beberapa plasma nutfah pada hutan produksi di Kalimantan Barat*. (Evaluation of several genetic resources on selected production forests in West Kalimantan). Kerjasama Kantor Menteri Negara KLH & PSL Univ Tanjungpura, Jakarta.

JACOBS, M. 1988. *The tropical rain forest*. Springer Verlag, London. 295 pp.

LEIGHTON, M. 1990. *Ecological interactions between plants and vertebrates: Comparison among rainforest habitats*. A report on several research topics of Gunung Palung Projects.

MENTERI PERTANIAN. 1982. *Penunjukan areal hutan di Wilayah Propinsi dati I Kalimantan Barat sebagai kawasan hutan (TGHK)*. (Assigned forest areas for forest land use in West Kalimantan Province). SK. No. 757/Kpts/Um/10/1982, Jakarta.

STEENIS, C.G.G.J. van. 1971. Plant conservation in Malaysia. *Bulletin du Jardin Botanique National de Belgique*. 41:189-202.

TIM RSTRP-KALBAR. 1991. *Rencana Struktur Tata Ruang Propinsi Kalimantan Barat Tahun 2005*. (Land use planning of West Kalimantan province towards 2005). Pemda Tkt. I. Kal. Barat.

WHITMORE, T.C. 1985. *Tropical rain forests of the Far East*. 2nd ed. Clarendon Press, Oxford. 352 pp.

Socio-economic values and tropical rain forest conservation

PETER EATON

Universiti Brunei Darussalam, Bandar Seri Begawan 2028, Brunei Darussalam

ABSTRACT. Many threats to tropical forests and their biodiversity result from social and economic factors. Policies to conserve the forests depend on being able to demonstrate their economic value and on social measures involving local community participation. Research to calculate the total economic value of the forests has involved estimating the direct use value of forest products, the indirect values from environmental benefits and services, and option and existence values. Threats to the forest have endangered the social and economic welfare of many indigenous groups; their survival and the conservation of the forests can often be achieved by reinforcing traditional systems of land tenure and environmental management.

KEY WORDS: Borneo, conservation, national parks, New Guinea, tropical forest, valuation.

INTRODUCTION

Global concern over issues such as the depletion of tropical rain forests and the preservation of their biodiversity formed a major part of the agenda of the United Nations Conference on Environment and Development held in Brazil in 1992. Products of the conference resulted in a Convention on Biological Diversity and a statement of principles for the management of forests. These and other significant documents such as Agenda 21 and the Declaration on Environment and Development have emphasised the need for the sustainable development, conservation and rational uses of these forests. Whether the good intentions, rhetoric and compromises of the conference will provide the foundations for effective policies and actions still remains to be seen, but one point of general agreement in the discussions there and elsewhere is that most of the problems associated with the tropical rain forests are social or economic in their origins. If, for example, one analyses the reasons for deforestation in Southeast Asia, immediate causes which are usually identified are commercial logging, shifting cultivation and conversion of forest to other types of land use. However, these in turn are largely a result of other problems such as excessive population growth, unequal patterns and terms of trade between industrial and developing countries, and the financial operations of political and economic elites. In addition, there are often inappropriate systems of land tenure which may both fail to protect the customary rights of indigenous forest dwellers and also force the landless and hungry into clearing forest on marginal land basically unsuited to long-term agriculture.

In the analysis of such problems and the formulation of policies to solve them, the social sciences necessarily have a key role to play. However, it should be emphasised that in issues of tropical rain forest management and conservation there is no place for "intellectual marking of territory" (Machlis 1992, p.2) by any one discipline, but instead there is a need for cooperation and sharing of knowledge, not only between social and natural scientists, but also involving foresters, protected area managers and the indigenous people who are directly dependent on the resources of the forest.

Policies to conserve tropical rain forests depend on being able to demonstrate their economic value to governments, their communities and rural populations. In the long term, to be successful they also depend on the support and participation of these local communities.

D. S. Edwards et al. (eds.), Tropical Rainforest Research – Current Issues, 417–423.

In this paper I shall discuss some of the methods used by economists in attempting to measure economic values of the tropical rain forests, and then go on to consider measures for local involvement in their conservation.

THE TOTAL ECONOMIC VALUE OF TROPICAL RAIN FORESTS

Attempts to assess the total worth of tropical forests have generally been based on both marketable and non-marketable values. The most comprehensive calculations include direct and indirect use values, and in addition what are termed option, bequest and existence values. Direct uses involve the consumption of forest products, including use for recreation, research and education. Indirect uses include vital environmental functions such as watershed protection and nutrient cycling. In addition, the assessment of total economic value should encompass option values, the amount individuals and society are willing to pay for future uses of the forests, and bequest values which relate to the amount the present generation is willing to pay to preserve use for their descendants and future generations. Finally, allowance should be made for existence values that represent how much individuals (or collectively society) are prepared to pay for an area to be preserved, even though they are not going to use it themselves.

Methods of calculating these different environmental values vary and some are necessarily more subjective and imprecise than others. In the case of commercial timber and forest products it is possible to calculate a market value. For subsistence production and consumption it may be possible to estimate substitute values from these products or similar goods bought commercially. Changes in productivity, the cost of defensive expenditure, replacement costs, and the use of shadow projects that provide substitute environmental services, can all be used as means of evaluating environmental changes and resource values. Surrogate markets can be used to estimate the value people place on forest conservation areas; entry fees, foregone earnings and travel costs providing an indication of how much people are prepared to pay to visit these areas. Option, bequest and existence values are more difficult to reliably assess and often rely on contingent valuation techniques by which people indicate how much they are willing to pay to prevent degradation to an environmental asset.

Examples of the economic valuation of environmental assets provide persuasive arguments for conservation. In the case of the Amboseli National Park of Kenya it as been estimated that each lion was worth US$27,000 a year in tourist earnings, and that the earnings of the park averaged about US$40 per hectare, far more than could be produced from agriculture (Western 1984). For a tropical rain forest reserve in Costa Rica research using travel cost

Table 1. Classification of the total economic value of tropical forests. (Sources: Munasinghe 1992, Pearce 1990).

DIRECT USE	INDIRECT USE	OPTION VALUE	EXISTENCE VALUE
Sustainable timber products	Nutrient cycling	Future Uses	Value from knowledge of continued existence
Non-timber products	Watershed protection	Biodiversity	Heritage
Medicine	Air pollution reduction	Conserved habitats	Endangered species
Plant genetics	Micro-climatic functions		
Recreation	Carbon store		
Education			
Research			

Table 2. Valuation Methods Relevant to Tropical Rain Forests.

Market value of products and services
Market value of surrogate products
Changes in productivity
Defensive expenditure
Replacement costs
Shadow projects
Travel costs
Contingent valuation - willingness to pay

Table 3. Financial returns from different forest uses, US$/ha (Source: Peters *et al.* 1989)

Non.timber harvesting - fruit and latex	- 6330
Clear-cutting timber harvesting	- 1001
Plantation timber harvesting and pulpwood	- 3184
Cattle ranching	- 2960

methodology suggested a value for the reserve of US$1,250 per hectare (Tobias & Mendelsohn 1991). The change-in-productivity technique was used in a study of Amazonian forest in Peru (Peters *et al.* 1989); the results of this study can be seen in Table 3 and are interesting in that they show that the financial returns from harvesting non-timber products, mainly wild fruit and latex, are greater than alternative land and resource uses such as clear felling, forest plantations, and conversion to cattle ranching.

Valuation methods have direct relevance to benefit-cost analysis of conservation and development projects. A study was made of the Khao Yai national park in Thailand (Dixon & Sherman 1990) that calculated benefits from tourism, wildlife preservation, and watershed protection against management expenses and foregone opportunity costs from other types of land use such as logging and agriculture. The conclusion of this study was that although the park was 'socially beneficial', government intervention was needed for resource protection as the directly appropriable benefits from exploitation would outweigh the diffused benefits and returns from protection if the area was left to private ownership and market forces.

Here, in Brunei Darussalam, a national park is being developed in the Ulu Temburong area where the Kuala Belalong Field Studies Centre is located. It is possible to assess the potential benefits from this park under a number of different headings which include biodiversity and species conservation, watershed protection, research and education, recreation and tourism, the use of forest products on a sustainable basis, and the provision of local employment and income.

Not all the types of resource use listed in Table 4 are compatible, and this poses problems for park planning and management. This can be partly solved by a system of zoning which restricts certain uses to particular areas, separating for example strictly conservation and research uses from recreation and tourism. A zoning plan which has been proposed for the park provides for an intensive use administrative area near the confluence of the Temburong and Belalong rivers, a medium intensity use recreation area for outdoor activities along the Temburong river, a low intensity wilderness zone, a scientific research and education area in the catchment area of the Belalong river, and security and restricted zones near the international boundaries.

Brunei is fortunate in that it has been able to conserve a large proportion of its forests, and at present there are few economic pressures to exploit its timber resources. Elsewhere in Southeast Asia the arguments for commercial logging have often proved more persuasive than those for conservation areas. The economically quantifiable benefits listed in support

Table 4. Potential benefits from the Ulu Temburong National Park.

1.	Preservation of biodiversity, species and ecosystems.
2.	Preservation of ecological processes.
3.	Watershed protection.
4.	Scientific research.
5.	Recreation and tourism.
6.	Environmental education.
7.	Local employment and income.
8.	Regional/district development.
9.	Sustained local use of forest products.
10.	Aesthetic, spiritual and cultural.

of logging have included the timber industry's contributions to the gross domestic product, foreign earnings and government revenue. If, for example, we look at Brunei's neighbouring East Malaysian states, we find that during the past twenty years royalties and taxes from the timber industry in Sarawak and Sabah have formed a large proportion of these states' government revenues. It can, however, be argued that much of the income from the exploitation of timber resources has not been recovered by the states, and has not been adequately redistributed among the rural indigenous populations, whose forest resources have often been destroyed or degraded. There is, for instance, considerable illegal logging in both states which does not come into the national accounts or contribute to revenue. Even using the official statistics, the proportion of the profits or resource rent which goes to the state is much lower than we have been led to believe. One study of resource rent recovery (Vincent 1990) gave a figure of only 18.4 per cent for Sarawak and 46.2 per cent for Sabah. This and other analyses (Repetto & Gillis 1988) would seem to indicate that much of the economic benefit from logging in Southeast Asian countries has been enjoyed by a very small section of the population rather than spread among the community as a whole.

Conventional national accounting, the compilation of national income and gross domestic product, has not in the past made any allowance for the destruction or depletion of natural resources, or for the defensive expenditures necessary as a result of environmental degradation. These have been pioneered by the governments of France, Norway and Canada (Pearce *et al.* 1989), but have also been applied to developing countries and the loss of tropical forests. Estimates of the depletion of forest, soil and petroleum resources between 1971 and 1984 have shown that, although the gross domestic product was calculated to have grown at an average annual rate of 7.1 per cent during this period, if allowance had been made for the loss of natural resources then the average annual growth rate would have been only 4 per cent (Repetto *et al.* 1989). This type of national accounting which measures losses of natural capital is necessary if governments are to plan for the sustainable use of tropical rain forests.

A form of direct use of the forests which is frequently underestimated by economists is that of non-timber forest products. In Borneo these have always been of great significance to the indigenous people, both as a vital part of the subsistence economy and as the basis for early trade. The great range of plant and animal products has been an indication of both the rich biodiversity of the forests and the traditional knowledge and skills which have been applied to their utilisation.

The variety of forest products was recorded in many of the early descriptions of Borneo and its inhabitants (Hose & McDougall 1912, Low 1948). The role of the forest in the subsistence economy has also featured in the work of more recent anthropologists, for example the studies of the Iban by Freeman (1955) and Sandin (1967).

Ethnobotanical studies have spanned the boundaries between the natural and social sciences, showing that rural communities still use a large number of forest products for food, handicrafts, construction, medicines and ceremonies. A survey of four Iban longhouses in the Pantu area of Sarawak (Pearce *et al.* 1987) identified 142 plant species as being collected for food and other purposes. Research into traditional medicines has provided further evidence of the utilisation and potential value of the forests. A study of three Kenyah villages on the Apoh Kayan plateau of Kalimantan found 213 different plant species that the local people used for medical purposes (Leaman *et al.* 1992) . Another study of one village in Central Kalimantan identified 107 medicinal plants as being in general use (Riswan *et al.* 1992). In the Gunung Mulu area of Sarawak 38 species of plants were identified as being used for medicinal purposes by the Penans (Kedit 1982). The tropical rain forests also serve as the habitat for a variety of wildlife species, some of which are important as a source of food and significant to the local economy. Their value is again difficult to quantify, but one study in Sarawak (Caldecott 1986) has estimated the total worth of the wildlife harvest as being between M$83 million and M$115 million each year.

LOCAL INVOLVEMENT IN TROPICAL RAIN FOREST CONSERVATION

The importance of non-timber products to rural indigenous communities living in or near the forests provides them with an important motive to attempt to conserve forest resources, and also to maintain and protect their rights to use them. Experience has shown that efforts to preserve the forest by establishing national parks, reserves or other types of protected area are unlikely to gain support if these rights are ignored and excluded. Conservation often proves more successful if local people are involved, if they feel it is to their advantage and that their traditional rights are being protected. It has also been found that traditional forms of land tenure and resource management may prove more effective in protecting the environment than introduced laws and institutions. The World Commission on Environment and Development recognised these principles when it advised: "The recognition of traditional rights must go hand in hand with measures to protect the local institutions that enforce responsibility in resource use" (WCED 1987, pp.115-6).

In Southeast Asia local involvement and the protection of traditional rights have not always been a feature of conservation policy. In Thailand and Indonesia the rapid expansion of national parks and reserves has often resulted in conflict with resident populations and competing land use demands. The Gunung Leuser Park in Sumatra is an example of continuous encroachment through illegal logging and farming (Wells *et al.* 1990). The forests of the Khao Yai Park were reduced by 10 per cent during the first twenty-five years of the park's existence (Dixon & Sherman 1991, p.100); further north in Thailand it has been estimated that 43 per cent of the Doi Inthanon Park has been affected by clearing for agriculture and fuelwood collecting (Dearden *et al.* 1992).

More success has been achieved where parks have been shown to be economically beneficial to local people, especially through providing employment and income from tourism. This seems to have been a factor in creating support among Kadazan villagers for the Kinabalu Park in Sabah (Jacobson 1991), and has helped to create more favourable local attitudes to the Khao Yai Park (McNeely 1988, pp.153-5). The creation of linkages between protected areas and local economies has often helped to promote both rural development and conservation, as happened in the case of the Dumoga-Bone Park in Sulawesi when it was perceived that the protection of the forests was ensuring a continuous supply of water for irrigation projects in neighbouring areas (Wells *et al.* 1990).

Although certain traditional rights to hunt and collect forest products may be allowed in parks, there has generally been a reluctance in Southeast Asia to encourage conservation based on customary tenure. Where indigenous groups have attempted to protect forest resources this has often led to conflict with the government, as in the case of the Penans, Kayans and Kelabits who blockaded the logging roads in Sarawak (Sahabat Alam Malaysia 1989, Hong 1987). In order to find a network of protected areas based on traditional tenure, it is necessary to examine conservation in Papua New Guinea, a country on the periphery of the Southeast Asia region.

In Papua New Guinea 97 per cent of all land is held under customary systems of tenure, under which rights to use land and forest resources depend on membership of the kinship group or clan. Traditional rules and communal controls have in the past helped to protect resources by excluding outsiders, imposing seasonal bans to prevent overhunting, and protecting certain species and areas by taboos (Morauta *et al.* 1982). Over three-quarters of the country is still covered by forest, and its products and wildlife remain important to the subsistence economy. Social and economic changes have, however, placed increased pressure on these resources as have technological innovations, such as the replacement of the stone axe by the chainsaw, and the bow and arrow by the shotgun.

National parks on state-owned land have not proved very successful in Papua New Guinea. Their development and expansion have often been hindered by land disputes and compensation claims (Eaton 1982, 1985, Gare 1986). Only a relatively small area, 127 square kilometres, has been reserved for parks, and even these are subject to encroachment. Customary tenure has proved a constraint to the national parks system, but it has provided the basis for other types of protected area, in particular those known as wildlife management areas. These are established by rural communities on their own customary land; village representatives form management committees, designate the boundaries, and draw up regulations designed to solve local problems of wildlife depletion. These rules vary according to the needs of the particular area, but generally include prohibitions on hunting with firearms, and destruction of the forest. Complete or seasonal protection is provided for certain valued but endangered wildlife species; hunting by outsiders is either not allowed, or subject to licences, royalties and quotas. Wildlife management areas have varied in their effectiveness, but have proved popular with the customary landowners who see them as a means of formalising their traditional rights and ensuring that they retain the power to retain the power to manage their own local resources. During the period from 1975 to 1992, eighteen wildlife management areas were established, plus five other conservation areas on customary land. Their total area is 10,529 square kilometres, over sixty times that protected under the country's national parks legislation.

Similar community participation has proved effective in the western or Indonesian part of New Guinea (Craven *et al.* 1992), but has not yet been a feature of conservation programmes in other parts of Southeast Asia, where there often remains a need to reinforce rather than destroy local systems of resource management. If we wish to conserve biological diversity, we may find that one of the most effective means is by supporting local institutions that also preserve cultural diversity.

LITERATURE CITED

CALDECOTT, J. O. 1986. *Hunting and Wildlife Management in Malaysia.* World Wildlife Fund of Malaysia, Kuala Lumpur.

CRAVEN, I., PURBA, P., SETIONO, D., WOMISWOR & D. & THAMRIN, S. 1992. *Community Involvement in Protected Area Establishment and Management in Irian Jaya.* Paper presented at IVth World Congress of National Parks and Protected Areas, Caracas.

DIXON, K.A. & SHERMAN, P.B. 1990. *Economics of Protected Areas*. Earthscan, London.

EATON, P. 1982. Land Tenure and National Parks. Pp. 223-226 in Morauta, L., Pernetta, J. & Heaney, W. (eds). *Traditional Conservation in Papua New Guinea*. Institute of Applied Social and Economic Research, Port Moresby.

EATON, P. 1985. *Land Tenure and Conservation: Protected Areas in the South Pacific*. South Pacific Commission, Noumea.

EATON, P. 1986. *Grassroots Conservation: Wildlife Management Areas in Papua New Guinea*. University of Papua New Guinea, Port Moresby.

FREEMAN, J. *Report on the Iban of Sarawak*. Government Printing Office, Kuching.

GARE, N. 1986. The Making of a National Park. *Parks* 11:13-18.

HONG, E. 1987. *Natives of Sarawak*. Institut Masyarakat, Penang.

HOSE, C. & MCDOUGALL, W. 1912. *The Pagan Tribes of Borneo*. Frank Cass & Co., London.

JACOBSON, S.K. 1991. Resident Attitudes about a National Park in Sabah, Malaysia. Pp.250-262 in West, P.C. & Brechin, S.R. (eds). *Resident Peoples and National Parks*. University of Arizona Press, Tucson.

KEDIT, P. 1982. An Ecological Study of the Penan. *Sarawak Museum Journal* XXX, 51:225-279.

LEAMAN, D.J., YUSOF, R. & ARNASON, J.T. 1992. Kenyah Medicinal Plants: Beyond the Inventory. Pp. 456-459 in Ismail, G., Mohamed, M. & Omar, S. (eds). *Forest Biology and Conservation in Borneo*. Yayasan Sabah, Kota Kinabalu.

LOW, H. 1848. *Sarawak; its inhabitants and productions*. Richard Bentley, London.

MCNEELY, J.A. 1988. *Economics and Biological Diversity*. International Union for the Conservation of Nature and Natural Resources, Gland.

MORAUTA, L., PERNETTA, J. & HEANEY, W. 1982. *Traditional Conservation in Papua New Guinea*. Institute of Applied Social and Economic Research, Port Moresby.

MUNASINGHE, M. 1992. Biodiversity Protection Policy: Environmental Valuation and Distribution Issues. *Ambio* 12:3.

PEARCE, D. 1990. *An Economic Approach to saving the Tropical Forests*. London Environmental Economics Centre, London.

PEARCE, D., MAKANDYA, A. & BARBIER, E.B. 1989. *Blueprint for a Green Economy*. Earthscan, London.

PEARCE, K., AMAN, V. & JOK, S. 1987. An Ethnobotanical Study of the Iban Community of the Pantu Sub-district, Sri Aman, Division 2, Sarawak. *Sarawak Museum Journal* XXXVII, 58:223-270.

PETERS, C., GENTRY, A. & MENDELSOHN, R. 1989. Valuation of an Amazonian Rainforest. *Nature* 339:665-6, 227-236.

REPETTO, R., MAGRATH, W., WELLS, M., BEER, C. & ROSSINI, F. 1989. *Wasting assets: natural resources in the national income accounting*. World Resources Institute, Washington.

REPETTO, R. & GILLIS, M. 1988. *Public Policies and the Misuse of Forest Resources*. Cambridge University Press, Cambridge.

RISWAN, R., MAHYAR, U.W., SANGAT-ROEMANTYO, H. & BURLEY, J.S. 1992. Ethnobotany of several medicinal plants in Harowu village, Central Kalimantan. Pp. 441-455 in Ismail, G., Mohamed M. & Omar, S. (eds). *Forest Biology and Conservation in Borneo*. Yayasan Sabah, Kota Kinabalu.

SAHABAT ALAM MALAYSIA. 1989. *The Battle for Sarawak's Forests*. Penang.

SANDIN, B. 1967. *The Sea Dayaks of Borneo before the White Rajah Rule*. Macmillan, London.

TOBIAS, D. & MENDELSOHN, R. 1991. Valuing Ecotourism in a Tropical Rainforest Reserve. *Ambio* 20:91-93.

VINCENT, J.R. 1990. Rent Capture and the Feasibility of Tropical Forest Management. *Land Economics* 66:212-223.

WELLS, M., BRANDON, K. & HANNAH, L. 1990. *People and Parks: Linking Protected Area Management with Local Communities*. World Bank, Washington.

WESTERN, D. 1984. Amboseli National Park: Human Values and the Conservation of a Savanna Ecosystem. Pp. 93-100 in McNeely, J. & Miller, K. (eds). *National Parks: Conservation and Development*. Smithsonian Institution, Washington.

WORLD COMMISSION ON ENVIRONMENT AND DEVELOPMENT (WCED). 1987. *Our Common Future*. Oxford University Press, Oxford.

Impact of rainforest destruction - the Siberut Island case, Sumatra, Indonesia

F. GOELTENBOTH *and* K.H.TIMOTIUS

Universitas Kristen Satya Wacana, Fakultas Sains dan Matematika, Jl. Diponegoro 52-60, Salatiga 50711, Indonesia

ABSTRACT. Indonesia is covered with about 143,969,000 ha of tropical forest, stretching from Aceh in Northern Sumatra to Irian Jaya on the island of New Guinea. About 13,000 islands, 6,000 inhabited, form the archipelago with its very distinct biogeographical regions. One group of islands in the Wallacea region, the four Mentawai Islands off the shore of Western Sumatra have a very unique ecosystem with an array of endemic species of flora and fauna rarely found anywhere else. This is due to the isolation of these islands from mainland Sumatra and the rest of the Sunda shelf for the past 500,000 years, and the relative stability of the islands ecosystem until recent days. Over 65% of the animal species and 15% of the plant species are thought to be endemic. The major island, Siberut, has an area of 448,000 ha, 90% of which is covered by Lowland Dipterocarp Rainforest, 3% by Riverine Forest, and 5% by Sago Palm Swamp Forest. The high endemism rate and the unique culture of the indigenous population of about 24,000 people (18,000 still living their traditional lives with a sedentary form of permanent garden agriculture) were reasons for the incorporation of Siberut Island into the Man and Biosphere Program of the United Nations in 1981. Since then a continuous struggle has been going on to find the optimal way to reconcile biosphere necessities and development. Main impacts come from logging companies who hold 256,000 ha of concession area; plans for the establishment of palm oil plantations; transmigration plans for about 10,000 families from Sumatra and Java; government resettlement schemes for the indigenous population ; and steadily increasing tourism.

KEY WORDS: Mentawai Islands, Wallacea, endemic flora and fauna, conservation, exploitation.

INTRODUCTION

The island of Siberut, off the shore of Western Sumatra is the largest of the four Mentawai Islands (Figure 1). It is well recognized that the island belongs to one of the most outstanding environments of the globe (Ave *et al.* 1990, McNeely 1978, Mitchell 1982, Shefold 1980, Whitten *et al.* 1979, Whitten *et al.* 1980, Whitten *et al.* 1981).

About 65% of the recorded animals (Tables 1 & 2) and 15% of the recorded plant species are endemic. Since the mid-Pleistocene, Siberut has been essentially an oceanic island, and its fauna and flora has evolved in isolation from the dynamic evolutionary events on the Sunda Shelf. The evolutionary processes on this island have led to a dynamic equilibrium resulting in a very high number of endemics. Long isolation has allowed the survival of relicts of an early Indo-Malayan fauna and flora and these forms are of great importance to studies of the evolution of the modern fauna and flora of South East Asia (Mitchell 1982).

Since 1981 a part of Siberut Island has been identified as a National Biosphere Reserve under the Man and Biosphere Program of UNESCO. Despite this, the island is under increasing pressure due to various logging operations and other ongoing development processes.

General Facts

Siberut lies within 0^0 55' to 2^0 00' South and 98^0 31' to 99^0 30' East. It is the largest of the four Mentawai Islands, being 112 km in length and 40 km in width, and with an overall

D. S. Edwards et al. (eds.), Tropical Rainforest Research – Current Issues, 425–433.

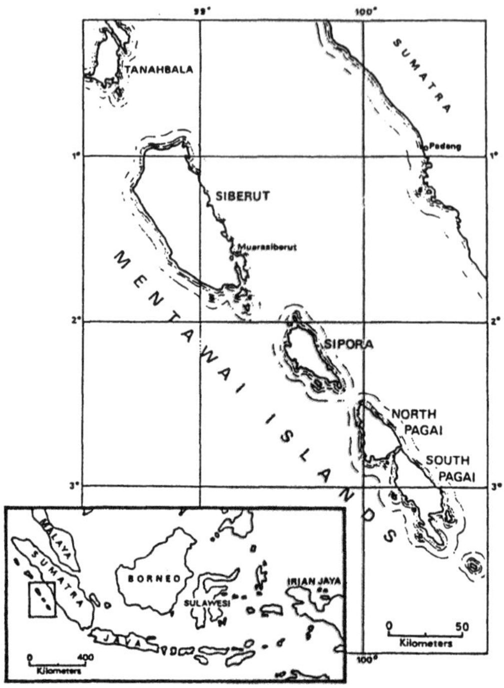

Figure 1. Location map of Siberut Island (after Whitten *et al.* 1980).

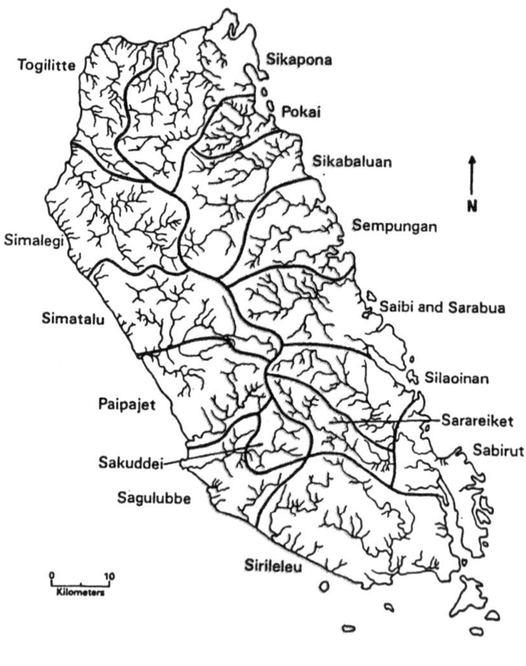

Figure 2. Drainage map of Siberut Island with linguistic regions occupying mostly an entire river catchment basin (after Whitten *et al.* 1980).

Table 1. Endemism of Mentawai mammals (after Whitten *et al.* 1980).

	Level of endemism			
Species	Genus	Species	Subspecies	None
Insectivores				
Tupaia glis siberu			+	
Ptilocercus lowii continentis				+
Bats				
Pteropus hypomelanus enganus				+
Pteropus vampyrus malaccensis				+
Cynopterus shinx angulatus				+
Macroglossus sobrinus fraternus			+	
Emballonura monticola				+
Hipposideros galeritus				+
Myotis muricela abbotti				+
Kerivoula harwickei engana				+
Megaderma spasma ssp.			?	
Pangolins				
Manis javanica				+
Squirrels				
Callosciurus melanogaster		+		
Sundascirus fraterculus		+		
Laricus obscurus		+		
Iomys sipora		+		
Petinomys hageni lugens			+	
Hylopetes sipora		+		
Rats				
Chiropodomys karlkoopmani		+		
Rattus tiomanicus mentawi			+	
Rattus sabanus siporanus			+	
Rattus surifer pagensis			+	
Civets				
Paradoxurus hermaphroditus siberu			+	
Hemigalus derbyanus sipora			+	
Otters				
Aonyx cinera			+	
Deer				
Cervus unicolor oceanus				+
Primates				
Macaca pagensis		+		
Presbytis potenziani		+		
Hylobates klossii		+		
Simias concolor	+			

area of about 448,000 ha. It is a hilly island with the highest peak at 384 m above sea level.

Siberut was uplifted as a relative flat surface with some sedimentary beds having been overturned and dipping at up to 90^0. Exposures are very scarce, consisting of tuffs, sandstones, and claystones of the Sagulubek Formation. The rather regular sedimentary formation is of a young geological age, and erosion rates have been high, resulting in the

Table 2. Endemic subspecies of birds (after Whitten *et al.* 1980).

Species	Character difference from mainland or other island forms		
	Smaller	Larger	Colour
Honey Buzzard *Pernis apivorus ptilorhynchus*	+		+
Crested Serpent Eagle *Spilornis cheela sipora*	+		+
Red Cuckoo-Dove *Macropygia phasianella elassa*			+
Blue-Rumped Parrot *Psittinus cyanurus pontius*		+	
Chestnut-breasted Malkoha *Phaenicophaeus curvirostris oenicaudus*			+
Green Broadbill *Calyptomena viridis siberu*			+
Ashy Drongo *Dicrurus leaucophaeus siberu*	+		+
Spangled Drongo *Dicrurus hottentotus viridinitens*		+	+
Black-naped Oriole *Oriolus chinensis mentawi*			+
Dark-throated Oriole *Oriolus xantonotus mentawi*			+
Magpie Robin *Copsychos saularis pagiensis*		+	
Black-naped Monarch *Hypothymis azurea leucophila*			+

development of a rugged landscape with many relatively short rivers and streams (Figure 2). The hills rise steeply almost without interruption from river terraces. As a result, a very complex drainage pattern can be found on Siberut, with no clear island divide. The dominant soil type is red-yellow podsolic of tropudults with dystropepts and troporthents on eroded sites. The larger rivers are usually 30-40 m wide up to one kilometer inland from the coast, and are navigable throughout the year. Eight catchment areas are identified, varying in size from 110-550 km^2 with a calculated mean runoff of 7-35 m^3 sec^{-1}. Siberut has extensive areas of hillrocky tuffaceous sedimentary plains, extensive coalescent estuarine or riverine plains, and some minor valley floors within the hill region.

The river system is fed by a wet equatorial climate with no distinct dry season. Annual rainfall is 3364 mm, temperature oscillates between 20° and 32° C, humidity is 82-86%, mean wind velocity is 24-34 km, and mean annual sunshine is 1290 hours. According to the Koppen System the climate can be described as tropical, rainy, isothermal with a hot summer, no dry season, little seasonal temperature changes and a bimodal rainfall distribution. In every month, more than 50% of the days have rain. The high rainfall on non-resistant soil results in an ongoing formation of silt terraces in the lowlands of the east coast in particular. Subsidence has also occurred along the east coast and is reflected in a very irregular coastline with bays, capes, islets, and coral reefs. The west coast is rather straight and lined with broad sandy beaches.

Natural vegetation and fauna

The natural vegetation of Siberut is tropical rainforest with some distinctive features (Figure 3).

1. Primary Dipterocarp Forest. This type of forest is confined to the high ridges and hills. In comparison to other Dipterocarp forests the genus *Dipterocarpus* is represented in greater numbers than the genus *Shorea*. *Palaquium* and *Hydnocarpus* species emerge from the 30-35 m high canopy. Large woody climbers and epiphytes are not abundant and the ground layer is relatively sparse with many saplings and a low herb layer. Fan and fishtail palms are absent and the rattans usually low and stemless.
2. Primary Mixed Forest. This type is found on the slopes and low hills with many tree species of the families Myristicaceae, Euphorbiaceae, Dilleniaceae and Dipterocarpaceae. Legumes, in contrast to Sumatra, are rare. The commonest emergent tree genera are

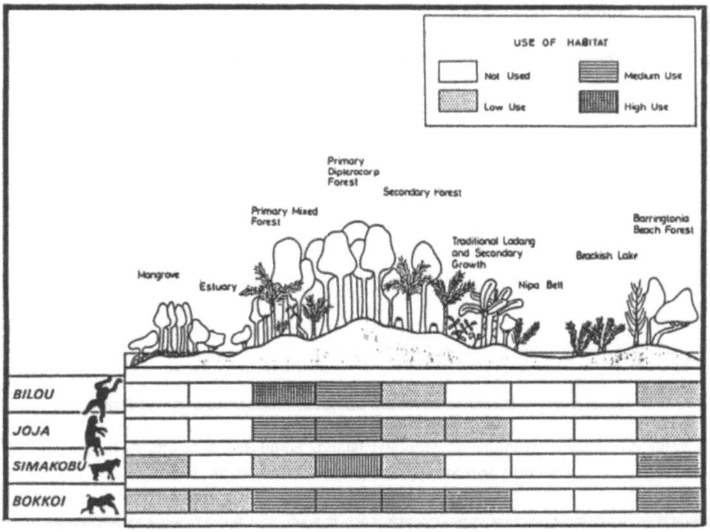

Figure 3. Different habitats of Siberut Island and their use by four endemic primates (after Whitten *et al.* 1980). Bilou: Kloss Gibbon - *Hylobates klossii*; Joja: Mentawa Langur - *Presbytis potenziani*; Simakobu: Pig-tailed Langur - *Simias concolor*; Bokkoi: Mentawai Macaque - *Macaca pagensis*.

Shorea, *Dipterocarpus*, *Dialium*, *Pentan* and *Durio*. The forest is rich in climbers and epiphytes, the ground vegetation dense with palms dominating. The canopy height is 15-30 m with fewer emergents than in the Dipterocarp forest.

3. Freshwater Swamp Forest. This has a specialized and limited tree flora dominated by *Terminalia phellocarpa*. The ground layer is composed of feather palms, rattans, pandans, and aroids. The soil is persistently wet and studded with pneumatophore roots.
4. Mangrove Forest. This type is only found on the shoreline of the Eastern coast. Usually *Rhizophora* species are replaced by taller *Bruguiera* trees where the substrate is exposed at low tide. The habitat is very specialized and the stands are composed of very few species. Estuaries of large rivers are fringed with *Nypa fruticans*.
5. *Barringtonia* Forest. This forest type is found on the west coast, with pure stands of *Casuarina equisetifolia* and tree-shrub associations of *Barringtonia*, *Hibiscus*, and *Pandanus*. *Eugenia grandis* and *Calophyllum inophylloides* can be found on headlands.

The major botanical collections on Siberut were made by Kloos in 1924 and described by Ridley in 1926 (Ridley 1926). Besides the fact that 15% of the recorded 330 plant species are endemic, it is remarkable that plant species and genera which are also found elsewhere in South East Asia often have quite peculiar characteristics on Siberut. *Xanthophytum villarii* (Rubiaceae), otherwise known only as a small tree from the Philippines, is a tall tree on Siberut, while *Phaeomeria minor* (Zingiberaceae) is the smallest in the genus. In Siberut, *Metroxylon sagu* is extraordinarily large with a trunk 18 m in length and 1.5-2.0 m in girth, whereas on average they reach only 9.5 m in length and 0.75-1.20 m in girth in Malaysia (Whitmore 1975).

The importance of the Siberut forests as a plant genetic resource is highlighted by the fact that on the smaller islands each plant species tends to fill a broader niche than on the mainland or larger islands. Species are usually adapted to a wider range of environmental conditions and are likely to be more suitable for silviculture or cross-breeding than the more narrowly specialized forms. One very important fruit tree originating from forests in Siberut is the durian (*Durio* spp.).

Table 3. Forest plants utilised by man (after Whitten *et al.* 1980).

Category	Number of species
Fruit and root crops	31
Seed grains	3
Spices	2
Animal forage	7
Medicine	10
Fibre	6
Resin	5
Gums	5
Insecticides	4
Total	73

Table 4. Endemic primates on Mentawi Islands compared with other islands with endemic primate species (after Whitten *et al.* 1980).

Island(s)	Area (km^2)	Number	Ratio Area/Primates
Mentawai	7,000	4	1,750
Sumatra	473,370	1	473,370
Sulawesi	189,037	7	27,005
Kalimantan	725,478	4	181,369
Madagascar	587,042	19	30,897
Taiwan	35,962	1	35,962

Of the 330 plant species Kloos collected in 1924, 73 (22%) represent genera of economic importance to man (Table 3). The forest of Siberut is clearly essential for the continuing existence of its rich endemic wildlife. Of 31 terrestrial species of mammals, 28 are dependent on forest (Whitten *et al.* 1980). All four of the endemic primates (Table 4) are dependent on forest to some extent. Rattan fruits are a significant food item for 3 of the 4 species and one such rattan (*Calamus manan*) is only found in mature forest. A similar pattern of dependence exists for the 107 species of birds.

This interrelationship is also true for many of the rainforest plants which are dependent on animals for seed dispersal, pollination and insect control. Thus the forest of Siberut, as all other natural forests, comprise a complex interacting system of plants and animals which are dependent on each other for their continued existence.

The main force reducing forest diversity on Siberut is modern man.

Human effects on the biosphere of Siberut

Five rather clear categories can be identified:

1. Undisturbed areas composed of forest showing negligible effects. About 70% of the Western part of the island belongs to this category, but virtually all these areas have been given out as timber concessions since the 1970's.
2. Moderately disturbed areas where human effects have noticeably altered the appearance of the forest. For example: areas around settlements, areas of commercial rattan collection, and mangrove areas.
3. Very disturbed areas where traditional systems of agriculture have been practiced. These are found mostly near navigable rivers and flat swampy areas dominated by Sago and Nipa palms.
4. Totally disturbed areas which include planned government and missionary villages, and

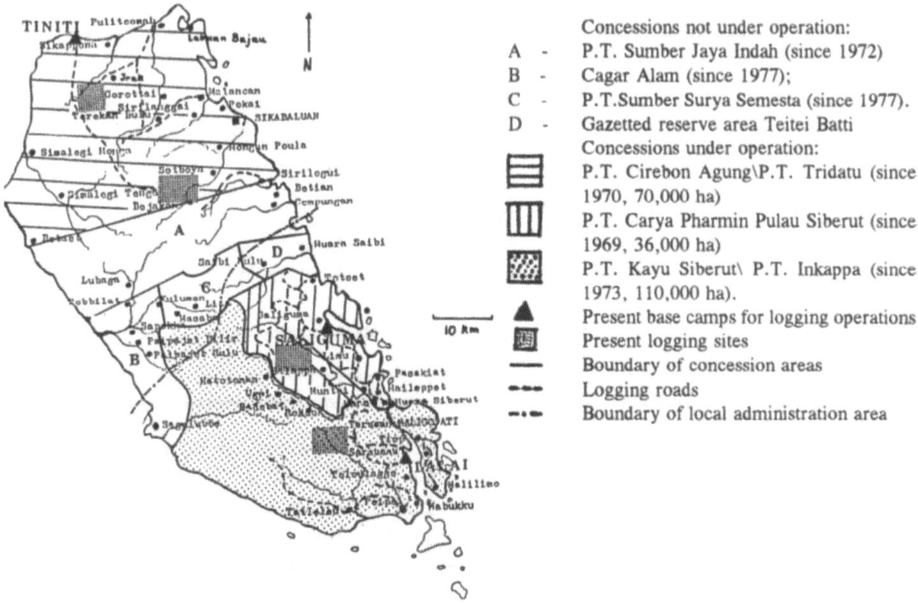

Figure 4. Timber concessions and concessionaires (status as at January 1993).

Figure 5. Heavily disturbed areas due to logging activities (Status as at January 1993).

their surroundings. For administrative reasons the traditionally scattered populations have been relocated in newly created, mostly coastal, resettlements. This practice is prone to create ecological wastelands through which the exchange of plants and animals is greatly hampered.

Table 5. Major impacts of logging operations on the environment and the population.

East coast	
Coral islands	Destruction by silt due to high erosion rate
Shallow sea	Coral reef destruction due to siltation
Mangroves	Destruction of fish nurseries; erosion by sea
Estuaries	Loss of habitat; loss of species
Primary Mixed forest Primary Dipterocarp forest	Changes of habitats; grassland areas split forest into 'islands' too small for maintenance of some species; increased erosion; depletion of various resources
Secondary forest	Increased erosion; reduced fertility
Brackish lake areas Nipa belt Barringtonia beach forest	)) Loss of habitat and associated species)
West coast	

5. Areas disturbed by logging, including all those where commercial logging takes place (Figures 4 & 5). At present 256,000 ha, more than 50% of the total area of Siberut Island, is logged by 3 major companies. This occurs despite the fact that major parts of Siberut have been declared or are needed as Biosphere reserve areas.

Due to inappropriate practices by contractors and sub-contractors, the extraction of 10% of the timber results in about 65% destruction of the forest. The felling of one major tree and its extraction may result in the damage of about 0.2 ha of forest. The logging roads on soft clay soil, up to 60 m wide, result in extremely high erosion rates. The logging does not benefit the Siberut people (Table 5); less than 3% of the work force is from Siberut itself.

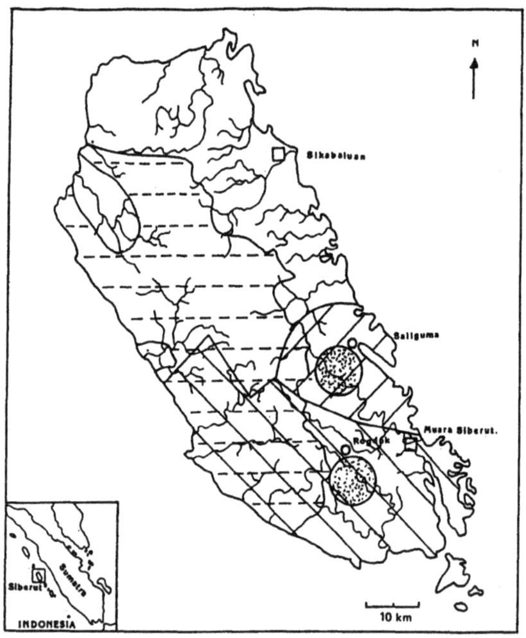

Figure 6. Proposed area for the planned National Park (198,000 ha); the two timber concession areas under operation until the end of 1994; the two recent resettlement areas for the local people. After Momberg 1994, unpublished.

CONCLUSIONS

The long isolation of Siberut has allowed a relict fauna to survive and new forms to evolve. This gives the island a very high rate of endemism, and offers a natural laboratory for the study of island biogeography. If logging is allowed to continue, these endemic forms of animals and plants will be gone forever.

Traditionally, conservation was assured by cultural sanctions and a complex system of taboos. Any method of conservation has to take into consideration the traditional systems of balanced exploitation by recognizing the rights of the indigenous people. Present logging practices must be seen as a threat to both the livelihood and ancient culture of the Proto-Malayan people and to the widely varied endemic fauna and flora.

At the very least, an extended totally protected area of about 190,000 ha is needed to guarantee the continuation of some of the unique fauna and flora (Mitchell 1982) (Figure 6). At present only 90,000 ha lies within gazetted reserve areas. Transmigration settlements, palm oil plantation schemes, uncontrolled tourism, and the ongoing resettlement programme are major threats to the uniqueness of this island. Use of the island for the transmigration of up to 10,000 families from Java and Sumatra would be disastrous. In addition, the lack of agreement between timber producers and forest conservation authorities on the safe exploitation of the forest in the buffer zones around the reserves will result in a loss for world heritage, the indigenous people, and the scientific community.

LITERATURE CITED

AVE, A. & SUNITO, S. 1990. *Medicinal plants of Siberut.* World Wildlife Fund Report, Gland, Switzerland. 186 pp.

MCNEELY, J.A. 1978. Siberut: conservation of Indonesia's Island Paradise. *Tiger Paper* 5(2):16-19.

MITCHELL, A.H. 1982. *Siberut Nature Conservation Area West Sumatra Management Plan 1983-1988.* World Wildlife Fund Report, Bogor, Indonesia. 139 pp.

RIDLEY, H.N. 1926. Spolia Mentawiensia - The flora of the Mentawai Islands. *Kew Bulletin* 1926:56-94.

SHEFOLD, R. 1980. *The Siberut Project.* Survival International Reviews: 4-12.

WHITMORE, T.C. 1975. *Tropical Rainforests of the Far East.* Oxford University Press, Oxford.

WHITTEN, A.J., WHITTEN, J.E.A. & HOUSE, A.P.N. 1979. Solution for Siberut. *Oryx* 15:166-169.

WHITTEN, A.J. & WHITTEN, J.E.A. 1980. *Saving Siberut: A conservation masterplan.* World Wildlife Fund Report, Bogor, Indonesia. 134 pp.

WHITTEN, A.J. & SARDER, Z. 1981. Masterplan for a tropical paradise. *New Scientist* 91:230-235.

WHITTEN, A.J., DAMANIK, S.J., ANWAR, J. & HISYAM, N. 1984. *The ecology of Sumatra.* Gadja Mada University Press, Yogyakarta, Indonesia. 583 pp.

Higher-order categories in Brunei Dusun Ethnobotany: the Folk-Classification of Rainforest Plants

JAY H. BERNSTEIN

Department of Sociology & Social Anthropology, Eliot College, University of Kent at Canterbury, Canterbury, Kent CT2 7NS, United Kingdom

KEY WORDS: Brunei Darussalam, Borneo, ethnobotany, Dusun, human-rainforest interactions.

INTRODUCTION

For more than 100 years botanists, agricultural scientists, and foresters working in island Southeast Asia have collected indigenous names for plants (e.g. Burkill 1935, Corner 1940, Ochse 1931, Quisumbing 1951). The reasons for this are plain. There needs to be some provisional way of identifying species new to science, scientists need to communicate with local people about plants, and various government departments require a convenient set of names for efficient administration. Unfortunately, it is also hazardous to use names and other isolated fragments of local knowledge about plants outside of the ethnobotanical context from which they are drawn (Ellen, this volume). Terms and associated data have not always been collected carefully, sometimes displaying linguistic and cultural ignorance. In the case of Brunei, Ashton's (Ashton 1964, also Pukul & Ashton n.d.) use of indigenous terms for trees keyed to scientific terms is open to confusion. Native terms are used as if they were vernacular synonyms for species names, but there is an unexplained overlap of names. In some cases where there are no known vernacular names new ones are suggested.

This paper presents a preliminary report on the overall structure of Brunei Dusun folk-botanical classification system, and may go some way towards illustrating some issues and problems of which rainforest biologists should be aware.

Brunei is an excellent location for an ethnobotanical study seeking to lay bare general principles of classification. The tropical rainforests of Southeast Asia have one of the highest biodiversities on earth, and in Brunei, through good fortune and effective planning, they are well-preserved. The cultures of the Austronesian peoples who have lived in Borneo for over two thousand years have co-evolved with the rainforest environment, and their survival has depended upon knowledge of the forest (Ellen, in press). Bornean groups have always relied on forest products, not only for food but to obtain materials needed for medicine, clothing, and other aspects of technology. The extraction of forest products, both for consumption and for trade, has long been an essential element of the economies of traditional Bornean populations, even those who define themselves first and foremost as 'farmers' (King 1993, pp. 77-102). With rapid socio-economic change occurring throughout Borneo (Cleary & Eaton 1992) traditional patterns of rainforest use have undergone massive change, and are now in decline.

The Area and its Inhabitants

The field site for the project reported here is the Dusun village of Merimbun, located in the Rambai *mukim* (administrative sub-district) in Tutong district. Tutong, one of Brunei's four districts, is the traditional homeland of the Dusun, who are among seven ethnic and linguistic groups constitutionally recognized as native Bruneians. Hence they are classified

D. S. Edwards et al. (eds.), Tropical Rainforest Research – Current Issues, 435–450.

for government purposes as 'Malay', and for this reason there are no contemporary statistics on the number of Dusun. A well considered estimate (Antaran 1993, p. 19) is that there are 5,000 non-Muslim Dusun in Brunei. It is clear, however, that the Dusun population is decreasing as the result of marriages to Malays and Chinese. Furthermore, traditional Dusun language and culture is not being effectively reproduced (Kershaw 1992). In large part, this is due to social and economic mobility: with a large public sector economy, Bruneians are abandoning traditional economic pursuits to join labour forces away from their villages. As this occurs, traditional social formations, and the cultural knowledge which sustains them, decline.[1] Another crucial factor in the transformation of Brunei's traditional indigenous peoples is conversion to Islam.

Traditional Dusun culture revolves around the cultivation of dry rice and fruit orchards. Values relating to these activities are expressed in ritual and belief, for example in the shamanic *temarok* ceremony and in bird omenism. Other major economic pursuits include hunting, fishing, and the collection of various forest products. Until recent years rubber cultivation and tapping also contributed to the Dusun economy.[2]

Merimbun village is situated at a large lake (*tasek*), and this is a source of fish for all villagers. It is also a bathing spot and a source of potable water when other supplies are depleted. Like most of Brunei, it is also surrounded by forests. The forests in this region are of mixed dipterocarp type, and are the source of many of the economic products that provided the subsistence of the traditional Dusun.

The Merimbun area has been made accessible by metalled roads leading into the village, and has been developed as a recreation site by the provision of bridges, pavilions, and toilets. On Sundays and public holidays it is a favoured spot for picnickers and nature-lovers. It is considered that the area has been degraded ecologically by this development by increased water pollution due to the use of motor boats, among other reasons. Heavy use of walking paths and the construction of roads has also disturbed and even destroyed local habitats (Wong 1992). Villagers note that fish are less plentiful than in bygone days, and that they now have to walk increasingly further to reach game-hunting grounds. But it is the rapid social, cultural, and economic change, rather than environmental degradation, that is leading to what seems the inexorable decline of forest use and ethnobotanical knowledge. Much of the folk botanical lexicon may no longer be in use in 20 years time as Malay (and English) supplant the Dusun language.[3]

A Framework for Ethnobiological Classification

Folk-classification differs in several important respects from the scientific, biotaxonomic classification of plants which is primarily a reflection of the position of a plant in evolution. In scientific taxonomies, reproductive and genetic criteria are highly significant: classifications are based to a great extent on the structure of reproductive parts (flowers, fruits, or spores). Folk classification, on the other hand, is relatively unconcerned with these matters, though it does reflect empirical criteria - particularly morphology, ecology, and (in the case of animals) behaviour - that are used in scientific classification. However, these are never entirely independent of other cultural criteria, as the present analysis will demonstrate.

Berlin (1992) suggests that plant classifications everywhere display evidence of certain universal regularies. He claims that the reason non-literate peoples have elaborated extensive domains of ethnobiological knowledge - and why these are strikingly similar in structure and content - is "human beings' inescapable and largely unconscious appreciation of the inherent structure of biological reality" (Berlin 1992, p. 8). The identity of a plant is immediately apprehended, due not to cultural logic but to the perception of 'discontinuities in nature'. For Berlin, folk classification relies upon the "natural system" of classification, in which plants and animals are categorized in accordance with "their overall morphological structure, or morphological plan" (ibid., p. 9).

Berlin claims, additionally, that living things are classified in terms of the following ranks, listed below from most inclusive to most specific.

(1) **kingdom:** a taxon with one member including all taxa of lesser rank.

(2) **life form:** a taxon containing a small number of highly distinctive morphotypes based on the recognition of gross morphological structure and ecological adaptation.

(3) **intermediate:** a taxon grouping together a small number of generic taxa, most commonly members of life-form taxa, showing marked perceptual similarities.

(4) **generic:** a taxon naming the smallest fundamental biological discontinuities easily recognized in any particular habitat.

(5) **specific:** a taxon partitioning the generic taxa into two or more members, differing in a few ways.

(6) **varietal:** a taxon further dividing specific taxa.

In this model, generally, taxa at all ranks except for those of kingdom and intermediate are lexically labelled; taxa at those ranks are usually unlabelled (covert), and some life-form taxa may also be unlabelled. Another axiom of Berlin's model is that terms at the generic rank are primary, meaning that they "are precisely the names of common speech" (Berlin 1992, p. 53). Psychologically, they are more salient than names at other ranks.[4]

The significance of Berlin's theory is that it challenges the idea that folk classifications of plants bear no resemblance to biotaxonomic classifications, but rather are determined to a great extent, if not completely, by cultural values, culture-specific logic, or the structure of language.

In this paper I use Berlin's model as a framework to discuss my data on one aspect of Dusun ethnobotany: the grouping of generically-named plants into more inclusive, 'higher-order' categories. I use the terms "life-form", "intermediate" and "generic" accepting the model of a ranked order they presuppose. I agree with Berlin that categories are predominantly constructed on the basis of their biological distinctiveness, and not because of independent cultural valuation and classification systems. At the same time I share the view of some of his critics, such as Ellen (1986, 1994) and Morris (1993), that categories derived from criteria that are neither morphologic nor phenotypic may intrude into the 'natural system of classification', even though such categories are indistinct. For this reason I include in my analysis higher-order categories elicited in the course of fieldwork that do not fit comfortably in the Berlin scheme. I conclude not only that higher-order categories occur at different ranks, but that they are weighted differently, and that it is possible to differentiate the most salient categories from less salient ones.

THE PROJECT AND ITS FIELD METHODS

The University of Kent - Brunei Museum project on human-rainforest interaction has among its aims the documentation of the current significance of rainforest resources and the ecological parameters and opportunities presented by rainforest in the lives of rural Brunei Dusun. By extension, it seeks to further our understanding of other comparable parts of Southeast Asia. The project also aims to contribute to knowledge about human impacts and dependency on rainforest (indigenous and commercial), and the realistic valuation of rainforest as an economic resource. As part of the research it was necessary to obtain data on people's knowledge and uses of specific rainforest plants. This was accomplished primarily through the collection of voucher specimens.

Plant collection was undertaken in the Merimbun area for four months in 1992 and three months in 1993. Four principal informants were engaged in this work. All were in their late fifties or older, with an extensive knowledge of forests, having spent most of their lives engaged in traditional subsistence activities, including hunting. Normally, only one informant collected plants on any one trip; however, on certain occasions two informants were used simultaneously. At the time of collection the informant would provide the Dusun name of the plant, the name of the location and the kind of habitat in which it was found, information about its uses, and any other information about the plant. Optionally at first, but later more regularly, the informant was asked to categorize the plant. Each evening, following collection, plants were preserved for further processing. Specimens were sometimes shown to other informants to obtain their identifications, before they were removed for processing.

Five hundred and thirty-five herbarium specimens of various kinds were collected, for which 426 different names (exclusive of synonymous or alternative names for identical plants) were provided. Collections were made in uncultivated areas, and almost no cultivated plants were included.[5] These specimens are currently being identified by the Royal Botanic Gardens at Kew and the Forestry Centre in Brunei. At the time of writing, about 300 of the specimens have been identified. A further 132 additional plant names were collected where no specimens were taken, through inventories enumerating each plant found within circumscribed areas of forest. For this exercise two informants named every plant they could recognize. Most plants were identified at the generic rank, while some were identified at a sub-generic rank; those plants for which generic names were unknown were identified at a more inclusive rank. In the course of the exercise, involving the identification of many thousands of plants, only one plant could not be identified to any category.

Other plant names were elicited in the course of interviews in which informants were asked to name plants similar to those given in prompts. The names of plants were put to several informants, consisting mainly of people not involved in plant collection work. They were asked to state what kind of plant it was and give any other information they could about it.

Another technique used was to present informants with sets of cards containing the names of plants which they were then asked to sort into groups that were similar in some way. For illiterate informants the names of plants were read aloud. They were then asked what the plants so grouped had in common, and how they contrasted with those in other groupings.

NOMENCLATURE AS EVIDENCE FOR CATEGORIZATION

The names of plants provide a major key to the ordering of an ethnobotanical domain, though not an entirely reliable one (Conklin 1962). There is no overall name for all plant life which would correspond with the kingdom rank. From the names of plants themselves the following list of category names may be drawn up: **akau** (vine or liana), **bulu'** (bamboo), **kayu** (tree), **kulat** (mushroom), **usak** (alternatviely **bunga**) (flower), **umbus** (vegetable), and **uwai** (rattan). These terms are used regularly in Dusun as the first element in a polynomial plant name (e.g., **akau gamuri**, **kayu baru, kulat kamanci**). These are all higher-order categories, groupings of generically named plants. They are the most inclusive groupings within the un-named domain 'plant'. These names, among others, were elicited when informants were asked, given a certain plant, what sort of a plant it was. How is it possible to determine which of these terms represent culturally salient categories corresponding with Berlin's "life-form" rank?

The specification of this term is difficult. It indicates, as Wierzbicka (1985) states, "a category of categories" (p. 189), "a kind including many different kinds" (p. 192). **Usak** (flower) and **umbus** (vegetable) are excluded on grounds that they are not completely distinctive: a plant can belong simultaneously in one of these categories and in some other group. **bulu'** (bamboo) is associated with other, non-life-form categories, though it does not overlap with any of them. For this reason it does not seem distinctive enough to be a life-form, despite the fact that it is, in a way, a category of categories. The remaining categories, **akau, kayu, kulat,** and **uwai** are treated here as life-forms.

The distinction between trees and vines is general though not universal in folk-classification (Brown 1977), and these are extremely common features of the forest environment. The focal perceptual difference between a tree and a vine is immediately apprehendable, and it is doubtful that there is much cultural variation in the construction of these categories in clear cases. Berlin (1976, p. 385) considers them (along with herbs) to represent "such distinct perceptual discontinuities that their recognition may constitute a substantive near-universal in prescientific man's view of the plant world."

The distinction between a tree and a vine is based on morphology (stem growth pattern), but is not recognized in scientific biological taxonomy. Plants whose stem growth pattern is intermediary between that of a "tree" and a "vine" must be included in one of these categories, and the basis of such a plant's categorization seems arbitrary in terms of morphology.

Rattans (climbing palms) are a distinctive category in Dusun terms, but they meet at least some of the specifications for 'vines' as well. While recognizing the similarity between **akau** and **uwai**, the Dusun do not consider **uwai** to be a kind of **akau**. In conducting my plot inventories I found that informants would initally identify rattans as vines at first, but then correct themselves when I repeated the name to them. It appears that the focal definition of **akau** excludes rattans. One non-palm was identified at the time of collection by a highly reliable informant as **uwai asu-asu** (JHB 280),[6] though I questioned him as to whether it really was **uwai**, and in my field notes for the specimen I wrote : "Sort of a lake grass, said to be a rattan. Not found on dry land. Useless, except for binding." The specimen was subsequently identified as *Flagellaria indica*.[7] I have also collected, with another informant, **uwai asu-asu bukid** (JHB 377, not yet identified), which like **uwai asu-asu** was thornless, though I did not question that it was a rattan.

Of 210 kinds of plants collected that were categorized as **kayu** (some ambiguously so; see below), the names of only ten were prefixed by the life-form name; one name included the term alternatively; one name ended with the term.

Turning to the **akau** category, of 71 kinds of plants collected, the names of 32 were habitually prefixed by the word **akau**, two alternatively used the prefix, and one included the life form name as the second term.

In the **uwai** category all but one of 16 kinds contained the life-form name **uwai** as a first term. In the **kulat** category all 13 kinds contained the word **kulat** as the first term.

It will be seen that in the case of **kayu**, the category term is rarely obligatory, for **akau** it is commonly obligatory, for **uwai** it is nearly always obligatory, and for **kulat** it is always obligatory. This fact underlies the high salience of **kayu** and the decreasing salience of the other categories, and supports Brown's (1977) theory that 'tree' is the first life-form term to emerge in an evolutionary sequence in any language.[8]

In other categories, as well, the category name appears as part of the name: **usak** (alternatively **bunga**) ('flower') and **bulu'** ('bamboo').[9] There are several other kinds of plant, but their category-labels are not similarly used. While these are higher-order groupings, they do not correspond to the 'life-form rank' as defined. The **usak** category seems roughly equivalent to 'herbaceous plant', which is often a life-form (Brown 1988), but the Dusun term is not used consistently to denote this. The criteria for the category seem not to be purely morphological but to be influenced by what Wierzbicka (1985) points out is not exactly "function" or "utility" but more generally a *relation to people*. Some woody plants are called **usak,** and this is probably due in part to their size. Some, but by no means all, plants in the **usak** category are "decorative plants." Finally, the **usak** category is not culturally salient. Only 13 specimens were placed primarily in this category, though some marginal plants, difficult to place in *any* life-form category, were placed here, as well as in one or more other categories. Furthermore, in an enumeration of plants in forest plots, two informants were more frequently unable to provide names at the generic rank for these than for other life-forms; in many cases **usak** were said not to have a name. **bulu'** is represented by no more than five different kinds, and is not considered to be completely distinct from some other kinds of plants.

It must be noted that viewing **kulat, uwai,** and possibly **akau** as life-forms violates Berlin's third principle of ethnobiological nomenclature, which holds that life-form and generic taxa should have primary names, while subgeneric taxa should (with exceptions) have secondary names (Berlin 1992, p. 29).

Berlin (1992, p. 27) distinguishes between primary and secondary plant (and animal) names. A primary plant name may be simple (e.g. oak) or complex (e.g. skunk cabbage, forget-me-not). To give an example of a secondary name, and explain the difference between such a name and a complex primary name, Berlin (1992, p. 28) writes:

> "*Sugar **maple*** can be shown unambiguously to be a secondary plant name in that (a) one of the name's constituents, *maple*, is also the name of the taxon that immediately dominates the category labelled by the expression *sugar maple*, and (b) *sugar maple* occurs in a contrast set whose members are also laeled by names that include that same constituent (*red **maple**, Norway **maple**, Oregon **maple**, vine **maple**, . . . n **maple***). In contrast, complex primary names such as *tulip tree* or *catfish* do no show this distributional pattern. These expressions occur as the names for taxa in contrast sets some of whose members are labelled by simple primary names such as *hickory, ash,* and *poplar* (all kinds of *tree*), or *bass, crappie*, and *carp* (all kinds of *fish)*"

According to the third principle of ethnobiological nomenclature mentioned above, any category name that habitually appears in a taxon, as is the case in **uwai** and **kulat,** should not be a life-form but a genus. My difficulty in applying this principle to Dusun ethnobotany is that it is at odds with the definition of a life-form as a kind including 'many different kinds' (though it must be admitted that such a vague criterion begs the question). Furthermore, the category prefix **akau** appears in about a half of generically-labelled vines. There seems to be no difference between the categorizing function of the term **akau** and that of **uwai** or **kulat**.

No ethnobiologist appears to have recognized rattans as a life-form (Brown 1977); Bartlett (1940) even presents his data on rattans as exemplifying the concept of folk-genus. But to see **uwai** as a generic we would have to accept the premise that the primary names of all 19 kinds of **uwai** are not generic but specific.[10] The generic level taxon would be empty, for there is no focal, generic, unmarked **uwai.** This contradicts Berlin's axiom that generic taxa are primary in any folk-biological system of nomenclature (Berlin 1992, pp. 52-101). The main evidence that **uwai** is not a folk-generic in Dusun is that it contrasts with **kayu** and **akau**, at the life-form rank. If **uwai** were a folk-generic it would presumably be included in some other life-form, probably as **akau**.

The same scepticism about the life-form status of rattans holds for **kulat** (fungi). Fungi are completely distinct from all other life-forms; they are infrequently encountered in forests. Some informants said that mushrooms growing out of the ground could be put into the more inclusive category of **sakot**, roughly translated as 'weed' (see below). However, **kulat** meets a fundamental criterion of a life-form: it is a kind that includes many different kinds. Unlike folk-specifics fungi differ not in a few distict ways but in many general ways.

Life-Forms

I will argue that **kayu** is the most salient plant life-form. As I have indicated, far more forest plants in that category were collected than in any other, meaning that they are encountered and recognized more often than other kinds of plants.

That trees are the most prominent and frequently-occurring feature of the forest environment is obvious enough, and this is reflected at the level of folk terminology in our analysis of two randomly selected patches of land, both on hilly slopes, one categorized as *entalun* ('forest', never previously cleared in known memory) and the other categorized as *gapu'* (secondary growth, cleared about 20 years ago). In enumerations made by one informant, in both the *entalun* and *gapu'* plots, 13 of the 20 most frequently mentioned plant-names were in the **kayu** category, followed, again in both cases, by 4 in the **akau** category. Considering the *entalun* survey only, this informant provided the names for 103 kinds of **kayu**, 26 kinds of **akau**, 8 kinds of **uwai**, and 21 names for all other kinds of plants, not counting plants he could not identify or remember the names for. A second informant (who surveyed an adjacent area half the size the one surveyed by the first informant) named 114 kinds of **kayu**, 19 kinds of **akau**, 10 kinds of **uwai**, and eight kinds of other plant.

As mentioned above, the difference between **kayu** and **akau** is not so much phylogenetic as morphological. Plants in both categories have many of the same uses: edible fruit and/or leaves; medicines, poisons, and antidotes; and numerous specialized uses, especially for leaves. The fruit of vines is rarely important in the trade economy compared with those of trees. But it is the role of trees in providing wood for building things that gives trees their unique cultural, social, and economic precedence in the plant world. The stereotypical purpose of vines, on the other hand, is as binding material. This purpose is not, however, unique to the **akau** category, but is shared with rattans (**uwai**). While this is a reason to reject a purely utilitarian explanation of life-form categories (such as that offered by Randall & Hunn 1984), it does seem plausible that utilitarian factors play some role in their relative salience.

The categories **kayu** and **akau** are semantically oppositional and dialectical, with **kayu** being primary. While about half of the kinds of **akau** are referred to habitually as **akau X**, very few trees are referred to habitually as **kayu X** rather than simply **X**.[11] Some terms occurred in both **akau** and **kayu** categories:

Akau	Kayu
1. **Akau melia'** (JHB 155) *Dichapetalum*?	**Melia'** (JHB 275) *Diospyros*
2. **Akau Sengkarai** (JHB 321) *Agelaea borneensis* (Hook.f.) Merr.	**Sengkarai** *Polyathia* sp. (Brunei checklist)
3. **Akau tis** (uncollected)	**Tis** (JHB 149) *Galearia fulva* (Tul.) Miq.
4. **Akau terakang** (JHB 277) *Oxyceros*	**(kayu) terakang** (uncollected)
5. **Bagu akau** (JHB 137) *Gnetum*	**Bagu kayu** (JHB 147) *Gnetum*

The fact that the life-form **akau** is marked in three cases seems to substantiate the claim that constituents of the **kayu** category are psychologically and culturally primary *vis-à-vis* those of the **akau** category (Brown 1984, cf. Berlin 1992, pp. 27-28). The use of the category term following the generic term in the case of **bagu** is explained by the premise that while the other plants are different things, **bagu akau** and **bagu kayu** are two kinds of one thing, and hence the category names are used as adjectives. Another example of this structure is found in the names of two kinds of cassava (*Manihot utilissima*), e.g. **ubi kayu** ('woody cassava') and **ubi bogo** ('creeping cassava').

An exception to the generalization that the **kayu** form is primary because it is unmarked *vis-à-vis* genera in other life-forms is **kayu tungod** (JHB 212, *Saurauia*), which contrasts with **tungod** (JHB 174, *Pentaphragma*), a 'grass' in the **telasai** category, the salience of which is low.

There is another instance of a term being identified nomenclaturally as both **kayu** and **akau**:

Akau piantok (JHB 264) (unidentified at time of writing)
Kayu piantok (JHB 126) Rutaceae?
Piantok (JHB 165) Celastraceae?

All of these plants are usually categorized as **akau** rather than **kayu**. Informants did not group it together with other **kayu**. As one informant said, **kayu piantok** "is considered **kayu**, but in reality it is **akau**. It is sliced and made into chopsticks (*supit*)". It seems to be a marginal type. In field notes both **akau piantok** and **kayu piantok** are described as 'vines', while **piantok** is described as a small plant. In this case plants that are different both phylogenetically and morphologically but are used in the same way occupy a single category.

A similar anomaly is **kayu sangga** (JHB 247, *Schefflera*), categorized by all informants as **akau**.

The categories **kayu** and **akau** are sufficiently complex and rich that informants knowledgable about forests are able to categorize them according to various dimensions such as hardness and size. Many trees and vines are associated with specific, distinctive purposes, such as the manufacture of a certain object or a role in a certain ritual. **Uwai** (rattans), on the other hand, is a much smaller category. With the exception of **uwai pios** (JHB 13, *Calamus comptus* J. Dransf.), which is used in a rice planting ceremony, rattans have only five contemporary uses: binding, weaving, frame-building, edible fruit, and edible shoots.[12]

Another phylogenetically focused group, that I consider to meet the criteria for a life-form, is **kulat**. Unlike **kayu, akau,** and **uwai**, they are not ubiquitous in the forest, and they are relatively insignificant culturally and economically.

Fungi are unmistakably different from all other kinds of living things, but are regarded by Dusun as plants. A certain number are edible, and one or two varieties may fetch high prices, but they are not an important part of either Dusun diet or economy, or of the forest landscape. An indicator of the very specific utility of **kulat** (as food) is that my informants were unable to provide generic names, for most inedible mushrooms lack generic names, even though it was recognized that they vary one from the other. Inedible mushrooms may be called **kulat andi kanon** (inedible mushroom) or **kulat raat** (bad mushroom).[13] However, **kulat jelundong** (JHB 289) (literally 'shade mushroom') is generically named, and the category so identified is even subdivided further: **kulat jelundong purak** (white shade mushroom) and **kulat jelundong gapu'** (secondary growth shade mushroom). None of the edible mushrooms collected were identified at a sub-generic rank.

An alternative possibility is that **kulat** is not a life-form but a folk-generic unaffiliated with any life-form (see discussion of 'unaffilated generics' below). By this analysis edible mushrooms as well as inedible **kulat jelundong** are marked off from generic **kulat** at the rank of folk-species. I resist this conclusion because it seems to me that the differences between the various kinds of inedible fungi are perceptually recognized as different genera without being coded.[14]

Intermediate categories

I have identified life-forms through category names that appear as proper parts of generic rank taxa. Some generically-named plants fall outside all life-forms, but are grouped in other categories. Berlin (1992) properly calls these categories 'intermediate', since they occupy a rank between generic and life-form.

Among Dusun ethnobotanical categories are those that may be glossed as 'grasses', 'gingers', 'bananas', and 'ferns', as well as various kinds of palms. These do not constitute life-forms as they are not labelled by any single over-arching term. For example, several kinds of ginger (including those called **encalongon** [JHB 41, *Plagiostachys crocydocalyx*], **kunyit** [*Curcuma*], **sagang** [JHB 43, *Etlingera ?punicea*], and **sumbang** [JHB 39, *Hornstedtia reticulata*]) were said to be sorts of **tumid**, while others (those called **terabak**)

Table 1. Higher-order Dusun plant categories (excluding cultivated plants).

Life-forms
kayu (tree)
akau (vine)
uwai (rattan)
kulat (fungus/mushroom)
Intermediates
Licuala plams (**silad-benjiru-ukang**)
Grasses (**kumpau-telasai-rumput**)
Ferns (**gerajai-paku-limputong-engkubuk-kuban**)
Gingers (**tumid-lingkuas-layoh**)
Bananas (**punti-rutai-binci-encarawan-powow**)
Problematic higher-order categories
usak (flower)
sakot/sakot tanah ('weed')
raun (leaf)
umbus/sancam (vegetable)

were categorized as kinds of **lingkuas**. It is possible that other gingers, called **layoh** (**layoh lamatai** JHB 372, *Globba brachyanthera* K. Schum. var *rubra* R.M. Smith; **layoh entalun** JHB 467, not yet identified), are not included in either of these groupings. While there is a clustering around focal members of these groups, the categories are also reciprocally related to one another. Although there is no term for a single category uniting all gingers, such a grouping is recognized by Dusun. The idea of covert categories in ethnobotanical classification, as proposed by Berlin, Breedlove & Raven (1968) is now generally accepted, but identification has proved to be a controversial matter (e.g. Berlin 1974, 1992, Brown 1974, Ellen 1993, Hunn 1977, Taylor 1984).

Another set of plants that do not fit into any other higher-order grouping but are related to each other are *Musa* spp. ('banana') plants. True bananas (*Musa sapientum*) are called **punti**, but several other genera are found in the wild: **rutai**, (JHB225, *Musa gracilis*) **binci**, **encarawan**, and **powow** (all uncollected).

'Grasses' seem a plausible candidate for life-form status, as Dusun **kumpau**, can be glossed by *rumput*, the Malay word for 'grass'.[15] However, in the course of research I found that several plants categorised by Dusun informants as **rumput**, and which I considered 'grasses', were said *not* to be **kumpau. Kumpau** occurs as a folk-generic name for *Paspalum conjugatum* Berg. (Graminae) (JHB 316), but also refers to an overarching category of grasses each of which have separate folk-generic names. Most grasses are included under this overarching cover-term, although tall grasses in both the Gramineae and Cyperaceae families are excluded. **Telincim** (JHB 230, *Paramapania*, and JHB 318, *Mapania sumatrana* (Miq.) Benth. ssp. *sumatrana*) and **telasai** (e.g. **telasai biasa**, JHB 219, *Scleria*), both Cyperaceae, were said *not* to be **kumpau**. A Gramineae, **kumpai** (JHB 298, no generic identification), as well as two Cyperaceae grasses, **purun**, (JHB 287, *Lepironia articulata*) and **tarupuk** (JHB 347, *Phragmites karka* (Retz.) Steud.) were classified as **telasai. Kumpau** and **telasai,** therefore, seem to suggest two prototypes around which similar folk genera may be clustered conceptually.

Similarly, ferns are classified as **engkubuk**, **gerajai** (alt. **derajai, rajai**), **gerintik, limputong, paku,** and **kuban**, without there being an overall Dusun term for 'ferns'. They are not grouped together as a life-form, though there is some reciprocal labeling, e.g. **gerajai** classified as a kind of **limputong**. Informants classify them as **rumput** or **sancam** ('vegetable'). One informant, asked to classify **engkubuk** (JHB 223), was able to reply only: "not **kayu**, not **akau**."

The final covert intermediate cluster includes palms of the *Licuala* genus and possibly an *Areca*. The three *Licuala* are **benjiru** (*L. spinosa*,[16] cf. JHB 80), **silad** (JHB 60, 61, *L. orbicularus*), and **ukang** (JHB 168). These were categorised reciprocally: **silad** was said to be a kind of **benjiru** and vice versa; there was no overall category name. Another palm, **landai tiwow** (JHB 181, *Areca minuta*), was also said by two informants at the time of collection to be a kind of **ukang**; another informant subsequently classified it as **kayu**. Other informants who had heard of this plant could not classify it at all.

Two other categories of palm are clustered around salient, focal members. The first grouping consists of palms similar to **pinang**, (*Areca catechu*, the betel-nut palm). As with the grass **kumpau** and **telasai** (see above), **pinang** is a folk generic that is the focal member of a larger grouping. Five palms of this kind have been collected. One is **pinang derato**[17] (JHB 465, not yet identified), the binomial directly indicating that it is folk-species of the folk-genus **pinang**. Another is **landai tiwow** (JHB 181, *Areca minuta*), the name of which means "peacock's tail-feathers". However, four other palms in this group have uninomials, including **barang** (JHB 216, *Pinang salicifolia*) and **raring** (JHB 278, *Cyrtostachys renda*).[18] Such names do not subdivide a recognized folk-genus. Sub-generic taxa should, according to Berlin, be labelled with *secondary names*: "linguistically complex expressions, one of whose constituents indicates a category superordinate to the form in question" (Berlin 1992, p. 28) - unless one taxon is the prototype of the genus or is of "major cultural significance" (ibid., p. 30). The betel-nut palm (*Areca catechu*) is without doubt highly significant culturally throughout Southeast Asia, but **barang** and **raring** are decidedly *in*significant.

Another palm category has as its focal member **luba** (uncollected), and includes **tiad** (JHB 182, *Eugeissonia minor*), **bidang** (*Borassodendron borneense* [19]), **ramok** (JHB 161, *Arenga undulatifolia*) and **iyul** (uncollected).

Problematic higher-order categories

Four or five supra-generic categories used by the Dusun in classifying plants have some of the characteristics of life-forms, but seem very different in other ways. They appear to be able to contrast with life-forms, and are treated like life-forms by Dusun informants who, when given a plant name and asked to classify it, provided the category terms to be discussed. They differ, however, in two important respects: they are not based on morphological or phenotypic stereotypes or indeed on any kind of visual cognitive prototype, and they are used idiosyncratically. Some members of these categories may belong in other categories. Others are placed in these categories by default, as they do not belong in any of the main groups. The categories are labelled (1) **usak** (Malay: **bunga**) (2) **sakot** or **sakot tanah**, (3) **sancam** (Malay: **sayur**) or **umbus**[20], and (4) **raun** (Malay: **daun**). These groupings occupy much of the same classificatory space as a comprehensive 'herbaceous plant' life-form, which does not exist in Dusun. However, none of these categories can clearly be described as life-forms since they do not specify rank.

The term **usak**, besides meaning 'flower', is used in Dusun in the same way as other life-form terms. Thus, if a Dusun informant is asked what kind of plant **X** is, he or she is able to respond: '**usak**'. This term seems to mean 'decorative plant', 'shrub', or 'small herbaceous plant'. However, it overlaps with other categories, including **kayu**: a woody plant such as **usak bila pinggan,** (JHB 8, 192, *Ixora*, and 91, *Pavetta*?, both Rubiaceae, plus 378, 385, and 416, unidentified) is classified primarily as **usak**, though it is also considered **kayu**. Usually a low shrub, it can grow up to four meters high. An attribute of **usak** is that it is close to the ground. A plant classified primarily as **usak** may be optionally included in other categories. The reverse is not the case. **Usak** as a plant category seems in some ways to imitate a life-form, but it may also include growth stages of plants which when mature are obviously members of other, more clearly-defined, life-forms.

The word **sakot** refers to any detritus lying around that cannot be identified. It may refer to rubbish. In the case of plants the term **sakot tanah** (or **sakot bumi**) is often used, suggesting a literal translation of 'ground detritus' (or 'earth detritus'). More idiomatically, it seems similar to English 'weed': a useless or even undesirable, relatively small plant. In Malay the term **rumput** generally has this connotation. And indeed the boundary between the 'grasses' and **sakot** is permeable. **Miau-miau** (JHB 37, *Lophatherium gracile* Brogn.) is categorized by some as **sakot**, and by others as **kumpau** or **rumput**. But plants lexically classified elsewhere may also be placed into this group, e.g. **bunga rampai** (JHB 152, *Dianella*) and **akau belan** (JHB 48, *Merremia borneensis*). Grasses, such as those in the **telincim** group are categorized as **sakot**. **Galugut** (JHB 116, moss) is put in this category, as are those mushrooms that grow out of the ground. A few informants put **basulok** (JHB 345, *Senna alata* (L.) Roxb.) in this category. There is considerable variation between informants as to which plants they classify as **sakot**. But the category is not haphazard, nor is it a "special purpose" category (Berlin *et al.* 1966, p. 274), as it does not have any single, 'special' purpose. Nor, on the other hand, can all plants described in this way be said to be useless, annoying, or disvalued, as the term 'weed' implies, since a number have medicinal or other uses. **Sakot** seems to be less a category with a clearly defined focus than a residue including culturally less important species that tend to be small and not noticed as individual plants but rather as clusters or clumps.[21] **Usak**, in contrast, groups plants in ways closely resembling those of a life-form, and it may perhaps be considered a 'pseudo-life-form'. Unlike a true life-form it lacks an ethnobiological focus and does not specify rank.

The categories **sancam** ('vegetable') and **raun** ('leaf') are different. They are also residual, used for plants that do not belong in the main groupings. The terms are most often used for **lamba** (JHB 72, *Curculigo vilosa* (Kurz) Merrill). This is a big leafy plant with no stalk. Informants varied in how they classified it. The generic category **lamba** is quite complex, and not all its consituents are edible. **Lamba gabuk** ('macaque lamba') (JHB 19, *Curculigo racemosa* Ridley), like **lamba** (alternatively called **lamba biasa**, 'ordinary lamba'), is an edible member of the family Hypoxidaceae. **Lamba-lamba**[22] (JHB 146, 160, *Corymborkis*) and **lamba entalun** ('forest lamba') (JHB 198, 392, *Niewiedia*) are both orchids, and are not eaten.

Eight other plants were classified as **sancam** or **sayur**.[23] For example, one informant categorized the shrub **kelupak manuk** ('chicken kelupak') (JHB 33, not yet identified) as **sayur**, saying that it was neither **kayu** nor **akau**.

Besides the **lamba** group, Araceae plants such as **latu** (including **raun latu,** JHB 199, *Homalomena cordata* (Houtt.) Schott) are classified as **raun**. Allocating a plant to the category of **raun** indicates that an attempt to put it into any other overarching category would be unsuccessful, but that it is possible to single out its outstanding characteristic. Like **usak**, **raun** is the term for a part of a plant. As for **sancam** (vegetable), it appears to be a functional category, like English 'toy' (Wierzbicka 1984), defined in terms of the use of its fleshy parts. Wierzbicka argues that such functional categories should not be treated as collective groupings. By this argument **sancam, umbus**, or **raun** may not be plant categories at all, and plants such as **lamba** and **kelupak manuk** may be 'unaffiliated generics' (see below).

Unlike the categories of folk genera and life-forms, the objects grouped together under the heading of **usak, sancam, raun, sakot,** or **umbus** lack any "inherent physical nature or 'essence'" (Atran 1993, p. 198) uniting them. However, the categories *are* natural in that they appear in natural discourse; **sakot** in particular is used contrastively with **kumpau**, **kayu**, **akau**, and the others as a category grouping together various genera. These problematical categogies are not exclusively contrastive: a plant could belong primarily to a category such as **sakot** and secondarily to a life-form. Plants occupying the fuzzy zone between **kayu** and **akau**, such as **kayu piantok** (JHB 126 Rutaceae?, also **piantok,** JHB 165, Celastraceae?), or **nunuk** ('strangling figs', e.g. JHB 445, not yet identified) cannot similarly be included in both categories.

Unaffiliated generics?

Three kinds of plant are unaffiliated to any of the life-forms, intermediate-rank categories, or the problematical categories just described. Like life-forms they are 'kind of kinds', contrasting with other such groupings. Unlike life-forms, there are only a few members of each grouping, and these are closely related biologically, rather than "cross-cut[ting] biologically natural groupings of organisms" (Berlin 1992, p. 167). In the case of the first two categories to be considered, it is awkward to consider them as folk genera. Berlin (1992, pp. 171-176) has addressed the problem that unaffiliated generics "cannot easily be removed from the contrast set they share with full-blown life-form taxa" (ibid., p. 174). If one accepts Berlin's ethnobiological rank system, one must agree with him that one cannot have one's cake and eat it too (Berlin 1992, p. 175) by placing a category in both life-form and generic ranks. If Berlin's schema is correct, then a taxon may occupy only one ethnobiological rank. Rank, however, is distinguished from 'taxonomic level' (ibid., p. 64), which refers to structural position within a hierarchy as well as cognitive aspects.[24]

Bamboos are uncommon in the Merimbun area. Only two kinds of bamboo were collected, **bulu' baluy** (JHB 21, Graminae, JHB 430, not yet identified) and **bulu' embiling** (JHB 431, not yet identified). I also collected Dusun terms for four other bamboo: **bulu' aur, bulu' batong, bulu' gana**, and **bulu' ranap.**[25] Linguistically, **bulu'** suggests a folk-genus. But unlike other folk-genera sub-divided and labelled at the folk-specific rank, there is no unmarked, or 'regular' (*biasa*), 'genuine' (*bonor*) **bulu'.** Also, informants did not classify **bulu'** into any life-form or other higher-order category. Rather, they classified members of this group simply as **bulu'.** Thus it seems that **bulu'** could be described as a life-form. Some informants considered **bulu'** to be affiliated with certain palms, such as **tiad** (JHB 182, *Eugeissonia minor*) and **raring** (JHB 278, *Cyrtostachys renda*).

A clearer example of an unaffiliated generic is **bamban**, members of the Marantaceae family, including **bamban** (JHB 289, *Donax*), **bamban aig** (JHB 249, *Schumannianthus* prob. *dichotomus* (Roxb.) Gagnep.), and **bamban batu** (JHB 517, not yet identified). They are affilated with, and may be compared to, **bulu'** in that they have a similar general structure and may be used in the same way (constructing fish-traps and similar object), but in terms of their basic categorization they stand alone in the **bamban** category.

CONCLUSION

The goal of this paper has been to provide a provisional outline of Dusun ethnobotanical classification in terms of its most inclusive categories. From this study we can see how the Dusun order knowledge of the forest environment with which they have interacted. Through the collection of forest plants of all kinds, along with sociocultural information about those plants, a series of general, inclusive, higher-order categories emerge. By analyzing these it can be seen that they do not have the same cognitive or logical status. I have interpreted my data using as a starting point the concepts and terminology developed by Berlin (1992). The results lead me to conclude that Dusun general ethnobotanical categories are not symmetrical but lopsided, and may shade into each other, and that non-taxonomic categories of the sort Berlin would exclude are an *intrinsic* feature of Dusun ethnobotanical classification, and provide only ambiguous evidence for completely separate cross-cutting 'special purpose' classifications.

In the social and psychological process of classification there is a yearning to find symmetry, but if we agree with Berlin's premise that ethnobiological classification follows nature's plan, the appearance of asymmetry should not be surprising - natural discontinuities are not evenly distributed. My data show that the general rules set forth by Berlin for recognizing categories at various ranks hold in the main, but become unworkable in certain cases.

The Dusun evidence shows that higher-order ethnobotanical categories do not have equal weight. Life-forms are more salient than other categories, but even the life-forms show differential salience. On the basis of nomenclatural evidence, prominence in the environment, and knowledge about different kinds, **kayu** is the most salient life-form, followed by **akau**, **uwai**, and **kulat**. As life-form salience decreases, categories are more easily treated as folk-genera. Likewise, some polytypic unaffiliated folk-genera have some characteristics of life-forms. In marginal cases, deciding whether a category is a highly polytypic unaffiliated generic or a minor life-form is almost a matter for subjective judgment, taking into account number of members and degree of biological focus. Bulmer (1974, p. 23) calls attention to this problem, and ultimately dispenses with Berlin's concept of life-form and indeed the whole rank system. And Ellen's (1993, p. 116) finding for Nuaulu ethnozoology holds true Dusun ethnobotany: "there is a gradation of inclusive categories, some highly salient culturally and widely shared, others - decreasingly - less so." Ellen, too, is moved to question the existence of life-forms and the reality of ethnobiological rank. But I find that these concepts ring true in the main: they provide a helpful framework in which to make sense of most of Dusun ethnobotanical classification. I agree with Hays (1983, p. 593) that while Berlin's system requires "slight modifications. . . critics' insistence on the need for a radically different approach in folk biology is unwarranted."

Categories of intermediate rank, constructed around prototypes, include various plants not included in life-forms. Other categories based on selected characteristics rather than overall morphology may be used as life-forms in classifying plants. These heterogeneous, non-taxonomic categories intrude in the 'natural' system of classification which Berlin says is paramount.

Anomalous plants may be categorized marginally in several groups, though never in more than one major life-form. **Patagar anak** (JHB 100, *Schizaea dichotoma* Swartz), for example, is ambiguous, being woody but tiny, and growing straight out of the ground. It has been classified as **kayu**, **usak**, and **sakot**. Similarly, **badoi** (JHB 256, *Costus globosus* Bl.) is classified variously as **usak**, **ukit**, **uwai**, and **tumid**. It may be ascribed to one of these categories on the basis of its use, for example, as a minor food, or as binding material; or because of the way it grows low to the ground. Where perceived features do not result in a conclusive, unambiguous category, informants seeking to categorize a plant must consider use. Thus, a plant may be categorized with rattans (**uwai**) on the grounds that it is used to manufacture the carrying basket, *saging*. Certain plants may not be fully accommodated in any larger groupings, even though they may have affinities with plants in other categories.

ACKNOWLEGMENTS

The research described in this report was made possible by U.K. Economic and Social Research Council award R000 23 3028 to Roy Ellen for a study of "The Ecology and Ethnobiology of Human-Rainforest Interaction in Brunei". Travel funds to attend the RGS/UBD conference were provided by the British Academy. I thank the Brunei Museum (particularly the Ethnography Section) for its collaboration, and The Royal Botanic Gardens, Kew, and the Department of Forestry in Brunei for their assistance. Bantong Antaran of the

Brunei Museum collaborated in the collection of data and provided helpful suggestions on this paper. Above all I thank Roy Ellen not only for critically reading several earlier drafts of the paper but also for his guidance and support through all stages of the study.

LITERATURE CITED

ANTARAN, B. 1993. *The Brunei Dusun: An Ethnographic Study*. M.Phil. thesis, University of Hull.

ASHTON, P.S. 1964. *A Manual of the Dipterocarp Trees of Brunei State*. Oxford University Press, London.

ATRAN, S. 1993. Folk biological cognition. *Current Anthropology* 34:195-198.

BARTLETT, H.H. 1940. The concept of genus. I. History of the generic concept in botany. *Bulletin of the Torrey Botanical Club* 67:349-362.

BERLIN, B. 1972. Speculations on the growth of ethnobotanical nomenclature. *Language in Society* 1:51-86.

BERLIN, B. 1974. Further notes on covert categories and folk taxonomies: a reply to Brown. *American Anthropologist* 76:327-331.

BERLIN, B. 1976. The concept of rank in ethnobiological classification: some evidence from Aguaruna folk botany. *American Ethnologist* 3:381-399.

BERLIN, B. 1992. *Ethnobiological Classification: Principles of Categorization of Plants and Animals in Traditional Societies*. Princeton University Press, Princeton, NJ.

BERLIN, B., BREEDLOVE, D.E. & RAVEN, P.H. 1966. Folk taxonomies and biological classification. *Science* 154:273-275.

BERLIN, B., BREEDLOVE, D.E. & RAVEN, P.H. 1968. Covert categories and folk taxonomies. *American Anthropologist* 70:290-299.

BROWN, C.H. 1974. Unique beginners and covert categories in folk biological taxonomies. *American Anthopologist* 76:325-327.

BROWN, C. H. 1977. Folk-botanical life-forms: their universality and growth. *American Anthropologist* 79:317-342.

BROWN, C.H. 1984. *Language and Living Things: Uniformities in Folk Classification and Naming*. Rutgers University Press, New Brunswick, NJ.

BULMER, R.N.H. 1974. Folk biology in the New Guinea highlands. *Social Science Information* 13:9-28.

BURKILL, I.H. 1935. *A Dictionary of Economic Products of the Malay Peninsula*. Crown Agents for the Colonies on behalf of the Governments of the Straits Settlements and Federated Malay States, London.

CONKLIN, H.C. 1962. Lexicographical treatment of folk taxonomies. In Householder F. W. & Saporta S. (eds). *Problems in Lexicography*. Indiana University Research Center in Anthropology, Folklore, and Linguistics.

CLEARY, M. & EATON, P. 1992. *Borneo: Change and Development*. Oxford University Press, Singapore.

CORNER, E.J.H. 1940. *Wayside Trees of Malaya*. 2 vols. Government Printing Office, Singapore.

DRANSFIELD, J. 1992. *Rattans of Sarawak*. Royal Botanic Gardens, Kew, Surrey.

EDER, J. 1988. *On the Road to Tribal Extinction: Depopulation, Deculturation, and Adaptive Well-Being among the Batak of the Philippines*. University of California Press, Berkeley.

ELLEN, R. 1986. Ethnobiology, cognition and the structure of prehension: some general theoretical notes. *Journal of Ethnobiology* 6:83-98.

ELLEN, R. 1991. Grass, grerb or weed: a Bulmerian meditation on the category *monote* in Nuaulu plant classification. In Pawley A (ed.). *Man and a Half: Essays in Honor of Ralph Bulmer*. Uniprint, Aukland.

ELLEN, R. 1993. *The Cultural Relations of Classification: An Analysis of Nuaulu Animal Categories from Central Seram*. Cambridge University Press, Cambridge.

ELLEN, R., in press. Modes of subsistence and ethnobiological knowledge: between extraction and cultivation in Southeast Asia. In Warren & Slikkerveer (eds). *Pithecanthropus Centennial, Vol. C*. Pithecanthropus Foundation, Leiden.

ELLEN, R. 1994. Review of Ethnobiological Classification, by B. Berlin. *Man*.

ELLEN, R. (this volume). Putting plants in their place: Anthropological approaches to understanding the ethnobotanical knowledge of rainforest populations.

ELLEN, R. & BERNSTEIN, J., in press. *Urbs in rure*: cultural transformation of the rainforest in Brunei. *Anthropology Today*.

HAYS, T.E. 1983. Ndumba folk biology and general principles of ethnobotanical classification and nomenclature. *American Anthropologist* 85:592-611.

HUNN, E.S. 1977. *Tzeltal Folk Zoology: The Classification of Discontinuities in Nature*. Academic Press, New York.

KERSHAW, E.M. 1992. *Final shifts: some why's and hows of Brunei Dusun convergence on Malay*. Paper presented at Borneo Research Council 2nd Biennial International Conference, Kota Kinabalu.

KING, V.T. 1993. *The Peoples of Borneo*. Basil Blackwell, Oxford.

MORRIS, B. 1993. Review of Ethnobiological Classification, by B. Berlin. *American Anthropologist* 95:733-734.

OCHSE, J.J. 1931. *Vegetables of the Dutch East Indies*, English edition. Department Landbow, Buitenzorg.

PERANIO, R.D. 1972. The Bisaya. In LeBar F. M. (ed.). *Ethnic Groups of Insular Southeast Asia, Volume I: Indonesia, Andaman Islands and Madagascar.* Human Relations Area Files, New Haven.

PERANIO, R.D. 1977. *The Structure of Bisaya Society: a Ranked Cognatic Social System.* Ph.D. thesis, Columbia University.

PUKUL, H. & ASHTON, P. n.d. *A Check List of Brunei Trees.* Brunei Forestry Department, Bandar Seri Begawan.

QUISUMBING, E. 1951. *Medicinal Plants of the Philippines.* Technical Bulletin no. 16, Department of Agriculture and Natural Resources, Manila.

RANDALL, R. & HUNN, E.S. 1984. Do life forms evolve or do uses for life? *American Ethnologist* 11:329-349.

RIDLEY, H.N. 1925. *The Flora of the Malay Peninsula.* L. Reeve, Ashford, Kent.

TAYLOR, P. 1984. "Covert categories" reconsidered: identifying unlabeled classes in Tobelo folk biological classification. *Journal of Ethnobiology* 4:105-122.

TAYLOR, P. 1990. *The Folk Biology of the Tobelo People: A Study of Folk Classification.* Smithsonian Contributions to Anthropology, no. 34. Smithsonian Institution Press, Washington.

WIERZBICKA, A. 1984, Apples are not a "kind of fruit": the semantics of human categorization. *American Ethnologist* 11:313-328.

WIERZBICKA, A. 1985. *Lexicgraphy and Conceptual Analysis.* Karoma Publishers, Ann Arbor.

WONG, M. 1992. *The Future of Tasek Merimbun.* Unpublished Position paper, Brunei Museum, Bandar Seri Begawan.

ENDNOTES

1. Compare Eder (1987) on 'deculturation' - the loss of cultural and social identity. See also Ellen & Bernstein, in press.

2. The only complete ethnographic description of the Brunei Dusun is Bantong Antaran's thesis (Antaran 1993). However, these people are very closely related - culturally, socially, and linguistically - to the Bisaya of Sarawak (see Peranio 1972, 1977). The main difference between the two groups is a political one: they occupy territories in different states.

3. Comparative studies have shown that specialised terms such as those for plants are among the last to disappear from a lexicon. However, in the Dusun case that knowledge of the plant world needed to preserve this part of the lexicon is itself disappearing in the younger generation, not only because traditional subsistence is giving way to wage employment, but also because parents now tend to speak to their children in Malay rather than Dusun (Kershaw 1992).

4. Wierzbicka (1985) challenges the universality of this claim, offering as counter-evidence the primacy of higher-level (viz. life-form) terms in English - especially **bird**, **tree**, and **fish**. The case, she says, is the same in other European languages. She suggests this may be due to the fact that native English-speakers do not live close enough to nature for various kinds of some plants and animals to be important. See also Berlin (1972).

5. On cultigens and cultivars see Antaran 1993, pp. 71-79. The author provides Dusun terms for 17 types of common vegetable, 32 types of seasonal fruit tree, and 12 types of non-seasonal fruit. He also provides names for 15 varieties of rice, and discusses the cultivation of tobacco and sago.

6. JHB followed by a number refers to a specimen in the herbarium of the University of Kent - Brunei Museum Tasek Merimbun project. Sets are held at the Brunei Forestry Centre, the Royal Botanic Gardens, Kew, and the University of Kent. At the time of writing only 55 percent of specimens have been identified. I indicate genus and species name if available, providing family names only in instances where I wish to show a contrast.

7. *F. indica* is also categorized as a rattan in Malaysia, Java, and Sumatra (Burkill 1935, pp. 1024-1025; see also Bartlett 1940, p. 353).

8. The Dusun ethnobotanical data do not support the rest of Brown's evolutionary theory, in which 'grerbs' (grasses plus herbs) emerge as life-forms before 'vines', and in which herbaceous plants and bushes but not rattans are life-forms.

9. Because of its lack of complexity, **buluh** is considered an unaffiliated generic rather than a life-form, and is described under the section heading, 'Unaffiliated generics?'.

10. In rattans as in other plant classes, my herbarium specimens and other data are not by any means complete. Dransfield (1992) has documented, for Sarawak (Brunei's neighbour), 102 species of rattans in 13 genera.

11. This points to a distinction between uninomial (one-word) and binomial (two-word) nomenclature. When the life-form term in not included in a binomial plant name, the first term is as a rule the generic name and the second name is a sub-generic modifier. Dusun ethnobotanical classification below the genus level will be discussed in another paper.

12. Before the introduction of metal scrapers rattan was also used to scrape cassava tubers.

13. It is unclear from my study that other Dusun communities also lack generic names for inedible mushrooms. It may also be the case that women's ethnomycological knowledge exceeds that of men.

14. Cf. Taylor's (1990, p. 64) discussion of 'residue' and 'residual taxa'.

15. For interesting discussions of the ethnobiological rank or status of 'grass' see Brown (1984, pp. 20-21) and Wierzbicka (1985, p. 190).

16. The identification of a specimen in the Brunei Checklist Project, Royal Botanic Gardens, Kew.

17. Derato is a god in Dusun cosmology. The nuts of this palm are believed to be used and preferred by Derato. Hence this palm is 'Derato's pinang'.

18. Two other palms, uncollected, were also included in this grouping:- **dudor** and **uwat endali'**.

19. Specimen in the Brunei Checklist Project Database.

20. I provisionally consider these terms to be synonymous, but it is possible that some distinction exists between plants grouped as **sancam** and those grouped as **umbus**.

21. Ellen (1991) has considered the classificatory status of a very large but heterogeneous collection of plants, called 'monote', among the Nuaulu. Like Dusun **sakot**, they are a residual grouping and lack focal members. Ellen considers and ultimately rejects the term 'weed', with its underlying criterion of dis-utility, as a gloss for monote.

22. The use of reduplication in this case seems to mean 'like **lamba**, or **lamba**-ish.'

23. Additionally, two plants are identified in their names as **umbus**, which also means 'vegetable': **umbus kenawai** (JHB 353) and **umbus tunjong** (JHB 132), neither identified as yet.

24. Berlin considers level to be "trivially easy to define," while the definition of 'rank' is "quite slippery" (1992, p. 65). However, he never explains exactly what he means by level.

25. The staff of the Royal Botanic Gardens, Kew, have provided me with a listing from their Brunei Checklist Project Database, still in progress at the time of writing. The Kew database includes the following Dusun terms for bamboos.
Buloh baden (alt. **badan**) *Dinochloa*
Buloh gana *Gigantochloa*
Buloh betung (alt. **abang**) *Gigantochloa* aff. *levis*
Buloh nanap *Schizostachyum latifolium*
The Checklist also includes two specimens identified, though not specifically in the Dusun language, as **buluh balui** (*Gigantochloa* and *Gigantochloa* sp. nov.).

The Socio-Economic Benefits of Indonesian Forest

MUBYARTO

Faculty of Economics, Gadjah Mada University, Jalan Kaliurang, Bulak Sumur Complex, Yogyakarta 55200, Indonesia

KEY WORDS: forest products, timber concessions, exploitation, rural development, legislation.

INTRODUCTION

Since the beginning of the New Order Government, forest products, especially timber, has become the most important export product of Indonesia after oil. Some 60 million ha of Indonesian production forest (out of a total of 144 million ha), has been contracted out to 579 concessionaires, most of them in East and Central Kalimantan (255 concessionaires or *Hak Pengusahaan Hutan* (HPH), 24 million ha), Maluku and Irian Jaya (64 HPH, 10 million ha), and Riau-Jambi (96 HPH, 9 Million ha). The revenue from forest exploitation has been so large that these regions have experienced a timber bonanza, giving a degree of prosperity to the majority of their population. In East Kalimantan, the economic impact has been so dramatic that people from other regions have flooded the province to take part in enjoying the bonanza.

According to Manning (1971, p.56) "The boom has brought a tremendous influx of people into the region - Bandjarese from South Kalimantan, Bugis fishermen, and farmers and townsfolk from Makassar, under-paid and under employer East Javanese, and Malaysians from Sabah and Sarawak. Within the region too there has been a substantial movement of labour into the timber and associated trades. Dajak Farmers and hunters have left their traditional pursuits to join bandjir kap teams and earn anything from Rp 15,000-25,000 to as high as Rp 100,00-150,000 per month. The wide variation in wages results from the fact that extraction of timber is almost totally dependent on a heavy rainfall (bandjir) which swells rivulets so that logs can float downstream and government officials have left prestigious positions to take up jobs in timber companies where they may earn anything from 15,000 to 45,000 rupiahs per month. The great variety of imported goods in the night markets, the flash of Japanese motor scooters down the narrow streets of Samarinda, the provincial capital, and the constant traffic of speedboats or tugs towing rafts of logs down the Mahakam and returning with supplies for the isolated timber camps all attest to the new economic prosperity brought about by the boom."

Legislation issued by the Indonesian Government to facilitate forest exploitation has included the Foreign Investment Law (1967), the Domestic Investment Law (1968), the Basic Forestry Law No. 5/1967, and Government Regulations No. 22/1967 and No. 21/1970. The influence of all this legislation has been very effective in facilitating large scale investment. So much so that after 20 years of forest exploitation (since 1970) there is now growing concern about the sustainability of our tropical forest. The free hand and generous economic stimuli given to HPH holders have resulted in these concessionaires becoming uncontrollable in maximising their profits.

In response to this behaviour, the government is currently working on a variety of arrangements to appropriate a greater share of the forest economic rent and to find ways to

D. S. Edwards et al. (eds.), Tropical Rainforest Research – Current Issues, 451–456.

force the HPH holders to help rural development in the concession areas. These are the so-called social forestry programmes which have become one of the more ambitious aspects of the next five year plan (1994-1999).

SOCIO-ECONOMIC BENEFIT: THEORY AND PRACTICE

Theoretically, forest exploitation should provide multiple benefits to many people of the region where the forest is located, and to the national economy as a whole. The benefits should be in the form of increasing supply of forest products, increasing income and foreign exchange, and of course increasing employment.

In practice, however, forest exploitation has rarely resulted in such benefits. Initially, foreign investors found difficulties in getting skilled labour for forest exploitation, forcing them to bring manpower from their own countries. By 1971 there were 4800 foreign labourers, mostly from Malaysia and the Philippines, working especially in East Kalimantan.

Early in the boom period, the Government became aware of the possible conflict between the short term objective of income and profit maximisation and the long term objective of forest sustainability, and entrusted the State forest corporation Perhutani to hold monopoly power. This monopoly, accompanied by government control of foreign investment did not last long, due perhaps to inefficiency and weak management. In 1970 the joint venture between Perhutani and a Japanese company was dissolved and Perhutani suffered a loss of US$ 10 million. Eventually a Government Regulation (PP No. 21/1970) was applied. Thus began a new phase in forest exploitation, where the Government gave great freedom to private companies, national as well as foreign, to operate, even going so far as to freeze the customary right of forest utilization to avoid possible problems for the concessionaires.

The stipulation to freeze the local rights to collect forest products clearly indicated that the Government 'sided' with the investors, allowing the companies to operate without difficulties, and with the understanding that the government should collect large royalties which could be used for national development. It was also expected that the forestry companies would create employment for local people.

Government regulations did anticipate possible dangers of deforestation as a result of such encouragement. Thus article 8 stipulated that in principle HPH holders (the forest companies) are allowed to do only selective logging and that they are required to practice forest conservation and reforestation.

This stipulation, however, is seldom implemented, mainly because there are not enough foresters to enforce the practice in the field. In theory, the companies should submit an annual working plan and a five year working plan to which they should adhere. But again because of the lack of supervision, local forestry representatives are usually forced to approve all the plans without question. It is therefore understandable that there is a great danger of corruption at the various levels of the forest bureaucracy.

THE FOREST AND THE PEOPLE'S ECONOMY

Article 33 of 1945 Constitution formulated by the then economist Vice-President of the Republic of Indonesia emphasizes the people's aspect of the economy. This aspect has always been the basis of state guidelines, law, and other legislation. The present National Development Plan (*Garis Besar Haluan Negara*) of 1993 gives even stronger emphasis on developing the people's economy. In recent years, it has become clear that local people in some provinces which are rich in forest resources have not received direct benefits. Equity in wealth distribution, which is supposed to be the country's official policy (within the framework of the trilogy of development), does not seem to be easily implemented in the regions.

The New Order Government which in 1966 inherited an almost bankrupt economy where poverty was widespread, had few alternatives when designing a recovery programme. Because economic growth was a prerequisite for further development, the Government chose a growth-oriented policy in all sectors, and for this it turned to forest and petroleum products (oil) to finance the economic recovery programme. The Government first introduced the Foreign Investment Act, followed by the Basic Act on Forestry, both in 1968, and in the following year the Government issued the Domestic Investment Act.

In order to facilitate forest exploitation, the Basic Forestry Act (No. 5/1967) anticipated possible problems of the forestry companies *vis a vis* local people possessing customary rights. In one of the articles (No. 17), the customary right is recognized but exercising such rights must not "disturb" companies holding concessionary rights granted by the Minister of Agriculture/Forestry in Jakarta.

This 'neutral' stipulation, however, was changed drastically in Government Regulation No. 5/1970, which explicitly froze the customary rights of the local community, in order for forest exploitation to operate without impediment. Local people who had been dependent for their livelihood on forest resources for generations, were forced to sacrifice their sovereignty to the incoming investors in order for the latter to help the Government earn much-needed foreign exchange. This resulted in perpetuating local people's poverty, and of course exacerbated the income and social gaps, even during the time when the country as a whole was achieving impressive economic growth.

Now in the face of the second long term National Development Plan which gives priority to equity programmes, there must be radical change in the development strategy, emphasizing the alleviation of poverty. In other words the new strategy must not be solely growth oriented but more equity oriented to achieve social justice for all Indonesian people.

The Government (i.e. the Ministry of Forestry) has indeed started equity programmes, the most recent of which has been the HPH Bina Desa in which the HPH holder must help local rural development. The programme is new and faces many problems, especially because the profit orientation of the HPH holders makes it difficult for them to operate useful rural development programmes. It is also apparent that the programmes rely too much on the 'ingenuity' of HPH holders. In reality the HPH holders in turn rely too much on the regional representatives of the Ministry of Forestry.

Socio-economic and cultural problems, and also that of shifting cultivation, can be solved only through correcting the many different acts (laws) concerning forests and forest exploitation. Above all, the two acts (No. 5/1967 and Government Regulation No. 21/1970) which are not helping the development of the people's economy must be modified.

The people's economy is a self-reliant economy. Sustainable development must be based on the development of the people's economy, especially in the rural areas. It is this rural people's economy which should become the basis of national economic resilience. And in the forestry sector, shifting cultivators are the principle actors. The shifting cultivators must be protected and must be encouraged to become permanent farmers. Such considerations are included in the social forestry programme already formulated in the 1993 State Guidelines.

THE SHIFTING CULTIVATORS AND FOREST SQUATTERS

In its overall effort to sustain the forest, the Government faces difficulties in pinpointing who is responsible for damaging the forest. The easiest thing is to accuse the shifting cultivators because, based on their name, they are destroying the forest to grow food crops. Thus the only way to control them is to force them to stop the practice and to settle them as farmers.

The Ministry of Forestry has made an effort to do this through training cadres of settled farmers (KANITAP).

In 1989, it was estimated that over one million families practiced shifting cultivation, covering an area of 7.3 million ha, so that each family of five would cultivate 6.9 ha. Even though the Government has attempted to reduce shifting cultivation, there is strong evidence that the area affected keeps increasing. An intensive study has shown that forest damage (estimated to be expanding at 1.3 million ha per year) is not caused solely by shifting cultivators but is also the result of deliberate logging for short term gain. It is suspected that the HPH holders, through close cooperation and under the protection of powerful individuals may be involved in such illegal logging because the 'stolen' timber is much cheaper. It is reported that application of strict controls on timber cutting in Samarinda for example, quickly increased the timber price in Balikpapan to a 'normal' level.

Native local people involved in shifting cultivation have indeed carefully practiced sustainable forest management, and are likely to become productive settled farmers if seeds, fertilizer, and farm implements can be purchased cheaply or obtained on credit. On the other hand Government action is needed to discourage the illegal forest squatter, rather than those exercising traditional rights. At the same time the illegal logging must be stopped.

The Ministry of Forestry is the most suitable agency to take responsibility for ensuring the sustainability of the forest. In the 1993 State Guideline, it is stipulated that:

1. Forest exploitation and forest product utilization is to be organized, jointly with the local Government, in such a way as to guarantee the highest revenue for the State, in line with the interests of the people living in and around the forest area.
2. Forest exploitation must result in minimal damage and utilization of the forest must be conducted within the framework of the law.
3. Forestry extension must be aimed at upgrading forestry personnel and increasing the participation of the local community in the form of social forestry.

FOREST ECONOMIC RENT

The economic aspect which has attracted much debate among writers is the 'inequitable' sharing of forest economic rent among the HPH holders, the Government, and the local community. Following a national seminar on the economic aspects of forest exploitation held in October 1992, the journal Prisma (No. 6, 1992) focussed on this problem.

The problem was the subject of a Doctoral Dissertation submitted to London University, and a summary of this research was presented in BIES in July 1979. It was shown that during 1973/1974, the average economic rent collected by the Government was as low as 30% of the fob price, while the portion received by the timber company was more than two times greater (63% of the fob price) (Ruzicka 1979).

It is astonishing that 13 years later, there had been no improvement in the government share, which even dropped to 8% in 1989 and 17% in 1990 (Mubariq 1992). In absolute terms the 1989 economic rent in terms of stumpage value should have been US$ 99.24 per m^3 of log, so that for the total timber production of 31,215,000 m^3 the total economic rent would have been US$ 3,098,000,000. However, in the case of logs which were not exported (export was banned completely in 1985) the economic rent, based on the stumpage value was US$ 29.66 per m^3, so that the uncollected economic rent amounted to US$ 13.66 or 46%. Thus the domestic price of timber, which at that time was only 55% of the fob price, decreased the stumpage value of logs to only one third (31%).

It is now clear that the failure of the Government to collect the economic rent represented a 'hidden subsidy' to the logging industry, resulting in the domestic price of timber being lower than it should have been. This in turn created an excessive demand for logs. The great demand for logs on the domestic market pushed up the price and this resulted in excessive logging. Based on this economic analysis, if the Government desires to encourage sustainable tropical forest logging in Indonesia, there is a need to collect a greater share of the total economic rent which should be used by the Government to finance reforestation.

In recent years the Government has encouraged the development of forest plantations (*Hutan Tanaman Industri*, HTI) using the funds collected from the Reforestation Fund (DR). The development of HTI will of course reduce the pressure of demand on the natural forest, especially for the supply of pulp and paper factories. In 1991/92 it was reported that HTI areas reached 516,038 ha, consisting of 207,058 ha in Sumatra, 165.331 ha in Kalimantan, 117,417 ha in Java, and 26,232 ha in other regions (Soeharto 1992). That HTI developed faster in Sumatra than in Kalimantan is an indication that the remaining natural forest area is still much larger in Kalimantan. In addition, most of the pulp and paper factories are located in Sumatra, primarily in Riau and North Sumatra.

There is a danger that encouraging the HTI policy may accelerate deforestation. This is likely because there is a provision that allows transfer of HPH concessions to HTI. This means that the requirement for selective logging can be changed to clear felling in order for concession holders to become eligible to apply for HTI concessions. The HTI holder is then eligible to receive government capital support of up to 46% of the total investment; the remainder can be financed by credit from the reforestation fund at zero interest rate. Under this scheme, there is a danger that these incentives will encourage the creation of fictitious areas of HTI to obtain subsidies, which may be used for other investment projects. At the same time the HTI becomes a 100% government project owned by private companies.

RECOMMENDATION TO THE HPH HOLDERS

The HPH holder is a business entity and has a 'mission' to earn profit. Since sustainable harvesting requires an allocation of funds, this means reducing possible profits. The Government must work out how to allow the company to earn sufficiently attractive profits but at the same time set aside sufficient funds to sustain the forest.

In addition, since the Government desires each company to absorb as many labourers as possible, there is a need for other incentives to achieve this latter objective, for example by offering an incentive for every 100 or 1000 labourers employed. It is in this context that the social forestry approach requires the HPH holder to involve more and more local participation in forest exploitation, or, looking at the problem from the other side, that the company must pay more attention to rural development. If cooperation can be improved between the company and thc local people, not only can government supervision be reduced, or made easier, but community participation will also be made more effective.

At the present time, relationships between the HPH company and the local people are not cordial, the causes of which can be traced back to government regulations. In China, reforestation is the responsibility of the whole village community, so that the individuals who actually carry out the reforestation are controlled directly by the village community. If this system can be applied in Indonesia, it would increase community participation in forest sustainability. However, in order to do this the Government and the Parliament must take the necessary steps to change some forest legislation. Government regulations which separated the HPH company and the local people must be abolished and replaced with regulations emphasising the social forestry approach stipulated in the 1993 State Guidelines. Of course this must be implemented in the context of increasing regional autonomy and decentralization related to the implementation of Law No. 5/1974 on government in the regions.

On the other hand, the HPH holder and the forest industries must realize the need to apply this social forestry approach. The HPH company needs to establish a special section and to appoint special personnel responsible for applying social forestry approach. The need for a sustainable forestry management programme is urgent since the target date to achieve sustainability, set down by the International Tropical Timber Organisation (ITTO), of which Indonesia is a member, is the year 2000. Some people consider that Indonesia will be able to satisfy the ITTO criteria in 1995. However, if we examine present forest conditions in some regions, one cannot be so optimistic.

Companies will need to set aside some portion of their profit to contribute to the achievement of sustainable forest management. They will have to pay more attention to rural development. It must be noted that many of the HPH holders do not really know what to do because of the multitude of regulations issued by the Ministry, which change too quickly. It is recommended that the HPH companies should be more active in proposing policies to the Ministry.

CONCLUSIONS

The socio-economic benefits of the Indonesian forest are now being questioned because of the demand from the people to improve living conditions. Sustainable forest management must be achieved in order to guarantee a decent standard of living for the next generation.

ITTO set the target of sustainable forest management by the year 2000 and Indonesia is optimistic that this target can be met. However, not all of our working methods are sufficiently good. In fact much of the legislation must be modified to accommodate the needs of a wider sector of society.

The 1993 State Guidelines explicitly mentioned that forest destruction must be stopped and that forestry companies should pay higher royalties to the Government. If these funds can be paid to the Government early every year, then the Government is in a much better position to implement sustainable forest management.

LITERATURE CITED

MANNING, C. 1971. The Timber Boom. *Bulletin of Indonesian Economic Studies* VII(3).

MUBARIQ, A. 1992. Rente Ekonomi dalam Eksploitasi Hutan Tropis. *Prisma* 6.

RUZICKA, I. 1979. Rent Appropriation in Indonesia Logging: East Kalimantan. *Bulletin of Indonesian Economic Studies* XV(2).

SOEHARTO. 1992. *Pidato Kenegaraan* (Presidential State Address 1992). Pp. 242-243.

Putting plants in their place: Anthropological approaches to understanding the ethnobotanical knowledge of rainforest populations

ROY ELLEN

Department of Sociology and Social Anthropology, Elliot College, University of Kent at Canterbury, Canterbury, Kent CT1 7NS, United Kingdom

INTRODUCTION

Over the last decade or so ethnobotany has assumed a scientific prominence previously denied it. It is endorsed by institutions with a high international profile (Kew, the Royal Geographical Society, WWF, UNDP, UNESCO), has a market value placed upon it by foresters, agronomists, development advisers and pharmacologists, and has become pivotal in preserving the cultural identity and knowledge of indigenous peoples whose traditional way of life is under threat (Posey 1990). Ethnobotanical knowledge has, therefore, become both economic commodity and political slogan. This is particularly true with respect to the plant knowledge of rainforest peoples, as these peoples are often those with the highest media profile. However, in our eagerness to exploit a product and to demonstrate its usefulness, there has been a tendency to oversimplify what ethnobotany entails and just how it can be useful. I argue in this paper that we must not be narrow-minded or simplistic in our conception of ethnobotanical knowledge, and that to take anything less than a broad culturally-contextualised approach may miss the point of the relevance of indigenous knowledge altogether.

In an historical review of the subject published a little over a decade ago, Richard Ford (Ford 1978, p. 29) says of ethnobotany that it represents a common discourse but lacks a unifying theory. That this is still the case is in part due to the historical differences between what we might call biological ethnobotany and anthropological ethnobotany.[1] The first operates within a bioeconomic paradigm. At its narrowest this has been simply the study of plants used by indigenous people (whoever they might be). Data thus provided have often been no more than species identifications against vernacular names, and lists of uses. In its most recent version this approach has become heavily identified with the discovery of new applied botanical and phytochemical possibilities, and with the economic valuation of rainforest (Peters *et al.* 1989); and so becomes virtually synonymous with what is widely understood as the remit of economic botany. Although the global significance of such work should not be underestimated, bare lists are of limited application (see e.g. Alcom 1981a, p. 228).

The second kind of ethnobotany operates primarily within a cultural-linguistic paradigm, is concerned with indigenous rules and categories and is particularly evident in the work of Harold Conklin, Brent Berlin and those who have been influenced by them in the US; and in the work of those within a Francophone tradition closely associated with what is now the Laboratoire d'Ethnobiologie-Biogéographie in Paris, and which finds much of its intellectual legitimation in the writings of Claude Lévi-Strauss. Although there is some overlap between

[1] This is similar to (though by no means identical with) other distinctions which are sometimes drawn, such as between different 'levels' of ethnobiological enquiry (Bulmer 1974).

D. S. Edwards et al. (eds.), Tropical Rainforest Research – Current Issues, 457–465.

the data and findings of the two kinds of ethnobotany, it is the second which - especially when situated in systematic botanical investigation - comes closest to what Ford (*ibid.*, p. 44) believes best exemplifies what ethnobotanists should be doing: namely, the *direct* investigation of the relations between plants and humans, placing plants in their total cultural context.

I wish to explain here why it is vital to adopt a 'progressive contextualist' (Vayda 1983) approach, beginning at the species level and working outwards; an approach which locks specific local knowledge within increasingly more general but denser culturally-relative paradigms, and which links indigenous ecological know-how to general subsistence and social behaviour. This is not - nor should it be - privileged or esoteric anthropological methodology, but a strategy of practical relevance to botanists, foresters and all those working in the development field.

The main weakness of the first kind of ethnobotany, then, is one of definition: that for many it has become simply that empirical knowledge embodied in fixed folk traditions which can be of value to global science. A pharmacologist looking at ethnobotany in this sense thus sees potential drugs, a botanist scientifically-unrecorded species, food scientists new foods, materials scientists new materials (for example, milled coconut fibre as a substitute growing medium for peat), and the Body Shop some new politically-correct cosmetic. Such an approach tends to reduce ethnobotany to partial unconnected bits: or to put it another way, knowledge is transformed into information. In the process much of potential value is lost. Rather than generating selected bits of information in a framework determined by the quite specialised requirements of conventional biological science and taxonomy, we should be focussing on connected systems of local knowledge, informed by an understanding that such knowledge is intrinsically highly variable between individuals and populations, fundamentally situational, and dynamic. My remaining remarks - which put more flesh on these bare assertions - fall under four headings: (a) species-focussed empirical knowledge, (b) knowledge of the rainforest as an entity, (c) the interconnectedness of social and mundane knowledge, and (d) domains of ethnobotanical knowledge defined functionally. I shall conclude by returning to the question of variability and change.

SPECIES-FOCUSSED EMPIRICAL KNOWLEDGE

By species-focussed knowledge I mean those data which an investigator will primarily elicit with a single voucher specimen as the unit of analysis, or multiple specimens of the same 'natural kind'. This will include all names which informants give to a particular specimen, the terminology used for its parts, knowledge of variation in form, its reproductive cycle, seasonal and geographic distribution, growing conditions; in fact, all information that a field botanist would ordinarily seek concerning a particular species being investigated. The secondary data, *secondary* only in some putative logical order of elicitation, refer to cultural uses, and practices with respect to the specimen type. These include how it is located, collected, prepared for use, protected and cultivated, including of course its recognised effects: material, nutritional, medicinal or symbolic; and evaluations of the plant type and particular specimens of the plant. As Hays has stressed (Hays 1974, 1982), it is really only when we examine our data in this way that we can see what constitutes a 'use', and that one use is by no means functionally equivalent with another. A third dataset at this level consists of associated ecological knowledge. This includes knowledge of the identity and activity of predators and diseases, the effects which the plant has on the habitat in which it is found and on associated flora, and the effects they have on the plant. Rich data in this area has emerged over the years with respect to specific cultigens, for example the important work of Paul Richards on folk understandings of *Zonocerus* infestations of cassava in southern

Nigeria (Richards 1979, 1985). More recently, and of more concern to rainforest specialists, Darrell Posey (Posey 1988) has shown how Amazonian Kayapó maintain buffer zones between gardens and forest which contain plants with nectar-producing glands on their foliage which have the effect of drawing away aggressive ants and parasitic wasps from crops. Then there are the effects of different patterns of rainfall, predation and human extraction; the role of other organisms in the dispersal of seeds, and the use made of the plant by other non-human organisms as food. Nuaulu, for example, in the lowland rainforests of Seram in eastern Indonesia, are well-informed on many species not because they are directly of use to humans, but because they represent the food of animals which they hunt, particularly cassowary, pig and deer. In other words, plants have to be understood as part of the web of forest life, not simply in isolation.

Although such knowledge as discussed here can be presented in terms of a checklist for the species as a whole, it is important to realise that most folk-botanical knowledge differs fundamentally from conventional scientific knowledge in not being organised abstractly within some convenient general-purpose classification, but rather with respect to particular contexts, defined perhaps in terms of different subsistence activities (hunting, planting trees, collecting, artefact manufacture, burning swidden trash, or whatever). Moreover, how that knowledge is apprehended by people will be determined by culturally relative coordinates of sense perception which sometimes deviate sharply from the expectations of scientifically-trained personnel; for example, the significance of olfactory and textual stimuli compared with the purely visual. Thus, much knowledge is inaccessible except via a research strategy which allows a multi-focal approach; and if investigators additionally wish to appreciate how local people make key subsistence decisions, they must attend to the categories of knowledge locally applicable (Ellen 1982). To do all this requires a general understanding of the underlying principles of ethno-botanical knowledge (e.g. Berlin *et al.* 1974, Conklin 1954), but goes beyond the more formal systematical aspect of folk classification, to what we might call folk-autoecology (rather, that is, than folk-synecology). But, then, that ethnobotany and the study of indigenous knowledge of particular plants needs to be ecological has been recognised for over half a century (Carter 1950, Jones 1941), even if the implications have seldom been followed through, or its relevance fully understood.

Species-focussed knowledge is, additionally, variable between individuals, and so may not be fully documented until a number of people have been queried regarding the same species, or indeed the same individual has been interviewed on the same subject on a variety of occasions. Moreover, since such knowledge is situational, the optimal conditions for finding out are with respect to growing specimens, at different stages in the life-cycle, in contexts where there is a predetermined subsistence goal, such as cutting bush for swiddens or collecting fungi for food. Just as knowledge is not fixed in its contemporary distribution, so it also changes through time: the result of generations of trial-and-error testing, extensive experimental evidence, enormous individual specialist and collective experience; new plants moving in and out. This does not mean that mistakes are never made, but simply that daily routine pragmatic considerations act as a major incentive for identifying effectiveness. As in science as a whole, there may be bad folk science; but there is also a wealth of good science (Richards 1993). It is this which pharmaceutical firms are now taking advantage of, but they and us also need to appreciate *how* that experience has been gained.

KNOWLEDGE OF THE RAINFOREST AS AN ENTITY

Ethnobotanical studies of rainforest peoples have uniformly demonstrated an impressive breadth of knowledge of plant species, in addition to a *depth* of knowledge of particular significant species. A recent attempt to collate data on the total inventories of plant categories for different subsistence populations (Brown 1985, p. 44) shows strikingly how rainforest populations have repeatedly been found to yield much longer lists than populations living in other environments, lists which may consist of between 800 and 2000 items at the upper end. Clearly, to some extent this reflects the subsistence necessity of those who extract from such environments, and also the strong tendency for non-significant and non-useful species to go unnamed or to be lumped within larger generic categories. Though it is still debated whether or not human populations could ever have entirely survived on rainforest plant matter, other than when they can rely on palm starch (Headland 1987), breadth of knowledge is undoubtedly a key part of any adaptive strategy.

In addition to this, it has become clear that systematic encyclopaedic knowledge is situated within folk-models which reflect an ability to connect observations at the species level with informed perceptions about forest structure and dynamics. What constitutes 'forest' is, of course, something we might expect to vary cross-culturally, but even if we restrict ourselves to focal shared meanings it is clearly a complex categorical construction. Thus, the Nuaulu generic term *wesie* (Ellen 1993) is anything but uniform or empty in the way they perceive, understand and respond to it. It is more like a mosaic of resources, and a dense network of particular places each having different material values. In this sense it is much like the modem scientific modelling of rainforest as a continuous aggregation of different biotopes and patches, varying according to stages in growth cycles, degree of regeneration, underlying geology, altitude, geography and natural contingency. In such a view, simple distinctions between 'secondary' and 'primary' begin to look pretty academic. Seventy-eight percent of the 272 forest trees identifiable by Nuaulu have particular human uses, and it is through their uses that they are apprehended (Ellen 1985), wherever they are found. The picture is similar elsewhere. Carneiro (Carneiro 1988, p.79) reports that, on average, Panare, Tembe, Urobu and Chacobo peoples of the Amazon basin use at least two-thirds of the tree species growing in the forests (see also Carneiro 1978). Such peoples, like the Indonesian Nuaulu, being forest-fallow swidden cultivators, also have sensitive understandings of how forest changes as a consequence of different soils, selective extraction, cutting and burning; and of the regrowth stages following abandonment. It is this knowledge which permits such strategies - despite persistent rumours - to be self-sustaining (Dove 1983a, b), and which underlies its deliberate application in a way which assists the recovery of degraded areas (Shanley 1991). The classic description of this knowledge is still that by Conklin (Conklin 1957) for the Hanunóo of Southeastern Mindoro.

THE INTERCONNECTEDNESS OF SOCIAL AND TECHNICAL KNOWLEDGE

Empirical knowledge of plants, as I have referred to so far, does not exist apart from a broader socially-informed understanding of the world, in some kind if hermetically sealed vacuum from which other aspects of culture are excluded. Detailed knowledge of plant reproduction or symbiosis may, for example, comfortably co-exist with beliefs about the world which have not been empirically tested in a conventional scientific sense. Everything is seen as connected through chains of mutual causation to give rise to a complex notion of nature. Indeed, it is often plausibly argued that rainforest peoples have cosmologies which in certain respects anticipated the systems view of the world which underlies modem ecology; as argued, for example, by Reichel-Dolmatoff (1976) for the Tukano of the northwest Amazon. Such a view is also reconstructable for the Nuaulu, and we can infer their conceptualisation of forest and attitudes towards it from systematic ethnobiological data, from human subsistence practices with respect to it, from the rules of land tenure, from general statements about forest, from rules relating to the extraction of resources and sanctions consequent upon their infringement, indirectly from myths, stories, taboos, and so on. From these we can distill four general characteristics which, taken together, summarise the most important aspects of Nuaulu conceptual engagement with 'forest'. The first characteristic has already been discussed, namely that forest is never experienced as homogeneous. The remaining three are that forest is a complex categorical construction, that there is an inner connection between history, identity and forest, and that forest is not morally neutral.

Although uncut forest is recognised by the Nuaulu as a single entity (*wesie*), it contrasts in different ways with other land types depending on context. It may contrast with *wasi* (owned land, which may sometimes display very mature forest growth), emphasising a jural distinction; with *nisi* (garden land), emphasising human physical interference; or with *niane* (village), emphasising landforms: empty as opposed to well-timbered space, inhabited (dwelt) as opposed to uninhabited space, untamed as opposed to tamed space, all with various symbolic associations and practical consequences for Nuaulu consumers. Although there are no Nuaulu words for either 'nature' or 'culture', it is in the various and aggregated senses of *wesie* that the Nuaulu come closest to having such a term, and from which the existence of an abstract covert notion of 'nature' can reasonably be inferred.

The values which Nuaulu attach to forest are thus multi-faceted and differential, simultaneously materially useful and culturally meaningful. And in the same way that the material uses to which forest is put must be understood in specific and local terms, so too the social implications. While the Western conception of environment is something which is 'opposed' to people, or some kind of medium in which we *dwell*, and which is therefore bounded; the Nuaulu conception of environment is not as a space in which they hang, but much more like a series of fixed points to which particular clans and individuals are connected. These points are objects in an unbounded landscape linked to their appearance in myths; use of land is at every turn inseparable from specific sacred knowledge, sometimes mutually contradictory and obscure, though never absent.

The undeniable effect of merging practical usefulness, mythic knowledge and identity in the construction of the category *wesie* is to give it a moral dimension. That is, there are right and wrong ways in which to engage with forest which arise in part from the specific social histories of parts of it, but also from its intrinsic mystical properties. Forest is unpredictable, dangerous and untamed, and various attempts are made to control it. This is reflected in the inferential symbolic opposition between 'nature' and 'culture' evident in most ritual, in the specific rituals conducted prior to cultivating forest, in the charms which are used to protect travellers in the forest, in the prohibitions on certain behaviours and utterances while in the forest, in the correct ritual disposal of its products.

The practical implications of the interconnections between the social and the environmental can be very important, and it is often the case that subsistence practices triggered by cultural beliefs (for example, linked to prohibitions) appear to regulate resources. It is in the context of all this that we must understand Nuaulu ritual restrictions (*sasi*) on harvesting certain forest products at particular times. Certainly, the *effect* of all of these things may well be to conserve resources and maintain biodiversity, and in particular cases people may consciously do so. But we should not confuse effect with purpose. Much of what appears to be 'ecological balance' amongst rainforest peoples is either illusory or simply a beneficial function of low population densities and benign subsistence practices. However, when governments and other agencies interfere and seek to introduce 'rational' measures to conserve resources, ignorant of local cultural representations of the forest, their purposes may be meaningless to local peoples, as Richards (Richards 1992) has shown for Mende living on the edges of the Gola in Sierra Leone. Similarly, governments acting with the best of intentions may interfere with cultural regulators (purposeful or inadvertent) which are often more sensitive, and in the long term, more effective (e.g. Morauta *et al.* 1982).

FUNCTIONALLY-DEFINED DOMAINS OF ETHNOBOTANICAL KNOWLEDGE

All human knowledge is represented and articulated not only through the systematic lens of general-purpose classifications, such as those offered by science, but also through the domains constructed around specialised applied knowledge. I have already noted how specifically 'botanical' knowledge is embedded in the particular contexts in which plants are encountered, but knowledge can also be presented in terms of functionally-defined abstractions, such as in the context of medical or veterinary knowledge, specific craft specialisations (such as house building), and so on. Indeed, many peoples who might otherwise deny formal knowledge of ethnobotany, may in fact have a considerable practical working knowledge known to them under other headings (see e.g. Brown 1992, Telban 1988). Thus, the unit of analysis (in this case most usually the voucher specimen) can always be organised into arrays which demonstrate significant discontinuities other than those found in classifications based predominantly on morphology.

VARIATION AND CHANGE

Ethnobotanical knowledge and practice must, therefore, be understood contextually, but it must also be recognised as being intrinsically variable and subject to change. One of the problems with the kind of narrowly-conceived ethnobotany which I have been criticising is that it conceals much of this. We need to take into account intra-cultural variation, often hidden in over-generalised accounts which imply that one respondent is as good as any other with respect to a particular subject. We now have convincing demonstrations that knowledge may vary qualitatively and quantitatively according to crucial social variables, such as gender, and even within groups defined in terms of a single key social indicator (e.g. Hays 1974). We also need to take into account variation between different rainforest populations, reflecting for example different modes of subsistence (Brown 1985), though it may be artificial to separate out populations on the basis of their apparent degree of interaction with forest, degree of acculturation or integration into the market. I take a broad view of what constitutes a rainforest population, including all those (such as peasants) interacting with the rainforest (Denslow & Padoch 1988). Knowledge may pass between superficially different groups which are in contact or who extract from similar biotopes. We must be wary of inventing a category of 'traditional' peoples whose knowledge is regarded as somehow pristine and superior, however much the temptation.

Narrow conceptions of the ethnobotanical enterprise also sustain a non-dynamic view of particular human populations. In the popular imagination, peoples of the tropical rainforest have hitherto been remote, isolated, living in more or less the same place, unchanging, living 'in harmony' with their surroundings. In fact, we now have plenty of evidence to the contrary. Ironically perhaps, it is ethnobotany which has provided much of this, in terms of the movement of plant species between isolated points on all sides of the globe. The migration of the main plant domesticates is well-documented, though we know less of the impact this has had on the forests themselves. Certainly, human settlement has introduced many varieties of cultivated aws into the lowland forest of Seram in the Nuaulu area, both deliberately and inadvertently. For example, *Tectona grandis* and *Toona sureni* are now established, though they were probably introduced during the seventeenth century; and the ornamental *Delonix regia* cannot have been planted before the nineteenth (Ellen 1985, p. 563). And the picture is no different as far as foraging populations are concerned (e.g. Fox 1952), where a nomadic lifestyle has served well to distribute certain humanly-valued trees widely. An excellent example of this is the role of the Mbuti of the Ituri forest of Zaire (Ichikawa 1992) in the propagation of genera such as *Canarium* and *Landolphia*. Many rainforest peoples selectively fell trees, protect valued species, replant the tops of wild *Dioscorea* tubers, transplant palm suckers, on a sufficiently systematic basis for us to speak of 'rainforest management'. Alcorn (Alcorn 1981b, p. 410) reports for the Huastec of Mexico's Sienna Madre that 63 percent of 800 wild species documented had uses and 25 percent were actively manipulated. The selective extraction of wild species, strategic burning, and swiddening at optimal conditions may combine to give rise to distinctive patches and new opportunities for colonisation; and there are numerous examples - such as those provided by the Kayapó (Posey 1988, p. 89) - of the deliberate preservation of corridors of mature forest between plots as some kind of biological reserve. The cumulative effect of this in some areas must be considerable, and persistent human interventions may be presumed to have co-evolutionary consequences. Such disturbances not only improve the rainforest as a human resource base but contribute in significant ways to its structural patchiness and biodiversity. They also signal, along with the gratuitous destruction of vegetation for short-term gain, such as the felling of entire trees to catch a single arboreal mammal (Ellen 1986, Rambo 1985), that the practical impact (if not the cosmological representation) of even the demographically smallest human group need not always be benign. Many tropical forest peoples have complex histories of movement, no better illustrated than from the intricate ethnic mosaic of northern Borneo (see e.g. Freeman 1970). Trade and exchange has existed for centuries between upriver peoples, including remote foraging populations, linking them to peoples of the forest fringes, the estuaries, and ultimately the global economy (Dunn 1975, Hoffman 1984).

We now have a considerable body of ethnobotanical knowledge and a growing volume of literature pertaining to rainforest peoples (Conklin 1972); for southeast Asia alone there is a critical mass (Barrau (n.d.)). But the data are of variable quality, ranging from the fragmentary notes of field botanists, the systematic inventories of an active generation of colonial officials (e.g. Burkill 1935), and detailed long-term studies within an anthropological context (e.g. Friedberg 1990, Revel 1990, see also bibliography in Brown 1985). The priority now, as I see it, is to organise such information into accessible databases and recognise the differential quality of our data (some of which is shoddy, much of which is inadequate). But perhaps even more importantly we need to shift our focus from abstracted empirical data on particular plants to contextual accounts. Indigenous peoples have perceived, interacted with and made use of tropical rainforest in historically diverse ways, and it is clearly important for those making recommendations in the fields of conservation and sustainable forest management to take this into account.

ACKNOWLEDGEMENTS

This paper is a by-product of an ESRC funded project (R000 23 3088) on 'The ecology and ethnobiology of human-rainforest interaction in Brunei'. I would like to thank Jay Bernstein for his comments, and am additionally grateful to the British Council for the award of a travel grant to make attendance at the UBD-RGS conference possible. Perhaps I should explain - at the risk of seeming an imposter - that my own empirical work has primarily focussed on ethnozoology. This paper might have had such a focus, and still dealt with essentially similar issues, though I judge that current debates concerning the role of indigenous rainforest knowledge assume that this is largely a matter of ethnobotany.

LITERATURE CITED

ALCORN, J.B. 1981a. Factors influencing botanical resource perception among the Huastec: suggestions for future ethnobotanical enquiry. *Journal of Ethnobiology* 1:221-230.

ALCORN, J. B. 1981b. Huastec noncrop resource management: implications for prehistoric rainforest management. *Human Ecology* 9:395-417.

BARRAU, J. (n.d.). *An ethnobotanical guide for anthropological research in Malayo-Oceania*. UNESCO Science Cooperation Office for Southeast Asia, (Paris).

BERLIN, B., BREEDLOVE, D. & RAVEN P. 1974. *Principles of Tzeltal plant classification: an introduction to the botanical ethnography of a Mayan-speaking people of the highland Chiapas*. Academic Press, New York.

BROWN, C. H. 1985. Mode of subsistence and folk biological taxonomy. *Current Anthropology* 26:43-53.

BROWN, K. 1992. *Medicinal plants, indigenous medicine and conservation of biodiversity in Ghana* (GEC 92-36). Norwich: CSERGE Working Papers, UEA.

BULMER, R. N. H. 1974. Folk biology in the New Guinea Highlands. *Social Science Information* 13:9-28.

BURKILL, T. H. 1935. *A dictionary of the economic products of the Malay peninsula (2 volumes)*. Crown Agents for the Colonies, London.

CARNEIRO, R. 1978. The knowledge and use of rainforest trees by the Kuikuru Indians of central Brazil. In Ford, R.I. (ed.). *The nature and status of ethnobotany*. Anthropological Papers 67, Museum of Anthropology, University of Michigan, Ann Arbor, Michigan.

CARNEIRO, R.L. 1988. Indians of the Amazonian forest. In Denslow, J. S. and Padoch, C. (eds). *People of the tropical rainforest*. University of California Press, Berkeley.

CARTER, G.F. 1950. Ecology-Geography-Ethnobotany. *Scientific Monthly* 70:73-80.

CONKLIN, H.C. 1954. *The relation of Hanunóo culture to the plant world*. Unpublished Ph.D. dissertation. Yale University, New Haven, Connecticut.

CONKLIN, H.C. 1957. *Hanunóo agriculture, a report on an integral system of shifting cultivation in the Philippines*. Food and Agricultural Organization of the United Nations, Rome.

CONKLIN, H.C. 1972. *Folk classification: a topically arranged bibliography of contemporary and background reference through 1971*. Department of Anthropology, Yale University, New Haven, Connecticut.

DENSLOW, J. and PADOCH, C. (eds) 1988. *People of the tropical rainforest*. University of California Press, Berkeley.

DOVE, M.R. 1983a. Forest preference in swidden agriculture. *Tropical Ecology* 24:122-142.

DOVE, M.R. 1983b. Theories of swidden agriculture and the political economy of ignorance. *Agroforestry Systems* 1:85-99.

DUNN, F.L. 1975. Rainforest collectors and traders: a study of resource utilization in modern and ancient Malaya. *Monographs of the Malaysian Branch of the Royal Asiatic Society* 5.

ELLEN, R.F. 1982. *Environment, subsistence and system: The ecology of small-scale social informations*. Cambridge University Press, Cambridge.

ELLEN, R.F. 1985. Patterns of indigenous timber extraction from Moluccan rain forest fringes. *Journal of Biogeography* 12:559-587.

ELLEN, R.F. 1986. What Black Elk left unsaid: on the illusory images of Green primitivism. *Anthropology Today* 2:8-12.

ELLEN, R.F. 1993. Rhetoric, practice and incentive in the face of the changing times: a case study of Nuaulu attitudes to conservation and deforestation. In Milton, K. (ed.). *Environmentalism: the view from anthropology*. ASA Monograph 32. Routledge, London.

FORD, R.I. 1978. Ethnobotany: historical diversity and synthesis. In Ford, R.I. (ed.). *The nature and status of ethnobotany*. Anthropological Papers 67. Museum of Anthropology, University of Michigan, Ann Arbor, Michigan.

FOX, R.B. 1952. The Pinatubo Negritos: their useful plants and material culture. *Philippine Journal of Science* 81:173-414.

FREEMAN, J.D. 1970. *Report on the Iban*. LSE Monographs 41. Athlone, London.

FRIEDBERG, C. 1990. Le Savoir Botanique des Bunaq. Percevoir et Classer dans le Haut Lamaknen (Timor, Indonésie). *Mémoires du Muséum National d'Histoire Naturelle, Botanique* 32. Muséum National d'Histoire Naturelle, Paris.

HAYS, T. E. 1974. *Mauna: Explorations in Ndumba ethnobotany*. Ph-D. dissertation, University of Washington, Seattle.

HAYS, T. E. 1982. Utilitarian/adaptationist explanations of folk biological classification: some cautionary notes. *Journal of Ethnobiology* 2: 89-94.

HEADLAND, T. N. 1987. The wild yam question: how well could independent hunter-gatherers live in a tropical rainforest ecosystem. *Human Ecology* 15: 463-492.

HOFFMAN, C. 1984. Punan foragers in the trading networks of southeast Asia. In Schrire C. (ed.). *Past and present in hunter-gatherer studies*. Academic, London.

ICHIKAWA, M. 1992. Traditional use of tropical rainforest by the Mbuti hunter-gatherers in Africa. Pp. 305-317 in Itoigawa, N., Sugiyama, Y., Sackett, G. P. and Thompson, R. K. R. (eds). *Topics in primatology: behaviour, ecology and conservation*. University of Tokyo Press, Tokyo.

JONES, V. H. 1941. The nature and scope of ethnobotany. *Chronica Botanica* 6:219-21.

MORAUTA, L., PERNETRA, J. and HEANEY, W. (eds) 1982. *Traditional conservation in Papua New Guinea: implications for today* (Monograph 16). Institute for Applied Social and Economic Research, Papua New Guinea.

PETERS, C. M., GENTRY, A. H. and MENDELSOHN, R. 0. 1989. Valuation of an Amazonian rainforest. *Nature* 339: 665-6.

POSEY, D. 1988. Kayapó Indian natural-resource management. In Denslow, J. S. and Padoch, C. (eds). *People of the tropical rainforest*. University of California Press, Berkeley.

POSEY, D. A. 1990. Intellectual property rights: what is the position of ethnobiology? *Journal of Ethnobiology* 10: 93-98.

RAMBO, A. T. 1985. *Primitive polluters: Semang impact on the Malaysian tropical rainforest ecosystem*. Anthropological Papers 76, Museum of Anthropology. University of Michigan, Ann Arbor, Michigan.

REICHEL-DOLMATOFF, G. 1976. Cosmology as ecological analysis: a view from the rain forest. *Man (N.S.)* 11: 307-318.

REVEL, N. 1990. *Fleurs de Paroles: Histoire Naturelle Palawan*. Volumes I and 2. Peeters/SELAF, Paris.

RICHARDS, P. 1979. Community environmental knowledge and African rural development. *IDS Bulletin* 10: 28-36.

RICHARDS, P. 1985. *Indigenous agricultural revolution: ecology and food production in west Africa*. Hutchinson, London.

RICHARDS, P. 1992. Saving the rain forest? Contested futures in conservation. In Wallman, S. (ed.). *Contemporary futures: perspectives from social anthropology*. ASA Monographs 30. Routledge, London.

RICHARDS, P. 1993. Natural symbols and natural history. In Milton, K. (ed.). *Environmentalism: the view from anthropology*. ASA Monograph 32. Routledge, London.

SHIRLEY, P. 1991. *Recovery and utilisation of ethnobotanical knowledge: the Rio Capim, Eastern Amazonia. Final Report: Ethnobotanical exchange between Asia and Amazonia*. UNDP, New York.

TELBAN, B. 1988. The role of medical ethnobotany in ethnomedicine: a New Guinea example. *Journal of Ethnobiology* 8: 149-169.

VAYDA, A. P. 1983. Progressive contextualization: methods for research in human ecology. *Human Ecology* 11: 265-281.

Valuing a tree: the ethics of environmental evaluation

DAVID H. BENNETT[1]

Aboriginal and Torres Strait Islander Commission

ABSTRACT. The paper looks at four sets of arguments for valuing trees: economic, instrumental, deeper ethical, and cultural. On the economic level, I consider material and functional values of trees. On the instrumental level I look at the arguments of John Passmore for the preservation of trees (or any part of the environment) on a use basis. I also look at an additional argument that rainforest lock in deadly viruses. On the deeper ethical level, I give the example of Aldo Leopold's extension argument and Deep Ecology. On the cultural level, I consider an idealised traditional relationship between Australian Aborigines and their environment.

KEYWORDS: economic, instrumental, land ethic, deep ecology, Australian aborigines, value.

There are a number of ways to value a tree or a forest. Four sets of arguments for valuing trees that I should like to discuss briefly are economic, instrumental, deeper ethical, and cultural.

ECONOMIC VALUE

The first set of arguments are economic arguments. Perhaps one of the simplest ways to value a tree or forest is economically. A usual way of valuing a tree economically is for its material value, such as woodchips or timber. In the case of woodchips the value can be determined by some formula like:

$$V = H + P + T + R + X + M$$

where Value (price per tonne) = Harvesting + Processing + Transporting + Royalties + Taxes + Profit Margin or some similar set of variables, where value, expressed in dollars (or yen or marks, etc.) per tonne, equates to the costs of obtaining a tonne of woodchips plus a margin for profit. Considerations such as labour, are built into the variables of harvesting, processing, transporting, and so on. What should be noted is that there is no variable in this equation for value of the tree or forest. The value of a tree or forest is an externality to the calculation. The tree is not an object of economic consideration, let alone of ethical consideration.

This is not so in the case of trees used for timber. A timber equation would include a factor for the type of tree or type of forest

$$V = H + P + T + R + X + M + L$$

where Value (price per tonne) = Harvesting + Processing + Transporting + Royalties + Taxes + Profit Margin + Timber Type.

Here the tree enters the equation, because factors such as the quality, origin, colour, hardness, scarceness, and end purpose for the timber, among other things, impact on the price per unit. So a plank of teak (*Tectona grandis*) will cost more than the same quantity

[1]Present address: Native Title Section, Land, Heritage and Environment Branch, Aboriginal and Torres Strait Islander Commission, P.O. Box 17, Woden, ACT 2606, Australia.

D. S. Edwards et al. (eds.), Tropical Rainforest Research – Current Issues, 467–475.

of Monterey pine (*Pinus radiata*). But considerations such as these do not take into account such factors as durable wood products being carbon prisons or carbon sinks, which remove and keep out of the atmosphere at least some of the 5.5 billion tonnes per year of anthropogenic carbon emissions. That requires another way of determining the economic value of a tree, besides its *material value*.

Another and more interesting economic analysis is the *functional value* of a tree. One such economic functional analysis estimates the value of a single mature tree at $196,250 (in Singapore dollars).

According to Professor T.M. Das of the University of Calcutta "a tree living for 50 years will generate $31,250 worth of oxygen, provide $62,000 worth of air pollution control, control soil erosion and increase soil fertility to the tune of $31,250, recycle $37,500 worth of water and provide a home for animals worth $31,250. This figure does not include the value of the fruits, lumber or beauty derived from trees".[2]

Here it is the services that a tree provides that determine its value. This method is crude, of course. It indicates only an average functional value for a hypothetical tree. The functional value of a given tree of a given species might be more or less than this amount. The point remains that the economic value of a tree cannot be reduced to its materials. A living tree has values that it does not retain when converted to chips or lumber. It will no longer remove carbon from the atmosphere even if it may continue to imprison carbon it trapped while alive. Even a standing dead tree will have functional values that it will cease to have when converted to chips or lumber. It will no longer provide shelter, food, and protection to non-human species, independent of any value it may have for humans. And this point can be taken in the opposite direction. Just as the total value of a single tree is not the value of one aspect, so the value of many trees is not merely the sum total of each of those trees. A forest is more than a collection of trees. It is an ecosystem. A system whose material and functional values exceed the value of its component parts, including not just trees, but undergrowth, epiphytes, animal species, the relationships among these and other elements, and so on. A forest is not just a stand of trees. Its value is not just a stand of trees. A mere stand of trees is a tree plantation.

But there other ways, besides economic, of valuing the services that a tree provides. Also there are other services, instrumental services, that a tree provides.

INSTRUMENTAL VALUE

The second set of arguments are instrumental arguments. Instrumental value construes nature and the environment simply as a means to ends. Instrumental value is attributed to an item that is in some way of value to or useful to another, usually to human ends and values. But the ends are not necessarily economic ends. The ends may be ethical. Instrumental value arguments are usually arguments for the ethical value of a thing.

There are a variety of instrumental arguments that may be used to argue for the value of a tree. The Australian philosopher, John Passmore, while he does not argue specifically for trees, lists economic, scientific, recreational, renewal, aesthetic, and posterity varieties. To take the last argument first, the '*Posterity Argument*' holds that the present generation has a duty to hand over the world "to our successors in a better condition" (Passmore 1974, p.91). Part of handing over the world in a better condition (from the ecologic, as opposed to the economic, perspective) is not to destroy species or let them perish through contributory

[2] Poster from the Singapore Zoological Gardens, *The Value of a Tree*, based on *Update Forestry*, Michigan State University.

negligence and thereby deprive posterity of the opportunity to observe, study, and otherwise have the experience of the diversity enjoyed by the present generation. For example, this is the last generation with a chance to preserve pristine rainforests. If this generation does not do it, no other can. With a 'Posterity Argument' Passmore is not calling for passive restraint on the part of humans in their dealings with the environment. For him, handing over the world in a better condition means "transforming the world into a civilised state" (Passmore 1974, p.178). Betterment through transformation may mean the preservation of some rainforest, for example, but with the major responsibility to other humans, it could mean the destruction of parts of the environment considered harmful or 'un-civilised'. Civilisation has too often been on a collision course with environmentalism. But in the words of Rachel Carson, "The question is whether any civilization can wage relentless war on life without destroying itself, and without losing the right to be called civilized?" (Carson 1962, p.99).

Yet, Passmore attempts to soften the harsh impact of this line of reasoning by a second argument: the *'Gene Pool Argument'*, or as another Australian philosopher, Bill Grey (formerly William Godfrey-Smith) labels it, the '*Silo Argument*' (1979). As humans utilise the plant and animal resources of the world they modify them through cultivation and domestication, but a modification for one purpose, say breeding cattle to produce more beef or greater drought resistance, can cause undesirable side effects, say less hybrid-vigour. Therefore, it is in humanity's own interest to keep wild stocks as a reservoir or gene pool.

Passmore holds "a purely economic argument will suffice to establish at least a *prima facie* case against the clearing of wilderness [and] the destruction of species" (1974, p.102). But obviously the 'Gene Pool' or 'Silo Argument' is reasonable in the case of trees only for preserving sample wild counterparts of cultivated species that produce fruit, blossoms, timber, resins, or other economically valuable products. Also it could be objected that a *prima facie* case does not hold if it is more economic to destroy a wilderness or species for immediate gain. "Uncontrolled timber extraction, slash and burn agriculture and large scale aggressive plantation development are destroying genetic materials" (Tinggal 1992, p.143). These same practices have placed some rainforest species on the edge of extinction and some rainforests on the edge of annihilation.

To protect species that are not currently held to be valuable, Passmore appeals to the '*Rare Herb Argument*'. "A species often turns out to be unexpectedly useful, a tropical plant to contain pharmacologically valuable substances" (Passmore 1974, p.102). Grey claims of this argument, "The rare herb argument is a sound argument in favour of preservation on the grounds of *prudence*, but it is not, on the face of it, a *moral* argument" (Godfrey-Smith 1980, p.31). If the argument is effective, it need not be a moral argument. But prudence has not prevented the destruction or near destruction of rainforests and the myriad of species that inhabited them. In Australia alone "up to 75 per cent of rainforest has been lost" by direct or indirect human intervention since 1788, the establishment of the first white colony (World Commission on Environment & Development 1990, p.44). Hopefully the 'Rare Herb Argument' could be applied to argue against the obliteration of additional species; but it says nothing of the numbers, conditions, or manner under which these reservoirs are kept and could permit the decimation of these species to plantation populations, botanic gardens, or token specimens.

Related to the 'Rare Herb Argument' is the '*Lock Argument*'. The 'Lock Argument' is not one suggested by Passmore. If the 'Rare Herb Argument' holds that there are potentially beneficial substances locked in rainforests that may one day prove valuable, as an example, for their ability to prevent or cure diseases then the 'Lock Argument' holds that there are potentially dangerous substances locked in rainforests. Fifty percent of the world's biological diversity is in rainforests.

"The rain forest, being by far the earth's largest reservoir of both plant and animal species, is also its largest reservoir of viruses, since all living things carry viruses. Just how large the tropical reservoir of viruses is no one knows, but here is one way to consider the question. The earth is estimated to contain between three million and thirty million species of plants and animals. Most of the species are fungi, insects, and non-insect arthropods, such as ticks and mites, and the bulk of them live in tropical forests. Viruses often adapt to one or two species. For example, human beings carry more than a hundred different cold viruses that are adapted almost exclusively to the human host. If we suppose that every species carries one virus exclusively adapted to it, then there may be from three to thirty million strains of viruses. Possibly the number of virus strains is much larger than that - perhaps a hundred million - but nobody has ever tried to count them.
When an ecosystem suffers degradation, many species die out and a few survivor-species have population explosions. Viruses in a damaged ecosystem can come under extreme selective pressure. Viruses are adaptable: they react to change and can mutate fast, and they can jump among species of hosts. As people enter the forest and clear it, viruses come out, carried in their survivor-hosts - rodents, insects, soft ticks - and the viruses meet *Homo sapiens*" (Preston 1992, p.62).

Exactly how many viruses and diseases are lurking in rainforests may not be the most important question, when the virulence of some that have already emerged is considered. Not how many there are, but how lethal they are, may be the question. Among those viruses suspected to have come from rainforest is H.I.V. "The human immunodeficiency virus looks like an example rather than a culminating disaster. As lethal viruses go, H.I.V. is by no means nature's preëminent display of power" (Preston 1992, p.62). Apart from H.I.V., there is Ebola Zaire, "a slate wiper in humans" (Preston 1992, p.59). It kills 88 per cent of the people it infects. At least 16 other lethal viruses have also emerged (Preston 1992, p.62).[3]

Thus to destroy a rainforest is to break the 'lock'. If the viruses remained with their hosts and if rainforests were of no other value, then the current rapid rate of their destruction (3 hectares per minute) could be a preventative. But the viruses do not remain with their hosts and rainforests do have other values besides 'locking away' lethal viruses. The emergence of the viruses appears to be a natural consequence of the ruin of the tropical biosphere. This says nothing of the threat these viruses pose to other species.

Prudence would demand that humanity not knowingly endanger itself or other species. *Prudential arguments* are pleas for prudent behaviour and are typically grounded in human interests.[4] Prudential arguments are arguments encouraging humans (whose Latin binomial, *Homo sapiens*, even carries the idea of wisdom in it) to exercise wisdom, although mainly the wisdom of protecting human interests. These arguments have a powerful appeal because they pander to human interests, but they do not acknowledge the value of the environment or parts of it, such as a tree, except so far as they promote human well-being or survival.

Another complementary argument to the 'Rare Herb Argument' is the '*Laboratory Argument*'. The 'Laboratory Argument' states the obvious thought that unless possible sources of pharmacologically valuable substances are preserved for experimentation then the diversity of humanity's environmental laboratory is diminished. As with the 'Silo

[3] Some of the emerging viruses are Lassa, Rift Valley, Oropouche, Rocio, Q fever, V.E.E., Guarnarito, Ross River, Monkeypox, Dengue, Chikungunya, Hantaan, Machupo, Junin, Mokola, Duvenhage, Le Dantec, Kyasanur Forest virus, Semliki virus, Crimean-Congo, Sindbis, O'nyong-nynong, Marburg and Ebola.

[4] These interests may be quite oblique, amounting for instance to merely thinking that something (valuable) is there.

Argument', the thrust of the 'Laboratory Argument' is that biological modification and experimentation by humans requires natural species as norms, as base lines and as fresh resources. Grey states: "This is the argument that wilderness areas provide vital subject matter for scientific inquiry which provides us with an understanding of the intricate interdependencies of biological systems, their modes of change and development, their energy cycles, and the sources of their stabilities. If we are to understand our own biological dependencies, we require natural systems as a norm, to inform us of the biological laws which we transgress at our peril" (Godfrey-Smith 1979, p.311).

In addition to the environment providing material resources, Passmore also holds "that there is refreshment as well as enjoyment to be found in wandering through wild country. (Not only recreation but re-creation; it renews one's sense of proportion)" (Passmore 1974, p.107). This can be split into two arguments: the '*Recreation Argument*', where humans use wilderness, such as a rainforest, as their gymnasium and the '*Cathedral Argument*' where humans draw spiritual revitalisation from wilderness.

But the instrumental argument upon which Passmore builds much of his overall case for a stronger adherence to existing ethics and ethical principles rather than the development of a new ethics is the '*Aesthetic Argument*'. Passmore holds that, "to justify action against the beauty-destroying polluter; at most what is needed is a strengthening of existing moral principles" (Passmore 1974, p.56). The 'Aesthetic Argument' is not limited to protecting the environment against polluters. As another Australian philosopher, H.J. McCloskey observes, "we hold the beautiful and mysterious in nature in awe and wonder, see much as precious, and view with moral shock and even moral outrage the wilful destruction of beautiful species" (McCloskey 1983, p.34). The 'Aesthetic Argument' conveys the concept of 'beauty-destroying', which Passmore finds is a "more explicitly Western tradition that it is wrong unnecessarily to destroy - a principle embodied in the concept of 'vandalism'" (Passmore 1974, p.124). This argument places the moral onus on the destroyer. It is intended to protect the environment from "'wanton' destruction where no defence of the destruction can be offered" (Passmore 1974, p.124).

Passmore states that these arguments should be enough to preserve the environment and give value to a tree. But they have certainly failed on the first count. As is apparent from just one statistic, Edmund O. Wilson graphically captures the American attitude towards tropical rainforests, "More money is spent in the bars of New York City in two weeks than is spent annually on tropical research" (Gordon & Suzuki 1990, p.52). If human interests change and humans, at least in the dominant Western and Western styled societies, placed even less value on the environment than they already do, then these arguments fail as well. These arguments still do not place value on the tree itself, for itself. For that a different value needs to be recognised, an intrinsic value.

DEEPER ETHICAL VALUES

The third set of arguments are deeper ethical arguments. These are arguments that recognise that instrumental values are not enough to protect the environment or rainforest or a tree. These deeper ethical arguments recognise the value of natural items for their own sake.

While instrumental arguments can point people in the correct direction as far as saving the environment, they are only a first step. Further steps are needed. One further step is Aldo Leopold's land ethic. He recognised that items in the natural environment have value-in-themselves as well as or despite any value they may have for humans. He summed this up in the maxim, "A thing is right when it tends to preserve the integrity, stability, and beauty of the biotic community. It is wrong when it tends otherwise" (Leopold 1966, p.262). He calls for an extension of ethical consideration to the land, or more broadly and accurately, extending the ethical community to include the ecosphere. Leopold sought to make the ethical and biotic communities co-extensive. His extension principle is summed up as, "The land ethic simply enlarges the boundaries of the community to include soils, waters, plants, and animals, or collectively, the land" (Leopold 1966, p.219).

The value of a tree on Leopold's land ethic is the value of one being among many in the ethical community. Thus moral consideration is extended to trees. This does not necessarily mean that a tree is extended equal moral consideration to say a human in every case, nor does it necessarily mean that a tree looses out to a human in every case.

But further steps are possible. It is often maintained that what makes the difference between something that is valuable merely because it is of use and something that is valuable in-itself is that the latter has a *telos* or goal of its own. This is not simply that a tree passes from a seed to a sapling to maturity to death. This is closer to the idea that Aristotle put forward about the virtue of an item is its *telos* or that for which it is intended. This is the difference between instrumental value and intrinsic value. A thing may have both instrumental value and intrinsic value. This is generally held to be the case with human beings. This means recognising that a tree's functions of 'carbon sink' and 'oxygen pump' are more than simply instrumental values but a 'virtue' of the tree.

A still further step is to take a deeper environmental perspective such as Deep Ecology. "Deep Ecology is a normative and policy- and lifestyle-oriented theory" (Sylvan 1985, p.43). Like the science of ecology, which deals with the relationships between organisms and their environments, Deep Ecology is concerned with the place of organisms in their environment, but it is bound up with value judgments about the organisms, their environments, and the relationships between them, and with the wisdom of living by the principles revealed by ecology, and not, as with the science of ecology, with doing experiments that reveal the principles. In a sense, Deep Ecology uses ecology to overcome the embarrassment of science at not being able to convert knowledge into wisdom and state what is most needed - an appreciation of how to value the environment and ecological relationships.

The argument from Deep Ecology is that while humans are entitled to protect their own vital needs, they are not entitled to pollute the habitats of other species for non-vital reasons. In seeking to protect their short-term vital needs humans often damage systems that will meet their long-term needs. Long-term viability is part of determining a vital need. The human pollution of the environment is imprudent because it affects humans, but it is also wrong of humans to ignore that they share the environment with other species that have a claim to its use and an interest in its quality. Pollution modifies habitats and directly kills other species. Once again humans need to remember that they need other species more than other species need them. Humans are more likely to miss rainforests than rainforests are likely to miss humans.

Resource depletion is another form of human interference. In an interview, Arne Naess, founder of Deep Ecology, observed, "The shallow ecology movement talks only about resources of mankind, whereas in deep ecology we talk about resources for each species" (Bodian 1982, p.10). While the most common concern is with the depletion of non-renewable resources -i.e., there are finite quantities of fossil fuels such as coal and petroleum available and once they are depleted they cannot be replaced (not, at least, in human time spans) - there is also a significant and growing concern about what have been considered renewable resources. The depletion of the systems by which renewable resources are produced would be an eco-catastrophe for all species. Supposedly renewable resources are being depleted faster than they can be replaced. A case in point is the circular connection exists on Christmas Island with Abbott's Booby, nesting trees, and guano mining.

> "The conservation of Abbott's Booby provides a classic example of a conflict between the preservation of a scientifically interesting species - which occurs nowhere else and

whose value is difficult to access - and the utilisation of valuable resources. There are surface phosphate deposits derived from bird dropping over countless years. The main breeding areas of the Abbott's Booby coincide with areas rich in phosphate" (Ovington 1978, p.106).

The mining of the phosphate requires removing the trees in which the Boobies nest and thus destroys the breeding habitat of the Boobies, which are the source of the phosphate. The collection of the resources destroys the source. The source is depleted along with the resource. Here the trees are a vital resource for the Boobies and the guano is a vital resource for the trees. The value of the trees 'is difficult to assess', but the value of the loss of the trees means the loss of the Boobies and the guano industry.

Another more tragic example is the depletion of rainforests. The 'lungs of the world' are burnt out of existence. Vast resources are wasted by the burning of the rainforests to create new grazing lands.

But if the value of things in themselves is recognised then while a tree may still be used, it must be used to meet vital needs rather than mere convenience. This leads to the fourth form of argument, cultural values and the way, in this case, that Australian Aborigines both use and respect trees.

CULTURAL VALUE

A fourth set of arguments are cultural arguments. A tree can form part of a culture's identity. As Bill Neidjie, an Aborigine from Kakadu in the far north of Australia, says:

"When you cut tree,
it pump life away,
all the same as blood in my arm
If you feel sore....
headache, sore body,
that mean somebody killing tree or grass
You feel because your body in that tree or earth.
Nobody can tell you,
you got to feel it yourself." (Neidjie 1985, p.82)

A valuable lesson that can be learned from Aborigines is respect for the environment. Aboriginal traditions can help Western societies realise a role in the environment compatible with ecological principles, or as Leopold said, compatible with the continuing "integrity, stability and beauty of the biotic community". An Aboriginal viewpoint is very much like Leopold's, in that, it is concerned with the maintenance of the whole. No part can exist without the whole. Human beings are like any other species in this regard - they are part of the whole. As part of the whole, they are interrelated with the other parts. An appeal can be made for the maintenance of the other parts in order to maintain humans. What Aborigines have accepted is the concept that everything in the environment is connected to everything else. The connections may not be direct and they may be very weak, but they are nonetheless connected. It is almost too obvious to remark that if Western societies could continue to manipulate the world without any serious environmental repercussions to themselves, they would. Yet if it is accepted that the part cannot exist without the whole, then preserving the whole is in the interest of each part. Insofar as humans are a part of the whole it is in their best interests to preserve, protect or nurture the whole. If humans recognise that they are part of the environmental/ecological whole and not separable from the whole nor outside the whole, then the other parts can be seen as valuable for maintaining the whole and thus maintaining human interests. If human preservation is subordinated to environmental preservation on the grounds that human preservation is not possible without environmental preservation, then human preservation becomes instrumental to environmental preservation. This is what Aborigines practiced. This is stepping into the web of life as a participant rather than standing outside of it as a voyeur.

In summary, Aboriginal individuals feel a unity with and responsibility for specific objects or phenomena or species, but this bond with these specific elements does not preclude use of those objects or phenomena or species. The relationship is one of spiritual unity and mundane utility. Thus a tree is more than a resource, for those with a special relationship to a given species of tree, it is part of their very being.

CONCLUSION

We profit more from trees than just their dollar value. They perform vital functions from which we profit. They are part of the same ecosystems of which we are a part and they and we are interconnected functionally and ecologically. To realise the value of a tree we must look beyond its dollar value, we must take lessons from older cultures who realised the interconnectedness of the environment.

While traditional Aboriginal practices and attitudes cannot contribute so much to the specific content of changes, they can provide inspiration for general changes in context for Western societies. For example, Aborigines have a generalised attitude towards the environment that can be specifically applied. This attitude can be likened to the maxim of 'Think globally, act locally'. Aborigines practiced a form of bioregionalism. Ways in which the context of Western practices and attitudes can be changed include but are not limited to acknowledging that the distinctions between humans and other species (us/them) are less sharp than it is usually made out to be and that therefore humans are subject to the same ecological pressures and limitations as other species in their dealings with the environment; that a life simple in means can be rich in ends and that such a life can lower human impact on the environment; that Western societies should be encouraged to 'dwell in situations of inherent value'; that the relationships between humans and other species is not solely a matter of the other species functional utility to humans; that Western societies can better live in accord with their environment by accepting ecological principles and living according to the character of their particular bioregion. "We might observe what we have been doing badly to perpetuate our own welfare at the expense of diminishing the welfare of other species. Then we might consider Aboriginal approaches to grasp what, more elementally, we might better have pursued - not in imitation of Aborigines, but symbolically of their attitudes towards the maintenance of their environment and their participation in it and tempo of life. We might consider Aboriginal approaches that would permit us to be happier creatures and give us and other species a greater hope of ecological survival" (Bennett 1991).

LITERATURE CITED

BENNETT, D.H. 1991. Animal Rights and Aboriginal Concepts. Pp.53-69 in Croft, D.B. (ed.). *Australian People and Animals in Today's Dreamtime: The role of comparative psychology in the management of natural resources.* Praeger Press, New York.

BERNDT, R.M. 1974. *Australian Aboriginal Religion.* E.J. Brill, Leiden.

BERNDT, R.M. 1979. *A Profile of Good and Bad In Australian Aboriginal Religion. Australian and New Zealand Theological Review* 12:17-32.

BODIAN, S. 1982. Simple in Means, Rich in Ends: a conversation with Arne Naess. *Ten Directions* 7:10-12.

CARSON, R. 1962. *Silent Spring*. Houghton Mifflin, Boston.

GODFREY-SMITH, W. 1979. The Value of Wilderness. *Environmental Ethics* 1:309-19.

GORDON, A. & SUZUKI, D. 1990. *It's a Matter of Survival*. Allen and Unwin, Sydney.

LEOPOLD, A. 1966. *A Sand County Almanac*. Oxford University Press, New York (First published 1949).

McCLOSKEY, H.J. 1983. *Ecological Ethics and Politics*. Rowman and Littlefield, Totowa, N.J.

NEIDJIE, B., DAVIS, S. & FOX, A. 1985. *Kakadu Man...Bill Neidjie*. Mybrood, Sydney.

OVINGTON, D. 1978. *Australian Endangered Species: mammals, birds and reptiles*. Cassell Australia, Stanmore, NSW.

PASSMORE, J. 1974. *Man's Responsibility for Nature*. Duckworth, London.

PRESTON, R. 1992. *Crisis in the hot zone: a reporter at large*. New Yorker, 26 October 1992.

ROSE, D. B. 1988. Exploring an Aboriginal Land Ethic. *Meanjin* 3:378-387.

STANNER, W.E.H. 1960. On Aboriginal Religion. *Oceania* 30:245-278.

STANNER, W.E.H. 1979. *White Man Got No Dreaming: essays 1938-1973*. Australian National University Press, Canbera.

STREHLOW, T.G.H. 1970. Geography and the Totemic Landscape in Central Australia: a functional study. Pp. 92-140 in Berndt, R.M. (ed.). *Australian Aboriginal Anthropology: modern studies in the social anthropology of the Australian Aborigines*. University of Western Australia Press for the Australian Institute of Aboriginal Studies, Nedlands, W.A.

SYLVAN, R. 1985. *A Critique of Deep Ecology*. Australian National University, Department of Philosophy, Canberra (Discussion Papers in Environmental Philosophy, 12).

TINGGAL, SERUDIN. 1992. *Brunei Darussalam Fruits in Colour*. Universiti Brunei Darussalam, Brunei Darussalam.

WORLD COMMISSION ON ENVIRONMENT AND DEVELOPMENT 1990. *Our Common Future*. Australian edition. Oxford University Press, Melbourne.

Extraction and Tropical Rain Forest Conservation in Eastern Brazil

ROBERT A. VOEKS[1]

Department of Geography, California State University, Fullerton, CA 92634, USA

ABSTRACT. Management and extraction of non-timber products represents an alternative to destructive exploitation of Latin America's tropical forests. Well adapted to subsistence level economies, extraction has nevertheless yet to prove itself ecologically sustainable when exposed to outside market forces and economically competitive when confronted with more destructive land uses. By means of a case study, this paper examines two objectives of the extraction option: (1) Can extraction remain economically competitive over the long term? and (2) Can management for extractive products contribute to the conservation of tropical forest biodiversity? Piassava palm (*Attalea funifera* Mart.) is endemic to the tropical rain forests of Bahia, Brazil. Its fibre has represented a commercially important product since the 16th century, first in the fashioning of anchor ropes and later in the manufacture of brooms and brushes. Destructive harvest methods during Brazil's Imperial period resulted in near extirpation of the species. Privatization of the royal forests led to more sustainable harvest and management techniques. Fibre extraction now represents an economically sustainable land use. Piassava-rich lands are managed as long-fallow, second growth forests. An inventory 0.3 ha of managed piassava land revealed the presence of 52 tree and 19 shrub species, most of which also inhabit primary forests. By serving as a temporary repository of primary forest species, long-fallow forest managed for extractive products subsidizes the conservation of tropical biodiversity. These results suggest that (1) extraction of native forest products can compete economically with more destructive land uses, particularly if the land is of marginal value; (2) provided that sustainability incentives exist, private ownership is not necessarily inconsistent with the successful management of extractive resources; (3) relatively high species richness can be maintained as part of commercial extractive systems; and (4) secondary forests may represent the best candidates for the development of extractive enterprises.

KEY WORDS: non-timber products, Piassava palms, *Attalea funifera*, fibres, conservation, sustainable management.

INTRODUCTION

Sustainable management and extraction of non-timber products offers a promising alternative to destructive exploitation of Latin America's tropical forests. Under ideal conditions, extractive enterprises provide meaningful employment for native and other forest-dwelling people, protect a large measure of the biodiversity and contribute to the local and regional economy. Indigenous land use practices, employing a combination of short and long fallow swiddens, sparing and transplanting of valuable species, and simple extraction, represent time-tested procedures for sustainable forest exploitation (Bennett 1992, Denevan *et al.* 1985, Dufour 1990). Aside from protecting biological values (Chapin 1992), subsistence management and extraction methods provide the framework within which sustainable agrosystems can be developed for non-native peoples (Alcorn 1984, Altieri & Merrick 1987, Clay 1988). The establishment of reserves around the collection of fruit, nuts, fibre, and latex seems a logical extension of the sustainable extraction philosophy. If not a panacea for the problem of rain forest degradation (Browder 1990), extractive reserves at least represent a notable improvement over current practices (Allegretti 1990, Fearnside 1989).

[1]Present address: Department of Geography, Universiti Brunei Darussalam, Bandar Seri Begawan 2028, Brunei Darussalam.

D. S. Edwards et al. (eds.), Tropical Rainforest Research – Current Issues, 477–487.

As a model for regional development and conservation, petty resource extraction exhibits several serious shortcomings. First, extractive activities retain a dubious reputation in Latin American history (Anderson 1990). Brazil's forested regions witnessed a succession of boom-and-bust extractive cycles during the colonial period, resulting in the near extinction of valuable plant and animal species. Policymakers, as a result, continue to view extractive enterprises as both agronomically regressive and environmentally destructive (Alvim 1989). Second, as a serious development alternative, extraction must compete favorably with other land uses. Small-scale studies on the collection of latex, fruits, and nuts (Peters *et al.* 1989) and forest medicinals (Balick & Mendelsohn 1992) suggest that petty extractive values exceed those of forestry and cattle ranching activities when considered over the long term. Secondary forests appear particularly desirable, as they are noted to be hot spots for useful forest products (Grenand 1992, Kohn 1993). Coupled with the potential benefits of protecting future pharmaceuticals, crop genetic resources, as well as less tangible environmental values (Myers 1984, Randall 1991), extraction would seem, at least in principle, a reasonable alternative to present practices. In reality, extractive systems seldom persist when confronted with the short-term economic benefits of more destructive forms of land use. Third, considerable biodiversity is maintained within extractive systems that are subsistence based, such as those practiced by Amazonian Indians. It is unclear, however, whether the social mechanisms that encourage environmental conservation in subsistence economies will persevere as extractive products become increasingly responsive to outside market forces.

This paper considers the viability of non-timber forest extraction by means of a case study from the Atlantic rain forests of eastern Brazil. With a history reaching back over four centuries, the collection of piassava palm fibre (*Attalea funifera* Mart.) represents one of the longest running extractive activities in the New World tropics. The exploitation of this palm is examined in terms of meeting two goals of tropical forest conservation and development: (1) Is it economically sustainable over the long term?, and (2) Does it conserve a relatively high level of biological richness?

PIASSAVA BIOLOGY AND GEOGRAPHY

Piassava palm is endemic to the Atlantic coast of Bahia, Brazil (Bondar 1942, Silva and Vinha 1982), where it is distributed from roughly 130-170°S latitude (Figure 1). Reaching 15 to 20 m in height, the species is easily recognized in the field due to the presence of erect 10 m fronds covered with droopy leaflets. The valuable sheath fibre is 3-5 m long and is available for harvest soon after the fronds have sloughed off the terminal meristem. Flowering and fruiting occurs throughout the year. Outcrossing is effected by sap beetles (*Mystrops* sp.) and weevils (*Phyllotrox tatianae*) (Voeks 1985). Individual palms change gender during their life cycle, producing first staminate and later pistillate flowers (Voeks 1988b). Fruits average 300-400 gm, and contain edible kernels.

The regional climate is a Köppen Af tropical rain forest type. Mean annual rainfall ranges from 1800-2100 mm, with no pronounced dry season. Mean monthly temperatures reach a low of 21°C in August and a high of 25°C in February (Milde 1983). Soils grade from oxisols in the interior zones to spodosols nearer the coast. Responsive to this edaphic transition, the vegetation changes from tall, broadleaf evergreen formations, *mata higrófila*, to shorter, heath-like forests, *restinga*. Piassava is facultatively restricted to sandy spodosols, although it readily colonizes oxisolic areas that have been deforested (Voeks 1990).

The Atlantic coastal forests of Brazil may represent the most speciose tropical forests in the world (Anon. 1993). The vegetation is characterized by largely buttress-free, broadleaf

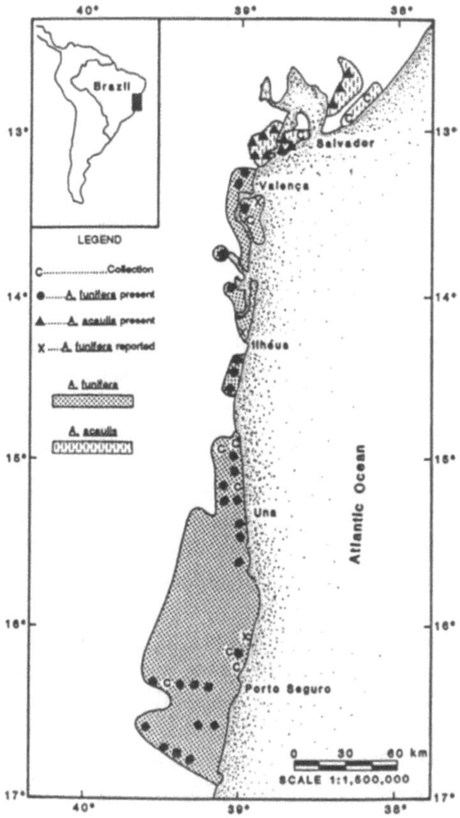

Figure 1. Distribution of piassava palm (*Attalea funifera*). North of the city of Valença, *A. funifera* is replaced by its shrub form, *A. acaulis*, from which sheath fibre is also extracted.

evergreen trees with relatively heavy loads of lianas and epiphytes (Mori *et al.* 1983). The high percentage of endemic tree species in the Bahian forests suggests lengthy geographical isolation from Amazonian formations (Mori *et al.* 1981). With over 95% of the biome converted to agricultural and urban landscapes, these forests represent one of the most threatened tropical biotas on Earth (Fonesca 1985, Mori 1989, Myers 1988).

FIBRE EXTRACTION: 1587-1987

The presence of piassava palm and its probable indigenous use was recorded within a few days of the Portuguese discovery of Brazil. Ship scribe Pero Vaz de Caminha reported that "the entire beach is covered with palms" and, later, that the huts of the natives were covered with fibre (Braga 1968, p. 33). Since coconut palms were yet to be introduced to eastern South America, Caminha was most likely describing piassava, which is the most numerous native palm at the site (Andrade-Lima 1984).

The first commercial use of piassava was recorded in 1587 by Gabriel Soares de Sousa, a sugar plantation owner. Able to withstand long periods of immersion in sea water, piassava was employed on local ships as a cable material (Sousa 1938). A century later, in his 1699 passage through Bahia, Captain William Dampier similarly reported that piassava cables were "strong and long lasting" and were "let to hire to European ships" (Masefield 1906, p. 388). During the 18th and 19th centuries, piassava became the cable material of choice for Atlantic shipping (Lapa 1968, p. 92). The Count of Atouguia noted their value to the local shipping industry in 1751 (Atougia 1925, p. 186), and by 1757 piassava cables were being exported to foreign countries (Caldas 1951, pp. 221-2). Prince Maximilian von Wied-Neuwied, in his 1816 passage through Bahia, described the fashioning of cables by African slaves (Wied-Neuwied 1940, p. 224). In 1820 Spix and Martius (1928, p. 122) noted that piassava rope making represented the principal occupation of the pacified coastal Indians. But while some touted cable making as a "lucrative branch of commerce" for the region (Henderson 1821, p. 314), the economic contribution of this cottage industry to colonial Bahia was in fact negligible (Bahia 1771-1777, Bahia 1789-1796, Camara 1789, Vilhena 1969).

The utility of piassava fibre as a broom and brush material was discovered in the mid-1800's. Employed as dunnage to pack Brazilian sugar, piassava fibre often accumulated as rubbish on foreign docks. An enterprising dock worker brought some to a British brush maker who, it is reported, fashioned the first piassava fibre broom (Kiddier 1927). By the 1840s, piassava brooms were well known in Europe and the Americas. In Cuba, they were referred to as "cleaning angels" (Amaral 1925, pp. 89-90). Sir William Hooker (1849, p. 122) commented that "few have walked the streets of London without remarking that of late years these streets are . . . kept peculiarly neat and clean, by the stiff fibres of a new material for making brushes and brooms . . .". As the value of piassava fibre to the broom and brush industry grew, its utility as a cable material apparently fell out of favour (Dodge 1897).

Responsive to the burgeoning international market, piassava fibre extraction became the principal industry in 19th century southern Bahia, surpassing sugar, cotton, and coffee in value (Silva Campos 1981). The British vice-council to Bahia noted that piassava was "... much procured and exported to Europe" (Wetherell n.d., p. 115). By the 1880s, nearly nineteen thousand tons were reaching the foreign markets, most arriving in England, France, Germany, and the United States (Relatorios 1867-1940). "In the future", predicted one contemporary observer, piassava "promises to be . . . one of the great sources of receipts for the province" (Carigé 1882, p. 39).

Bahia's 19th century fibre boom was short lived. Sustained harvest and export of ten to twenty thousand tons per year during the 1880s was followed by a precipitous decline in the 1890s (Figure 1). For the next six decades, annual collection figures stabilized at one to three thousand tons, and exports fell to a fraction of their former importance. Faced with dwindling supplies and increasing costs, broom manufacturers in Europe and North America switched to less costly African fibres (Kiddier 1927, Moraes 1911).

The piassava fibre crash was associated with over-exploitation of the resource. Until 1888, much of Bahia's forested landscape belonged to the Imperial government. Administration of the royal woods and their resources was delegated to the various relevant municipalities. In order to attain access rights to piassava and other non-timber products, local entrepreneurs were required to apply for a license and pay a user fee to the municipal government. In their letters of application, fibre merchants carefully outlined the precautions that would be taken to protect the forests and, in particular, how the fibre would be collected in a non-destructive manner. Such assurances were largely unnecessary, however, as municipal authorities lacked the resources to properly oversee their vast holdings. Moreover, although forest regulators were well aware of ongoing forest abuses, user fees represented an important source of revenue for the municipalities (Letters 1864-1889).

Incentives to harvest piassava on a renewable basis existed at neither the harvest nor the regulatory level. Fibre merchants, as a result, collected the fibre by the most destructive means possible. A municipal judge complained that exporters were "causing total destruction because . . .they cut the whole tree in order to collect the fibre" (Letter 1881). Another observer noted that "the present mode of obtaining [fibre] is to cut the tree down . . . a very foolish proceeding, considering the time that the fallen nuts take to germinate and grow into bearing trees" (Booth 1889, p. 239). This "destructive method of harvest", according to a provincial employee, was carried out "because those with licenses have nothing to lose, since the forests belong to the state" (Letter 1885).

Given these destructive extraction methods, 5% of the entire piassava population was destroyed on an annual basis (Webering 1937). By the 1890s, piassava was commercially extinct as an export commodity. As a poorly regulated extractive product subject to intense international demand and inappropriate modes and levels of exploitation, piassava supplies declined rapidly. Its demise may have been exacerbated by labour shortages associated with the abolition of slavery and the inability of the region to attract immigrant workers (Leao 1982, p. 315). There is, however, little evidence linking piassava harvest to dependence on slave labour. Extraction was and continues to be carried out by *caboclos*, mixed descendants of African, Amerindian, and Portuguese ancestry. In any case, by the end of the century, another boom-and-bust extractive cycle had run its course.

After decades of quiescence, piassava production began to rebound in the mid-twentieth century. Total collection doubled during the 1950s and by 1967 had exceeded the highest figures recorded in the 1880s. The 1987 harvest exceeded 66 thousand tons, 3.5 times higher than the fibre boom years (Figure 1). Wholesale values climbed from US$300,000 in the 1950s to over US$16,000,000 in the 1980s. Unlike the earlier peak production years, however, the recent upsurge in production has been accompanied by decreasing dependence on export markets. With over 99% of the raw material destined for broom and brush manufacturers within Brazil, piassava exports have slumped to 400 tons per year, barely 2% of their 1880s total. Declining interest in piassava, according to European importers, is due to the relatively high cost of Bahian fibre and the popularity of monofilament fibres (Voeks 1987, pp. 208-210).

Improvements in fibre management and harvest appeared as fibre-rich lands passed from public to private control. Formation of the Brazilian Republic in the latter part of the century led to the opening of the royal woods for settlement. Fibre-bearing forests that were not already in private hands were settled by locals and incoming colonists. Some were incorporated into cacao-based latifundia. Most, however, appear to have been occupied by peasant farmers.

Occupation of piassava-rich forests represented anything but a land rush. Rather, with nutrient-poor, sandy soils, heath forests were viewed as being of marginal agronomic value. Piassava was associated with neither quick wealth nor social status. It was and continues to be perceived as a 'poor man's' pursuit. With a strong incentive to protect the value of their properties, peasant smallholders initiated the non-destructive collection methods that are in place today.

Under current practices, piassava fibre is collected on a yearly cycle. Rather than killing the palms, as was done previously, fibre collectors climb the trees and cut away from the nearly senescent outer fronds with a machete. After being stripped from the fronds, the fibre is hauled to a common cleaning area and manually separated. The *borra* fibre, a papery material, is sun dried, woven into thatch, and sold as roofing material for fashionable homes and beach cabanas. The more valuable strand fibre is transported to fibre merchants in the nearest city. After being cleaned, sorted into grades, and cut, the fibre bales are forwarded to broom and brush manufacturers throughout Brazil (Voeks 1988a).

Piassava is frequently managed as part of a long-fallow system. Every thirty to sixty years the vegetation is cut and burned. Only piassava individuals are spared. Manioc is planted during the first and second year after the burn so that some economic return is realized prior to the palms achieving harvestable status. Following the fire, previously established piassava seedlings vigorously resprout through the charred remains. In five to eight years, the palm fronds have grown large enough to produce harvestable fibre. Within 20 years, the forest is numerically dominated by adult piassava (Voeks 1988c). Given the present mode of harvest and management, piassava palm fibre has made the transition from destructively exploited export product to sustainably managed regional resource.

BIOLOGICAL RICHNESS

Piassava's transition from destructive to sustainable management during this century means that two of the objectives of extractive development are being met, that is, creating employment for the local population and contributing to the regional economy. Nevertheless, piassava palm is often managed by fire. Every half century or so, the heath forest biodiversity is converted to ash and atmospheric gas. Given this level of disturbance, is the management of this extractive resource consistent with the goals of protecting biological diversity?

This question was examined by enumerating the tree and shrub species in a 0.3 ha (25 m x 120 m) plot. Slashed and burned in the 1930s, the plot is managed as long fallow piassava forest. All trees and palms with a dbh of 5.0 cm or greater were marked with aluminum tags. Voucher specimens were collected over an 18 month period (1984-86). Shrubs were also collected, although the large number of individuals was not recorded. Over 90% of the collections were made during flowering or fruiting. Thus, although many specimens have yet to be identified to species, I am confident that most collections identified as distinct species will withstand further scrutiny. Vouchers are permanently housed at the herbarium, Centro de Pesquisas do Cacau, Itabuna, Bahia.

Enumeration results are listed in Table 1. A total of 52 tree and palm species (or morphospecies) distributed among 22 families were recorded. Nineteen shrub species were distributed among 11 families. Among the trees, the most speciose families included the Myrtaceae (6), Rubiaceae (4) and Myrsinaceae (4). The most species-rich shrub families included Melastomataceae (5), Rubiaceae (3), and Euphorbiaceae (2).

Numerically dominated by piassava (85 of 185 individuals), the plot nevertheless retains several characteristics in common with old growth tropical forest. For example, fully 60% of the tree species are represented by a single individual. Excluding piassava, the 51 tree species were represented on average by less than two individuals. Further similarities between this second growth forest and old growth are evidenced by calculating Simpson Ds and Shannon-Wiener H' diversities. Including piassava in the calculations, Ds = 0.78 and H' = 3.90, figures that are diverse by temperate standards but not by primary tropical forest standards. Excluding piassava from the calculation, however, yields figures of Ds = 0.98 and H' = 5.35. Comparing these latter figures with those reported from old growth rain forests (cf. Bongers *et al.* 1988, Knight 1975, Uhl & Murphy 1981) indicates that this plot is diverse even by primary forest standards.

The high species diversity exhibited by this second growth plot is not an isolated occurrence. Comparable results have been reported for a plot near this study area (Passos 1984), as well as others in northern Amazonia (Saldarriaga *et al.* 1986) and Malaysia (Kochummen & Ng 1977). In their summary of secondary tropical forests, Brown & Lugo (1990) report that species richness often reaches that of old growth forest within 80 years of disturbance. In some cases, secondary forest accumulates more woody species than the forest it replaced.

Table 1. Species Enumeration of 0.3 ha Second Growth Plot. T=tree, P=palm, S=shrub, NC=no collection, VC=various collections.

Species	Abund-ance	Life form	Coll. No.	Species	Abund-ance	Life form	Coll. No.
Anacardiaceae				**Melastomataceae**			
Tapirira guianensis Aubl.	1	T	108	*Clidemia capitellata*	-	S	27
Annonaceae				var. *dependens*			
Annona salzmannii Alph. DC.	1	T	65	*Henriettea succosa* DC.	-	S	8
Annona sp.	1	T	45	*Miconia albicans* Steud.	-	S	60
Aquifoliaceae				*Miconia amoena* Triana	2	T	82
Ilex sp.	4	T	94	*Miconia pileata* DC.	-	S	17
Ilex sp.	5	T	54	*Ossaea retropila* Triana	-	S	46
Ilex sp.	1	T	63	**Mimosaceae**			
Arecaceae				*Abarema filamentosa* (Benth.)	-	S	71
Attalea funifera Mart.	85	TP	VC	Pittier			
Attalea humilis Mart.	1	TP	NC	*Albizia pedicellaris* (DC.)	1	T	68
Bactris sp.	-	SP	NC	ined.			
Geonoma blanchetiana	1	TP	29	**Moraceae**			
Wendland ex Drude				*Ficus* sp.	1	T	9
Boraginaceae				**Myrsinaceae**			
Cordia nodoso Lam.	1	S	61	*Conomorpha oblongifolia* A.	1	T	21
Cordia superba Cham.	1	T	66	DC.			
Cordia sp.	3	T	34	*Cybianthus gardneri* (A. DC.)	1	T	18
Chrysobalanaceae				G. Agostini			
Couepia impressa ssp.	1	S	53	*Myrsine umbellata* Mart.	2	T	41
cabraliae Prance				*Rapanea* sp.	1	T	5
Hirtella bahiensis Prance	1	T	72	**Myrtaceae**			
Licania lamentanda	1	T	28	*Myrcia* sp.	4	T	96
Clusiaceae				*Myrcia* sp.	1	T	51
Clusia sp.	2	T	10	*Myrcia* sp.	2	T	109
Clusia sp.	2	T	43	?	1	T	69
Symphonia globulifera L.	1	T	44	?	4	T	73
Erythroxylaceae				?	6	T	59
Erythroxylum mattos-silvii T.	1	S	58	**Nyctaginaceae**			
Plowman				*Pisonia* sp.	2	T	85
Euphorbiaceae				?	1	T	47
Chaetocarpus sp.	2	T	30	?	-	S	121
Croton selowii Baill.	-	S	2	**Piperaceae**			
Pera sp.	5	T	89	*Ottoma* sp.	-	S	40
Sebastiania sp.	-	S	31	**Polygonaceae**			
?	1	T	35	*Coccoloba* sp.	1	T	66
Fabaceae				*Coccoloba* sp.	1	T	84
Andira sp.	1	T	102	*Coccoloba* sp.	3	T	97
Platymiscium blanchetii	1	T	19	**Rubiaceae**			
Benth.				*Psychotria erecta* (Aubl.)	-	S	111
Swartzia apetala Raddi.	1	T	12	Standl. et Steym.			
Flacourtiaceae				*Psychotria jambosioides*	-	S	52
Casearia commersoniana	1	T	78	Schlecht			
Cambess.				*Psychotria mapourioides* DC.	1	T	7
Casearia sp.	1	T	83	*Psychotria* sp.	6	T	77
Icacinaceae				*Salzmannia* nitida DC.	-	S	3
Emmotum affine Miers.	1	T	80	?	1	T	57
Lauraceae				?	1	T	NC
Persea sp.	2	T	55	**Sapindaceae**			
?	2	T	62	*Cupania oblongifolia* Mart.	2	T	11
Malpighiaceae				**Sapotaceae**			
Byrsonima sericea DC.	9	T	79	?	1	T	88
Marantaceae				**Solanaceae**			
Stromanthe porteana A.Gris.	-	S	20	*Solanum polytrichum* Moric.	-	S	1

Table 2. Species Diversity: Old and Second Growth Forest. Simpson Index: Ds = 1-(Σni(ni-1)/N(N-1). Shannon-Wiener Index: H' = (N log N - Σni log ni)/N.

Simpson Ds	Shannon-Weiner H'	
0.93	4.80	Knight 1975 (Panama)
0.97	5.86	
0.95	5.38	
0.97	5.71	
0.95	5.27	
0.94	3.63	Bongers *et al.* 1988 (Mexico)
0.92	4.80	Uhl & Murphy 1981 (Venezuela)
0.96	5.40	
0.96	5.30	
0.93	5.00	
0.78	3.90	This study - study site + piassava
0.98	5.35	This study - study site - piassava

The biological value of long-fallow forest management is further evidenced by comparing the species list (Table 1) with a list of the trees and shrubs known to inhabit primary forests in the region (Mori *et al.* 1983). The majority of trees and shrubs enumerated in the second growth plot are seen to represent components of old growth tropical forest. Examples include the trees *Tapirira guianensis*, *Annona salzmannii*, *Cordia nodosa*, and *Swartzia apetala*. Several locally endemic old growth species were also recorded in the plot, including *Couepia impressa* ssp. *cabraliae*, *Hirtella bahiensis*, and *Erythroxylum mattos-silvii*. It is also noteworthy that nearly all the species in the plot were found to be reproductively active. Thus, rather than representing a destructive form of land use, long-fallow forest management appears to create fluctuating second-growth repositories for the conservation of primary forest biodiversity.

Piassava fibre extraction appears to be compatible with the maintenance of biodiversity values. By discouraging less profitable and more destructive land uses, fibre extraction subsidizes the existence of the remaining members of the forest community. Heath forest owners, with a vested interest in maintaining the biological integrity of their property, represent ideal land stewards. As an adjunct to biological preservation in parks and reserves, extraction should be considered a serious conservation alternative.

SUMMARY

Extraction of non-timber forest products represents one of the mix of development options for tropical rain forests. In the case of piassava fibre extraction, local employment of forest dwellers is realized, species-rich second growth forests are conserved, the region produces an economically valuable product, and the country achieves self-sufficiency in a commodity that otherwise might need to be imported.

Although it is difficult to generalize from a single example, these results at least suggest the following. First, extraction of native forest products can, under some circumstances, compete economically with more destructive land uses. In the case of Bahia's fibre-rich heath forests, the marginal agronomic value of sandy spodosolic soils has allowed piassava extraction to dominate the economy for over four centuries. Given the relatively low production of extractive systems, marginality may prove to be a prerequisite to long term sustainability. Second, private ownership of native forests is not necessarily inconsistent with the successful management of extractive resources. In the case of piassava, transfer of property from the state to the private sector was associated with the transition from destructive to sustainable practices. This does not mean that collective ownership or customary rights management are necessarily less desirable than private ownership. Rather, serious incentives for resource protection at the grass roots level appear to be essential to long term management of extractive resources. Third, species richness figures from disturbed forests are high, sometimes as high as old growth forests. Periodic disturbance of old growth forests, in this case to enhance the population status of a particularly valuable species, may be consistent with the goals of biodiversity preservation. Finally, secondary forests are increasing in Latin America at the expense of old growth. Learning to sustainably exploit these already disturbed tropical ecosystems might relieve some of the pressure to remove the increasingly rare primary forest.

ACKNOWLEDGEMENTS

Field work was funded in part by a Fulbright Exchange Fellowship. The logistical support supplied by the Centro de Pesquisas do Cacau, Itabuna, Bahia is gratefully acknowledged. The project could not have been accomplished without the assistance and advice of A. M. de Carvalho, L. A. Mattos e Silva, T. S. Santos and S. G. da Vinha. I also thank Glyn Davies for constructive criticism of the manuscript.

LITERATURE CITED

ALCORN, J. B. 1984. Development policy, forests and peasant farms: reflections on Huastec-managed forests' contribution to commercial production and resource conservation. *Economic Botany* 38:389-406.

ALLEGRETTI, M. H. 1990. Extractive reserves: an alternative for reconciling development and environmental conservation in Amazonia. Pp. 252-264 in Anderson, A. (ed.). *Alternatives to deforestation.* Columbia University Press, New York.

ALTIERI, M. A. & MERRICK, L. C. 1987. In situ conservation of crop genetic resources through maintenance of traditional farming systems. *Economic Botany* 41:86-96.

ALVIM, P. de T. 1989. Tecnologias apropriadas para a agricultura nos trópicos umidos. *Agrotrópica* 1:5-26.

ANON. 1993. New in the botanical book of records: highest tree diversity in the world. *Field Notes from the New York Botanical Garden* 2:1-2.

AMARAL, L. 1925. A exploraçao de fibras na Bahia (1843). *Anais do Arquivo Público da Bahia* 14:89-90.

ANDERSON, A. B. 1990. Extraction and forest management by rural inhabitants in the Amazon estuary. Pp. 65-85 in Anderson, A. (ed.). *Alternatives to deforestation.* Columbia University Press, New York.

ANDRADE-LIMA, D. de 1984. A botânica da carta de Pero Vaz de Caminha. *Rodriguesia (Rio)* 36:5-8.

ATOUGIA [to the Royal Court, 1751] 1925. Letter to the Royal Court. *Anais do Arquivo Publico da Bahia* 14:185-86.

BAHIA [to Portuguese Arsenal] 1771-1777. Notices of shipments of piassava anchor ropes. *Inventários dos documentos relativos ao Brasil* II: 261, 269, 318, 345, 369. Centre of Bahian Studies, Salvador.

BAHIA [to Portuguese Arsenal] 1789-1796. Notices of shipments of piassava anchor ropes. *Inventários dos documentos relativos ao Brasil* III:112, 115, 117, 249, 263, 282, 283, 290, 291, 309, 311, 312, 316, 317, 319, 322, 348, 350, 357, 383, 390. Centre of Bahian Studies, Salvador.

BALICK, M. J. & MENDELSOHN, R. 1992. Assessing the economic value of traditional medicines from tropical rain forests. *Conservation Biology* 6:128-130.

BENNETT, B. C. 1992. Plants and people of the Amazonian rainforests. *BioScience* 42:599-607.

BONDAR, G. 1942. A piassaveira e outras palmeiras Attaleianeas na Bahia, *Instituto central de fomento econômico da Bahia, boletim no. 13.* 73pp.

BONGERS, F., POPMA, J., MEAVE DEL CASTILLO, J. & CARABIAS, J. 1988. Structure and floristic composition of the lowland rain forest of Los Tuxtlas, Mexico. *Vegetatio* 74:55-80.

BOOTH, W.S. 1889. Bahia piassava. *Kew Bulletin of Miscellaneous Information* 34:327-42.

BRAGA, R. 1968. *Pero Vaz de Caminha: carta a el Rey Dom Manuel.* Editora Sabiá, Rio de Janeiro.

BROWDER, J. O. 1990. Extractive reserves will not save the tropics. *BioScience* 40:626.

BROWN, S. A. & LUGO, A. E. 1990. Tropical secondary forests. *Journal of Tropical Ecology* 6:1-32.

CALDAS, J. A. 1951. *Notícia geral de toda esta Capitania da Bahia desde o seu descobrimento até o presente ano 1759*. Reprint edition, Tipografia Benitiana Ltd.

CAMARA, M. F. 1789. Ensaio da descriçao fízica, e econômica da Comarca dos Ilhéus na América. *Memórias Econômicas da Academia Real das Ciências de Lisboa*, pp. 304-350.

CARIGE, E. 1882. *Geografia, physica e política da província da Bahia*. Impresa Econômica, Bahia.

CHAPIN, M. 1992. Coexistence of indigenous peoples and the natural environment in Central America. *Research and Exploration* Spring (special map supplement), 16 pp.

CLAY, J. W. 1988. *Indigenous peoples and tropical forests*. Cultural Survival, Inc. Cambridge, MA.

DENEVAN, W. M., TREACY, J. M., ALCORN, J. B., PADOCH, C., DENSLOW, J. & PAITON, S. F. 1985. Indigenous agroforestry in the Peruvian Amazon: Bora Indian management of swidden fallows. Pp. 137-155 in Hemming, J. (ed.). *Change in the Amazon Basin. Vol. 1., man's impact on forests and rivers*. Manchester University Press, Manchester.

DODGE, C. R. 1897. *A descriptive catalogue of useful fiber plants*. U.S. Department of Agriculture Report No. 9.

DUFOUR, D. L. 1990. Use of tropical rainforests by native Amazonians. *BioScience* 40:652-659.

FEARNSIDE, P. M. 1989. Extractive reserves in Brazilian Amazonia. *BioScience* 39:387-393.

FONESCA, G. A. B. 1985. The vanishing Brazilian Atlantic forest. *Biological Conservation* 34:17-34.

GRENAND, P. 1992. The use and cultural significance of the secondary forest among the Wayapi Indians. Pp. 27-40 in Plotkin, M. & Famolare, L. (eds). *Sustainable harvest and marketing of rain forest products*. Island Press, Washington D.C.

HENDERSON, J. 1821. *A history of the Brazil, comprising its geography, commerce, colonization, aboriginal inhabitants, etc*. Longman, Horst, Reese, Orme & Brown, London.

HOOKER, Sir Wm. 1849. Piaçaba; fibre and fruit of the coquilla nut, *Attalea funifera*, Mart. *Hooker's Journal of Botany and Kew Garden Miscellany* 1:121-23.

IIBGE (Instituto Brasileiro de Geografia e Estatística) 1946-1982. *Anuário estatístico do Brasil*. Conselho Nacional de Estatística, Rio de Janeiro.

KIDDIER, W. 1927(?). *The brushmaker and the secrets of his craft: his romance*. Jonathan Cape, London..

KNIGHT, D. H. 1975. A phytosociological analysis of species-rich tropical forest on Barro Colorado Island, Panama. *Ecological Monographs* 45:259-284.

KOCHUMMEN, K. M. & NG, F. S. P. 1977. Cited in Whitmore, T. C. 1984. *Tropical Rain Forests of the Far East*. Clarendon, Oxford.

KOHN, E. O., in press. Some observations on the use of medicinal plants from primary and secondary growth by the Runa of eastern lowland Ecuador. *Journal of Ethnobiology*.

LAPA, J. R. 1968. *A Bahia e a corrida da India*. Nacional, Sao Paulo.

LEAO, S. 1982. *The evolution of agricultural land use patterns in the state of Bahia, Brazil*. Ph. D. dissertation, University of Western Ontario.

LETTER [Judge José Augusto Beirbosa Coelho to Senator Conselheiro Joao Lustosa? da Cunha Paranagua] 13 October 1881. Colonial Section, Agriculture-Piassava. Public Archive of Bahia, Salvador.

LETTER [Commissao de Mediçao de Terras to Jessé Martins Sampaio] 2 September 1885. Colonial Section, Agriculture-Piassava. Public Archive of Bahia, Salvador.

LETTERS 1864-1889. Colonial Section, Agriculture-Piassava. Public Archive of Bahia, Salvador.

MASEFIELD, J. 1906. *Dampier's voyages, by Captain William Dampier*. Dutton, New York.

MILDE, L.C.E. 1983. *Estudo de precipitaçao diária: regimes pluvio-métricos e modelos de distribuiçao para a regiao cacaueira do sudeste da Bahia*. Unpublished Master's Thesis, Federal University of Paraíba.

MORAES, P. de, 1911. A piassabeira: a extinçao dos piassabaes naturais e a sua necessária cultura. *Chácaras e Quintais* 3:15-16.

MORI, S. A., BOOM, B. M. & PRANCE, G. T. 1981. Distribution patterns and conservation of eastern Brazilian coastal forest tree species. *Brittonia* 33:234-45.

MORI, S. A., BOOM, B. M., CARVALHO, A. M. de & SANTOS, T. S. 1983. Southern Bahian moist forests. *The Botanical Review* 49:155-232.

MORI, S. A. 1989. Eastern extra-Amazonian Brazil. Pp. 427-454 in Campbell, D.G. & Hammond, H.D. (eds). *Floristic inventory of tropical countries: the current status of plant systematics, collections, and vegetation*. New York Botanical Garden, New York.

MYERS, N. 1984. *The primary source: tropical forests and our future*. Norton, New York.

MYERS, N. 1988. Threatened biotas: "hot spots" in tropical forests. *The Environmentalist* 8:187-208.

PASSOS, J. R. 1984. *Estudo comparativo da vegetaçao em tres estágios da sucessao e a determinaçao da area minima de amostragem para inventario floresa na Estaçao Ecologica do Pau-brasil*. Centro de pesquisas do cacau. Unpublished manuscript.

PETERS, C. M., GENTRY, A. H. & MENDELSOHN, R. O. 1989. Valuation of an Amazonian rainforest. *Nature* 339:655-656.

RANDALL, A. 1991. The value of biodiversity. *Ambio* 20:64-68.

RELATORIOS. Associaçao Comercial da Bahia. 1867-1940. Wilcke, Picarde Co.

SALDARRIAGA, J., WEST, D. C. & THORP, M. L. 1986. Cited in Brown, S.A. & Lugo, A.E. (1990). Tropical secondary forests. *Journal of Tropical Ecology* 6:1-32.

SILVA, L. A. & VINHA, S. G. da 1982. A piassaveira (*Attalea funifera* Mart.) e a vegetaçao associada no Município de Ilhéus, Bahia, *Centro de pesquisa do cacau, boletim técnico no. 101*, Ilhéus, BA, Brasil.12p.

SILVA CAMPOS, J. 1981. *Crônica da Capitania do Sao Jorge de Ilhéus*. Nacional, Rio de Janeiro.

SOUSA, G. S. de. 1938. *Tratado descritivo do Brasil em 1587*. n.p.

SPIX, VON & VON MARTIUS, C. F. P. 1928. *Atraves da Bahia: excertos da "obra reise in Brasilien"*, translated by da Silva, P. & Wolf, P. Imprensa Official, Bahia.

UHL, C. & MURPHY, P. G. 1981. Composition, structure and regeneration of a tierra firme forest in the Amazon Basin of Venezuela. *Tropical Ecology* 22:219-237.

VILHENA, L. dos S. 1969. *A Bahia no século XVIII*. Volume 1. Editora Itapoa, Salvador.

VOEKS, R. A. 1985. Preliminary observations on the reproductive ecology of the piassava palm (*Attalea funifera*). *Anais, Academia Brasileira de Ciências* 57:524-525.

VOEKS, R. A. 1987. *A biogeography of the piassava fiber palm (Attalea funifera Mart.) of Bahia, Brazil*. Ph.D. dissertation, University of California, Berkeley.

VOEKS, R. A. 1988a. The Brazilian fiber belt: harvest and management of the piassava fiber palm (*Attalea funifera* Mart.). *Advances in Economic Botany* 6:262-275.

VOEKS, R. A. 1988b. Changing sexual expression of a Brazilian rain forest palm (*Attalea funifera* Mart.). *Biotropica* 20:107-113.

VOEKS, R. A. & da VINHA, S. 1988c. Fire management of the piassava fiber palm (*Attalea funifera*) in eastern Brazil. *Yearbook, Conference of Latin Americanist Geographers* 14:7-13.

VOEKS, R. A. 1990. Edaphic limitation of a Brazilian rainforest palm: the role of energy allocation and competition for sunlight. *Physical Geography* 11:154-171.

WEBERING, V. A. 1937. Notas sobre a piassaveira. *Bahia Rural* 5(49/50): 1828-30.

WETHERELL, J. n.d. *Brasil apontamentos sobre a Bahia: 1842-57*, trans. M.P. do Rio Branco, S.A. Artes Gráficas, Salvador.

WIED-NEUWIED, Prince Maximilian. 1940. *Viagem ao Brasil*. Editora Nacional, Sao Paula.

Geographic information systems in rainforest ecosystem research: the Sarawak pilot model

EBERHARD F. BRUENIG *and* SONJA CSOMOS

University of Hamburg, Leuschnerstr. 91, D-21031, Hamburg, Germany, and Forest Department Sarawak, Wisma Sumber Alam, Petra Jaya, 93660 Kuching, Sarawak, Malaysia

THE DEVELOPMENT OF CONCEPT

In 1963, a 20 ha permanent ecological research plot, now Forest Department Research Plot (RP) 146, was established in an area of Kerangas and Mixed Dipterocarp forest in Sabal Forest Reserve, West Sarawak. All trees over 10 cm dbh were enumerated and the topography sketched in 20 x 20 m recording units (exactly: chain-squares). The trees were not mapped because we did not realize at the time the great heterogeneity of soils and species associations. The conventional traditional textbook analyses of species richness, tree species diversity (evenness of mixture), and general growing stock patterns and densities in the sixties were easily determined (Weiscke, 1982). However, we encountered problems with determining associations between tree species, between species and soil features, and between patterns and assumed dynamic trends. The chain-square grid was too coarse to catch the fine fabric of soil and vegetation heterogeneity which varies at a scale of metres and even less. Another disadvantage was that plant locations within the 20 x 20 m square were not recorded. This made it impossible to analyse species associations and cohort dynamics among the smaller trees. Unless dynamic cohorts and neighbourhood conditions can be determined and observed in time series of remeasurements, any measurements of growth, nutrient stocks and cycles and hydrological conditions will be difficult to interpret. The shifting-mosaic dynamics have proved to be so strong and the time-series of patterns of the shifting mosaics among the smaller trees so turbulent (Poker 1992) that research into processes which are influenced by the autecological properties and physiological state of the trees at the point of measurement is a shot in the dark unless the fine-grained spatial information is available.

Consequently, in our next ecosystem study in the seventies, the International Amazon Ecosystem MAB-pilot project at San Carlos de Rio Negro, we established and demarcated a coordinate grid covering the 20 ha research area. This was intended (Bruenig 1977, 1978) to provide a permanent frame of reference for the coordination and linking of research activities by different scientists and different disciplines. This purpose was achieved but a further deficiency emerged. Research into the interactions between stationary plants and mobile animals and humans required larger areas and observations outside the 20 or 50 ha vegetation research plots. The coordinate grid had to be extended accordingly far beyond the boundaries of the natural-ecological ecosystem research area.

This was done in the eighties in the two tropical natural forest and restoration ecosystem studies in Guangdong (Xiaoliang barren coastal ecosystem) and Hainan (Bawangling evergreen tropical forest ecosystem) which were part of the Chinese-German Coordinated Ecological Research Programme (CERP). The purpose was to:

- establish a coordinate grid (1 dm resolution) for the natural ecosystem study as in San Carlos de Rio Negro
- extend the grid into the landscape beyond to include agricultural and degraded lands

D. S. Edwards et al. (eds.), Tropical Rainforest Research – Current Issues, 489–494.

areas under native customary rights of the indigenous Li people, and village ecosystems
- to coordinate and link the data sources of research with those of practical forestry, agriculture, infrastructural and general economic and social development
- to secure safety and integrity of the research area by means of integrating the research area into the landscape ecosystem which not only widened the scope for research, but also made local authorities and planning agencies better aware of the existence of the area and also of research as a source of information of value to themselves.

OVERALL OBJECTIVES

The ultimate objectives for establishing the spatial information system for ecosystem research are to:
- improve the efficiency of collecting and utilizing research data
- improve inter-disciplinary analyses of associations and cause-and-effect relationships
- provide a stable spatial reference for longterm time series
- improve the security and integrity of the research area against abuse by scientists and encroachment from outside
- provide a means to prevent data and projects to be lost and forgotten
- stop the notorious habit of scientists to re-invent the wheel and to repeat work by making it impossible for newcomers not to recognize existing data and knowledge from past and ongoing projects.

At the same time it was hoped that spatial linkages and easier identification and access would lead to better integration of forestry and general land-use practice and would eventually reduce:
- misallocation of sites to unsuitable uses
- failures to match species with natural and economic site conditions
- application of obsolete or unsuitable techniques
- wastage of time, money, natural resources and opportunities as a result of ignorance of existing data sets and knowledge

REMAINING PROBLEMS

While scores of diverse proprietary brands of geographic information systems have recently flooded the market, none has yet met our needs adequately. Also, none has been or is sufficiently user-friendly that it could be efficiently used for research and application in a tropical rainforest country. Improvements of the quality of ecosystem research by the use of efficient spatial (geographic) information systems are urgently needed. This applies at all levels of the ecosystem hierarchy (Grossmass in Bruenig *et al.* 1986):
- natural ecosystem level (forestry, landscape)
- technological ecosystem level (forest management and conservation)
- economic ecosystem level (local, regional and national economy, tactical policies)
- socio-political ecosystem level (formulation and decision of strategic policies).

A dramatic increase in demand for spatial information systems which provide links between research and practice and between the various sectors involved in land-use analysis, planning and monitoring has arisen from the decision and pledge by the member states of the International Tropical Timber Organisation (ITTO) to achieve sustainability of forest management by the year 2000 (ITTO 1991), to which Sarawak has particularly committed itself. This target will not be met unless an efficient information system is in place on time, that means immediately. But information systems which adequately answer our specific needs of forest ecosystem research, forest management and forest and environmental conservation at all levels are not yet available.

METHODOLOGICAL APPROACHES

The problem has been attacked by our team (E.F. Bruenig, Jackson Lau, Steven Chan, Sonja Csomos, Stephen Price) at three levels (Figure 1):

(1) Research

- advancement and improvement of the Sabal - San Carlos - Bawangling concept for GIS ecosystem research
- establishing operational data linkages of the ecosystem GIS for basic scientific research in natural rainforest ecosystems with existing or evolving special research GIS (e.g. genetics, autecology silviculture, growth and yield, protection utilization, biodiversity) and higher-level ecosystem research GIS (socio-economic, anthropological, socio-political).

(2) Practice of Management, Economics, Sociology

- Developing operational linkages between existing sectoral GIS (e.g. between agriculture, forestry, infrastructure, industrial development).

(3) Integrated Multi-Sectoral Information System (IMSIS)

- This is a new development of a comprehensive GIS with a holistic structure which integrates the sectoral GIS of research and practice which we developed over the past three years in the form of an integrated multi-sectoral information system (Figure 1). The primary objective is to provide a solid, sound and reliable data and information base for technical planning and monitoring of forestry and agricultural land use and for general economic and social development. A further objective is to link scientific research, application-oriented experimentation and practice within an efficient, comprehensive information system. IMSIS is particularly suited to prevent or solve conflicts arising out of ignorance or disregard of conflicting interests and intentions, of conservation needs, native customary rights, local attitudes, local spiritual and cultural requirements. It was a condition that it should be flexible and applicable to the different levels of local and regional forest management planning and monitoring and of regional, state-wide and national land-use and economic planning and monitoring. IMSIS is designed to incorporate any proprietary brand of GIS which a user may have already installed or may want to use.

IMSIS in combination with data from remote-sensing provides graphic and non-graphic data for:

- information of past and present states (statistics, areas, etc.)
- information on existing sectoral plans and ongoing planning projects in the area covered
- analysis of relationships and interactions between various sectors (mutual supports, conflicts, synchronization of operations)
- analysis of sensitivities and dynamics of the sectoral systems involved.

Combination with dynamic system analysis and simulation permits the assessment of possible future states of ecosystems (Bossel & Bruenig 1992, Bruenig *et al.* 1986). IMSIS can, if required and wished by government, also be used as an instrument for exchange of data with international regional data stores of regional networks or with global data stores (Figure 1).

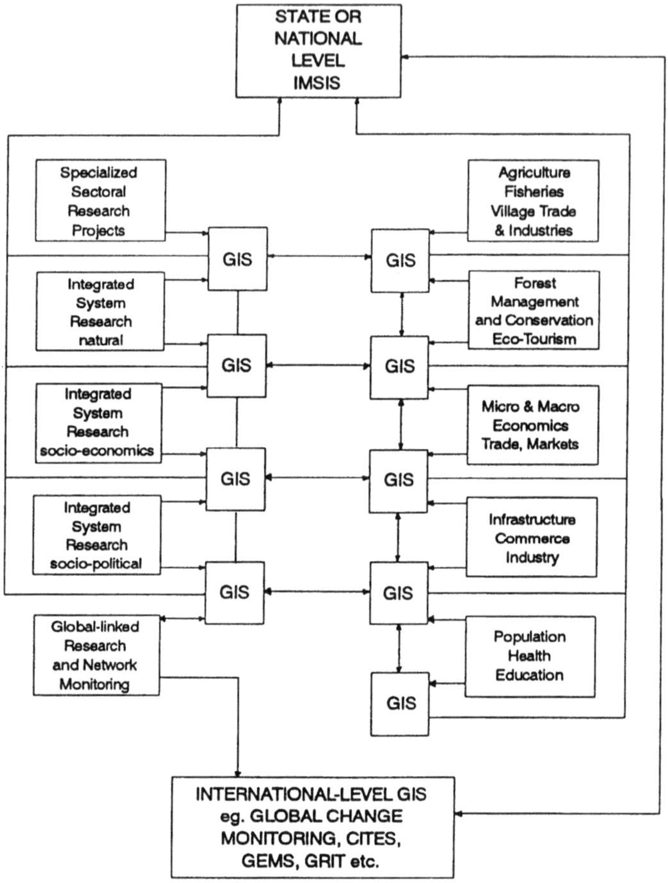

Figure 1. The structure of the harmonized and integrated geographical information system complex serving research and practice of sustainable forest management, land-use and socio-economic development and monitoring. The various GIS of local and specialised research projects (left) are linked directly among themselves. They may also be linked directly to corresponding GIS in practice (right) if desired. The more efficient analysis and synthesis function is supplied by the comprehensive Integrated Multi-Sectoral Spatial Information System (IMSIS). IMSIS supplies landuse , socio-economic and socio-cultural information for the identification and assessment of capacities and opportunities, the analysis of potential conflicts and sources of instability and for long-term monitoring and evaluation of research and practice.

ACTIVITIES TO SOLVE THE PROBLEM

The development of the GIS technology for individual projects of scientific ecosystem research is an open-ended, ongoing process. The standards for data collection and processing in rainforest ecosystem research are well established since the inception of the IBP and MAB programmes (Bruenig 1977). Adherence to these standards assures compatibility of data, but leaves the user freedom to decide which GIS he wants or develop to suit particular needs.

IMSIS, as a new concept and technique, has been developed to the state of an operational prototype which can be used on a PC or work station. The next phase is field testing in a suitably heterogeneous forested and agricultural region of Sarawak. This will then lead to full operationality for use in research and forestry practice in 1994. The ecosystem research GIS for individual ecosystem research projects will simultaneously be further advanced, based on the experiences from recent field work in Sabal Forest Reserve, RP 146. The plot is now being re-established and extended southwards to cover a research area of approximately 600 ha Kerangas and Mixed Dipterocarp forest, including the remaining 12 ha of the original 20 ha plot.

CONCLUSION

Rainforest ecosystem research and tropical forest research has, until now, been rather inefficient and its results have been largely ineffective in influencing forestry practice and the general public. Forestry practice has notoriously and generally fallen short of targets and is specifically yet far from the ITTO Target 2000. Among the major reasons for this are the lack of knowledge of research done and its results, lack of communication and coordination between forest research and practice, and between forestry and the other sectors of the socio-economic, social and political systems. Underlying these shortcomings, which are crucial obstacles to holistic approaches to sustainable development, is the notorious failure of governments to support the development and installation of comprehensive, integrated and multisectoral user-friendly spatial information systems which would benefit themselves, the private sector and the public. An efficient, user-friendly GIS for rainforest ecosystem research is essential to improve research quality and the application of results to management and policy practice. IMSIS is therefore essential for advancement and improvement as an instrument of integration and a vehicle of communication between ecological science, the practice of forest management and conservation, and the practice and politics of landuse.

The still existing wide gap between information needs and supply has most recently again been confirmed by the 14th Commonwealth Forestry Conference in Kuala Lumpur. The final recommendations of the conference state "that new developments in fields such as biotechnology, geographical information technology and valuation of non-market benefits (of forests) be promoted" by Commonwealth Governments in their very own interest.

ACKNOWLEDGEMENTS

The authors are thankful to the German Research Foundation (DFG), the Federal Ministry for Research and Technology and the Forest Department, Sarawak, for supporting the ecosystem research projects which provided the opportunities for developing the ecosystem GIS and the higher-level IMSIS. Special thanks are due to Miss Julia Philip for processing the manuscript, to Dr. T. Schmaltschinski and Mr. William Then for computer-drafting.

LITERATURE CITED

BOSSEL, H. & BRUENIG, E.F. 1992 *Natural Resource Systems Analysis. Ecological and Socio-economic Systems and Sensitivity Analysis for the Conservation and Management of Forest Ecosystems and Natural Resources in South-East Asia.* Feldafing, Food and Agriculture Development Center (ZEL), DSE, Dok.Nr. 1655 A/a-TK 78-300-89.

BRUENIG, E.F. (ed.) 1977. *Transactions of the International MAB-IUFRO Workshop on tropical rainforest ecosystem research.* Chair of World Forestry, Hamburg University. Special Report No. 1. Pp. IV + 363.

BRUENIG, E.F. 1979. The International Amazon Rainforest Ecosystem Pilot Project at San Carlos de Rio Negro: Vegetation Classification and Structure. Pp. 67-100 in Bruenig, E.F. & S. Adisoemarto, S. (eds). *Transactions of the 2nd International MAB-IUFRO Workshop on Tropical Rainforest Ecosystem Research, Jakarta, 21.-25.10.1978.* Chair of World Forestry, Hamburg University, Special Report No. 2.

BRUENIG, E.F., BOSSEL, H, ELPEL, K.-P., GROSSMANN, K.-D., SHCNEIDER T.W., WANG, ZHU-HAO & YU, ZUO-YUE 1986. *Ecological-socioeconomic system analysis and simulation: A guide for application of system analysis to the conservation, utilization and development of tropical and subtropical land resources in China*. Hamburg, Inst. World Forestry. Feldafing, ZEL-DSE; and Bonn, German MAB Nat. Comm. Mitt. 24. Pp. XV + 388.

BRUENIG, E.F. & SYNNOTT, T.J. 1977. An outline for a basic interdisciplinary research programme with emphasis on a common approach to research methodology and measurement techniques. Pp. 265-339 in Bruenig, E.F. (ed.). *Transactions of the international MAB-IUFRO Workshop on tropical rainforest ecosystems research*. Chair of World Forestry, Hamburg-Reinbek, Special Report No. 1. Revised Edition, 1978.

ITTO 1991. Beyond the guidelines - an action program for sustainable management of tropical forests. ITTO Technical Sevices No. 7. Pp. VI + 187.

WEISCKE, A.S. 1982. *Structure and functions of forest ecosystem: structural comparison between Kerangas and Caatinga*. Unpublished thesis, Chair of World Forestry, Faculty of Biology, University of Hamburg. Pp. III + 83 (in German).

POKER, J. 1992. *Mosaikdynamik in tropischen Regenwald, (Mosaic dynamics in the tropical rainforest)*. Unpublished thesis, University of Hamburg, cyclostyled. Pp. 220.

The creation of a geographical information system for the Kuala Belalong Field Studies Centre

JOHN T. WILLS

Department of Geographical and Environmental Sciences, University of Huddersfield, Queensgate, Huddersfield HD1 3DH, United Kingdom

ABSTRACT. Whilst the applications for Geographical Information Systems (GIS) are expanding, their use in tropical rainforest research is relatively undeveloped. At the conclusion of many scientific expeditions and projects, the data produced becomes dispersed between a large number of individuals and institutions, resulting in some information being under-utilised or even lost. Apart from the traditional monograph, many projects involving large numbers of scientists have also lacked the means to illustrate their research findings visually. GIS offers a means to display, store and manipulate geographically referenced data. Recording data in this way ensures its long term survival and use, allowing for frequent update, the revisiting of sample sites and an expansion in the uses of the information for scientists and non-scientists alike.This paper details the aims, methods and problems encountered in the creation of a GIS for The Kuala Belalong Field Studies Centre in Temburong District, Brunei Darussalam, as part of the UBD/RGS Brunei Rainforest Project 1991-92. Topographical data relating to the project area was digitised and results of the scientific research added as they became available in the form of database files linked to geographical points or areas. Problems were encountered with the accuracy of locational data and in converting some data into a suitable form for input into the GIS. A number of suggestions are given to aid the creation of future GIS's.

KEY WORDS: tropical rainforest, GIS, Brunei Darussalam, Borneo.

INTRODUCTION

"*A Geographical Information System is an organised collection of computer hardware, software, geographic data and personnel designed to efficiently capture, store, update, manipulate, analyse and display all forms of geographically referenced information*" (ESRI 1990). Within a GIS distinct classes of data are stored in different layers (see Burrough 1985, p.20). If the data layers within a GIS relate to the same geographic area, they can be overlaid in any combination and it becomes possible to perform spatial analysis on them.

Many commonly used computer programmes such as Minitab and Autocad can handle simple geographic or spatial data but are not considered to be Geographical Information Systems. Whilst modern Computer Aided Design (CAD) systems are capable of placing different types of geographic features on individual layers and linking these layers with attributes in a database, they are merely graphic systems (Cowen 1988). Automated mapping systems are capable of more complex linkages between geographic and display data, but are still *"restricted to the functions of data retrieval, classification, and automatic symbolisation.......a computer mapping system still is not a GIS"* (Cowen 1988, p.1552). To distinguish itself from computer mapping systems, a GIS must be capable of performing the following four main tasks (Scholten & Lepper 1991):

1: Storing, managing and integrating spatially referenced data:
- spatial data relating to points, lines or areas;
- descriptive data relating to the attributes of the points, lines and polygons.

2: Spatial query - for example, searching for the areas in which a particular species is found.

D. S. Edwards et al. (eds.), Tropical Rainforest Research – Current Issues, 495–502.

3: Geographic analysis - for example, statistical analysis of the relationship between data sets on tree girth and altitude.
4: Displaying data in the form of high quality maps.

AIMS

The primary aim of this project was to create a GIS which could be used as an educational tool for students and members of the public visiting the field centre, as well as being useful to visiting scientists to illustrate the research work previously conducted. It was hoped that map print-outs would form a useful means of information display for publications and that the data contained in the GIS could be used for detailed spatial analysis by other researchers.

METHODOLOGY

Hardware & Software configurations
This project used a vector based GIS: PC ARC/INFO version 3.4D on an IBM compatible 386 PC. Digitising was done on a Summagraphics Microgrid II digitiser, whilst maps were plotted on a Hewlett Packard HP7475 plotter. A Toshiba 2200 8086 laptop was used for database entry in the field. Some data was also transferred to a raster based image processing and GIS package: IDRISI version 3.0, to produce digital elevation models, slope angle and slope aspect maps.

Topographical and related data
The first stage in establishing a GIS is the creation of an accurate topographical representation of the world's surface. To this end, 1:50000 scale Ordnance survey maps of the project area detailing contours, rivers, streams and the Rectified Skew Orthographic (RSO) grid system were digitised. After discussions with project members, data layers for the paths, sub-camps and project boundary (based on contour interpretation and estimation) were created by drawing in on the Ordnance Survey maps and re-digitising (Figure 1).

To up-date the existing maps and provide accurate positions for specific locations a Trimble "Transpack" Global Positioning System (GPS) was used. GPS is a US Department of Defence satellite-based navigation system which began development in the late 1970's. The system consists of twenty-four satellites in high, geosynchronous orbits arranged in six orbit planes. Each orbit plane which contains four satellites is inclined to the equator an angle of 55 degrees. This means that all areas of the earth have at least four satellites visible at any one time. Ground-based GPS receivers intercept signals from the satellites to calculate the user's position (Cross 1991). Before using the GPS it is necessary to select the appropriate map projection for the area of the world in which the fixes are to be taken; in Brunei's case this is the Timbali projection. This enables the GPS to convert the locational information from World Geodetic System 1984 (WGS-84) coordinates into the correct latitude and longitude for different map projections (Brown 1992).

Existing data
After completing the topographic coverages a search was made for existing data relating to the project area. Much of Brunei has been covered by detailed geological and soil surveys, but due to its remoteness the project area has not been included in these surveys, the only data available being of a very coarse resolution. The "Temburong Renewable Resources Study" (Anderson & Marsden 1983) was found to be the only comprehensive dataset available. This survey covered the whole country and provides detailed information

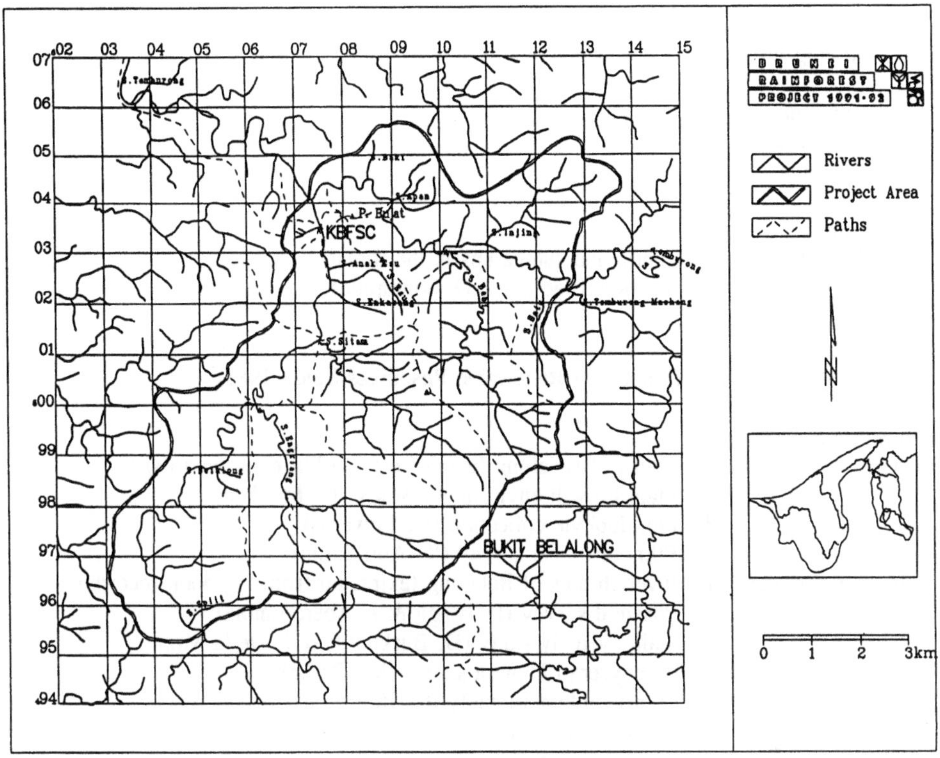

Figure 1. Example of the output from the GIS. Included are data layers of rivers, paths, the project boundary and the sub-camps.

on vegetation type in Southern Temburong district. The relevant 1:50000 maps were digitised and exist as a separate data layer.

Project data

With over seventy scientists conducting research in the project area, the volume of data produced was substantial and spans a wide range of scientific disciplines. To enable a variety of users to make effective use of the system it was therefore necessary to record the data produced in a concise and understandable form.

Some data was received in the form of written reports from scientists who had completed their field studies whilst the remaining scientists were interviewed. Once it was decided that a particular research project would require GIS input, a specific request for data was made: for example, the points or areas sampled and the species found there. This data was entered into specific subject database files containing spatial data (e.g. species present) and attribute data (e.g. an identification number linking the data to a particular sampling point or area). Some files relate to the work of individual scientists while others are the combination of results from two or more scientists working in the same field. The files were modified and updated as required.

Each database file was stored as a different data layer to allow clear viewing of the results and to avoid confusion. There are six background coverages relating to basic topology and the project's infrastructure. Twenty four coverages relate to the scientific results. This series of coverages accurately reflects the complexity of the real world in a clear, intelligible way.

RESULTS

The following information was entered into the GIS:

Topographic data
1 - Project boundary
2 - Contours
3 - Rivers and streams
4 - Paths
5 - Pondoks (the field centre and sub-camps in the forest)
6 - Grid reference system (Rectified Skew Orthographic)

Existing data
1 - Vegetation Type (Temburong Renewable Resources Study 1983)

Project data

A: Plants	1 - Plants of Belalong (extracted from log book records)
	2 - Plant growth environment study (D. Mitchell)
	3 - Pteridophyte collection (D. Edwards)
	4 - Ethnobotanical survey (K. Abu Salim)
B: Invertebrates	1 - Invertebrates of Belalong (extracted from log book records)
	2 - Phasmid survey (P. Bragg & I. Abercrombie)
	3 - Stalk eyed fly study (D. Burkhardt & I. De la Motte)
	4 - Brachyceran Diptera study (M. Drake)
	5 - Microlepidoptera study (G. Robinson)
	6 - Woodlice study (J. Medway)
C: Fish	1 - Fish survey (S. Choy)
D: Herps	1 - Frog survey (C. Reading)
	2 - Reptiles of Belalong (I .Das & log book records)
	3 - Amphibians of Belalong (I. Das & log book records)
E: Birds	1-4 - Birds of Belalong (C. Mann, D. Wells & log book records)
F: Mammals	1 - Mammals of Belalong (J. Charles & log book records)
	2 - Bat survey (C. Francis)
	3 - Mammal transect survey (L. Bennett)
G: Physical data	1 - Waterlevel records of S. Belalong (manual records)
	2 - Rainguage records at KBFSC (manual records)

Because these coverages relate to the same base area it is possible to overlay them in any desired combination and perform spatial analysis. For example, it is possible to show which species are present in any sampling point or area on any of these coverages, and vice versa in which sampling points or areas a particular species is found.

DISCUSSION

Spatial accuracy

The only maps available for the project area were at a scale of 1:50000, based on aerial surveys flown some thirty years ago. The absence of more up-to-date maps is due to the remoteness of the area, although a new aerial survey was flown in 1992 and will be used to produce 1:10000 maps. Some of the scientific results produced relate to very specific locations and it was therefore felt that the detail of the base map should be supplemented to provide accurate locations at a larger scale. This is not to say that the original 1:50000 maps are inaccurate, rather that they are not of a sufficiently large enough scale to identify points only a few metres apart accurately.

As previously mentioned, a GPS was used to fix the positions of the permanent scientific plots, project sub-camps, the Field Studies Centre and streams not recorded on the Ordnance Survey maps. The latitude-longitude positions produced by the GPS were converted into RSO grid references with the help of the Land Survey Department in Brunei, by using a computer programme.

It was found that in open locations unobscured by the forest canopy three or more satellites could be tracked and the GPS used in two dimensional or three dimensional mode. If the height of a given location was known, the best results were obtained by using the GPS in two dimensional mode so that locational accuracy could be concentrated on the two horizontal dimensions. The Trimble handbook states that an inaccuracy of five metres in the figure entered for height will lead to a position error of approximately 15-20 metres, although in the author's experience this error was sometimes considerably larger.

Under closed canopy or in semi-obscured locations it was necessary to use the GPS in two dimensional mode permanently, as the physical obstruction of the trees limited the number of satellites 'visible' and caused considerable position drift. By taking several readings it was possible to obtain a good position cluster, of which the centre theoretically provides the best location. Whilst some results exhibited a high degree of clustering, the positions cannot be assumed to be totally reliable. The altimeter used had to be calibrated at a point of known height, with the height of other points being derived by the altimeter applying the relationship between increased altitude and decreasing air pressure (and vice versa). Tropical climates are noted for their unstable bodies of air and large diurnal changes in air pressure (Briggs & Smithson 1985). Thus, an altimeter which relies solely on air pressure to determine height will produce different readings at the same point over time. Consequently, position readings from the GPS taken over time at the same point show considerable drift, in line with the findings on height-position error in the previous section. Whilst readings taken at the same time may show a high level of clustering, the central point cannot therefore be taken as the optimum location, unless the height value entered is known to be accurate. To assess the effect of altitudinal error, readings were taken using altimeter height and observer's estimated height (estimated from a given point on the Ordnance Survey map). There was frequently a discrepancy between the two height values of up to 25 metres, with the resultant position readings varying by up to 250 metres (see Figure 2).

GPS is therefore highly accurate in two dimensional mode where the height is accurately known, although in closed canopy areas some persistence may be required for the machine to track sufficient satellites and position drift may result from tree obstruction. It was more difficult to obtain positions in valley side locations than in valley bottoms and on ridge tops, although this will depend on the position of satellites at the time the fix is taken. As an indicator of accuracy, the machine gives its own measure of the precision of each fix, displaying +/-30 metres to +/-1000 metres with each reading. More accurate positions can therefore be obtained by recording only those with the highest accuracy rating. Inaccuracies in the height value entered will lead to considerable errors in position location to such an extent that the positions may not be valid. This problem appears to be the biggest stumbling block to the further use of GPS in forested areas. Whilst it is suggested that the GPS would be most useful in open areas such as desert or tundra, methods for lifting the GPS aerial above the canopy, for example by balloon or telescopic pole, coupled with improved altimeter accuracy could increase its usefulness in mountainous or forested areas.

The GPS data are currently being analysed to assess the precise relationship between the number of readings taken and the degree of clustering. This analysis will show whether or not the assumptions about clustering hold true.

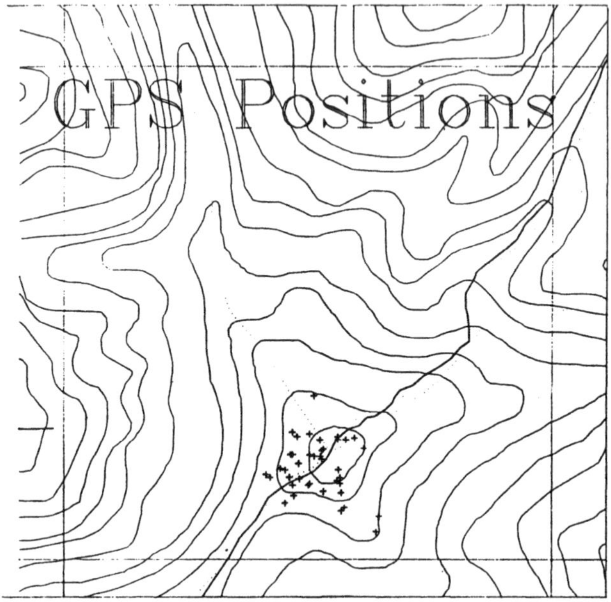

Figure 2. Distribution of GPS readings recorded from Bukit Belalong. Grid square = 1000 m x 1000 m.

The lack of spatial data

The majority of Geographical Information Systems relate to areas for which considerable spatial data already exists; for example, data on land-use classes, geology or soil type. With the exception of the vegetation type data and base maps, no detailed spatial data for this area existed before the Brunei Rainforest Project took place. Whilst considerable spatial data now exists as a result of the project, this relates to buffered areas around paths and streams (for 16 coverages with streams buffered at 15 m and paths at 60 m) and points or sampling stations (7 coverages). As research in the project area continues and more data become available, it may be possible to extrapolate them to produce species location maps or species density predictions. However, this process is problematic, especially where existing species data are poor and the species habitat parameters wide. Where the habitat parameters are narrow, species presence estimates are more feasible. For example, in the case of a particular fish (a species confined to small spatial areas) a strong correlation was found to exist between stream order and altitude for existing data on the species of fish found at thirty-nine sampling stations in the project area.

Data unification

Ideally the data pertaining to one subject area (e.g. plants) would be combined as a single coverage to make the storage and presentation of data more logical. However, this would necessitate the construction of complex database files and would only be possible where all of the data had a comparable degree of locational accuracy and related to approximately the same geographical area. For example, it might not be desirable to combine the data on Microlepidoptera which relate to precise sampling locations with the other datasets on Invertebrates which relate to much broader sampling zones.

The utility of Arc/Info

Arc/Info operates by using a large number of individual commands. Some of these are very similar, have complex suffixes and can take a considerable time to learn. Arc/Info proved to be an excellent platform for digitising, display and analysis, easily meeting the requirements of this project. The map output was of a very high quality and it was possible to produce clear, detailed maps with relative ease. Arc/Info uses Dbase files to store data; for this GIS, Dbase III+ was used. This version of Dbase has a limited number of fields. In the case of the bird data, because there were over 200 species found in the project area it was necessary to create four separate coverages. This is not essentially a fault with Arc/Info and could be rectified by transferring the data into Dbase IV.

SUGGESTIONS FOR EFFECTIVE GIS CREATION

The following comments stem from the author's GIS experience and are relevant to the creation of future GIS's:

Spatial accuracy

Ideally the area covered by this GIS would have been mapped in detail before the scientific studies began, thereby increasing the accuracy of the data produced. For example, data could then be expressed as species 'X' present at a particular grid reference. This would greatly increase the value of the GIS for detailed analysis and modelling work: maximising spatial accuracy increases the effectiveness of a GIS.

Data protocols

The establishment of data protocols prior to the start of the project would have simplified data entry into the GIS. For example, a GIS data entry form could have been created for project members to complete with their final reports, thereby reducing the time needed to talk to all members individually. Creating data protocols formalises the system of data input and ensures that all data are compatible with the requirements of the GIS.

Following closely defined aims

The primary aim of this project was to provide a GIS for educational use. The planned uses of a system will influence the data included, the levels of accuracy required and the data format. Unless the aims are outlined in advance, the final system may not meet the user requirements. It is therefore essential that all GIS's are created with specific aims in mind.

Ease of use

It is essential that GIS's are user-friendly to operate, to enable the maximum number of people to use the system with the minimum training and specialist knowledge. To this end, the completed datasets for the Belalong GIS were entered into ARCVIEW, a display/analysis package produced by the makers of ARC/INFO.

CONCLUSIONS

This project has created a large number of coverages and provides a good overview of the area's topology and the research work conducted. In view of the educational aims of this GIS, the problems of spatial accuracy were not of paramount importance. Whilst the effectiveness of the GPS was limited due to physical obstruction and altimeter problems, its use in forested areas would be greatly increased by following the suggestions outlined. Arc/Info proved to be a very effective package and had no significant limitations. Maximising spatial accuracy, establishing data protocols, following closely defined aims and ensuring ease of use were found to be very important factors in creating an effective GIS.

ACKNOWLEDGEMENTS

I thank the Land Survey Department of Brunei Darussalam for converting the GPS positions into RSO grid co-ordinates (in particular Dy. Mas Suria Watia bte Hj Watib) and Brunei Shell Sdn Bhd for offering to provide a Landsat TM image of the project area. I am grateful to all Brunei Rainforest Project members for providing their data and co-operation, and to all of my sponsors, Universiti Brunei Darussalam and the Royal Geographical Society for making my work in Brunei Darussalam possible.

LITERATURE CITED

ANDERSON, J. & MARSDEN, D. 1983. *The Temburong Renewable Resources Study*. Forestry Department, Bandar Seri Begawan, Brunei Darussalam.

BRIGGS, D. & SMITHSON, P. 1985. *Fundamentals of Physical Geography*. Unwin Hyman, London.

BROWN, A. 1992. The GPS Coordinate system explained. *GIS Europe* 1:62-63.

BURROUGH, P. A. 1986. *Principles of Geographical Information Systems for Land Resources Assessment*. Clarendon Press, Oxford.

COWEN, D. 1988. GIS versus CAD versus DBMS: What Are the Differences? *Photogrammetric Engineering and Remote Sensing* 54:1551-1555.

CROSS, P. A. 1991. GPS for GIS's. *Mapping Awareness & GIS Europe* 5:30-34.

ENVIRONMENTAL SYSTEMS RESEARCH INSTITUTE (ESRI) 1990. *Understanding GIS: The ARC/INFO Method*. ESRI, Redlands, California.

SCHOLTEN, H. J. & LEPPER, M. J. C. De 1991. The benefits of the application of Geographical Information Systems in public and environmental health. *World Health Statistics Quarterly* 44.

Tropical rain forests in Sri Lanka: characteristics, history of human impact, and the protected area system

W. ERDELEN

Department of Biogeography, University of the Saarland, Postfach 1150, D-66041 Saarbrücken, Germany

ABSTRACT. Natural forests cover about 20% of the tropical island of Sri Lanka. However, distribution in the major hygroclimatic zones is biased towards the drier parts of the island. Forest coverage is about 30% in the dry zone, but only 9% in the wet zone where Sri Lanka's rain forests occur. They contain the most diverse communities of the island with the highest concentration of endemic species. Over the past centuries rain forests have been reduced to small and more or less isolated fragments. To a large extent this has been a result of human dispersion and concomitant impact on natural landscapes. Most significant changes took place during the British colonial time (1796-1948) and after independence (1948). Present population trends in the wet zone of Sri Lanka tend towards exacerbating impacts on the remaining rain forests. Therefore, the existing protected area system should be extended with emphasis on remaining rain forest fragments.

KEY WORDS: forest communities, vegetation history, forest reserves, conservation.

INTRODUCTION

Tropical rain forests have been a central focus of international conservation activities such as FAO's Tropical Forestry Action Plan and the International Tropical Timber Agreement. Another major and related concern has been the preservation of global biodiversity (e.g. WCMC 1992). As the most complex of our extant ecosystems, both with regard to their plant and animal communities, rain forests have played a key role in this context. Although research on tropical forests has a long tradition, our knowledge about their structural and functional characteristics is still relatively poor. Up to the 1960s the tropics had been considered a region with a past of constant environmental conditions. Later evidence clearly documented that many tropical rain forest ecosystems have experienced drastic changes in geographic extension and species composition, particularly during the Quaternary (overviews in Flenley 1979, Goudie 1985). In addition, the mosaic character of rain forests (Remmert 1991) and their intrinsic (patch) dynamics (Pickett & White 1985) have changed our views of the 'constant' tropical environment. Moreover, substantial evidence has shown that alternative interpretations if not changes in ecological paradigms may be required to explain structural and functional characteristics of tropical ecosystems (overview in Waide & Lugo 1992).

A drastic reduction in extent of tropical forests, especially in the neotropics, has taken place over the last decades. This also applies to Sri Lanka where, particularly in the wet zone lowlands and in the montane areas, severe human impacts have occurred. The rain forests were already fragmented in the middle of the last century (Ishwaran & Erdelen 1990). Immediate measures are required to preserve these remnants and to extend the protected area system with emphasis on forest fragments in the wet zone. This paper describes characteristics of the rain forests of Sri Lanka as well as the history of human impacts and gives an overview of the protected area system.

D. S. Edwards et al. (eds.), Tropical Rainforest Research – Current Issues, 503–511.

GEOGRAPHIC BACKGROUND

Sri Lanka is a tropical island, located southeast of the Indian peninsula, between longitudes E 79°39' and 81°53' and latitudes N 5°54' and 9°52' and with an area of 65,610 km^2. The island comprises three peneplains with average altitudes of 30 m, 500 m, and 1500-1800 m, respectively (Adams 1929, Domrös 1976). The central highlands in the southern part of the island comprise the second and third peneplain (Figure 1). These highlands consist of three different geographic units, the Knuckles mountains in the northeast (highest peak: Gombaniya, 1906 m), the central massif (Pidurutalagala, 2524 m), and the Sabaragamuwa ridges in the southwest (Beralagala, 1386 m).

The dominating climatic factors are the southwest monsoon (May-September) and the northeast trade wind (November-February). The southwest monsoon causes most of the precipitation in the southwestern parts of Sri Lanka. During the northeast trade wind, the northern and eastern regions receive their maximum precipitation. Convectional rains dominate during the intermonsoon periods. Based on seasonal and spatial differences in rainfall, Sri Lanka is subdivided into two major country-specific hygroclimatic zones, the wet zone with mean annual rainfall above 1904 mm and the dry zone with means below 1904 mm (Domrös 1974, Figure 1).

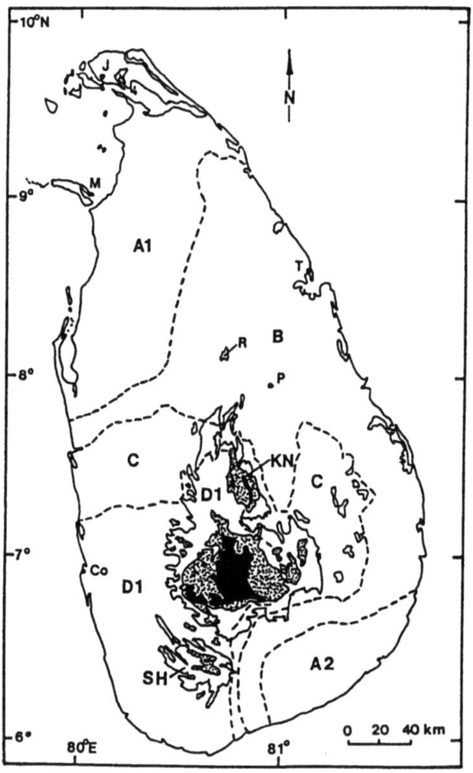

Figure 1. Relief and major vegetation types of Sri Lanka (after Erdelen 1989). KN = Knuckles mountains, R = Ritigala, SH = Sabaragamuwa hills. A1, A2 = Tropical thorn forest, B = semi-evergreen forest, C = intermediate forest, D1 = rain forest below 900 m, D2 (stippled) = submontane rain forest (900-1500 m), D3 (black) = montane rain forest (above 1500 m). Towns: Co = Colombo, J = Jaffna, M = Mannar, P = Polonnaruwa, T = Trincomalee.

Temperature, relative humidity and day length show little seasonal variation. However, average annual temperatures, which are about 27°C at sea level, decrease about 0.50-0.65°C per 100 m altitudinal difference. Monthly temperature means vary only little, but daily temperature maxima and minima differ between 5 and 10°C. Relative humidity is high throughout the year, annual means are between 77% (Trincomalee) and 85% (N'Eliya) and vary on a regional and, more markedly, daily scale (see Domrös 1974 & 1976 for details).

THE FOREST COMMUNITIES OF SRI LANKA

Distribution of Natural Vegetation

The natural vegetation of the island, particularly the forest communities, has been classified into various units as summarized, for instance, by Greller & Balasubramaniam (in press). In the National Atlas of Sri Lanka (Somasekaram 1988), 11 major formations are distinguished, including seral stages such as tropical savanna forests and grasslands which in most cases represent secondary vegetation (Pemadasa 1984), as well as edaphically controlled formations such as mangroves. For this overview a subdivision of the island into 7 zones with 6 types of natural vegetation seems sufficient (Figure 1; for more details see Gaussen *et al.* 1968). These include tropical thorn forests (or monsoon scrub jungle, *Manilkara-Chloroxylon* series,) in the driest parts of the island (NW and SE) with 5-7 dry months and an annual precipitation (a.p.) of about 1000-1400 mm, semi-evergreen forests (*Chloroxylon-Berrya-Vitex-Schleichera* series), mainly in the northeastern, eastern and southern parts (3-6 dry months, a.p. 1400-2000 mm), an intermediate type between semi-evergreen forests and rain forests (*Filicium-Euphoria-Artocarpus-Myristica* transitional series), in the intermediate zone of the island (1-2 dry months, a.p. 1900-2000 mm), and the rain forest communities in the southwestern part and the hills (no dry season, a.p. over 2000 mm; Figure 1). The latter consist of lowland forests (*Doona-Dipterocarpus-Mesua* series, sea level to about 900 m), submontane forests (*Doona-Calophyllum-Syzygium* series, 900-1500 m), and montane forests (*Syzygium-Gordonia-Michelia-Elaeocarpus* series, above 1500 m). Floristic and physiognomic characteristics of these formations are discussed, for instance, in Balasubramaniam *et al.* (in press), Gaussen *et al.* (1968), Gunatilleke & Ashton (1987), Jayasuriya *et al.* (in press) and Werner (1982, 1984).

The Rain Forests of Sri Lanka

Typical rain forests or "tropical moist forests" (*sensu* WCMC 1992) occur only in southwestern Sri Lanka and in the hill region, i.e. in the wet zone (Figure 1). This zone represents the only area of aseasonal wet climate between western Malesia and eastern coastal Madagascar (Ashton & Gunatilleke 1987). Even the most humid regions of South India experience a relatively dry phase (Legris & Viart 1961, Pascal 1984) which may explain the lack of endemism in certain taxa such as the dipterocarps in this region (Gunatilleke & Gunatilleke, personal communication).

The rain forests of Sri Lanka are not only the most diverse and species-rich of the natural forests but they also contain the highest proportions of endemic plant and animal taxa (Baldwin 1991, Erdelen 1989, Gunatilleke & Ashton 1987). Amongst angiosperm plants the dipterocarps are the most characteristic elements of these forests. Of the 56 species occurring in Sri Lanka, 55 are endemic (Balasubramaniam 1985, but see also Ashton 1980). A reexamination of endemism in the lowland wet zone revealed that endemism in the woody flora, as calculated from 9 sites sampled, was 68% (Gunatilleke *et al.* 1987). Moreover, a recent study of the understorey plants in Sinharaja forest has also documented not only their high species richness but also their high percentage of endemics, almost 50% of all woody and non-woody plants (DeZoysa *et al.* 1988).

Studies on the forest communities have shown that in Sri Lanka 15 floristic regions may be distinguished (Gunatilleke & Gunatilleke 1990). Excluding the coastal and marine belt, 11 (79%) of the remaining 14 regions are located in the wet zone and the hills (Ashton & Gunatilleke 1987). Even within these floristic regions tree communities may vary considerably in physiognomy and species composition. For instance, studies of aerial photographs of Sinharaja forest have shown that the tree communities do not only vary with altitude but have also a very pronounced mosaic character (Merritt & Ranatunge 1959, Erdelen, in press). Though the best studied of the Sri Lankan rain forests, its internal dynamics are still poorly understood. Only recently studies have been launched on the phenology of major tree taxa, their pollination system, and reproductive biology (e.g. Dayanandan *et al.* 1990, DeZoysa *et al.* 1991).

Many plant and animal species have very restricted geographic ranges within the island. This has been documented for instance for ***Solenostemon elongatus*** (Labiatae) and other endemics confined to Ritigala hill in the northern part of the island (Figure 1; for details see Jayasuriya 1984) or the endemics restricted to Hindumkanda, an isolated hill in the southwestern lowlands (Gunatilleke & Gunatilleke 1984). These very localized geographic distribution patterns have not adequately been considered with regard to conservation of the forest communities in Sri Lanka, particularly those of the wet zone. Restricted geographical distribution is one feature which makes species susceptible to extinction, another is low population density. The latter has been shown to apply for many of the better known tree taxa and also for the only rain forest understory community studied more in detail, i.e. that of Sinharaja forest (DeZoysa *et al.* 1988). For instance, according to Gunatilleke *et al.* (1987) in the 9 study sites mentioned above, 98% of the woody endemic subcanopy and canopy species and 85% of the understory species are either rare, vulnerable or endangered in terms of the IUCN Red Data Categories. This heterogeneity in plant and animal distribution and the fact that many species may be characterized as naturally already prone to extinction becomes an even more important factor to be considered for conservation measures with respect to human population growth in the island.

The History of Forest Cover in Sri Lanka

The impact of man on extension and successional stages of natural forest communities in Sri Lanka has been an important environmental factor since the Holocene. During prehistoric times, the centers of human population were the southwest (wet zone) and the peripheral areas of the central hills (Figure 2). In the transitional phase between prehistory and protohistory, human settlements shifted towards the dry zone, mainly to the north-central parts of the island. Population density increased between 250 BC and 1235 AD, the culmination period of Singhalese culture. During the Late Historical Phase, beginning in the 13th century, the political, economic and cultural centres shifted from the north-central, eastern and southeastern parts of the island towards the lowland areas of the wet zone, the central hills, and the extreme northern part. This process was coupled with the decline of the high cultures in the dry zone and the beginning of the colonial phases, i.e. the Portuguese, the Dutch, and the British periods. In particular during the British period (1796-1948) impacts on natural forests in the southwestern parts and the hills were severe and extensive. Since Sri Lanka's independence in 1948, human impacts have reached new scales. High population growth triggered shrinkage and fragmentation of the remaining natural forests. Natural forest cover has been drastically reduced during this century. In 1900 forest cover was still 70%, by 1948 it had already been reduced to 50% and now it amounts to about 20% (Preu & Erdelen 1992). This process has not equally affected the different vegetation types. As a result, extension of areas under natural forest is presently highly skewed towards the dry zone where semi-evergreen forest is the most widespread forest type. Wet zone lowland and hill forests have been reduced to small fragments.

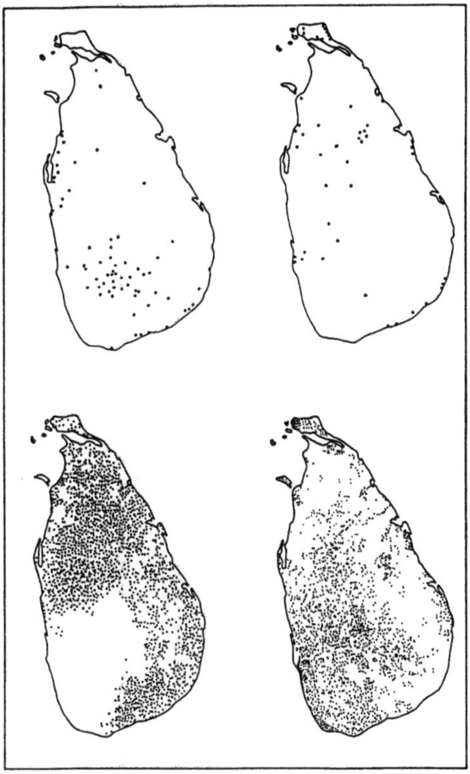

Figure 2. Human settlement patterns in Sri Lanka during the prehistoric phase (top left), the protohistoric and early historic phase (top right), the Anuradhapura and Polonnaruwa period (bottom left) and at present (bottom right). Based on archeological maps, data on irrigation systems ("tanks"), and actual population densities (data from Somasekaram 1988).

At present the natural forest cover amounts to about 30% in the dry zone, but only 9% in the wet zone. About two thirds of Sri Lanka's population are already living in this latter zone. Extrapolating from present figures, population density in the wet zone will be about 860 persons per km^2 by the end of this millennium (Erdelen 1993). It remains to be seen what percentage of the formerly widespread rain forests will be left by that time.

SRI LANKA'S PROTECTED AREA SYSTEM

Sri Lanka has the highest biodiversity per unit area of all Asian countries (Baldwin 1991). The existing extent of protected areas amounts to almost 8000 km^2 or about 12.2% of the total land area (IUCN 1990). An additional 0.4% of the area is proposed for reserves.

The protected area system consists of National Parks (N= 10, Ruhuna and Yala East combined), Strict Natural Reserves (3), Nature Reserves (2), one Jungle Corridor (Nilgala), and Sanctuaries (49). In addition, international designations have been assigned to the Hurulu Forest Reserve (Biosphere Reserve), Bundala Sanctuary (Ramsar Wetland), and Sinharaja (Biosphere Reserve, World Heritage Site). Their assignment to the IUCN management categories and other details are given in IUCN (1990). Despite this extensive

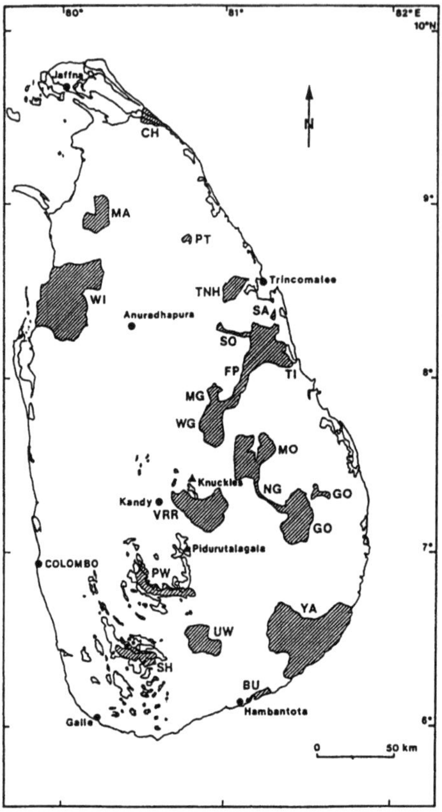

Figure 3. Protected area system of Sri Lanka. Simplified after Collins *et al.* 1991. Only wet zone forests shown. Sanctuaries: BU = Bundala, CH = Chundikulam, MA = Madhu Road, PT = Padaviya Tank, PW = Peak Wilderness, SA = Seruvila-Allai, TNH = Trincomalee Naval Headworks, VRR = Victoria-Randenigala-Rantambe. Jungle Corridor: NG = Nilgala. Nature Reserves: MG = Minneriya-Giritale, TI = Tirikonamadu. National Parks: FP = Flood Plains, GO = Gal Oya, MO = Maduru Oya, SO = Somawathiya Chaitiya, UW = Uda Walawe, WG = Wasgomuwa, WI = Wilpattu, YA = Yala. World Heritage Sites: SH = Sinharaja.

area coverage the different vegetational zones are not equally represented. Most of the protected areas are located in the dry zone (Figure 3). In other words, the regions with highest biodiversity and concentration of endemics are under-represented. This also applies with regard to the size distribution of the different categories of protected areas (Figure 4). Only two of the protected areas are over 1000 km^2 in extent, i.e. Wilpattu and Yala National Park. The only larger tracts relevant for rain forest conservation are Sinharaja (c. 90 km^2) and the Peak Wilderness Sanctuary (c. 224 km^2).

Except for the two international biosphere reserves mentioned, 47 forest reserves have been declared National Biosphere Reserves (IUCN 1990) within the national man and biosphere program (details in Bharathie 1979). Most of these reserves contain forests degraded to various extent, but would nevertheless form an important component for the establishment of better connectivity between the elements of the protected area system. Of particular importance in this regard would be national MAB plots located in the wet zone. Integration of these areas into the existing protected area system, however, would require considerable strengthening of links between the two responsible bodies, the Department of Wildlife Conservation and the Forest Department of Sri Lanka.

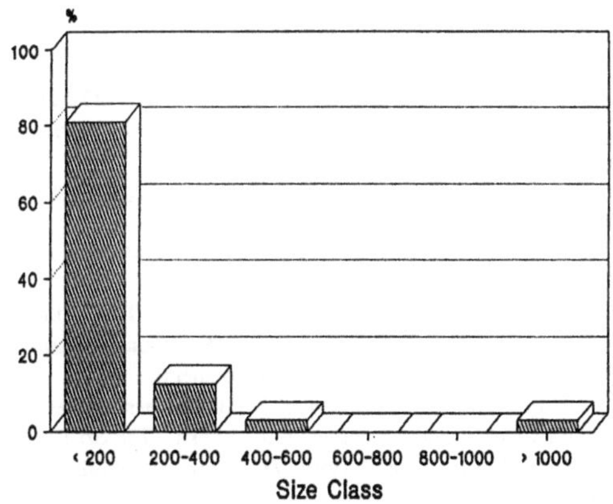

Figure 4. Size distribution of protected areas in Sri Lanka (N = 64). Size classes in km^2. Based on data in IUCN (1990).

CONCLUSIONS

Although our knowledge of the flora and fauna of Sri Lanka is relatively good compared to that of other Asian countries, there is still a clear need for further research on species composition and species distribution, in particular of understorey plants and invertebrate animals. As already pointed out (Erdelen 1993), studies on population viability and vulnerability of target species for conservation have not yet been carried out. Furthermore, some wet zone species have not been recorded since the last century and may have become extinct by now. One major ecosystem of this zone, the swamp forests, formerly distributed in areas now under rice cultivation, has been reduced to a single locality.

Data on actual forest cover, even as used in the most recent publications, date back to 1986 or even earlier (e.g. Somasekaram 1988). Present trends in human population growth underscore the need for more updated information on natural forest cover in the island. This could easily be accomplished by means of interpretation of satellite imageries.

In Sri Lanka, conservation activities in the future should focus on effectively protecting remaining forests in the more humid parts of the island which harbour not only highest species diversities but also the highest numbers of endemics. To accomplish this, updated information on actual forest cover and an ecological monitoring system are urgently required.

The need for new organizational frameworks for the sustainable use and conservation of tropical ecosystems has repeatedly been pointed out (e.g. Erdelen & Ishwaran 1991). First steps towards developing cooperative programs and launching new initiatives to improve the environmental situation in Sri Lanka have already been undertaken (overview in Erdelen & Preu 1990). This has been but a starting point, particularly with regard to the increasing pressure on the remaining natural landscapes in the island.

ACKNOWLEDGEMENTS

I am grateful to the Deutsche Forschungsgemeinschaft (German Research Foundation) for sponsoring my participation in the rain forest conference in Brunei Darussalam. I thank the organizers for inviting me to contribute. I appreciate the comments of Ralf Peveling and an anonymous reviewer on an earlier draft of the manuscript.

LITERATURE CITED

ADAMS, D. 1929. The geology of Ceylon. *Canadian Journal of Research* 1:425-511.

ASHTON, P.S. 1980. Dipterocarpaceae. Pp. 364-423 in Dassanayake, D.S. & Fosberg,F.R. (eds). *Handbook to the Flora of Ceylon I.* Oxford & IBH Publishing Company, New Delhi.

ASHTON, P.S. & GUNATILLEKE, C.V.S. 1987. New light on the plant geography of Ceylon. I. Historical plant geography. *Journal of Biogeography* 14:249-285.

BALASUBRAMANIAM, S. 1985. Tree flora of Sri Lanka. Pp. 58-67 in Jayatilleke, A. (ed.). *Proceedings of the International Conference on Timber Technology*, Moratuwa, Sri Lanka.

BALASUBRAMANIAM, S., RATNAYEKE, S. & WHITE, R., in press. The Montane Forests of the Horton Plains Nature Reserve. In Erdelen, W., Preu, C., Ishwaran, N. & Madduma Bandara, C.M. (eds). *Ecology and Landscape Management in Sri Lanka.* Margraf Publishers, Weikersheim.

BALDWIN, M.F. (ed.) 1991. *Natural Resources of Sri Lanka.* Keells Business Systems Ltd., Colombo.

BHARATHIE, SRI K.P. 1979. Man and biosphere reserves in Sri Lanka. *The Sri Lanka Forester* 14:37-40.

COLLINS, N.M., SAYER, J. & WHITMORE, T. (eds) 1991. *The conservation atlas of tropical forests. Asia and the Pacific.* Macmillan Press, London.

DAYANANDAN, S., ATTYGALLA, D.N.C., ABEYGUNASEKERA, A.W.W., GUNATILLEKE, I.A.U.N. & GUNATILLEKE, C.V.S. 1990. Phenology and Floral Morphology in Relation to Pollination of Some Sri Lankan Dipterocarps. Pp. 103-133 in Bawa. K.S. & Hadley, M. (eds). *Reproductive Biology of Tropical Forest Plants.* UNESCO, Paris.

DEZOYSA, N.D., GUNATILLEKE, C.V.S. & GUNATILLEKE, I.A.U.N. 1988. Diversity of understorey vegetation in the Sinharaja rainforest. *The Sri Lanka Forester* 18:121-130.

DEZOYSA, N.D., GUNATILLEKE, C.V.S. & GUNATILLEKE, I.A.U.N. 1991. Comparative Phytosociology of Natural and Modified Rain Forest Sites in Sinharaja MAB Reserve in Sri Lanka. Pp. 223-233 in Gómez-Pompa, A., Whitmore, T.C. & Hadley, M. (eds). *Rain Forest Regeneration and Management.* UNESCO, Paris.

DOMRÖS, M. 1974. *The Agroclimate of Ceylon. A Contribution Towards the Ecology of Tropical Crops.* Geoecological Research 2. Steiner Verlag, Wiesbaden.

DOMRÖS, M. 1976. *Sri Lanka. Die Tropeninsel Ceylon.* Wiss. Länderkunden 12. Wiss. Buchgesellschaft, Darmstadt.

ERDELEN, W. 1989. Aspects of the Biogeography of Sri Lanka. Pp. 73-100 in Schweinfurth, U. (ed.). *Forschungen auf Ceylon III.* Steiner Verlag, Wiesbaden.

ERDELEN, W. 1993. Conservation of Tropical Forest in Sri Lanka - Problems and Issues. Pp. 123-144 in Balakrishnan, M., Bie, S.W. & Borgstrom, R. (eds). *Tropical Ecosystems. A Text Book of Tropical Ecology and Conservation.* Norwegian Centre for International Agricultural Development, Norway.

ERDELEN, W., in press. The Relevance of Scale and Hierarchy Concepts for Landscape Management - A Sri Lankan Perspective. In Erdelen, W., Preu, C., Ishwaran, N. & Madduma Bandara, C.W. (eds). *Ecology and Landscape Management in Sri Lanka*, Margraf Publishers, Weikersheim.

ERDELEN, W. & ISHWARAN, N. 1991. Information Inputs to Regulate Use and Conservation of Tropical Ecosystems - The Need for New Organizational Frameworks. Pp. 195-202 in Erdelen, W., Ishwaran, N. & Müller, P. (eds). *Tropical Ecosystems - Systems Characteristics, Utilization Patterns, and Conservation Issues.* Margraf Publishers, Weikersheim.

ERDELEN, W. & PREU, C. 1990. Background and Outcome of an International and Interdisciplinary Symposium on "Ecology and Landscape Management in Sri Lanka". *GeoJournal* 22:215-217.

FLENLEY, J.R. 1979. *The equatorial rain forest: a geological history.* Butterworths, London.

GAUSSEN, H., LEGRIS, P., VIART, M. & LABROUE, L. 1968. *Explanatory notes on the vegetation map of Ceylon.* Government Press, Colombo.

GOUDIE, A. 1985. *Environmental Change.* Clarendon Press, Oxford.

GRELLER, A.M. & BALASUBRAMANIAM, S., in press. Physiognomic, Floristic, and Bioclimatological Characterization of the Major Forest Types of Sri Lanka. In Erdelen, W., Preu, C., Ishwaran, N. & Madduma Bandara, C.M. (eds). *Ecology and Landscape Management in Sri Lanka.* Margraf Publishers, Weikersheim.

GUNATILLEKE, C.V.S. & ASHTON, P.S. 1987. New light on the plant geography of Ceylon II. The ecological biogeography of the lowland endemic tree flora. *Journal of Biogeography* 14:295-327.

GUNATILLEKE, C.V.S., GUNATILLEKE, I.A.U.N. & SUMITHRAARACHCHI, B. 1987. Woody endemic species in the wet zone lowlands of Sri Lanka and their conservation in botanic gardens. Pp. 183-194 in Bramwell, D., Hamman, O., Heywood, V.B. & Synge, H. (eds). *Botanic Gardens and the World Conservation Strategy.* Academic Press, London.

GUNATILLEKE, I.A.U.N. & GUNATILLEKE, C.V.S. 1984. Distribution of endemics in the tree flora of a lowland hill forest in Sri Lanka. *Biological Conservation* 28:275-285.

GUNATILLEKE, I.A.U.N. & GUNATILLEKE. C.V.S. 1990. Distribution of floristic richness and its conservation in Sri Lanka. *Conservation Biology* 4:21-31.

ISHWARAN, N. & ERDELEN, W. 1990. Conserving Sinharaja - An Experiment in Sustainable Development in Sri Lanka. *Ambio* 19:237-244.

IUCN (INTERNATIONAL CONSERVATION UNION) 1990. *IUCN Directory of South Asian Protected Areas.* Gland.

JAYASURIYA, A.H.M. 1984. Flora of Ritigala Natural Reserve. *The Sri Lanka Forester* 16:61-156.

JAYASURIYA, A.H.M., GRELLER, A.M., BALASUBRAMANIAM, S., GUNATILLEKE, I.A.U.N., GUNATILLEKE, C.V.S. & DASSANAYAKE, M.D., in press. Phytosociological Studies of Mid-Elevational (Lower Montane) Evergreen Forests in Sri Lanka. In Erdelen, W., Preu, C., Ishwaran, N. & Madduma Bandara, C.M. (eds). *Ecology and Landscape Management in Sri Lanka.* Margraf Publishers, Weikersheim.

LEGRIS, P. & VIART, M. 1961. *Bioclimates of South India and Ceylon.* Institut Francais de Pondichéry. Traveaux de la Section Scientifique et Technique 3:165-179.

MERRITT, V.G. & RANATUNGA, M.S. 1959. Aerial photographic survey of Sinharaja forest. *The Ceylon Forester (New Series)* IV:103-156.

PASCAL, J-P. 1984. *Les forets denses humides sempervirentes des Ghats Occidentaux de l'Inde. Ecologie, structure, floristique, succession.* Institut Francais de Pondichéry. Traveaux de la Section Scientifique et Technique 20. Pondichéry.

PEMADASA. M.A. 1984. Grasslands. Pp. 99-131 in Fernando, C.H. (ed.). *Ecology and biogeography in Sri Lanka.* Martinus Nijhof, The Hague.

PICKETT, S.T.A. & WHITE, P.S. (eds) 1985. *The Ecology of Natural Disturbance and Patch Dynamics.* Academic Press, New York.

PREU, C. & ERDELEN, W. 1992. Geoecological Consequences of Human Impacts on Forests in Sri Lanka. Pp. 147-164 in Goldammer, J.G. (ed.). *Tropical Forests in Transition: Ecology of Natural and Anthropogenic Disturbance Processes.* Birkhäuser Verlag, Basel.

REMMERT, H. 1991. The Mosaic-Cycle Concept of Ecosystems - An Overview. Pp. 1-21 in Remmmert, H. (ed.). *Tropical Forests in Transition: Ecology of Natural and Anthropogenic Disturbance Processes.* Ecological Studies, Volume 85. Springer-Verlag, Berlin.

SOMASEKARAM, T. (ed.). 1988. *The National Atlas of Sri Lanka.* Survey Dept., Colombo.

WAIDE, R.B. & LUGO, A.E. 1992. A Research Perspective on Disturbance and Recovery of a Tropical Montane Forest. Pp. 173-190 in Goldammer, J.G. (ed.). *Tropical Forests in Transition: Ecology of Natural and Anthropogenic Disturbance Processes.* Birkhäuser Verlag, Basel.

WCMC (WORLD CONSERVATION MONITORING CENTRE) 1992. *Global Biodiversity: Status of the Earth's living resources.* Chapman & Hall, London.

WERNER, W.L. 1982. The Upper Montane Rain Forests of Sri Lanka. *The Sri Lanka Forester* 15:119-135.

WERNER, W.L. 1984. *Die Höhen- und Nebelwälder auf der Insel Ceylon (Sri Lanka).* Trop. subtrop. Pflanzenwelt 46. Steiner Verlag, Wiesbaden.

A Conflict over Natural Resources between Highland and Lowland Populations in Thailand

WARANOOT TUNGITTIPLAKORN[1] *and* BENGT-ERIK BENGTSSON[2]

Asian Institute of Technology, Natural Resources Programme, GPO Box 2754, Bangkok 10501, Thailand

ABSTRACT. The ongoing conflict between the highlanders and the lowlanders at Mae Soi, Jomthong, Chiang Mai, was explored to identify its causes. The study shows that perceptions about the conflict by both groups are at variance. Lowlanders perceive the conflict to be environmental, i.e. due to deforestation, water reduction, and contamination. Highlanders attribute social, political and economic causes to the conflict. The study also finds a high dependency on the forest for fuel wood, fodder and building materials, and a low dependency on the forest for food products. It can be concluded that environmental effects of the agriculture practices in the hill areas have reduced both the forested area and its quality. Water reduction has occurred but it is not possible to single out one particular group as responsible for this change. Cash crops introduced among the hill tribes to replace opium poppy cultivation need pesticides (both organochlorines and organophosphates are used), but pesticide concentrations found in downstream river water in this study are estimated to be within safe limits for drinking purposes. The conflict is both environmental and socio-economic. Resolving the conflict thus requires a broader perspective than resolving environmental problems alone.

KEY WORDS: environmental damage, deforestation, water reduction, pesticide contamination, socio-economic conflict.

INTRODUCTION

Increasing conflict between different interest groups over natural resources has become apparent in Thailand in recent years. This study highlights a prevailing natural resource management problem, the environmental conflict between highlanders and lowlanders. By using one of the most widely publicized cases, the study examines the nature of the conflict from the perspective of the local people. The aim of this study is to have a better understanding of the social and environmental aspects of conflicts over resources in northern Thailand.

Of over seventeen million hectares of northern Thailand, more than two-thirds is upland (>300 m <1000 m above sea level [asl]) or highland (>1000 m asl), which contain most of the country's remaining forests. The highland forests are the origins of tributaries of major rivers. While the majority of the population is concentrated in valleys, the uplands are home for over five hundred thousand ethnic minorities, widely known as Chao Khao or hill tribes. These geographical and demographical characteristics create a situation susceptible to conflict over natural resources, especially control over headwaters.

The situation in Mae Soi reflects such problems and has developed into a severe crisis.

Some data in this article are from "Highland-Lowland Conflict Over Natural Resources: A Case of Mae Soi, Chiang Mai, Thailand", in Society and Natural Resources, Vol. 8, pp. 279-288. Taylor & Francis. 1995.

[1]Waranoot Tungittiplakorn is currently a PhD candidate at the University of Victoria, Geography Department, Victoria, Canada.

[2]Bengt-Erik Bengtsson is an Associate Professor at the Institute of Applied Environmental Research, Stockholm University.

D. S. Edwards et al. (eds.), Tropical Rainforest Research – Current Issues, 513–522.

The lowlanders demanded that Hmong hill tribes move out of the headwaters forest. The Hmong's agriculture practices allegedly destroy the forest which in turn reduces stream flow that feeds agriculture downstream. Supported by a conservationist monk and an affiliated foundation, lowlanders marched up the mountain and erected a 12-kilometre fence, so limiting access to the forest and arable land. Protests were staged and letters written. Finally, the cabinet in 1989 announced that the Hmong of Pakluay village would be relocated. Hmong villagers, however, argued that they had cooperated with the government in crop replacement and reforestation programmes and were sceptical about the resettlement scheme. They appealed to the Royal Northern Project, the Department of Public Welfare and the Military. These are the agencies that have most contact with the hill tribes. The conflict widened the gaps not only between highlanders and lowlanders but also between the agencies involved. Unable to reach a consensus, government agencies have stalled the resettlement programme.

METHODS

One hundred and fifty people in the lowlands and highlands were interviewed. We selected the respondents by simple random sampling, using the voter list as a sampling frame. The villagers were from four villages identified by the district officer as most actively involved in the conflict. Interviews were done both in northern dialect and in Hmong language. Some Hmong women refused to be interviewed. In the Hmong culture, contacts with outsiders are more the realm of the men. In such cases we selected a replacement. We also talked to staff of various agencies including the Tribal Research Institute, the Thai-Norwegian Highland Development Project, the Forestry Department, the United Nations Program for Drug Abuse Control, Agricultural Extension Office, Health Office, District Office, and the Dharmmanart Foundation for Conservation and Rural Development. These interviews were informal, using a few structured questions as guidelines. We compared the result of the interviews between highlanders and lowlanders regarding dependency on forest and water and their perceptions of the conflict .

The Health Officer from Jomthong hospital and the District Agricultural Extension Officer have collected water from the Mae Soi and other streams routinely since 1990. The samples were generally sent to the laboratory at the Ministry of Health. On 18 November 1992, however, an additional set of water samples were taken from three sites around the Hmong village for analysis at the Environmental Engineering laboratory of the Asian Institute of Technology (AIT), Bangkok. Water from Mae Tia stream was collected one kilometre downstream from the Hmong village and two and three kilometres upstream. Two other samples were collected from the Inthanon area, where intensive cultivation of similar crops occurs.

Table 1. Distribution of Respondents, Mae Soi, Chiang Mai, Thailand, 1992.

Location	Village Name	Total No. of Households in village	Number of Respondents	Households represented (%)
Lowland	Nong Khan	166	34	20.48
	WangNamyad	128	46	35.94
	Huay Ha	90	17	18.89
Highland	Pakluay	82	53	64.63
Total		466	150	32.19

The water samples that were sent to AIT were analyzed according to the following procedure. One litre from each water sample was extracted three times with Hx: DCM, 1:1. The combined organic phase from each sample was evaporated, dried with sodium sulphate and the extracts concentrated to 1 ml. The extracts were injected into a Hewlett Packard 5890 gas chromatograph (GC) with a mass selective detector and analyzed for pesticides and PCBs. The column was 50 m x 0.12 mm Chrompac, CP-Sil 8CB with film thickness 0.25 μm. The carrier gas was helium and the flow 0.5 ml/min. The oven was held at 60°C for two minutes and raised by 20^0 min^{-1} to 220°C. The temperature was then raised by 3^0 min^{-1} to 280°C. The injector and detector temperatures were 200°C and 300°C respectively. Scan mode was used.

The Study Area

Mae Soi is a subdistrict located south of Jomthong, approximately 80 km south-west of Chiang Mai city. The four villages under study were Nong Khan, Wang Namyad, Huay Ha and Pakluay. The first three are lowland villages found at the elevation 270-300 m asl. Pakluay is a highland village located at 1410 m asl (Figure 1).

The Pakluay cultivation area is adjacent to the headwaters of three streams: Mae Soi, Mae Tim and Mae Tia. The Mae Soi stream is central to the dispute between the lowlanders and the highlanders. Orographic precipitation allows Pakluay to depend largely on rain for agriculture; however farmers who own land adjacent to the streams transport water through pipes to their fields. The quantity of water in these streams was "little" and in the dry seasons many of them dried up completely (Anon 1987).

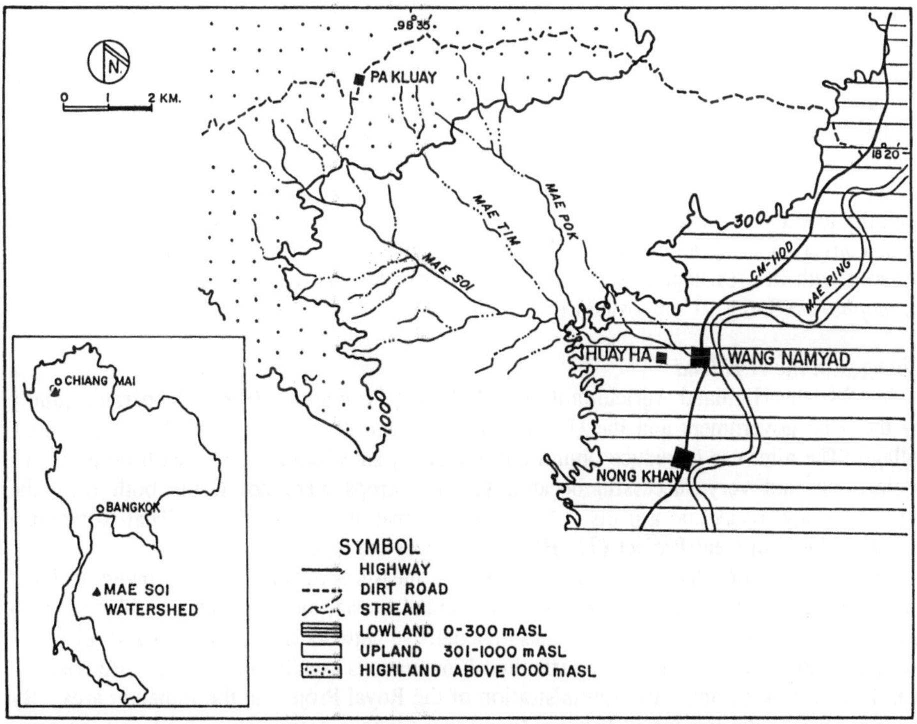

Figure 1. The study area, Tam bon Mae Soi, Chiang Mai Province, Northern Thailand.

Traditionally, the Hmong are pioneer swiddeners who regularly migrate when productivity of their lands decline. Opium poppy and swidden rice were their two main crops. The village was settled approximately 25 years ago. The current land shortage does not allow them to move to a new area. They cultivate secondary forest, rotate their fields and even cultivate the grassland. Pakluay used to raise a few hundred livestock. However, because of the theft of the cattle and horses during the early 1980's, they have abandoned this activity. Pakluay shifted from opium and rice to vegetable growing in the mid-1980's.

The lowland rice farmers depend on the Mae Soi stream for their paddy. The villagers manage irrigation communally. The majority of the people, however, engage in upland rain-fed cultivation of such cash crops as garlic, red onion, soil bean, peanut and chili. In the past, tobacco was an important crop, as evidenced by the presence of tobacco curing plants still in the area. Wood milling and charcoal making were also important means of making a living in Wang Namyad and Huay Ha, but are now greatly reduced. Approximately 300 cattle were raised in 1991, with the upland forest providing grazing grounds. Fruit trees, especially longans, have become popular in recent years. Most lowlanders are landless or near landless. There have been major land transactions in the area, and a large part of the land has been transferred to outsiders.

Natural forest in the lowlands, like any other place in Thailand, has long been cleared. Around the late 1930's, teak was being extracted from the valley by the Bombay-Burma Company. Following the path of the British company, the remaining hardwood species were logged by Thai timber companies. Around the early 1970's, defective and poor quality trees were removed to be used as fuel wood to cure tobacco leaves. After the tobacco company concession ended, villagers took over the forest and harvested young trees for charcoal burning. By 1977 the forest in Mae Soi was cleared (Maxwell 1992). What remains is a remnant of the original deciduous forest along the Mae Soi stream. This area was considered sacred by the villagers in the past, though many people have now clearly abandoned their belief in the spirits of the grove. This is reflected in the extent of utilization of the area. The upland at the base of the mountain is dominated by bamboo which is a common pioneer in the process of forest regeneration. Further up, at 700-1000 m asl, the deciduous forest prevails with evergreen species along the water course (Maxwell 1992).

Pine and evergreen hardwood dominate the highland. This area has been less disturbed than the lowland and original species still exist. However, a large area of the highland has been converted to agricultural land by highland villagers. The old fields left fallow are covered with fire resistant weeds such as *Imperata cylindrica* L. (Poaceae) and *Eupatorium adenophorum* Spreg. (Compositae) (Maxwell 1992).

Changes in the Highland

In 1981 the Highland Agricultural and Marketing Programme (HAMP) operated jointly by the Thai government and the United Nations began their work in the Pakluay Hmong village. The aim was to reduce opium cultivation by substituting it with cash crops. This project was not very successful because the new crops were not viable both from the productive and economic aspects. After the termination of HAMP, the Thai-Norwegian Highland Development Project (TN-HDP) took over the task.

Around 1985 the Thai government stepped up measures against opium growers (raiding fields and arresting farmers). This had a substantial impact on the farmers and led them to gradually change crops. The Hmong has an extensive network through clan-based relations. According to some villagers, their agricultural innovations usually come from those relatives who live in villages under the administration of the Royal Project in the Inthanon area. By the mid 1980's, the Hmong in Pakluay found that cabbage grew well in their environment. Most important of all, however, was the presence of marketing outlets for cabbage in Jomthong. TN-HDP followed up the innovation by providing seeds, pesticides and fertilizers to farmers. From 1985 onward, Pakluay has been transformed into a cabbage growing village.

Cabbage has brought significant income into the village. The 1986 survey conducted by the Department of Land Development estimated the profit to be 1,700-12,000 baht per rai. It also reported the per capita income of Pakluay villagers in 1986/7 to be -27 to 3,263 baht (Tansiri & Chondamrongkul 1987). The figure for northern rural villagers was 826 baht in 1988 (Anon 1993). Nevertheless, cabbage requires a tremendous amount of input. Most Hmong farmers today claim that they are indebted to lowland merchants because they bought inputs for cabbages on credit.

RESULTS

Dependency on Forest

Both highlanders and lowlanders depend heavily on the forest for firewood, building materials and fodder (see Table 2). The highlanders use vegetation from the forest to supplement maize for their pigs. The lowlanders also gather firewood from the forest and take lumber from the local forest. However, lowland villages are closer to alternative sources of building materials and fuel, thus depending on the forest for these products to a lesser degree than the hill tribes.

Food obtained from the forest is only supplementary (Table 3). The Hmong hunt more frequently. They also gather medicinal plants more than the lowlanders. Mae Soi villagers in the lowlands collect bamboo shoots and mushrooms, which they sometimes sell. They also gather vegetables and ant's eggs occasionally. It was not possible to obtain data about the biomass of these products in this study. The significance of forest as a source of food is small. Most people from both groups buy food from local vendors. Respondents from both areas stated that forest resources have been greatly reduced in recent years.

Dependency on Water

Figure 2 shows the periods when water is needed in the highlands and the lowlands. The most critical period of water shortage is from April to the beginning of May. The rain has not yet arrived and water is needed both in the highlands and the lowlands. The highlands need water in May mainly for the vegetable seedlings. The villagers draw water from the stream and transport it through pipes to their seedlings beds.

The lowlands need water for the second cropping of rice in April and May. This is a recent phenomenon facilitated by 'electrical irrigation', a new irrigation scheme in which water is pumped from the Ping river. This should not affect the Mae Soi stream but may decrease the available cattle grazing lands. Cattle are also affected by the water shortage in the Mae Soi stream.

Table 2. Percentage of respondents obtaining resources from the Mae Soi forest, Thailand 1992. n: number of respondents; * significant at 95% confidence level using Chi-square test of association.

Resources	Highlands		Lowlands	
	n	% of respondents	n	% of respondents
Firewood*	53	100	73	63.0
Fodder	51	43.1	45	40.0
Building materials*	52	100	54	75.9

Table 3. Forest Food Collection Frequency in 1991/92, Mae Soi. Thailand. 1992. (Multiple Responses); n: number of respondents; * significantly different at 95% confidence level using Chi-square test of association.

Food Resources	Highland (n = 51)				Lowland (n = 37)			
	mean	min	max	S.D.	mean	min	max	S.D.
Mushroom*	0.8	0	7	1.6	2.1	0	15	3.3
Bamboo shoot*	1.2	0	5	1.4	6.9	0	100	17.9
Animal*	5.4	0	98	19.6	0.0	0	0	0.0
Honey	0.0	0	0	0.0	0.1	0	3	0.6
Vegetables*	0.0	0	0	0.0	1.4	0	20	3.5
Ant's eggs*	0.0	0	0	0.0	1.4	0	10	2.8
Medicinal plants*	6.1	0	97	17.7	0.6	0	9	1.7

Longan has become very popular in the lowland because of the demand from abroad. As a result, many villagers are turning their upland gardens into fruit orchards. However, longan must receive adequate amounts of water before fruiting (May-July). The quality of the fruit largely depends on the amount of water the trees receive. Smaller fruits cannot be exported and fetch a very low price, thus villagers who can afford pumps take water from the stream in great quantities. Because of the increase in longan in the areas, water shortage is likely to escalate.

There are also some complaints about the reforestation project which withdraws water from the stream during the critical period. The reforestation project is carried out by the conservation NGO in cooperation with the Royal Forestry Department. This has resulted in some villagers having a negative attitude towards this project.

As summarized by Shinawatra (1985), various studies suggest that cultivation in the highland reduces stream discharge. Her analysis of run-off, in the dry season, of major rivers in Jomthong reveals that the volume of water was reduced from 109.26 million cubic metres per year during 1921-1951 to 68.5 million cubic metres during 1970-1980. However, it is difficult to single out a particular activity carried out by a particular group of people as being responsible for this change.

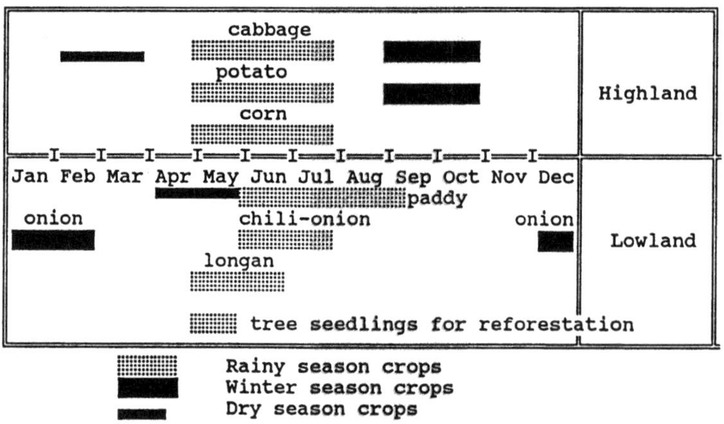

Figure 2. Agricultural calendar for highlanders and lowlanders of Mae Soi, northern Thailand.

When asked how long the water in the Mae Soi streams had been reduced in volume, 71 per cent of the lowlanders indicated 3-6 years. This contrasts with the length of time the conflict in Mae Soi has existed. The earliest evidence that we gathered indicated the year 1981 as marking the beginning of the conflict. The conflict has thus lasted for over a decade. Water shortage may not be the original cause of the conflict.

Results of Water Sample Analysis

No peaks on the gas chromatograph that corresponded to standard solutions of pesticides (HCB, Aldrin, Dieldrin, Lindane, Mirex, op-DDE, pp-DDD, pp-DDT) or PCBs (IU-28, IU-52, IU-101, IU-118, IU-138, IU-153, IU-180) could be detected. The detection limit for the contaminants listed above in Scan mode was less than 0.5 ug/L river water.

In order to detect even lower concentrations of contaminants, water extracts were also analyzed with a GC equipped with an electron capture detector. From the 1 ml extracts dissolved, 0.2 ml was taken out, diluted to 1 ml and injected into a Hewlett Packard 5890 GC. The instrument settings were the same as described under Methods with the exception that nitrogen was used as carrier gas.

The extracts were compared with standard solutions of pesticides and pentachlorobiphenyls (PCBs) but no pesticides or PCBs could be detected. The approximate detection limit was 0.006 ug/L river water but no contaminants could be detected. This means that no trace of pesticides and PCBs could be found at any of the sampling sites near Mae Soi or Khun Klang Hmong villages. To fully investigate pesticide pollution of river water, the sampling should be carried out at all stages in the agricultural cycle.

Villagers in Pakluay claimed that in 1970 they were the first Hmong to apply chemical fertilizers to poppy fields. They also claimed they were pioneers in the use of herbicides in the poppy fields around 1976-7 (Prangkhiaw 1984). When they shifted to cabbage cultivation they began using pesticides.

Several water analyses have been carried out to evaluate the water quality from agricultural fields in the highlands. The first was done by TN-HDP and Office of National Environment Board in February 1990. The Jomthong agricultural extension office and the district hospital also tested water samples in January 1991 and January 1992. All the results indicated that the levels of pesticide were within safe limits. Our water analysis was carried out in response to a request from the hospital and the agricultural extension staff. The aim was to compare the results of the AIT laboratory and those of the Ministry of Health laboratory, but the Ministry water samples were lost and no comparison was possible.

The results of the water analyses only provide some indication, not proof, that there are no pesticides being released from the highland fields. The water samples were all collected at the same season, thus not taking into account any possible seasonal variations. Secondly, there seem to be no control points or consistency in the selection of areas being sampled.

Potato, a secondary crop to cabbage in the highlands, needs more pesticides than cabbage. Some farmers reported having to apply the chemicals every five days. Fields that are used continuously also need more pesticides. Winter and dry season cabbages need more pesticides than rainy season crops.

The lowlanders used pesticides most commonly in their onion and garlic fields. They indicated that onions needed pesticides every 3-4 days. Moderate amounts of pesticides were used on other crops. The surveys done by the agricultural extension and the hospital in Jomthong indicated that pesticides are used most frequently in the lowlands in December and January. This corresponds to the onion growing period.

Deforestation

The document obtained from the district office indicated that, since they settled in the area around 1975, the Pakluay villagers have cleared at least 320 ha. of forest. After the lowlanders fenced off part of the forest, the Hmong abandoned half of their fields. The Department of Land Development reported that the area belonging to Pakluay in 1987 was 161.5 ha. It reported that the cultivated area covered 139.5 ha. (see Table 4). The size of the Pakluay land has become quite stable since 1987, although, critics of the Hmong believe that these figures are gross underestimates (Kaye 1990).

Nature of the conflict

Table 5 shows the divergence of the highlanders' and the lowlanders' perceptions of the causes of the conflict. The lowlanders perceived the conflict as largely environmental, caused by the Hmong's destruction of forest and the contamination of the river. The highlanders attributed the conflict to more diverse causes. Most of them believed that the conflict stemmed from the incident when a cattle thief (incidently a lowlander) was killed in the Hmong village in 1984. About one-third of the interviewees mentioned the success of their cash crops to be the cause of the lowlanders' resentment. An equal proportion of the highland respondents thought the conflict originated from the lowlanders' concern over the forest. Other reasons cited were ethnic difference and the intervention by the conservationist monk.

Table 4. Pakluay Cultivation Area (in hectares). Source: Jomthong district 1990 *Tansiri, B. and P. Chondamrongkul 1987. **Jomthong district 1991. nr = not reported; HAMP: Highland Agricultural and Marketing Programme; TRI: Tribal Research Institute; TN-HDP: Thai-Norway Highland Development Project; ONCB: Office of Narcotic Control Board; PLO: Provincial Land Office.

Year	Total	Poppy	Potato	Cabbage	Fallow	Reported by
1983	nr	70.72	nr	nr	nr	HAMP
1984	nr	78.08	nr	nr	nr	HAMP
1985	nr	47.20	29.28	nr	nr	TRI
1986	345.76	1.44	20.96	37.92	nr	TN-HDP/ONCB
1987	161.44	0	28	59.84	nr	TN-HDP/DLD*
1988	148.32	0	36.64	90.88	20.80	TN-HDP/ONCB
1989	nr	0	37.60	93.92	nr	TN-HDP/ONCB
1990	160.96	0	129.60		31.36	PLO**
1991	nr	0	139.44		nr	PLO**

Table 5. Perceived Causes of Conflict, Mae Soi, Chiang Mai, Thailand.1992. (Multiple Responses)

Perceived causes	Highland (n = 41)		Lowland (n = 81)	
	n	% of Resp.	n	% of Resp
Deforestation by the Hmong	12	28.6	60	81.4
Pesticide contamination by the Hmong	1	2.4	12	16.0
Murder of the lowland thief	22	54.8	0	0.0
Racial issue	10	23.8	1	1.2
Economic competition	13	30.9	0	0.0
Intervention by the conservationist monk	8	19.0	5	6.2
Others	4	9.5	5	6.2

DISCUSSION

The pressure to use the resources of the highlands and lowlands is high. The most serious threat to the highlands comes from the expansion of agricultural land. The promotion of cash crops in the hills have undesirable effects in terms of decreasing forested land. However, Pakluay's lands seem to have been stable since 1989, although if the Hmong continue their cultivation of the same area, the soil productivity is likely to decrease. The Hmong are already beginning to realise the effect of soil degradation. They spend a great proportion of their income buying fertilizers and pesticides to maintain productivity of their land.

Pakluay villagers still use open fires for cooking and heating. This method consumes a large amount of wood, although in some houses more efficient clay stoves have been used to save fuel. Such innovations should be followed up by government or non-government agencies concerned about forest destruction.

One obstacle against initiating constructive measures in the highlands is the lingering indecision on whether to move the Hmong village. Agencies involved express reluctance to invest any energy and resources in conservation if the village is to be relocated. However, there is little possibility of finding a suitable area for relocation in the lowlands. The government may resort to forced eviction, which is disruptive to the Hmong and also to the existing population of the destined site. The situation is likely to remain a stalemate.

The conflict in Mae Soi involves both environmental and socio-economics. The conflict evolved around the forest and water issues. The absence of a management plan over resource use has resulted in over-exploitation both of the highlands and the lowlands. The lowlanders have become environmentally conservative as a result of the campaign by an outside NGO and a conservationist monk. The Hmong, existing outside the Nation's and Buddhism's realm, are largely excluded from this campaign. The work of international projects in the highlands has caused some resentment by the lowlanders, who think they are being bypassed. Currently, the Hmong are better off economically while the Mae Soi people are mostly poor and landless. This situation is partly the result of environmental effects as a result of forest destruction in the highlands. Equally, if not more important, the economic substandard in the lowlands originates from larger political and economic pressures, such as land speculation, that has occurred throughout the nation in the last few decades.

CONCLUSION

It can be concluded that environmental effects of agriculture in the highland have reduced the forested area and its quality. The remnant forest is probably not extensive enough to remain viable, and major conservation measures are needed to protect these remaining tracts of forest.

The reduction in water flow has been confirmed from various sources. However, it is not possible in the present study to single out deforestation as being responsible for this change.

Cash crops introduced to the highlanders need pesticides. The pesticide concentrations according to other studies are too low to be considered hazardous for drinking purposes and too low to be detected by the sensitive analytical methods used in the present study. A more rigorous study is needed to address this question.

The study reveals that highlander and lowlander perceptions of the conflict are at variance. Lowlanders perceive the causes of the conflict as mainly environmental. The Hmong have allegedly caused forest destruction, water reduction and contamination of the water. Highlanders suggest more diverse factors, such as economic factors and ethnic pressures, compounded by local incidents.

Differences in perceptions amongst the lowlanders and highlanders concerning the causes, seriousness and possible resolution of the conflict are serious obstacles to resolving the present situation. The apparent environmental conflict is linked to social, economic and political conflict. Simply addressing environmental issues may not solve the problem.

ACKNOWLEDGEMENTS

The project was financially supported by a scholarship from the Rockefeller Foundation and research grants from the Canadian International Development Agency (CIDA) and the Swedish International Development Authority (SIDA). We would like to thank the staff of the various agencies we interviewed, and especially the villagers of Pakluay and Mae Soi, whose kind cooperation made this study possible. Finally, Ms Charlotta Lindberg, AIT, is thanked for her analysis of the water samples.

LITERATURE CITED

Anonymous 1993. *Basic Information Regarding Women and Society in Northern Thailand*. Faculty of Education, Chiang Mai, Thailand.

Anonymous 1987. *Report of Soil Survey and Land use Planning for the Thai-Norwegian Highland Development Project, Ban Pakluay (Mahin Luang 1,2,3) and Doi Lium Plot, Huay Manao Area, Amphoe Jomthong, Changwat Chiang Mai*. Department of Land Development, Bangkok (in Thai).

Kaye, L. 1990. Of Cabbage and Cultures. *Far Eastern Economic Review*, 13 December 1990.

Maxwell, J.F. 1992. *Vegetation of the Mae Soi Conservation Area, Jomthong District, Chiang Mai Province*. Department of Biology, Faculty of Science, Chiang Mai University (mimeograph.)

Prangkhiaw, J. 1984. *Summary of Socio-Economic Survey, Pakluay, Mae Soi, Jomthong, Chiang Mai* (in Thai).

Shinawatra, B. 1985. *Highland-Lowland Inter-relationships in Northern Thailand: A Study of Production, Distribution and Consumption*. PhD dissertation, Michigan State University, Ann Arbor.

Tansiri, B. & Chondamrongkul, P. 1987. *Report of Soil Survey and Landuse Planning for the Thai-Norwegian Highland Development Project, Ban Pa Kluay (Mahin Luang 1,2,3) and Doi Lium Plot, Huay Manao Area, Amphoe Jomthong, Changwat Chiangmai*. Department of Land Development, Bangkok (in Thai).

Rattan inventory: determining plot shape and size

M. C. STOCKDALE *and* H. L. WRIGHT

Oxford Forestry Institute, Department of Plant Sciences, University of Oxford, Oxford OX1 3RB, United Kingdom

ASTRACT. Increased recognition of the importance of non-timber forest products has led to a need to develop inventory methods for these species. One of the most valuable non-timber forest products are rattans, the stems of climbing palms belonging to the sub-family Calamoideae and the basis of a furniture industry worth an estimated U.S.$ 6.5 billion per annum in international trade. This paper discusses the methodology for determining plot shape and size for the inventory of rattan. This methodology involves selecting a desired level of precision (e.g. a sampling error of 20%) and finding the plot shape and size which minimises the cost, in terms of time, of achieving it. Rectangular plots, oriented parallel to the direction of the slope, are more cost efficient than square plots; however, there are limits to rectangularity, as plots which are very long and narrow are prone to errors in boundary identification and area estimation. Thus, a minimum plot width of 5 m is recommended. The most cost efficient plot size ranges from 0.0025 ha to 0.025 ha, and is positively correlated with the desired sampling error, the total area under inventory, and the plot rectangularity. It also varies with the parameter(s) being assessed. The sample size is then determined by the plot size.

KEY WORDS: Palmae, Calamoideae, plot sampling, cost efficiency, Brunei Darussalam.

INTRODUCTION

Inventory methods for tree species have been well developed and documented (e.g. F.A.O. 1973, Husch *et al.* 1972, Loetsch *et al.* 1973, Philip 1983). Increased recognition of the importance of non-timber forest products has led to a need to develop inventory methods for these species (Williams 1991). Rattans are one of the most valuable non-timber forest products; their durable stems are the basis of a furniture industry worth an estimated U.S.$ 6.5 billion per annum in international trade and are responsible for providing approximately half a million jobs (Caldecott 1988, Manokaran 1990).

Rattans, belonging to the sub-family Calamoideae (family Palmae), comprise approximately 600 species from 13 genera, and range from the forests of southern China through to Australia and Fiji, with largest concentrations in Peninsular Malaysia, Borneo and New Guinea, as well as perhumid equatorial Africa (Uhl & Dransfield 1987). These climbing palms vary in growth form; some species occur as solitary stems whilst others produce suckers, resulting in clumps of stems.

Rattan inventory, as in other resource inventories, requires decisions to be made on the sampling pattern or design, the nature of the sampling units used, the population and parameters to be included, and the most appropriate methods of measurement (Philip 1983). This study focuses upon the nature of the sampling unit, which may take the form of a point or a plot. Decisions about sampling units very much depend upon the nature of the population under study.

Point sampling involves the measurement of distances from a point to individual plants, from which the number of plants per unit area may be calculated. This method has the advantage of being quicker than sampling with plots, but if the assumption of random distribution is violated, the density of a population will be underestimated (Grieg-Smith 1983). Point sampling was not tested in this study as rattans often grow in dense thickets;

D. S. Edwards et al. (eds.), Tropical Rainforest Research – Current Issues, 523–533.

for example, Tandug (1978) found the frequency distribution of rattan in a plot in the Philippines to be non-random, and best fitted by the negative binomial function. A study (Stockdale, in prep.) of the frequency distribution of rattan in Brunei and Indonesia will further investigate this issue.

Plot sampling may use plots of a variety of shapes and sizes. Inventories - where the aim is to enumerate as precisely as possible the nature and extent of the resource - may require different plot shapes and sizes from those required for ecological studies or growth modelling (Vanclay, in press). In an inventory, it is desirable that plots maximize within-plot variation in order to minimise between-plot variation. With plots for ecological studies or growth modelling, however, within-plot homogeneity may be preferable. In addition to precision, cost is an important consideration in the selection of plot size and shape as, at the level of large-scale inventories, an inefficient sampling method can waste much time and money.

Larger plots incorporate greater within-plot variation than smaller plots, but they tend to involve greater cost for the same precision (Philip 1983). As far as plot shape is concerned, a rectangular plot is better than a square or circular plot at representing the variance of a population, if the population has a linear trend and the long axis of the plot is oriented parallel to this trend (Freese 1961). However, the disadvantage of rectangular plots, compared to circles and squares, is that their lower perimeter to area ratios make them liable to serious bias from errors in identification of borderline plants and area estimation (Philip 1983). Circular plots are not recommended for inventories in mixed tropical forests where the undergrowth is dense, making it difficult to walk in an arc around a central point (Alder & Synnott 1992). Thus, only rectangles and squares, of a range of sizes, were compared in this study.

Plot sizes in rattan inventories have ranged from 5 m by 5 m (Tandug 1978) to 250 m by 250 m (Sharma & Bhatt 1982) or 10 m by 400 m (Siswanto & Soemarna 1988). Plot shapes have been circular (Tomboc 1988), square (Sharma & Bhatt 1982, Siswanto & Soemarna 1988, Tandug & Lasmarias 1984) and rectangular (Siswanto & Soemarna 1988, Tandug & Lasmarias 1984, Zieck 1967). Only Tandug's (1978) study, however, has tested the relative efficiency of different plot shapes and sizes for rattan inventory; her results will be discussed subsequently.

The optimal plot shape and size for an inventory can be determined by two related approaches (Arvanitis & O'Regan 1967, Taafe 1979, Tandug 1978, Zeide 1980). One approach is to set a desired level of cost (in this study represented by the time taken to conduct the inventory) and choose the plot shape and size which maximises precision. The second is to set a desired level of precision (in this study represented by the sampling error) and choose the plot type which minimises cost. This study used the latter approach.

METHODS

The work was conducted at a 1.5 ha study plot near the Kuala Belalong Field Studies Centre in the Temburong District of Brunei, as part of the University of Brunei Darussalam/Royal Geographical Society's Brunei Rainforest Project. The long axis of the 300 m by 50 m plot extended from just above the Belalong river floodplain (60 m a.s.l) to two-thirds of the distance up the east ridge (210 m a.s.l.). The plot consisted of five parallel lines spaced 10 m apart, which were marked with tagged poles at 10 m intervals. Thus the study plot was subdivided into one hundred and fifty subplots of dimensions 10 m by 10 m, as shown in Figure 1. The terrain in the plot encompasses small ridges, valleys and plateaus, with slopes ranging from 0° to 60°. The vegetation of this area is lowland mixed dipterocarp forest (U.B.D./ R.G.S. 1991).

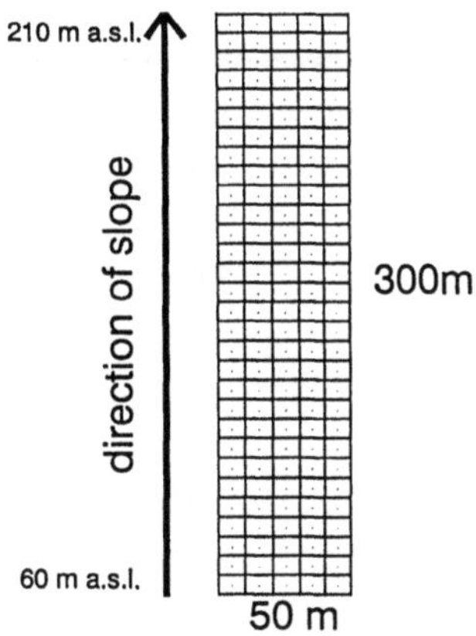

Figure 1. Grid map of 1.5 ha study plot.

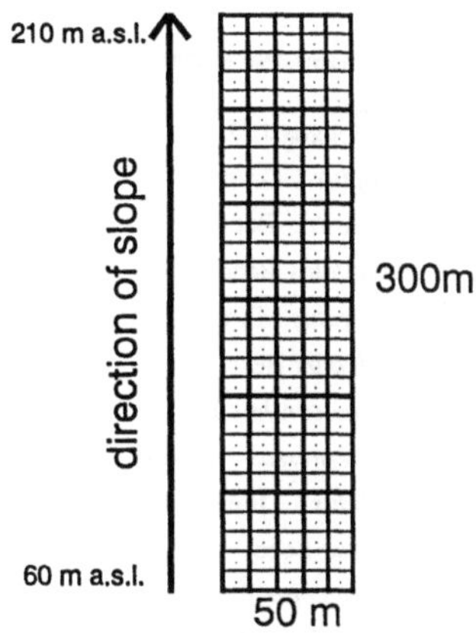

Figure 2. Grid map of 1.5 ha study plot, overlaid with 30 plots of rectangularity = 5 and size = 0.05 ha.

A 100% enumeration of all rattan species in the study plot was conducted by a team of three people. One person recorded the data and walked up the marked line. The other two people, on either side of the line, used 2.5 m poles to determine whether the central points of rattan clumps were within 5 m of the line. Only rattan clumps mature enough to have developed at least one stem were enumerated. Data recorded for each 10 m x 10 m subplot were, for each species, the number of clumps and the number of stems, as well as time taken to enumerate each subplot. On separate occasions, time taken to demarcate the tagged poles and to walk between all poles in both directions was also recorded. Data on species are vouchered by herbarium material housed in the Forestry Department of Brunei Darussalam and the Royal Botanic Gardens, Kew, United Kingdom.

RESULTS

Enumeration

The results of a 100% enumeration of the 1.5 ha plot are listed in Table 1. There were twenty-one species of rattan from four genera. All species together comprised 343 clumps (228 clumps/ha) and 794 stems (529 stems/ha).

Determining plot shape and size: the coefficient of variation

The study plot grid was overlaid with all possible combinations of plot shapes and sizes which could completely cover the plot. An example is shown in Figure 2, using a plot shape or rectangularity (measured as the ratio of the long axis to the short axis) of 5 and a plot size of 0.05 ha. The plots were always oriented so that their long axes were parallel to the direction of the slope, thus encompassing the maximum amount of topographical variation. Plot shapes ranged from rectangularities of 1 to 30, and plot sizes ranged from 0.01 ha to 0.75 ha.

For each plot shape/size combination, the mean number of clumps/ha and stems/ha were calculated, together with their co-efficients of variation. The coefficient of variation is a measure of variation and is calculated as the ratio of the standard deviation to the mean, normally expressed as a percentage. As the total area covered was the same for each plot

Table 1. Results of 100% enumeration of rattans (family Palmae: sub-family Calamoideae) in a 1.5 ha plot in Temburong District, Brunei Darussalam.

Species	No. Clumps	No. Stems
Calamus amplijugus J.Dransf.	6	16
Calamus comptus J.Dransf.	2	2
Calamus conirostris Becc.	18	30
Calamus flabellatus Becc.	92	210
Calamus laevigatus Mart. var. *laevigatus*	2	2
Calamus muricatus Becc.	9	9
Calamus pogonacanthus Becc.	23	124
Calamus praetermissus J.Dransf.	6	24
Calamus sordidus J.Dransf.	18	88
Calamus zonatus Becc.	2	7
Ceratobolus concolor Bl.	14	43
Daemonorops didymophylla Becc.	1	1
Daemonorops korthalsii Bl.	7	7
Daemonorops longipes (Griff.) Mart.	4	4
Daemonorops oxycarpa Becc.	13	13
Daemonorops periacantha Miq.	105	105
Korthalsia debilis Bl.	4	20
Korthalsia echinometra Becc.	3	13
Korthalsia hispida Becc.	8	48
Korthalsia furtadoana J. Dransf.	4	14
Korthalsia rostrata Bl.	2	14
Total: 4 genera, 21 species	343 (228 per ha)	

shape/size combination, the means were also the same; however, the coefficients of variation varied with plot type.

Multiple regression analysis found the percentage coefficients of variation, or CV, of both clumps/ha and stems/ha estimates to be more significantly related to both plot shape and size than to plot size alone:

Clumps/ha:

$$LOG_{10}\ CV = 3.18 - 0.13\ LOG_{10}\ SHAPE - 0.28\ LOG_{10}\ SIZE \quad (1a)$$
$R^2_{adj}=0.95$, $F=144.9$, $P < 0.001$

$$LOG_{10}\ CV = 2.98 - 0.30\ LOG_{10}\ SIZE \quad (1b)$$
$R^2_{adj}=0.85$, $F=77$, $P < 0.001$

Stems/ha:

$$LOG_{10}\ CV = 3.74 - 0.24\ LOG_{10}\ SHAPE - 0.19\ LOG_{10}\ SIZE \quad (1c)$$
$R^2_{adj}=0.64$, $F=13.6$, $P < 0.01$

$$LOG_{10}\ CV = 3.37 - 0.23\ LOG_{10}\ SIZE \quad (1d)$$
$R^2_{adj}=0.34$, $F=8.2$, $P < 0.02$

Other workers who have conducted tree inventory studies and examined the relationship between the coefficient of variation and plot size (but not plot shape) have also found equations of the generalised form: $LOG_{10}CV = h - g\ LOG_{10}SIZE$ to be significant. Their

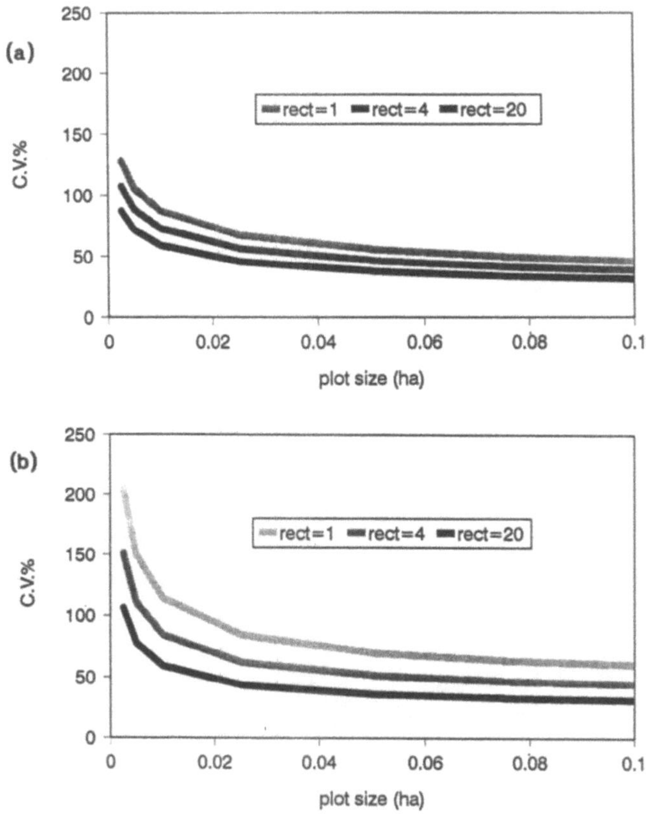

Figure 3. The coefficient of variation (C.V.%) vs. plot size and shape, for: (a) clumps/ha and (b) stems/ha.

values for g ranged from 0.25 (Freese 1961, Prodan 1968) and 0.27 (Taafe 1979, Tardif 1965) to 0.44 (Zeide 1980), depending upon the degree of heterogeneity of the forest. A value for g of 0.5 indicates a random distribution. Thus the value for rattan clumps, 0.30, is comparable to that of some tree species, and the value for rattan stems, 0.23, is indicative of its more clumped distribution.

Figure 3 depicts this relation between the coefficient of variation and plot shape and size, for (a) clumps/ha and (b) stems/ha. The variation between plots is greater for estimates of stems/ha than for clumps/ha, although this difference is less marked at higher plot sizes and rectangularities. Plots of larger size or of higher rectangularity have lower variation between plots. The coefficient of variation can thus be predicted for a given plot size and shape, and following this, the sample size for a given precision calculated.

Determining plot shape and size: the precision

One way of expressing the precision of an estimate is as the sampling error. This is calculated as the ratio of the confidence interval to the mean; thus if there is a sampling error of 20%, the confidence interval is 20% of the mean, at the given level of probability (Philip 1983).

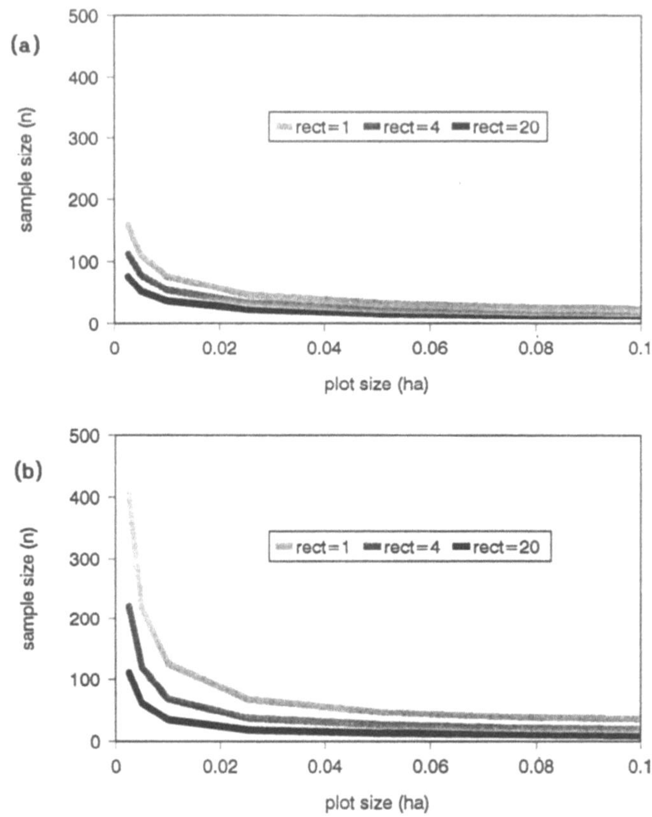

Figure 4. The sample size (n) required for a sampling error (E%) of 20% (p=0.95) vs. plot size and shape, for (a) clumps/ha and (b) stems/ha.

The number of plots (n) of a certain shape and size required for a chosen percentage sampling error (E) can be calculated from the equation:

$$n = (t^2 CV^2) / E^2 \qquad (2)$$

where t is the student's 't' statistic and the coefficient of variation (CV) is calculated from equations (1a) and (1c).

Figure 4 displays, for the parameters of (a) clumps/ha and (b) stems/ha, the number of plots needed to obtain a sampling error of 20%, at a 95% probability level, for a range of plot shapes and sizes. More plots are needed for estimating stems/ha than for estimating clumps/ha. Looking at shape, the higher the rectangularity of the plots, the fewer are necessary for the same sampling error. Looking at size, the larger the plots, the fewer are necessary to achieve the same precision. All of these plot number/shape/size combinations depicted in the figure are equal in terms of precision; the optimal plot number/shape/size combination is that which achieves this precision for the least cost.

The cost

The cost was approximated as the total time taken to conduct the inventory:

$$T_{total} = T_{travel} + T_{enumerate} \qquad (3)$$

which is divided into time to travel between plots (T_{travel}) and time to delineate and enumerate the plots ($T_{enumerate}$) (Gambill *et al.* 1985, Taafe 1979).

The time to travel between plots:

$$T_{travel} = T_1 (nA)^{1/2} \qquad (4)$$

is the product of the mean walking time per unit distance (T_1), and the total minimum distance between n randomly placed plots in a total area A, which is approximated by imagining a square grid of area A, covered by a systematic sample of n points. The distance between any two points will be $(A/n)^{1/2}$; this value is then multiplied by the total number of points (n) (Arvanitis & O'Regan 1967, Zeide 1980).

The time to delineate and enumerate the plots:

$$T_{enumerate} = T_2 (nP) \qquad (5)$$

is the product of the mean time to delineate and enumerate a plot per unit area (T_2), and the total area delineated and enumerated, calculated by multiplying n plots by their plot size (P) (Bonilla 1969, Tardif 1965).

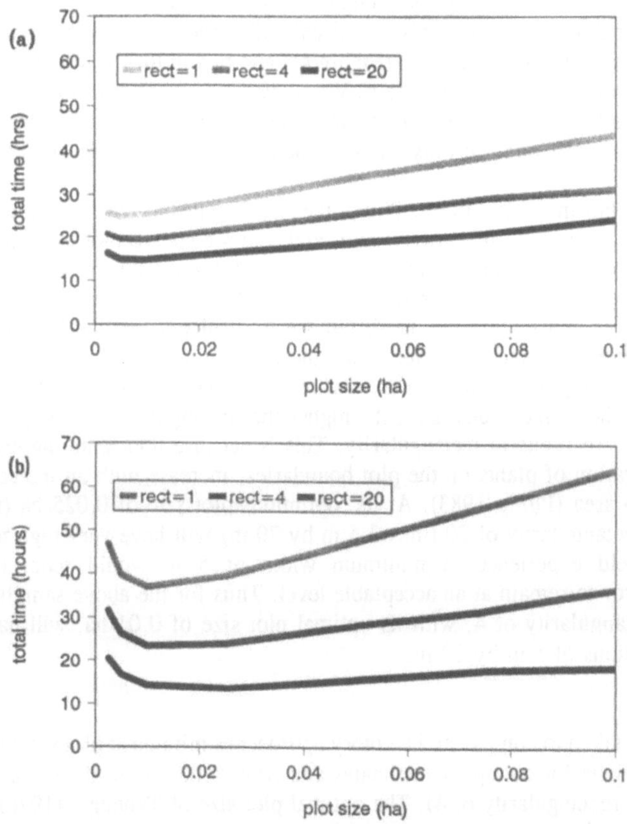

Figure 5. The total time taken to conduct an inventory of an area (A) of 200 ha at a sampling error (E%) of 20% (p=0.95) vs. plot size and shape, for: (a) clumps/ha and (b) stems/ha.

Figure 5 displays, for the parameters of (a) clumps/ha and (b) stems/ha, the total time taken to inventory an area of 200 ha with a sampling error of 20%, for a range of plot shapes and sizes. For the same precision, enumerating stems is more costly than enumerating clumps. Looking at shape, the higher the rectangularity of the plot, the more efficient it is in terms of time. Looking at size, the most efficient plot sizes are those at which total time taken approaches a minimum value. These optimal plot sizes vary with shape; the higher the rectangularity of the plot the larger the optimal plot size.

The above figure indicates optimal plot size for a specific sampling error (20%) and total area (200 ha). Tables 2 and 3 list the optimal plot sizes (and their corresponding sample sizes) for a range of sampling errors and total areas, for clumps/ha and stems/ha respectively, with rectangularity held constant at 4. The optimal plot size, which determines the sample size, is shown to be positively correlated to both the sampling error and total area. The optimal plot sizes for estimates of clumps/ha, ranging from 0.0025 ha to 0.01 ha, are less than those for stems/ha, which range from 0.005 ha to 0.025 ha.

DISCUSSION

Plot shape

In this study, plots with the highest rectangularity were more cost efficient than plots of lower rectangularities (Figure 5). This is because fewer rectangular plots than square plots are needed for the same sampling error (Figure 4), and this in turn is because the higher the rectangularity of the plot, the lower the C.V., or variation between plots (Figure 3). It is likely that there is less variation between rectangular plots because their orientation (parallel to the direction of the slope) permits them to encompass more of the variation within the plots.

A study of rattan inventory in the Philippines by Tandug (1978) compared rectangular plots with square plots, both of various sizes, and found a square plot to be most efficient. No mention was made of the orientation of the rectangular plots, thus it is possible they were randomly oriented and did not incorporate as much of the overall variation as they might have done. Another explanation for the difference in results is that the Philippine site may be flatter than the Brunei site; in flatter areas, the rectangularity and orientation of the plot may not have as strong an effect as in areas where the topography is sharply dissected.

In theory, for the Brunei study area, the higher the rectangularity of the plot, the better; in practice, there are limits to rectangularity. This is because non-sampling errors, such as incorrect enumeration of plants on the plot boundaries, increase with an increasing ratio of plot perimeter to area (Philip 1983). At its 'optimum' plot size of 0.025 ha (for E=20%, A=200 ha), a rectangularity of 20 (thus 3.5 m by 70 m) will have very high non-sampling errors. From field experience, a minimum width of 5 m would seem to allow the non-sampling error to remain at an acceptable level. Thus for the above sampling error and total area, a rectangularity of 4, with its optimal plot size of 0.01 ha, will have the more workable dimensions of 5 m by 20 m.

Plot size

The total time taken to conduct an inventory approaches minima at plot sizes ranging from 0.0025 ha to 0.01 ha for clumps/ha estimates and from 0.005 ha to 0.025 ha for stems/ha estimates (with a rectangularity of 4). The optimal plot size of Tandug's (1978) study, 0.01 ha (with a rectangularity of 1), is within this range. In Tandug's study, clumps/ha were estimated; unlike the present study, this included the enumeration of seedlings.

The reason that the cost (in terms of time) approaches a minimum is that for plots smaller than the optimal size, travel time becomes the dominant component of the total time taken,

Table 2. Optimal plot size (ha) and its sample size (n) for estimating clumps/ha (rectangularity = 4).

Total Area (ha)	Sampling Error (E%)				
	10	15	20	25	30
50	0.0025 (n = 442)	0.0025 (n = 197)	0.0025 (n = 113)	0.005 (n = 50)	0.005 (n = 35)
100	0.0025 (n = 442)	0.0025 (n = 197)	0.005 (n = 78)	0.005 (n = 50)	0.005 (n = 35)
200	0.0025 (n = 442)	0.005 (n = 136)	0.01 (n = 53)	0.01 (n = 35)	0.01 (n = 25)
400	0.005 (n = 300)	0.01 (n = 92)	0.01 (n = 53)	0.01 (n = 35)	0.01 (n = 25)

Table 3. Optimal plot size (ha) and its sample size (n) for estimating stems/ha (rectangularity = 4).

Total Area (ha)	Sampling Error (E%)				
	10	15	20	25	30
50	0.005 (n = 472)	0.01 (n = 121)	0.01 (n = 68)	0.01 (n = 44)	0.01 (n = 30)
100	0.005 (n = 472)	0.01 (n = 121)	0.01 (n = 68)	0.01 (n = 44)	0.025 (n = 16)
200	0.01 (n = 272)	0.01 (n = 121)	0.01 (n = 68)	0.025 (n = 24)	0.025 (n = 16)
400	0.01 (n = 272)	0.01 (n =121)	0.025 (n = 37)	0.025 (n = 24)	0.025 (n = 16)

since much larger numbers of these plots are necessary to achieve the same sampling error (Figure 4). For plots larger than optimal size, the delineation and enumeration time becomes increasingly significant, as an increasing fraction of the total area must be delineated and enumerated to achieve the same sampling error (Figure 4).

The optimal plot size, which determines the sample size, is positively correlated with the chosen sampling error, the total area to be inventoried, and the rectangularity (Figure 5, Table 2). Thus, a decision about desired sampling error, plus external limits such as the total area to be inventoried and a minimum plot width of 5 m (which will determine the rectangularity), will enable an inventory planner to determine an optimal plot size.

The parameter also has an effect upon optimal plot size; the most efficient plot size for estimates of the parameter clumps/ha is smaller than that for estimates of stems/ha, all else being held constant (Figure 5, Tables 2 and 3). According to F.A.O. (1973), when several parameters are measured in the same inventory, the most important parameter should be selected for optimisation. If several parameters are equal in importance, the parameter with the greatest variance should be used. On both counts, it is better to use the plot sizes optimal for estimating stems/ha; it is the more important parameter in terms of assessing economic potential of an area, and it also has the higher variance (Figure 3).

CONCLUSIONS

In the sharply dissected Brunei site, rectangular plots, oriented parallel to the direction of the slope, are more cost efficient than square plots for the inventory of rattan. As errors in boundary identification and area estimation increase with the degree of rectangularity, however, a minimum width of 5 m is recommended. The most cost efficient plot size ranges between 0.0025 ha and 0.025 ha, and can be selected from Tables 2 or 3 after determining the desired sampling error, the total area under inventory, and the parameter(s) to be estimated.

ACKNOWLEDGEMENTS

This research was funded by grants from the Canadian International Development Agency and the Dennis Curry Charitable Foundation. We would like to thank Judith Stockdale, Impin Lassah and Simpurai Anggat, for their hard work. This research was undertaken as part of the Brunei Rainforest Project, organised by the University of Brunei Darussalam and the Royal Geographical Society. Many thanks to the corporate patrons, Royal Brunei Airlines, The Baring Foundation, DICAM, GreenCard Trust, Hongkong Bank, Morgan Grenfell, and Nomura-NIMCO for their commitment to the Kuala Belalong Field Studies Centre. Thanks also to Dr. Peter Kanowski and Dr. John Dransfield for commenting on drafts of this paper.

LITERATURE CITED

ARVANTIS, L. G. & O'REGAN, W. G. 1967. Computer simulation and economic efficiency in forest sampling. *Hilgardia* 38:133-164.

BONILLA, J.A. 1970. Plot size. The size most useful for mensurational surveys. *Silvicultura* 19:15-38.

CALDECOTT, J. 1988. Climbing towards extinction. *New Scientist* 118 (1616):62-66.

FREESE, F. 1961. Relation of plot size to variability. An approximation. *Journal of Forestry* 59:679.

GAMBILL, C.W., WIANT, H.V., Jr., & YANDLE, D.O. 1985. Optimum plot size and BAF. *Forest Science* 31:587-594.

GRIEG-SMITH, P. 1983. *Quantitative plant ecology* (3rd edition). Studies in Ecology Vol. 9. Blackwell Scientific Publications, Oxford.

F.A.O. 1973. *Manual of forestry inventory with special reference to mixed tropical forests*. F.A.O., Rome.

HUSCH, B., MILER, C.I. & BEERS, T.W. 1972. *Forest Mensuration* (2nd edition). The Ronald Press Company, New York.

LOETSCH, F., ZÖHRER, F. & HALLER, K.E. 1973. *Forest Inventory, vols I and II*. BLV Verlagsgesellschaft, München.

MANOKARAN, N. 1990. The state of the rattan and bamboo trade. *RIC Occasional Paper* No. 7.

PHILIP, M. S. 1983. *Measuring trees and forests*. Division of Forestry, University of Dar Es Salam.

PRODAN, M. 1968. *Forest Biometrics*. Pergamon Press, Oxford.

SHARMA, S. K. & BHATT, P.M. 1982. An assessment of cane potential of Baratang Island in South Andaman Forest Division. *Indian Forester* 108: 270-282.

SISWANTO, B. E. & SOEMARNA, K. 1988. Metode inventarisasi rotan di K. P. H. Pontianak, Kalimantan Barat (Rattan inventory method in Pontianak Forest District, West Kalimantan). *Bulletin Penyelidikan Hutan (Forest Research Bulletin)* 503:1-11.

TAAFFE, K.E., Jr. 1979. Computing optimum plot size for wildland inventories. *Resource Inventory Notes* BLM-23:8-15.

TANDUG, L. M. 1978. Sampling method for inventory of Philippine rattan and its distribution. *Sylvatrop Philippine Forestry Research Journal* 3:155-170.

TANDUG, L. M. & LASMARIAS, V.T. 1984. *Determination of the most appropriate sampling design for the inventory of Philippine rattan*. P.C.A.R.R.D.-I.D.R.C. National Integrated Research Program on Rattan. Terminal report (unpublished). 3 pp.

TARDIF, G. 1965. Some considerations concerning the establishment of optimum plot size in forest survey. *Forestry Chronicle* 41:93-102.

U.B.D./R.G.S. 1991. Information and joining instructions for European members (2nd edition). University of Brunei Darussalam/Royal Geographical Society Brunei Rainforest Project 1991-1992.

UHL, N. W. & DRANSFIELD, J., 1987. *Genera Palmarum: a classification of palms based on the works of Harold E. Moore, Jr.* The L. H. Bailey Hortorium and the International Palm Society, Lawrence, Kansas.

VANCLAY, J., in press. *Growth models for tropical moist forest*. C.A.B. International.

WILLIAMS, T. 1991. *Research needs for bamboo and rattan to the year 2000.* Unpublished manuscript of the Tropical Tree Crops Program of the International Fund for Agricultural Research, Washington D.C., U.S.A.

ZIECK, J. F. U. 1967. Minor forest products investigation - to investigate the potential for the economic exploitation of rattans in the territory of Papua and New Guinea. *Research Notes U11-2.* Forest Products Research Centre, Port Moresby.

ZEIDE, B.B. 1980. Plot size optimisation. *Forest Science* 26:251-257.

Conservation of biodiversity in Kenya's indigenous forests: biological currencies used in management planning

GLYN DAVIES

Kenya Indigenous Forest Conservation Project, P.O. Box 71493, Nairobi, Kenya

KEY WORDS: endemism, endangered species, habitat-dependence, indicator species, national parks.

INTRODUCTION

With the recent up-surge of interest in conserving biological diversity, especially through effective conservation of the world's tropical forests, there has been a considerable increase in the funding focussed on forest management and a change in the way development agencies now carry out economic appraisals of natural biological systems. A wealth of new literature has been produced over the past 5 years outlining the different levels at which economic and financial analyses must be applied (e.g. Brown & Moran 1993, Pearce 1991), and it is an appropriate time for biologists to make a greater contribution and so allow economic models to be refined further.

In this context, one issue of applied tropical forest research which needs to be addressed more vigorously is the need to present a clearer set of biological guidelines to use in evaluating different areas of forest. Whenever biologists discuss forests, mention is invariably made of a particular species of interest, often because it is rare and/or endemic. In other words, it is not just the number of species which makes an area important for biodiversity conservation, but there are biological qualities which are used when assessing biological value. These are distinct from the economic values of 'direct use', 'indirect use' and 'option' (*sensu* Pearce 1991). For example, Collar & Stuart (1985) compiled a book on the threatened birds of Africa and related islands in which the importance of an area for bird conservation was based on the number of rare species found within. Indeed, Collar (1986), when reviewing the conclusions of the African Bird Red Databook, emphasised how few rare species were currently, or potentially, of any economic value, but that the risk of extinction was the main concern.

The primary aim of this paper is to describe biological qualities of species which are used when developing conservation strategies. The magnitude of these qualities varies between species, and so they can be used as 'currencies' when calculating the overall value of an assemblage of species within a particular forest. Field data on tree/shrub, butterfly, bird and mammal species from four of Kenya's indigenous forests, have been compiled, and using these biological currencies the relative importance of each forest for biodiversity conservation has been assessed.

KENYA'S INDIGENOUS FORESTS

Kenya covers an area of 582,644 km^2 which encompasses about 10,700 km^2 of lake area, including two large lakes: Turkana (6,405 km^2) and part of Victoria (3,785 km^2). Rainfall is relatively low, with very few areas receiving 2,000 mm per annum - the minimum

D. S. Edwards et al. (eds.), Tropical Rainforest Research – Current Issues, 535–546.

Figure 1. Agroclimatic zones with potential for forest cover in Kenya.

requirement for rain forest (Richards 1952). However climatic conditions are sufficiently wet, with rainfall spread over two seasons, to support forest vegetation over about 15% of the country (Figure 1). In these areas the ratio of average annual rainfall to average annual potential evaporation exceeds 50% (Kenya Soil Survey, 1980).

Within these wetter parts of Kenya, forest cover a decade ago amounted to 1.255 million ha (Doute *et al.* 1981). Almost all of this lies within the 2.2 million ha of gazetted Forest Reserves (1.7 million ha) plus proposed Forest Reserves (0.5 million ha). As is also apparent, there are many areas of gazetted Forest Reserve which contain no forest vegetation.

The areas in which indigenous forest occurs also have the highest potential for agriculture, and support about 80% of Kenya's 24-30 million people, a population which is expanding at about 3.7% p.a. Not surprisingly, closed-canopy indigenous forests only occupy about 2% of the country because many areas have been cleared for agricultural development, some have been burnt by pastoralists in the dry regions, and since the 1950's about 10% of the forest estate has been converted to industrial plantations mostly comprising exotic trees. In addition to total loss of forest cover, some areas of forest have been degraded through over-exploitation of forest products: timber extraction; charcoal and fuelwood collection; carving wood and polewood cutting; hunting and trapping of ungulates (Davies 1992).

As a consequence of these pressures the condition of Kenya's forests has been deteriorating and the British Overseas Development Administration has given support to the

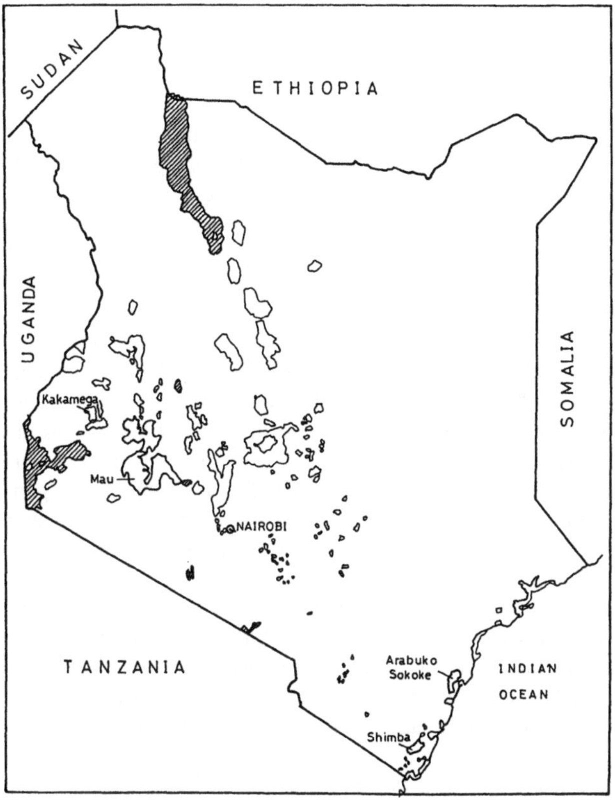

Figure 2. Gazetted forests of Kenya, showing those surveyed during the KIFCON project.

Kenya Government to establish a Kenya Indigenous Forest Conservation Project (KIFCON Project). The main aim of this project is to assist the Ministry of Environment and Natural Resources to develop a strategy for improved conservation management of Kenya's indigenous forests. This has initially required multidisciplinary surveys to be conducted in different forest areas, and this paper reports the results of Biodiversity Surveys carried out in four forests (Figure 2, Table 2) by survey teams of the National Museums of Kenya, working under contract to the Forestry Division/KIFCON Project.

Throughout this paper, therefore, a national perspective will be taken, since it is forest management within Kenya which is the context of the development programme technical surveys. One of the required outputs of the Biodiversity Surveys is to identify priority forests, and priority areas within forests, which require protection to ensure against extinction of plant or animal species. Furthermore, this biological information must be presented in a form that allows evaluation of species protection in relation to forest-use by rural communities, commercial enterprises etc. To develop such a multiple-use strategy for forest management, the biological information must be understood by planners who may have no biological training, and it must be applicable across a broad range of forests.

BIOLOGICAL CURRENCIES

The recent report on Global Biodiversity, compiled by the World Conservation Monitoring

Centre expands on many of the issues noted below, and discusses them more fully (Groombridge 1992).

Biodiversity

Throughout the environmental literature the conservation of global biodiversity is cited as a fundamental aim, yet there is seldom a clear understanding of what 'biodiversity' means. This, in part, is because biological diversity (= biodiversity) refers to the total variety of living organisms on earth (Stuart *et al.* 1990) and can be considered at three levels: a) the gene; b) the species, of which the individuals are products of genetic and environmental inter-play, and; c) the ecosystem, which is composed of species interacting with one and other and with the non-biological components of their environment.

There is therefore ample scope for confusion, both in understanding and discussion, depending on the level at which biodiversity is considered. A confusion exacerbated by the direct linkages between the three levels - changes at any one level will affect the other two.

In this paper, biodiversity will be considered at the level of the species, accepting current taxonomic nomenclature. This is a natural way to consider whole-animal variation, although it does have the short-coming of failing to take full account of the genetic differences between species. Some species may be more closely related and hence share more genetic characters than others, so biodiversity is greatest when distantly related species are combined (Groombridge 1992).

Furthermore, only four taxonomic groups will be considered in this paper: vascular plants (especially woody shrubs and trees), butterflies, birds and mammals. These groups are relatively well-studied in Kenya and there are reasonably accurate species lists for several forest areas. Information on trees and shrubs is an immediate requirement for assessing the vegetation of a forest, since these larger plants are cut/collected most frequently. Data on different birds species was sufficiently detailed from short surveys to be analyzed for an assessment of forest quality in terms of supporting rare or forest-dependent species etc. The effects of hunting, however, do not show up in avifaunal surveys so larger mammals, especially the ungulates, were also surveyed.

Information from other groups (e.g. amphibians/reptiles, other invertebrates, non-vascular plants and microbes) was very sketchy, and could not be improved upon during the brief KIFCON surveys, largely because of the great seasonal changes in insect populations and detectability of the herpetofauna. Information on butterflies was supplied by Dr Steve Collins, who has collected extensively in Kenyan forests, and gives distributional data which were not seasonally biased.

Throughout Kenya, the number of tree and shrub species, calculated from a total of 1,532 named species in the East African Herbarium in 1980 and adding about 10% for new specimens collected in the last decade, is estimated at 1,700 species. For all vascular plants, a figure of about 6,000 species is the current estimate (C.Kabuye, personal communication). For butterflies the current estimate is 875 species (S.Collins, personal communication) and for birds 1,079 (L.Bennun, personal communication). A mammal checklist is currently being drawn up, with best estimates at the present time of 305 species, but only 119 species if rodents, shrews and bats are excluded.

Rarity

Loss of biological diversity, through extinction of species, is often a gradual process in which a species becomes increasingly rare until, eventually, there are too few individuals to sustain the population. Species threatened with extinction have always attracted conservation attention, and IUCN developed a set of categories for threatened species in decreasing order of rarity: endangered, vulnerable and rare. Recently there has been much debate about these categories, especially since there is an inevitable lack of information on most threatened species.

A recent proposal has been made to improve the threatened species categories (Mace & Lande 1991). For given populations or sub-populations (of reproductive individuals) the following definitions have been offered:

Critical: 50% probability of extinction within 5 years or 2 generations, whichever is longer

Endangered: 20% probability of extinction within 20 years or 10 generations, whichever is longer

Vulnerable: 10% probability of extinction within 100 years.

The threatened species referred to in this paper probably all fit the above three definitions, although there is very little information on rates of reproduction to confirm which category applies, a problem encountered all over Africa (Andre *et al.* 1992). All that is known, or in some cases inferred, is that the population is small and declining in Kenya.

Habitat-dependence

Not withstanding the many causes of species extinctions, whether through chance events (e.g. storms) or direct causes (e.g. hunting), a central issue for forest conservation in any country is the loss or degradation of forest habitats. Species which are dependent on forest habitat are those which are being lost globally, and there is a need to focus attention on these even though many can live at high population densities where there is suitable habitat. If the forest in an area is lost or degraded, the total number of species may increase as a result of invasion by savanna-preferring species. This would not constitute an improvement in biodiversity conservation, however, since it merely indicates replacement of forest species by savanna species many of which are already abundant outside forest environments.

A forest-dependent species is one which attains its highest population densities in forests which have been little-modified by human activities. Ornithologists in Kenya have divided forest species into two sub-categories: **F** species which are forest-specialists that only occur where there is dense multi-storeyed forest, and **f** species which are forest-generalists that occur on forest edges and in forest gaps (Lewis & Pomeroy 1989, Pomeroy 1989). In one Kenyan afromontane forest it was the populations of **F** species which declined in areas greatly modified by past timber extraction (Bennun & Waiyaki 1992). A similar trend applies to **F** birds in the largest coastal forest (J.Fanshawe, personal communication).

Forest-dependent mammals include antelopes (e.g. red duiker) which are rare outside forest habitat and primates that do not occur in unforested areas (e.g. colobus monkeys). Forest-dependent butterflies are largely those whose larvae feed on forest plants.

For the category of forest-dependence to have any meaning, it is necessary to define 'forest'. The most obvious definition, which is followed in this paper, is that of White (1983) in his review of African vegetation:

"a continuous stand of trees; canopy varies in height from 10m to 50m and usually consists of several layers; crowns of individual trees interdigitate or overlap, and are often inter-laced with lianes. A shrub layer is normally present being densest where the tree canopy is most open. The ground layer is sparse or absent, generally lacking grass cover."

In the wettest parts of Kenya, where slopes and soil are suitable, forest vegetation is easily identified using this definition. There are two main problems, however: first, there is a gradual transition from forest formations to more open canopy woodland formations as conditions become drier, making it difficult to define a cut-off point between forest and woodland vegetation in a given area; second, interspersed in many forest areas are grassy glades which obviously do not constitute forest vegetation but form an integral part of forest area and cannot be ignored.

Figure 3. Floristic zones in Kenya (after White 1983). Key: 2 - Guineo-Congolian rainforest (drier type); 11a - lowland forest/secondary grassland mosaic; 16a - coastal mosaic/Zanzibar-Inhambane; 16b - coastal mosaic/forest patches; 19a - Afromontane forest and bamboo.

Endemism

The degree to which a species is restricted in its geographical range can also contribute to its rarity. An endemic species is one which is restricted to a particular area, varying in size from a mountain, to a country, to a continent. In a recent assessment of global bird conservation, ICBP (Bibby *et al.* 1992) developed the concept of restricted-range (r-r) species for land birds whose breeding range has never occupied more than 50,000 km^2. On the basis of two 2 r-r species occurring in an area it is designated an Endemic Bird Area (EBA) and by focussing on EBA's there is no chance of double counting species which occur in different countries, for example, but still within the same EBA.

By focussing on different biogeographical regions in Kenya in this paper (see below), this problem of double counting is largely avoided.

BIOGEOGRAPHY OF KENYAN FORESTS

Following the phytogeographical zones to which White (1983) has assigned African vegetation formations, Kenyan forests occur in three floristic zones: coastal, afromontane and western (Figure 3).

A narrow coastal belt, running inland approximately 30 km from the shore-line, is the northern part of the Zanzibar-Inhambane region which stretches from northern Mozambique to southern Somalia (White 1983). This region contains many isolated forest blocks which have been separated from one and other for protracted periods and therefore the species within each show high levels of endemism, and because they are frequently small, high levels of rarity also. Furthermore, the coastal area can be divided into two sub-regions north and south of the Mombasa-Kilifi area.

The bulk of the country's forests are in the afromontane region, a region which stretches from the highlands of Guinea and Liberia in the west to Ethiopia in the east and the Drakensburg Mountains in the south. There are close relationships between plant species in all these forests, although the region has been further sub-divided into seven afromontane systems (White 1978); within Kenya all the forests are in the Imatong-Usambara system. There is a further distinction between forests on the west side of the Rift Valley, which have low levels of plant endemism, and east-of-Rift forests which have higher levels both amongst plants (Beentje 1990) and birds (Lewis & Pomeroy 1989).

Finally, there is a small area in the Lake Victoria basin which is the eastern-most extension of the Guineo-Congolian forests that are distributed from the Congo basin to coastal west Africa. Although the area occupied by this type of forest in Kenya was probably once much more extensive, it is now represented by a single small forest, Kakamega.

By selecting forests from these different biogeographical regions, the species richness patterns which are described below, especially for plants and birds, refer to many different species. When more forests are considered, many from the same biogeographical region, care will have to be taken to ensure against 'double counting' species which occur in more than one forest.

BIODIVERSITY IN KENYAN FORESTS

Biogeographic Regional Overview

Lists have been drawn up of Kenyan species which are threatened at an international level, for trees and shrubs (Beentje 1988), birds (Fanshawe & Bennun 1991) and mammals (IUCN 1990). Excluding species found in savanna, dry bushland and high moorlands, the distribution of threatened forest species can be extracted from these lists and related to the different biogeographical zones in Kenya (Table 1).

From a botanical perspective, almost half of the threatened trees and shrubs are found in coastal (38%) or riverine (9%) forests, with some overlap between the two areas. Similarly,

Table 1. Internationally threatened forest species in different regions of Kenya. Sources: Beentje 1988, Fanshawe & Bennun 1991, IUCN 1990.

	Trees/Shrubs	Birds	Mammals
Number of threatened species	159	71	9
% forest species in different regions:			
Coast	38%	27%	55%
Taita Hills	11%	4%	0%
Riverine	9%	3%	22%
Afromontane	12%	14%	22%
Western	0%	6%	0%

Table 2. Altitude, rainfall, gazetted area and vegetation cover of four indigenous Forest Reserves. Forest types after Beentje 1990.

	Arabuko/Sokoke	Shimba Hills	Mau	Kakamega
Altitude (m a.s.l.)	50	150	2,150-2,500	1,600
Rainfall(mm/annum)	600-1000	1,150	1,000-2,000	2,000
Gazetted area(ha)	42,000	22,000	126,550	21,600
Vegetation:				
grass	+	3,000	11,346	4,250
shrub	13,300	7,800	23,800	1,500
forest	27,400	10,400	57,000	12,440
bamboo	-	-	32,100	-
Forest Types	4	6	4(+)	1

over a quarter of the threatened Kenyan birds are found therein, and 55% of mammals (5 species) also occur in coastal forests or associated grasslands.

The afromontane forests are moderately important for mammals (2 species), plants and birds, but the western forests hold few threatened birds, no threatened mammals and no threatened trees/shrubs.

The Taita Hills, and adjacent Kasigau, occur in an outlier patch of the afromontane region near Usambara mountains of Tanzania and cover only a small area (<1,000 ha), yet are of surprisingly great importance for flora and birds. They account for 11% of Kenya's threatened plants and 4% of threatened birds, including endemic species/subspecies.

This regional overview immediately shows the importance of coastal forests for the conservation of biodiversity within Kenya, irrespective of the numbers of species found in different forests. However, as indicated below, when nationally rare species and forest-dependent species are considered, forests in other regions assume greater importance.

Focal forests

In the course of KIFCON Project work four forests have been surveyed for fauna and flora (Table 2). These forests are located in all three main biogeographical regions. At the coast, Arabuko-Sokoke (42,000 ha) and Shimba Hills National Reserve with the two contiguous Mkongani Forest Reserves (22,000 ha) are located north and south of the Mombasa-Kilifi biogeographical subdivision respectively. The West Mau, Southwest Mau and Transmara Forest Reserves are contiguous and cover 126,550 ha in the west-of-Rift afromontane region, and finally Kakamega (21,600 ha) occurs in the western region. Within these forests, all species which are considered rare in Kenya, irrespective of their abundance elsewhere in Africa, are noted in the comparisons.

As expected, the largest numbers of rare trees were in the coastal forests; Shimba having 19 species compared with Sokoke's 11, Mau's 9 (including 3 new KIFCON records for the East African Herbarium) and 7 nationally rare species in Kakamega (Table 3). The overall number of species was also very high in the coastal forests, especially considering their small size - Shimba is probably the most botanically rich forest in Kenya. Mau and Kakamega are of much lower importance in terms of conservation of higher plant biodiversity.

The butterflies show a similar trend, with the highest species numbers at the coast. Sokoke has many more species than Shimba, but Shimba has more rare species. Kakamega has a similar number of rare species and is exceptionally rich in forest-dependent species, although overall species richness does not reach the levels of the coastal forests.

Kakamega forest comes into its own when birds are considered. For such a small forest, the 351 species that have been recorded is extraordinarily high. Indeed, Kakamega is unique

Table 3. Threatened, forest-dependent and total species counts for four forests. (e) = estimate based on species collected to date.

	A/Sokoke	Shimba Hills	Mau	Kakamega
PLANTS				
Threatened trees/shrubs	11	19	9	7
All trees/shrubs	127	350	100	93
Vascular plant species (e)	208	400	273	135
BUTTERFLIES				
Threatened	6	15	3	13
Forest-dependent	20	24	20	67
All	369	299	240(e)	263
BIRDS				
Threatened	18	13	9	15
Forest-specialist	20	18	40	75
Forest-generalist	63	37	60	124
Non-forest	32	30+	12	152
Aquatic	11	?	5	?
All	126	85	117	351
MAMMALS				
Threatened	5	4	6	2
Forest-dependent	7	8	7	9
All	31	28	28	20+

in continental terms as a forest in which the lowland species of the Guineo-Congolian forests intermix with those of the upland species of the afromontane zone (Collar & Stuart 1988).

Like the butterflies, a large proportion of the Kakamega birds are forest-dependent, and there are 15 species threatened within Kenya. To this can be added the large number of savanna birds found in long-standing forest glades and anthropogenic grasslands of more recent origin, of which 54 are grassland dependent and 10 are rare in Kenya. The importance of forest glades for these species stems from the scarcity of uncultivated or ungrazed grasslands outside protected forest areas. This observation introduces a new category of forest-dependent species, namely 'forest-grassland dependent' birds!

The coast forests support many rare birds: 18 in Sokoke, of which 5 are considered international restricted-range species, and 13 in Shimba of which 2 are r-r species (Bibby *et al.* 1992). There are also a moderate number of forest species and some open-country species in the drier forest formations such as the *Brachystegia* woodland of Sokoke.

Mammals, excluding shrews, bats and rodents about which little is known, have been all but eliminated from the small Kakamega forest by intensive hunting pressure from the adjacent densely populated communities' hunters. The coastal forests fare better, and include the savanna-dwelling sable antelope in the Shimba grasslands. However, it is the large Mau forest area which has the greatest importance for mammal conservation because the large area supports viable populations of some of the rarest larger species (e.g. Bongo and Yellow-backed Duiker).

Overall, therefore, the coastal forests, with their high degree of endemism are important for fauna and flora alike, but are especially important for their high plant species richness and rare bird species. The Mau forest has even lower levels of plant species endemism than reported from the east-of-Rift afromontane forests (Beentje 1990), and remarkably few plant species considering the huge area involved, perhaps reflecting the negative effects of high altitude on plant species richness. But it is very important for mammal conservation. Kakamega is important for bird and butterfly species conservation.

Interestingly, and contrary to general continental trends (Bibby *et al.* 1992, Groombridge 1992), forests which are important for one vertebrate group are not necessarily important for another. Kakamega is very important for birds but of little importance for mammals; Mau is important for supporting populations of rare mammals but only supports birds which are widely distributed in other afromontane forests.

Comparing the total number of species in each forest weights the results in favour of groups with the greatest numbers of species (i.e. plants and butterflies). To avoid this bias, the following three species categories were considered: a) threatened, b) forest-dependent, and c) all species, and each forest was assigned a rank (1 to 4), for the four taxonomic groups for which data are presented (Table 3). The number of times each forest scored the lst or 2nd ranks (i.e. highest numbers of species) was summed and the most 'important' forest assessed (maximum score = 12). The results were: Shimba (8), Sokoke (7), Kakamega (6) and Mau (4). Shimba's importance stems from its plant and butterfly species richness, Sokoke's from plant, bird and mammal species and Kakamega from birds and butterflies.

Taking only threatened species, and using the same ranking procedure, the order changes slightly (maximum score = 4): Sokoke (3), Shimba (2), Kakamega (2), Mau (1). If the actual numbers of threatened species are compared, the order reverts to that of the previous exercise: Shimba (51), Sokoke (40), Kakamega (37) and Mau (27), with Sokoke and Kakamega changing places if the Kakamega rare grassland birds are included.

Since threatened species are of more immediate concern than nonthreatened species, the latter ranking gives a useful priority list for forest conservation planning at a national level. However, it must not be forgotten that there are a further 12 large forest reserves, and over 100 smaller forest reserves to be assessed.

PLANNING AND POLICY ISSUES

Assessing biodiversity priorities

Have we developed a set of biological currencies which can be used to answer the question "What forests in Kenya are most important for biodiversity conservation?", a question that the Government of Kenya, donor agencies and NGOs are all asking.

The species level comparison works well, at least for the taxonomic groups considered, and is an easy concept for nonbiologists to grasp even if biologists debate the definition! The taxonomic groups considered also provided enough information with which to compare threatened, forest-dependent and total species counts despite the incompleteness of inventory work in all four forests.

Comparison could be made between biogeographic regions, which is an essential requirement. In this case, since all four forests occurred in different biogeographical regions and many species were unique to each forest, Shimba, Sokoke and Kakamega are all considered priority forests for conservation intervention.

There are other large forests which are also important: Mt Kenya, Aberdares, Mathews Range etc., and prioritising conservation intervention in these, which occur in the same biogeographical region, will require a more detailed assessment of shared species.

Two other factors will have to be introduced into the comparisons: viability of populations, and other values of the forest. As indicated for the Mau forest, its large size is particularly important for maintaining viable populations of larger mammals, which is central to maintaining forest biodiversity in the long-term. Information on species' abundance is essential to allow estimation of the minimum area of a particular habitat required to support 200-2000 breeding individuals. The calculation must, however, be sensitive to different species population dynamics - 2000 elephant hardly occur in any forest areas!

Biodiversity value may not be the only, or even the main reason for conserving or protecting an area of forest. Environmental importance, such as for controlling water quality, needs to be factored into the comparative calculations along with the more conventional use values of forest resources.

Using 'indicator' species

Given the very large numbers of animal and plant species in forests, and the lack of information on distribution or abundance for many of them, comparison of different areas has relied on using a sub-section of the biological community about which most population information is available. The use of such 'indicator' species has become standard practice with preliminary analyses showing that, at least at continental level comparisons, there is a good relationship between the areas of high endemism for different vertebrate groups (e.g. Bibby *et al.* 1992, Collar & Stuart 1988).

However, the broad scatter of data points which form a correlation at the continental scale can cease to apply within a smaller national framework. In the four Kenyan forests, bird species tend to be fewer in forests with most plant species. There was no obvious relationship between bird and butterfly species numbers. Kakamega is very important for birds but unimportant for mammals, and the converse applies in Mau.

The use of indicator species in analyses of biodiversity must be done with great caution, therefore, especially if there are a few small areas scattered across more than one biogeographical zone. In particular, the invertebrates account for the majority of the species-level biological diversity in forests, yet there is a remarkable lack of information on the populations of all but a few groups (e.g. termites, butterflies); only information on butterflies could be used in these analyses.

In general, time and resources are very limited for biodiversity surveys for management planning, as opposed to in-depth research, and methods need to be developed that allow information to be gathered quickly (ca. 7 day) and analysed quickly (ca. 1 month), from several forests, or several areas of the same forest. Standard survey methods could ideally be developed for several invertebrate 'indicator' groups which would allow a fuller assessment of the biodiversity importance of different forests. Development of these methods should be done at a few well-studied sites where rapid appraisal survey results can be calibrated with known population estimates from more thorough research.

Economic assessment of biodiversity

From the point of view of economic analyses, the existence value of those species which will cease to exist soonest should be higher than for those which are abundant and less likely to be lost. This obviously applies irrespective of the increased value of commercial species, whose supply of individuals or products declines while demand is constant or increasing. Furthermore, rare species should be more valuable whether they are attractive to policy makers or not; 'sexy' species which attract funding are still more valuable if they are rare than if they are common.

This separation of species according to their abundance or rarity has yet to be included in economic analyses of existence value (e.g. Brown & Moran 1993). Part of the problem is that biologists must supply economists with the appropriate information and secondly, giving economic value to species stumbles on the unresolved question of "What is a species worth?". Perhaps it is better to retain the biological currency, and give high value to threatened species without trying to reduce the value to economic currencies.

ACKNOWLEDGEMENTS

I would like to thank Universiti Brunei Darussalam for the invitation to participate in this conference, and the Royal Geographical Society for assistance with my travel costs. The work presented entirely reflects the results of KIFCON field surveys carried out by scientists and technicians from the National Museums of Kenya, and I thank them for the field data which have been included in this paper. In addition, Dr Steve Collins (Monsanto Ltd) supplied a checklist of butterflies. The KIFCON Project and Overseas Development Administration (UK) gave permission and funds to allow me to attend this conference.

LITERATURE CITED

ANDRE, H. M., LEBRUN, P. & NOTI, M. I. 1992. Biodiversity in Africa: a plea for more data. *Journal of African Zoology* 106:3-15.

BEENTJE, H.J. 1988. Atlas of the Rare Trees of Kenya. *Utafiti* 13:71-124.

BEENTJE, H.J. 1990. The forests of Kenya. Proceedings of the twelfth plenary meeting of aetfat. *Mitteilungen Institut fur Allgemeine Botanik Hamburg* 23a:265-286.

BENNUN, L.A. & WAIYAKI, E.M. 1992. *Ornithological Surveys of the Mau Forest Complex.* Unpublished KIFCON/NMK Report, Nairobi.

BIBBY, C.J., COLLAR, N.J., CROSBY, N.J., HEATH, M.F., IMBODEN, C.H., JOHNSON, T. H. , LONG, A. J., STATTERSFIELD, A.J. & THIRGOOD, S.J. 1992. *Putting biodiversity on the map: priority areas for global conservation.* International Council for Bird Preservation, Cambridge.

BROWN K. & MORAN D. 1993. *Valuing Biodiversity: The Scope and Limitations of Economic Analysis.* Conference proceedings on the Convention on Biological Diversity: National Interests and Global Imperatives. ACTS/UNEP, Nairobi.

COLLAR, N.J. 1986. Species are a measure of man's freedom: reflections after writing a Red Data Book on African birds. *Oryx* 20:15-19.

COLLAR, N.J. & STUART, S.N. 1985. *Threatened birds of Africa and related islands.* ICBP/IUCN, Cambridge.

COLLAR, N.J. & STUART, S.N. 1988. *Key forests for threatened birds in Africa.* ICBP, Cambridge.

DAVIES, A.G. 1992. Deforestation: East and West Africa. Pp. 23-34 in *Global Environment concerns: Impact on Kenya.* Kenya Museum Society, Nairobi.

DOUTE, R., OCHANDA, N. & EPP, H. 1981. *A forest inventory of Kenya using remote sensing techniques.* KREMU, Nairobi.

FANSHAWE, J. & BENNUN, L.A. 1991. Bird Conservation in Kenya: Creating a national strategy *Bird Conservation International* 1:293-319.

GROOMBRIDGE, B. ed. 1992. *Global Biodiversity: status of the world's living resources.* Chapman and Hall, London.

IUCN 1990. *Red list of threatened species.* International Union for Conservation of Nature and Natural Resources, Gland.

KENYA SOIL SURVEY 1980. *Exploratory soil map and agroclimatic zones of Kenya.*

LEWIS, A. & POMEROY, D. 1989. *A bird atlas of Kenya.* A.A.Balkema, Rotterdam.

MACE, G.M. & LANDE R. 1991. Assessing Extinction Threats: Toward a Reevaluation of IUCN Threatened Species Categories. *Conservation Biology* 5:148-157.

PEARCE, D.W. 1991. An Economic Approach to Saving the Tropical Forest. In Helm, D. (ed.). *Economic Policy Towards the Environment.* Blackwell, Oxford.

POMEROY, D. 1989. *The making of inventories: an introductory survey of methods for trees, shrubs, mammals and birds.* Unpublished report, University of Makarere, Uganda.

RICHARDS, P.W. 1952. *The tropical rain forest.* Cambridge University Press, Cambridge.

STUART, S.N., ADAMS, R.J. & JENKINS M.D. 1990. *Biodiversity in Sub-saharan Africa and its Islands: Conservation, management and sustainable use.* IUCN & The World Conservation Union.

WHITE, F. 1978. The afromontane region. Pp. 463-513 in Werger, M.J.A. (ed.). *Biogeography and ecology of southern Africa.* Junk, The Hague.

WHITE, F. 1983. The vegetation of Africa. UNESCO, Paris.

INDEX TO GENERA AND SPECIES

Genera and species listed in appendices are not indexed

GENERAL INDEX

MIX
Papier aus verantwortungsvollen Quellen
Paper from responsible sources
FSC® C105338

If you have any concerns about our products,
you can contact us on
ProductSafety@springernature.com

In case Publisher is established outside the EU,
the EU authorized representative is:
Springer Nature Customer Service Center GmbH
Europaplatz 3, 69115 Heidelberg, Germany

Printed by Libri Plureos GmbH
in Hamburg, Germany